International Conference on Automation, Computing and Communication (ICACC-2022)

ITM Web of Conferences Volume 44 (2022)

Nerul, India
7 - 8 April 2022

Editors:

Mukesh D Patil
Vishwesh A Vyawahare

ISBN: 978-1-7138-5246-9

Printed from e-media with permission by:

Curran Associates, Inc.
57 Morehouse Lane
Red Hook, NY 12571

Some format issues inherent in the e-media version may also appear in this print version.

For additional information, please contact EDP Sciences – Web of Conferences
at the address below.

EDP Sciences – Web of Conferences
17, Avenue du Hoggar
Parc d'Activité de Courtabœuf
BP 112
F-91944 Les Ulis Cedex A
France

Phone: +33 (0) 1 69 18 75 75
Fax: +33 (0) 1 69 28 84 91

contact-edps@webofconferences.org

Additional copies of this publication are available from:

Curran Associates, Inc.
57 Morehouse Lane
Red Hook, NY 12571 USA
Phone: 845-758-0400
Fax: 845-758-2633
Email: curran@proceedings.com
Web: www.proceedings.com

TABLE OF CONTENTS

Preface

The third International Conference on Automation, Computing and Communication (ICACC-2022) was organized by Ramrao Adik Institute of Technology, D. Y. Patil deemed to be University, Nerul, Navi Mumbai, India, between April 7-8, 2022. The conference was held in online mode due to COVID19 post pandemic scenario.

ICACC-2022 was organized with a motivation to provide a platform for researchers working in the areas of automation, computing and communication to showcase their research and to create a strong community which thrive on the frontier of technology.

The conference received an overwhelming response from academicians, industrialists and the research community. We received 197 papers among three different tracks: Automation, Computing and Communication. All submissions undergone the quality and similarity check before the review process. The quality submissions were reviewed by the experts from academia and industry. Only 87 submissions with originality of work were accepted (approx. 44 % acceptance ratio). The accepted papers were from a broad spectrum of fields like control systems, signal and image processing, AI and ML applications, intelligent systems, Deep Learning, Blockchain, virtual reality and many more. The keynote speakers and authors presented their work in an online mode. All the plenary and technical sessions were interactive and brainstorming.

We would like to take this opportunity to express our deep sense of gratitude towards our management, Hon. Dr. D. Y. Patil (Founder President), Hon. Dr. Vijay Patil (President), and Hon. Mrs. Shivani Patil (Managing Trustee), Dr. Shirish Patil (Hon. VC) and Dr. Vandana Mishra (Hon. Pro-VC), D. Y. Patil Deemed to be University for encouragement and motivation to plan and conduct this conference. We put on record our sincere thanks to the members of the international programme committee and technical programme committee, keynote speakers, reviewers, session chairs, session coordinators, authors and participants for making this conference a grand success. Thanks are due to our colleagues in the institute for supporting us. We would like to devote special thanks and appreciation to the organizing team for their tireless efforts and meticulousness to make this event successful. We would also like to acknowledge the cooperation and support by the teams at Easychair conference management system and MATEC, Web of Conferences. We wish pleasant reading and a fruitful impact of ICACC2022 proceedings on your future innovations and research.

We look forward to welcoming all in ICACC-2023.

Prof. Mukesh D. Patil, Conference Chair, ICACC-2022

Prof Vishwesh A. Vyawahare, Convener, ICACC-2022

April 2022.

Statement of Peer review

In submitting conference proceedings to *Web of Conferences*, the editors of the proceedings certify to the Publisher that

1. They adhere to its **Policy on Publishing Integrity** in order to safeguard good scientific practice in publishing.
2. All articles have been subjected to peer review administered by the proceedings editors.
3. Reviews have been conducted by expert referees, who have been requested to provide unbiased and constructive comments aimed, whenever possible, at improving the work.
4. Proceedings editors have taken all reasonable steps to ensure the quality of the materials they publish and their decision to accept or reject a paper for publication has been based only on the merits of the work and the relevance to the journal.

Title, date and place of the conference

International Conference on Automation, Computing and Communication, April 7-8, 2022
Ramrao Adik Institute of Technology, Nerul, Navi Mumbai, India

Proceedings editor(s):

Dr. Mukesh D Patil,
Dr. Vishwesh A Vyawahare

Date and editor's signature

April 25, 2022

Analysis of CSTR Temperature Control with PID, MPC & Hybrid MPC-PID Controller

Mayur Kumavat[1], *Sushil* Thale[2]

[1]Fr. Conceicao Rodrigues Institute of Technology, Department of Electrical Engineering, India.
[2]Fr. Conceicao Rodrigues Institute of Technology, Department of Electrical Engineering, India.

Abstract. This paper presents an analysis of the continuous stirred tank reactor (CSTR) temperature control with the Proportional-Integral-Derivative (PID) Controller, Model Predictive Controller (MPC) and Hybrid-Model Predictive Controller-Proportional Integral Derivative Controller (MPC-PID). It is the main goal of this project to find a suitable improvement strategy for the system's stability and accuracy to be more stable. By creating a model, the control system is implemented for all the above mentioned control methods and so comparative analysis is carried out to find the best control method for CSTR. Simulation data inspector is used to compare the performance of different types of control systems: PID, MPC and MPC-PID. It has been observed that the hybrid MPC-PID has a more effective control action than a PID controller; with some tuning, the MPC controller can maintain the temperature within a reference or set point range. The control and simulation toolbox is used to construct the model predictive control method in LabVIEW platform. The performance of controllers is measured in terms of settling time, rise time and percentage of overshoot. Finally, a comparative overview of PID, MPC and Hybrid MPC-PID controllers on system performance is presented and discussed.

1 Introduction

A Continuous Stirred Tank Reactor (CSTR) is a system with an exothermic reactor that exhibits highly nonlinear behavior, for example, the abundance of stable conditions [1]. Fig. 1 below shows the system of CSTR. Continuous Stirred Tank Reactors are widely used in the chemical plants and food industries. To test new control algorithms, CSTR is a challenging plant with process variables like reactor temperature and manipulation variables like coolant flow rate. Real reactors, on the other hand, can be dangerous and expensive to operate. As a result of the heating system's high reliance on thermal induction, maintaining precise system control is never easy. Since it's a dynamic, ever-changing system, it is difficult to realize proper control of the system.

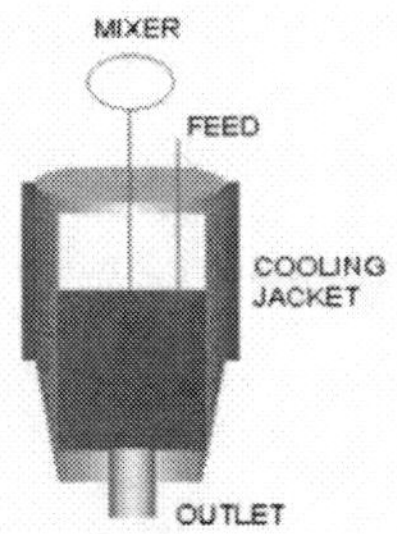

Fig. 1. CSTR System

The most critical aspect of any control system is its design. There are numerous controller kinds, which might be either traditional or intelligent. Model Predictive Control (MPC) is extensively regarded as the industry standard for resolving some of the most difficult control problems encountered in the processing industry. MPC is a very effective optimal control method capable of performing well for nonlinear systems Complex non-linear systems can pose a challenge for basic PID controllers. Control mechanisms for the CSTR have been developed in this study. PID controllers are well-suited for non-linear process control. Instinctively hybrid MPC-PID systems have maximum applications when the plant goes through transition or shows non-linear behaviour and when the plant structure is unknown [2].

The cstr in this project works on chemical and thermal energy and is a composite and dynamic system so the outlet temperature of the liquid in the reactor is constantly changing. The reactor outlet temperature of the liquid should be between 100°C to 300°C when it is taken for further processing.

2 PID controller

Feedback controllers such as proportional-integral-derived (PID) controllers are widely employed in a wide range of control applications. Because its primary goal is to keep the process as close as feasible to the desired set point while minimising errors, a PID controller analyses the error between its input process variable and that target value before taking appropriate action to bring the process' error down as low as possible. Actions are taken based on current errors, summation of mistakes, and error conversion rate. The Proportional term replies based on current error.

Using these three parameters together, the process can be fine-tuned to achieve the desired outcome. PID's equation is presented below [3].

$$u(t) = K_P e(t) + K_I \int_{t0}^{t} e(t)\,dt + K_D \frac{de(t)}{dt} \quad (1)$$

The PID controller returns a Process Variable (PV), an average of its three corrective terms such as Proportional, Integral and Derivative.

$$PV(t) = P_{out} + I_{out} + D_{out} \quad (2)$$

3 MPC controller

Model predictive control is becoming even more important and widespread because of its ability to handle nonlinear multivariable process models and constraints on inputs, states, and output. It uses open loop restricted optimization of limited horizon criteria for receding horizon control. Fig. 2 below describes the receding horizon control approach [4]. Due to the inherent complexity of the chemical process it is usually problematic to obtain a precise model.

System dynamics with nonlinear dynamics can only be accurately predicted by using linear dynamic models in the MPC control scheme. System models that can forecast future output values on a defined time horizon, known as the prediction horizon, are explicitly used in model predictive control.

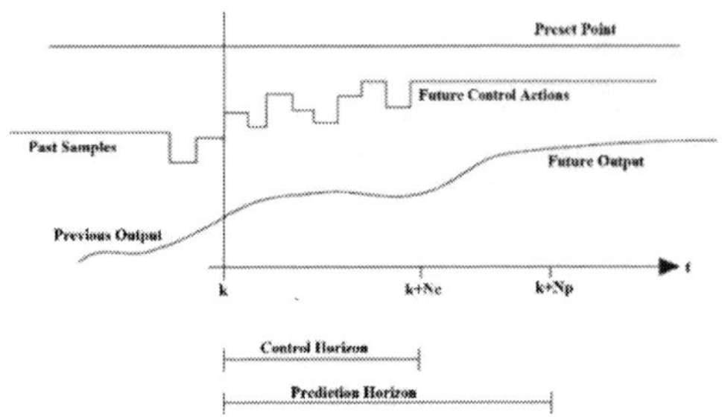

Fig. 2. Receding Horizon Control

The control algorithm is as follows:

1. When a new value of the controlled variable y (t+k) is collected at each sample, it is reported on the prediction interval P (k = 1,... P). This estimate is based on the control variable u (t + k)'s instantaneous values on the control horizon of k = 1,...... M, where M<=P.
2. r (t+k) is the projection of the intended projector to be followed by the system response as a reference point.
3. u (t + k) is calculated in a way that reduces the cost function of future actions.
4. The plant is subsequently subjected to the optimal control, and the yield is measured. The early phases of the model for subsequent iterations are derived from these data of plant conditions.

The Receding Horizon Strategy dictates that the steps 1 to 4 must be performed at each sample.

4 Mathematical model of CSTR

The rate of reaction of non linear dynamic model can be given as,

$$\Re_x = kC_x = C_x k_0 exp\left(\frac{-A_e}{RT_r}\right) \quad (3)$$

where $\Re_x$ = Consumption rate of reactant X

F = Flowrate of feed

C_x = Concentration of reactant X in reactor

V_r = Volume of Reactor

k_0 = Preexponential factor

A_e = Activation energy

T_r = Reactor Temperature

T_c = Coolent Temperature

ΔH = Heat of reaction

Four nonlinear ordinary differential equations form dynamic model of reactor and jacket [5].

Total mass balance (kg/s):

$$\frac{d(V_r,\rho)}{dt} = \rho_0 F_0 - \rho F \quad (4)$$

Component A balance (kmol A/s):

$$\frac{d(V_r,C_x)}{dt} = F_0 C_{xf} - FC_x - V_r \Re_x$$

$$= F_0 C_{xf} - FC_x - V_r kC_x \quad (5)$$

Reactor energy balance (J/s):

$$\frac{d(V_r,\rho H_c,T_r)}{dt} = \rho_0 H_{c0} F_0 T_0 - \rho H_c F T_r - \lambda V_r \Re_x - Q \quad (6)$$

The heat transfer rate of a jacket water system with a jacket temperature T_j is affected by the temperature gradient between the jacket and the surrounding water.

$$Q = UA_j (T_r - T_j) \quad (7)$$

The dynamic model of jacket is

$$\frac{d(V_j,\rho j\ Hcj,T_j)}{dt} = F_j \rho_j H_{cj} T_c - F_j \rho_j H_{cj} T_j + Q \quad (8)$$

For a given density and heat capacity, time derivative in Eqs. (4), (5) and (6) can be taken out of the equations and replaced with the physical parameters (8). The variables V_r and V_j can be eliminated from the derivatives if the reactor volume and the jacket volume remain constant. As a result, equation (4) is reduced to,

$$F_0 = F \quad (9)$$

The differential equations (5) and (6) reduce to the following:

$$\frac{dC_x}{dT} = \frac{F}{V_r}\left(C_{xf} - C_x\right) - K_0 * e^{\frac{-A_e}{RT_r}} * C_x \quad (10)$$

$$\frac{dT}{dt} = \frac{F}{V_r}\left(T_f - T_r\right) + \left(\frac{-\Delta H}{\rho H_c}\right) K_0 * e^{\frac{-A_e}{RT_r}} * C_x + \frac{U*A}{V\rho H_c}\left(T_c - T\right) \quad (11)$$

Where, $\lambda = \frac{-\Delta H}{\rho H_c}$. In order to mimic the dynamic response of CSTR, we will employ these two basic equations. Without controllers, the system will behave in a "open loop" manner. There is no change in flow of cold water in this situation. Using the "close loop" method, a temperature controller regulates the flow of cold water into the reactor to keep it at a constant temperature.

To solve these two equations, it is necessary to specify all the parameters and variables except two (C_x and T_r). Fig. 3 shows the Simulink model for CSTR system derived from eqn (10) and eqn (11). Table I gives the parameter values for CSTR Model [6].

Table 1. Parameter values of CSTR model

Parameters	Description	Values
F/V_r	Flow Rate per Unit Volume	0.83
K_0	Preexponential Factor	10e15
$(-\Delta H)$	Heat of reaction	6000
A_e	Activation energy	-69815
ρHc	Density times Heating Capacity	500
T_f	Feed Temperature	312
C_{XF}	Feed Concentration	8
UA/V	Heat Transfer per Unit Volume	145
R	Reaction Rate	8.314

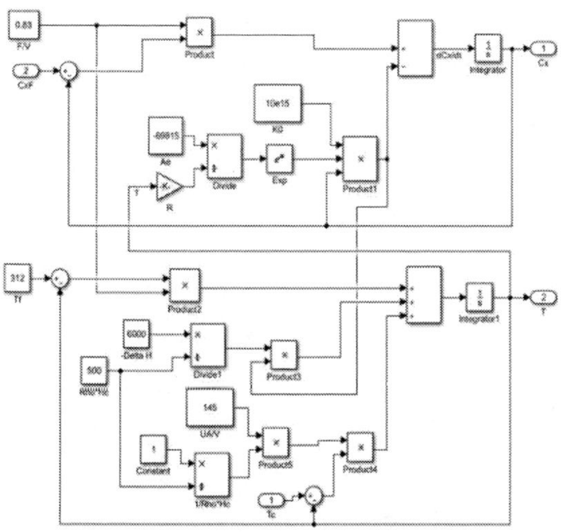

Fig. 3. Simulink Model of CSTR

5 Simulation & analysis

A controller is a tool for regulating a dynamic system so as to accomplish a desired behavior. The goal is to reduce undershoots and overshoots from the system and generate output signals from systems that are close to the set point value and maintain the accuracy and stability of the system. Control systems for CSTR are designed using a combination of PID, MPC, and MPC-PID.

5.1 Temperature Control using PID Controller

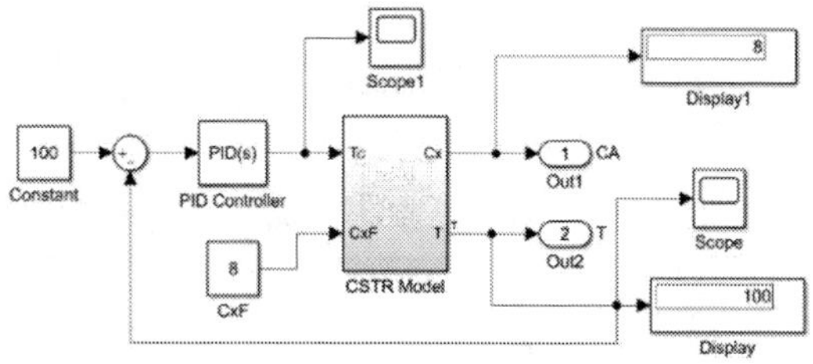

Fig. 4. Temperature Control using PID Controller

Fig. 4 above shows a Simulink model of a PID controlled CSTR system. The desired temperature is set at a set-point of 100°C. The output of the system is given in Fig. 5.

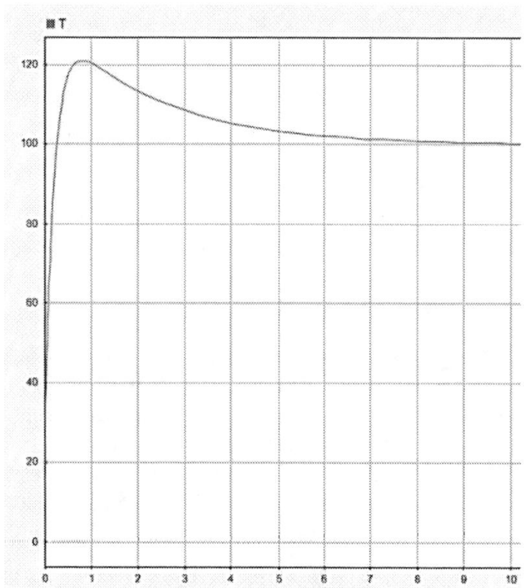

Fig. 5. Output of PID Controller

The figure above shows the output response of a PID controlled CSTR system, where we can observe the output temperature equal to the desired temperature and give good accuracy as well as an overshoot of about 21% which is fine but can be / should be reduced for more stability.

https://doi.org/10.1051/itmconf/20224401001

5.2 Temperature Control using MPC Controller

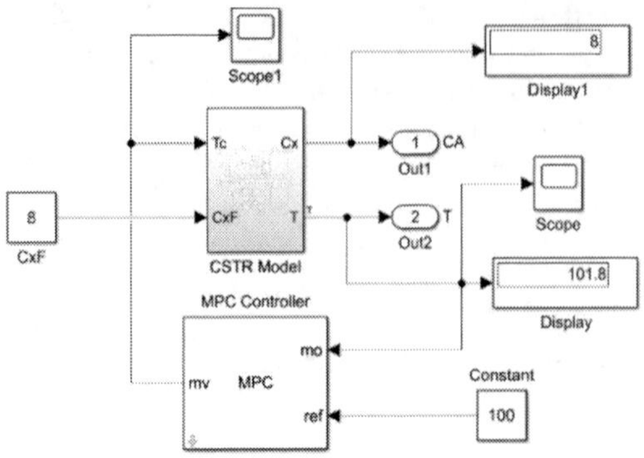

Fig. 6. Temperature Control using MPC Controller

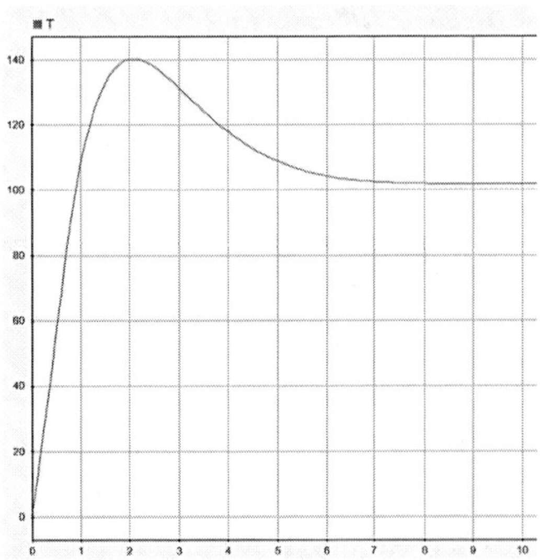

Fig. 7. Output for MPC Controller

Fig. 6 shows a Simulink model of the MPC controlled CSTR system. The desired temperature is set at a set-point of 100°C. The output of the system is shown in Fig. 7. In Fig. 6, we can observe that the output temperature is close to the desired temperature. This gives a difference of about 1.8°C which is acceptable but not completely accurate. Also, we can observe an overshoot of about 40% which should be reduced for better stability.

5.3 Temperature Control using Hybrid MPC-PID Controller

Fig. 8 states the flow chart of the system flow when designing the proposed Hybrid MPC-PID controller [7].

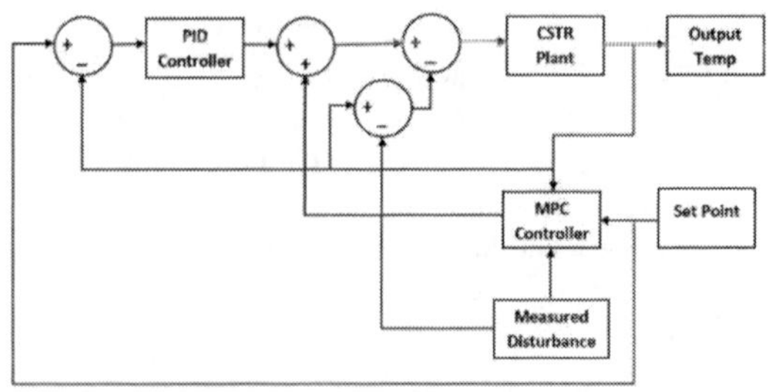

Fig. 8. Block diagram of Hybrid MPC-PID Controller

Use the model to predict future plant responses, which are utilised to fine-tune control signals. This new control signal is generated by an internal model, which is based on the prediction horizon (Np) and the system control horizon (Nc). Further tuning and refinement is required before controller performance is considered satisfactory.

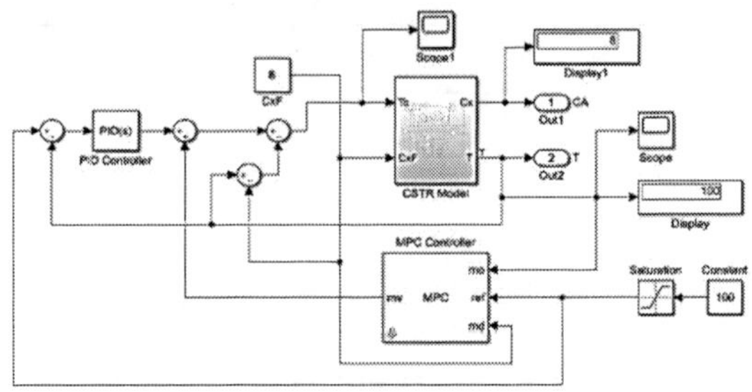

Fig. 9. Temperature Control using Hybrid MPC-PID Controller

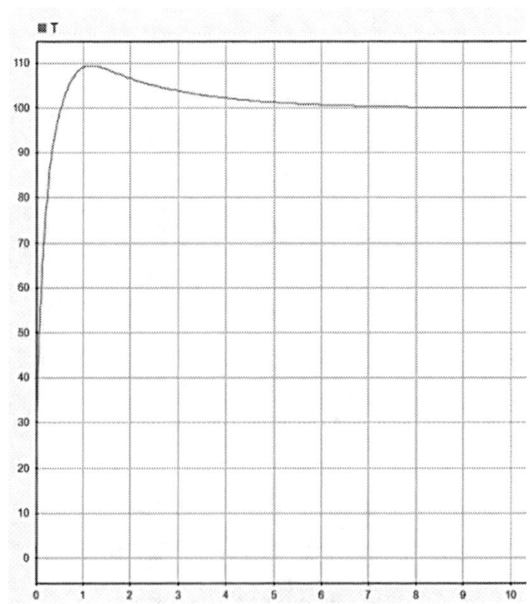

Fig. 10. Output of Hybrid MPC-PID Controller

In Fig. 9, a Simulink model for our proposed hybrid MPC-PID controller is shown. Where the output of the PID and

https://doi.org/10.1051/itmconf/20224401001

MPC controller is fed together to the CSTR system so that the error is minimized. The desired temperature is set at 100°C. The output response of the hybrid MPC-PID controlled CSTR system is shown in Fig. 10.

In Fig. 10, we can observe that the output temperature value is equal to the desired temperature value and given good accuracy as well as we can see that the overshoot is reduced to 9% which is superior to the other two controllers we have seen and discussed previously. So the proposed controller gives better stability than others, as well as if we are concern about rise time and settling time, the given controller gives satisfactory results.

5.4 PID Controlled CSTR in LabVIEW

Fig. 11 below shows the Front Panel of PID Controlled CSTR.

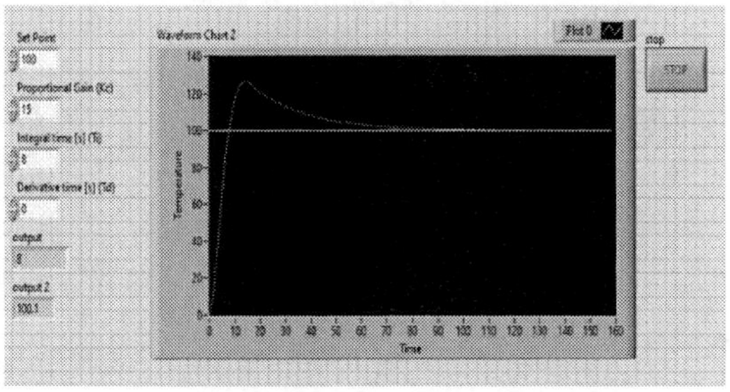

Fig. 11. Front Panel of PID Controlled CSTR

The front panel was made with LabVIEW, and it has a place for you to enter all of the PID parameters that were chosen in the design. As we can see in the fig, for the given set-point of 100°C we can observe the real time output of 100.1°C with overshoot of around 24 to 25% and some minor fluctuations.

Fig. 12 below shows the Block Diagram of PID Controlled CSTR. As we have discussed earlier in the MATLAB simulation and design of CSTR model, the model of CSTR is presented here in LabVIEW using Control and Simulation Loop with same those equations. In Fig. 12 as we can see, the model is controlled using PID Controller with T_c (Coolant Temperature) as the process variable.

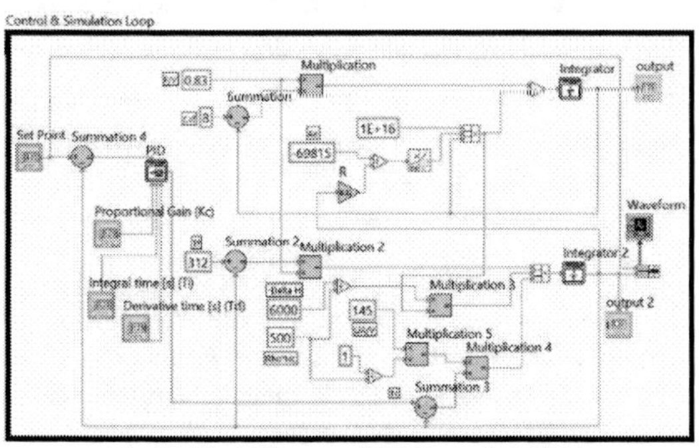

Fig. 12. Block Diagram of PID Controlled CSTR

5.5 MPC Controlled CSTR in LabVIEW

Fig. 13 below shows the Front Panel of MPC Controlled CSTR. The front panel was made with LabVIEW, and it has a place for you to enter all of the MPC parameters that were chosen in the design.. Linearized state space model is obtained by converting transfer function into state space by using CD Convert to State Space Model as shown in Fig 14.

The Transfer Function,

$$G(S)=\frac{0.29s+0.2407}{s^2+1.95s+0.9296} \tag{12}$$

was obtained after the linearization of model in Matlab.

As can be seen, the MPC was made to have 10 samples for the prediction and 10 samples for the control. In Fig. 13, you can see how controller output constraints and cost function parameters can be set up.

As can be seen in the Fig 13, for the given set-point of 100°C we can observe the real time output of 100°C with overshoot of around 48% with no fluctuations and maintaining good stability.

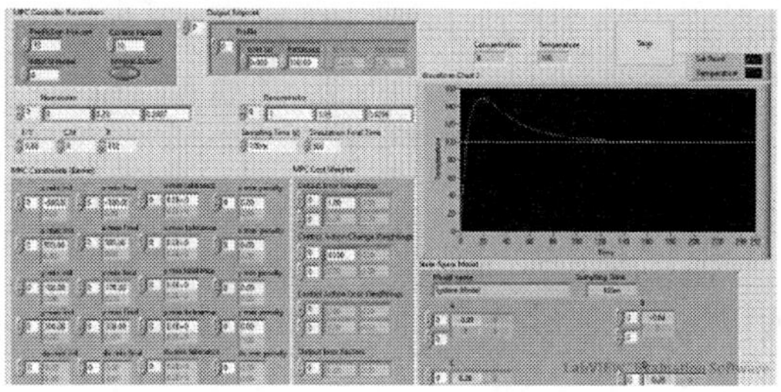

Fig. 13. Front Panel of MPC Controlled CSTR

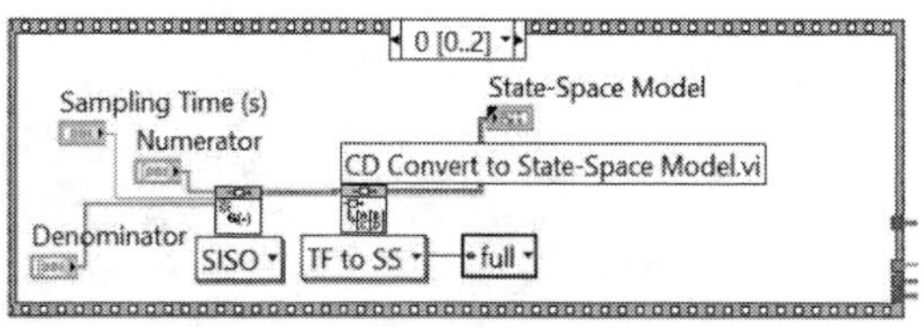

Fig. 14. Block Diagram of MPC Controlled CSTR

Fig. 14-16, shows the Block Diagram of MPC Controlled CSTR.

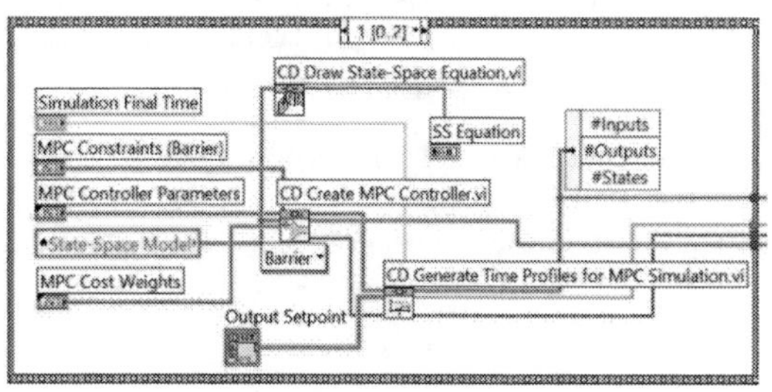

Fig. 15. Block Diagram of MPC Controlled CSTR

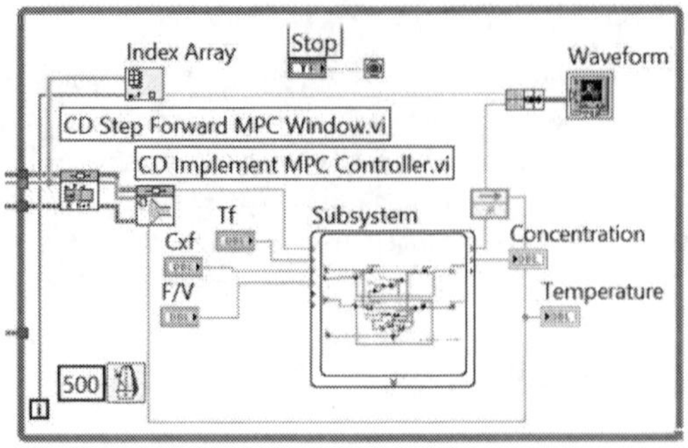

Fig. 16. Block Diagram of MPC Controlled CSTR

In Fig. 15 and 16, the VI has two parts. The first part is to make sure the system is ready for MPC by getting all of the user's MPC parameters. The second stage makes sure that the MPC scheme's control actions work.

5.6 Hybrid MPC-PID Controlled CSTR in LabVIEW

Following figures illustrates the proposed Hybrid MPC-PID control scheme for controlling the CSTR temperature in LabVIEW. Fig. 17 below shows the front panel of Hybrid MPC-PID Controlled CSTR.

As we can see in the Fig. 17, for the given set-point of 100°C we can observe the real time output of 100.5°C which further reaches to 100°C smoothly and steadily by the time. With overshoot of around 7 to 8% which is lesser than the conventional PID controlled and MPC controlled methods by maintaining its stability and accuracy to a great extent.

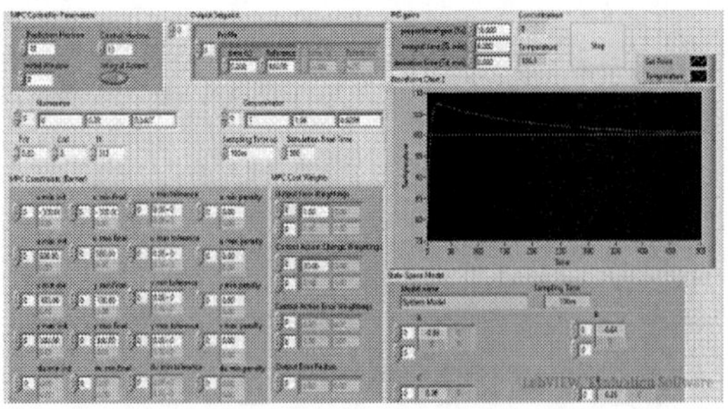

Fig. 17. Front Panel of Hybrid MPC-PID Controlled CSTR

Fig. 15 & 16 shows the block diagram of MPC scheme that is explained already in previous method that further connects with PID scheme in fig 18.

The Controlled output of MPC and PID together is given to the CSTR plant/model so as to maintain the temperature of reactor as shown in Fig 18.

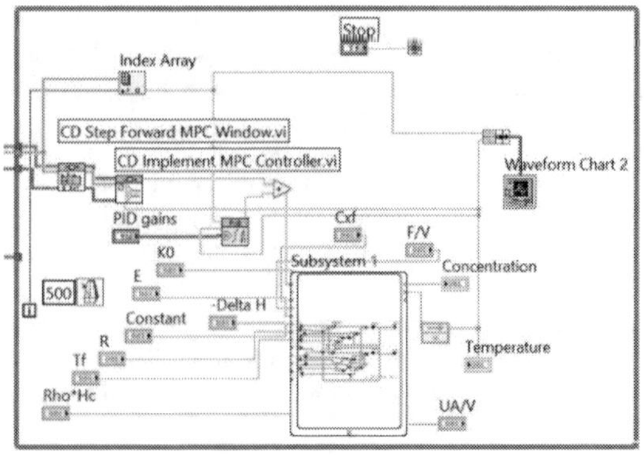

Fig. 18. Block Diagram of Hybrid MPC-PID Controlled CSTR

6 Results & conclusion

To analyze the performance of the proposed controller, the system is simulated using MATLAB / SIMULINK. The controller parameter values for the simulation are given in the table below:

Table 2. Results

Controllers	Setpoint	Feed Concentration	Feed Temperature	% Overshoot	Peak Value	Peak Time	Rise Time	Settling Time
PID	100	8	312	21	121	0.8	0.2	7
MPC				40	140	1	0.6	7
MPC-PID				9.3	109	0.65	0.5	6.48
PID	100	58	362	26	126	0.7	0.18	6.5
MPC				57	157	1.8	0.6	3.5
MPC-PID				14	114	0.8	0.3	4.5

The primary goal of this analysis was to integrate an MPC and a PID controller in order to manage the outlet liquid temperature of a CSTR and to pick the optimal model using a system identification technique. MPC and PID controllers significantly enhanced performance in our analysis. While the PID controller assisted in maintaining a reasonable level of liquid temperature in the CSTR, the system remained unstable. In order to improve our system's performance and stability, we incorporated an MPC controller together with a PID controller.

The PID controller's design and implementation can be modified to eliminate overshoots and undershoots and to ensure that the output is near to the target set point value.

The MPC controller assisted us in achieving optimal performance for our system by minimising overshoot, rising time, and settling time. MPC can aid in the development of more effective control measures for non-linear or dynamic systems whose reaction is not stable.

The primary goal of the hybrid MPC-PID design approach and implementation was to determine whether the control system could be enhanced by keeping a stable temperature level. The Hybrid MPC-PID scheme's results demonstrate that performance of the system can be increased with minor modifications. The controller's efficiency is determined by the temperature of the CSTR outflow liquid.

All these schemes are also implemented in LabVIEW. The system has minor fluctuations with PID scheme in real time while MPC has comparatively higher overshoot but Hybrid MPC-PID controller has stable output with comparatively lower overshoot of around 7-8%. Only thing is system requires more settling time in real-time system. For the creation and implementation of MPC controllers, the control and simulation toolbox includes pre-built MPC blocks.

By changing the values of parameters such as feed concentration and feed temperature for the same given system, PID and Hybrid MPC-PID controllers give better results while MPC controller give sluggish and incomplete responses.

References

1. Mrs. M. Shyamalagowri, Dr. R. Rajeswari, "Model Predictive Control Design for Nonlinear Process Control Reactor Case Study: CSTR (Continuous Stirred Tank Reactor)", *IOSR Journal of Electrical and Electronics Engineering (IOSR- JEEE).*, vol. 7, issue. 1, pp. 88-94, 2013.
2. Vivek Agrawal, Manish Gupta, Umesh Gupta , Rahul Saraswat, "A model predictive controller using multiple linear models for Continuous stirred tank reactor (CSTR) and its implementation issue", *Fourth International Conference on Communication Systems and Network Technology.*, pp. 1001-1005, 2014.
3. Liuping Wang, "*PID Control System Design and Automatic Tuning using MATLAB/Simulink*". UK: John Wiley & Sons, 2020.
4. Lars Grüne, Jürgen Pannek, *Nonlinear Model Predictive Control Theory and Algorithms.* Switzerland: Springer, 2017.
5. William L. Luyben, *Chemical Reactor Design and Control.* New Jersey: A John Wiley & Sons, 2007.
6. Rahul Upadhyay, Rajesh Singla, "Analysis of CSTR Temperature Control with Adaptive and PID Controller (A Comparative Study)" *IACSIT International Journal of Engineering and Technology.*, vol. 2, no. 5, pp. 453-458, 2010.
7. Md Mafizul Islam, Md Abdul Salam, "Modelling and Control System design to control Water temperature in Heat Pump" *Master's Program in Electrical Engineering.*, pp. 1-60, 2013.

Energy Efficient Li-Ion Battery Based Electric Bicycle

Ajay Narkhede[1], *Aditya* Jagtap[1], *Sumit* Makne[1], and *Santosh* Gadekar[1]

Department of Electrical Engineering

JSPM'S Bhivarabai Savant Institute of Technology and Research, Wagholi, Pune

Abstract - This paper presents the energy efficient electric bicycle based on Li-ion dry type battery. Since last 50 years the fossil fuels reserves are continuously decreasing. So, it becomes very necessary to find out the solution to use alternate energy resources in reliable and efficient way. The increase demand for IC engine-based vehicles creates the pollution problems in many cities of India as well as in world. Hybrid electric vehicle is a best alternative to overcome the problem of air pollution and noise pollution. The PMDC Motor is used to drive the electric bicycle. The PMDC Motor eliminates the extra converter required if the motor used is of DC type. LI-ion battery is used for delivering high power storage capacity. A 24V controller is used for controlling inputs and outputs. It consists of one switch which on/off all system over battery. It is hybrid type bicycle which can operate on battery and pedalling. The hybrid nature of the bicycle makes it more energy efficient and eco-friendly.

Keywords: *Electric Bicycle, PMDC Motor, Li-Ion Battery, Controller, E-Bike, Eco-Friendly Vehicle*

1. Introduction

The main motive to layout the electric bicycle is to conquer the trouble with the pollution and with the financial system. Destiny for electric bicycle is the pleasant technical utility as solution for the higher world and upcoming generation. The E-bicycle is a battery-operated vehicle this is very comparatively cheap with low renovation price and less pollutants. E motorcycles are an attractive opportunity to each conventional bicycles and conventional cars, offering an environmentally pleasant, a laugh, efficient and handy manner to travel. E-bikes are pushed with the assist of battery that's coupled with electric motor. E- bicycle is the plug-in electric bike with two wheels. The electricity on which this bike works is saved in a rechargeable battery which drives the motor. Now a days these motorbikes are manufactured at a very large scale. Typical parts used in E-bike are PMDC motor, battery, controller, throttle, chain set [1]. A brushless DC motor uses electronic commutators and sensors rather than the mechanical commutation process, which is used in a brushed DC motor. The use of permanent magnet DC motor reduces mechanical losses and hence efficiency is improved. A permanent magnet DC motor has high dynamic response, better speed-torque characteristics high efficiency and more reliable as compared to a brushless DC motor [2]. A lithium-ion batteries allow an excellent thermal design to meet this requirement. As a result, a compact power source system can be constructed that can be expected to improve vehicle mount ability [3]. Through its high potential, and high energy density and capacity, this battery type has already contributed to improving our lives, and arguably will continue to do so in the years to come. However, battery development is very daunting and challenging in general, and perhaps particularly so when it comes to lithium-based cells [4]. Controller is just like a brain of the e-bicycle which controls power from battery bank to the different parts of the system i.e., motor, and accessories like headlight, backlight, horn and indicators and the power from charging ports to the battery [5].

2. Functional Block Diagram

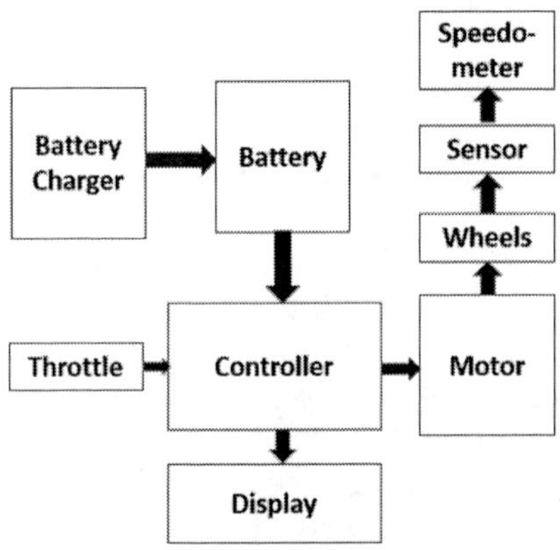

Fig.1: Block Diagram

Fig.1 shows the Functional Block diagram of Electric bicycle. When power supply using the battery is provided then the state of controller changes from OFF state to ON state. A controller can be called as the brain of the whole system as it carries out all the input and output signals which are essential for the functioning of the system. Controller receives throttle signals and transfer output signal to motor. It is a PMDC motor which operates on principle, "whenever a current carrying conductor placed in a magnetic field it experiences a force". Motor is attached to

 https://doi.org/10.1051/itmconf/20224401002

wheel shaft of bicycle by using gearing and chain method. DC motors are in particular popular in high power and precise servo applications due to their reasonable cost and ease of control [6]. Throttle works as an accelerator. By using throttle accelerating torque of the motor increases and motor current increases. The power flow works in collaboration with power provided by the rider from the pedals [8]. Here, electromagnetic sensor is placed in between wheel spoke and wheel support frame which is connected to LCD display which Tracks the speed and distance are totally based on its stability in nature. It can check and verify the speed and distance, have many functions, average speed, maximum speed, time of riding, odometer, time clock, distance, scan, maintenance alarm etc. There is another display which displays the battery capacity in percentage and the battery voltage(8-70V) and is connected directly to the battery. The weight of the electric bicycle is approximate 10kg, it have around double the power output and the need to be pedalling at hills and road with high inclination [8].

2.1. Table 1: Specifications of components

Sr. No.	Name of Component	Specification
1.	PMDC Motor	24V,10A, 250W, 3800RPM
2.	Controller	DC 24V, 21A, 250W
3.	Li-Ion Battery	(24 to 28.80) V, 10Ah

2.1.1 PMDC Motor

The Permanent magnet DC motor is an energy efficient motor. It has very low mechanical losses, it is compact in size and easy to available in market for reasonable price. Due to its compact nature, it is easy to mount that motor on the bicycle. The direction of motor can be clockwise or anti-clockwise by reversing the motor connections. The no load speed of the PMDC motor is 3800 rpm; the rated speed of the motor is 3000 rpm.

2.1.2 Controller

The controller works as a monitor of electric bicycle, it monitors the operation of all the components are used. It consists of connections for motor, accelerator, brake, battery, battery charging, brake light and power lock.

2.1.3 Battery

Compared with traditional battery technology, lithium-ion batteries charge faster, last longer and have a higher power density for more battery life in a lighter package. Lithium-ion battery types can also be moulded into different shapes or unique spaces, making them ideal for high capacity and low power applications. The battery can be charged up to maximum 28.8V. A long life with full capacity for up to 2500 charge cycles and has low maintenance

3. Results

3.1 Time to Voltage Graph:

3.1.1. No load conditions

Table 2: Observation Table of no-load condition

Time in Min	*Voltage in Volts*
45	28
30	27
25	26
18	25
15	24
10	23
0	22

Table.2 shows the performance of battery voltage with respect to time, for No-load condition. Maximum charged voltage of battery is 28V and battery gets fully discharge at 22V.

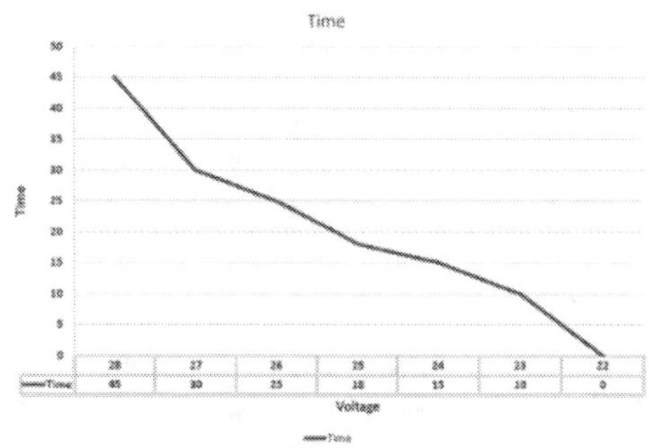

Fig.2: Time to Voltage Graph

In Table.2, the graph of time to voltage is shown, it represents the time in minutes on the Y-axis and voltage in volt in X-axis.
From the above Table.2 and fig.2 it is observed that the battery is discharged at 22V. Voltage is directly proportional to the time, as the voltage drops at the same instant the time of run also decreases. For a drop of 1 volt, it takes maximum time of 45 minutes and minimum time taken to 1Volt drop is 10 minutes. In fig.2, time taken to drop the voltage from 28V to

https://doi.org/10.1051/itmconf/20224401002

27V is 45 minutes and from 27V to 26V is 30 minutes respectively. Similarly, it takes 25 minutes to drop the voltage from 26V to 25V, it takes 18 minutes to drop the voltage from 25V to 24V and 15 minutes for the voltage drop from 24V to 23V. After dropping to 23V the battery will no longer perform its operation until it is charged again. So 23V is the minimum voltage required for battery to function.

3.1.2 On load condition

Table 3: Observation table of On-load condition

Time in min	Voltage in volt
40	28
28	27
22	26
15	25
10	24
5	23
0	22

In Table.3, the performance of battery voltage with respect to time, for No-load condition. Maximum charged voltage of battery is 28V and battery gets fully discharge at 22V.

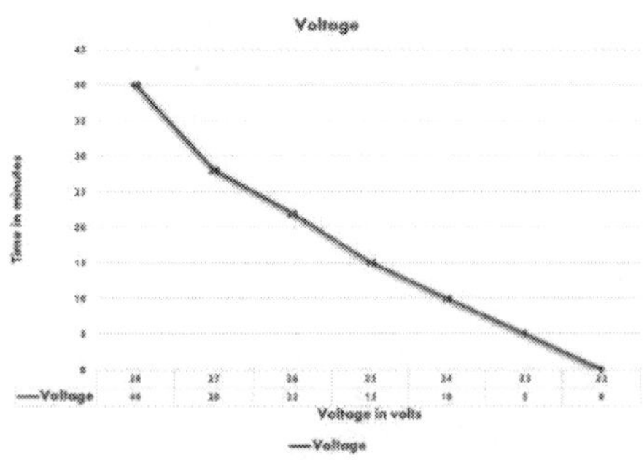

Fig.3 Time to Voltage Graph

In Table.3, the graph of time to voltage is shown, it represents the time in minutes on the Y-axis and voltage in volt in X-axis.
From the above Table.3 and fig.3 it is observed that the battery is fully discharged at 22V. Voltage is directly proportional to the time, as the voltage drops at the same instant the time of run also decreases. For a drop of 1 volt, it takes maximum time of 40 minutes and minimum time taken to 1Volt drop is 5 minutes. In fig.3, time taken to drop the voltage from 28V to 27V is 40 minutes and from 27V to 26V is 28 minutes respectively. After 23V battery will not perform any operations. So, 23V can be called as the threshold voltage of the battery.

3.2 Observation Table:

3.2.1 Range for variable loads.

***Table 4*: Observation table of bicycle performance**

Sr. No.	Person/kg	Distance/charge/km
1	40-50kg	40km
2	55-65kg	35km
3	65-75kg	30km
4	>75kg	< 30km
5	2 seats	25km

The Table 4 represents the range of the electric bicycle with respect to different loads. It gives a clear idea about how many kilometres it can be driven for a particular load or weight on it. For example, for a weight ranging between 40 kg to 50 kg the bicycle runs for an average distance of 40 kilometres. As the weight on the bicycle increases the range of it significantly decreases.

4. Conclusion

The idea of low-cost electric bicycle based on PMDC motor with Li-Ion battery is successfully presented. The results suggests that the PMDC motor can be the best alternative for light load applications. Also, the size of the battery, motor and their mounting position makes this electric bicycle aerodynamic in nature which assist the speed obtained by bicycle. The project has been able to design, analyse and assemble the electric bicycle which can achieve a top speed of 35 km/hr. For an average weight that is for the weight ranging between 55 kg to 65 kg the electric bicycle can travel for an approximate distance of 35 kilometres on a single charge.

References

[1]. S. Matey, A. Prabhu, "Design and Fabrication of Electric Bike" International Journal of Mechanical Engineering and Technology- Vol. **8** Issue 3- March (2017).
[2]. A. Jain, P. Sarkar, K. Siddique, "Controlling of Permanent Magnet Brushless DC Motor using Instrumentation Technique" International Journal of Advance Engineering and Research Development- Vol. **2** Issue 1- (2015).
[3]. O. Shimamura, T. Abe, K. Watnabe, Y. Ohsawa, H. Horie "Research and Development Work on Lithium-ion Batteries for Environmental Vehicles" The World Electric Vehicle Association Journal, Vol. **1**, (2007)
[4]. J. Godenough, M. Whittinghum, A. Yoshino "Li-Ion Batteries" Kungl. Vetenskaps-akademein (2019).
[5]. A. kumar, A. Dewangan, V. Dange, A. Agrawal "Eco-Friendly E-Bicycle" International Journal of Recent Technology and Engineering (IJRTE) ISSN: 2277-3878, Volume-**8** Issue-5, January (2020).

[6]. M. Ruderman, J. Krettek, F. Hoffmann and T. Betram, "Optimal State Space Control of DC Motor," Proc. of the 17th World Congress of the Int Federation of Automatic Control, pp. 5796-5801, July 6-11. (2008).
[7]. A. Parker, A. Alan, "The electric powered assisted bicycle; a clean vehicle to reduce oil dependence and enhance the mobility of the elderly" International Conference on Sustainability Engineering and Science, Auckland, New Zealand, July (2004).
[8]. A. Pauldanie, R. Jerin Sam, A. Nabisha, S. Nithya, "Design and Fabrication of Hybrid Electric Bike", International Journal of Research Publication and Reviews Vol. **2** Issue 7, (2021).

Design and Implementation of Voice Controlled Wheelchair using MATLAB

Kaushal Balu Karande
Department of Electronics Engineering
Ramrao Adik Institute of Technology,
DY Patil deemed to be university,
Nerul, Navi Mumbai, Maharashtra
kaushalkarande24@gmail.com

Jagruti Dilip Zope
Department of Electronics Engineering
Ramrao Adik Institute of Technology,
DY Patil deemed to be university,
Nerul, Navi Mumbai, Maharashtra
jagrutizope27@gmail.com

Sakshi Somani
Department of Electronics Engineering
Ramrao Adik Institute of Technology,
DY Patil deemed to be university,
Nerul, Navi Mumbai, Maharashtra
sakshi.somani@rait.ac.in

Balu Bhusari
Department of Electronics Engineering
Ramrao Adik Institute of Technology,
DY Patil deemed to be university,
Nerul, Navi Mumbai, Maharashtra
balu.bhusari@rait.ac.in

Abstract: **Most physically challenged people to rely on others for their motion and the number of such people is increasing day by day. Manual Wheelchairs are quite difficult to operate by physically challenged people as they require force for the motion. To overcome this drawback, conventional wheelchairs or powered wheelchairs are available in the markets. Even if they are meant to design for the comfort of users they are quite expensive, so not every physically challenged can afford them. To reduce the percentage of use of a manual wheelchair and to give physically challenged people a better experience and comfort in an affordable manner we have proposed this model. Our proposed model is based on neural networks and the backpropagation algorithm is used as a medium to train the Artificial neural networks which are trained by the user's voice command so it uses voice recognition as acontrol mechanism. The proposed prototype is been trained with five voice commands such as (Forward, Backward, Left, Right, and Stop). These voice commands are successfully identified and converted to text. The following text is further written on target hardware wirelessly by using a Bluetooth module for the motion of DC Motors to control the directions of a wheelchair. The entire prototype is been designed on Matlab R2020a and Arduino UNO as target hardware. The main aim of our proposed work is to develop a smart wheelchair that overcomes the problems of existing systems and should be easy, impressionable, and convenient for physically challenged people.**

Keywords: Voice Recognition, Neural Network, Back Propagation Algorithm, Linear PredictionCoefficient, Matlab, and Arduino UNO.

I. INTRODUCTION

A Wheelchair is defined as a device that is used to assist physically challenged people. It can be driven either manually or automated using control systems. Manual Wheelchairs are quite difficult to operate by physically challenged people as they require force for the motion. To overcome this drawback conventional wheelchair-poweredwheelchairs are available such as joystick-based, eye-controlled, brain-controlled, gesture-controlled, and manymore. Even though these powered wheelchairs are *designed* for user's comfort but economically they are quite costly and have complex logic to operate. Also, there is an increase in the demand for a wheelchair in the Present andFuture markets. Due to this, there are a lot of developments for automating the wheelchair. To contribute to this development we introduce a cost-efficient and reliable smart voice-controlled wheelchair which makes use of a backpropagation algorithm to train neural networks and anArduino Uno board to control the motion. Speech Recognition techniques enable the machines to identify theuttered words and convert them into understandable text. Inour proposed work a speaker-dependent speech recognitionsystem is used to build up a smart powered wheelchair. This speech recognition technique is implemented using neural networks. The input from the user is recorded in .wav format. To train these user input data, a Backpropagation algorithm is used. Also, Linear Prediction Coefficient (LPC) is a feature extractor employed by us at the input which extracts the pattern and helps the neural networks totrain faster. The training phase is completed only when a minimum error is achieved. After that, the model is validated i.e. recognized voice is converted into text and is sent to the Arduino board where it matches the initial and previously trained data and based on that it recognizes the command and executes the instructions. A Bluetooth module is used to send the data wirelessly from system to board. Thus the main aim of our proposed work was to develop a comfortable, cost-efficient, and versatile model of the wheelchair so that more physically challenged people would benefit from it.

II. RELATED WORK

The goal of developing a smart wheelchair is to provide cost-efficient and reliable solutions for automating the movements of physically impaired people. Before developing this thought we have gone through various works related to wheelchair automation. Some of the important methods are mentioned as Dilok Puanhvuan(2014) proposed a work in which the patients can drive thewheelchair in two ways either manually or based on eye movement. The user can start or stop the system using eye closing protocol. The wheelchair can automatically travel to a pre-mapped location by safely avoiding and detecting the obstacles [10]. But the system comes with a drawback which causes stress on the human eye. Tao Lu (2015) develop a motion control system for an intelligent wheelchair based on hand gesture recognition. An automatic gesture segmentation algorithm was developed to identify individual gestures in a sequence and the Hidden Markov Model (HMM) is used for the hand gesture model training [9], But this method is proven to be complex for people with disabilities in wrist movements. Sumet Umchid (2018) developed a wheelchair that can operate byusing voice commands through the given input [2]. The proposed wheelchair was developed using a voice recognition module, DC motors, and a microcontroller. In addition to this their proposed model was embedded with ultrasonic sensors for detecting obstacles and for applying immediate brakes when an object is suddenly detected in front of it. This developed system achieved a high accuracybut failed in terms of high cost due to a wide variety of hardware components. Mohammed Faeik Ruzaij (2018) developed an intelligent wheelchair-using speech recognition system to control the movement of a wheelchair in different directions by using voice commands and also the simple movement of the patient's fingers with keypad control [5]. But the main problem in the proposed work was to handle an inaccurate response inthe noisy environment and it can become difficult for the user to move in such an environment.

III. SYSTEM DESIGN

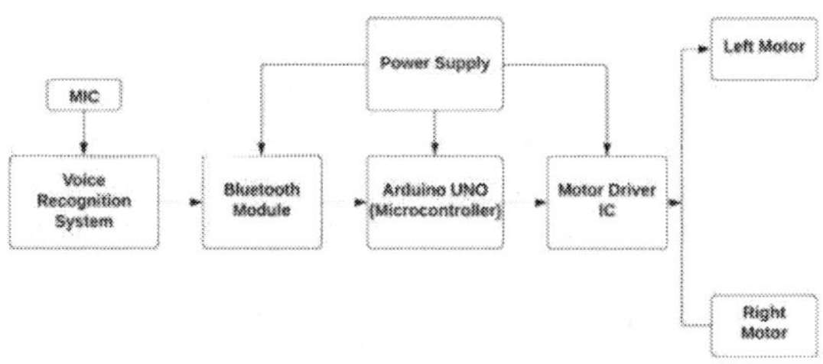

Fig 1. Block Diagram of Proposed System

The above diagram is a Block Diagram of the proposed system. There are three main components of the Voice Recognition System (VRS), Bluetooth module, and Arduino UNO as a Micro-controller. Others are Motor Driver IC, Micro-phone, Left and Right DC Motor. The first is a microphone is used to capture the voice command and is given to the Voice Recognition System as input. Voice Recognition System (VRS) is a model based on Deep Learning and Signal Processing. The main Function of VRS is to identify the user voice command and passed it to the Controller through Bluetooth [14]. Next is Arduino UNO as Micro-Controller has main functionality is to control all hardware-related components such as Motor Driver IC, Left Motor, and Right Motor. Arduino UNO is used as a computational-processing device. It is connected to both motors through driver IC. Last is Bluetooth Module acts as the communication medium between Microcontroller and Voice Recognition System (VRS). It is connected to Arduino as it requires the power supply and Processing Power [12]. The workflow in the proposed system is an audio signal as the input given to the Voice Recognition System (VRS) after the identification of the voice command, a Bluetooth connection is established between the micro-controller and VRS, and the output of VRS is given to the micro-controller which has a set of instructions to control both motors through Driver IC which is based on the output of Voice Recognition System.

IV. SYSTEM IMPLEMENTATION

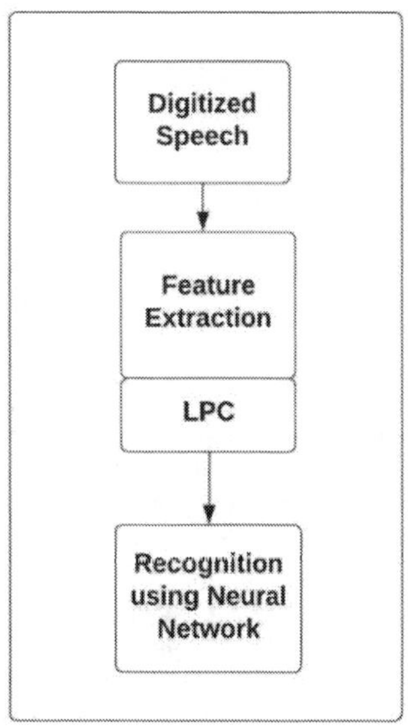

Fig. 2 Flow Diagram of Flow diagram of an Automatic Speech Recognition System

To design a prototype that can think like a human brain is a difficult task. Many Machine Learning algorithms had been developed to build a system that can replicate a human brain. One of the most popular techniques among them is artificial neural networks. The ANNs are the sequence of the algorithms used to train themselves i.e. systems to recognize different patterns that may be in image or audio form. Different types of algorithms can be used to implement the neural network. In the proposed work, Back Propagation Algorithm and LPC (Linear prediction coefficients) feature extraction is used. The back Propagation Algorithm is the mechanism to train neural networks [11]. This algorithm is used to compute gradient descent by randomly initializing the weights. It is also called the backpropagation ion of errors. It mainly consists of three layers input layer, hidden layer, and output layer as shown in figure 3. Simple calculations like vector addition, multiplications, and partial differentiation are computed to predict the output. BPA first starts training by random weight and outputs in a feed-forward way and then starts calculating the error in an inverse way from The last node to the first node. Everything in the initial stage is an assumption and based on the biases and weights are set. Each output node is computed using the sigmoid function to fit the output in the range of 0 to 1. The main reason for deploying the backpropagation algorithm in our proposed work is they are quick, straightforward, and simple to program. In this Speech Recognition system, we are sampling the speech signals to 8 kHz.

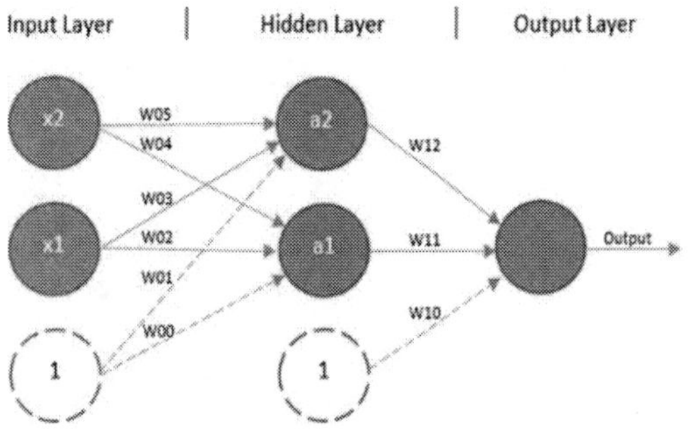

Fig. 3 Structure of Backpropagation Algorithm

This signal is fed as an input to feature extractors to extract the different patterns from voice samples. The voice samples are stored in wav format with 8 bits per sample, 1 channel, and 30 words per command. In total, we are making use of 150 voice samples for training our neural networks. Back Propagation algorithm is deployed on Matlab to train the neural networks. Our proposed model has 13 nodes in the input layer due to the feature vector size of 13 obtained from the LPC algorithm. As we are classifying a total of 5 classes in the output the number of nodes in the output layer is 5. For indicating the input phase, every output will be set to 0, and only one output is kept to 1. We choose 6 numbers of nodes for the hidden layer. In this algorithm, we are making using two matrices D and X as input. Matrix X represents the vector-matrix of sample input. It is a K X M matrix where K denotes no input samples and M denotes the size of the feature vector. D is a dummy matrix used to assume the output in the same pattern. It is also a K X M matrix where K denotes the no of input samples and M denotes the number of output classes to classify i.e. 5. A weight matrix is designed as a sequence of vectors (w0,w1,w2,..,wn). This matrix is present between each layer. We initially randomize these weights from random values in the range of -1 to +1.

All the activation vectors (x1, x2, xn) and net vectors (n1,n2, nn) are set to 1, and all the weight vectors to zero. Both matrices are K X M matrices. i.e. 150 X 5 matrices. From the net value, the output is calculated using the tan(h) sigmoid function. The prediction between the actual and assumed output values must be equal. This error value is computed by summing all the output errors. We had continued the training process until the minimum error of0.01 is achieved. The entire dataset is passed from all networks for and through 3000 times. A total of 150 voice samples were recorded from users. Out of 5 classes, each class is allotted 30 words per class. This input is fed to the neural networks where an LPC extractor is used to extract the feature. The network is trained using both feed-forward as well as a backpropagation method. After training the network, user voice samples which are taken by microphone tested with trained audio samples, and voice recognition take place. After the voice recognition is done successfully the recognized word is converted to text as shown in figure 5.

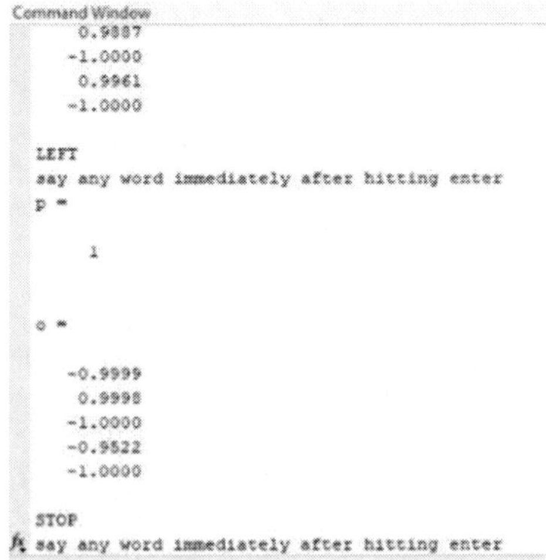

Fig. 4 Speech to text on MATLAB ENVIRONMENT

This text is then transferred to the target hardware in form of letters for eg (F, B, L, R, and S) wirelessly using a Bluetooth module. The Arduino code is written using the Arduino IDE platform where logic is assigned to rotate the motor based on voice commands. In the proposed system, Arduino UNO is used as a micro-controller to control the peripherals of the system. Motor Driver IC is used to provide sufficient current for the operation dc motor. The DC motor is controlled based on the user's voice command. The Bluetooth Module acts as communicating medium between the neural network and micro-controller.

V. EXPECTED RESULT & IMPACT

Based on the entire process of implementing a prototype of a voice-controlled wheelchair has been implemented. The prototype of the circuit and model are shown in the images below.

Fig. 5 Implementation of Proposed Circuit Model

Fig. 6 Proposed model of Voice Controlled Wheelchair

VI. CONCLUSION

This work tackled the speech recognition problem by applying the Backpropagation algorithm to neural networks. This proposed method has a lot of contributions to making physically challenged people's life comfortable. The goal behind this project is to contribute a simpler way of developing an Automatic Voice Controlled Wheelchair System by overcoming many traditional methods. This project contributes a novel technique that eliminates the use of complicated algorithms, high costs, and huge datasets. The system is designed in such a way that the user should feel comfortable without giving stress to any of his body parts. While working on this prototype several research objectives have been decided by us to make the system more efficient. In future work, research planning is to deploy the entire Matlab code directly to the target hardware using the Matlab coder. This will yield us many benefits such as the elimination of the Bluetooth module and controlling system. Thus a low-cost and accurate autonomous model can be implemented that can make the operation easier. Thus, this paper concludes that in this work, a design is provided that is efficient in helping physically disabled people without putting their strengths and efforts to pull the wheelchair, by commanding it on their voice.

VII. REFERENCES:

1) P. P. Dutta et al., "Design and Development of Voice Controllable Wheelchair," 2020 8th International Conference on Reliability, Infocom Technologies and Optimization (Trends and Future Directions) (ICRITO), 2020, pp. 1004-1008, doi:10.1109/ICRITO48877.2020.9197765.
2) S. Umchid, P. Limhaprasert, S. Chumsoongnern, T. Petthong, and T.Leeudomwong, "Voice Controlled Automatic Wheelchair," 2018 11th Biomedical Engineering International Conference(BMEiCON), Chiang Mai, 2018, pp. 1-5,doi:10.1109/BME iCON.2018.8609955.
3) S-.Y. Cho, A. P. VINOD and K.W.E. CHENG, Towards a Brain-Computer Interface Based Control for Next-Generation Electric Wheelchairs", 2009 3rd International Conference on Power Electronics Systems and Applications, Digital Reference: K210509123.
4) Rini Akmeliawati, Faez S. Ba Tis, Umar J. Wani., Design and Development of a Hand-glove controlled Wheel Chair", 2011 4th International Conference on Mechatronics (ICOM), 17-19 May 2011, Kuala Lumpur, Malaysia, pp. 1-5.
5) Mohammed Faeik Ruzaij and S. Poonguzhali, "Design and implementation of a low-cost intelligent wheelchair," 2012 International Conference on Recent Trends in Information Technology, Chennai, Tamil Nadu, 2012, pp. 468- 471, DOI:

10.1109/ICR-TIT.2012.6206809.

6) Umar J. Wani. Rini Akmeliawati, Faez S. Ba Tis,

May2011, Kuala Lumpur, Malaysia, 2001.

7) Zhang-fang Hu1 Lin Li2. and Yuan Luo3 Yi Zhang4 Xing Wei5, \A Novel Intelligent Wheelchair Control Approach Based On Head Gesture Recognition", 2010 International Conference on Computer Application and System Modeling (ICCASM 2010), pp. V6 157-163.

8) T. Lu, "A motion control method of intelligent wheelchair based on hand gesture recognition," 2013 IEEE 8th Conference on Industrial Electronics and Applications (ICIEA), Melbourne, VIC, 2013, pp. 957-962, DOI: 10.1109/ICIEA.2013.6566505.

9) D. Puanhvuan, S Khemmachotikun, P. Wechakam, B. Wijam, and Y. Wongsawat, "Automated navigation system for eye-based wheelchair controls," The 7th 2014 Biomedical Engineering International Conference, Fukuoka, 2014, pp. 1-4, doi:10.1109/BMEiCON.2014.7017444.

10) D. Puanhvuan, S Khemmachotikun, P. Wechakam, B. Wijam, and Y. Wongsawat, "Automated navigation system for eye-based wheelchair controls," The 7th 2014 Biomedical Engineering International Conference, Fukuoka, 2014, pp. 1-4, doi:10.1109/BMEiCON.2014.7017444.

11) S. R. Avutu, D. Bhatia and B. V. Reddy, "Voice Control Module for Low-Cost Local-Map Navigation Based Intelligent Wheelchair," 2017 IEEE 7th International Advance Computing Conference (IACC), Hyderabad, 2017, pp. 609- 613, DOI: 10.1109/IACC.2017.0129.

12) S. U. UPase, "Speech recognition based robotic system of a wheelchair for disabled people," 2016 International Conference on Communication and Electronics Systems (ICCES), Coimbatore, 2016, pp. 1-5, DOI: 10.1109/CESYS.2016.7889851.

13) 12) S. A. M. S. Sheikh and D. R. Rotake, "An evolutionary approach for smart wheelchair system," 2015 International Conference on Communications and Signal Processing (ICCSP), Melmaruvathur, 2015, pp. 1811-1815, DOI: 10.1109/ICCSP.2015.7322836. 13) B. Borges, A. Chandra, R. Kalantri, S. Gupta, G. Dsilva and S. Rajguru, "Android Controlled Wheelchair," 2018 First International Conference Secure Cyber Computing and Communication (ICSCCC),Jalandhar, India, 2018, pp. 1-7, doi:10.1109/ICSCCC.2018.8703345.

Design and development of a hand-glove controlled wheelchair," 2011 4th International Conference on Mechatronics (ICOM), 17-19

14) 14) S. Soma, N. Patil, F. Salva, and V. Jadhav, "An Approach to Develop a Smart and Intelligent Wheelchair," 2018 9th International Conference on Computing, Communication and Networking Technologies (ICCCNT), Bangalore, 2018, pp. 1- 7, DOI: 10.1109/ICCCNT.2018.8494050

Electric Heavy Vehicle Transportation Highway (By Catenary System)

Suyash Dhayarkar, *Suraj* Sontakke, *Vipul* Shinde, and *Nilam* Ghuge

Department Of Electrical Engineering, JSPM's Bhivarabai Sawant Institute of Technology & Research, Pune, India.

Abstract This paper provide a review of the latest work in the sector of electric vehicle. The transport Sector massively depends on Non-Renewable Sources. Continuous release of harmful substances in the surrounding by the vehicle should be confined and restore with alternative transportation. Electric Highway is one of efficient Solution. It is modern technology of electrical road systems. It is a technique in which electricity is taken from power grid through dynamic Pantograph which is attached with overhead transmission line. Thus, EV's with a combination of Electric Highway can charge the battery in dynamic motion and reduce the time of recharging of a vehicle. Shortly, the paper introduces some features of EV, their limitation and one of the effective alternate modern solution considering the well-being of the environment.

1. INTRODUCTION

As electric vehicle boost the automobile sector but having some demerits like the Electrical heavy vehicle is not preferable for long distance travelling and it requires more charging time of the battery , also more battery capacity is required [3,5,10]. So the main aim of the paper proposal is to overcome these demerits . For that we are making the working model . The system is used in this model is similar to the system used in Electric train ,i.e. catenary system (In that system the current is collected from the overhead transmission line with the help of pantograph and gives to the motor and vehicle will runs) [2].Vehicles like trucks and hybrid trucks can instantly charge it's battery using the power line [2]. The sensors inside the trucks takes for an electric contact line then truck automatically lift it's pantograph and contact with electric line or else if there is no such sensor the driver manually set the position. The pantograph transfer these energy to electric motors as well as the charge the battery .When there is no power line the truck can runs on battery . Due to E-highway zero emission possible reduces CO2 emission [13] .Currently the E-highway are more flexible to truck transport as these are the major transports vehicle for goods and services [1,4] . Even the existing roads can be upgraded with E-highway structure because of this project crude oil imports can be reduce and save the cost of fuel [1,4].

2. OBJECTIVE

- EV's are able to travel for long distances [5].
- No extra Charging time is required.[9,14].

3. LITERATURE SURVEY

1. The world's originally zapped street (June 2016) - which charges the batteries of electric vehicles and trucks as they roll over it - has opened close to Stockholm, Sweden [4,10]. The street extends 1.2 miles and is important for an administration driven arrangement to zap almost 12,500 miles of roads and roadways the nation over [10].

2. In August 2017, Siemens Mobility was commision by the province of Hesse to fabricate an upward contact line for charged cargo transport on a ten-killometer stretch of expressway. With this field trail, the E-Highway is being tried on a public parkway in Germany interestingly.

3. The ELISA (ELektrifizierter, Innovativer Schwerverkehr auf Autobahnen that means zapped, inventive weighty vehicle traffic on motorways) was the primary show to be initiated in May 2019. This venture, drove by the Road Authority Hessen Mobil, is situated between the urban areas of Darmstadt and Frankfurt on motorway A5 [4,6,10].

4. PROPOSED SYSTEM

An Electric Road System (ERS) in which electrical energy is transferred during motion from the road to the electric vehicle for both propulsion and charging of battery which is a technology concept with great potential for reducing dependency on conventional sources and increased energy efficiency in the transport sector [1,9,12] as shown in fig.10.ERS is stated as dynamic power transfers from the road to the vehicle when the vehicle is in running condition and could be achieved through different power transfer technologies from the road to the vehicle [1,6]. In this paper we have used the Catenary System technology on which Electric train runs , however we have used Two Line AC System (230 AC). This power is collected through Pantograph and Rectifier is used to convert 230 AC into 12 volt DC (Fig 7) and by utilising this electrical Power, battery is charged and able to drive the vehicle [2,7] as shown in fig 1. In this technology we can used small capacity battery (12V-7AH lead acid battery)(Fig.8) hence overall weight as well as initial cost of the vehicle is also reduced. It reduces the Static charging time of the electric vehicle and increase the range of travelling distance of heavy vehicle. When Pantograph is disconnected or any obstacles in overhead system does not affect the operation of the vehicle because change over takes place and battery will be ON state and further drive the Vehicle. The Model works on the following three conditions that is

1)If the Supply is ON that time vehicle will run and battery will charge simultaneously [6,9].

2)If the supply is Cut OFF that time vehicle will working on battery supply only [6,9].

3)If the vehicle is on standstill position that time battery will be charging [9].

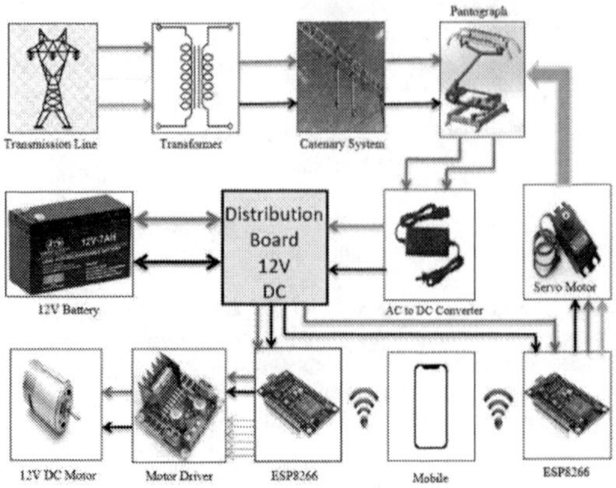

Fig 1. Wiring diagram.

4. METHODOLOGY

In our project we have given 230V AC to overhead Transmission line which is taken through supply. When pantograph is connected to transmission line, current is transfer to the rectifier (to convert AC current to DC current) and then provide to motor of the Vehicle [2,7]. Simultaneously current is also provided to battery for charging. It means at the same time vehicle is running on the supply as well as battery is charging [6,10]. We have used change over that is Switch for operational changes. Whenever the overhead supply is turned off, battery will be on and drive the vehicle without any delay in the motion of the vehicle [6]. For the operation of vehicle we have used Rover Controller which is control by software in the mobile as shown in fig 2. Vehicle wheels movements in either forward or reversed direction and it left or right directions is controlled through mobile app. Similarly the operation of pantograph (i.e. lifting up and down) is done through software on mobile app. We have provide indicator on the vehicle which gives indication if pantograph is connected or not. We have used battery indicator which also gives voltage reading to indicate amount of battery charge in case of pantograph is in operational state and vice-versa (Fig.9).

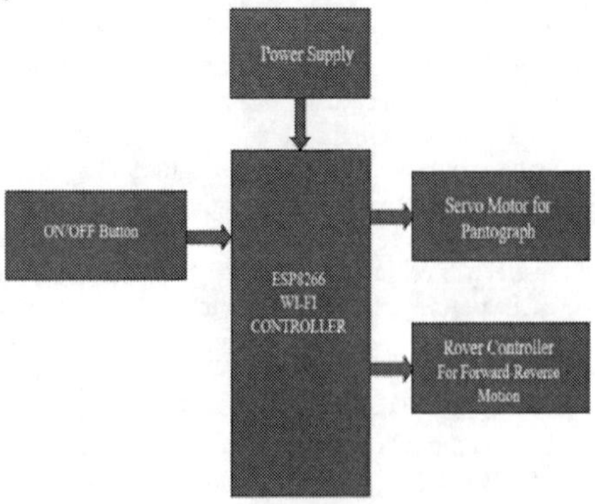

Fig 2. Block diagram .

5. COMPONENT DESCRIPTION

5.1 PANTORGAPH

Used to collect current through overhead line and given to further for operation. Pantograph directly transfer the current from the overhead transmission line to electric motor of the truck [2].

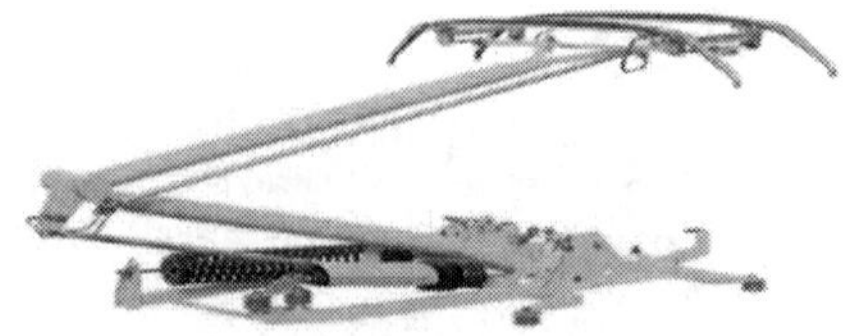

Fig 3. Sample Pantograph

5.2SERVO MOTOR

This Tower Pro MG996R Digital Metal Gear High Torque Servo Motor (180 Degree Rotation)features metal equipping coming about in extra11kg slowing down force in a little bundle. The outfitting and engine have likewise been moved up to work on dead data transmission and jogging. The unit comes total with 30cm wire and 3 pin 'S' type female header connector that fits most collectors, including GWS, Cirrus, Corona, Berg, Spektrum and Hitec.

Specifications for Digital Metal Gear High Torque Servo Motor of Tower Pro MG996R

- Operating voltage: 4.8 V - 7.3 V
- Rated Current : 500 mA – 900 mA (6V)
- Stall Current : 2.5 A (6V)
- Temperature range in degree celcius :[0 °C – 55 °C]
- Control System: Analog
- Operating Angle: 180degree
- Required Pulse: 900us-2100us
- Operating speed: 0.18 s/60° (4.8 V), 0.14 s/60° (6 V)
- Stall torque: 9.9 kgf·cm (4.8 V), 10.9 kgf·cm (6 V)

Fig 4. Servo Motor.

5.3 ESP8266

ESP8266 is a computer on a chip. It is an integrated chip that is usually a part of an embedded system [8,11]. It is a self contained, independent and yet function as a tiny, dedicated computer [8,11]. It also supports IOT Applications due to built-in Wifi [8,11] (Fig.5).

ESP8266 SPECIFICATION

- Microcontroller: Tensilica 32-bit RISC CPU Xtensa LX106
- Operating Voltage: 3.3V
- Input Voltage: 7-12V
- Digital I/O Pins (DIO): 16
- Analog Input Pins (ADC): 1
- UARTs: 1
- SPIs: 1
- I2Cs: 1
- Flash Memory: 4 MB
- SRAM: 64 KB
- Clock Speed: 80 MHz
- USB-TTL in view of CP2102 is incorporated locally available, Enabling Plug n Play .
- PCB Antenna Small Sized module to fit insightfully inside your IoT projects .

Fig 5. ESP2866 Microcontroller .

5.3 ROVER CONTROLLER

Rover control is used to move robot in forward or reverse direction or in left or right direction for moving to robot in desired direction rover is used [11].The pin connection of Rover Controller and ESP8266 (Fig. 6).

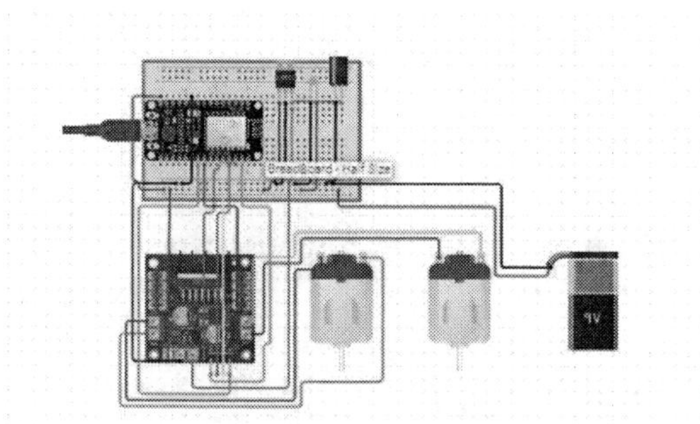

Fig 6. Rover controller

5.4 RECTIFIER

It takes current from overhead transmission line (230 V AC) and converts to 12V-1.7 AH DC and provides to the motor of the vehicle.

Fig 7. 12 V-1.7A Adapter

5.5 POWER SUPPLY

Provide power to E-Vehicle to drive the motor as well as to Arduino of pantograph. When pantograph is connected to overhead line , battery is in recharging condition.[9,10,11]

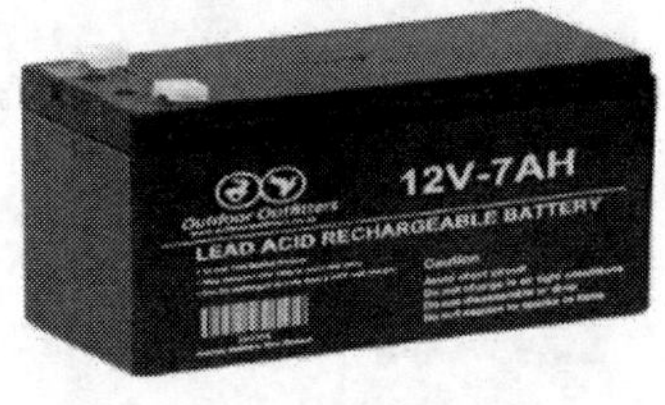

Fig 8. 12V-7AH Lead Acid Rechargeable Battery

5.6 BATTERY PERCENTAGE INDICATOR

This Indicator indicates the amount of battery charged as well as Voltage Rating.

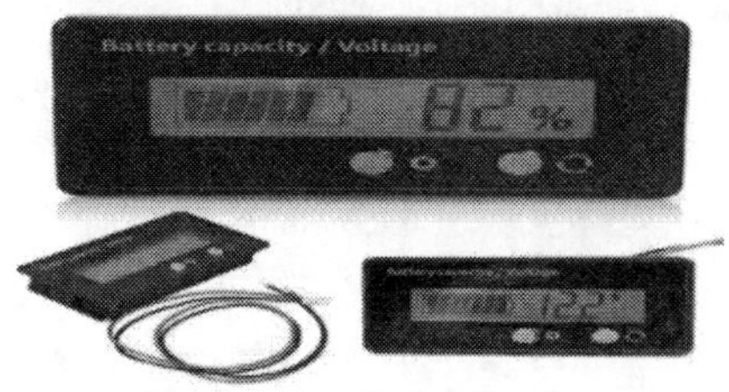

Fig. 9. Battery Percentage Indicator

6. RESULT , DISCUSSION AND PERFORMANCE ANALYSIS

This module provides a successful result which is obtained from proposed method. Vehicle is in operation due to overhead flowing current and simultaneously charge the battery. Indicators are used to show mode in which vehicle is running i.e, on battery or without battery on direct supply. Precautions are taken to avoid contact of AC and DC Equipment.
Hence, Heavy Vehicle long transportation is easily accessible and no extra charging time required for battery charging.

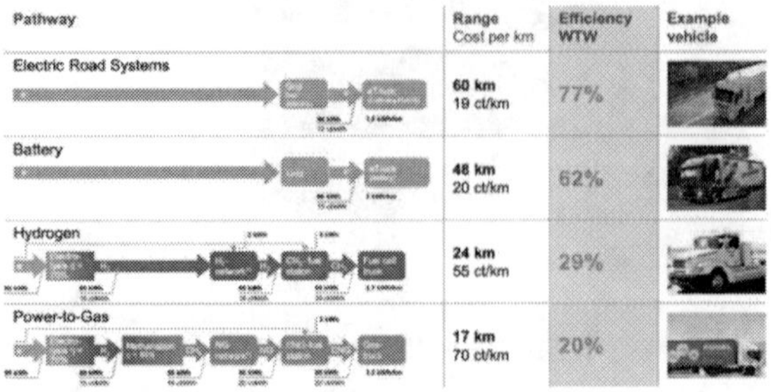

Fig. 10. Efficiency Of Vehicels

Fig. 11. Actual Model

7. CONCLUSION

Electric Highway framework not just works on the advantage of diminishing the unsafe Green house gases outflows into the air yet in addition works on the country's economy by its real System. This cutting edge energizing expressway is two times as rational as gas powered motors. Aside from this, it is a silent framework which turns out as an eco-accommodating strategy in transportation area. Contrast with that of Non-Renewable petrol based transportation , running expense is less in E-Highway Catenary System that is power utilization is less.

The world's first electrified road that recharges the batteries of cars and trucks driving on it has been opened in Sweden. Germany has presented the main electric expressway on its motorway framework to re-energize cross breed trucks as they drive. In Rail technology used in Germany, the path of vehicle cannot be changed while in catenary System we can change the lane in dynamic motion by taking pantograph downward. Also, the fault detection and fault recovery in catenary system is easy as compared to rail technology. During monsoon, the rail technology is less effective and not safe as water is clog on the surface of track.

The thought behind the e-expressway framework is to expand the effectiveness of jolted trucks and expanding the country's economy by lessening the utilization of non-renewable energy source. Thus, here we conclude that Electric heavy vehicle able to travel long distances as well as charging time of vehicle is reduced.

REFERENCES

1. Sundelin, Hakan; Gustavsson, Martin G. H.; Tongur, Stefan. [IEEE International Conference on Electrical Systems for Aircraft, Railway, Ship Propulsion and Road Vehicles & International Transportation Electrification Conference (ESARS-ITEC) - Toulouse, France] International Conference on Electrical Systems for Aircraft, Railway, Ship Propulsion and Road Vehicles & International Transportation Electrification Conference (ESARS-ITEC) - The maturity of electric road systems. , 1–5. doi:10.1109/ESARS-ITEC.7841380 (2016).

2. K.Chandra Mouli; N.Pannirselvam; V.Anitha, D.V.V.Nagasai pardhu. International Journal of Engineering and Advanced Technology (IJEAT) ISSN: 2249 – 8958, Volume-8, Issue-2S2, (January 2019).

3. Navidi, Thomas; Yue Cao, ; Krein, Philip T. [IEEE Power and Energy Conference at Illinois (PECI) - Urbana, IL] IEEE Power and Energy Conference at Illinois (PECI) - Analysis of wireless and catenary power transfer systems for electric vehicle range extension on rural highways., ,1–6. doi:10.1109/PECI..7459224 (2016).

4. Taljegard, Maria; Thorson, Ludwig; Odenberger, Mikael; Johnsson, Filip. IEEE Transportation Electrification Conference and Expo, Asia-Pacific (ITEC Asia-Pacific) - Harbin, China IEEE Transportation Electrification Conference and Expo, Asia-Pacific (ITEC Asia-Pacific) - Electric road systems in Norway and Sweden-impact on CO<inf>2</inf> emissions and infrastructure cost. , 1–6. doi:10.1109/ITEC-AP.2017.8080779 (2016).

5. Bischoff, Joschka; Marquez-Fernandez, Francisco J.; Domingues-Olavarria, Gabriel; Maciejewski, Michal; Nagel, Kai. IEEE 6th International Conference on Models and Technologies for Intelligent Transportation Systems (MT-ITS) - Cracow, 6th International Conference on Models and Technologies for Intelligent Transportation Systems (MT-ITS) - Impacts of vehicle fleet electrification in Sweden – a simulation-based assessment of long-distance trips. , 1–7. doi:10.1109/MTITS.2019.8883384 (2019).

6. Alakula, Mats; Marquez-Fernandez, Francisco J. IEEE Transportation Electrification Conference and Expo, Asia-Pacific (ITEC Asia-Pacific) - Harbin, China IEEE Transportation Electrification Conference and Expo, Asia-Pacific (ITEC Asia-Pacific) - Dynamic charging solutions in Sweden: An overview.1–6.doi:10.1109/ITEC8080781 (2017).

7. Kokkosis, A.; Barbarosou, M.; Charitopoulos, A.; Paraskevas. [Institution of Engineering and Technology 8th Mediterranean Conference on Power Generation, Transmission, Distribution and Energy

Conversion - Cagliari, Italy 8th Mediterranean Conference on Power Generation, Transmission, Distribution and Energy Conversion - A low voltage DC Power supply for fluorescent lamps with battery protection unit. ,58–58.doi:10.1049/cp.(2012).

8. Mesquita, Joao; Guimaraes, Diana; Pereira, Carlos; Santos, Frederico; Almeida, Luis [IEEE 23rd International Conference on Emerging Technologies and Factory Automation (ETFA) - Torino, Italy (2018.9.4-2018.9.7)] IEEE 23rd International Conference on Emerging Technologies and Factory Automation (ETFA) - Assessing the ESP8266 WiFi module for the Internet of Things. , 784–791. doi:10.1109/ETFA.2018.8502562 (2018).

9. Gautham Ram Chandra; Venugopal, Prasanth; Bauer, Pavol (2017). [IEEE 2017 International Symposium on Power Electronics (Ee) - Novi Sad, Serbia] International Symposium on Power Electronics (Ee) - Future of electric vehicle .charging.,,17. doi:10.1109/PEE.8171657(2017).

10. David Wenander; Philip Abrahamsson; Francisco J. Márquez-Fernández; Mats Alaküla AEIT International Conference on Electrical and Electronic Technologies for Automotive (AEIT AUTOMOTIVE).2022IEEE.DOI:10.23919/AEITAUTOMOTIVE52815.2021.9662755 (07 January 2022).

11. Sundaram, S. M. Proposal for establishing an Electric Highway to show the way for future of personal mobility. 2015 IEEE International Transportation Electrification Conference (ITEC). doi:10.1109/itec-india.2015.7386881(2015).

12. Aksjonov, Andrei; Nedoma, Pavel; Vodovozov, Valery; Raud, Zoja; Petlenkov, Eduard. [IEEE International Conference on Mechatronics (ICM) - Ilmenau, Germany] IEEE International Conference on Mechatronics (ICM) - Speed Control of Electric Vehicle Propulsion with Autotuning at Changeable Driving Modes and Road Conditions. , 584–589. doi:10.1109/ICMECH.2019.8722909 (2019).

13. Singh, Archit ; Hasselgren, Björn ; Madani Larijani, Hatef "Electric Road Systems- A feasibility study investigating a possible future of road transportation" by Archit Singh Master of Science Thesis EGI_-090, MSC, KTH Sustainable Energy Engineering, Energy and Environment, SE-100 44 STOCKHOLM(2016).

14. Liimatainen, Heikki; van Vliet, Oscar; Aplyn, David (2019). The potential of electric trucks – An international commodity-level analysis. Applied Energy,236(),804–814. Doi:10.1016/j.apenergy.(201 8.12.017).

ITM Web of Conferences **44**, 01005 (2022) https://doi.org/10.1051/itmconf/20224401005
ICACC-2022

Coal Mine Safety Monitoring and Alerting System with Smart Helmet

Mangesh Rudrawar[1*], *Shivam* Sharma[2**], *Madhuri* Thakur[3***], and *Vivek* Kadam[4****]
[1,2,3,4] Department of Instrumentation Engineering, Ramrao Adik Institute of Technology, Navi Mumbai, Maharashtra, India

Abstract. Traditional monitoring systems in coal mines are difficult to install, hazardous, and difficult to power. Because of the complexity of the mining environment and the wide range of operations performed incoal mines, it is vital to monitor and maintain the parameters in the background t increase the efficiency and safety of mineworkers. As a result, traditional monitoring methods cannot be relied on to ensure coal workers'safety. This research represents a ZigBee-based wireless monitoring system using a smart helmet. The presented wireless monitoring system is capable of detecting and transmitting critical parameters in coal minessuch as methane gas, high temperature, humidity, and fire. In an emergency, this monitoring system transmitsdistress signals. A buzzer will sound if emergency conditions are detected, and the monitored variables will be displayed on the user interface machine. Moreover, the Parameters are wirelessly transmitted to the controlroom, allowing people to determine the safety situation of the mine. This model is easily reprogrammable. Experiments have demonstrated the system's reliability and stability.

Keywords - Coal mine Safety, Monitoring system, IoT, ZigBee, Smart helmet.

1 Introduction

An accident that occurs while mining minerals or metals is known as a mining accident. Each year, tens of thousands of miners die in mining accidents, most of which occur in underground coal mining, while accidents also happen in mining [1]. Due to rock strata that are plain, typically incompetent rock, presence of CH4 gas, and coal powder, Coal mining is considered significantly more dangerous than hard rock mining. Most deaths nowadays occur in underdeveloped nations and rural areas of wealthy countries when safety precautions are not properly implemented [2].

As a result, it is critical to maintaining track of circumstances that might contribute to an accident to protect human safety. This project introduces a device used to keep track of the conditions within a coal mine, which might pose a threat to human life. The sensors andZigbee are employed for monitoring in this system. This system also incorporates a smart helmet with a panic button that each worker may activate individually. All sensors are linked to specific applications and send an alert signal when it exceeds a particular value. The alarm message is sent to the control room to take the necessary action quickly. If coal mine accidents are not adequately controlled, it may result in massive human fatalities. A monitoring system has been built to record the readings ofdangerous gases and risky working conditions. The information gathered is subsequently forwarded to professionals for analysis and action.

Over the years, it has been observed that coal mine contributes significantly to the country's rapid economic and social development. Scholars from around the world have undertaken substantial research to improve the degree of safety. Coal mines are an essential source of energy for human growth. Development. Mining failures may be exploited to increase security in the industry. Industrial series of failures or flaws are typically the cause of an accident. There are a number of practical measures available. Initiatives involving mines, such as the construction of safety lamps, can significantly impact. Law and creating a self-contained coal mine safety monitoring system are both in the works. The trajectory of China coal mine accidents over the last ten years were investigated, and the human elements involved in these accidents were investigated using multi-dimensional statistical analysis. The number of significant coal mine accidents and the number of people killed were constantly reducing, but occasional death accidents still accounted for most deaths. Human factors accounted for 94.09 percent of the causes of these incidents, with the willful violation, mismanagement, andflawed design accounting for 35.43 percent, 55.12 percent, and 3.54 percent, respectively [3].

A round penetrating synthetic pulse radar device to probe through a coal pillar in search of hidden structures or abnormalities was employed. To produce a velocity picture of the inside of the pillar, direct matrix inversion was performed. The reconstructed picture revealed the presence and position of a substance with a low velocity. A clay vein in the cores corresponded to the low-velocity material shown in the tomogram [4]. Technology alone will not be adequate to address the difficulties for a substantial number of individuals

* e-mail: man.rud.rt18@rait.ac.in, **e-mail: shi.sha.rt18@rait.ac.in, ***e-mail: mad.tha.rt18@rait.ac.in,
****e-mail: vivekkaddam@rait.ac.in

The dangers of coal mining; proper education and training arealso essential for mine safety [5].

The research was conducted for safety monitoring of mine roof integrity and dangerous gases in coal mines, fibre optic sensors have been developed and implemented. The inherent safety, multi-location, and multi-parameter monitoring features of the FOS-based mine danger detection system are unmatched. They might be utilized to create expert systems for the early identification and prevention of mining hazards [6]. Underground mines are prone to serious safety incidents. The development of safe mining technologies is critical. This problem can be solved with smart mining. One of the most important components in smart mining is the unmanned electric locomotive [7].

The rest of the paper is organized as follows: In section2 methodology of the proposed system has been explained. In section 3, a brief description of the system is given. Section 4 comprises of working of the monitoring system and smart helmet. Output result analysis and discussion have been specified in section 5, and the end section tells us about conclusions and the future scope of the project.

2 Methodology

The Arduino microcontroller is used to detect and monitor variables in a coal mine. Live readings are provided by the temperature sensor, humidity sensor, IR flame sensor, and gas sensor. A microcontroller and a transceiver are connected to all of these sensors. The data is sent to the microcontroller, and communication between the gateway and the specific node is done via Xbee WPAN IEEE 802.15.4. As mentioned previously, the data is sent to the control room via the XBee protocol. In an abnormal situation, an alert message is sent to the system, which is also displayed on an LCD screen connected to Arduino at the coalfield's entrance. A buzzer is also programmed and controlled with the help of Arduino, which activates at any abnormal reading detected by the above sensors.

A smart helmet is also computed with a microcontroller which simply has a push button connected to it. The microcontroller is also comprised of a ZigBee transceiver. This measure has been taken to maximize the safety of workers in coalfield. Whenever a worker needs medical attention or has any discomfort, a panic button can be used, which transmits a message in the control room about an emergency so that medical attention can be given to that worker.

3 Description of System

3.1 Block diagram

As shown in Fig.1, This block diagram is of a prototype that includes some sensors that would be needed in a coalfield and are connected to a microcontroller to detect environmental conditions. The Arduino Uno microcontroller was used and connected to LM35

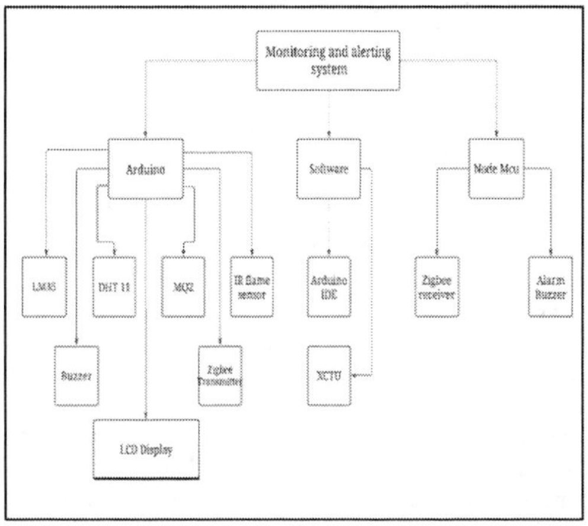

Fig. 1: Block Diagram Of Coal Mine Safety And MonitoringSystem.

(Temperature sensor), DHT11 (Humidity sensor), Mq2 (Gas sensor), and an IR flame sensor. With the help of these sensors, continuous readings of temperature, humidity, and gases present in the coalfield can be taken, as well as air density, and we've programmed the Arduino board in such a way that any abnormality in the parameters as mentioned above will trigger an alert, and the buzzer will be activated.

A 16x2 LCD display was added, which will display all of the readings. Workers will be able to monitor the live circumstances of the coalfield and take necessary action using this LCD, which will be installed at the coalfield's entrance. To create safe and dependable communication to the control room, the ZigBee Protocol has been employed. The environmental parameters of the coalfield are communicated with the aid of the XBee transmitter and receiver, which are coupled together with the help of XCTU software so that necessary measures may be done in time in critical situations.

Fig.2 below is a block diagram. Another safety gadget is a smart helmet equipped with a push-button, buzzer, and a ZigBee transmitter. At any sudden point of time, if any worker feels uneasiness or needs medical attention, that worker can use the push button, which will trigger the buzzer and transmit an emergency message to the controlroom via ZigBee.

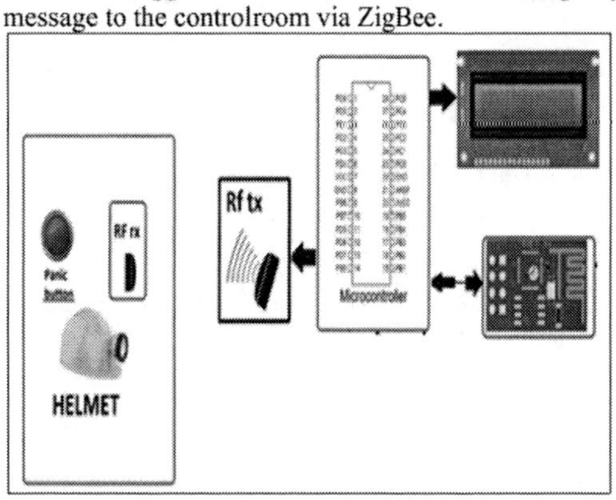

Fig. 2: Block Diagram of Smart Helmet

3.2 Sensors Specifications

This prototype is intended to monitor and measure environmental factors such as temperature, gas concentrations such as CH4, CO2, and humidity, as well as to detect inflammation or fire. Table 1 shows the sensors utilized in the prototype and their parameters. The LM35 sensor measures temperature since it is small and accurate to within 3°C [8]. An MQ2 gas sensor is used to detect gas concentration with an accuracy of 85%, while a DHT11 sensor is used to measure humidity with an accuracy of ±5% RH. Another sensor used in the prototype is an infrared flame sensor, which is used for fire detection and has a higher sensitivity and better accuracy than other types of flame sensors. These sensors are long-lasting, can withstand the harsh environmental conditions of the coalfield, and are cost- effective.

For wireless M2M and IoT networks that are low-cost and low-power, Zigbee is a wireless technology that is based on standards. In comparison to a WIFI network, ZigBee provides significantly lower data rates and use a mesh networking protocol to eliminate hub devices and build a self-healing architecture. This allows for the potential mixing of implementations from different manufacturers; however, it has been changed and improved. Zigbee is a standard protocol with a modest data rate and consumes less power [9].

Table 1: Specification table for selected sensors

Sr no.	Parameters	DHT11 Humidity sensor	MQ2 Gas sensor	LM35 Temperature sensor	IR Flame sensor	ZigBee Module
1	Sensing variable/Uses	Humidity	Gas concentration	Temperature	Fire	Transceiver
2	Operating Voltage	3.5V to 5.5V	5V	+4V to 30V	3.3V to 5V	2.1V to 3.6V
3	Operating Current	0.3 mA	800 mW	60 micro-A	15 mA	45 mA
4	Output Voltage	5.5 V	0 to 10 V	10 mV	Digital output (0 or 1)	_
5	Operating Temperature Range	0 to 50 degrees Celsius	-20 to 70 degree Celsius	-55 to 150 degrees Celsius	-25 to 85 degrees Celsius	-40 to 85 degree Celsius
6	Output/sensing Range	20% - 90%	300 to 10000 ppm	-55 to 150 degrees Celsius	760 to 1100 nm	4000 ft

3.3 Flowchart

Fig. 3 shows the operational flow of the monitoring system. Begin by connecting the prototype circuit to a power source. Now we'll have to keep track of all of the sensors' values attached to the Arduino UNO. Set a threshold value for each sensor after evaluating the readings and determining when environmental circumstances become aberrant.

The condition will be true if the sensor output is greater than the defined threshold value, and a buzzer will sound, readings will be displayed on the LCD screen with an alert message, and data will be transferred through Zigbee protocol to the control room. The Buzzer will not turn on if the sensor value is less than the defined threshold value, but the sensor readings will be shown on the LCD and relayed to the control room. This is when the flow of the procedure comes to a stop.

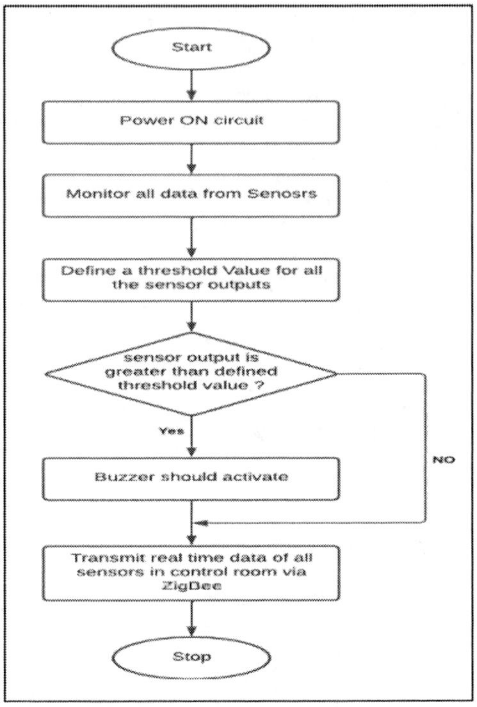

Fig. 3: Flow Chart For Controller Programming

4 Working

Arduino Uno is the central component, and it sends the information on the internet via the Zigbee module. DC power supply is used for supplying voltage to the circuit.This is the circuit diagram for the smart helmet's monitoring and alerting system. This system will be builtusing Arduino, ZigBee modules, and sensors such as the DTH11 humidity sensor and the IR flame sensor, among others. The sensors we'll use will detect changes in the coal miner's environment, and variables like temperature, gas concentration, and fire detection will be continuously monitored. It will also have an extension that will connect to an LCD display, which will be placed at the entrance tothe coal minefield. Also, the data which will be recorded by sensors will be transmitted via the Zigbee module to the control room so that proper actions can be taken effectively in minimum time. Fig.4 shows the representation of the circuit diagram designed.

A push-button, a buzzer, and a ZigBee transmitter will be included in another safety device, the smart helmet. If any worker becomes uneasy or requires medical attention at any time, that worker can use the push button to activate the buzzer and send an emergency message to the controlroom via ZigBee.

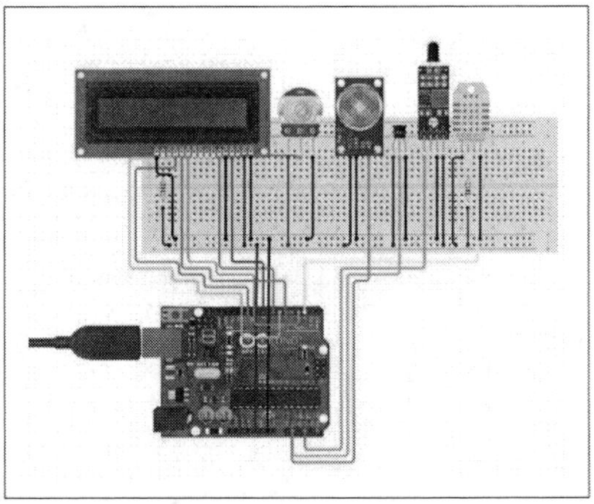

Fig. 4: Circuit Diagram of Coal Mine-Side Monitoring System.

5 Result and Discussions

IoT concepts and sensors are used to create a small, effective, and cost-efficient prototype. The temperature was successfully measured using LM35 sensors during the testing phase, and the output was satisfactory. The DTH11 sensor was also used to monitor humidity accurately. The MQ2 gas sensor detected gas concentrations in the area and successfully displayed a warning message on the screen. The IR flame sensor also detected fire and displayed an alert message informing the user of its presence. For wireless communication, the Zigbee Protocol was developed and tested in various locations and distances. Wireless communication was functional, and Zigbee delivered the best results and was fully functional. Prototype in coal mine can be seen in Fig.5, and prototype of control room can be seen in Fig 6. Smart helmet, on the other hand, was working correctly and displaying a help message on the screen.

During an emergency, the smart helmet communicates signals to the control room. The smart helmet is fitted with a push-button and a ZigBee module, as illustrated in fig.8. When the amount of carbon dioxide or methane in a coalfield rises, miners begin to feel dizzy. In this case, they can hit the panic button on their helmet, which alerts the workers in the control room and effectively reduces the time it takes for individuals to be rescued.

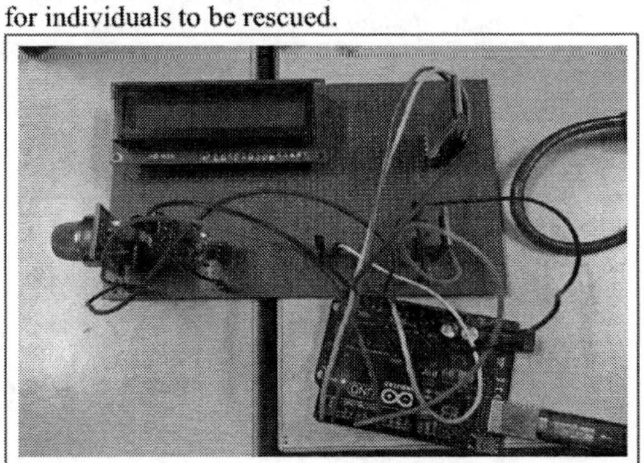

Fig. 5: Prototype Of Monitoring System in Coal Mine

Fig. 6: Prototype Of Monitoring System in Control Room.

The result obtained via the monitoring system has been analyzed and represented in Table 2 and Table 3, respectively, whereas results obtained are shown in fig 7. Whereas Table 2 describes the output of the monitoring system when conditions are under control, compares the output from all the sensors with their respective thresholds set and shows the hardware changes and their current state. Whereas table 3 describes output in an emergency situation and what hardware changes will take place in that situation. Fig.7. shows the actual readings obtained from the monitoring system prototype. Arduino UNO is programmed so that the readings of all the sensors are displayed on the user interface at an interval of 5 seconds. When the threshold value has crossed, warnings like 'high gas concentration', 'close fire', 'high humidity' and 'high temperature' are displayed.

Table 2 : Result analysis of monitoring system

Sr no.	Parameter	Threshold value set	Sensor reading	Hardware changes	remark
1	Temperature (degree Celsius)	25	21.09	Buzzer off	Safe condition
2	Gas (PPM)	600	518.2	Buzzer off	Safe condition
3	Humidity(%RH)	45	33	Buzzer off	Safe condition
4	IR flame sensor (digital output)	0	0	Buzzer off	Safe condition

Table 3 : Result analysis of monitoring system

Sr no.	Parameter	Threshold value set	Sensor reading	Hardware changes	remark
1	Temperature (degree Celsius)	25	31.05	Buzzer on	unsafe condition
2	Gas (PPM)	600	694.1	Buzzer on	unsafe condition
3	Humidity(%RH)	45	48.3	Buzzer on	unsafe condition
4	IR flame sensor (digital output)	1	1	Buzzer on	unsafe condition

```
temperature: 21.09 °c
gas density: 561.00
Humidity = 33.5
* Close Fire *
temperature: 21.14 °c
gas density: 561.00
Humidity = 33.5
* Close Fire *
temperature: 21.19 °c
gas density: 563.00
Humidity = 33.5
* Close Fire *
temperature: 21.14 °c
gas density: 561.00
Humidity = 33.5
```

Fig. 7: Result of Monitoring System

Fig. 8 shows the final prototype image of the smart helmet. During an emergency, the smart helmet communicates signals to the control room. The smart helmet is fitted with a push-button and a ZigBee module.When the amount of carbon dioxide or methane in a coalfield rises, miners begin to feel dizzy or Any feeling of uneasiness in health or injury in the body, they can hit the panic button on their helmet, which alerts the workers in the control room and effectively reduces the time it takes for individuals to be rescued as help will reach fasterdue to quick action.

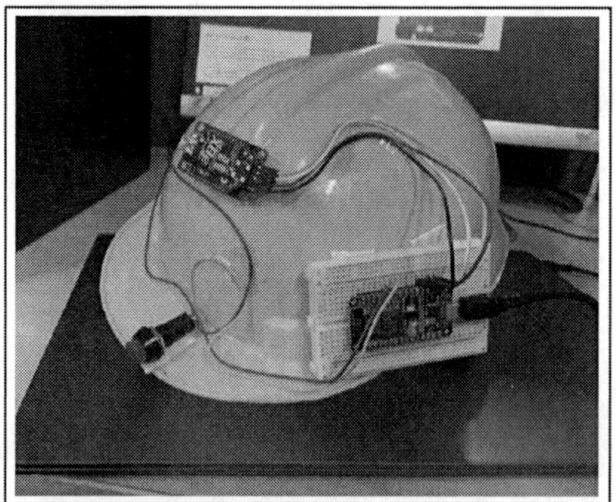

Figure 8: Prototype of Smart Helmet

Table 4 does the analysis of results transmitted through ZigBee. A push-button can have only two possible outputs, 0 or 1. When the push button is pressed, the output will be 1, and so the buzzer will get activated, whereas if the push button is not pressed by default, the output will be 0, and the buzzer will be in the deactivatedstate.

Table 4: result analysis of smart helmet

Sr.no	Sensor reading	Hardware changes	remark
1	0	Buzzer OFF	Safe condition
2	1	Buzzer ON	emergency

6 Conclusion

This paper presents a Zigbee-based coal mine monitoring system. The proposed system is used to monitor the subsurface characteristics of a coal mine and will aid in the prevention of mine disasters. ZigBee technology features a simple and adaptable networking concept, is tiny in size, and consumes little power. Additionally, the proposed system addresses all of the problems associated with wired connections. As a result, this is an advancement above the traditional coal mine safety system. Furthermore, a distress signal from the smart helmet is useful in abnormal environmental conditions or when a person needs medical attention. This system has applications in securities of industries as well. It is easy to upgrade and modify further as innovations arise.

References

[1] Y. Shi, J. Chen, J. Hao, J. Bi, M. Qi and X. Wang, "Statistical Analysis of Coal Mine Accidents of China in 2018," *2019 Prognostics and System Health Management Conference (PHM-Qingdao), (2019).*

[2] Ericsson, M., Löf, O. Mining's contribution to national economies between 1996 and 2016.

[3] Hong Chen, Hui Qi, Ruyin Long, Maolong Zhang, Research on 10-year tendency of China coal mine accidents and the characteristics of human factors, *Safety Science, Volume 50, Issue 4, (2012).*

[4] Dennen, R S, and Stroud, W P. Radar hazard Detection in a coal structure. *United States : N.p., (1991).*

[5] Hebblewhite, Bruce. (2009). Mine safety–through appropriate combination of technology and management practice. *Procedia Earth and Planetary Science. 1. 13-19. 10.1016/*j.proeps.2009.09.005, (2009).

[6] T. Liu et al., "Advances of optical fiber sensors for coal mine safety monitoring applications," 2013 International Conference on Microwave and Photonics (ICMAP), (2013).

[7] Jiang Y, Li Z, Yang G, Zhang Y, Zhang X. Recent progress on smart mining in China: Unmanned electric locomotive. *Advances in Mechanical Engineering. March 2017.*R. K. Kodali, T. Devi B. and S. C. Rajanarayanan, "IOT Based Automatic LPG Gas Booking and Leakage Detection System," 2019 *11th International Conference on Advanced Computing (ICoAC),(2019).*

[8] T. H. Nasution and L. A. Harahap, "Predict the Percentage Error of LM35 Temperature Sensor Readings using Simple Linear Regression Analysis," (2020) *4rd International Conference on Electrical, Telecommunication and Computer Engineering(ELTICOM),(*2020).

[9] V. Kadam, S. Jadhav, M. Parihar and A. Karande, "Development of wireless embedded automation system for a batch process," (2015) *5th Nirma University International Conference on Engineering(NUiCONE),(2015).*

IOT Based Wireless Data Technology Using LORA and GSM

VEDANT GANDHI[1,*], ADITI SARDAR[2], PARIDHEE WANI[3], RAKESH BORASE[4], and JAYANAND GAWANDE[5]
[1,2,3,4,5]Ramrao Adik Institute of Technology, Navi Mumbai, India

Abstract. Traditional agriculture is converting into smart agriculture due to the bulge of the IoT (Internet of things). An agriculture services platform is developed to support environmental monitoring and to improve the efficiency of agriculture management. Contemporary the Internet of things (IoT) is one of the highest promising application areas in information technology for forthcoming products and services. And, the agriculture field is changing expeditiously pointing to the future of automated and embedded systems with a bunch of sensors to monitor and curb the flourishing plants in a way to profit associated with it. However, one of the major issues of IoT is still a conversion between devices, notably on long-range. LoRa is lately accepted as auspicious communication technology, due to its properties such as long-range, two-way communication, and low cost. In this work, the LoRa technology is applied in the agriculture sector for making long-distance, low-cost communication. This work includes sensing and monitoring the parameters such as level, temperature, and humidity and also controlling the said parameters with the help of a remote unit. The proposed system is based on a high-performance microcontroller, integrated temperature and humidity sensor on real-time data analysis. The use of Data collection, pattern classification and apply strategic analysis then control execution for the final result.

Keywords: IoT, LoRa , Wi-fi, GSM

1. Introduction

Because of the rapid expansion of the Internet and the Internet of Things (IoTs), a variety of useful service applications are now being employed in a variety of industries. Smart agriculture is a new agricultural information and communication technology that was developed in recent years to suit the needs of farmers for data gathering, signal processing, data analysis, and equipment control. This paper presents a wireless sensor network and LoRa communication technology-based agricultural service platform [1,2]. To overcome the problem of communication failure and save energy, the work employs LoRa as a network transmission interface. To support environmental monitoring and increase farm management efficiency, a smart agriculture service platform is being created [3].

The purpose of this paper is to incorporate Internet of Things (IoT) awareness and communication technology into a smart agriculture platform [4]. Sensors of diverse types have their accuracies measured, and these sensors are then merged into a multi-function sensor component. Then, multi-functional sensor components and LoRa wireless network components are combined [5]. The goal of this work is to design and build a smart sensor network platform for agricultural applications.

Many people in our modern world make extensive use of technology, and as a result, they complete their tasks quickly and effectively. However, many rural farmers are still unable to use technology. In addition, due to an increase in the world population, the need for food is increasing. At the time, IoT (Internet of Things) was tending to create particular agricultural approaches in order to boost food output. Farmers can also acquire helpful information about moisture and soil requirements, among other things. As a result, we're putting in place this smart agricultural system for them. Farmers can use this system to check the temperature, humidity, and soil moisture of their farm, which is done using various IOT sensors such as DHT and soil moisture, as well as regulate various components such as motors, and so on. This system is incredibly straightforward to use and operates entirely on wireless technology. To use this system, the farmer must install transmitter modules at various locations throughout his field, as well as a receiver in his home that is connected to the server. The farmer can now use the internet or a mobile application to monitor and control the system. As previously stated, this system is built utilising wireless technology, and we currently have a

* Corresponding author: vedantgandhi18@gmail.com

number of wireless protocols such as BLE (Bluetooth Low Energy), Wi-Fi, and cellular, among others. However, these technologies are not appropriate for IoT sensor nodes, which must broadcast data over a long distance while utilising little power and at a cheap cost [6,7]. Apart from that, LoRa technology is quite important in agriculture. That's why we're relying on LoRa technology, which allows us to send data or information across vast distances at a low cost and without the need for an internet connection. Long Range is the abbreviation for LoRa [8]. LoRa is a wireless technology that transmits data in both ways over great distances. LoRa typically has a range of 15 to 20 kilometres and may operate for a year on a single charge. In India, the LoRa Technology licencing frequency band is 865 MHz to 867 MHz[9].

2. METHODOLOGY

Agricultural data is currently collected using a variety of sensors, including soil moisture, light intensity, and air temperature and humidity. However, because such sensors lack an integrated computing platform, they can only perform environmental monitoring and data collecting; as a result, the data collected cannot be efficiently exploited or processed. An agricultural intelligence platform is being investigated and developed to overcome this challenge. The objective of this work is to create a smart agricultural platform that combines communication and multi-functional sensing. The intelligent agriculture platform may be used to collect environmental data in agricultural areas and communicate it to remote computers using LoRa or another IoT strategy, such as sending data over the internet.
The goal of a smart agricultural system that uses LoRa wireless technology is to monitor and regulate the field's temperature, humidity, moisture, and motor via a website and an Android mobile application that uses GSM technology. Without an internet connection, field data will be transmitted using LoRa technology.

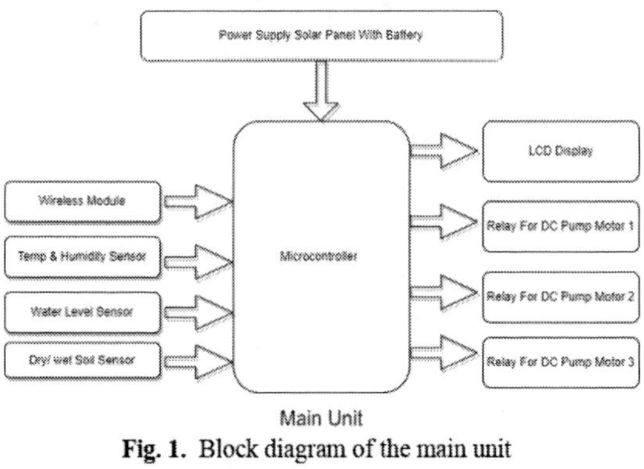

Fig. 1. Block diagram of the main unit

The main device, which includes a microprocessor, power supply solar panel with battery, wireless module ie lora RA-02, temperature and humidity sensor, water level sensor, dry and wet soil sensor, and temperature and humidity sensor. On the main unit, we've also attached a lcd display, three relays, and a dc motor pump. We attached a wireless module, the LORA RA-02, switches, and a microcontroller to the remote control unit. GSM MODULE 800I, battery, lcd display, buzzer, and led The LORA RA-02 is a wireless data monitoring and control device.

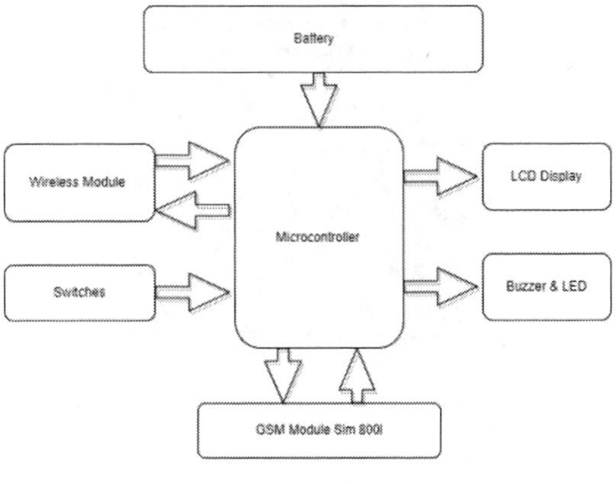

Fig. 2. Block diagram of remote module

The design is split into two parts: the main unit and the remote unit.
The major goal of the initiative is to make agricultural production easier.
Farmers can handle their agricultural fields in remote places where there is a lack of power and energy with the help of this concept. The major goal of this paper is to make farming easier at a low cost and to make the process work without the use of electricity.
Manpower is being conserved.
Quality and efficiency have both improved.
Working in a hostile atmosphere is a must. As the temperature of the field rises, the relay will activate and display the value on the LCD screen; motor 1 will activate and sprinkle water, lowering the temperature value.
The motor, relay, and buzzer light will switch on when the water level in the sensor falls below a specified level. In order to fill the tank to the desired level, the motor will deliver water to the field. When the dry wet soil sensor determines that there is insufficient water in the soil, motor 3 is activated and water is supplied to the soil.

The LORA RA-02 will be installed on the main unit as well as the remote unit.

https://doi.org/10.1051/itmconf/20224401006

3. EXPERIMENTAL SETUP

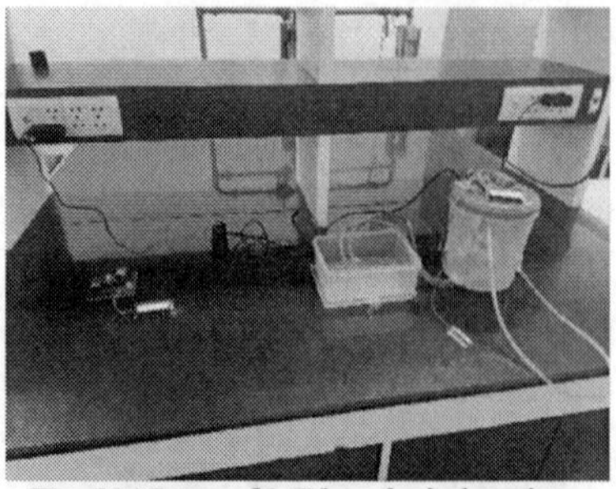

Fig. 3(a) Setup of IoT based wireless data technology using LoRa and GSM

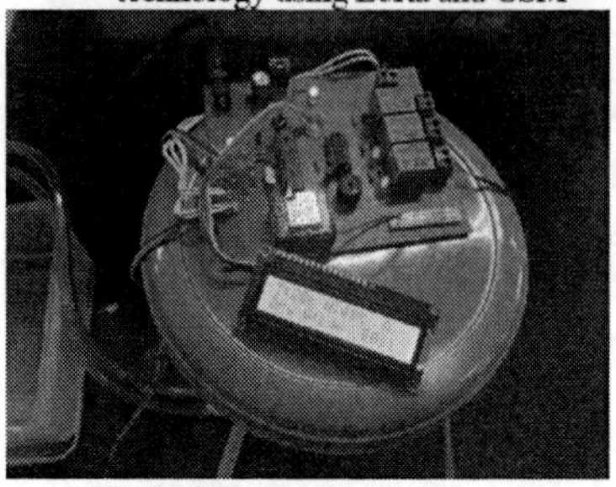

Fig.3(b) Controller Unit

The controller unit includes a microprocessor, power supply, battery, wireless module ie lora RA-02, temperature and humidity sensor, water level sensor, dry and wet soil sensor, and temperature and humidity sensor as shown in figure 3(b). On the main unit, we've also attached a lcd display, three relays, and a dc motor pump. We attached a wireless module, the LORA RA-02, switches, and a microcontroller to the remote control unit. GSM MODULE 800I, battery, lcd display, buzzer, and led The LORA RA-02 is a wireless data monitoring and control device.

Fig.3(c) Remote unit

The remote unit consist of LORA RA-02, microcontroller, GSM module 800I, LCD Display, 4 Switches to control the system in manual mode as shown in figure 3(c). Also, there is a switch connected for the real data to be uploaded on the web page.

The design is split into two parts: the main unit and the remote unit. The major goal of the initiative is to make agricultural production easier. Farmers can handle their agricultural fields in remote places where there is a lack of power and energy with the help of this concept. The major goal of this paper is to make farming easier at a low cost and to make the process work without the use of electricity. Manpower is being conserved. Quality and efficiency have both improved. Working in a hostile atmosphere is a must. As the temperature of the field rises, the relay will activate and display the value on the LCD screen; motor 1 will activate and sprinkle water, lowering the temperature value. The motor, relay, and buzzer light will switch on when the water level in the sensor falls below a specified level. In order to fill the tank to the desired level, the motor will deliver water to the field. When the dry wet soil sensor determines that there is insufficient water in the soil, motor 3 is activated and water is supplied to the soil. The LORA RA-02 will be installed on the main unit as well as the remote unit.

The farmer will be able to control the field using a remote module and a GSM SIM 800I. In the main unit there is a switch for auto mode and manual mode if the farmer turn on the manual mode he can control and analys the real time data within 1km of range. As we can see, there are buttons on the website for mannual and auto mode; in auto mode, we can see the live status of our project; in manual mode, if the operator is away from the field, he can change the setting to manual mode and control the field; in manual mode, we can see the live status of our project; in manual mode, we can see the live status of our project; in manual mode, we can see the live status of our project; in manual mode, we can see the live status with the help of buttons on the remote control and the web page By upgrading antenna design, the LoRa module has extended its range even farther. For the unique wireless sensor network architecture, the modify web-page design for real-time data analysis. In a different approach, the number of parameters increases as the number of sensors increases, resulting in real-time data with great accuracy.

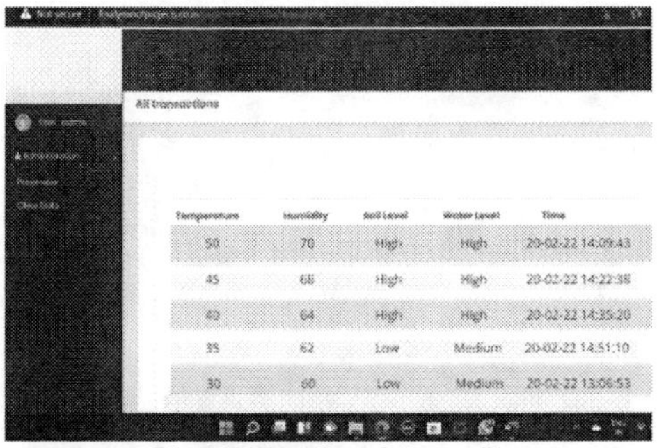

Fig.4(a) Real time data reading

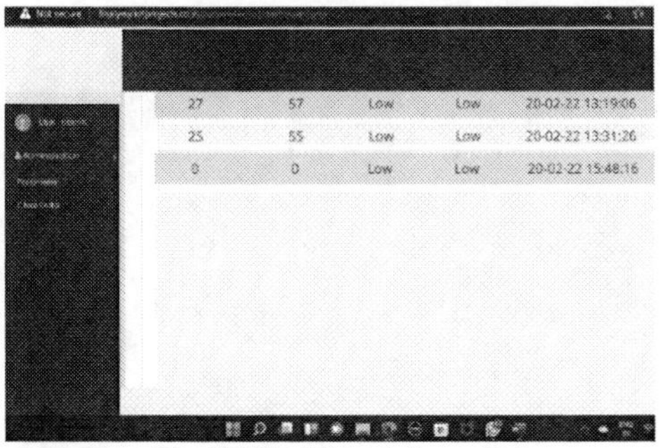

Fig.4(b) Real time data reading

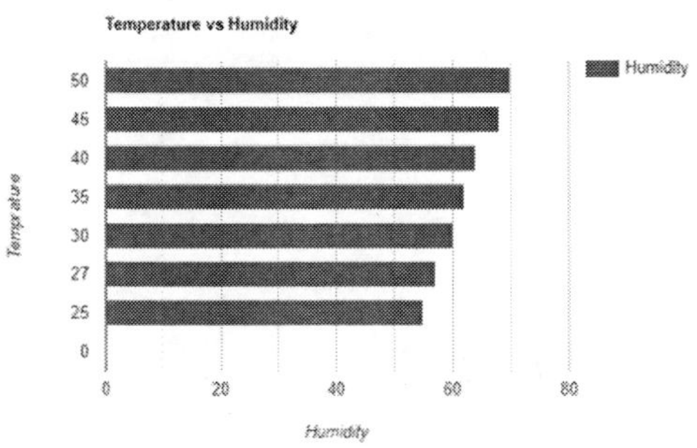

Fig.4(c)Graph of temprature vs humidity

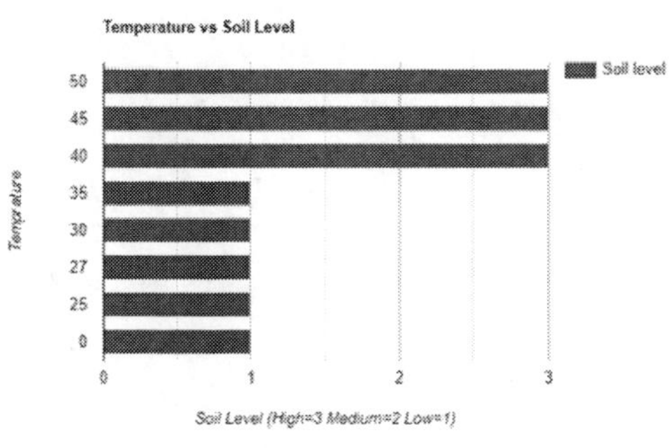

Fig4.(d)Graph of temprature vs Soil level

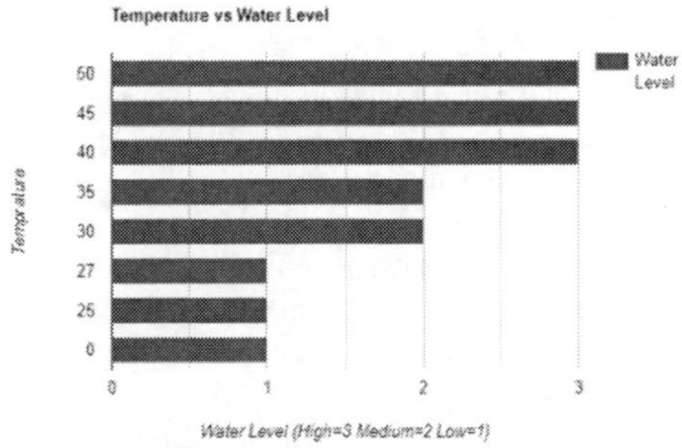

Fig4.(e)Graph of temprature vs water level

Fig5.(a)IoT website

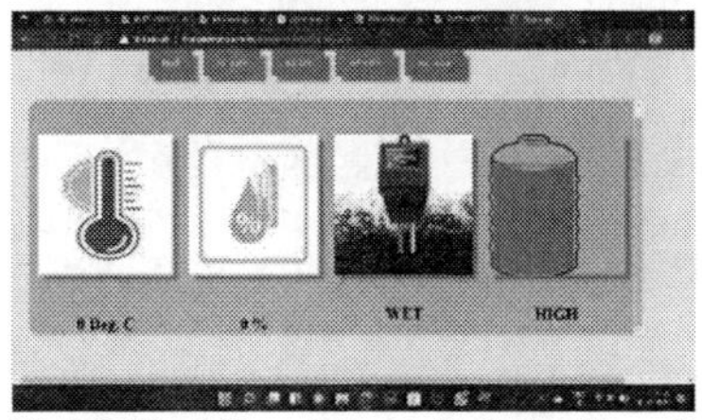

Fig5.(b)Real time data analysis dashboard

4. CONCLUSION

In the proposed work, the LoRa technology is effectively utilised for agriculture application for making long-distance, low-cost communication. The results shows the efficacy of proposed work that includes sensing and monitoring of the parameters such as level, temperature and humidity and also controlling of the said parameters with the help of remote unit. The proposed system is based on a high-performance microcontroller, integrated temperature and humidity sensor on real-time data analysis. The key benefit of the proposed work is that we can control and analyse it with the help of real-time data that will be shown on website. This study can be expanded in the future as more data is collected, resulting in more accurate and exact results through the use of algorithms. For smart monitoring systems, wireless sensor networks offer a lot of potential. WSNs are extremely appealing for smart agricultural systems because of their ability to quickly construct low-cost, simple end devices that can be deployed in the field.

References

1. Yi-Wei Ma1 and Jiann-Liang Chen, "Toward Intelligent Agriculture Service Platform with LoRa-base Wireless Sensor Network" IEEE International Conference in 2018
2. P. P. Ray, M. Mukherjee and L. Shu, "Internet of Things for Disaster Management State- one of the-Art and Prospects," IEEE Access, vol. 5, pp. 18818-18835, 2017.
3. Christopher Brewster, Ioanna Roussaki, Nikos Kalatzis, Kevin Doolin, and Keith Ellis "IoT in Agriculture: Designing a Europe-Wide Large-Scale Pilot" IEEE Communications Magazine September 2017.
4. Liwei Geng!!", Tingting Dong Hebei, Agricultural University, baoding, China "An agricultural monitoring system based on wireless sensor and depth learning algorithm" iJOE journal Conference in 2017.
5. S. Persia, C. Carciofi and M. Faccioli, "NB-IoT and LoRA connectivity analysis for M2M/IoTs smart grids applications," Proceedings of the AEIT International Annual Conference, pp, 2017
6. O. Georgiou and U. Raza, "Low Power Wide Area Network Analysis: Can LoRa Scale?," IEEE Wireless Communications Letters, 2017.
7. J. Haxhibeqiri, A. Karaagac, F. Van den Abeele, W. Joseph, I. Moerman and J. on Hoebeke,"LoRa indoor coverage and performance in an industrial environment: Case study,"Proceedings of the IEEE International Conference Emerging Technologies and Factory Automation, 2017.
8. Manikandan .S.V1, Jayapriya .P2, "Precision Agriculture Using Wireless Sensor Network System: Opportunities and Challenges" International Journal Of Engineering And Computer Science ISSN: 2319-7242, 2016.
9. R. Fatchurrahman, H. Putra, I. Joyokusumo, M. H. Habib, I. Imawati and S. P. Hadi, "Light sensor selection of Wi-MoLS (wireless modern light sensor) based on analytic hierarchy process (AHP)," Proceedings of the International Conference of Industrial, Mechanical, Electrical, and Chemical Engineering, pp. 152-155, 2016.

Mayday – Automated Home Security System using IOT

Vinay Bhongale[1], *Ashlesha Jagtap*[2], *Ajit Bhandarkar*[3] and *Nilima Dongre*[4]

[1]Department of Information Technology, R.A.I.T., Nerul, India, vin.bho.rt18@rait.ac.in
[2]Department of Information Technology, R.A.I.T., Nerul, India, ash.jag.rt18@rait.ac.in
[3]Department of Information Technology, R.A.I.T., Nerul, India, aji.bha.rt18@rait.ac.in
[4]Department of Information Technology, R.A.I.T., Nerul, India, nilima.dongre@rait.ac.in

Abstract: The rise in theft, break-ins, accidents which caused due to negligence at home. Now-a-days both parents are working, leaving their kids and elderly people at home, as all make mistake kids and elderly people too tend to make some mistake which may cause accidents. It is at great concern to develop a Security and Safety System which can detect these threats and try to notify the owner so that he can prevent it well in time. The Main aim of our project is to provide Security and Safety to the user by notifying the user if any hazardous activity is taking place in the presence or absence of the user. The hazardous activities may include opening of Door, Gas Leakage in kitchen, fire in the house (due to any reason like short-circuit, negligence, etc.). It can also help to avoid wastage of electricity. It will notify the user when someone opens the door, Recognize the face of person and send the data to the owner in form of alert. If the visitor is an intruder, owner can take necessary steps to stop the person. It will also notify about accidents in home like fire, gas or smoke. Face recognition will be used to segregate between authorized user and intruder.
Keywords Security, Safety, Face Detection, Automation, IOT, Home Security Automation

1 Introduction

The safety and security is an attractive research topic due to its wide application areas in day to day life of human. In this era of automation, the surveillance system need to be fully automated and free of human intervention in order to maintain transparent and error free system [1]. Now-a-days, both parents go to work leaving kids or elderly people at home. Kids and elderly people tend to forget things such as leaving the house unlocked and leaving electrical appliances on which results in wastage of electricity and sometimes that may even result in accidents.

Every 3 minutes, a burglary, robbery or a break-in is taking place in India. In 2018, 70% of all thefts in India are home thefts while only 30% are digital thefts. In India, from 2001 to 2014 there were over 3 lac deaths due to fire, In 2015 2200 deaths and 2016 11000 deaths respectively. Top 3 reasons for fires are I). Negligence in Cooking. ii.) Heating Equipment's. iii.) Faulty Wiring.

Most of the time people tend to forget things such as keeping the door open, leaving the electrical appliances on, household appliances on which may lead to different hazardous activities such as fire, robbery, gas leakage, which may cause damage to the property and can be dangerous to one's own health. Few reasons like cylinders in kitchen are not properly monitored due to which gas leakage can take place in whole kitchen which may lead to harm.

Therefore, a system is required to maintain and avoid these hazardous activities which may cause harm to one's own life. So, a system is designed which will monitor all such activities and notify the user accordingly so that they can take proper preventions and measures. This is achieved by implementing various modules in system such as the very first module Intruder Detection System which will detect any suspicious person/activity and if the system finds any suspicious person trying to break in then simply the system will block the entry and it will be brought to notice of owner. The next module in the system is about detecting gas leakage/fire alert and this module is achieved by using MQ2 sensor which will detect the hazardous gas and its concentration. If it is near to threshold value accordingly alert will be given to the owner. The last module is to notify user if he/she leaves any electrical appliances on and leaves the home, so this can be notified to the owner immediately in order to prevent from any accident that can take place.In next section, we have reviewed various work done on our topic. In methodology section, we have mentioned the process of execution. In upcoming sections we have mentioned results and conclusion.

2 Literature survey

Home Security system [1] was designed that detects the opening of Door and sends a notification on Smartphone, Owner can control the alarm from his smartphone. This will help in giving information about visitor to the home owner which will increase security.

IoT-based smart home security system [2] was created, for alerts and door access control via a smart phone. The system allows for remote door access and voice alerting by smartphone, as well as receiving a taken image of a person at the door as an email alert.

The goal of the Gas Detection and Monitoring System [3] is to develop a system that can detect and monitor a variety of gases in the environment, as well as their ppm concentrations. When the gas concentration exceeds a specified threshold, it sounds an alarm and sends a text message to a mobile phone via the GSM module. If the user is unable to react in time and the gas concentration rises to a harmful level, the valve will be turned off instantly to prevent further damage.

According to review done by Mrs. S. S. Sankpal [4] on different technologies used in Security System most of the system are automated to avoid any errors or negligence caused by humans.

Fig 1 is a low-cost GSM/GPRS based wireless Home Security System [5]. This system has three kinds of wireless security sensors, which are mainly the door security nodes, nodes for infrared security and nodes for fire alarm. These all sensors will help to alert whenever security is compromised. GSM will digitize and compress the data and then it will send down to a channel. It consists of two microprocessors, the first one is connected to the GSM/GPRS module and also to the wireless transceiver module. It is also connected to the lcd screen and also to the led's and the buzzer. The second microprocessor is connected to the keyboard where the user will enter the password. And the whole module is connected to the power supply. House, Residential Complex and Small Offices are the major applications of this system.

In fig 2 secure door system [6], it uses human tracking method with the help of laser range finder. This laser range finder is also called laser telemeter. The objects which are at distance can be determined with the help of laser beam. It consists of automatic door, the laser range finder, the computer, an identification device and also the camera alarm dual door. The laser range finder is connected to computer with the help of USB. When the trespasser enters the region of interest, the camera captures the human moment and also counts the number of people in the region of interest.

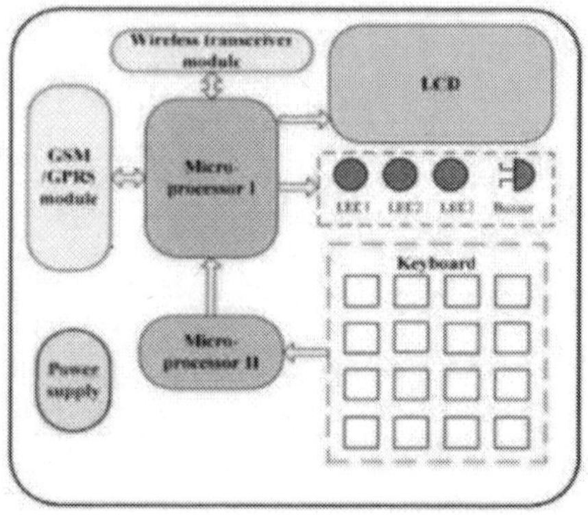

Fig 1: GSM/GPRS based wireless Home Security System[5]

If the person is not authenticated then the camera will capture the face ofthe trespasser and also the alarm will ring. If the person is authenticated then the door will open and he will be allowed to enter. This system is basically designed to provide high level security and help find any trespasser/intruder who wants to enter in authenticated region. This system was designed in a cost-efficient manner

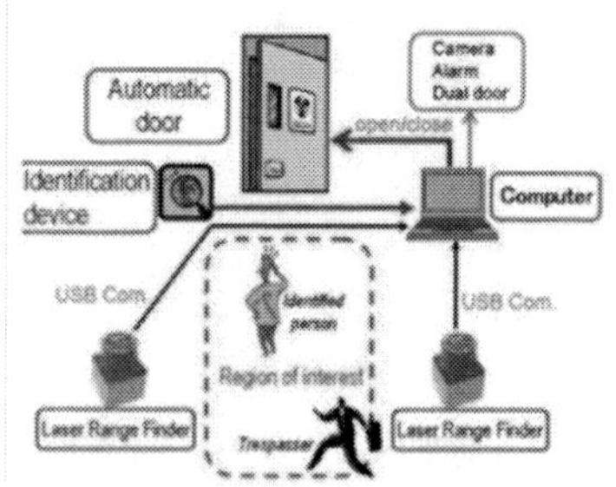

Fig 2: Human Tracking Method with Laser Range Finder[6]

The Raspberry-Pi home security system [7] is a system which consists of different sensors together called a sensor unit, Camera and the servo motors which together represents Camera Unit. Then it also consists of MCU which has Raspberry-Pi3 connected to webserver which is then connected to pc and smart device. The sensor unit consists of Ultrasonic sensor01, Ultrasonic sensor02, IR sensor. This ultrasonic sensor measures the distance of an object with the help of ultrasonic sound ways. IR sensor is Infrared sensor which sense some aspects of surrounding.

https://doi.org/10.1051/itmconf/20224401007

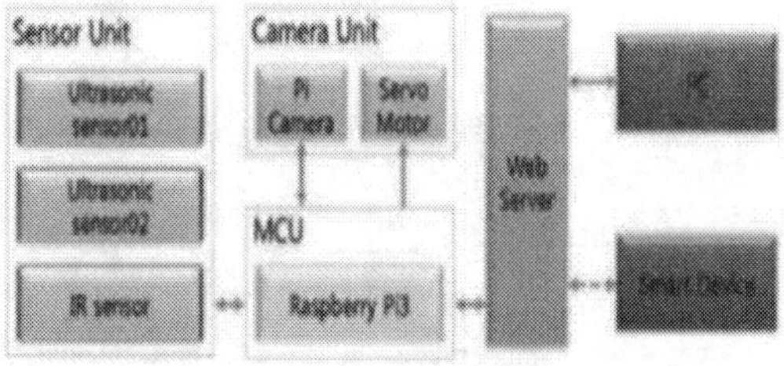

Fig 3: Raspberry-pi based home security system[7]

3 Methodology

As a person opens the door, camera starts to capture and using face recognition the individual would be categorized as a member of the house or an intruder an alert notifying the same will be sent to the user. In case of fire, gas leakage or smoke user will be notified at the early phases with all the alerts light status and room temperature will be sent. Whenever user leaves the house, he can check the light status and room temperature, if mistakenly user has left lights on or the AC's on, he can simply go in and turn it off. Images of all members would be stored in database for face recognition.

To implement Security and Safety of the user by notifying the user if any hazardous activity is taking place in the presence or absence of the user. The hazardous activities may include opening of Door, Gas Leakage in Kitchen, Fire in the house (due to any reason like short circuit, negligence, etc.). This will be achieved using sensors and other modules. It can also help to avoid wastage of electricity. It will notify the user when someone opens the door, recognize the face of person and send the data to the owner in form of alert in our application.

Visitor or User can be verified in fig 4 by password based authentication or with the help of Face Recognition. Based on these two above categories, a result will be generated depending upon which we can see that whether the user is an authorized user or an unauthorized user or an intruder. An alert message will be sent to the owner notifying that the visitor or Intruder is trying to open the door.

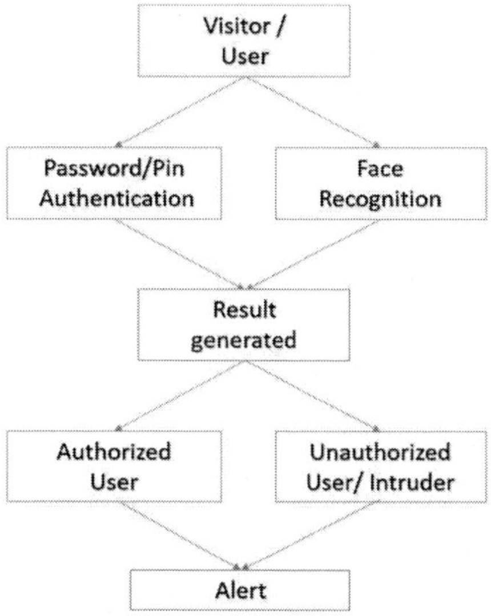

Fig 4: Authenticating the visitor

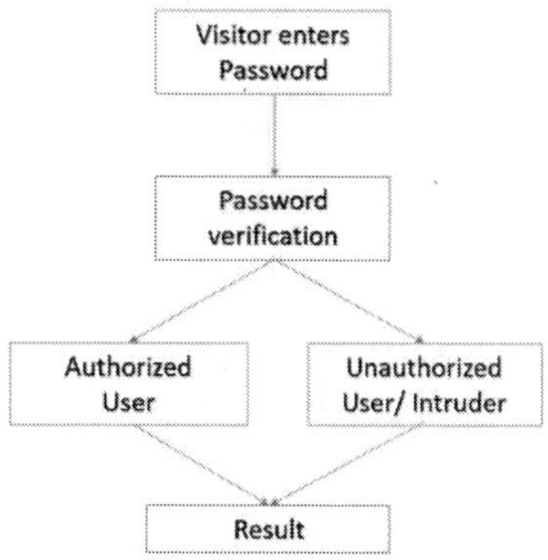

Fig 5: Password Based Authentication

When visitor arrives on the door he can enter his password. The system will verify the password and depending upon the password the user will be categorized.

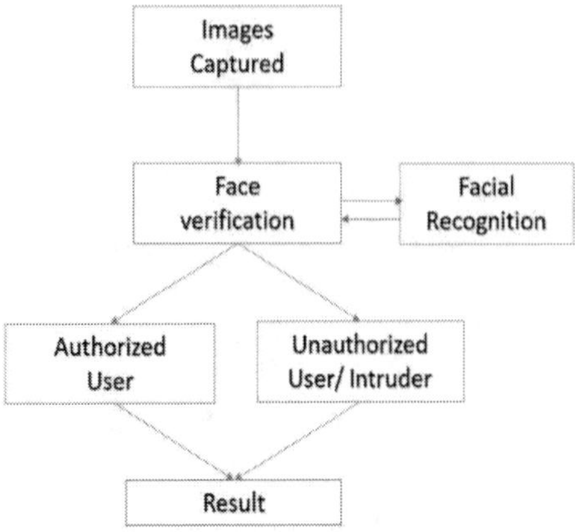

Fig 6: Face Recognition based Authentication

When visitor arrives at the door, the images of the visitor get captured. As soon as the images get captured, face verification is done with the help of facial recognition model against the images stored in dataset. As a result, depending upon this, the visitor is categorized as an authorized or an unauthorized user/intruder.

Face recognition model is created using computer vision and deep learning with haar cascade classifier[11] and local binary pattern classifier. Other model developed using deep learning models and convolution neural network (CNN) is most accurate with images. The CNN is widely used for image classification. The CNN in further depth can be used to get more accurate models.[12] The network architecture of model consists of convolutional layer followed by batch normalization and ReLU.

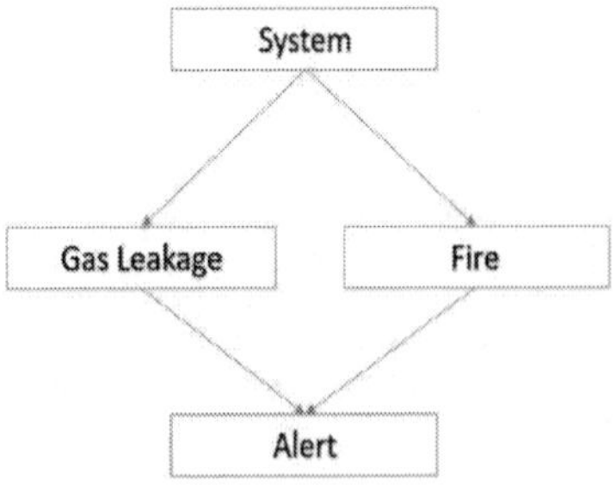

Fig 7: Gas Leakage & Fire detection

With the help of gas leakage and fire detection module system will generate the alert and notify the member's or owner well in time so that proper caution will be taken. In case of gas leakage, when there is gas leakage below the threshold a warning is sent to the user. As soon as the leakage of the gas goes above the threshold then alert is sent to the owner.

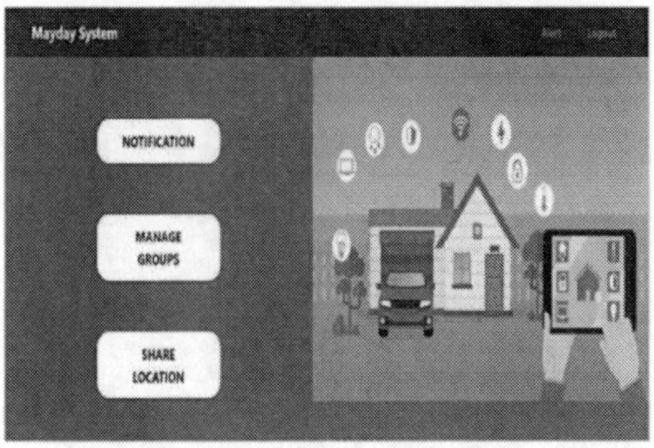

Fig 8: Home-Page of the website

The above figure shows us the webpage of the MayDay system. In this, we can see the notification, manage groups and share location buttons. On clicking the notification button, the user can see all the notification received by the system. On clicking the manage group button, the user can view or manage different groups in which he/she is part of. And on clicking the share location button the user can share the location to other members of the group.

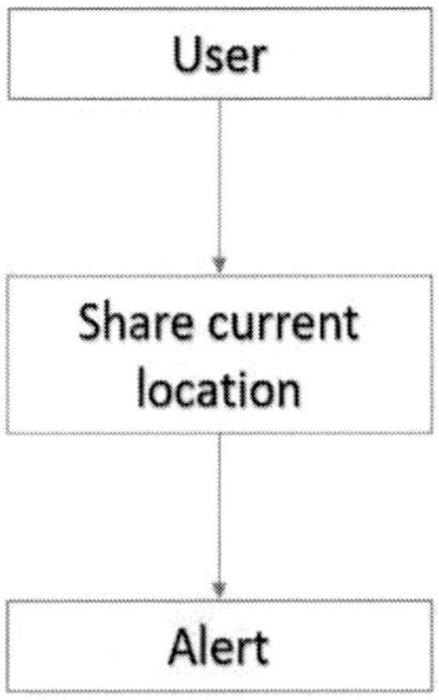

Fig 9: Share Location

Whenever the user clicks on the share location button, the location of the user along with the username gets shared to all the groups he/she is present in. The other group members receive an alert which contains the current location where the user is stranded.

In the fig 10 we can see that, we can add members to our group. On writing the username and clicking the add member button, the user can add any member to the group. Also, we can see who all are the participants of the group, and can also view the member's profile.

 https://doi.org/10.1051/itmconf/20224401007

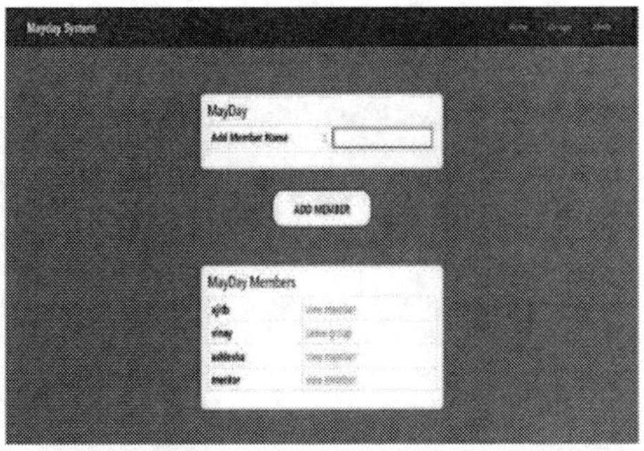

Fig 10: Manage Group

4 Result

The Proposed Method solves the problem of Security and Safety of the user in a Residential Complex, working environment and minimize the problem occur due to intruder and other accidents caused due to negligence.

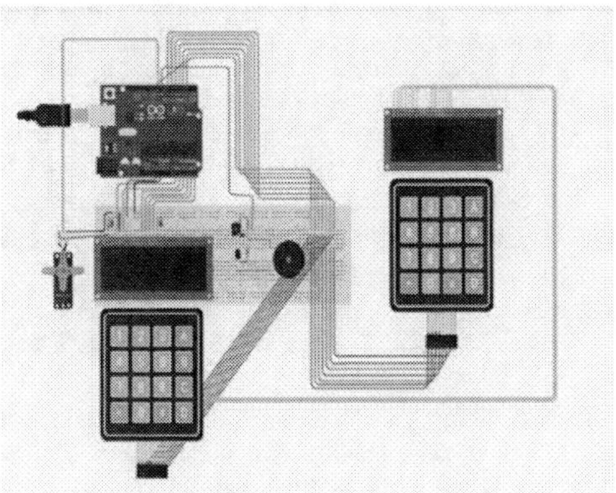

Fig 11: Password Validator Module

In figure 11 we can see that, we have to enter the password in the keypad while entering and exiting through door.

In figure 12 we can see that the different sensors used in this system. These sensors collect the data and system processes the data collected to send the alert to the user. There is the mq2 sensor, PIR sensor, lm35 sensor which collects the data and system generates the alert and sends it, to the user during the time of the crisis.

In figure 13 we can see that, the alerts are being sent by the system whenever there is gas leakage, door unlocking, or there is fire the place. We can also see that; we receive the location send by the user in our group.

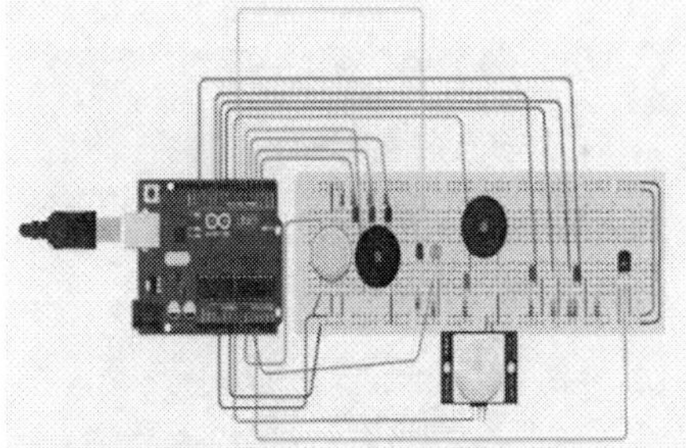

Fig 12: Sensor Module

MayDay

User	Alert	Lgt	Tmp	Date/Time
SYSTEM	GAS LEAKAGE	on	34	2022-02-26 21:26:27
Gauri	Door Unlocked	on	36	2022-02-26 21:24:42
SYSTEM	GAS LEAKAGE	off	17	2022-02-26 21:24:04
SYSTEM	FIRE	off	39	2022-02-26 21:23:57
vinay	Shared location rait, Mumbai 400059		0	2022-02-02 11:50:35
ajtb	Shared location Thane, Mumbai 400059		0	2022-02-01 22:53:22
SYSTEM	FIRE	on	56	2022-02-01

Fig 13: Alert Page

In fig 13 we can see that the alerts such as the door unlocked, gas leakage, fire and the location of the user are sent to the people along with the status of the lights, temperature and the date along with time. Thus, the user is notified along with the status of these things keeping the user safe and secure from different threats and intruder.

Therefore, the user who uses this system is constantly updated of all the current status of his/her home thus keeping the property safe.

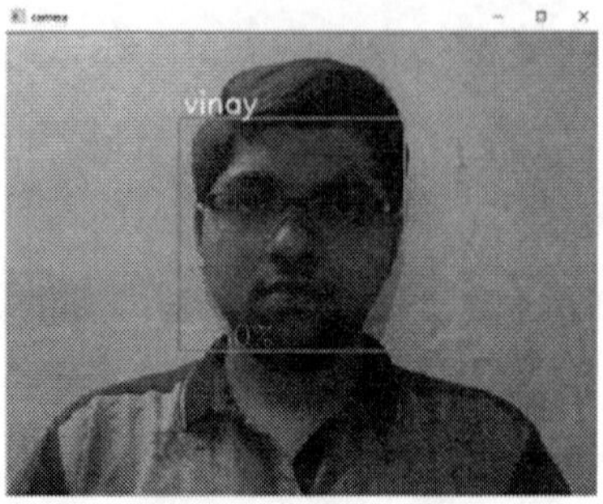

Fig 14: Face Recognition

In the above image, we can see that when a person opens the door, the camera starts to capture the image of the person and checks whether the person is authorized or not, depending upon the accuracy of the facial recognition of the person.Depending upon this the person can be recognized as an intruder or the authorized visitor. Thus, the burglary or the break-ins can be prevented and the house can be safe.

4 Conclusion

The Proposed Method solves the problem of Security and Safety of the user in a Residential Complex, working environment and minimize the problem occur due to intruder and other accidents caused due to negligence. Thus, by developing MayDay System it provides Security and Safety to the user by notifying the user if any hazardous activity is taking place in the presence or absence of the user. As user will be notified during/prior to the action of incident he will have a greater chance to stop/ prevent it. In future the system would be enhanced to make an accurate activity detection and classification model.

References

1. Anitha A (2017), "Home security system using Internet of things", IOP Conference Series Materials Science and Engineering volume 263, Accessed on November 11 2017.
2. G Sowjanya, S Nagarjuna (2016)," Design and implementation of door access control and security system based on IoT", International Journal of Engineering Research & Technology, Accessed on January 8 2016.
3. P. Nagendra Siva Kumar, Dr. A. Kamala Kumari," Design of Gas Detection and Monitoring System using IoT", IJERT VOLUME 08, Issued on November 11 2019.
4. Prof. Mrs. S. S. Sankpal, Ms. Vijaya Prakash Phalle," Review on Home Security System", International Journal on Recent and Innovation Trends in Computing and Communication ISSN 2321-8169, Volume 6, Issued on 3 March 2018.
5. YanboZhao ; ZhaohuiYe "A low cost GSM/GPRS based wireless home security system" IEEE Transactions on Consumer Electronics.
6. Jae Hoon Lee; Yong-Shik Kim; Bong Keun Kim; Ohba,K. ;Kawata,H.; Ohya,A.; Yuta,S. "Security Door System Using Human Tracking Method with Laser Range Finders" , IEEE- International Conference on Mechatronics and Automation 2007. ICMA 2007.
7. ArbabWaheed Ahmad, Naeem Jan, Saeed Iqbal, Chankil Lee "Implementation of ZigBee-GSM based Home Security Monitoring and Remote Control system" 978-1- 61284-857-0/11/ (92011 IEEE).
8. D.Pavithra, Ranjith Balakrishnan," IoT based monitoring and control system for home automation", Volume 8, Issued on 1 April 2015.
9. Dr Joga Rao Ganta, "IoT based web-controlled home automation using Raspberry-pi", International Journal of Scientific Research in Science, Engineering and Technology, Issued on 22 March 2019.
10. Kumar Mandula, Ramu Parupalli, E Magesh," Mobile based Home Automation using Iot", IOP Conference Series Materials Science and Engineering, Issued on 1 December 2015.
11. R. Padilla, C. F. F. Costa Filho and M. G. F. Costa, "Evaluation of Haar Cascade Classifiers Designed for Face Detection", International Journal of Computer, Electrical, Automation, Control and Information Engineering Vol:6, No:4, 2012.
12. Rondik J. Hassan & Adnan Mohsin Abdulazeez,"Deep Learning Convolutional Neural Network for Face Recognition: A Review", International Journal of Science and Business Vol:5, Issue:2, 2021.
13. Gaikwad Pranay, Gabhane Jyotsna, Golait Snehal, "A survey based on Smart Homes system using Internet-of-Things", Internation conference on computation of power energy, information and communication, Issued on 1 April 2015.

Design and Implementation of Smart Home Network using Cisco Packet Tracer

Ananya Mishra[1], Jay Ghayar[2], Ritika Pendam[3], Shilpa Shinde[4]

[1,2,3,4]Ramrao Adik Institute of Technology, D Y Patil Deemed to be University, Navi Mumbai, India

Abstract: Technology has become a critical part of our daily activities and routine life in the present day scenario. The need of the hour is to devise a smart home that will be able to control the operation of electronic devices via a smartphone. This concept, while absurd to imagine, can be easily implemented efficiently using a package tracking software which includes IoT functions for control and simulation of a smart home network. IoT technology has had a huge boom in the tech industry in the recent past. IoT, better known as Internet of Things, can be applied to several real time scenarios, such as home automation, treatment, campus, office, etc. In this paper, we will be focusing on a safe home automation system that will be including modules such as: smoke detection, humiture monitoring, motion detection, temperature monitoring and solar panel controlling using smart battery. The aim of our research is to devise a simulation network of smart devices which will be in control of the end-user remotely and implement the concept of smart home automation. This project will make use of the Cisco Packet Tracer to monitor the IoT devices in the smart home network. Simulation results have shown that these smart devices can be connected in a network to the home portal and objects can be successfully monitored which proves that such a model can be used for real life implementation in the user's home.

Keywords—Smart Home, Automation, Internet of Things, Cisco Packet Tracer.

1. INTRODUCTION

The Internet of Things (IoT), which is made up of two components- web and things, is a system devised to interface everything to the web using remote sensor networks. This new and upcoming field has made ground breaking changes in the technical industry in the recent past. With the use of this technology in which billions of entities have sensors for measuring and determining their status, all linked by common or proprietary protocols over public or private networks, we are able to create a smart home network using the Cisco Packet Tracer software which has all inbuilt functions to provide the necessities needed to build the system. This Smart Home includes objects with specific functions designed to improve comfort, increase security and automate day-to-day activities, helping the user in monitoring the home environment with great ease. The most crucial advantage of devising a smart home network is to relieve users of the stress of having to individually operate components of their household. Through the IoT concept, the devices operate together by sharing consumer usage data with each other and follow an algorithm of automated actions as per the preference of the users of the system. Along with providing this comfort, these smart home networks aim to provide increased security through use of motion detectors and web cameras. This enables the users to monitor their homes without having the need to be physically present. The smart home system also aims to improve the consumption of energy by analyzing the conditions and controlling the electric network accordingly, switching off appliances when not in use.ent data.

2. MOTIVATION AND SCOPE

The simulation in our project is demonstrated through the use of Cisco Packet Tracer (Version 8.0.1), through which we have created a framework for a Smart Home automation system. The Cisco Packet Tracer is a system devised by the Cisco Academy to model networks without physical limitations. The interface of this application proves to be highly efficient as it provides the drag and drop function while configuring complex networks. It also operates as a hybrid network to combine real and virtual networks. The advantages of using the Cisco Packet Tracer application for the development of the project are –

• Practical IoT machine simulations and visualizations provided

• Planning, creation and customization of smart homes allowed

• Supply of array of smart objects for use during smart home creation

• Board for control of intelligent objects provided

• Exploration of concepts of IoT made easy for the user

• Sensor detectors provided

• Energy meter provided to keep track of user's energy consumption

• User given access to data in the form of readings through mobile application

The scope of the system would be to provide users with the opportunity to access, control and monitor their homes with just a few button clicks, making use of smart technology. Interfaces of sensors can be created in the homes using sensors to monitor motion, lighting, temperature and humidity among a few things and also enable toggling automation in the devices as per conditions provided by the user. Energy conservation, comfort of living, enhanced lifestyle and heightened security are some of the features that the network aims to provide. Integration of the system with these fields of home automation give us the opportunity to be able to extend its impact on a much larger scale by devising automation of offices and factories, too. This automation network sets a global standard for linking smart devices to the web through cloud technology. It enables the user to control appliances of the smart home among which are lighting, irrigation sprinklers, fire detection systems, solar energy panels and web cameras to name a few. The expandability of these smart devices to connect with other networks enables for ease of energy management and security solutions.

3. LITERATURE SURVEY

This section describes the methodology adopted for the literature review. These papers represent an exploration of the contributions that have already been made in the home automation field.

Title and Year	Author	Survey
[1]"Designing Smart Campus using Internet of Things", 2018.	Ahmed Abdi	In this paper, research was conducted for the creation of a smart university to utilize technology to design a campus where components were to be replaced with smart devices to provide automation in the workplace with the help of cloud technology. The smart office was designed using IoT gateways to provide a link between devices in the office and internet, making use of the cloud.
[2]"Enabling communication between wireless sensor networks and the Internet of Things - ACOAP Communication Stacks", 2014.	Alexandria Olaisi	In this paper the author states the relationship between IoT and wireless sensor networks from ACOAP transport network is described, proving that there is compatibility between network of things and remote sensing systems over protocols as studied. The work done made use of communication constrained protocol (COAP) for the implementation of a smart campus
[3]"Design and Implementation of a WiFi based Home Automation System", 2010.	Miluzzo Chodhury and A. T. Campbell	In this research, the work was primarily done on the smartphone sensor which is considered to be the most important sensor today. This sensor consisted of various modules such as measurement of temperature, pressure, humidity etc. They worked on making the smartphone interface for creation of smart devices linked to sensors which can be controlled through the smartphone.

4. EXISTING SYSTEM

The existing systems studied demonstrate the use of smart networking for applications in campus and offices. Our project aims to demonstrate smart networking applications in a home system. We will be making use of the Cisco Packet Tracer which provides several built-in features of smart home automation networks as opposed to Raspberry Pi and Arduino incorporated with NRF modules that were used by researchers previously. With the help of Cisco tools, authentication and validation of wireless connections are attained with ease and interactive environments are made use of for creation of virtual network worlds to understand concepts and protocols for explanation, experimentation and exploration of technologies of Network Engineering. While the previous systems have been more hardware focused, making use of microcontrollers, our system focuses majorly on software networking, making use of wireless network connections to automate the control of smart home devices.

5. PROPOSED SYSTEM METHODOLOGY

Various smart objects are connected to the Home Gateway with the use of sensors. The Packet Tracer provides us with a programming environment to control these smart devices. The control mechanisms are enabled through registration of the objects on the Home Gateway. The components of the IoT are able to directly register with the IoT service on the Home Gateway or the network database.

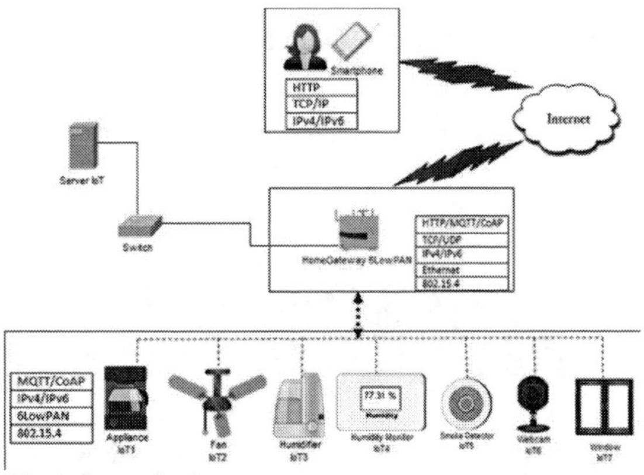

Fig. 1. System Design

HOME GATEWAY: The home gateway is equipped with 4 ethernet ports. It also has a wireless contact point on its channel 6 to enable us to make wireless links and connections through the SSID. The home gateway connects to the internet by making use of the WAN Ethernet port. This proves to be the interface which helps in the management of the automation network. Wireless connections and Ethernet cables aid in the control of devices both locally and remotely. Home gateway can be reached via its internal IP address as well as that on the internet. It also serves to be a DHCP server, thus assigning IP addresses to its connected devices. The list of devices that are registered and regulated by the gateway can be accessed and controlled.

The Algorithm of the Smart Home Automation System can be explained as follows

Step 1: The project is started on the Cisco Packet Tracer tool.

Step 2: Files with the extension .pkt can be opened and saved.

Step 3: Required components are added in the workspace, chosen from an array of inbuilt components provided by the Packet Tracer.

Step 4: Devices in the workspace are connected with the use of cables.

Step 5: Internet Service Router is set up and the devices are configured.

Step 6: The Home Gateway is added to the network.

Step 7: Smart Devices are connected to the wireless network.

Step 8: An end user device like the smartphone is added to the network.

Step 9: An application is developed to keep track and provide control of the smart devices.

Step 10: The project is completed and can be modified to meet the end user's needs.

As depicted, all the smart devices are connected to the Home Gateway with either wireless connections or through Ethernet Cables. The Home Gateway is connected to the Switch which allows it to set an IP address on the interface level. The IP addresses of the smart devices can be assigned through the Home Gateway. The Switch in turn is connected to the IoT Server which allows for linking of the smart devices in the LAN network which can be accessed by the user. Furthermore, the list of the connected devices can be linked with the smartphone to provide access to control the devices remotely through the smartphone by the end-user through the medium of the internet.

Fig. 2. Depicts the list of devices on the IoT Server that are connected to the network.

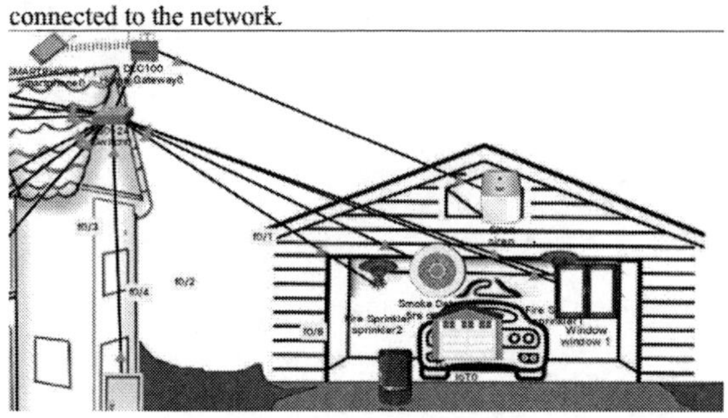

Fig. 3. Depicts the connection of the devices with the Home Gateway

FIRE DETECTOR: In the event of a fire, as the smoke level will rise, the siren will ring, the fire sprinkler will get switched on to sprinkle water and the window of the garage will open. the conditions upon which the fire detector will operate are as follows -

- If the level (L) detected by the fire detector is greater than 0.1 then the garage fire is set to be ON.
- When garage fire is ON, the Sprinkler will turn ON, Siren will turn ON, Window will OPEN and door lock will UNLOCK.
- If the level detected by the fire detector is lesser than 0.1 then the garage fire is set to be OFF.
- When garage fire is OFF, the Sprinkler will be OFF, Siren will be OFF, and Window will be CLOSED.

$$If\ L >= 0.1\ ;\ Status = True \qquad (1)$$
$$If\ L < 0.1\ ;\ Status = False \qquad (2)$$

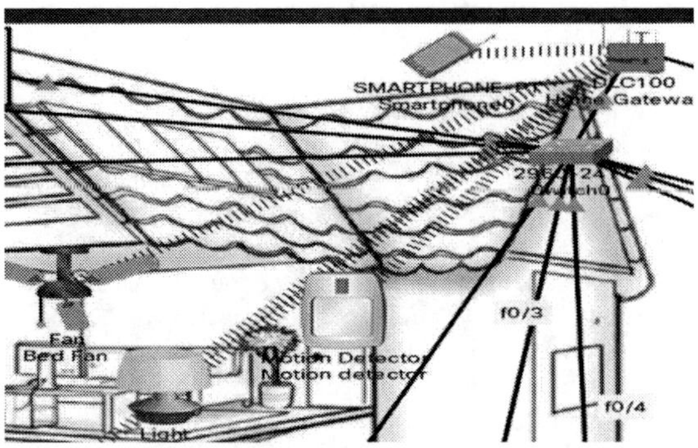

Fig. 4. Above shows the connection for the motion detector

MOTION DETECTOR: This sensor detects motion as soon as a person enters the room and the lights and fans switch on and off accordingly, thus promoting energy conservation when not in use. The conditions upon which the motion sensors will operate are as follows -

- Motion detector set to ON.
- If motion detector shows TRUE, fan status set to HIGH, and light status set to ON.
- If motion detector shows FALSE, fan status set to OFF, and light status set to OFF.

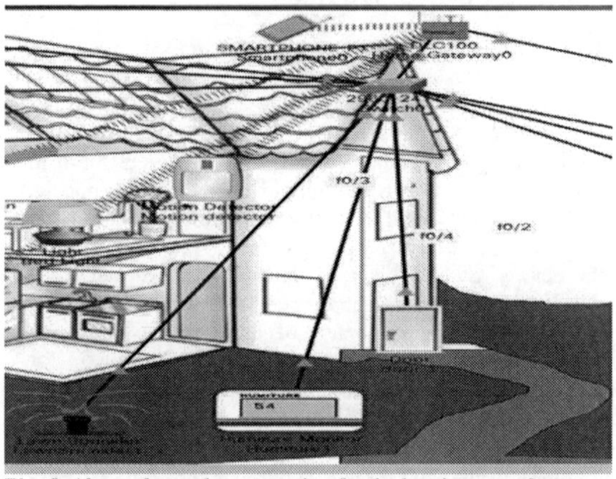

Fig. 5. Above shows the connection for the humiture monitor

HUMITURE MONITOR: This sensor detects the outdoor temperature and humidity of the environment and turns the lawn sprinkler on and off accordingly. The conditions upon which the humiture monitor will operate are as follows -

- If the outdoor temperature (t) is equal to or above 50 degrees, the lawn sprinkler will be turned ON.
- If the outdoor temperature is lesser than 50 degrees, the lawn sprinkler will be switched OFF.

$$If\ t >= 50\ ;\ Status = True \qquad (3)$$
$$If\ t < 50\ ;\ Status = False \qquad (4)$$

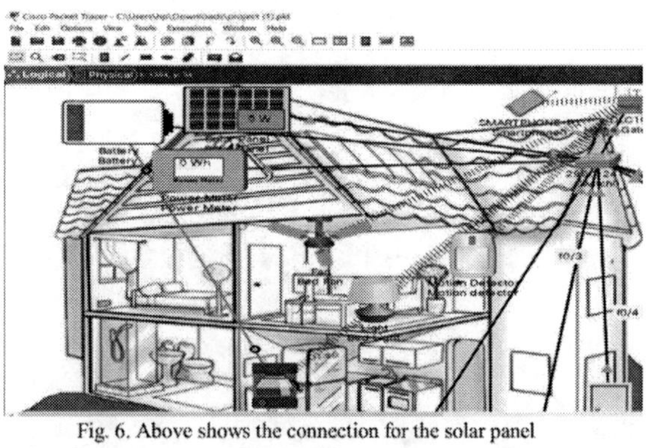

Fig. 6. Above shows the connection for the solar panel

SOLAR PANEL: This solar panel sensor detects the battery present in the solar panel and turns the coffee machine on and off accordingly. The conditions upon which the solar panel sensor will operate are as follows -

- If the available battery power (β) in the solar panel is greater than 60% then the coffee machine will be turned ON.

$$If\ \beta >= 60\ \%\ ;\ Status = True \qquad (5)$$
$$If\ \beta < 60\ \%\ ;\ Status = False \qquad (6)$$

6. RESULT ANALYSIS

Dataset taken from sensors is displayed through an Android application interface. The data is recorded through the use of Arduino connected to the sensors and the readings from the sensors are stored in the SQL database. These readings are reflected back to the user.

Fig. 7. Interface of the Android Application

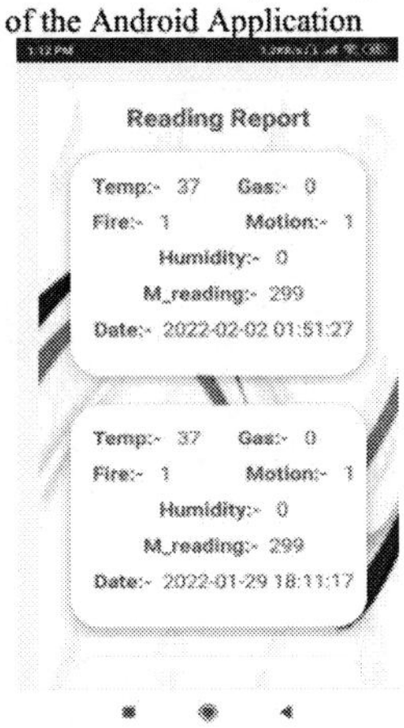

Fig. 8. Readings from the sensors are displayed to the user

	r_id	r_temp	r_humidity	r_fire	r_gas	r_meter_reading	r_motion	r_time
Edit Copy Delete	1	37	10	1	0	299	1	1643459656
Edit Copy Delete	2	37	0	1	0	299	1	1643459883
Edit Copy Delete	3	37	0	1	0	299	1	1643460008
Edit Copy Delete	4	37	0	1	0	299	1	1643460077
Edit Copy Delete	5	37	0	1	0	299	1	1643746687

Fig. 9. Readings from the sensors is stored in the SQL database

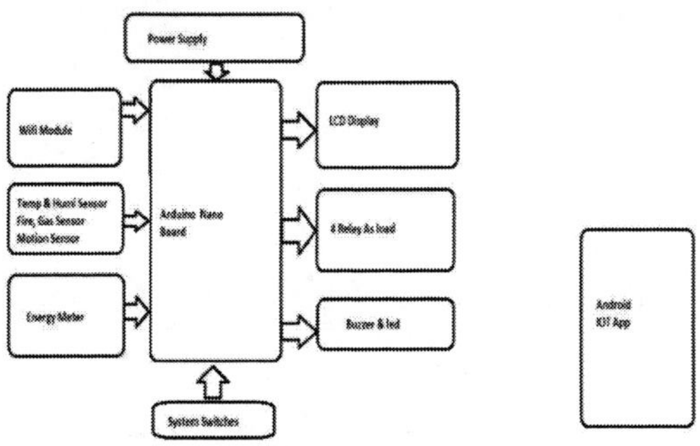

Fig. 10. Hardware model using Arduino connected to the sensors

The sensors will detect the change in environment and record the readings. These readings will be available through the interface of the app to the user. The user is able to generate a monthly report of the readings that are fetched from the database. Through this report generation module, the user is able to keep track of his/her energy consumption as well as observe the trends in the readings and therefore be informed.

A certain threshold value for each of the sensors is determined by the user. If the sensors detect a value above the threshold, alert mechanism is activated. The alert setting module on the application is equipped to send out a notification to the user as soon as the threshold value is crossed and keep sending out notifications until the readings go back to normal range. This ensures that the user is able to keep track and stay alert in case of any mishaps.

7. CONCLUSION

The idea behind this project is to explore the concept of home automation using the principle field of Internet of Things which enables us to interconnect gadgets with the assistance of web association. IoT has the power to enable 32 detection of the smart objects remotely which not only provide the user with enhanced comfort but also assist in natural resources conservation and a save in energy consumption. The question of increased security has also been covered by this Smart Home Network which provides security solutions with remote access. This gives rise to the future scope of this project which can be expanded to a large scale to cover multiple homes, office spaces and even entire industries as a whole.

Through integrating the Cisco Packet Tracer, we are able to provide a simulation which furthers the understanding and accuracy while developing a network for the Smart Home. The mobile application allows the user to have full access and control of the happenings in their environment and personalize the settings to tailor-suit their needs. Furthermore, the energy consumption module plays a vital role in promoting the need of the hour to use resources carefully and sparingly and cut costs that come along with wastage. The alert module of the mobile application will also prove to be successful in helping users avoid at-risk situations and thus enhance the quality of solutions for safety concerns regarding smart home automation.

8. ACKNOWLEDGEMENT

We take this opportunity to express our profound gratitude and deep regards to our professors, our institute and it's management for providing all necessary resources and helping us in all the possible ways. We also thank readers of this paper and all those who have directly or indirectly helped us in completion of this thesis.

9. REFERENCES

[1] "The Internet of Things in an Enterprise Context " in Future Internet-FIS 2008 Lecture Notes in Computer Science Vol. 5462 2008.

[2] "Smart Home System using IoT technologies" Computational and Information Sciences (CIS), 2010 Fifth Int. Conference on IEEE.

[3] Sri Reddy, P. Salim Akram, M. Adarsh Sharma, P. Aditya Sai Ram, R. Prithvi Raj, "Study and Analysis of Routing Protocols", International Journal of Emerging Trends in Engineering Research, Vol 8, No11, PP 454 – 460, 2021.

[4] K. Xu, Performance analysis of differentiated Quality of Service MAC in wireless local area networks (WLANs), Submitted to the Department of Electrical and Computer Engineering, Queen's University, Canada. September, 2003.

[5] Alexandria, Al-oisi (2014), "Enabling communication between wireless sensor networks and the internet of things- ACOAP communication stacks." International journal of Science and Engineering, 7-8 Vol.6.

[6] Ahmed Abdi (2018), "Designing Smart Campus Using Internet of Things." International Journal of Computer Science Trends and Technology (IJCST), Vol.6 PP 119-126.

[7] Miluzzo Chodhury and A. T. Campbell (2010), "Design and Implementation of a WiFi Based Home Automation System", pages 2177-2180.

[8] Arunvivek, J., Srinath, S., and Balamurugan, M. S. (2015). Framework development in home automation to provide control and security for home automated devices. *Indian J. Sci. Technol.* 8, 1–10.

[9] Berlo, A. V., and Allen, B. (1999). *Design Guidelines on Smart Homes: A COST 219bis Guidebook.* Brussels: COST, Commission of European Communities.

[10] Y. Upadhyay, A. Borole, and D. Dileepan, "MQTT based secured home automation system," in *Proceedings of the 2016 Symposium on Colossal Data Analysis and Networking (CDAN)*, pp. 1–4, IEEE, Indore, India, March 2016.

[11] S. Chattoraj, "Smart home automation based on different sensors and arduino as the master controller," *International Journal of Scientific and Research Publications*, vol. 5, no. 10, pp. 1–4, 2015.

Performance Assessment of Hetero-Junction Intrinsic Thin Film HIT Photovoltaic Module Using Machine Learning Methods

Prasad Kulkarni[1], *Tushar* Patil[1], *Aditya* Pandey[1], *Vishwesh* Vyawahare[1], *Dhiraj* Magare[1] *and Gajanan* Birajdar[1]

[1]Ramrao Adik Institute of Technology, Navi Mumbai

Abstract. A solar cell built of ultra-thin amorphous silicon and high-quality mono-crystalline silicon is known as a hetero-junction intrinsic thin film. It has a pyramid surface on the front that increases sunlight absorption. The operating environment has a significant impact on the performance of hetero-junction intrinsic thin-film photovoltaic modules with real I–V (current-voltage) characteristics. Changes in the environment have a significant impact on solar irradiation. Clouds also have a significant impact on the solar irradiation that a PV cell receives. In this project, we will use the Random Forest Regression machine learning algorithm to investigate the effects of sudden changes in environmental conditions on power output and module temperature of an HIT (Heterojunction with Intrinsic Thin Layer) module, where irradiance, temperature, and module efficiency parameters are taken into account when designing modules. The algorithm's output will be studied to gain a better understanding of performance variations as well as the behavior of the power output and module temperature when subjected to random influences induced by various environmental variables. The suggested algorithm is not restricted to a certain module technology or geographic location.

1 INTRODUCTION

PV cells, also referred to as solar cells, are electrical components that create electricity when they are exposed to photons, or light particles. The French physicist Edmond Becquerel gave the name to the photovoltaic phenomenon, which was discovered in 1839. PV cell modules started appearing on roofs towards the end of the 1980s. As the photovoltaic industry has flourished in the 21st century, the construction of vast solar farms has continually increased in photovoltaic capacity.

A photovoltaic cell is made up of several layers of materials, each with its own function. The specifically treated semiconductor layer is the most crucial layer in a solar cell. It has two separate layers (p-type and n-type) and is responsible for converting the Sun's energy into usable power via the photovoltaic effect. A layer of conducting material is present on both sides of the semiconductor, which "collects" the electricity generated. Note that the backside of the cell, which is shaded, can afford to have the conductor totally covered, whereas the front, which is illuminated, must use the conductors sparingly to prevent blocking too much of the Sun's energy from reaching the semiconductor. The anti-reflection coating is the final layer, which is only applied to the illuminated side of the cell. Reflection loss can be severe because all semiconductors are naturally reflective. To limit the amount of solar radiation reflected off the cell's surface, one or more layers of an anti-reflection coating can be applied. Solar cells can be grouped into arrays. Homeowners have installed solar cells in considerably smaller designs on their rooftops, referred to as solar cell panels or just solar panels, to replace or supplement their normal energy source.

Solar photovoltaic (PV) module temperature impacts the efficiency of PV modules, which increases the overall power output of the PV system [5]. PV modules lose output power when they heat up, and degradation accelerates as a result. There is a minor increase in short circuit current as the module temperature rises; nevertheless, the open-circuit voltage of the module drops dramatically, lowering the module's output power.

The wind is one of the most critical environmental elements that affects the temperature of the module, and whose effect is finally translated into the PV module's performance [3]. The PV module can be more efficient when the temperature is lower. As well as wind speed, the wind direction relative to the PV module's orientation determines what degree of temperature reduction the air circulation around the PV module causes. When comparing the variation of the module temperature with the ambient temperature and irradiance in parallel wind situations versus the perpendicular and low wind situation, it has been discovered that the variation of the module temperature is nonlinear and quite irregular with the ambient temperature and irradiance in parallel wind situations. The wind effect is more prominently noticeable at lower ambient temperatures, and module temperature has a very poor association with irradiance levels. Furthermore, when the irradiance is high, the parallel wind direction affects the module temperature more than the other two wind speeds. In addition, the area and period of contact between the module surface and the wind are reduced, lowering the overall cooling mechanism in a PV module. The data was measured using a Young wind sensor of model no. 05103 [14]. This model was developed considering how module temperature and several meteorological characteristics, including wind speed, wind direction, and in-plane irradiance, interrelate.

Although MPPTs (maximum power point trackers) or module-level power electronics (MLPE) such as microinverters or DC-DC optimizers can be connected to create arrays with the desired peak DC voltage and current capacity, this can be done with or without the use of independent MPPTs. In arrays with series/parallel coupled cells, shunt diodes help reduce shadowing power loss.

The data log from the National Institute of Solar Energy (NISE), Gurgaon, was analyzed in this paper. Experimental setup and data logging strategy has been described in [14].

2 LITERATURE SURVEY

Today, photovoltaic modules are commonly used. These modules are available in various shapes and sizes. Hetero-Junction Intrinsic Thin Film Module is one of them. They are the most modern and efficient photovoltaic module currently available. They are, nevertheless, still influenced by the conditions in which they labor. Wind speed, sunlight, weather, and other environmental conditions all have an impact on the module's performance. The temperature of the module is strongly affected by wind movement around it, which has an impact on the module's output power. Magare et al. [14] used the linear regression approach of machine learning to estimate module temperature and maximum power production in prior works such as "Wind Effect Modeling and Analysis for Estimation of Photovoltaic Module Temperature." However, because machine learning has advanced so much in recent years, researchers are continuously coming up with new ways to improve prediction. They are also always looking for methods to improve the module's efficiency in delivering power.

3 THEORY

3.1 Random Forest Regression

A random forest regression method is a regression technique utilizing ensemble learning as the supervised learning approach. A machine learning algorithm combines several different forecasts from different machine learning models to make more accurate predictions.

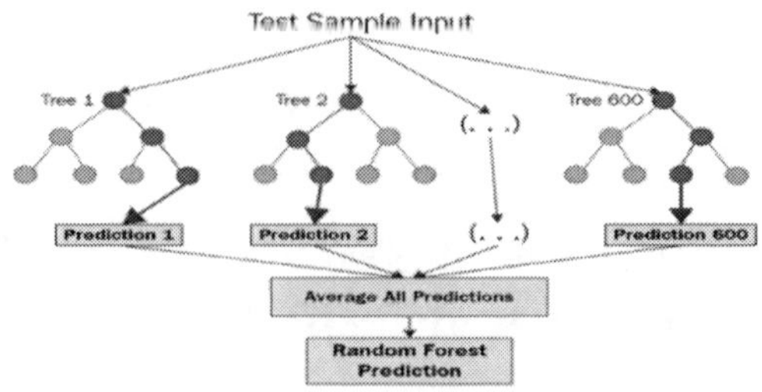

Fig. 1. Random Forest Regression

The structure of a Random Forest is depicted in fig 1. The trees run in a straight line with no contact. When a Random Forest is trained, many decision trees are constructed, and the mean of all the trees' predictions is derived. Let's go over the steps to acquire a better knowledge of the Random Forest algorithm:

• The training set is divided into n random data points.

• Analyze the k data points to create a decision tree.

• Steps 1 and 2 should be repeated for as many trees as N is required.

• Assign the new data point to the average of all predicted y values from all your N-tree trees for each new data point.

In terms of accuracy and power, Random Forest Regression stands out. It works well on a wide range of situations, including those with non-linear relationships. Because it exploits randomization on two levels, the ensemble of decision trees has a high level of accuracy.

• At each split, the algorithm selects a subset of features at random to be used as candidates. This avoids many decision trees from sharing the same set of features, which decorrelate individual trees.

• When generating splits, each tree takes a random sample of data from the training dataset. This adds a layer of unpredictability to the equation, preventing the trees from overfitting the data.

3.2 Working of PV Cell

A photovoltaic cell as shown in fig 2 is composed of semiconductor materials that capture photons from the sun and generate an electrical charge [1]. Solar radiation travels through the geosphere in the form of photons, elementary particles that travel at speeds of 300,000 kilometers per second. When photons impact a semiconductor material such as silicon, the electrons in its atoms are released, leaving a free space. The wayward electrons roam around aimlessly in search of a new "hole" to fill.

Electrons must, however, move in the same direction to generate an electric current. Silicon comes in two forms to do this. It has one more electron than silicon, so the side facing outward is doped with phosphorus atoms, whereas the side facing inward is doped with boron atoms, which have one fewer electron than silicon. With the sandwich constructed, the excess electron layer serves as the negative terminal (n), while the shortfall on the positive side serves as the positive terminal (p).At the connection between the two layers, an electric field is formed.

An electric field sweeps electrons to the n-side, while holes drift to the p-side, when they are stimulated by photons [1]. Electrical energy is transferred to the external circuit in the form of electrons and holes, which are directed to the electrical contacts on both sides. This results in a direct current. The top of the cell is coated with an anti-reflective material to reduce photon loss due to surface reflection.

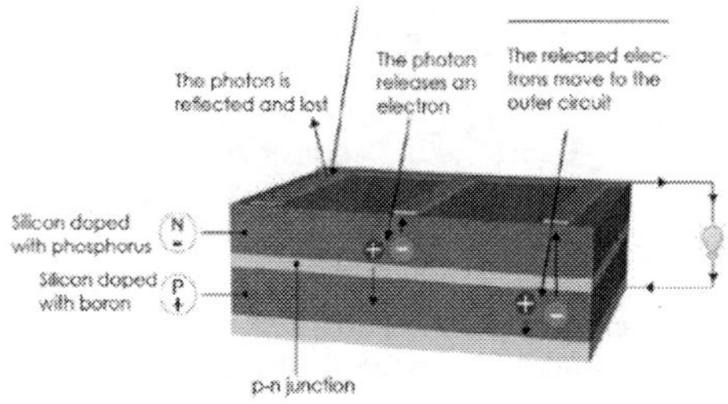

Fig. 2. Photovoltaic Cell

3.3 Photovoltaic Cell Efficiency

Because the visible range of electromagnetic radiation contains most of the energy in sunshine and artificial light, a solar cell absorber should be effective at absorbing radiation at those wavelengths. Solar cells are rated according to their energy efficiency based on how much electricity they produce compared to the amount of light they receive. In order to test the efficiency of the cell arrays, the cells are wired together into modules, which are then assembled into arrays. A solar simulator simulates optimal sunlight conditions including 1,000 watts (W) per cubic meter at 25 degrees Celsius with the panels positioned in front of it. Peak power, or the amount of electricity generated by the system, is a proportion of the solar energy received. The efficiency of a panel of one square meter that provides 200 W of electrical power is 20%. PV cells have a theoretical maximum efficiency of roughly 33%. The Shockley-Queisser limit is the name for this limit.

A solar cell's output, or how much electricity it can produce, is determined by several factors, including the solar radiation levels in the area, the efficiency of the cell, and the type of installation. In the Paris area, incident solar radiation is 1 megawatt-hour per square meter per year (MWh/sq.m/y), compared to 1.70 MWh/sq.m/y in southern France and approximately 3.0 MWh/sq.m/y in the Sahara Desert. In Paris, a solar panel with a 15% efficiency rating will create 150.0 kWh/sq.m/y, but in the Sahara, it will generate 450.0 kWh/sq.m/y [11].

3.4 Structure of HIT

Heterojunction with Intrinsic Thin-layer, or HIT, is an abbreviation for intrinsic thin-film heterojunction. Because Sanyo Corporation of Japan has sought for a registered trademark for HIT, it is also known as HJT or SHJ (Silicon Heterojunction solar cell). Three layers of photovoltaic material make up heterojunction solar panels as shown in fig 3. HJT cells combine crystalline and amorphous "thin-film" silicon technologies into a single device. The top layer of amorphous silicon captures sunshine and light that reflects off the lower layers before it reaches the crystalline layer. The middle layer, monocrystalline silicon, is responsible for converting most of the sunlight into energy. Finally, there is an amorphous thin-film silicon layer behind the crystalline silicon [16]. The remaining photons that pass through the first two levels are captured in this final layer. When these technologies are used together, more energy may be gathered than if they were used separately, with efficiency of 25% or higher. The basic structure of an HIT solar cell is shown in the diagram below, which is characterized by a pi-type a-Si: H film (film thickness 5-10 nm) on the light irradiation side and an n-type a-Si: H film (film thickness 5- 10nm) sandwiching a crystalline silicon wafer, forming transparent electrodes and collector electrodes on both sides to form an asymmetrical HIT solar cell.

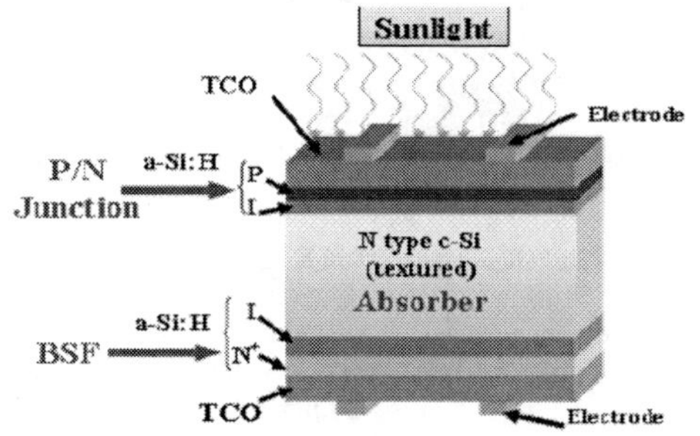

Fig. 3. Structure of HIT PV Cell

HIT combines the greatest features of crystalline silicon with those of amorphous silicon thin film to create a high-power hybrid cell that outperforms the industry's standard, PERC. There is the potential for significant cost reductions because the HIT manufacturing process requires four less steps than PERC technology.

3.5 Fabrication of HIT Cells

The fabrication order differs from one group to the next. The absorber layer of HIT cells is usually made of high-quality CZ/FZ grown c-Si wafers. The surface of the wafer is textured with alkaline etchants such as NaOH or (CH3)4NOH to generate pyramids of 5-10m height. After that, peroxide and HF solutions are used to clean the wafer. It is followed by the deposition of an intrinsic alkali Si passivation layer, which is usually done by PECVD or Hot-wire CVD. For deposition, silane gas (SiH4) is diluted with H2. The deposition temperature and pressure are 300°C and 0.1-1 Torr, respectively.

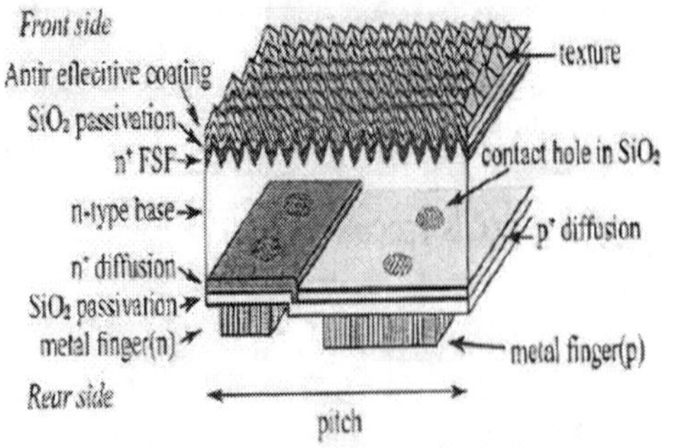

Fig. 4. Fabrication of HIT PV Cell

To avoid the production of faulty epitaxial Si, this stage must be precisely controlled. Deposition and annealing cycles, as well as H2 plasma treatment, were proven to offer excellent surface passivation. A-Si p-type layers are deposited by mixing phosphene gas with silane, whereas a-Si n-type layers are deposited by mixing diborane with silane. On a c-Si wafer deposited directly with doped a-Si layers, very poor passivation properties are observed. Dopant-induced defect formation in a-Si layers is most likely to blame. Because a-Si has a high lateral resistance, Bi-facial designs typically have front and back layers of Indium Tin Oxide (ITO) respectively that are transparent conductive oxides (TCOs)

4 DESIGN METHODOLOGY

This project proposes a prediction model using Random Forest Regression machine learning algorithm that can predict module temperature and maximum power output of a given photovoltaic module under different environmental conditions. A data log consisting of 18000+ data points of a whole year was used to train the model. Out of these, 80% data points have been used for training the model and 20% for testing.

The data set consists of parameters such as short circuit current (I_{sc}), open circuit voltage (V_{oc}), current at maximum power (I_{pmax}), voltage at maximum power (V_{pmax}), maximum power (P_{max}), module temperature (Temp_mid_avg), solar irradiance (G_t), wind speed (WS) and ambient temperature (T_{amb}). The parameters have been measured at an interval of 10 minutes. Out of these parameters, Temp_mid_avg and Pmax are dependent parameters. G_t, WS and T_{amb} are independent parameters for predicting temp_mid_avg. I_{sc} and V_{oc} are independent parameters for predicting P_{max}. Temperature coefficient of maximum power for HIT module is -0.33 %/℃ [14]. Linear regression machine learning algorithm has been used before, but it is a very basic algorithm and limited to only linear relationships. It also doesn't work efficiently with large data types. The PV module's maximum output power was measured and compared to the predicted maximum output power. Random Forest Regression was used to get the anticipated output power. Random forest uses ensemble technique which means that it combines multiple models.

4.1 Required Libraries

Various libraries have been used for achieving the prediction model. Pandas is used for data processing and manipulation. Sklearn is used for training and testing of the data. Matplotlib and seaborn are used for data visualization.

4.2 Data Visualisation

Graphs have been plotted to compare the various parameters. Using this comparison, correlations between the parameters were observed.

4.3 Training and testing

Random Forest regression model operates by making several decision trees at the time of training and the final output is given by the mean value of all the decision trees. It picks random data points from the data set and then the decision tree is made according to the picked data set . The number of trees are then chosen and the same process of picking random data points and making decision trees is continued.

5 RESULTS AND OBSERVATIONS

In order to predict the maximum power and temperature of the module, a prediction model was used on the data log from the National Institute of Solar Energy (NISE), Gurgaon. Detailed information about the experimental setup and data logging strategy is presented in [14]. The predicted values were compared to the actual values given in the data set by plotting the "predicted vs actual value" curves. The difference in the values, i.e., error was calculated. RMSE values for each of the dependent parameters was also calculated.

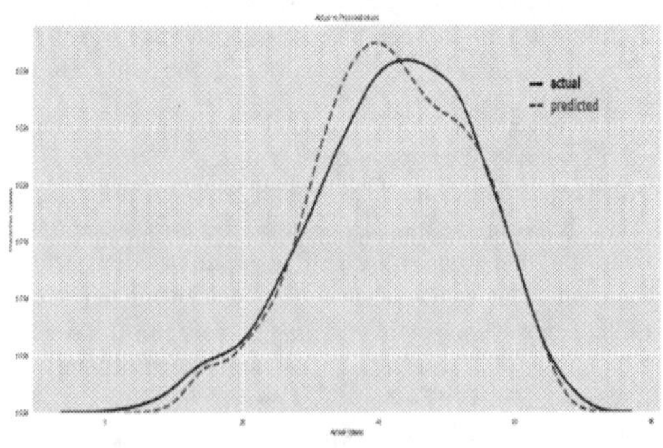

Fig. 5. Actual vs Predicted for Module Temperature using Linear Regression

Error Metric	Linear Regression (Module Temperature)	Linear Regression (Maximum Power)	Random Forest Regression (Module Temperature)	Random Forest Regression (Maximum Power)

RMSE	4.3	3.5	3.85	2.7

Table 1. RMSE values for different machine learning algorithms

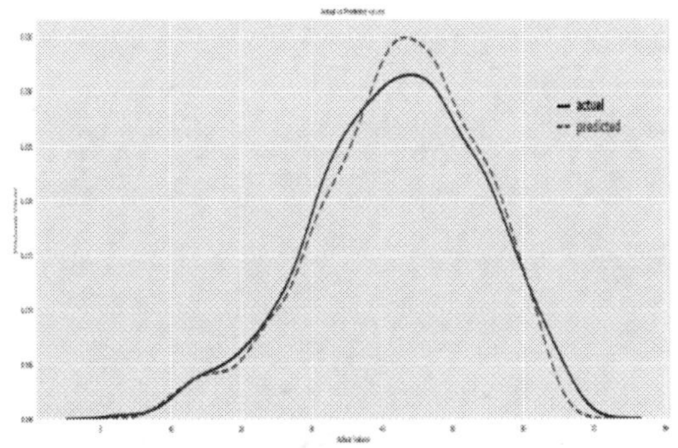

Fig. 6. Actual vs Predicted for Module Temperature using Random Forest Regression

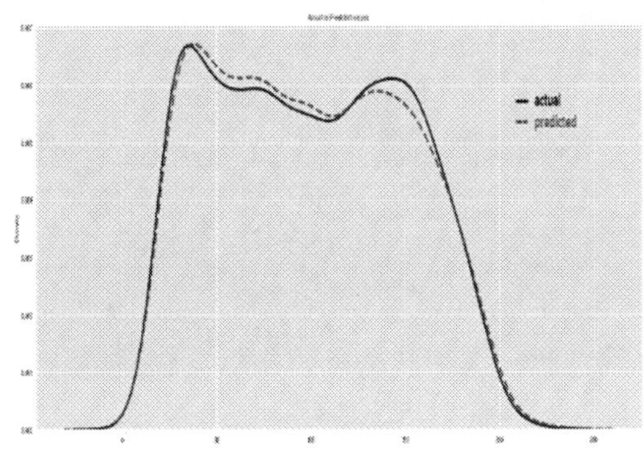

Fig. 7. Actual vs Predicted for Maximum Power using Linear Regression

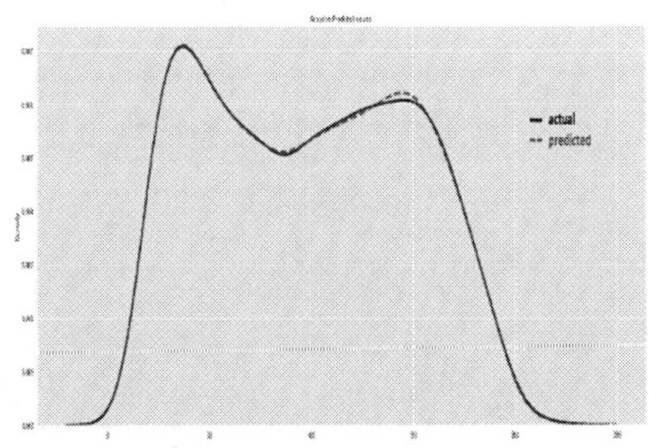

Fig. 8. Actual vs Predicted for Maximum Power using Random Forest Regression

As it is clearly observable from fig 5 and fig 6, random forest regression is much more accurate and efficient in predicting the module temperature than linear regression. Similarly, from fig 7 and fig 8, it can be observed that random forest regression is also more accurate and efficient in predicting the maximum power output of the module.

Temperature of the module has a non-linear relationship with its independent parameters, so the RMSE value is higher as compared to maximum power which has a linear relationship with its independent parameters. The RMSE values for linear and random forest regression is given in table . As it is apparent from the table, the random forest regression has a much lower RMSE value than linear regression. Thus, random forest provides much better prediction and is more efficient. The comparison of RMSE values is given in table 1.

6 CONCLUSION

The assessment of PV module temperature and maximum power was included in this project. With the machine learning model of random forest regression, a simple model for predicting PV module temperature and maximum power has been developed that considers wind speed and direction, in-plane irradiance, and module efficiency. Overall, the model's results show a good match between actual and predicted module temperatures and maximum power outputs for HIT PV cells. The methodology given in this project can be used to anticipate module temperatures of HIT PV cells under varying weather conditions with great consistency. The RMSE values of both linear regression and random forest regression were compared in this project. It was observed that linear regression has much higher RMSE value than random forest regression which makes the latter more accurate and efficient in predicting module temperature and maximum power. Actual vs predicted graphs were plotted as well and again random forest regression was more accurate.

7 REFERENCES

1. Green, M. A., (1982), Solar Cells: Operating Principles, Technology, and System Applications, Prentice Hall, Englewood Cliffs, NJ.
2. King, D. L., Boyson, W. E., and Kratochvil, J. A., (2002), "Analysis of Factors Influencing the Annual Energy Production of Photovoltaic Systems," 29th IEEE Photovoltaic Specialists Conference, New Orleans, LA, May 19–24, pp. 1356–1361.
3. Schwingshackl, C., Petitta, M., Wagner, J. E., Belluardo, G., Moser, D., Castelli, M., Zebisch, M., and Tetzlaff, A., (2013), "Wind Effect on PV Module Temperature: Analysis of Different

Techniques for an Accurate Estimation," Energy Procedia, 40, pp. 77–86.

4. Ross, R. G., Jr., and Smokler, M. I., (1986), "Flat-Plate Solar Array Project Final Report—Vol. VI: Engineering Sciences and Reliability," Jet Propulsion Laboratory, Pasadena, CA, Report No. DOE/JPL-1012-125.
5. Nordmann, T., and Clavadetscher, L., (2003), "Understanding Temperature Effects on PV System Performance," Third World Conference on Photovoltaic Energy Conversion, Osaka, Japan, May 11–18, pp. 2243–2246.
6. Duffie, J. A., and Beckman, W. A., (2006), Solar Energy Thermal Processes, 3rd ed., Wiley, Hoboken, NJ.
7. Bahaidarah, H., Rehman, S., Subhan, A., Gandhidasan, P., and Baig, H., (2015), "Performance Evaluation of a PV Module Under Climatic Conditions of Dhahran, Saudi Arabia," Energy Explor. Exploit., 33(6), pp. 909–930.
8. Jakhrani, A. Q., Othman, A. K., Rigit, A. R. H., and Samo, S. R., 2011, "Determination and Comparison of Different Photovoltaic Module Temperature Models for Kuching, Sarawak," IEEE First Conference Clean Energy and Technology (CET), Kuala Lumpur, Malaysia, June 27–29, pp. 231–236.
9. Bora, B., Sastry, O. S., Kumar, A., Renu, M., Bangar, M., and Prasad, B., (2016), "Estimation of Most Frequent Conditions and Performance Evaluation of Three Photovoltaic Technology Modules," ASME J. Sol. Energy Eng., 138(5), p. 054504.
10. IEC, (2009), "Procedures for Temperature and Irradiance Corrections to Measured I-V Characteristics," International Electrotechnical Commission, Geneva, Switzerland, Standard No. IEC 60891:2009.
11. IEC, (2011), "Photovoltaic (PV) Module Performance Testing and Energy Rating—Part 1: Irradiance and Temperature Performance Measurements and Power Rating," International Electrotechnical Commission, Geneva, Switzerland, Standard No. IEC 61853-1:2011.
12. Magare, D. B., Sastry, O. S., Gupta, R., Betts, T. R., Gottschalg, R., Kumar, A., Bora, B., and Singh, Y. K., (2016), "Effect of Seasonal Spectral Variations on Performance of Three Different Photovoltaic Technologies in India," Int. J. Energy Environ. Eng., 7(1), pp. 93–103.
13. Rosell, J. I., and Ibanez, M., (2006), "Modelling Power Output in Photovoltaic Modules for Outdoor Operating Conditions," Energy Convers. Manage., 47(15–16), pp. 2424–2430.
14. Magare, D., Sastry, O., Gupta, R., Bora, B., Singh, Y., and Mohammed, H. (December 22, 2017). "Wind Effect Modeling and Analysis for Estimation of Photovoltaic Module Temperature." ASME. *J. Sol. Energy Eng.* February 2018; 140(1): 011008.
15. E. Skoplaki, J.A. Palyvos, On the temperature dependence of photovoltaic module electrical performance: A review of efficiency/power correlations, Solar Energy, Volume **83**, Issue 5, (2009), Pages 614-624, ISSN 0038-092X
16. Haddad, Sofiane & Rabhi, Abdelhamid & Soukkou, Ammar & HASSAN ALI, Mohamed & El hajjaji, Ahmed. (2018). Experimental Analysis and Assessment of a HIT Photovoltaic Module

Smart Bus Tracking For Blind And Visually Impaired

Saloni Redij[1], *Ashwati* Rao[1], *Sagar* Anil Kumar[1], and *Ashish* Jadhav[1]*

[1] *Ramrao Adik Institute of Technology, D. Y. Patil Deemed to be University,Nerul, Navi Mumbai, Maharashtra, India.*

Abstract. Visually impaired and blind people (VIBP) face great difficulty finding public transport and bus stops due to their visual impairments. Over the past decade, several support systems have been prototyped and developed to address this problem. However, most of these are wearable prototypes that turn out to be grueling for them. In this project, we propose an AI-powered real-time bus navigation system that provides information about buses arriving at bus stops. The buses will be tracked using the GPS of the mobile phone installed with the driver-side application. The application will take audio input for the source and destination for the journey and provide the available buses from the nearest location of the user on that route as audio output. It will provide the ETA and location of the target bus. This app will coordinate with the driver's app and live notifications will be given to the driver for any visually impaired passenger.

Keywords– VIBP, AI voice assistant, real-time bus tracking, GPS, Google Maps.

1 Introduction

According to a WHO survey [1], it is estimated that there are 285 million people with visual impairments. 39 million of them have a visual impairment and 246 million have low vision. There are many challenges for people with visual impairments, especially when it comes to using public transportation[2]. In any unknown environment, navigation and orientation are traditional challenges. Buses are widely used for transportation in major cities. Most blind and visually impaired individuals depend on public transportation for their transportation needs, especially trains and buses, because public transportation is the only viable option for seeking education, work, and social networking [3][4].

In the limited environment they live in, they aren't able to perceive the outside world as well as others, and this is having a negative impact on their activities, such as in education or transportation since they rely on only their own intuition. Among other resources, it gives them access to employment, community resources, health care, and recreational activities [5]. Despite established public transportation infrastructure, it is not easy or comfortable for the visually impaired to use it. They are forced to settle for less-productive local employment opportunities due to unfriendly public transport systems not tailored to their needs. The majority of the visually impaired choose not to go to work far from home. This often makes them settle for less than what they could actually do.

Volunteer organizations such as helping hands provide important assistance to the visually impaired community[6]. They provide mobility training and other forms of assistance. It's very important for a visually impaired person moving into a new area to be able to get around with the help of a trained and knowledgeable guide. However, these solutions make the visually impaired dependent on travel aids, which makes it impossible to navigate outdoors without the assistance of a professional.

This paper presents a systematic and structured way to provide bus arrival information at an individual bus stop via a mobile application, which takes into account the voice of the visually impaired person when receiving the bus destination. The voice assistant will constantly

*e-mail: ashish.jadhav@rait.ac.in

assist and update the visually impaired person till they reach their destined bus station. The estimated arrival time of the desired bus at the bus station will be updated and informed in real-time. Once the person has boarded the bus he/she will be assisted and notified when they are about to reach their desired bus station so that they can stay alert and get down from the bus without any confusion.

2 Literature Survey

"An RFID Based System For Bus Location Tracking And Display" [7]. This paper uses an RFID system and RFID tags are embedded on buses and readers at alternative bus stops, to detect buses. The local government server receives information and displays all of this data at the bus stop. This system is not very economical to implement on a large scale since the RFID tags are expensive.

"Smart Bus Alert System for Easy Navigation of Blind" [8] proposes to provide a solution using a wireless sensor network (WSN). The ZigBee system is used to indicate that there are blind people at the bus stop. The voice module and APR9600 voice playback system are used. Since ZigBee has a range of 300-400 meters, this system can be an advantage[9]. In this system visually impaired people must carry a ZigBee mobile unit, this unit has a certain size and weight that make it difficult and troublesome to carry for them [10].

"Implementation of a Navigation Application for the Visually Challenged" [11]. Visually impaired people face many challenges while navigating both indoors and outdoors. Indoor navigation is provided with the help of Bluetooth Beacons and Indoor Atlas technology. This application provides voice assistance to users to navigate to their destination.

"Wireless Bus Identification System for Visually Impaired Person" [12],uses RF-based wireless communication to simplify movement for the visually impaired. The system consists of an RF module and ultrasonic sensors. The blind person activates the toggle switch which sends a signal to the bus driver who then helps them board the bus.

"Easy Navigation of Bus using Alert system"[13]. The bus alert system contains a microcontroller for processing bus route data and an RFID reader for reading bus routes and numbers at bus stops, as well as an LCD window to display the bus arrival status. The aim of this paper is to provide information on bus time and bus statuses. "Navigation of Blind People Using Passenger Bus Alert System"[14], ARM-7 is used to generate signals that are sent to to bus to indicate the presence of a blind passenger.

"BlindNavi: A navigation app for the visually impaired smartphone user[15]. Breadcrumb files are used. Based on the information on the breadcrumb that is given by the sensors, we can gain the desired location on the same. Despite its cost-effectiveness, the model lacks accuracy. "A radio network for guidance and public transport assistance of the visually impaired"[16], introduced a portable handheld device called "PAVIP transport system." It utilizes a PAVIP box in conjunction with the radio module installed on the device, which allows information to be exchanged between the vehicle and device. Using RFID chips, it sends route numbers, directions, and the next stop, but the disadvantages are that the signal may overlap with other signals.

"TYFLOSET" was proposed by Chechak J.[17], which combines a guiding system and a voice playback system. Using this system, the user is able to request driver information and the bus schedule while operating the acoustic system. The RAMPE model developed by Baudoin, G. et al. [18] is intended to provide an interactive mode of transport to the blind. But the vehicle does not interact directly with the user [19]. Using the Intermodal Transport Control System, Bischof et al. proposed the model, "NA VCOM," which utilizes the air interface. This model has the disadvantage of no direct communication between the user and the bus controller. It was hard to establish a new connection and sometimes hard to locate a new WLAN connection[20].

3 Problem Statement

For most visually impaired people today, public transport is the most convenient means of transportation. However, the existing infrastructure of bus transportation is inadequate for people who are visually impaired. There is no word on whether the bus they need is available. Considering the discomfort of traveling by bus, most of them avoid going to work at distant locations from their homes. It is therefore imperative that a system be developed for the purpose of easing the commute process of visually impaired people by bus. It should allow them to easily locate the bus that takes them to their destination. In order to make bus travel convenient, they should be informed when the bus is approaching and time of arrival.

4 System Design and Specifications

A system that is user-friendly and cost-effective is proposed to help visually impaired people navigate buses. This system consists of two major units- User Unit and Driver Unit.

1) User Unit This is the core of the system and the most important module. In order to activate this module's voice assistant, users need to tap the screen. All the bus details can be accessed from their smartphones. It is possible to track the location of a bus from any location. Users must ensure that their location services are active. Once the voice assistant is activated, the user can either ask for the nearest bus stop from their location or can ask for the bus number going to their desired destination bus stop.

2) Driver Unit This module can be used by authorized bus drivers. Before driving, they must activate their location services. Using a GPS device on the driver's mobile device, the server is automatically updated with the current latitude and longitude of the bus. The driver can start a new bus session or can select the bus from the list of already active buses. If the driver starts a new bus he can select the start and end of the bus from the given list. Whenever a visually impaired person gives a request for a particular bus the system sends a notification to all the active buses whose stops match with the given source(S1) and destination (D1) by the visually impaired user to indicate the presence of a visually impaired person along with their source and destination to the bus driver.

5 Methodology

Figure 1 shows the functioning of user unit and Figure 2 shows the functioning of driver unit. When the visually impaired person opens the application Google Maps API and GPS (Global Positioning System) fetches the real-time coordinates of the visually impaired person. Then by tapping on the screen the visually impaired person activates the AI voice assistant. The voice stream given by the user is converted to text via speech to text feature and is sent to the DialogFlow. DialogFlow agent handles the types of conversations and translates user audio input during a conversation to structured data. The user can either ask for the nearest bus stop from their location or for the bus number going to their desired destination bus stop. The nearest bus stop is calculated by finding the minimum distance between the location of the user and the bus stop. All the distances are calculated by using the open route service api and the Flutter Geolocator Plugin. If the visually impaired person doesn't explicitly give the starting point (S1) of their journey the system considers the nearest bus stop from the user's location as the beginning of the journey. Once the user has given a source and destination, the system looks for the matching destination in the list of bus stops of the upcoming active buses in the firebase cloud.

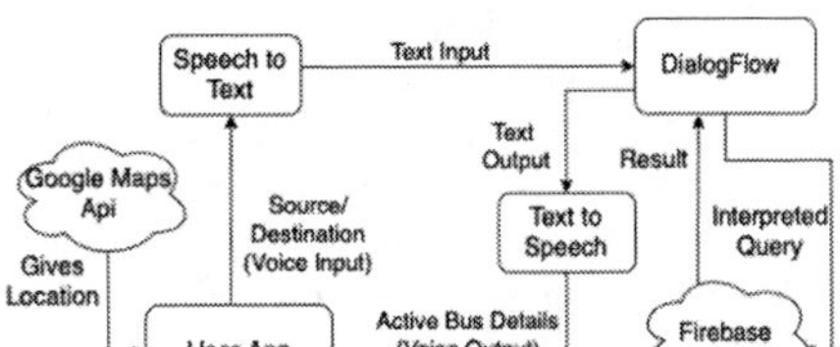

Figure 1: Functional Block Diagram of User Unit

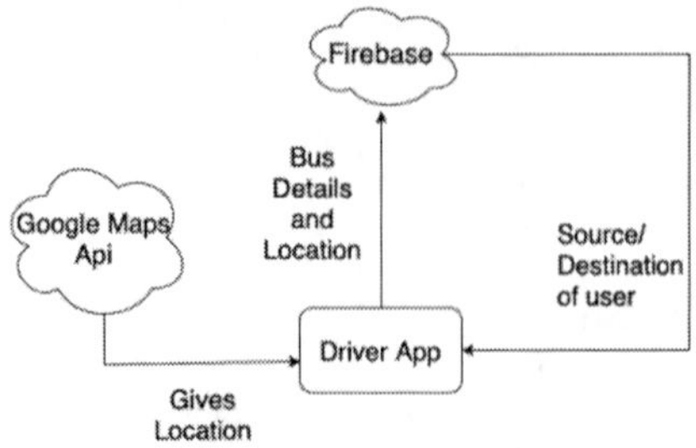

Figure 2: Functional Block Diagram of Driver Unit

If the destination matches, the system will announce the upcoming desired bus if the bus is within the range of 500m and 50m. If no information is given, it again asks for source and destination. When the target bus crosses S1 the AI assistant asks the user if they were able to embark the bus? If the user says no, then the next nearest upcoming bus for D1 is announced to the user. If the user says yes, then the application starts checking if the destination of the user has arrived. When the location of the bus matches with the location of destination, the AI assistant announces that the destination has arrived. The user can also access the location of the bus anytime during the wait and the journey by long pressing on the screen. On the driver side application of the system, the driver has to select the bus they will be driving and the start and end of the journey of that bus from the given list. The location of the bus is stored in the Firebase which will be updated continuously. Whenever a blind person gives a request for a particular bus the system sends a notification to all the active buses whose stops match with the given source(S1) and destination (D1) by the blind user. This way the system is able to indicate the presence of a blind person along with their source and destination to the bus driver so that they can wait longer than usual for the blind person to embark and disembark the bus. When the blind person is in the bus and their destination is about to arrive the notification that was given initially to all the active buses going to D1 are removed.

6 Result

An experiment was conducted to check if the interface was easy to use. A total of 85 students were blindfolded and were asked to interact with the application and were asked to rate their level of satisfaction on the scale of 1-10, 1 being the lowest and 10 being the highest. Figure 3 displays the overall results obtained.

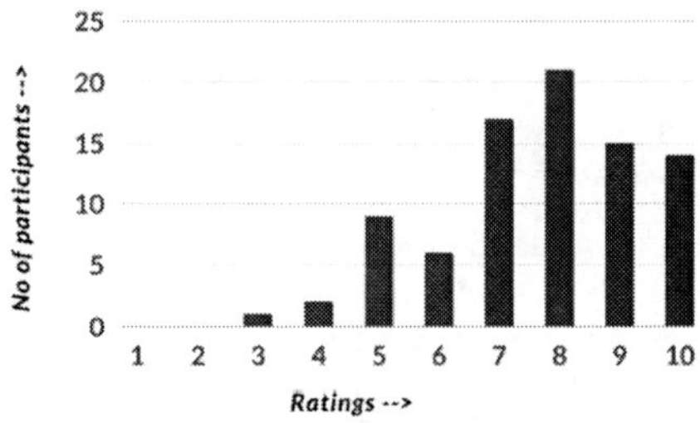

Figure 3: Statistics of survey responses

Based on the analysis of Figure 3, 86% of the participants have given a positive response, 11% gave a neutral response and 3% gave a negative response.

To test the functional efficiency of the proposed system, the prototype has been installed in the user's mobile and the driver's mobile with provided internet, in order to try all the possible solutions. A vehicle roamed through the roads of DY Patil campus in Nerul city for several days and at different times, for testing the functionality of the application. In accordance with the system model, this information is transferred via the internet to a Firebase server and then to the Android mobile application. Table 1 represents the interaction between users and the application on the end-users mobile application.

Table 1: Table of solutions and outcomes

Requirements Identified	**Solution Build**	**Outcome**
Finding nearest bus stop	U1: "Where is the nearest Bus stop?" A1: Nearest Bus Stop is *Name of the nearest bus stop*" U2: "Okay" A2: "You want to go from *Name of the nearest bus stop* to?"	Information about the nearest bus station is made available easily.
Finding bus	Case 1: The source bus stop is the nearest bus stop U1: " Take me from Datta Mandir(source) to Bahnani Chowk(destination)" A1: "Do you want to go from Mandir(source) to Bahnani Chowk(destination)" U2: "Yes" Case 2: The source bus stop is not the nearest bus stop U1: " Take me from Nerul Sector 21(source) to Bahnani Chowk(destination)" A1: "Do you want to go from Mandir(source) to Bahnani Chowk(destination)" U2: "Yes"	Helping to find the desired bus without assistance from any volunteer or any abled person.
Approaching bus	A1: Bus number N2 is within 500m of your bus stop A2:Bus number N2 is within 50m of your bus stop.	Alerts are given for the desired bus.
The bus has arrived at Source bus station	A1: Bus number N2 has arrived at your bus stop please board the bus	Alert is given so that the visually impaired person can board the bus.
The bus has left the user's starting bus stop	A1:Have you embarked the bus? U1:Yes A2:Distance from the destination is 1500m.	To make sure that the visually impaired person has boarded the bus, if not then details of the next nearest bus qualifying the user's input are made available to the user.
The Bus is about to approach to destination	Instance 1: The bus is 50m away from destination A1: You are about to reach your destination, prepare to disembark Instance 2: Bus is 25m away from destination A1: You are about to reach your destination, prepare to disembark	Alert the visually impaired about their approaching destination.
The bus has reached the destination of the user	A1: You have reached your destination, please disembark A2: Did you disembark? U1: Yes	Provide assistance to disembark the bus.

7 Conclusion

Using this proposed scheme, visually impaired individuals can independently travel by bus from their current location to their desired destination without encountering any obstacles. It is possible for users to get information on the nearest bus stops and when their desired bus will arrive through this application. This allows users to plan routes and itineraries accordingly. The advantages of this application are: (1)Easy to use (2) Users can travel independently. (3) Users don't need to carry another device for navigation.

8 Future Scope

The proposed system achieves the framework objectives, but there are still some improvements that can be made to make the system better. This system can be further enhanced to provide communication between the visually impaired person and their relatives and in case of any emergency, send the user's current location and destination to them. The system currently depends on the application to be installed by the driver on their smartphone. In the future, the software in the buses can be modified to include the current functions of the driver-side application. This will eliminate the need for the driver to explicitly carry a smartphone. The application interface presently focuses on the needs of visually impaired people. This system can be further extended for the use of all individuals.

References

[1] M. Markiewicz, M. Skomorowski, TST , (2010)

[2] S. Priya., IJRASET, (2018)

[3] B.Moulika, Y.R. Sankar, .K. Praveen, G.S. Surendra Babu, J. Engg Sci. **11**,(2020)

[4] S. S. Patil, V. Mareeswari, G. Amogh, M. S. Giridhar. ICSE, 353,(2019)

[5] K. Sasikanth, G. K. Vinay, K. Sudhakar, Y. Surya, M.S.S. Rao, IJRE **05**, 1 2018

[6] Dr.R.Thendral, A.Gayathri, P.Kalaivani, T.Kiruthika, IJERCSE **05**, (2018)

[7] M. Anu, D. Sarikha, G. K. Sai, J. Jabez, *Intl. Conf.and Cmpt. Tech.*,(2015)

[8] A. Holikatti, Dr. S. M. Kumar, IJANA, 508,(2016)

[9] D. Chander, M.V. Sireesha, IJESR **4**, 709,(2014)

[10] G. Lavanya; W. Preethy; A. Shameem; R. Sushmitha, ICCPCT, 798,(2013)

[11] S. Panicker, K. V. Maitreyi, M.Gonsalves, K. Gonsalves, S. Patil, S. Anand, IJETAE **07**, 137 (2017)

[12] A. Agarwal, K. Agarwal, R. Agrawal, A. K. Patra, A. K. Mishra, N. Nahak, *Odisha Intl Con. Elec. Pow. Engg, Commun. and Cmpt. Tech.*,(2021)

[13] S. A.Ghodake, M.Pallewar, R. Bandal, ICICET, 1,(2018)

[14] S. Gholap , G. Ekshinge , P. Naik, Prof.S.D.Chavan, IJSRP **05**, 189, (2015)

[15] Hsuan-Eng Chen, Yi-Ying Lin, Chien-Hsing Chen, I-Fang Wang, CHI,19,(2015)

[16] P. Korbel, P. Skulimowski; P. Wasilewski,HSI,485,(2013)

[17] J. Cechak, Intl. J. Electron. Telecommun. **57**,369,(2011)

[18] G.Baudoin, O.Venard, G.Uzan, A.Rousseau, Y.Benabou, A.Paumier, J.Cesbron, ITST, (2005)

[19] A. Meliones, D. Sampson,PETRA, **6(1)**,(2018)

[20] W. Bischof, E. Krajnc, M. Dornhofer, M. Ulm, ICCHP, 91, (2012)

https://doi.org/10.1051/itmconf/20224402002

Job Recommendation System Using Hybrid Filtering

Aneesh Mulay[1], Shriyash Sutar[1], Jiten Patel[1], Aditi Chhabria[1], Snehal Mumbaikar[1]
[1]Department of Computer Science and Engineering, Ramrao Adik Institute of Technology , DY Patil Deemed to be university
Nerul, Navi Mumbai, India

***Abstract*—As for today's era, recruitment can be considered as one of most difficult process to undergo for job seeking candidate. Many fresher candidates face issue while job recruitment process to undergo which field of interest. The proposed system will help the user to overcome this difficulties by matching their work experience,skills and other details with appropriate companies suitable for respective user. The system will also help experienced users in getting their intended job on the basis of their last job profile. The job recommendation algorithm developed is tedious nor complicated and will be using user-friendly approach to implement job search.The proposed system consist of user dataset with various attributes and company dataset with company details. The profile matching of user with the respective companies can be done using various recommendation algorithms such as content-based,collaborative and hybrid filtering. Since, the content-based and collaborative approach have their own disadvantages, so here implement hybrid filtering which overcomes the disadvantages of the content-based and collaborative filtering. The user can expect a well-proof recommendation from our model. The Project will focus of developing the job recommendation system using hybrid filtering. As for today's era, recruitment can be considered as one of most difficult process to undergo for job seeking candidate. Here, our job recommendation system comes in picture which neither is tedious nor complicated and makes use of user-friendly approach and helps user to accomplish the task easily.The project will also be focusing on developing the android application which will add a better user interface. The Android application will be user friendly and the user just have to fill in basic details such as his past years of experiences, project, internship, etc. That's it,the rest part of recommending the job to the users will be done safely by the recommendation model of this project.**

***Index Terms*—recommendations, content-based, similarity, jobs**

I. Introduction

Nowadays there is a rapid growth in Internet Technology, job seekers are releasing their own personal information whereas enterprises are continuously posting for jobs on the Internet. Because of this, there is a dramatic increase in availability of the job seeker's personal information and the recruiting information of various enterprises. Thus, the amount of such type of data keeps on increasing and when compared to the increase in data rate, there is not much increase in the utilization rate of this data or resources.

Given access to such huge amount of data with high veracity, an individual on his own may not be able to utilize this data in an efficient manner. This paper introduces Job Recommendation System that basically recommends jobs to users based on vast amount of information provided by the users and the huge amount of data regarding jobs that is available to us via various Internet Resources. The paper proposed using three types of filtering for providing recommendations - **Content Based Filtering, Collaborative Filtering and Hybrid Filtering**. Since, our job recommendation system uses multiple recommendation algorithms, the disadvantages or lack of efficiency of one algorithm is covered by another algorithm resulting in highly efficient recommendations.

The major objective of the paper is to build a model to recommend a job using **hybrid recommendation system** which is the combination of content-based filtering and collaborative filtering approach.The main motto is to make easy job search for users.

This recommendation depends on the user's past experiences as well as data from users with similar approach. The Recommendation model makes it easy for the users to get recommendation of various job profiles on basis of their past **experiences, projects, internships, skills** ,etc. The model will also help the experienced employees in recommending various job profiles based on their experience and skill based performance. The main reason being the freshers job recommendation approach as some of the students may get confused over various job profiles.

The system not only considers the experience factor of individual but also the skills and project developed to make the job recommendation more assuring from user's point of view. Hence, the user will not have any kind of uncertainty regarding the job posting recommended by our model.

Nowadays an enormous amount of data is available on the internet and Internet users can receive a huge amount of information. If the data volume or variety of data increases tremendously, then the individual user faces problems of excessive information, it can cause a problem to make the correct decisions.

People are often confused on what roles they fit in or where they should start their job search, especially younger people or graduates who are searching for their first job. For example, while the downsizing of a company, they usually let of the people with less experience. Such people with less experience in a particular field can face a problem of where they should start again or at what role are they supposed to fit in in a particular company. Considering that this issue is quite common for people with non-technical backgrounds. e.g.

Salesperson, marketing, etc.

To Resolve such type of problems, Job recommendation system comes in the picture.The Job Recommendation system can solve various problems by effectively finding user's probable requirements and select fascinating items from a vast amount of applicant information.

II. Literature Review

Pradeep Kumar Singh, Pijush Kanti Dutta Pramanik, Avick Kumar Dey, Prasenjit Choudhury [1] this paper provides a comprehensive study on the RS covering the different recommendation approaches, associated issues, and techniques used for information retrieval.

[2].This paper gives an overview of increasing data and explains users disadvantages to access useful recommendation informations. The paper depicts use of user's requirements, user's factors like location, music, shopping into recommendation system for giving possible recommendations.

Ravita Mishra, Sheetal Vikram Rathi [3] explains three types of recommendation algorithmsi.e Collaborative filtering, Content Based filtering, Content Based Filtering and Hybrid Filtering. It explains about various advantages and disadvantages of these algorithms.

Tanya V. Yadalam, Vaishnavi M. Govda, Vandhita Shiva Kumar, Disha Girish [4] explains about different methods such as Natural Language Processing, Cosine Similarity, Content Based Filtering,etc. Here, the paper depicts the use of Natural Language Processing for sentiment analysis on user feedback and also introduces encryption technique to handle data in a more secured manner.

Marwa Hussien Mohamed, Mohamed Helmy Khafagy, Mohamed Hasan Ibrahim [5] explains about Content Based Filtering and Collaborative Filtering. This paper proposed about various methods such as classification of data, cluster analysis, outlier detection, regression analysis, association analysis.

Greg Linden, Brent Smith, Jeremy York [6] gives a brief summary about customer to customer and item to item recommandations for amazon.com

Kunal Shah, Akshaykumar Salunke, Saurabh Dongare, Kisandas Antala [7] presents an overview of the field of recommender systems and describes the present generation of recommendation methods.

Gomez [9] introduce a business model while also building recommendation system. The paper also introduces the use of A/B test in recommendation system. It mentions the some isses while designing and interpreting A/B tests.

Thiengburanathum P, Cang S, Yu H [10] aims to build destination recommendation system. It introduces use of various algorithm to build recommendation model. The use of weighted hybrid and cascade hybrid methods is more depicted and used in the recommendation model.

Bell, R., Koren, Y., Volinsky, C [11] proposed the use of predicting the item raings using neighbourhood technique. The method consider using SVD to predict various user-item ratings based on the neighbourhood user-item rating dataset. The paper also highlights the imputation in order to fillin matrix entries from other algorithms

Karim, J. [12] have developed a Hybrid Recommender System Using Collaborative Filtering and Knowledge-Intensive Case-Based Reasoning.

E K Subramanian, Ramachandran. [13] proposed a Career Recommender System for Students based on the performance and marks obtained by them in various subjects.

Ivens Portugal, Paulo Alencar, Donald Cowan. [14] have introduced detailed information regarding different types of recommendation systems and the various machine learning algorithms associated with them. It also calculates the results of these systems and compared them using perfomance measures like precision, recall and f-measure.

III. Types Of Recommendation Techniques

A. *Content-based Recommendation System*

In Content-based Recommendation System, the final recommendations are generated based on user's profile data. This system provides the suggestion based on user's similarity with the items. Mainly, the concept of Term Frequency Inverse Document Frequency(TFIDF) is used in information retrieval and content-based recommendation system. TFIDF basically computes the frequency of words in respective documents.

1) Limitation of Content-based Recommendation System:

- Sparsity problem is the situation which concerns about insufficient data present in the dataset.
- Generally, the content based approach faces sparsity problem. Which means,this methods limits the recommendation only to user specifics.
- As the method only involves using the user related data, the dataset is insufficient as it does not involve rating given by the other users.
- The recommendation engine developed will not recommended anything besides user's interest.
- Hence this approach only helps to recommend the result based on user's interest and not based on other users preferences.

B. *Collaborative Filtering*

In Collaborative Filtering, historical data of users is used to make the recommendations. Based on the explicit ratings given by the users, the user to user similarity is calculated and then the corresponding items are recommended to the users.

1) Memory-based Collaborative Filtering:

- The idea behind implementing memory based collaborative filtering is to compute the similarity between different users based on their historical data.
- The approach works on ratings given by different users and then finally recommends the similar jobs to the users.

2) Model-based Collaborative Filtering:

- In Memory-based CF, SVD(Singular Value Decomposition) a machine learning algorithm is used to predict the user's ratings on unrated items.

 https://doi.org/10.1051/itmconf/20224402002

- In this technique, various algorithms can be applied but the most common and suitable approach would be matrix factorization model to apply SVD and reconstruct the rating matrix
- Finally, top recommendations for particular users and produced based on their predicted ratings.

3) Limitation of Collaborative Filtering:

- The recommendation system experiences cold start problem as it does not have any relevant past data of the users.The cold start problem is experienced in case of new user. When a new user register himself to the system, the proposed model does not know about his intereset and the user did not make any rating to the existing companies. Due to this scenario the system won't be able to recommend anything to the user.
- This approach uses more amount of data consisting of different perspective of different users.
- There is also sparsity problem which leads due to undefined similarity between different users.

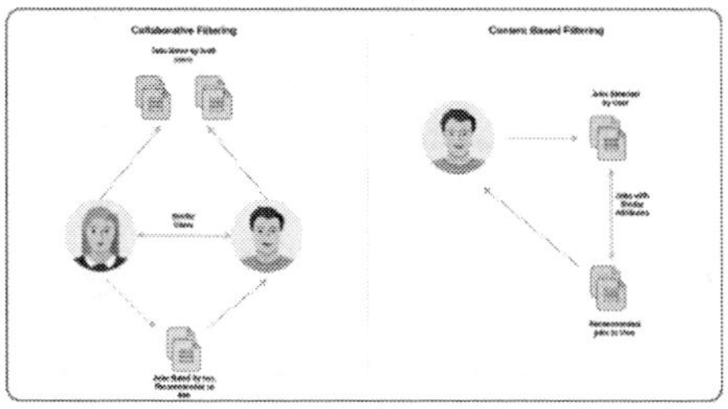

Fig. 1. Content-Based Filtering and Collaborative Filtering Recommendations

C. *Hybrid Filtering using Weighted Average Technique*

As explained in the above two approaches, both collaborative and content-based filtering techniques have their limitations. To resolve this, hybrid filtering techniques are used which is the combination of the above two mentioned approaches. In Hybird filtering using Weighted average technique, a weighted score is calculated using the results of final recommendations of both collaborative and content-based recommendations. The Hybrid Filtering helps in analyzing the results of recommendation systems when combined and when each recommendation system works alone.

IV. SYSTEM ARCHITECTURE

Initially to perform the content-based filtering approach, we need a dataset of set of companies. Hence we scraped the data from ambitionbox to get the companies dataset. After extraction, the paper depicts the use of TFIDF vectorization in the respective company dataset. With finding the words frequency, cosine similarity between each company attributes the similarity matrix is computed. Based on the similarity matrix Top-n content based recommendations are calculated.

For Collaborative Filtering, the paper follows the use of two approaches which are memory based approach in which user-user rating based similarity is calculated and model based approach which consist of deep learning techniques , matrix factorization. This paper shows the implementation using Singular Value Decomposition(SVD) to predict user ratings on unrated jobs. Both of these approaches provide Top-n recommendations .

The results from above approaches are combined using weighted average technique to form a Hybrid Recommendation System which in turn will provide the Top-n recommendations.

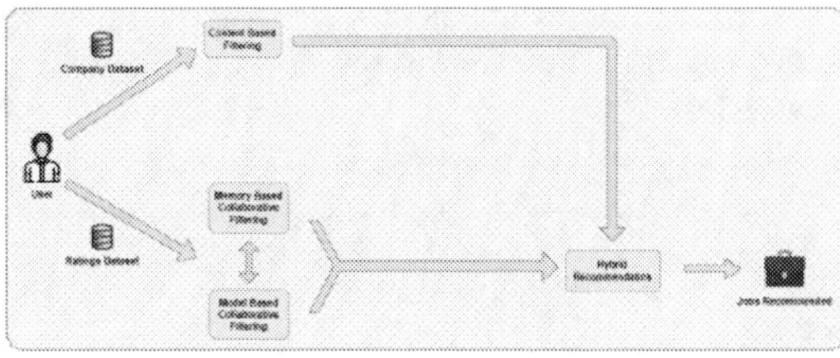

Fig. 2. System Architecture

V. EQUATIONS

A. *Cosine Similarity*

Cosine similarity is used to calculate the degree of similarity between two vectors in n-dimensional space. It is widely used in information retrieval.

$$sim(a.b) = \frac{a.b}{|a|.|b|} \tag{1}$$

B. *Weighted Average*

The weighted hybrid technique combines the result of both content-based and collaborative filtering techniques for comparison. This technique provides the results comparison when implementing both approaches combined and when each approach works alone.

$$W = \frac{\sum_{i=1}^{n} w_i X_i}{\sum_{i=1}^{n} w_i} \tag{2}$$

C. *Root Mean Square Error(RMSE)*

Root Mean Square Error is the square of all the errors. The use of RMSE is very common, and it is considered an excellent general-purpose error metric for numerical predictions.

$$RMSE = \sqrt{\frac{1}{n}\Sigma_{i=1}^{n}\Big(\frac{d_i - f_i}{\sigma_i}\Big)^2} \tag{3}$$

D. *Mean squared error(MSE)*

It measures the average of the squares of the errors—that is, the average squared difference between the estimated values and the actual value.

$$MSE = \sum_{i=1}^{n}(y_i - y'_i)^2 \tag{4}$$

E. Mean absolute error(MAE)

Mean absolute error (MAE) is a measure of errors between paired observations expressing the same phenomenon.

$$MAE = \frac{\sum_{i=1}^{n} |y_i - x_i|}{n} \tag{5}$$

F. Precision

Precision is calculated by true positive divided by sum of true positive and false positive. Precision is considered as positive predicted value.

$$Precision = \frac{TP}{TP + FP} \tag{6}$$

G. Recall

Recall is calculated by dividing true with sum of true positive and false negative. Recall is the fraction of relevant instances that were retrieved.

$$Recall = \frac{TP}{TP + FN} \tag{7}$$

H. F1 Measure

F1-measure is a measure of a test's accuracy. It is calculated from the precision and recall of the test, where the precision is the number of true positive results divided by the number of all positive results, including those not identified correctly, and the recall is the number of true positive results divided by the number of all samples that should have been identified as positive.

$$F1 = \frac{2 * TP}{2 * TP + FP + FN} \tag{8}$$

VI. RESULTS

A. Rating Dataset:

ID	userId	companyId	rating
0	1	1	4
1	1	3	4
2	1	6	4
3	1	47	5
4	1	50	5
5	1	70	3
6	1	101	5
7	1	110	4
8	1	151	5
9	1	157	5
10	1	183	5

B. Jobs Dataset:

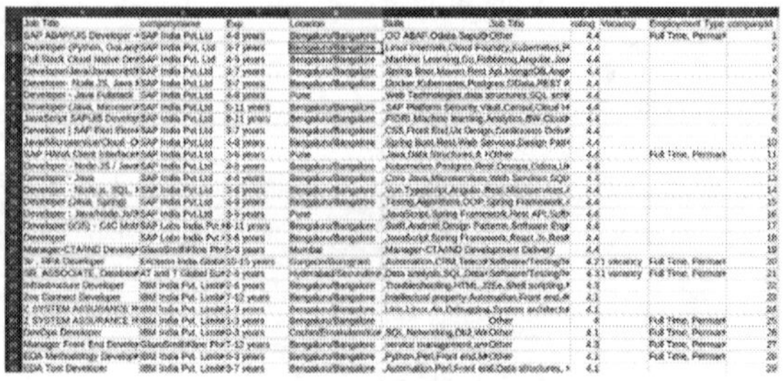

Fig. 3. Jobs Dataset 1.2

C. Final Results:

Company ID	ColllabRating	ContentRating	WeightedAvg
2593	0.2796	0.000	0.111
4690	0.2261	0.004	0.092
1612	0.2230	0.000	0.089
645	0.2295	0.000	0.355
506	0.3884	0.333	0.088
2283	0.2286	0.00	0.097
3655	0.2224	0.014	0.083
4428	0.2099	0.000	0.364
908	0.3674	0.333	0.08
750	0.2003	0.004	0.20

D. Evaluation with threshold

1) Content-Based Evaluation: For the evaluation phase of this study, we have used a threshold based approach in order to evaluate the relevance of the recommendations. First we select multiple random users and iterate over them, In each iteration we calculate their respective precision, recall and f-measure for a range of thresholds. In order to do that, we first calculate the threshold score. Then with the help of this threshold score we calculate the number of true positives, false positives and false negatives. Recommendations whose similarity scores are greater than the threshold score are considered as true positives and below that are considered as false positives. Recommendations whose similarity scores are equal to zero are considered as false negatives. With the help of these values, we calculate the precision, recall and f-measure for each threshold score of each user.

User 4590:

Threshold	Precision	Recall	Fmeasure
0.01	0.5654	0.5654	0.5654
0.02	0.553	0.5599	0.5564
0.03	0.5404	0.5542	0.5472
0.04	0.4734	0.5213	0.4969
0.05	0.4188	0.4907	0.4519
0.06	0.4046	0.4821	0.4399
0.07	0.3786	0.4655	0.4176
0.08	0.3708	0.4603	0.4107
...	...	...	...

User 4812:

Threshold	Precision	Recall	Fmeasure
0.01	0.7312	0.7312	0.7312
0.02	0.6616	0.71109	0.6854
0.03	0.5686	0.6790	0.6189
0.04	0.5406	0.66790	0.5975
0.05	0.5144	0.65679	0.57694
0.06	0.4294	0.61501	0.505712
0.07	0.3956	0.5954	0.4753
0.08	0.374	0.5818	0.4553
...	...	...	...

Then we calculate the average of these metrics at each threshold for each user.

User	Precision	Recall	Fmeasure
1721	0.3221	0.5204	0.3876
4846	0.2947	0.9607	0.3918
4542	0.577819	0.7478	0.6487
200	0.4247	0.5594	0.4797
793	0.3878	0.5328	0.4455
4590	0.3854	0.4635	0.4194
4812	0.3971	0.5765	0.4641
4730	0.3852	0.53801	0.4434
Average	**0.41112**	**0.6244**	**0.4768**

Then we calculate the average of these metrics at each threshold for all users as per the table below.

Threshold	Avg Precision	Avg Recall	Avg F-measure
0.01	0.7398	0.7479	0.7438
0.02	0.6555	0.7306	0.6894
0.03	0.6052	0.7143	0.6534
0.04	0.5441	0.6988	0.6059
0.05	0.5069	0.6853	0.5752
0.06	0.4522	0.6654	0.5276
0.07	0.4204	0.6496	0.4989
0.08	**0.3918**	**0.6362**	**0.4730**
...	...	...	...

Ideal Threshold: 0.08

The recommendation needs to be diverse in nature, selecting a high threshold would reduce the diversity of the recommendations whereas selecting a very low threshold would include almost everything which is more diverse. So we calculate the ideal balanced threshold value with the help of the average of all metrics that we found for each user.

2) Memory Based Collaborative: For User 5:

Threshold	Precision	Recall	Fmeasure
0.01	0.07061	0.1127	0.0868
0.09	0.0706	0.112	0.0868
0.11	0.0706	0.114	0.864
0.13	0.070	0.119	0.0861

For User 19:

Threshold	Precision	Recall	Fmeasure
0.01	0.0039	0.0067	0.0050
0.09	0.0039	0.0067	0.0050
0.11	0.0037	0.0067	0.0047
0.13	0.0037	0.0067	0.0047

Average of All Users:

Threshold	Precision	Recall	Fmeasure
0.01	0.115	0.161	0.134
0.09	0.0872	0.1266	0.1033
0.11	0.0784	0.115	0.093
0.13	0.07055	0.105	0.084
Avg	0.08	0.125	0.10

Average of Threshold:

Threshold	Precision	Recall	Fmeasure
0.01	0.109	0.151	0.126
0.09	0.0685	0.100	0.0816
0.11	0.0787	0.113	0.0930
0.13	0.0578	0.0863	0.0692
Avg	0.08	0.125	0.10

VII. Conclusion and Future Work

The proposed Job Recommendation System using Hybrid Filtering will be the most reliable medium for fresher candidates to get various job recommendation. The system will be also helpful for experienced users which will have no confusion or uncertainty from recommendation results. We will be using NLP for collecting various user feedback and then decide the efficiency of our model.

For better user experience this model can be embedded inside android application which will be done using flutter sdk for android development. Further, the user experience can be improvised by implementing this inside deep learning model.

References

[1] Pradeep Kumar Singh, Pijush Kanti Dutta Pramanik, Avick Kumar Dey, Prasenjit Choudhury. "Recommender Systems: An Overview, Research Trends, and Future Directions", 2021.

[2] Dr. Alka Singhal, Shivangi Rastogi, Nikhil Panchal, Shivani Chauhan, Shradha Varshney. "Research Paper On Recommendation System", Volume 9, Issue 8, August 2021.

[3] Ravita Mishra, Sheetal Vikram Rathi, Efficient and Scalable Job Recommender System Using Collaborative Filtering, Researchgate, 19 May 2020. Mishra Ravita, Rathi Sheetal, Efficient and Scalable Job Recommender System Using Collaborative Filtering. In: Kumar A., Paprzycki M., Gunjan V. (eds) ICDSMLA 2019. Lecture Notes in Electrical Engineering, vol 601. Springer, Singapore, 19 May 2020.
[4] Tanya V. Yadalam, Vaishnavi M. Govda, Vandhita Shiva Kumar, Disha Girish, Career Recommendation Systems using Content based Filtering, Published in 5th International Conference on Communication and Electronics Systems (ICCES), 10 July 2020.
[5] Marwa Hussien Mohamed, Mohamed Helmy Khafagy, Mohamed Hasan Ibrahim, Recommender Systems Challenges and Solutions Survey, Published in International Conference on Innovative Trends in Computer Engineering (ITCE), 2 February 2019.
[6] Greg Linden, Brent Smith, Jeremy York," Amazon.Com Recommendation", IEEE Computer Society, 2003.
[7] Kunal Shah, Akshaykumar Salunke, Saurabh Dongare, Kisandas Antala, Recommender systems: An overview of different approaches to recommendations, International Conference on Innovations in information Embedded and Communication Systems (ICIIECS), 2015
[8] Francesco Ricci, Lior Rokach, Bracha Shapira and Paul B. Kantor, Recommender Systems Handbook, Springer, 2010.
[9] Gomez- Uribe CA, Hunt N (2016) The netflix recommender system: algorithms, business value, and innovation. ACM Trans Manag Inf Syst (TMIS) 6(4)
[10] Thiengburanathum P, Cang S, Yu H (2016) An overview of travel recommendation system. In: IEEE 22th international conference on automation and computing
[11] Bell, R., Koren, Y., Volinsky, C.: Modeling relationships at multiple scales to improve the accuracy of large recommender systems. In: KDD 07: Proc. of the 13th ACM SIGKDD Int. Conf. on Knowledge Discovery and Data Mining, pp. 95104. ACM, New York, NY, USA (2007).
[12] Karim, J., 2014, Hybrid Systems for Personalized Recommendations, Research Challenges in Information Science (RCIS), 2014 IEEE Eighth International Conference, May 2014
[13] E K Subramanian, Ramachandran." Student Career Guidance System: Recommendation of a course". In " International Journal of Recent Technology and Engineering", Volume 7, Issue 6S4,2019.
[14] Ivens Portugal, Paulo Alencar, Donald Cowan." The Use Of Machine Learning Algorithms In Recommender Systems: A Systematic Review".
[15] Manish Kumar Singh, Dr. Dinesh Prasad Sahu." Research Aspects Of The System". In International Journal For Research In Applied Science Engineering Technology (IJRASET), Volume 5 Issue XI November 2017.
[16] Bhumika Bhatt, Prof. Premal J Patel, Prof. Hetal Gaudani. "A Review Paper On Machine Learning Based Recommendation System. In "International Journal Of Engineering Development And Research" Volume 2, Issue 4 ,2014.
[17] Kaveri Roy, Aditi Choudhary And J. Jayapradha." Product Recommendations Using Data Mining And Machine Learning Algorithms ". In ARPN Journal Of Engineering And Applied Sciences Vol. 12, No. 19, October 2017.
[18] Kaustubh Kulkarni , Keshav Wagh, Swapnil Badgujar, Jijnasa Patil , A Study Of Recommender Systems With Hybrid Collaborative Filtering, Volume: 03 Issue: 04 — Apr-2016.
[19] Priyanka." A Survey Paper On Various Algorithm's Based Recommender System ". In IOSR Journal Of Computer Engineering, Volume 19, Issue 3, (May - June 2017).
[20] Mukta Kohar, Chhavi Rana." Survey Paper On Recommendation System". In (IJCSIT) International Journal Of Computer Science And Information Technologies, Vol. 3 (2), 2012.
[21] Joeran Beel, Stefanlanger, Marcel Genzmehr, Bela Gipp, Corinna Breitinger, Andreas Nurnberger," "Research Paper Recommendation System: A Quantitative Literature Survey", International Journal On Digital Libraries (2015).
[22] N. Divya, S. Sandhiya, D.R. Anita Sofia Liz , P. Gnanaoli ,"A Collaborative Filtering Based Recommender System Using Rating Prediction", International Journal Of Pure And Applied Mathematics Volume 119 No. 10 2018.

Smart Surveillance System for Anomaly Recognition

Kunal Kamble[1,*], *Pranit* Jadhav[1,**], *Atharva* Shanware[1,***], and *Pallavi* Chitte[1,****]

[1]Ramrao Adik Institute of Technology, D Y Patil Deemed to be University, Navi Mumbai

Abstract. Situation awareness is the key to security. Surveillance systems are installed in all places where security is very important. Manually observing all the surveillance footage captured is a monotonous and time consuming task. Security can be defined in different terms in different conditions like violence detection, theft identification, detecting harmful activities etc. In crowded public places the term security covers almost all type of unusual events. To eliminate the tedious manual surveillance we have developed a smart surveillance which will detect an anomaly and alert the user and authority without any human interference. It is a very critical issue in a smart surveillance system to instantly detect an anomalous behaviour in video surveillance system. In this project, a unified framework based on deep neural network framework is proposed to detect anomalous activities. This neural network framework consists of (a) an object detection module, (b) an object discriminator and tracking module, (c) an anomalous activity detection module based on recurrent neural network. The system is a web application where user can apply for three different security services namely motion detection, fall detection and anomaly detection which is applicable for monitoring different environment like homes, roads, offices, schools, shops, etc. On detection of anomalous activity the system will notify the user and responsible authority regarding the anomaly through mail with an anomaly detected frame attachment.

1 Introduction

The term security covers various type of abnormal events like explosion, arson, burglary, assault etc. But some sorts of anomalies like violence detection is very difficult task to recognize in crowded place because it involves group activities. The anomaly analysis of a input stream involving crowd is very difficult because of large number of people are involved. Smart video surveillance systems are spatio-temporal and are capable of improving situational awareness. The goal is to explore the concepts of multi-scale spatio-temporal tracking through the use of real time video analysis and long term pattern analysis to encourage situational awareness. The aim is to presents a unique approach to recognize anomalies in intelligent video surveillance systems.We aim to solve the issues that arises in manual surveillance systems. We aim to recognize human activity and classify them as normal or one of the anomalies. We also aim to recognize anomalies that won't require human action interference. If a anomalous activity is detected the user and the responsible authority will get notified through mail along with an attachment of image in which the anomalous activity is detected. The application areas of Surveillance cameras are public places e.g. streets, traffic lights, banks, malls, schools, etc. in order to maintain public safety. However, the surveillance systems are not that efficient.

- This system aims on providing the user with 3 distinct types on anomaly namely motion detection, fall detection and anomaly detection.
- The system detect the anomaly and alerts the authorities but cam also be enhanced to discover a potential threat in advance for the incoming threat and hence, increasing the safety of people.
- The deep learning model is developed for only anomaly detection as motion detection and fall detection does not require an convolutional neural network to be trained on a dataset.
- Currently, the anomaly recognition model has been trained on different types of anomalous activities and normal activities. This can be broadened to integrate an enhanced variety of anomalies.
- The motion detection module doesn't require CNN model as it can be detected using only image preprocessing. The fall detection module only requires an object recognition framework to work hence, it saves a lot of time and results to quick response time.
- Video Surveillance systems, are being integrated with other industrial automation systems for intelligent application in fields like fire, access control, police services, manufacturing, transportation and many other fields so that surveillance cameras become a part of the Internet of Things.

Existing work focuses primarily on providing context-specific solutions on:

- Highways, traffic signals and main junctions

*e-mail: kunalkamble20k@gmail.com
**e-mail: pranit812jadhav@gmail.com
***e-mail: atharvamshanware@gmail.com
****e-mail: pallavi.chitte@rait.ac.in

- Apartment buildings, houses or other residential areas
- Crowd pull gatherings
- Different religious festival gatherings
- Inside office, school buildings, malls and shops

Resulting, insufficiency in the implementation of cctv cameras and an unfeasible ratio of human monitors to surveillance cameras. Maintaining a constant watch using surveillance cams is very tiresome for the operators. In real word situations it is necessary for intelligence to be visible. In real world scenarios response time taken for generation is highly important. In emergency situations like stampede prediction of certain behaviour, actions and movements is highly useful. The primary contribution of this paper is to offer a deep learning framework for identification of anomalous activities. A framework that utilizes YOLO network and Kalman filter as an object detector and discriminator model for the video frames, from which VGG-16 based CNN model extracts multiple features and finally a LSTM model trained on these features classifies whether the activity being performed or the general environment is anomalous or normal. The result obtained by this model reflects the significance of the dataset and provide scope for further work.

2 Literature Survey

Anomaly recognition involves identifying human activities that are carried out in an anomalous event. This involve identifying the objects of interests for feature extraction. The region of interests part of the video with some human interaction or objects that effect the environment and leads to the anomaly. Since the final results depends on the identification, the particular objects of interest are crucially important. There have been multiple attempts in developing a human activity or anomaly recognition model. In 2020, Shreyas D, Raksha S and Prasad B implemented an "anomalous human activity recognition system" [2], their methodology uses adaptive video compression method which compresses parts of the video that are irrelevant, retaining only the objects of importance. This method combines adaptive video compression and 3D CNN with contextual multiple scales based on the temporal features, to provide an accurate anomalous activity recognition system that works in real time. This method can recognize anomalous events with an accuracy of 80 percent. In 2019, Dinesh Jackson Samuel, Fenil E and Gunasekaran Manogaran implemented a "Real time violence detection framework for football stadium comprising of big data analysis and deep learning through bidirectional LSTM" [4], which uses a spark framework to process real time videos from different sources,the video frames are separated and the features of each frames are extracted by using Histogram of Oriented Gradients function. Then the frames are labeled as violent and normal, which are used to train the Bidirectional LSTM network for recognition of anomalous activities. 94.5 percent accuracy was achieved using this method. In 2018, Waqas Sultani, Chen Chen and Mubarak Shah implemented a "Real-world Anomaly Detection System in Surveillance Videos" [6] which followed a method based on videos which are considered normal and violent as bags, and segments of video as instances, an anomaly ranking model is automatically learned that predicts high anomaly scores for anomalous segments of videos. In 2017, Kwang-EunKo and Kwee-BoSim implemented a "deep convolutional framework for abnormal behavior detection in a smart surveillance system" [9], the proposed method discriminates and tracks objects of same type by using object recognition algorithm and Kalman filter based object entity discriminator. Developed a deep convolutional framework for extracting dynamic feature of human behavior from a video frame. The VGG-16 feature extractor used in this method is pose based CNN, which identifies different poses that are involved in anomalous activities. This method managed to achieve 89 percent of accuracy. In 2016, D. Xu, E. Ricci, Y. Yan, J. Song, and N. Sebe in "Learning deep representations of appearance and motion for anomalous event detection" [23], utilised deep learning-based auto encoders to develop a model of normal behaviours and reconstruction loss to detect abnormalities recently and proposed an unsupervised deep learning framework for building discriminative representations for video anomaly detection automatically. This method also known as double fusion scheme is based on learning the correlations between appearance and motion features which are combined to discover abnormal activities. In 2017, Cheng-Bin Jin, Trung Dung Do, Mingjie Liu and Hakil Kim implemented a "multilevel action descriptor" [5], which consists of three levels: posture, locomotion, and gesture level; each of which corresponds to a different group of sub-actions describing a single human action, for example, eating while sitting using appearance based temporal features with multiple CNN. Considering all the mentioned case studies for anomalous action recognition system, the viable approach to this problem is in implementing a anomalous recognition system using a CNN+RNN model. The regions of interest for anomaly detection involves different objects that affect change in the state of environment. For identifying region of interest, an object detection and recognition framework can be used. In 2020, A. Bochkovskiy, C.-Y. Wang, and H.-Y. M. Liao developed "YOLOv4: Optimal Speed and Accuracy of Object Detection" [1], which is an object detection framework. This network is trained on coco dataset for recognizing 72 different objects from images. YOLOv4 is a state-of-the-art detector which is faster in terms of FPS and more accurate than all available alternative detectors. YOLOv4 detects and recognized multiple object visible within the frame. It has a lot of features and out of them the features that improves the accuracy of both the classifier and and detector are selected. Training a machine learning model on dataset that includes data in form of videos requires implementing spatio-temporal tracking algorithm. This means the features extracted on single frame should be tracked throughout the sequence of frames. Hence it is necessary to implement a algorithm to track the object detected by YOLO network through entire sequence of frames. In 2012, Z. Fu and Y. Han implemented a algorithm, "Centroid weighted kalman filter for visual object tracking" [20]. It is a state estimation method which pre-

https://doi.org/10.1051/itmconf/20224402003

dicts the target state using the state model and combines the observation model to estimate the posterior probability density function of the state. Videos are sequences of images which are trained on a CNN network. Multiple examples of algorithm based on Convolutional Neural Network with impressive capabilities in computer vision problems have been developed and has managed to achieve good accuracy in tasks of object detection and recognition in the ILSVRC (ImageNet Large Scale Visual Recognition Challenge) 2012, ZF-Net, GoogLeNet, and ResNet were the best-performing entries to the ILSVRC in 2013, 2014, and 2015. Among which VGGNet was one of the top-performing models. In 2014, Karen Simonyan and Andrew Zisserman of the Visual Geometry Group Lab of Oxford University proposed "Very deep convolutional networks for large-scale image recognition" [18]. Even for video recognition, VGGNet is simple to use and has a high recognition rate.

3 Proposed Methodologies

The goal of this problem statement is to develop a application that takes input from surveillance footage from the user's surveillance system, acknowledge any anomaly occurred in the footage and notifies the user. Mistreat, arrest, arson, assault, explosion, brawling, gunshot, and hooliganism are among the abnormalities. There are two broad categories of activities that human beings engage in: normal activities and anomalous activities. An anomalous behavior is when an individual behaves differently from normal in a way that causes harm to him or to others. Mental discomfort usually causes such behaviors. Intensive research into the recognition of human activity and its applications has shed light on detection of anomalies. The contexts identified are listed as application areas. The proposed system for anomaly detection takes stream of data i.e. frames from the video stream from different sources and analyze them for anomalous activities. Anomalies can be categorized in different types ranging from a simple motion in an unrestricted areas to a fight or an explosion. The sequence of frames of videos are handled according to the different category of the anomaly. Here the anomaly detection has been divided into three main categories:

- Motion Detection
- Fall Detection
- Anomaly Detection

The work flow of the proposed system is illustrated in Fig 1. The proposed system is a web application which takes video stream as an input from multiple users. These users selected category for anomaly detection is stored into the database. The system takes stream of data from different sources as input, breaks it down into small video clips and are stored on a cloud storage. The video clips on the cloud are fetched sequentially. The video clips are converted into overlapped frames and are passed to the respective detection to be processed further. If any find of anomaly is detected the user is alerted.

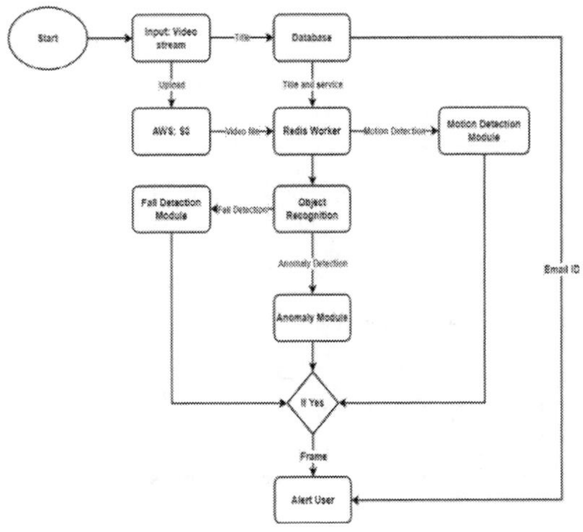

Figure 1. Proposed System activity flow

3.1 Motion Detection

Motion detection module detects any kind of motion captured and reports it. The motion detection module involve following steps:

1. Capture the video as a sequence of frames.
2. Read two frames from the sequence.
3. Get the difference between the current frame and the previous frame.
4. Apply Image manipulations like grayscaling and thresholding.
5. Finding Area of Contours to detect Motion.

3.2 Fall Detection

The fall detection module will recognize the human subject and the action of falling performed by the human. To recognize the human subject from the video an object detection framework is You only look one(YOLO) is used. The version of YOLO used in the proposed system is YOLOv4. YOLOv4 is a state-of-the-art detector which is faster in terms of FPS and more accurate than all available alternative detectors. YOLOv4 detects and recognized multiple object visible within the frame. It has a lot of features and out of them the features that improves the accuracy of both the classifier and and detector are selected. YOLOv4 recognize object within the frame and returns it's co-ordinates. The co-ordinates are used to determine the position of a person. Using these co-ordinates, height and width of the objects can also be retrieved. These co-ordinates are used for the fall detection of a human subject. Fall detection module involves following steps:

1. Capture the video as a sequence of frames.
2. Predict the person object using YOLOv4.

3. Retrieve the height and width of the object using the co-ordinates returned by YOLO.

4. Find the difference between height and width of the object between current and the previous frame.

5. If the height calculated in the current frame is smaller and the width is bigger than the person is considered as fallen.

3.3 Anomaly Detection

Anomaly detection is a machine learning framework Fig. 2 shows the flow of data through different modules for anomaly detection in a video.

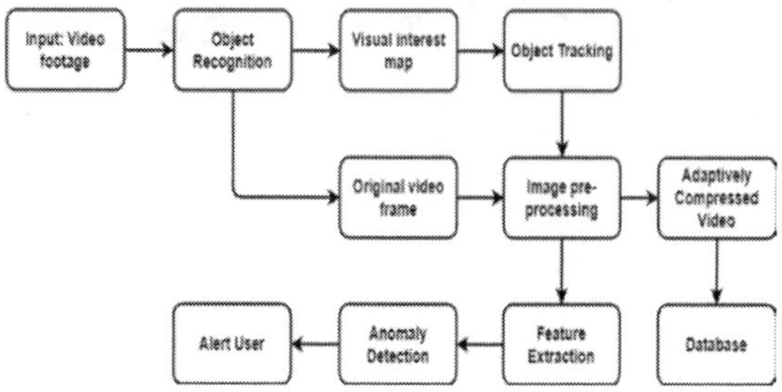

Figure 2. Anomalous Activity Detection

Input Module

This module takes video files as input. It converts video file into series of images and passes it to the next module.

Object Detection module

YOLOv4 object detection model is used for detecting different objects within each frame. It draws a bounding box around each object and mentions the object class. Following calculation shows the way each feature map is calculated:

$$W^1 = \frac{(W - F + 2P)}{S} + 1 \quad (1)$$

where W denotes the size of input image, F denotes size of receptive field, P denotes the number of pads, S denotes step size of stride, and W^1 denotes the size of resulted feature map. The input image is equally divided into a cell grid with S x S cells by the YOLO network. The bounding boxes B and $Pr(obj)$, the probability that each box will contain the object are calculated through the forward propagation of the network in each grid cell. The confidence score is computed using the probability of the object in the box and this reflects both the expectation of the object in the box and the extent of accuracy of the box prediction. The confidence score is depicted as $Pr(obj) * IOU^{truth}_{pred}$. Thus $Pr(obj) * IOU^{truth}_{pred}$ denotes that a value is computed by the intersection over union between two B, namely a box for ground truth (GT) box and a box for the predicted bounding box. If an object is not present in the predicted B the confidence score corresponds to 0. If the B and the GT box are completely matched the confidence score corresponds to 1.

Object tracking module

A YOLO object detector is able to detect an object extremely fast and accurately. However, the model is not able to discriminate between object of same type. The YOLO network detects all the specified objects within the frame discounting of their number. YOLO network unifies the object localizer and classifier into single CNN network to speed up the process of detecting objects. To develop a practical applications such as a smart surveillance system only object detection without discrimination is not enough. Hence, it is important to discriminate and track object entities among a group of the detected objects of same type.The centroid weighted Kalman filter based object entity discriminator not only discriminates the objects with same label but also tracks them through multiple frames is integrated in the system. For implementing an activity recognition it is necessary to discriminate each object and track them which is done by the object tracking module. The tracking region of the dynamic target is selected using background subtraction and the target position at the beginning of tracking is predicted by Kalman filter, and then centroid weighted method is utilized to optimize the target position for further enhancing tracking accuracy.

Image pre-processing module

In image pre-processing frames with bounding box are resized to (244, 244).

Adaptive Video Compression module

: The resized frames are combined together to form a compressed version of the video file and save it into the database.

Feature Extraction Module

Feature extractor is an important step in machine learning. Feature extraction decreases dimensionality by condensing a large number of input variables into a smaller number of feature groups. Even though the dimentionality is reduced this does not disturb the description of the original dataset. The feature extractor module use VGG. The Visual Geometry Group at the University of Oxford devised the VGG convolutional neural network model for image recognition, where VGG16 refers to a VGG model with 16 weight layers. The input layer accepts an image with dimensions of (224 x 224 x 3), while the output layer is a 1000-class softmax prediction. The feature extraction component of the model runs from the input layer to the last max pooling layer (labelled by 7 x 7 x 512), whereas the classification part of the model runs from the input layer to the last max pooling layer (labelled by 7 x 7 x 512). The weigths used in the feature extractor are trained on 14 million

Anomaly Detection Module

The next step is the analysis of the dynamics of anomalous behaviour based on RNN model. Feeding the output of a particular layer to the input in order to predict the output of the layer is the basis of the principle on which a RNN works. RNN accepts the current input data and previously received inputs in order to handle the sequential data. Due to the internal memory present in a RNN it is able to memorize the previous inputs. In RNN the data is cycled through a loop to the middle hidden layer. The input layer in RNN takes x as an input, processes it and passed it to the middle layer. Multiple hidden layers are present in the middle layer h and each has its own activation functions, weights and biases. Each layer generates out y. Fig 3 shows the working of RNN.

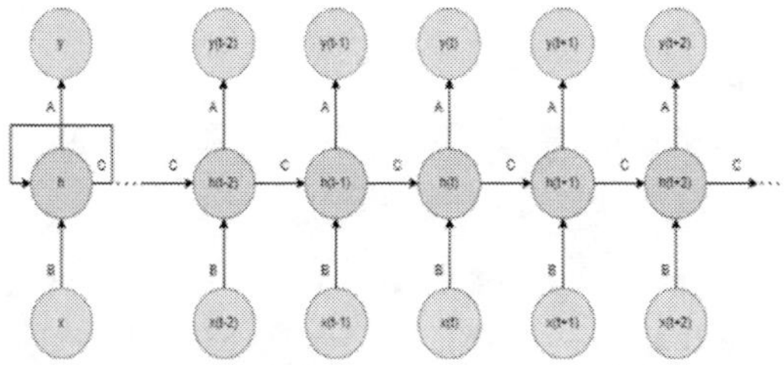

Figure 3. Working of RNN

The internal state of the network created by the architecture of RNN allows it to exhibit dynamic temporal behaviour. RNN uses backpropagation through time (BPTT) to perform weight update along with the time axis. However, the BPTT algorithm causes the RNN suffer through the Vanishing gradient problem. For an anomaly detection, it is assumed that it is necessary to identify the action sequence from the input which is also is assumed to be a sequence representing a series of actions. In an anomalous series of action the current always tend to be dependent on the previous or vice versa. The weight gradient gradually disappears in the learning process of a RNN model due to the increase in time interval of dependency. Long short-term memory networks (LSTMs) is a special type of RNN that are designed to avoid long term dependency problem. The LSTMs are suitable for usage in action recognition applications. By preventing gradient vanishing during the learning period, the structure overcomes the problem of long-term reliance. At every time step an input data sequence is received by the LSTM and the cell state and cell output vector is the result derived by each cell. Utilization of these two vectors is done at the next time step in the propagation of LSTM. The cell gives an output which is estimated as a class estimation which interprets the behaviour as normal or anomalous corresponding to the input data sequence as shown in the Fig 4.

The operation of a LSTM cell at time t is described. The output vector from the VGG-16 feature extractor is received as input X_t by each cell, along with the previous time's cell $t-1$, C_{t-1}, and the cell output h_1. The cell state updating procedure is developed in the following order based on these inputs: forget-input-output gate:

$$f_t = \sigma(W_{fh}h_{t-1} + W_{fx}X_t + b_f) \tag{2}$$

$$i_t = \sigma(W_{ih}h_{t-1} + W_{ix}X_t + b_i) \tag{3}$$

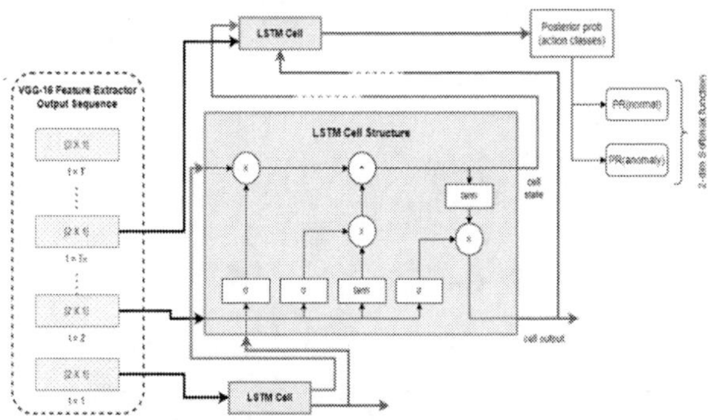

Figure 4. Anomaly detection process using LSTM with vector sequence recieved from VGG-16 feature extractor

$$o_t = \sigma(W_{oh}h_{t-1} + W_{ox}X_t + b_o) \tag{4}$$

h_t the final output and C_t the cell state of the corresponding cell at time t are obtained as follows:

$$C_t = f_t * C_{t-1} + i_t * tanh(W_{ch}h_{t-1} + W_{cx}X_t + b_c) \tag{5}$$

$$C_t = f_t * tanh(C_t) \tag{6}$$

The final output h_t from cell at time$t$ is the object class that is obtained from the LSTM. Based on the input given to the LSTM network at that time point along with output generated by that cell at previous time point the LSTM network outputs the result of behaviour class estimation i.e. normal or anomalous. The proposed method identifies ongoing activity from continuous streaming videos and enables in identifying whether the ongoing activity is normal or anomalous without completely perceiving the entire image sequence.

4 Results and Analysis

The primary objective of this proposed system is to develop a framework that is applicable in smart surveillance system for monitoring the well being of physically disabled or elderly people and also in infrastructure which require constant monitoring for commotion. A system capable of detecting normal and abnormal behaviour from the ongoing feature input generated from a video clip that of standard RBG image frame sequence has been developed. The hardware and software specifications of the environment in which the proposed system is developed and tested are as follows:

- Hardware specifications:
  - CPU: AMD Ryzen 5 3550H (8 CPUs, 2.1GHz)
  - GPU: NVIDIA GeForce GTX 1050 (3GB) CUDA enabled
  - Memory: 12GB
- Software specifications:

https://doi.org/10.1051/itmconf/20224402003

- Windows: 10
- CUDA: 10.1
- CUDNN
- OpenCV: 4.5
- Python: 3.7

4.1 Fall Detection

This module only utilizes YOLO object detection framework. The way object detection works is consider Fig. 5, which is a normal image, after giving this image as an input to YOLO network it gives Fig. 6 as an output.

Figure 5. Normal Image

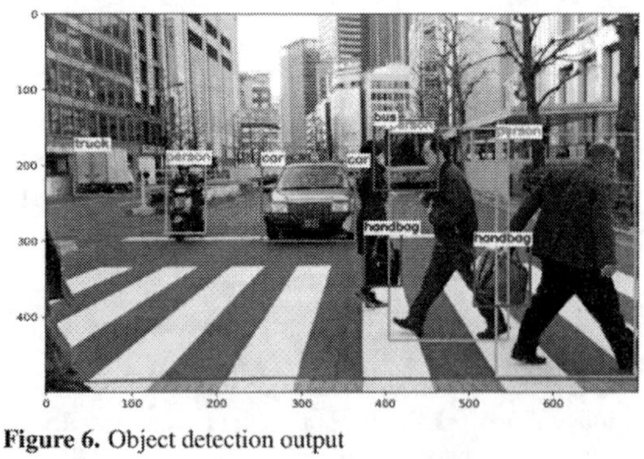

Figure 6. Object detection output

As it is seen in Fig. 6 that objects such as people, car motorcycle present in the Fig. 5 is displayed in bounding boxes. YOLO network returns the co-ordinates for the bounding box of the object along with the probability and class labels which is used to display the frames along with boxes and labels highlighting the objects within the image. The fall detection algorithm will takes sequence of images as output and will detect the person visible in the frames and whether the person is standing up or has fallen down. Fig. 7 shows a person standing upright, the YOLO network detects the person and determines the person is standing and is alright.

In Fig. 8 it is visible that the person is lying on the floor hence the algorithm detects that the detected person has fallen down and does the required action.

Figure 7. Person standing upright

Figure 8. Person fallen down

4.2 Anomaly Detection

The proposed method for anomaly detection is trained and tested on UCF Crime dataset. UCF-Crime dataset is a new large-scale dataset which consists of 128 hours of 1900 videos which are untrimmed real world surveillance footage captured on cctv cameras, with 13 realistic anomalies including arrest, shoplifting, arson, burglary, assault, explosion, vandalism, fighting, road accidents, robbery, abuse, shooting, stealing. These anomalies are selected because they have a significant impact on public safety. This categories present in the dataset is reduced to involving only arrest, arson, assault, explosion, vandalism, fighting, abuse, shooting, stealing which consists of 550 video clips. This is done to improve the computation with respect to the hardware specification. The above mentioned activities are considered as anomalous activities and the performance the methodology is evaluated. YOLO object detection and discriminator is implemented through Darknet library which is a open source deep learning library. Fig. 9 is the image in which the objects detected in Fig. 6 are discriminated fraom one another. Features are extracted from these frames which are used for training the VGG-16 feature extractor. VGG-16, a pre-trained CNN model based on large scale image dataset is used due to the dataset being small which is compatible for using a transfer learning approach. Caffe library, an open source deep learning framework is used in utilizing the VGG-16 feature extractor. Finally the feature vectors extracted from VGG-16 feature extractor is used for training the LSTM

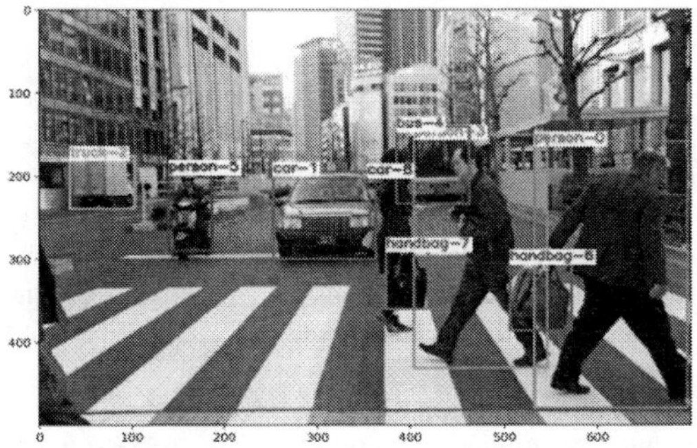

Figure 9. Object Discriminator Output

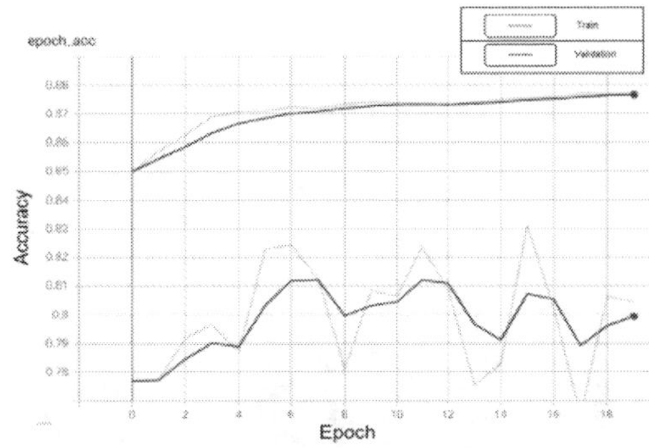

Figure 10. Train vs Validation Accuracy

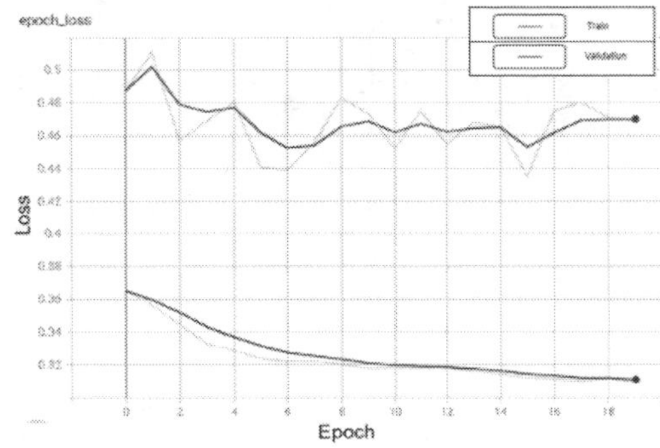

Figure 11. Train vs Validation Loss

network. For validation on the anomaly detection system the, precision and recall values are calculated.

The precision is measured as the ratio of the number of correctly identified Positive samples to the total number of Positive samples (either correctly or incorrectly). The precision of the model in categorising a sample as positive is measured. The precision is calculated in the following manner.

$$Precision = \frac{True_{positive}}{True_{positive} + False_{positive}} \tag{7}$$

The recall is determined by dividing the total number of Positive samples by the number of Positive samples accurately categorised as Positive. The model's ability to recognise Positive samples is measured by the recall. The higher the recall, the greater the number of positive samples found. The following is how recall is computed.

$$Recall = \frac{True_{positive}}{True_{positive} + False_{negative}} \tag{8}$$

The performance of the model across all classes is described by a metric called accuracy. It is useful when each class has equal importance. The ratio between the number of right guesses and the total number of forecasts is used to compute it. The accuracy is calculated as follows.

$$Accuracy = \frac{T}{T + F} \tag{9}$$

Where,

$$T = True_{positive} + True_{negative} \tag{10}$$

and

$$F = False_{positive} + False_{negative} \tag{11}$$

Accuracy measured during each epoch of training and validating the LSTM network in given in Fig. 10. Loss is also an important factor in training a LSTM model. The increase in loss causes decrease in the accuracy and vice versa. Hence, it is advised to keep loss to the minimum while training a machine learning model. Loss measured during each epoch of training and validating the LSTM network in given in Fig. 11.

The classification report generated for validation of the LSTM models is given in Table 1. The classification report consists of precision, recall, f1-score and support valued of both the classes along with the accuracy

Table 1. Classification report of LSTM model

	Precision	Recall	F1-Score	Support
Normal	0.81	0.75	0.78	147595
Anomaly	0.80	0.85	0.82	172345
Accuracy	-	-	0.80	319940
Macro avg	0.80	0.80	0.80	319940
Weighted avg	0.80	0.80	0.80	319940

Another metric for evaluating the performance of machine learning model is a confusion matrix. It is is *N* x *N* matrix where *N* is the number of target classes. It is a tabular summary of the number of correct and incorrect predictions made by a classifier. Since the LSTM model can predict between two target classes, hence $N = 2$. Table. 2 is the confusion matrix generated from the validation of the LSTM model.

Table 2. Confusion matrix of validation

	Predicted values	
Actual values	Normal	Anomaly
Normal	111124	36471
Anomaly	26142	146203

The way this model can be implemented is, consider Fig. 12, the frame is from the video clip of a protect taken place on 27 February 2022 in St. Petersburg, Russia. The video clip depict police brutality shown while arresting a protester, which is considered as a crime, hence it is recognized as an anomalous activity by the anomaly detection system.

Figure 12. Anomalous activity detection

On the other hand Fig. 13 is a frame from video clip taken from a cctv camera located on street-side. The video clip depicts people walking on the street and doing their job. There isn't any crime taking place in that clip, it is just people following their daily routine, hence it is recognized as an normal event by the anomaly detection system

Figure 13. Normal Acivity Detection

After training and analysis of the model, it has been observed that the proposed system can recognize normal and anomalous events with an accuracy of 80.43% with precision of 80% and recall of also 80%

5 Conclusion and Future scope

Due to the increase in crime rates, there is a great need for surveillance systems to be installed in all public spaces, such as residences, schools, and streets. However, simply installing surveillance cameras will not prevent criminal activity; there must also be a structure in place to give prompt assistance to victims of crime, as well as swift action against offenders. This is possible through continuous and careful monitoring of the surveillance footage which requires a lot of manpower. Smart surveillance system for anomaly recognition is a automated system which reads the surveillance footage from the cameras and alerts the user if any anomaly is found. As this system is automated there is no need of manpower to monitor the videos as it is done by the system itself. This system is helpful in many environments such as hospitals, schools, offices, highways, etc. The current system will detect whether the event is anomaly or not and will notify the user and authority that an anomaly has been detected. This system can be further updated to recognize the type of anomaly occurred and notify the specific authority to handle the anomaly. As this system is developed is developed in a low end environment, the dataset used to train the model is limited and also more complex algorithm require high computational power for training the model, hence training the model in high end environment will give better result than the current environment. In the current environment, actions performed in the video are processed and detected in real time. The YOLO and Kalman based combined network labels each object within the frame, vgg-16 CNN network extracts features from each frames, these features are labelled as normal and anomalous by the LSTM network. The system classifies the sequence of frames as normal and anomalous at an accuracy of 80%,

References

[1] A. Bochkovskiy, C.-Y. Wang, and H.-Y. M. Liao, "Yolov4: Optimal speed and accuracy of object detection," 23 2020.

[2] D. G. Shreyas, S. Raksha, and B. G. Prasad, "Implementation of an anomalous human activity recognition system," SN Computer Science, vol. 1, no. 3, p. 168, May 2020.

[3] G. Sreenu and M. A. Saleem Durai, "Intelligent video surveillance: a review through deep learning techniques for crowd analysis," Journal of Big Data, vol. 6, no. 1, p. 48, Jun 2019.

[4] D. J. Samuel R., F. E, G. Manogaran, V. G.N, T. T, J. S, and A. A, "Real time violence detection framework for football stadium comprising of big data analysis and deep learning through bidirectional lstm," Computer Networks, vol. 151, pp. 191–200, 2019.

[5] C.-B. Jin, T. D. Do, M. Liu, and H. Kim, "Real-time action recognition using multi-level action descriptor and dnn," in Intelligent Video Surveillance, A. J. R. Neves, Ed. Rijeka: IntechOpen, 2019, ch. 4.

[6] W. Sultani, C. Chen, and M. Shah, "Real-world anomaly detection in surveillance videos," in The IEEE Conference on Computer Vision and Pattern Recognition (CVPR), June 2018.

[7] Abdel, M., M, G., Rashad, H. and Zaied, A. N. H. (2018). A comprehensive review of quadratic assign-

ment problem: variants, hybrids and applications. Journal of Ambient Intelligence and Humanized Computing, 1-24.

[8] Huang W, Ding H, Chen G. A novel deep multi-channel residual networks-based metric learning method for moving human localization in video surveillance. Signal Process. 2018;142:104–13 (ISSN 0165-1684).

[9] K.-E. Ko and K.-B. Sim, "Deep convolutional framework for abnormal behavior detection in a smart surveillance system," Engineering Applications of Artificial Intelligence, vol. 67, pp. 226–234, 2018.

[10] Wang C, Yang H, Bartz C, Meinel C. Image captioning with deep bidirectional LSTMs and multi-task learning. ACM Trans Multimedia Comput Commun Appl. 2018;14:40.

[11] Zhang C, Tian Y, Guo X, Liu J. DAAL: deep activation-based attribute learning for action recognition in depth videos. Comput Vis Image Underst. 2018;167:37–49.

[12] Lee WK, Leong CF, Lai WK, Leow LK, Yap TH. ArchCam: real time expert system for suspicious behaviour detection in ATM site. Expert Syst Appl. 2018;109:12–24

[13] Dan X, Yan Y, Ricci E, Sebe N. Detecting anomalous events in videos by learning deep representations of appearance and motion. Comput Vis Image Underst. 2017;156:117–27

[14] Tsakanikas V, Dagiuklas T. Video surveillance systems-current status and future trends. Comput Electr Eng.

[15] Feng Y, Yuan Y, Lu X. Learning deep event models for crowd anomaly detection. Neurocomputing. 2017;219:548–56.

[16] Pang S, del Coz JJ, Yu Z, Luaces O, Díez J. Deep learning to frame objects for visual target tracking. Eng Appl Artif Intell. 2017;65:406–20 (ISSN 0952-1976).

[17] Zhou S, Shen W, Zeng D, Fang M, Zhang Z. Spatial–temporal convolutional neural networks for anomaly detection and localization in crowded scenes. Signal Process Image Commun. 2016;47:358–68

[18] Simonyan, K., Zisserman, A., 2014b. Very deep convolutional networks for large-scale image recognition, 2014b, aArXiv preprint arXiv:1409.1556.

[19] B. Hutchinson, L. Deng, and D. Yu, "Tensor deep stacking networks, IEEE Trans. Pattern Anal. Mach. Intell., vol. 35, no. 8, pp. 1944–1957, Aug. 2013.

[20] Z. Fu and Y. Han, "Centroid weighted kalman filter for visual object tracking," Measurement, vol. 45, no. 4, pp. 650–655, 2012.

[21] Şaykol E, Güdükbay U, Ulusoy Ö. Scenario-based query processing for video-surveillance archives. Eng Appl Artif Intell. 2010;23(3):331–45.

[22] M. Hasan, J. Choi, J. Neumann, A. K. Roy-Chowdhury, and L. S. Davis. Learning temporal regularity in video sequences. In CVPR, June 2016.

[23] D. Xu, E. Ricci, Y. Yan, J. Song, and N. Sebe. Learning deep representations of appearance and motion for anomalous event detection. In BMVC, 2015.

Hexagonal Microstrip Patch Antenna Design for UWB Application

Khushboo Singh [1*], *Sonal Patil* [1*], *Ashwini Naik* [1*], *and Sujata Kadam* [1*]

[1]Electronics and Telecommunication Engineering Department, Ramrao Adik Institute of Technology, Nerul, Navi Mumbai.

Abstract. Microstrip patch antennas is one of the most used antennas for wireless communication. Its key features include a limited bandwidth, low cost, and ease of manufacture. This paper describes about the design of microstrip patch antenna over the frequency range of 3.1 GHz to 10.6 GHz for UWB applications. The substrate material chosen is FR4, having a loss tangent of 0.02, dielectric constant of 4.4, and substrate thickness of 1.59 mm. This work presents design of microstrip antenna with hexagonal shaped radiating patch. An antenna is modelled and analysed using HFSS 2021 R2 software. The antenna structure provides return loss which is less than -10 dB and a VSWR less than 2 over the specified frequency range. The simulated results of proposed hexagonal shaped microstrip patch antenna provides a peak gain of 5.32 dB with radiation efficiency of 90.88 %. The planned antennas can be used for UWB applications.

Keywords. MSA, UWB, Gain and Radiation pattern.

1 Introduction

Wireless communication is one of the most rapidly evolving and dynamic technology areas within the field of communication. Wireless communications are now the frequently used in both human daily activities and industrial purposes. The most common wireless communication is a microstrip patch antenna. The key attributes of microstrip patch antennas are a cost feasibility, ease of fabrication and limited bandwidth.

A rectangular microstrip patch antenna of a I-shaped and dual C-slotted have been designed for 10 GHz. The breast cancer tumour is detected by the planned antenna which has bandwidth of 7.51dB, covers the frequency band of 4.9 to 7.89 GHz, has a return loss of less than -37dB, and has an antenna gain of a 2.45dB. The rectangular patch antenna is flexible and an efficiency of 87.1% were obtained [1]. The breast cancer detection system is used in compact UWB antenna have been designed. Here, the antenna has a frequency band of 3 to 12 GHz and an UWB of 120% for FCC band, with an overall size of 35mm x 20mm x 1.6mm. The suggested antenna features high gain, radiation patterns that are omnidirectional and good impedance matching [2]. The rectangular microstrip patch antenna for an UWB have been designed for microwave breast cancer detection and have evaluated the data mining in the classification and accuracy using random forest algorithm. Here, the rectangular MSA is designed over wide frequency range around 3 to 18 GHz. The random forest algorithm has been determined using breast cancer with a 94% accuracy, according to the classification results [3]. For UWB applications, a simple and compact microstrip-fed slot antenna have been proposed, which comprises of a circular patch as well as an open T-slot to implement a microstrip-line into slot transition over a broader frequency range. The observations and simulations show that such a simple structure, with its proposed system, from 2.5 to 12.5 GHz (bandwidth = 133.3%), yields a broadband impedance bandwidth with 10 dB return loss that can be adjusted by changing its parameters [4]. A new proximity-fed printed slot antenna with dual band notches for 3.5/5.5 GHz features have been designed and analysed. The designed circular microstrip patch antenna meets a return loss of -10 dB criterion over a frequency range of 2.7 to 17 GHz. To achieve dual band notched characteristics at 3.5 and 5.5 GHz, with an open ring slot has been etched off the circular patch as well as a -shaped slot has been etched off the microstrip feeding line [5]. A hexagonal monopole antenna which is relatively compact for UWB applications have been designed using 2 antennas which are fed to the single microstrip line over a frequency band of 4 to 14 GHz. The impedance bandwidth of the proposed antenna1 is approximately 111 %. The designed antenna2 has an impedance bandwidth for lower frequency is 31.58% and for higher frequency is 62.54%. A frequency notched band of 6.05 to 7.33 GHz was projected, while 6.22 to 8.99 GHz was observed, resulting in a single narrow band with axial ratio (1.435) have been achieved [6]. A slotted ground of three hexagonal MSA have been achieved using a circular polarization. The three hexagonal patch antennas are as simple patch, slotted ground patch, and parasitic element patch. The simple patch has a 2% impedance bandwidth, the patch with slotted ground has a 5.2% impedance bandwidth, and also the antenna including parasitic

* Corresponding author: khushis7891@gmail.com sonalpatil606a@gmail.com ashwini.naik@rait.ac.in sujata.kadam@rait.ac.in

element has a 6.35% impedance bandwidth. The patch with slotted ground has a measured axial ratio of 4.73%, while the patch with parasitic element has a measured axial ratio of 3.33%. Within the operating band, the suggested FSS provides a gain enhancement of roughly 3 dB [7]. A Novel modified Hexagonal microstrip antenna with curved partial ground plane have been proposed using UWB. Here, a modified hexagonal microstrip patch antenna has a 3.6 dB average gain (maximum gain ranging from 1.9 to 5.4 dB) and a relatively flat group delay and an impedance bandwidth of 3 to 11.3 GHz. The detected radiation pattern is observed as an omnidirectional [8]. The microstrip-fed hexagonal type UWB monopole with triple notched bands is used in the antenna. Two inverted T-shaped conductor backed planes, a microstrip feed line, and a regular hexagonal radiation patch with such a complementary split ring resonator (CSRR) placed inside an antenna underside make up the antenna. Here, the simulation and measurement result revealed that such a monopole UWB antenna can operate at frequencies ranging from 2.93 to 10.04 GHz with a return loss of -10 dB bandwidth, with the exception of 3 notched bands at 3.31 to 3.78 GHz, 5.33 to 5.77 GHz, and 7.24 to 7.72 GHz for dismissing WiMAX as well as downlink X-band satellite communications network signals [9]. A hexagonal microstrip patch antenna has been designed and simulated using breast cancer detection. Here, the antenna is mimicked by embedding a hexagon slot in the patch's centre, resulting in an impedance bandwidth of approximately 5 GHz [10]. A hexagonal and octagonal microstrip patch antenna has been implemented for UWB applications. For the designed antenna, the strip is built with a gap of d=3mm and length of 23mm between strip and ground planes to achieve matching. A geometry of hexagonal monopole is 10mm wide whereas octagonal has 9mm wide. The 2 printed monopole antennas have a total dimension of 45 x 60 x 1.6 mm and are etched on a FR-4 epoxy substrate. A Hexagonal strip monopole has a 5.5 GHz resonant frequency and a UWB impedance bandwidth around ($S11 < -10$ dB) of 1.54 - 9.41 GHz, whereas a octagonal strip monopole has a 5.5 GHz resonant frequency as well as a UWB impedance bandwidth around ($S11 < -10$ dB) of 1.3 - 5.65 GHz. Hexagonal has a VSWR of 1.52:1 over 2.09 GHz and Octagonal has a VSWR of 1.53:1 over 1.78GHz. Hexagonal has a bandwidth of 7.87GHz, whereas Octagonal has a bandwidth of 4.35GHz [11]. A hexagon shaped microstrip patch antenna has been designed and fabricated in three bands. The antenna is made up of a partial ground plane and hexagonal patch, which has the overall dimensions of 15 x 17 x 1.6 mm, therefore it works at three frequencies: 5.40, 6.76 and 8.82 GHz for TV satellite broadcasting, WLAN, WiMAX (5250 - 5850 MHz), 5.2 to 5.7 GHz for 5G Unlicensed band, 5.47 to 5.725 GHz for IEEE 802.11a, radar and weather monitoring [12].

2 Antenna Design Theory

2.1. Microstrip Patch Antenna (MSA)

Microstrip antennas are becoming increasingly important in the era of today's wireless communication. Antennas of various sorts have been developed for different wireless applications. The MSA is among the most suitable and attractive type of antenna compared to other antennas due to its light weight and smaller size. Microstrip patch antenna provides a numerous advantage, including conformability to mounting hosts, low profile, and ease of fabrication. Size reduction as well as bandwidth increase are also key design concerns for microstrip antennas within practical applications. These lightweight microstrip antennas are affordable, and durable. Their fabrication process is simple as it involves etching and photolithography, also simple feeding mechanism. The microstrip circuit elements are simple to utilise in an array or to combine with other microstrip circuit elements. Microstrip antennas are simple to integrate with MMICs and MICs. They are durable when installed on a device's hard surface. The antenna architecture and the feedline and matching network can be done at the same time. Microstrip patch antennas were born out of the desire to use printed circuit technology not just for transmission lines and circuit components, as well as for the radiating portions of an electronic system.

A metal patch and a ground plane are separated by a thin dielectric substance in a microstrip patch antenna. These antennas can be fed by using proximity feed, aperture coupled feed, microstrip line feed, and coaxial probe feed techniques. The energy is initially steered or linked to the region beneath the patch, which works as a resonant cavity with the open circuits on both of sides. Some of energy escapes the cavity and it is radiated into space, forming an antenna.

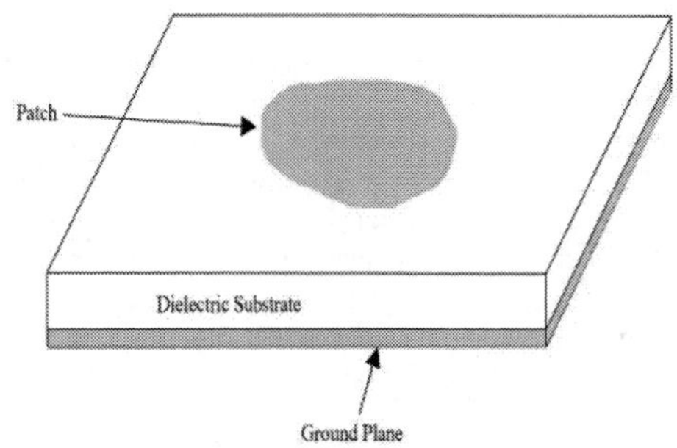

Fig. 1. Microstrip Patch Antenna

The figure 1 shows the basic structure of microstrip patch antenna. MSA can be made up of various shapes such as square, rectangular, circular, dipole, elliptical, circular ring, hexagonal and so on. Patch antennas are by far the most prevalent type of a micro strip antenna.

2.2 Design Theory of Microstrip Patch Antenna

Here, MSA is configured using hexagonal shaped radiating patch. The structure is designed to operate in the ultra-wideband frequency range of 3.1 to 10.6 GHz. The radius of Hexagonal Patch (a_h) is given by,

$$a_h = \frac{F}{\sqrt{\left\{1+\frac{2h}{\pi \varepsilon_r F}\left[ln\left(\frac{\pi F}{2h}\right)+1.7726\right]\right\}}} \quad (1)$$

Where,

$$F = \frac{8.791 \times 10^9}{f_o \sqrt{\varepsilon_r}}$$

The side length of hexagonal patch (l) is given by,

$$\pi a_h^2 = l^2 \frac{3\sqrt{3}}{2} \quad (2)$$

The height of hexagonal patch (L) is given by,

$$L = \sqrt{3}\, a_h \quad (3)$$

2.3 Proposed Structure

The MSA for ultra-wideband applications is proposed. The antenna structure consists of a dielectric substrate on one side which is a hexagonal shaped metallic radiating patch, and on the other side is a metallic ground plane. The substrate material chosen is of FR4. The relative permittivity, thickness and a loss tangent of FR4 substrate is 4.4, 1.59 mm and 0.02 respectively. Microstrip line model is used to feed the antenna. The antenna structure is modelled using HFSS 2021 R2 software. Ansys HFSS is a three-dimensional electromagnetic (EM) simulation software for developing and simulating high frequency electrical products like antennas. The antenna parameters are designed at resonating frequency of 6.85 GHz to cover frequency band range of 3.1 GHz - 10.6 GHz. The designed dimensions of hexagonal patch were obtained as radius of 10 mm, side length of 11 mm and height of 17.3 mm. The antenna structure is optimized to obtain return loss less then - 10 dB and VSWR less than 2. The geometry of proposed antenna structure and details of design dimensions are shown in figure 2 and 3 respectively. The optimized values of control parameters of HMSA are listed in the table 1.

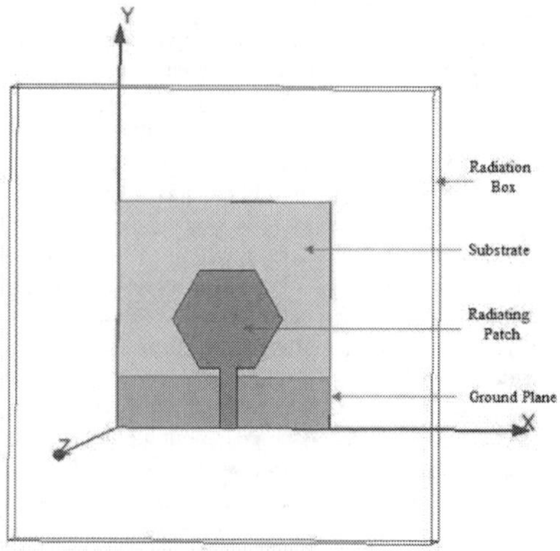

Fig. 2. Geometry of proposed antenna structure

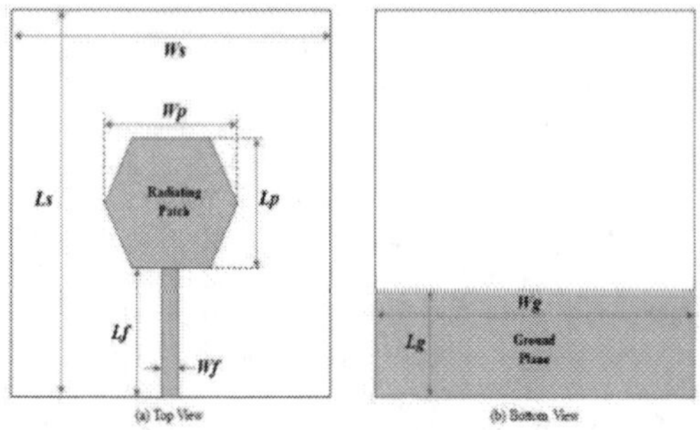

Fig. 3. Design details of dimensions of HMSA

Table 1. Parameters design details of HMSA

Parameter	Ls	Ws	Hs	Lg	Wg
Value (mm)	39.96	38.52	1.59	9	39.96

Parameter	Lp	Wp	Lf	Wf
Value (mm)	17.3	20	11.32	3

3 Results and Discussion

The performance of a hexagonal shaped microstrip patch antenna for ultra-wideband has been investigated by using HFSS software. The proposed antenna structure provides return loss less than -10 dB for the frequency range of 3.1 - 10.6 GHz as shown in figure 4.

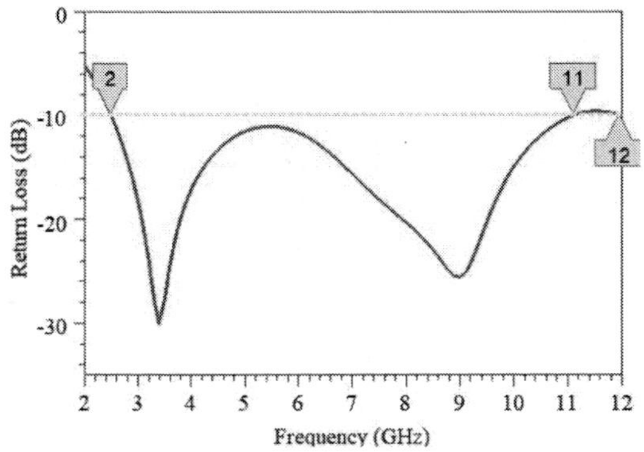

Fig. 4. Return loss verses Frequency of HMSA

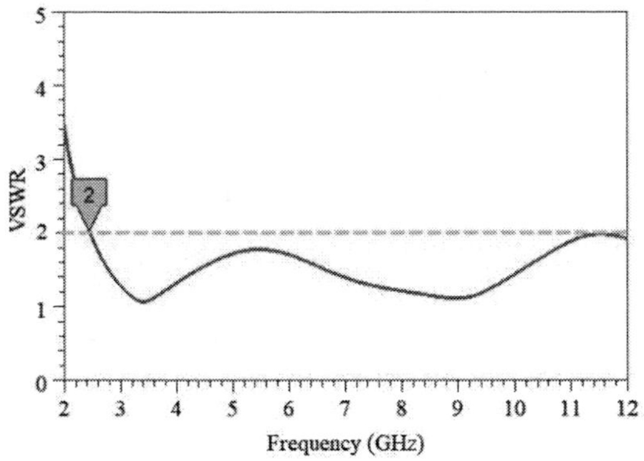

Fig. 5. VSWR verses Frequency of HMSA

Also, the VSWR of HMSA structure obtained is less than 2 for the frequency range of 3.1 - 10.6 GHz as shown in figure 5. The antenna structure provides the peak gain of 5.32 dB. The gain variations are shown in figure 6. The radiation efficiency obtained by the structure is 90.88 %. The H-plane and E-plane radiation patterns of HMSA at frequency 6.85 GHz are shown in figure 7.

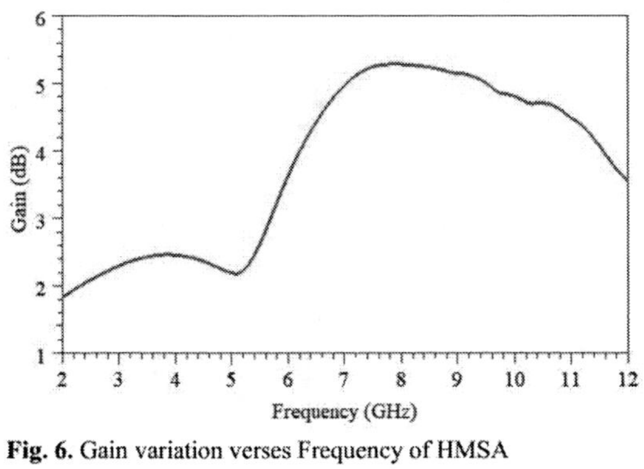

Fig. 6. Gain variation verses Frequency of HMSA

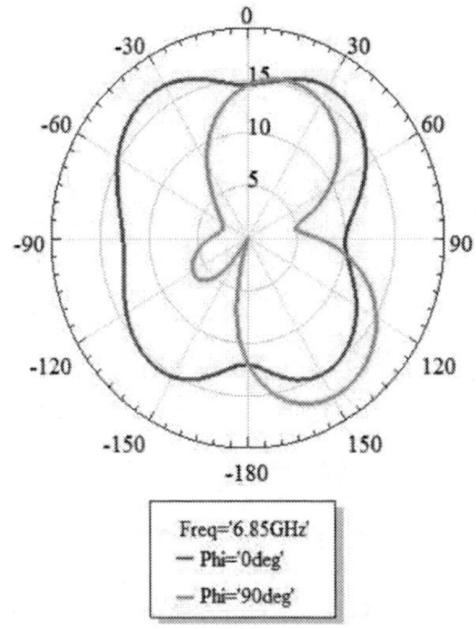

Fig. 7. E-plane and H-Plane Radiation Pattern of HMSA at 6.85 GHz

The simulated results of the proposed antenna structure are compared with the similar antennas present in the literature. The comparison is shown in table 2.

Table 2. Comparison with Literature

Ref. No.	Antenna Structure	Antenna Size (mm^2)	Gain (dB)
1	I-shaped dual C-slotted Rectangular MSA	0.5 x 6	2.45
2	Square ring shape antenna	35 × 20	5.2
4	Circular patch antenna with an open T-slot	30 × 35	4.1
8	Hexagonal MSA	30.4 × 35.4	3.6
Proposed Antenna	Hexagonal MSA	39.96 × 38.52	5.32

4 Conclusion

Here, the hexagonal shaped microstrip patch antenna is proposed. The antenna is made on a well-known FR4 substrate, which is frequently used to make printed circuit boards with a 1.59mm substrate thickness and a permittivity of about 4.4. The proposed antenna structure is simulated to operate in ultra-wideband frequency range covering a band from 3.1 to 10.6 GHz. The simulated result of hexagonal microstrip patch antenna provides the peak gain of 5.32 dB. The radiation efficiency of 90.88% is achieved. The designed antenna structure is useful for ultra-wideband applications.

Acknowledgment

The authors would like to thank our guide Dr. Ashwini Naik and co-guide Mrs. Sujata Kadam for their guidance and assistance to publish this paper. The authors are grateful to Dr. Mukesh D. Patil, Principal and also Dr. Chandrakant Gaikwad, Head of Electronics and Telecommunications Department at Ramrao Adik Institute of Technology for their support.

References

1. T. V. Padmavathy, P. Venkatesh, D. Bhargaval and N. Sivakumar, Design of I-shaped dual C-slotted rectangular microstrip patch antenna (I-DCSRMPA) for breast cancer tumor detection, Springer (2018)
2. I. Amdaouch, O. Aghzout, A. Naghar, A. V. Alejos and F. Falcone, Breast tumor detection system based on a compact UWB antenna design, *Electromagnetics*, **64**, 123-133 (2018)
3. E. A. Aydin and M. K. Keles, UWB rectangular microstrip patch antenna design in matching liquid and evaluating the classification accuracy in data mining using random forest algorithm for breast cancer detection with microwave, JEET (2019)
4. X. F. Zhu and D. L. Su, A study of a compact microstrip-fed uwb antenna with an open T-slot, *Electromagnetics*, **13**, 181-189 (2010)
5. M. M. M. Ali1, A. A. R. Saad and E. E. M. Khaled, A design of miniaturized ultra-wideband printed slot antenna with 3.5/5.5GHz dual band-notched characteristics analysis and implementation, *Electromagnetics*, **52**, 37-56 (2013)
6. A. A. Shaalan & M. I. Ramadan, Design of a compact hexagonal monopole antenna for Ultra—Wideband applications, Springer (2010)
7. N. Kushwaha and R. Kumar, Design of slotted ground hexagonal microstrip patch antenna and gain improvement with FSS screen, *Electromagnetics*, **51**, 177-199 (2013)
8. A. S. Ali and D. Mathur, A novel UWB modified hexagonal microstrip antenna with improved gain, IJCE, **3** (2015)
9. X. Gong, L. Tong, Y. Tian and B. Gao, Design of a microstrip-fed hexagonal shape UWB antenna with triple notched bands, *Electromagnetics*, **62**, 77-87 (2016)
10. A. Kahwaji, H. Arshad, S. Sahran, A. Garba and R. I. Hussain, Hexagonal microstrip antenna simulation for breast cancer detection, IEEE (2016)
11. M. L. Naidu, Dr. B. R. Rao and Dr. C. Dharamraj, Implementation of octagonal and hexagonal microstrip patch antennas for UWB applications, SSRG IJECE, **3** (2016)
12. P. P. Singh and S. K. Sharma, Design and fabrication of a triple band microstrip antenna for WLAN, satellite tv and radar applications, *Electromagnetics*, **117**, 277-289 (2021)
13. J. M. Rathod, Comparative study of microstrip patch antenna for wireless communication application, IJIMT, **1** (2010)
14. S. T and C Nandagopal, Analysis on microstrip patch antennas for wireless communication, IEEE (2015)

ITM Web of Conferences **44**, 02005 (2022) https://doi.org/10.1051/itmconf/20224402005
ICACC-2022

Toolkit Carrying Drone For Maintenance Of High Altitude Transmission Lines

Shivam Kadam, *Piyush* Dharmasare, *Shubham* Ganjale and *Nilam* Ghuge

Department Of Electrical Engineering, JSPM's Bhivarabai Sawant Institute of Technology & Research, Pune, India.

Abstract Research and analysis on drones is increasing rapidly nowadays in various fields of application. One of the major problem which arises in the drone application is its ability of lifting heavy loads when attached to it that requires proper design and selection of the required type of drone for its ability to carry out various missions including lifting a heavy load. One of the important type of drone that is widely used is hexacopter drone. This paper shows the design of hexacopter drone that lifts the toolkit which has to be delivered for maintenance of transmission lines at high altitude. It is designed with the help of calculation and analysis of the criteria and constraints by the aid of software. The calculations shows the ability of hexacopter drone of lifting a toolkit and delivering it to the location at higher altitude.

1. INTRODUCTION

Automated Aerial Vehicle or UAV acquired ubiquity in the universe of science and innovation in view of its different applications in different fields like hunt and salvage activity, farming tasks, ethereal photography, studying, reconnaissance reason, firefight, observing, and so forth [1,2]. Considering its process for lifting, the UAV are organized in two sorts, rotor wing and fixed wing. The wing rotor involve helicopter, tricopter, quadcopter, hexacopter and octocopter depending on the amount of stimulus motor. A critical examination has been done in this handle like independent helicopter control [1], disposition and elevation quadcopter development control [3, 4], tricopter displaying [5, 6], displaying and controlling hexacopter [7, 8], octocopter navigation control [9]. The main advantage of multi rotor is it do not require a runway to takeoff. It can perform vertical takeoff or landing (VTOL), it has the ability to hover and maneuver to perform a task. Different multirotors have different load lifting capability due to addition of extra motor however it becomes more difficult to control as the system becomes more nonlinear. Since our task is to deliver toolkit which can be called as a heavy payload, thus hexacopter and octocopter would be preferable which has six motors and eight motors respectively. Although octocopter can perform the same task as that of the hexacopter it costs more because it has two extra motors compared to hexacopter. Therefore we have selected hexacopter to deliver toolkit for maintenance of transmission lines at high altitude as it is further discussed in this paper.

2. DESIGN OF HEXACOPTER

2.1 Hexacoper configuration

Hexacoper consist of six motors which rotates the six propellers attached to them. The lifting movement of hecacopter utilizes the thrust produced by the six rotating propellers in combination of hexacopter frame. The hexacopter has two types of configuration based on the frame design which are plus configuration (+) and X-configuration (x) as shown in Fig. 1. The hexacoper has 6 levels of chance and which is affected by the rotational speed of every motor. Thusly both the casings will have different model of movement elements. In this we have utilized hexacopter with X arrangement (x). Fig. 1. Shows that each arm of the casing is associated with a DC brushless engine and has a propeller with fixed pitch. It makes the rotor to drive the wind stream downwards which creates the lift force. The rotor has two heading of pivot that is clockwise and counter-clockwise. Three rotors turns in clockwise bearing while other three pivots in counter-clockwise course fixed on the other hand. So we can obviously see the exclusively by the speed of motor pivot the unique movement of hexacopter gets affected.

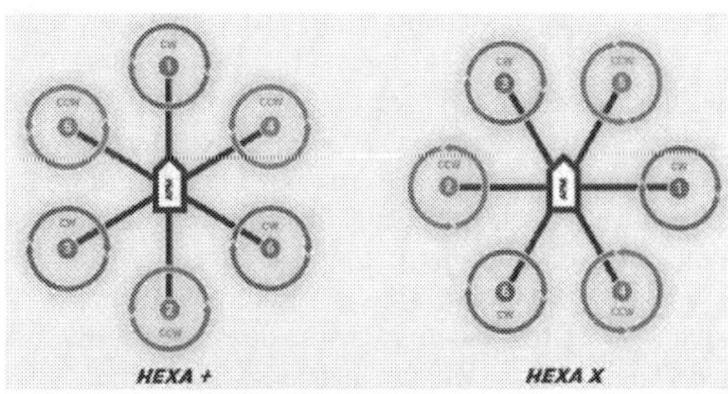

Plus configuration X- configuration

Fig. 1. Hexacopter configurations

https://doi.org/10.1051/itmconf/20224402005

Fig. 2. Toolkit caryying hexacopter frame.

Lightweight material should be used to construct the frame but it should be strong enough so that it can withstand its operational weight along with the weight of the toolkit that is to be carried. Accordingly it is very essential to pick the making material out of the edge admirably by examining its adaptability and strength. The plan of this casing has six equivalent long arms that are appended to the focal point of a multiplied hexagonal plate as displayed in Fig. 2.

2.2 Thrust and Motors

The motor is the main part as it is the impetus component of the hexacopter. It is critical to know the hard and fast weight to be lifted and the push expected to lift the hexacopter before concluding the constraint of the motor used in the arrangement. Calculation for determining the thrust per motor can be done using following equation:

$$\text{Thrust} = \frac{Total\ Weight\ X\ 2}{Number\ of\ Motors} \qquad (1)$$

Moreover, following equation can be used to calculate static thrust [11]:

$$T^3 = \frac{\pi}{2} D^2 \rho P^2 \qquad (2)$$

In this situation T represents Push, D represents measurement of the propeller, ꞵ is the air thickness and P is the force of engine. We can as certain worth of RPM from the KV consistent and the voltage utilized by the motor. For a hexacopter framework the really main thrust is given by the motor. Thus we use a BLDC motor. The advantage of this motor is that it does not use the brush and commutator which increases the efficiency as compared to normal DC motor also it produces high RPM (revolutions per minute). KV (rpm/volt) is the principle boundary of worry for BLDC engine. The boundary which expresses the extent of rpm increment for every unit of voltage that is utilized is KV. In this hexacopter we have used 1000 KV BLDC motors.

Fig. 3. BLDC Motor

2.3 Propeller

While selecting the propellers, pitch parameter is often noticed. The distance gushed by the liquid because of one turn of propeller's edges is called as pitch boundary. Assuming that the measurement and pitch of the propeller are huge the lifting power will be high however the turn of the motor will be more slow. Accordingly to lift a weighty burden the pitch and measurement of the propeller should be enormous. In this hexacopter model, to lift a tool stash we have utilized 1045 propellers where 10 is the width of the propeller in cm and 4.5 is the pitch of the propeller.

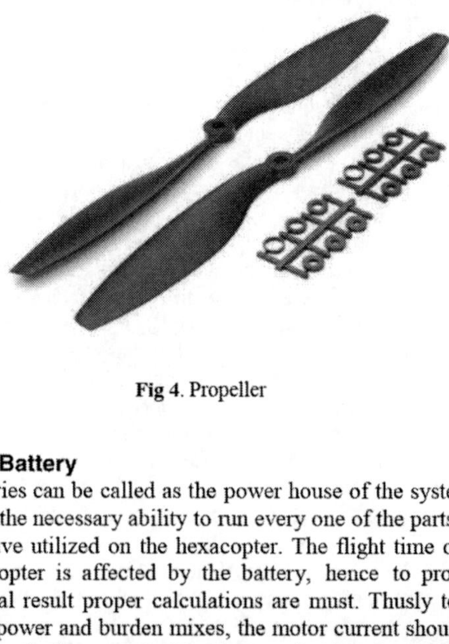

Fig 4. Propeller

2.4 Battery

Batteries can be called as the power house of the system. It gives the necessary ability to run every one of the parts that we have utilized on the hexacopter. The flight time of the hexacopter is affected by the battery, hence to produce optimal result proper calculations are must. Thusly to get ideal power and burden mixes, the motor current should be not exactly the current of the battery that is being utilized. Number of cells, release and limit are the most extensive boundaries while choosing the battery. The voltage of the battery in an unfilled not permanently set up by the amount of cells while how much current rating/current speed can be not totally settled by the delivery. How lengthy the battery can chip away at specific amperes is the limit of the battery.

https://doi.org/10.1051/itmconf/20224402005

For this hexacopter model we have used 2200 mah battery which has a discharge rate of 30C.

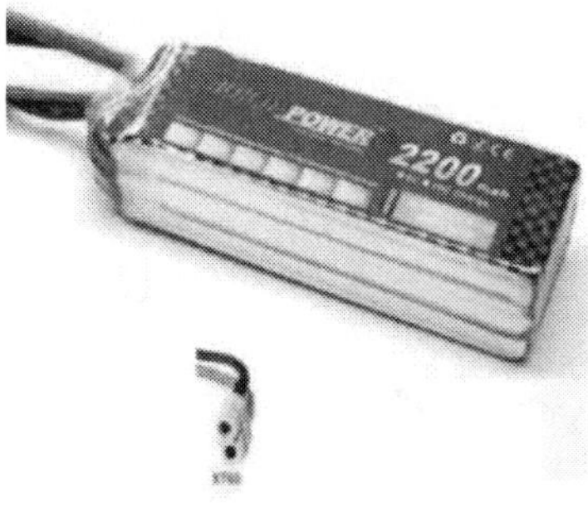

Fig 5. Battery

After assembling final drone is as shown below.

Fig 6. Drone

3. SOFTWARE

To make our toolkit carrying hexacopter drone ready to fly we have to do various calibration. Mission planner is the software that is used to do these calibration. It is like a ground control station. With the help of mission planner we can do accel calibration, compass calibration, radio calibration, ESC calibration, flight modes, etc. Mission planner is more like a dynamic control supplement for our drone.

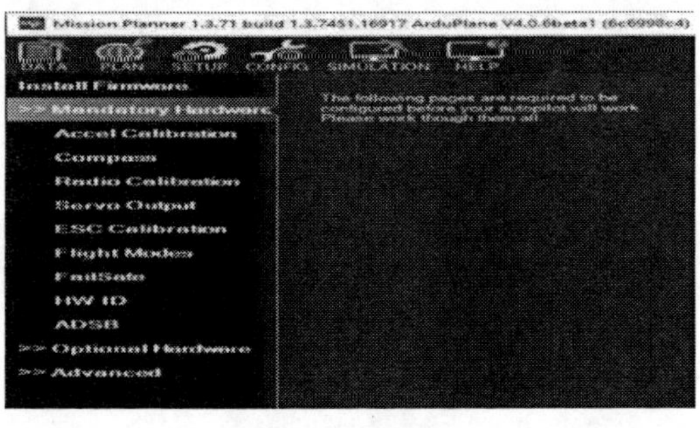

Fig. 7. Mission Planner

We can configure total 23 flight modes to our drone. There are 10 flight modes which are mainly used. However for this model and application we have equipped our hexacopter drone with three flight modes and they are stabilize mode, Alt hold mode and land.

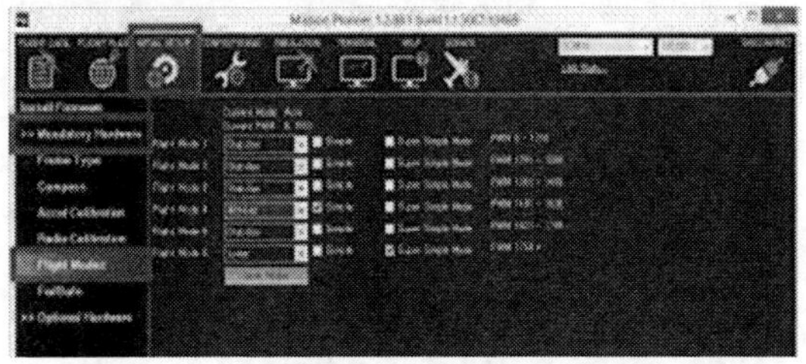

Fig. 8. Mission Planner

4. RESULT

The weight of each part is according to the accompanying:

Table No. 1

No.	Item	Unit	Weight (Grams)
1.	Frame	1	424
2.	Motor	6	390
3.	Propeller	6	42
4.	APM Flight Controller	1	80
5.	ESC	6	138
6.	APM Pix-Hawk Power Module	1	24
7.	Battery	1	170
8.	Shock Absorber	1	20
9.	Delivery Rope And Steel Hook	1	47
		Total	1335

With a full scale weight of 1335 grams , the push of each motor is determined as follows:

$$\text{Thrust} = \frac{Total\ Weight\ X\ 2}{Number\ of\ Motors}$$

Thrust = (1335 X 2) / 6

Thrust = 445 Grams

Thus, all six motors combined produce a thrust of 445 x 6 grams that is equal to 2670 grams. This much pushed will have the choice to lift Hexacopter which has a flat out weight of 1335 grams and on the off chance that the apparatus compartment that will be lifted is of 1000 grams, the by and large of full scale weight is 2335 grams. Consequently we have saved 335 grams for wellbeing. From the estimation, hexacopter can lift it and can fly with practically no issue and effortlessly.

5. CONCLUSION

The venture has option to plan, investigation and build toolbox conveying hexacopter drone that is fit for flying and lifting the tool compartment. With the battery provided, the drone can fly for a sufficient time period to perform the delivery of toolkit when fully charged. With the help of Flysky FS-i6S remote controller the drone can hover and can fly within the range of 1 km. The powerful motors of the hexacopter drone makes it capable of lifting a toolkit which has a weight between the range of 1000 grams to 1300 grams without any problem and with ease. As it is a tool stash conveying drone it can convey different payloads as well. So it has a future extension for conveying different things of weight under its functional cutoff points. It tends to be made do by adding more highlights like camera and GPS. With this highlights it be followed and controlled without any problem.

References

[1].S. d'Oleire-Oltmanns, I. Marzolff, K. D. Peter, and J. B. Ries, "Unmanned aerial vehicle (UAV) for monitoring soil erosion in Morocco," Remote Sensing, vol. **4**, pp. 3390-3416.(2012)

[2]. G. Belloni, M. Feroli, A. Ficola, S. Pagnottelli, and P. Valigi, "A mini UAV for security environmental monitoring and surveillance: Telemetry data analysis," in 'Workshop Proceedings of SIMPAR', pp. 426-433.(2008)

[3] A. Zulu and S. John, "A review of control algorithms for autonomous quadrotors," arXiv preprint arXiv:1602.02622, (2016)

[4] M. Heryanto, H. Suprijono, B. Y. Suprapto, and B. Kusumoputro, "Attitude and Altitude Control of a Quadcopter Using Neural Network Based Direct Inverse Control Scheme," 'Advanced Science Letters', vol. **23**, pp. 4060-4064, (2017)

[5] E. Servais, B. d'Andréa-Novel, and H. Mounier, "Ground control of a hybrid tricopter," in 'Unmanned Aircraft Systems (ICUAS)', International Conference on, pp. 945-950.(2015)

[6] M. Ramp and E. Papadopoulos, "On modeling and control of a holonomic vectoring tricopter," in 'Intelligent Robots and Systems (IROS)', IEEE/RSJ International Conference, pp. 662-668.(2015)

[7] B. Y. Suprapto, M. Heryanto, H. Suprijono, and B. Kusumoputro, "Altitude Control of Heavy-Lift Hexacopter using Direct Inverse Control Based on Elman Recurrent Neural Network," in 'Computer Modeling and Simulation', the Eighth International Conference, pp. 135-140.(2017)

[8] R. Baránek and F. Šolc, "Modelling and control of a hexa-copter," in 'Carpathian Control Conference (ICCC)', Thirteenth International Confrence, pp. 19- 23.(2012)

[9] M. J. Er, S. Yuan, and N. Wang, "Development control and navigation of Octocopter," in 'Control and Automation (ICCA)', Tenth IEEE International Conference, pp. 1639-1643.(2013)

[10] L. E. Romero, D. F. Pozo, and J. A. Rosales, "Quadcopter stabilization by using PID controllers," Maskana, pp. 175-186, (2016)

Recipe Recommendation System Using TF-IDF

Shubham Chhipa[1], *Vishal Berwal*[2], *Tushar Hirapure*[3] *and Soumi Banerjee*[4]

[1]*Department of Information Technology, Ramrao Adik Institute of Technology, Nerul, Navi Mumbai.*
[2] *Department of Information Technology, Ramrao Adik Institute of Technology, Nerul, Navi Mumbai.*
[3] *Department of Information Technology, Ramrao Adik Institute of Technology, Nerul, Navi Mumbai.*
[4] *Department of Information Technology, Ramrao Adik Institute of Technology, D.Y. Patil Deemed to be University, Navi Mumbai.*

Abstract - A Recipe Recommendation System is being proposed in this following paper. Food recommendation is a new area, with few systems that are focus on analysing and user preferences and constraints such as ingredients available at their side being deployed in real settings in the form of web application or mobile application [4]. The proposed model is a mobile application which allows users to search recipes using ingredients available at them including vegetables. For this work we have find a dataset which is a collection of Indian cuisines recipes and apply the content-based recommendation using Term Frequency – Inverse Document Frequency (TF-IDF) and Cosine Similarity [1]. This application gives the recommendation of Indian recipes based on ingredients available at them and allows users to filter out the recipes on course type, diet type, etc.

Keywords— Machine Learning, Recommendation System, TF-IDF, Cosine Similarity, bag-of-words, NLP, Flask.

1. Introduction

People make decisions related to food every day. They all think about what to eat, what to cook, how much nutritional value this food has, can this food make me healthy i.e., is it gluten free or sugar free, is it diabetic friendly or not and other questions. Recommendation systems help the user to make fast decisions in these complex information spaces.

Diet management systems, which have mostly replaced conventional paper-and-pen approaches, have gotten a lot of attention. These systems leverage educational information and services to persuade people to change their habits. Because of the popularity of these diet monitoring capabilities, these systems have a wealth of user preference data, which might be used to personalise interactive elements and boost user engagement with the system and diet programme. A meal recommender is one such tailored service that is great for educating diet. The nutritional properties of the food could be used by this recommender to inform its recommendations.

Recommender systems are one of the most widely used data science applications today. They are used to anticipate a user's "rating" or "preference" for a certain item. Almost every big IT firm has used them in some capacity. It's used by Amazon to recommend products to customers, YouTube to determine which video to play next on auto-play, and Facebook to suggest pages to like and people to follow [5].

For as long as people have existed, the challenge of what to cook/how to cook delicious food and make it nutritious has existed. For thousands of years, the only way to learn about amazing dishes and recipes was through trial and error or word of mouth, and neither approach took into account the nutritional content of a dish. [3]

Thinking about **"What to cook today?"** is a problem that every housewife and working women face. It seems to be easy decision but it requires some strategic and smart decisions when it comes to decide heavy or light food such that their family members digestion and health doesn't get disturbed and face acidity and gas issues and they also have to think about their budget.

The proposed application helps users to cook recipes from available ingredients at their end and filter the recipes based on course type, diet type, etc.

 https://doi.org/10.1051/itmconf/20224402006

2. Literature Survey

2.1. Recommendation of Indian Cuisine Recipes based on Ingredients-Nilesh, Vishal, Pritam, Dr. Madhu. [1]

They have created datasets by using web scraping and extract features form ingredients column using bag of words. Bag of Words contain the specific keywords for each recipe and for recommendation they have used cosine similarity between these keywords.

2.2. Recipe Recommender System Using Image Recognition of Food Ingredients – Mona Mishra, Yifan Gong. [2]

They have picked up twenty-two fruits and vegetables categories from ImageNet. They have created 5layer CNN model for image recognition. They have used Microsoft Azure for cleaning the dataset and then searching the ingredients and sends the response containing the recipes.

2.3. Healthy Recipe Recommendation using Nutrition and Ratings Models - Yew Siang Tang, Anita Hanzhi Zheng, Nicholas Lai. [3]

They have used dataset from Kaggle which contains 50,000 recipes which is scraped by allrecipes.com. They have created two models Nutritional Model and Ratings Model. They have used GNN for prediction of the recipes.

2.4. Personalized Food Recommendations - Jorge Imeida, Prof. Doutor Anibal Cavaco Silva, Technical University of Lisbon.[4]

They have used Yolp Content-based component and Yolp Collaborative Component. They have used Rochios Algorithm using FF-IRF for extracting weights.

3. Proposed System

This paper proposes a system which has greater accessibility as it is a mobile application and it is easy to use for users.

The existing system consists of approaches such as thinking what to cook today manually leading to stress, time consuming and worries about their budget for groceries. There are several cooking apps or websites available today, that are used to find recipes based on some keyword, like name of the food ingredient or type of cuisine, etc. These apps are mindful of the needs and interests of their users, but they fail at identifying their user's constraints, i.e., limited number of food ingredients. In such cases, users find themselves shopping for ingredients or they decide to substitute the missing ingredient with something else.

In the proposed system the user can explore recipes based on different course types such as lunch, dinner, breakfast, snacks, etc. as well as different diet types such as gluten free, sugar free, diabetic friendly, etc.
This system takes care of user's biggest constraints such as limited number of ingredients available at them.

The user have to enter or dictate the ingredients available at their end and then they have to click on see recipes button. Then app will make a request to the recommendation system passing the ingredients as a query parameters and then recommendation system sends the recipes that can be cooked using that ingredients to the app. Then app display all that recipes in a list format with the filter so that user can filter recipes based on their requirements.

Users can also create their shopping list for groceries in our app while cooking so that they don't procrastinate the task of making a list. This system also takes care of user's budget by creating multiple combinations of the ingredients and trying to give best suggestions / recipes so that their resources get used perfectly.

Fig. 1. displays the working of our system.

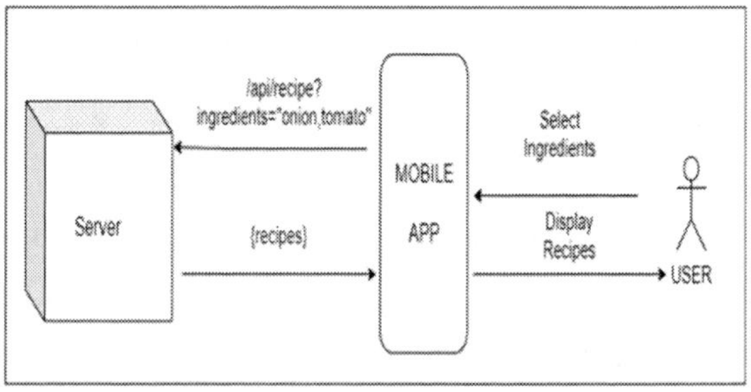

Fig. 1. Block Diagram of System.

4. Proposed Methodology

4.1 Dataset:

This paper have used the dataset available on the Kaggle. The name of dataset is **6000+ Indian Food Recipes Dataset** by Kanishk. This dataset was selected because it contains all the recipes which are of Indian Palette and it is created by scraping the archanaskitchen.com website.

Since this dataset does not contain images of a recipes then we have created a script that scrape all the images of the recipe from the same website and adds them to a new column Images in the dataset and created a new dataset.

https://doi.org/10.1051/itmconf/20224402006

The dataset can be used to answer a lot of questions related to Food Recipes such as serving sizes, time required to prepare a dish, most common ingredients, different cuisines, diets, courses and what not.

4.2 Pre-processing and Parsing of Ingredients:

There was a lot of redundant information in this dataset for example weights and measures such as tbsp, grams, etc are not going to add value to the vector encodings of the recipes. There are also other words in the ingredients that are useless to us. These are the words that are very common among all the recipes such as oil, fresh, leaf, or any other words.

To remove all this information from the ingredients column in the dataset we have created **ingredient_parser** function. At first we are removing all the punctuations from the ingredients using **maketrans** function from str library. Then we have lemmatize the ingredients using **WordNetLemmatizer()** from nltk to get root of the different words. After creating **ingredient_parser** function we have used lambda functions which makes it easy to parse all the ingredients.

Fig. 2. shows the flowchart of the preprocessing method.

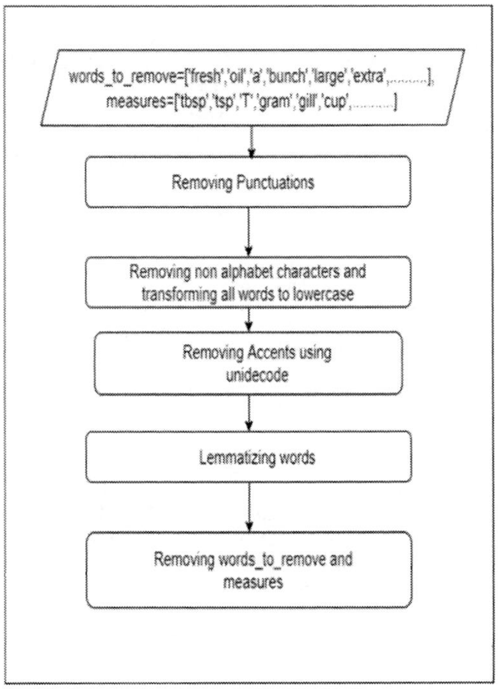

Fig. 2. Flowchart of Preprocessing Method.

4.3 Extracting Features

To encode each document (recipe ingredients) we have two choices NLP's bag of words and Term Frequency – Inverse Document Frequency (TF-IDF). Bag of Words performed ok but TF-IDF marginally out-performed it, so we opted for this instead.

The TF-IDF determines the relevance of a phrase by considering its importance across all documents. [6].

$$\mathbf{tfidf}(t,d,D) = \mathbf{tf}(t,d) \times \mathbf{idf}(t,D) \quad (1)$$

- t = term
- d = document
- D = document

$$\mathbf{tf(t,d)} = f_{t',d} \,/\, \sum_{t' \in d} f_{t',d} \quad (2)$$

$$\mathbf{idf(t,D)} = \log(\,|D| \,/\, |\{\, d \in D : t \in d \,\}|\,) \quad (3)$$

The paper have used Scikit-learn **TfidfVectorizer** method. We then saved the model and encodings using pickle in two different files.

4.4 Recommendation System

This paper have used content-based filtering which enabled us to recommend recipes to people based on the attributes i.e. ingredients.
To measure the similarity between the ingredients provided by user and ingredients of a recipe we have used Cosine Similarity.

Cosine Similarity is a metric that is used to measure how similar two items are and generate a scores based on the similarity [7]. By using the cosine similarity function we get the the scores of the matching recipes and then that scores are passed to **getRecommendation** function to rank these scores and output a pandas containing all the details of top N recommendations.
We have created an API for our model using Python web framework Flask [8] and hosted it on pythonanywhere.
Fig. 3. show the flowchart of the recommendation system.

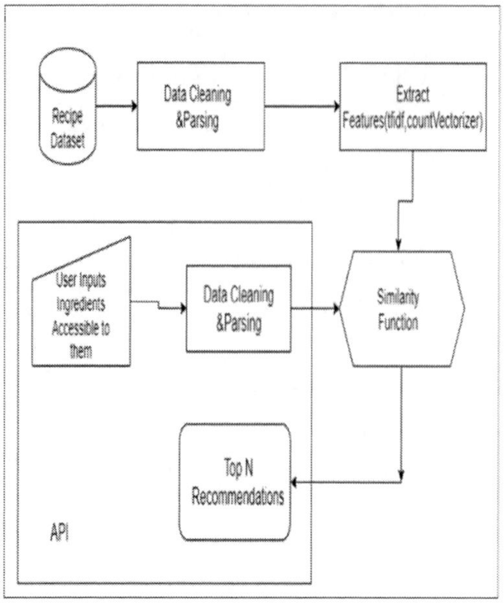

Fig. 3. Flowchart of recommendation system.

4.5 Developing An APP:

For developing an app for the users we have used React Native [9], which is open source cross platform UI Software framework developed by facebook.

This paper have also used React Native Paper [10] which is a UI library for react native and Redux for state management.

5. Results

In our evaluation we have tested the performance of our recommendation system by comparing the ingredients given by the users as input and the ingredients of the recipes that our recommendation system displays.

We have also taken the budget of the users into an account hence we create the powerset of all the ingredients given by the users to generate multiple combinations of different ingredients and then run our recommendation on each element or combination of a powerset.

Below is the table that shows the recommendations provided by our system on a certain input.

Users Input – Potato or Onion

Sr No	Recommendations
1	Aloo Ke Gutte Recipe
2	Konkani Style Batata Recipe
3	Parsi Style Chas Payelo Sakarkand Recipe
4	Niramish Aloo Recie
5	Cajun Style Potatoes Recipe
6	Spring Onion Recipe
7	Vaghrela Chawal Recipe
8	Odisha Style Aloo Recipe

Using Lunch Filter

Sr No	Recommendations
1	Aloo Ke Gutte Recipe
2	Niramish Aloo Recipe

Using Side Dish Filter

Sr No	Recommendations
1	Konkani Style Batata Recipe
2	Parsi Style Chas Payelo Sakarkand Recipe
3	Odisha Style Aloo Recipe

Users can also explore recipes based on different types of cuisines such as South Indian, North Indian, Italian, Chinese, Fusion.

On exploring Chinese cuisine type our recommendation system recommends recipes that are of Indo-Chinese taste. Our system recommends 48 recipes based on Chinese cuisines. Some of the recipes are shown below in the table.

Chinese Cuisine Recipes

Sr No	Recommendations
1	Idli Manchurian Recipe With Oriental Twist
2	Chilli Baby Corn Manchurian Recipe
3	Chilli Paneer Momo Recipe
4	Spicy Schezwaan Noodles

Users can also explore recipes based on their diet type such as Veg, Non Veg, Gluten Free, Diabetic Friendly and sugar free.

On exploring Diabetic Friendly Diet type our recommendation system recommends recipes that are diabetic friendly. Our system recommends 260 recipes based on Diabetic Friendly diet. Some of the recipes are shown below in the table.

Diabetic Friendly Recipes

Sr No	Recommendations
1	Masala Karela Recipe

2	Thalipeeth Recipe
3	Ragi Masala Roti
4	Kovakkai Podi Curry Recipe

All responses from our server is in JSON format and in app we have parsed and transformed these JSON.
Users will see list of recommended recipe in the flatlist format in app.

6. Conclusion

The Indian cuisines are broadly refreshing for its remarkable utilization of herbs and spices. Mostly Indian Cuisines are result of an innovation/jugaad done by various Indian housewives and working class women cause they were having the constraints of limited ingredients and keeping in mind of the health of their family members.

This paper presented a method for recipe recommendation using ingredients matching of recipes. In this paper, we tried to solve the burning problem of Indian housewives or working women that is **What to cook today?** Through this paper we have learned to build a recommendation system using TF-IDF method and understands how similarity scores gets calculated using cosine similarity.

7. Future Scope

The current system allows users to directly explore all the recipes without login, in future we are planning to add registration and login functionalities. If there will be login functionality then we will be able to personalise all the user experience by recommending recipes of their flavours preferences and we will be able to store the bookmarks and likes of the user.

The current system allows two approaches for taking input from users which is by selecting ingredients from list and other is by dictating the ingredients one by one. In future we are planning to add image recognition of vegetables so that users can directly scan the ingredients and search for recipes. [2]

References

[1]. Recommendation of Indian Cuisine Recipes based on Ingredients - Nilesh, V. Raman, P. Hazarika, Dr. M. Kumari.

[2]. Recipe recommendation system using Image recognition of Food Ingredients – M. Mishra and Y. Gong.

[3]. Healthy Recipe Recommendation using Nutrition and Ratings Models – Y. Tang, A. Zheng, N. Lai.

[4]. Personalised Food Recommendations – J. Almedia, D. Anibal.

[5].https://www.datacamp.com/community/tutorials/recommender-systems-python

[6]. https://towardsdatascience.com/tf-idf-explained-and-python-sklearn-implementation-b020c5e83275

[7]. https://towardsdatascience.com/using-cosine-similarity-to-build-a-movie-recommendation-system-ae7f20842599#:~:text=Cosine%20similarity%20is%20a%20metric,the%20items%20are%20100%25%20similar

[8]. https://flask.palletsprojects.com/en/2.1.x/

[9]. https://reactnative.dev/

[10]. https://callstack.github.io/react-native-paper/

Pragmatic Approach for Online Document Verification Using Block-Chain Technology

[1]Nilesh P. Sable, [2]Sachin R. Powar, [3]Queency Fernandes, [4]Nikita A. Gade, Akash B. Shingade

[1] Department of Information Technology, Bansilal Ramnath Agarwal Charitable Trust's, Vishwakarma Institute of Information Technology, Pune, India, drsablenilesh@gmail.com

[1, 2, 3, 4]JSPM's Imperial College of Engineering and Research Pune, India

ABSTRACT: As we all know, India has a plethora of universities, and many people graduate from them. Because it is possible for someone to falsify a degree, a secure based verification mechanism is required. It will be achieved in an existing system through the exchange of e-mails or postal mail; however, this is a time-consuming process that is insecure owing to human involvement. And, in order to solve these drawbacks, we implemented block chain technology into our system. When there is a security risk, block chain comes to mind. When it comes to data breaches, education is also not far away. For attackers, student data with little financial information has become a valuable commodity. At the same time, student verification is becoming a serious worry at educational institutions, which are being breached to generate phony identities and records. As a result, the greater the digitization of student information, the greater the need to protect student privacy.

Keywords: Block chain, Data Mining, SHA, Cloud computing.

I. INTRODUCTION

Any country's rapid expansion is dependent on its infrastructure, government regulatory policies, natural resources, skilled labor force, and a variety of other factors. The trained labor is one of the most essential variables in this. India has created multiple universities in response to the rising demand for trained workers in the industries, education, and health care sectors. According to official figures, India has a total of 1019 universities (as on UGC report dated 22-11-2021).

With the development in the number of institutions, enormous obstacles have arisen in the educational system as well as the university certification system. The issue of fraudulent degrees is a big problem that all universities, government sectors, public sectors, and private sector institutions are dealing with. Universities have their own verification system, but due to the lengthy process, it is not being used effectively. We provide an online certificate verification solution based on cloud computing that uses Block-chain technology to complete the verification procedure in a small amount of time.

Type of university	Number
State university	442
Deemed to be university	126
Central university	54
Private university	397

II. Block Chain

A block chain is a continually growing collection of information known as blocks. These blocks are encrypted and connected together. Each block contains a cryptographic hash of the previous block, as well as a timestamp and transaction data. The timestamp proves that the transaction data existed when the block was released, which is required to get into its hash. The timestamp confirms that the transaction data existed when the block was published in order to get into the hash. Because each block holds information about the previous one, they create a chain, with each new block strengthening the ones before it. As a result, block chains are resistant to data tampering since data in one block cannot be modified retrospectively without impacting the data in all previous blocks.

A. Structure

A block chain is a distributed, public digital ledger made up of information called blocks that is used to record transactions across multiple computers and make sure that no one block can be modified without impacting all subsequent blocks. This allows participants to review and audit transactions independently and for a low fee. A block chain database is run independently using a P2P network and a distributed time stamping server.

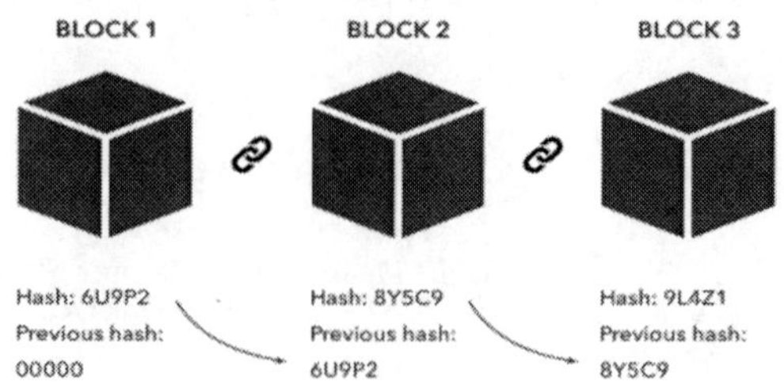

Figure 1: Block Structure

They are confirmed by extensive collaboration, which is fuelled by a sense of common good. This style of architecture encourages a consistent workflow with

 https://doi.org/10.1051/itmconf/20224403001

minimum data security issues among participants. The use of a block chain eliminates the ability of a digital item to be reproduced indefinitely. It ensures that each value of unit was transmitted only once, solving the long-standing problem of duplicate spending. A block chain is a process of exchanging value. A block chain can protect title rights by creating a record that requires offer and acceptance when set up correctly to describe the trade agreement.

III. ALGORITHMS

In the proposed system we have implemented several algorithms for solving easily and automate a solution to the problem .The SHA algorithm is implemented for the purpose of hash generation .To validate and explore system performance using consensus algorithm for proof of validation. P2P software allows "peers" (separate computer systems) to exchange files by connecting over the internet.

Secure Hash Algorithms (SHA) is a set of cryptographic methods for keeping data safe. It converts data using a hash function, which is a mechanism based on compression algorithms, modular additions and bitwise operations. After that, the hash function returns a fixed-length string with no similarity to the original. These approaches are meant to be a one-way function, which signifies that after they've been changed into hash values, it's very hard to alter them back in its original content.

When utilizing SHA to encrypt passwords, the server just has to keep track of a single user's hash value instead of the actual password. This is beneficial if an attacker hacks into the database and only discovers the hashed functions, not the passwords themselves. If they try to use the hash value as a password, the hash function will convert it to another string and prohibit them from obtaining the information.

The sha-256 technique was used to generate hashes in the proposed system. One of the most secure hashing functions available is SHA-256. The US government mandates its agencies to use SHA-256 to protect some sensitive data. While the specific details of how SHA-256 works are classified, we do know that it is made up of a Merkle-Damgard structure obtained from a one-way compression function that was made up of the Davies-Meyer structure from a specialized block cypher.

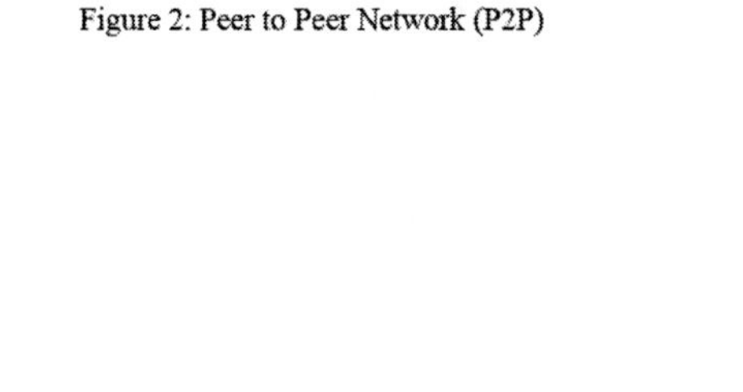

Figure 2: Peer to Peer Network (P2P)

P2P (peer to peer) is a decentralized and distributed communications methodology in which a group of devices or the nodes join to store and transport data, each functioning as a single peer. P2P communication takes place in this network without the use of a central administration or server, this means that all nodes have the same amount of power and do the same tasks.

B. CONSENSUS:

As we all know, block chain is a distributed decentralized network that allows for transparency, privacy, immutability and safety. Regardless the lack of a centralized system to authenticate and validate transactions, the Block chain believes every transaction to be completely safe and secure. It is only possible because of the consensus mechanism, which is an integral part of any Block chain platform.

Consensus Algorithm Each block added to the block chain is verified by all other nodes on the network to ensure that the node getting added is an authorized node. A consensus algorithm is used to complete this operation. They aid in the development of participant trust and network reliability. PoW, PBFT, and PoS are common consensus algorithms that are being used.

IV. PROPOSED MODULE

The system contains following modules:

Admin:

Admin is a system user who has access to data that has to be uploaded. An admin defines an access policy for accessing only those users with matching Enrollment and Name are given authorization to access the data.

Data User:

An authorized user who wishes to access data is referred to as a data user. The user creates an account with an Authority and receives a single Enrollment. If the Enrollment and Name satisfy an access policy connected with a document, the end user by entering proper Enrollment and Name, you will be able to check data and obtain access to educational database.

Distributed block chain: The block chain is a distributed ledger that represents the current state of the systems delegated access privileges. The Root User Authority and the Authorities (banking, medicine, marketing, etc.) manage permissions to interact with the block chain.

V. PROPOSED SYSTEM

As shown in Fig. 3, the suggested model's process is broken down into the following steps:

Step 1: University Admin uploads the documents.

Step 2: The uploaded document will be stored in database.

Step 3: A unique hash is created using the attributes: - admin name, email, contacts number, name of the student, student enrolment number, student name, system time, previous block and the time required for mining. The numbers of attributes are variable.

Step 4: The data with new hash value will be connected to the previous block.

Step 5: User with his unique id and password will login and request for the verification of his document.

Step 6: Admin upon receiving the request will verify if the user is the student of the particular university or not.

If yes then the request is approved and if not then the request is denied.

Step 7: The user's information is compared to the information recorded in the block chain. If the data is correct then document will be verified.

Steps 1 to 7 describe the whole data verification procedure and data flow. To eliminate data fraud, this will utilize secure cloud computing combined with block chain technology services.

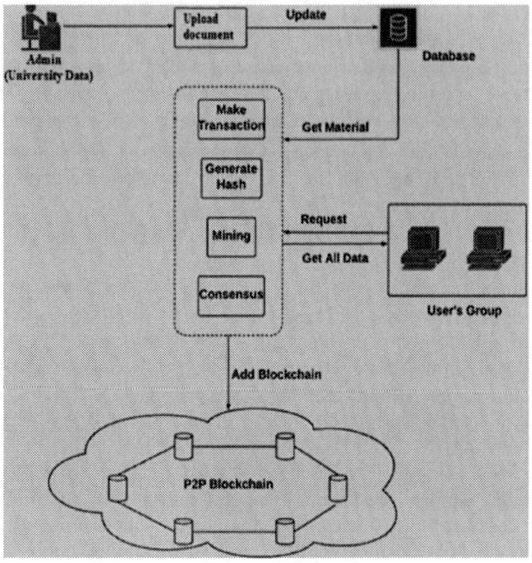

Figure 3: System Architecture

VI. CONCLUSION

We've developed an online certificate verification system that employs block-chain technology to enable online document verification. The online certificate verification system provides a cost-effective, simple to verify and maintain solution for introducing the cloud computing architecture to the online certificate verification system across all sorts of users with a single click. The block-chain concept is primarily being used to validate user credentials and offer data verification via a secure manner. Furthermore, depending on the needs of the system, this might be expanded to different types of data verification systems.

REFERENCES

[1] C. G. San Jose, "Document Security System Using Improved Hash Algorithm on Pre-processing Operation", Journal of Advanced Research in Dynamical and Control Systems, vol. 11, no. 11-, pp. 972-978, 2019. Available: 10.5373/jardcs/v11sp11/20193123.

[2] F. Kaspar, H. Rust, U. Ulbrich and P. Becker, "Verification and process oriented validation of the MiKlip decadal prediction system", Meteorologische Zeitschrift, vol. 25, no. 6, pp. 629-630, 2016. Available: 10.1127/metz/2016/0831.

[3] B. Houtan, A. Hafid and D. Makrakis, "A Survey on Blockchain-Based Self-Sovereign Patient Identity in Healthcare", IEEE Access, vol. 8, pp. 90478-90494, 2020. Available: 10.1109/access.2020.2994090.

[4] G. Dagher, J. Mohler, M. Milojkovic and P. Marella, "Ancile: Privacy-preserving framework for access control and interoperability of electronic health records using blockchain technology", Sustainable Cities and Society, vol. 39, pp. 283-297, 2018. Available: 10.1016/j.scs.2018.02.014.

[5] Y. Sharma, "A Survey On Privacy Preserving Methods Of Electronic Medical Record Using Block chain", Journal Of Mechanics Of Continua And Mathematical Sciences, vol. 15, no. 2, 2020. Available: 10.26782/jmcms.2020.02.00004.

[6] V. Saxena and S. Pushkar, "Risk Reduction Privacy Preserving Approach for Accessing Electronic Health Records", International Journal of Healthcare Information Systems and Informatics, vol. 16, no. 3, pp. 46-57, 2021. Available: 10.4018/ijhisi.20210701.oa3.

[7] Y. Park, Y. Kim and J. Shim, "Block chain-Based Privacy-Preserving System for Genomic Data Management Using Local Differential Privacy", Electronics, vol. 10, no. 23, p. 3019, 2021. Available: 10.3390/electronics10233019.

[8] G. George and L. Jayashree, "A survey on user privacy preserving blockchain for health insurance using Ethereum smart contract", International Journal of Information Privacy, Security and Integrity, vol. 5, no. 2, p. 111, 2021. Available: 10.1504/ijipsi.2021.120354.

[9] K. Gu, N. Wu, Y. Liu, F. Yu and B. Yin, "WPKI Certificate Verification Scheme Based on Certificate Digest Signature-Online Certificate Status Protocol", Mathematical Problems in Engineering, vol. 2018, pp. 1-19, 2018. Available: 10.1155/2018/7379364.

[10] Y. Yuan, J. Zhang, W. Xu and Z. Li, "Identity-based public data integrity verification scheme in cloud storage system via blockchain", The Journal of Supercomputing, 2022. Available: 10.1007/s11227-021-04193-6.

Plant Leaf Disease Detection Using Computer Vision Techniques and Machine Learning

Kalpesh Joshi[1,*], *Rohan Awale*[2], *Sara Ahmad*[3], *Sanmit Patil*[4], *and Vipul Pisal*[5]

[1-5]Department of Electronics Engineering, Vishwakarma Institute of Technology, Pune, India

Abstract. Agriculture production is extremely important in today's economy because disease development in plants is relatively common, early detection of disease in plants is critical in the agriculture field. The automatic finding of such early-stage disease detection is helpful as it decreases a great effort of supervising in large farmhouses of yields. Using digital image processing and machine learning algorithms, this paper presents a method for detecting plant disease. The disease detection is done on the yields' various leaves. The presented system for plant disease detection is simple and computationally efficient which requires less time for prediction than other deep learning-based approaches. The accuracies for the various plant and leaf diseases are calculated and presented in this paper.

1 INTRODUCTION

Developed countries are now putting more money into agriculture. Early detection of plant diseases is critical to avoid sufferers during the harvest. Different lesions on the leaf can be seen in the early stages of the plant disease. Physically detecting plant diseases is extremely difficult [1] It necessitates a significant amount of effort, knowledge of plant diseases, and an inordinate amount of time. The use of image processing aids in the accurate detection of plant diseases. The strategies for detecting plant diseases using their leaf pictures are presented in this paper. Image processing is a method of performing a few operations on a photograph to obtain a better photograph or extract a few useful records from it. A subset of Artificial Intelligence, machine learning (ML), works automatically or gives instructions to complete a task. Machine learning's main goal is to comprehend training data and fit that data into models that should be useful to people. Machine learning aids in the automatic detection of plant diseases [2]. It has aided good decision-making and forecasting due to the large amount of data generated. The color of the leaves, the amount of damage to the leaves, and the area of the unhealthy plant leaf are used to classify them. The system shown here is useful for monitoring large crop fields. The proposed solution for plant disease detection is less computationally expensive and takes less time to predict than other deep learning approaches because it uses statistical machine learning and image processing algorithms.

The remainder of the paper is divided into three sections. Section 2 shows similar work that has been done. Section 3 presents the proposed system. Section 4 elaborates on the experimental findings, and Section 5 brings the paper to a close.

2 LITERATURE REVIEW

Kothawale, et al. presented a system for detecting and classifying grape leaf disease using a support vector machine (SVM) classifier [3]. The leaf area is obtained using histogram thresholding, Grey Level Co-occurrence Matrix (GLCM) is employed for the feature extraction. They classified each image as healthy or affected and achieved an accuracy of 89.90%.

Pavithra, et al. implemented a method for rice leaf disease detection [4]. The shape and color-based features were extracted using Scale Invariant Feature Transform (SIFT). They have performed experimentation using SVM and Artificial Neural Network (ANN) classifiers and achieved accuracies of 95.2% and 92.2% respectively.

Islam et. al. presented a Multiclass Support Vector Machine-based method for detection of Potato Diseases[5]. Their experimentation performed on the public database 'Plant Village' yielded an accuracy of 95%.

* Corresponding author: rohan.awale18@vit.edu

https://doi.org/10.1051/itmconf/20224403002

Shima Ramesh et. al. implemented a technique using Random Forest for plant disease detection [6]. They created the dataset and extracted the features using the Histogram of an Oriented Gradient (HOG). The final classification was performed using different classifiers yielding maximum accuracy of about 70%.

Jagan Mohan et.al. presented a system for the detection and recognition of Paddy Plant Leaf diseases [7]. The disease was identified with an accuracy of 83.33 percent using Haar-like features and the Ada-Boost classifier. The SIFT feature and classifiers k-Nearest Neighbour (k-NN) and SVM are used to identify the paddy plant disease type. The accuracy rate for disease recognition was 91.10 percent and 93.33 percent, respectively.

John William et. al. identified three common diseases on rice plants using Back Propagation Artificial Neural Network [8]. The basic statistical and color features were extracted for detection of the disease using Neural Networks. They claimed an accuracy achieved of 100%.

Kusumo et. al. proposed a method for the disease detection of corn-plant [9]. In this technique, the features based on color, SIFT, speeded-up robust features (SURF), and object detectors such as HOG are utilized. They performed the performance analysis using various classifiers and achieved maximum accuracy of 87%.

H. Sabrol and S. Kumarn presented a decision tree classifier-based method for detecting tomato plant disease[10]. They demonstrated the detection of six basic diseases using statistical features, with a 78 percent overall accuracy.

3 IMPLEMENTATION

3.1 Dataset

The disease detection experiment was carried out with the help of a publicly available dataset called Plant Village, which was curated by Sharada P. Mohanty et al. [2]. As shown in Table 1, the dataset contains 87000 RGB images of healthy and unhealthy plant leaves divided into 38 classes.

Table 1. Specifications for Datasets

Plant	Disease Name & its Types	No. of images
Apple	Healthy	2008
	Scab	2016
	Black rot	1987
	Cedar apple rust	1760
Corn	Healthy	1859
	Cercospora leaf spot gray leaf spot	1642
	Common rust	1907
	Northern Blight	1908
Grapes	Healthy	1692
	Black rot	1888
	Esca (Black Measles)	1920
	Leaf blight (Isariopsis Leaf Spot)	1722
Potato	Healthy	1824
	Early blight	1939
	Late blight	1939
Tomato	Healthy	1926
	Bacterial spot	1702
	Early blight	1920
	Late blight	1851
	Leaf Mold	1882
	Septoria leaf spot	1745
	Spider mites Two-spotted spider mite	1741
	Target Spot	1827
	Tomato Yellow Leaf Curl Virus	1961
	Tomato mosaic virus	1790

3.2 Methodology

The presented system for leaf disease detection is implemented using major steps as shown in Fig. 1. The image acquisition step captures the image of the plant, then these images are preprocessed for making these images suitable for further process. The image segmentation step performs the localization of the disease pattern-like areas for further steps. The feature extraction step extracts the features related to the disease for classification and identification purposes.

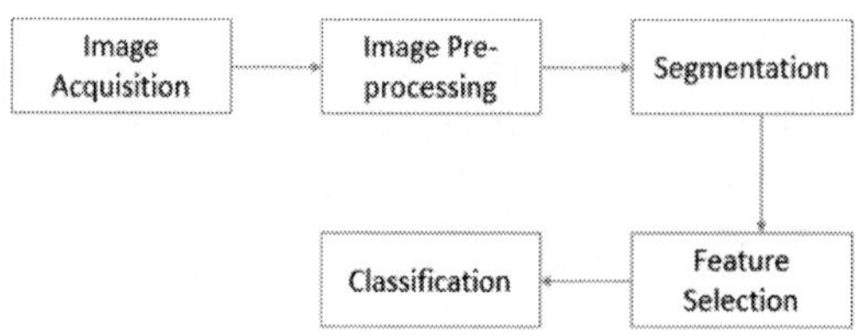

Fig. 1. Proposed System Flow

3.2.1 *Pre-processing*

In any computer vision-based system, data preprocessing is essential. The preprocessing steps for each image are depicted in Figure 2. Before extracting features, background noise should be removed to ensure precise results. The RGB image is converted to greyscale first, and then the image is smoothed with a Gaussian filter. After that, Otsu's thresholding algorithm is used to binarize the image. The morphological transform is then used on the binarized image to fill in the small holes in the foreground. Following foreground detection, a bitwise AND operation on the binarized image and the original color image yields the RGB image of the segmented leaf. The shape, texture, and color characteristics of the image are extracted.

Contours are used to calculate the area and perimeter of the leaf. The lines that run along the edges of objects that are all the same colour or intensity are known as contours. In an RGB image, the mean and standard deviation of each channel are calculated. Convert an image to HSV colour space and divide the total number of pixels in one channel by the number of pixels with hue (H) channel pixel intensities between 30 and 70 to find the amount of green colour. By subtracting the green colour part from 1, the non-green part of the image is calculated. The texture features were extracted from the grey level co-occurrence matrix after the colour features were extracted from the image (GLCM).The image's unique pixel relationship is represented by the GLCM. Extracting texture features from GCLM is a traditional method in computer vision. Contrast, Dissimilarity, and other characteristics were extracted from GCLM. Homogeneity, Energy, and Correlation are all words that come to mind when thinking about homogeneity.

The feature selection task is performed after all of the features from all of the images in the dataset have been extracted.

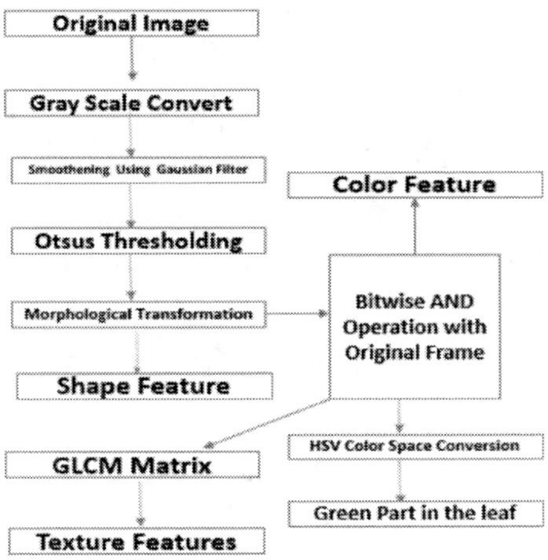

Fig. 2. Image Processing

3.2.2 *Feature Extraction*

The features characterize appropriate & refined qualities/evidence related to objects, distinguishing one object from other objects. The features are supportive of identifying objects and labeling the category tag to an object. The feature abstraction stage is most substantial in making the classification/recognition model and searching for the extraction of relevant attributes characterizing each class. The features used in the presented system are explained as follows.

i) Grey Level Co-occurrence Matrix (GLCM):

It is used in a series of "second-order" texture calculations to measure the combinations of picture element brightness values (grey levels) that occur in a picture. The matrices are used to figure out how pixels are related spatially. The GLCM technique is extremely sensitive to any changes in the images, such as rotation, scale, and so on. By observing repeating patterns, spatial distribution, colour arrangement, and intensity, we can visually distinguish the texture of an image.

ii) Shape Features:

For the calculation of the shape feature, we have used the contour. Contours are described as a curve ioining all the continuous points (along the boundary), ng the same color or intensity. The contours are a ul tool for shape analysis and object detection and gnition.

ßtatistical and Color Feature:

The standard statistical feature such as mean, ian, and standard deviation are utilized here. The r feature is extracted as the amount of green color in eaf. The RGB value of true Green color is (0,255,0). the actual images there is always variation in the ge color values due to various lighting conditions, low & even due to noise added by the camera while cing & subsequently processing the image. To ulate the amount of green color in an image, convert HSV color space and divide the total number of pixels ne channel by the number of pixels with hue channel pixel intensity between 30 and 70. racting the green color portion from One yields the green portion of the image.

3.2.3 *Classification*

The correlation matrix method is being used to select features. The correlation matrix can be used to predict how features are related to one another. In all machine learning problems, feature selection is critical. The features are selected based on the variables' correlation with the target variable. Figure 3 depicts the correlation of each variable with each other for the apple dataset. The F1 and F2 features have a high correlation (1), indicating that they are interdependent. As a result, one of them has been removed from the list (F2). Green channel mean, blue channel mean, green channel standard deviation, f4, f5, f6, f7, and f8 are less correlated features that will not contribute significantly to the model for apple disease prediction.

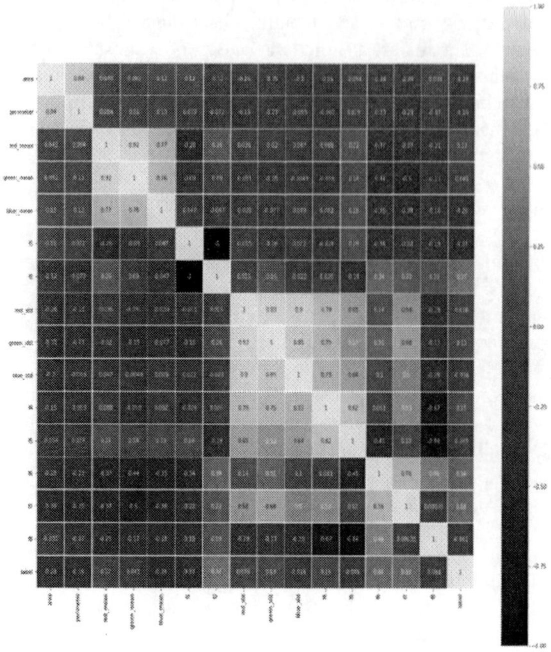

Fig. 3. Correlation map for apple dataset

3.2.4 Classical Model Development

For classification, the Random Forest algorithm is used. For regression, classification, and other tasks, a random decision forest is a mixed methods training technique. Several decision trees make up a random forest. In the case of a classification problem, Random Forest aggregates the results of all decision trees during training and outputs the class, while in the case of a regression model, it outputs the mean prediction. One of the main issues in decision tree algorithms is overfitting, which is overcome by using a random decision tree. Random forest is a classifier that uses many decision tree models to create an ensemble classifier.

4 RESULT & DISCUSSION

The proposed algorithm is tested on a database of 300 images of potato leaves taken from the publicly available dataset 'Plant Village,' which includes 100 healthy and 200 diseased leaves. The database was divided into two sets during the experiment: the training set, which contained 180 images (70 percent), and the testing set, which contained 120 images (30 percent). The accuracy, sensitivity, recall, F1-score, and precision of the presented system were calculated to evaluate its performance. The classification testing accuracy is 95 percent when the train-test split is 60 percent -40 percent. Furthermore, 5-fold cross-validation was used to make the model more robust, and 93.7 percent accuracy was achieved. Table 2 illustrates the performance indicators.

Table 2. Performance Measure Classification

Plant dataset	Accuracy	F1 score	Precision
Tomato	87%	0.93	0.91
Potato	98%	0.99	0.99
Grapes	95%	0.99	0.99
Corn	94%	0.98	0.97
Apple	91%	0.91	0.92

The graphical representation of the confusion matrix containing Tomato test data for 0: healthy, 1: Bacterial spot, 2: Early blight, 3: Late blight, 4:Leaf Mold, 5:Septoria_leaf_spot, 6: Spotted spider mite, 7: Target Spot, 8: Yellow Leaf Curl Virus, 9: the mosaic virus is shown in Fig 4. It depicts the visual representation of the presented algorithm's performance on the testing dataset. As shown in Fig. 5, the area under the ROC curve for our classification is 87 percent, indicating that our classification is extremely accurate.

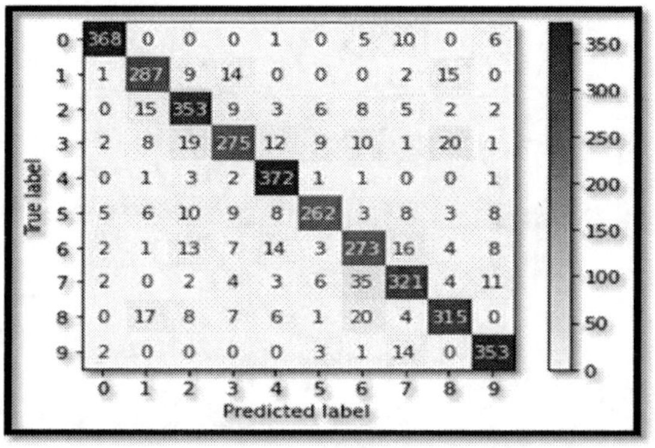

Fig. 4. Statistical data of confusion matrix on tomato dataset

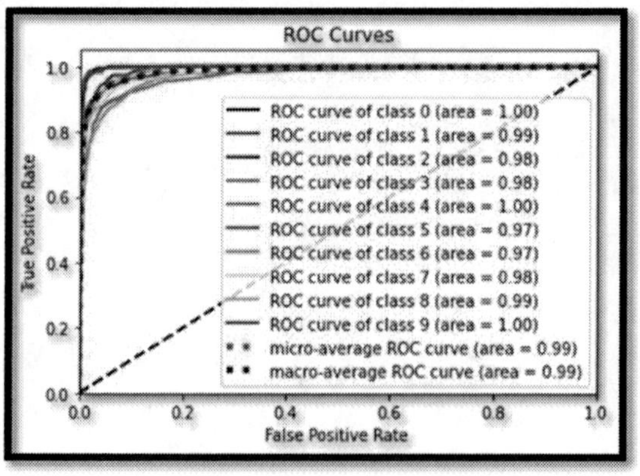

Fig. 5. ROC for tomato dataset

https://doi.org/10.1051/itmconf/20224403002

The graphical representation of the confusion matrix containing Potato test data for 0: healthy, 1: Early blight, 2: Late blight is shown in Fig.6. It represents a visual representation of our algorithm's performance on the testing dataset. As shown in Fig 7, the area under the ROC curve for our classification is 98 percent, indicating that our classification is extremely accurate.

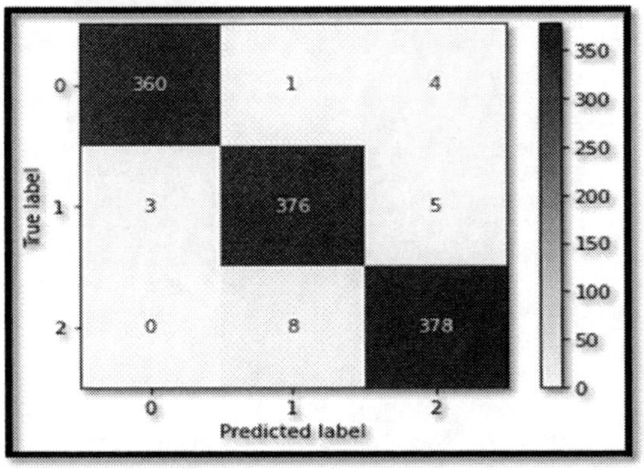

Fig. 6. Statistical data of confusion matrix on potato dataset

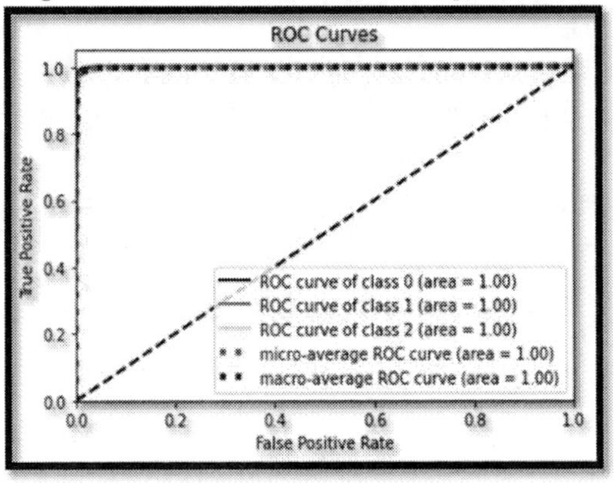

Fig. 7. ROC curve on potato dataset

The graphical representation of the Grape test data confusion matrix. for 0: healthy, 1: Black_rot,2: Esca(Black Measles), 3: Leaf blight (Isariopsis Leaf Spot) as shown in Fig. 8. The ROC curve achieved 95% as indicated in Fig. 9.

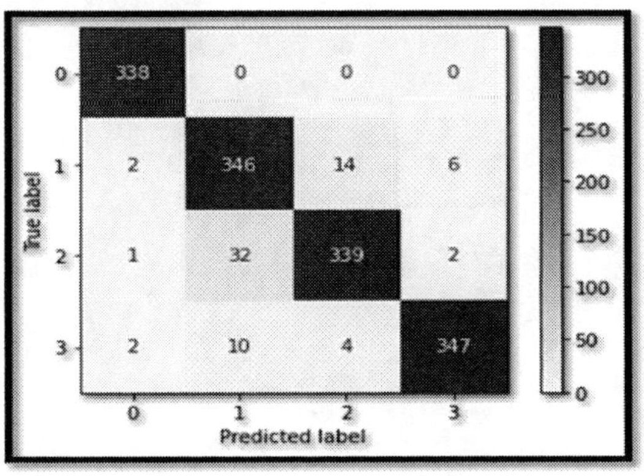

Fig. 8. Statistical data of confusion matrix on grapes dataset

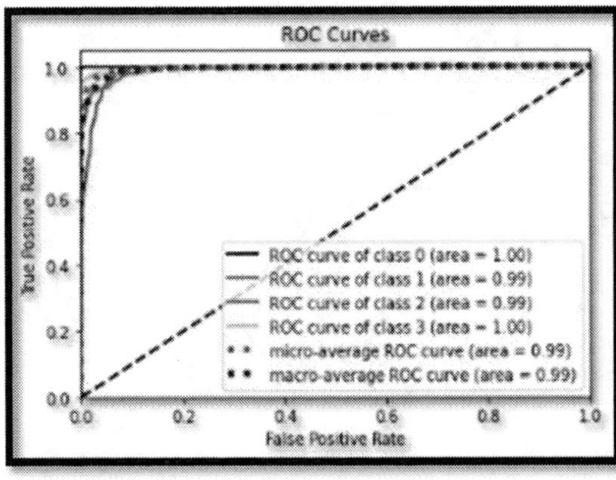

Fig. 9. ROC curve on the

The confusion matrix containing the Corn test data for which 0: healthy,1: Cercospora leaf spot Gray leaf spot, 2: Common rust, 3: Northern Leaf Blight is presented in Fig. 10. Fig. 11 shows that the realm under the ROC curve for our classification is 94%.

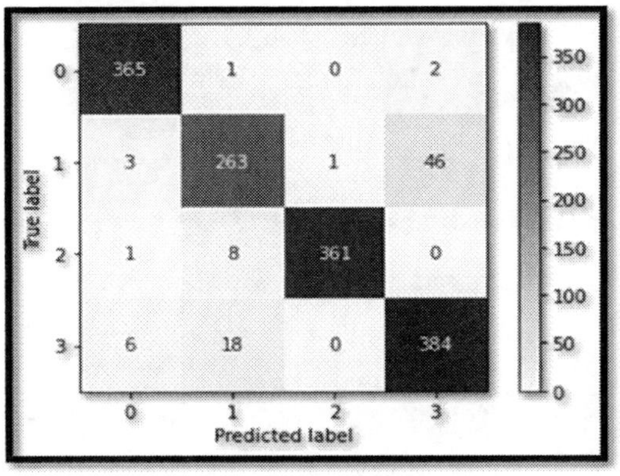

Fig. 10. Confusion matrix for Corn dataset

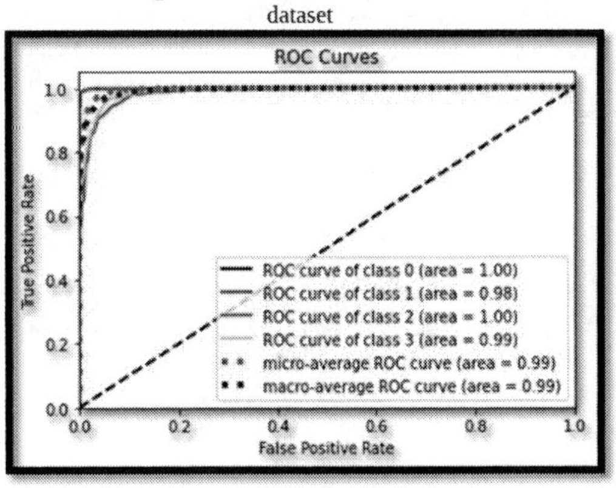

Fig. 11. ROC Curve for Corn dataset

The confusion matrix containing Apple test data for 0: healthy, 1: Apple scab, 2: Black rot, 3: Cedar apple rust is shown in Fig. 12. The realm under the ROC curve for our classification is 91% as seen from Fig. 13.

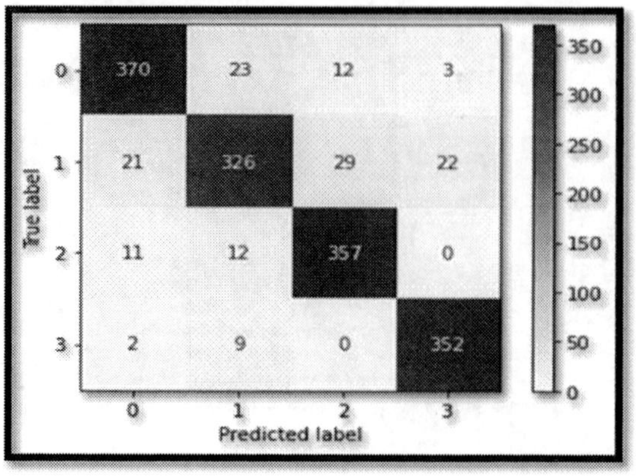

Fig. 12. Confusion matrix for Apple dataset

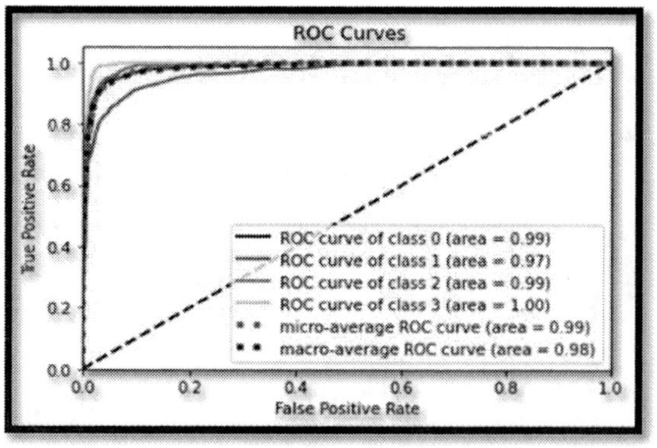

Fig. 13. ROC Curve for apple dataset

5 CONCLUSION

This paper describes a novel plant leaf disease classification technique that can be used for both automatic detection and classification of plant leaf diseases. As a result, related diseases for these plants were investigated. The best results were obtained with very little computational effort, demonstrating the efficacy of the proposed algorithm in the recognition and classification of leaf diseases. Another benefit of using this method is that plant diseases can be detected at an early or early stage. A 93 percent accuracy rate has been achieved. The research has real-world implications for crop disease classification, as it can be used to classify disease symptoms in fruits, vegetables, commercial crops, and other crops. The accuracy of the same can be further improved by implementing other classifiers.

REFERENCES

[1] E. David et al. (2020).: Global Wheat Head Detection (GWHD) dataset: a large and diverse dataset of high resolution RGB labeled images to develop and benchmark wheat head detection methods. https://arxiv.org/abs/2005.02162.

[2] SP . Mohanty, DP. Hughes and M. Salathe Using Deep Learning for Image –Based Plant Disease Detection .Front .Plant Sci. 7:1419 .DOI :10.3389/fpls.2016.01419(2016)

[3] Kothawale, Shweta S., S. R. Barbade, and P. P. Mirajkar. "Grape leaf disease detection using SVM classifier." International Journal of Innovative Research in Computer and Communication Engineering 6.4 (2018).

[4] S.Pavithra, A.Priyadharshini, V.Praveena and T.Monika Electronics and Communication Engineering, VSB Engineering College, Karur. "PADDY LEAF DISEASE DETECTION USING SVM CLASSIFIER." International Journal of Communication and Computer Technologies 3.1 (2015), 16-20.

[5] Islam, Monzurul, et al. "Detection of potato diseases using image segmentation and multiclass support vector machine." 2017 IEEE 30th Canadian conference on electrical and computer engineering (CCECE). IEEE, (2017).

[6] Ramesh, Shima, et al. "Plant disease detection using machine learning." 2018 International conference on design innovations for 3Cs compute communicate control (ICDI3C). IEEE, (2018).

[7] Mohan, K. Jagan, M. Balasubramanian, and S. Palanivel. "Detection and recognition of diseases from paddy plant leaf images." International Journal of Computer Applications 144.12 (2016).

[8] Orillo, J. William, et al. "Identification of diseases in rice plant (Oryza sativa) using backpropagation Artificial Neural Network." 2014 International Conference on Humanoid, Nanotechnology, Information Technology, Communication, and Control, Environment and Management (HNICEM). IEEE, (2014).

[9] Kusumo, B. Suryo, et al. "Machine learning-based for automatic detection of corn-plant diseases using image processing." 2018 International Conference on Computer, Control, Informatics and its Applications (IC3INA). IEEE, (2018).

[10] Sabrol, H., and S. Kumar. "Intensity-based feature extraction for tomato plant disease recognition by classification using the decision tree." International Journal of Computer Science and Information Security 14.9 (2016): 622.

[11] SE. Raza, G. Princ, JP. Clarkson, NM. Rajpoot Automatic detection of diseased tomato plants using thermal and stereo visible light images. PLoS ONE 10(4):1–20 (2015)

[12] MM. Hasan et al.: Detection and analysis of wheat spikes using Convolutional Neural Networks. Plant Methods. 14 DOI: 10.1186/s13007-018-0366-8. (2018).

[13] Wu Wei et al.: Detection and enumeration of wheat grains based on a deep learning method under various scenarios and scales. Journal of Integrative Agriculture 19:1998–2008. DOI: 10.1016/s2095-3119(19)62803-0, ISSN 2095-3119(2020).

[14] NGowda, S.N., Yuan, C (2019).: ColorNet: Investigating the Importance of Color Spaces for Image Classification. Computer Vision – ACCV 2018 Lecture Notes in Computer Science. 581– 596 (2019).

ITM Web of Conferences **44**, 03003 (2022)
ICACC-2022

https://doi.org/10.1051/itmconf/20224403003

Automated Lung Semantic Segmentation on X-Ray Using Convolutional Models

Aakash Ramesh[1*], *Aditya Bhatkhande*[2], Bhavesh Jadhav[3], Nipul Avdhuth[4], Kirti Rathi[5]

[1,2,3,4,5]Ramrao Adik Institute Of Technology, Electronics And Telecommunication Engineering, Navi Mumbai, Maharashtra, India

Abstract. Towards the culmination of 2019, an outburst of coronavirus disease 2019 (COVID-19) pandemic struck mankind, which was instigated due to severe acute respiratory syndrome (SARS) which originally transpired from Wuhan City, China. Myriad of people have acceded to this disease. The effects of the pandemic have been more severe on the populous nations of the world. In India, although over 1.5 billion vaccinations have been provided to the inhabitants, as of 21 February 2022, the pandemic has barely diminished, as seen in Figure 1.1. While restrictions are being somewhat relaxed, the chances of the ominous '4th wave' materialises large. In such scenarios, it is of supreme importance to have apparatus for rapid testing and diagnosis of the disease, to expedite a much faster process. This paper will give an insight of a model that can efficiently and precisely predict the presence of COVID-19 by using a CT scan of the lungs.

1 INTRODUCTION

The novel coronavirus baptised SARS-CoV-2 primarily broke out in Wuhan, China, in December 2019, and has since conquered most of the countries across the globe. By infiltrating the respiratory tracts, the virus produces respiratory syndromes. As of now, the pre-existing prognosis and diagnosis techniques are too tender-some, and are quite often, imprecise. Normally, the CT images are segmented physically but in case of Covid-19 contagion, the common assembly of the lung can be infected. Hence, manual semantic segmentation is not a feasible solution. Chest CT is evolving as a treasured diagnostic set-up for clinical administration of COVID-19 linked lung disease [10]. Lung CT image segmentation is a crucial step for lung semantic scrutiny, and is a preconditioned step to deliver a precise lung CT analysis. Therefore, in the proposed scheme we aim to employ CT (computer tomography) images. Our chief objective is to attain RGB images of the lungs of normal and affected patients, then perform necessary image analysis by applying segmentation, classification and regression to obtain the CT scan by means of DICOM images.

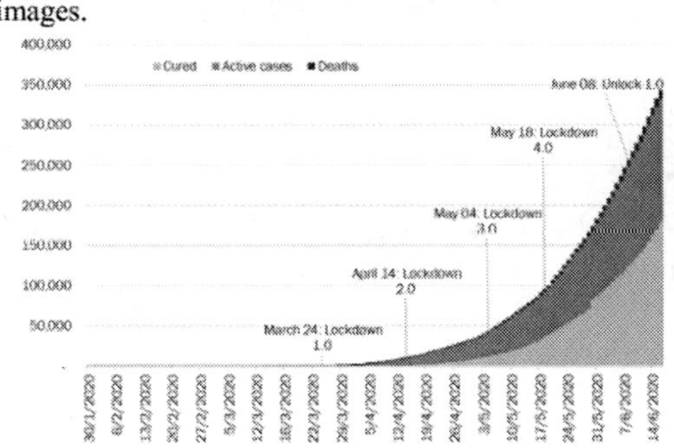

Figure 1.1: Total number of cases of Covid-19 in India [1]

Fig 1.2 characterizes the technique by which the patient is firstly diagnosed with covid-19 by using CT chest analysis which elasticities a visual quantification of lung lesions. The white portion in the image displays the affected area of the lungs. It cannot be detected via naked eyes, highlighting the need of CT analysis. If the affected area percent is above 50 percent the chances of cure is less and if it is between 26 to 50 percent the chances of treatment for patient is possible.

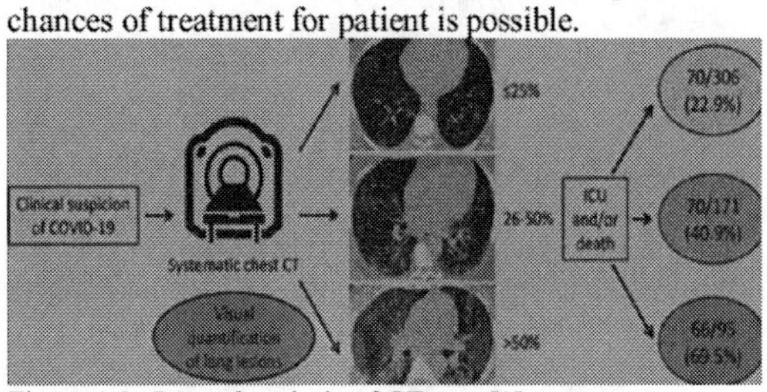

Figure 1.2: General analysis of CT scan [1]

This paper is prearranged as follows: Section 1 deals with the pre-existing mechanisms, and the requirement for a technique to enhance the same. Section 2 deals with the overall principal locations to look out for within a CT scan. Section 3 is concerned with the overall operation of the proposed model followed by Section 4 which provides with analysis of the results obtained. Finally, Section 5 puts forth the conclusion and possible modifications and upgradations of this model.

2 MEANS OF COVID-19 DETECTION IN LUNGS

Computed tomography (CT) of the chest uses premier x-ray devices to figure out flaws found persistent with other imaging scans, and to help diagnose the motive behind remote, chest pain, fever, shortness of breath, coughs, and other chest related symptoms. CT scanning is instantaneous, non-invasive, effortless and accurate. Since it can detect infinitesimal nodes in the lung, chest CT is especially competent for pinpointing lung cancer at its earliest, most curable stage. A CT scan generates images that can be reinvigorated across multiple

* Corresponding author: aakashramesh3007@gmail.com

dimensions and planes [3]. It can also produce n-dimensional images, which is the rudimentary qualification. Chest CT scans have a very crucial aspect in the prognosis, diagnosis, and detection of complications in COVID-19. The following principle arraignments are intermittently realized in the chest CT scans of COVID-19 patients :

- Ground-glass opacities (GGO) (bilateral, multifocal, subpleural, peripheral, posterior, medial and basal).
- Bronchovascular thickening (in the lesion).
- Traction bronchiectasis.
- Air space consolidation.

Ground glass opacities, i.e., GGO and consolidations are most regularly visualized in CT presentations, not enclosed to bilateral middle and lower lobe predominance and peripheral circulation (76– 85 percent). The disease is broadly multifocal, although single lesions can be visualized on preliminary imaging. This is sometimes seen about 4 days post infection. In a active patient's CT scan, the heart takes up the center (in white) and is enclosed by darker shades, which is the lung. But in a COVID-19 infected lung, the scan shows whitish regions over the lungs, which are injuries/scars subordinated on the "alveoli" because of the COVID virus. Around 4-8 days post infection, there will be expulsion.

In Figure 2.1, one can observe the GGO designated by white arrow(s), chiefly in the lower- left lobe, but sometimes also in the right-middle lobe. Lower right lobe consolidation with air bronchogram is labelled with a black arrow [3] of crazy paving with septal thickening on top of a background of GGO. There are also bilateral infiltrates that are predominantly posterior. A reverse halo is also visualized in the CT scans of patients

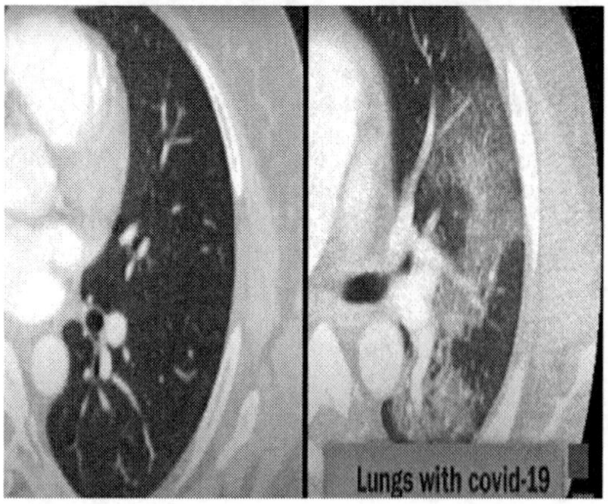

Figure 2.1: This is the chest CT scan of a covid positive pneumonia patient.[3]

diagnosed with COVID-19 with lower attenuation centrally, as seen in Figure 2.3. It is normally observed in patients who have COVID-19 with organizing pneumonia. Crazy paving and GGO is also observed.

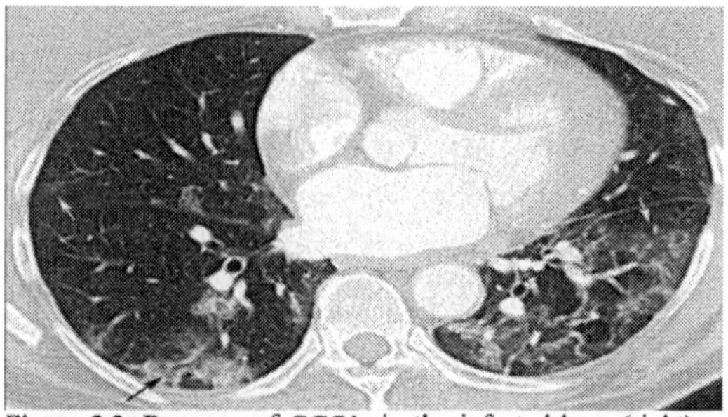

Figure 2.2: Presence of GGO's in the infected lung (right) as opposed to a healthy lung (left) [3]

About 13-14 days after the spread of the infection, a full-blown consolidation is realized. It is an infiltrate where vessels are not conspicuous. A reverse halo can also be visualized. It may be unilateral as well. When patients get really sick and get ARDS, Diffused Alveolar Damage (DAD) is detected and the trademark of DAD in pathology is surfactant annihilation. CT scans can aid in examination of COVID-19 in the appropriated ambience, and set a reference line for patients who are at risk for exposure, may benefit from suggesting alternate diagnosis and comorbidities in the vulnerable patients with COVID-19. Employing a CT scan is more preferred for our model, as they do not influence the contrast of the images, which can critically hinder with the precision of the proposed system. Since the images will be in .dcm file format, we will employ Micro-DICOM viewer for siphoning the same.

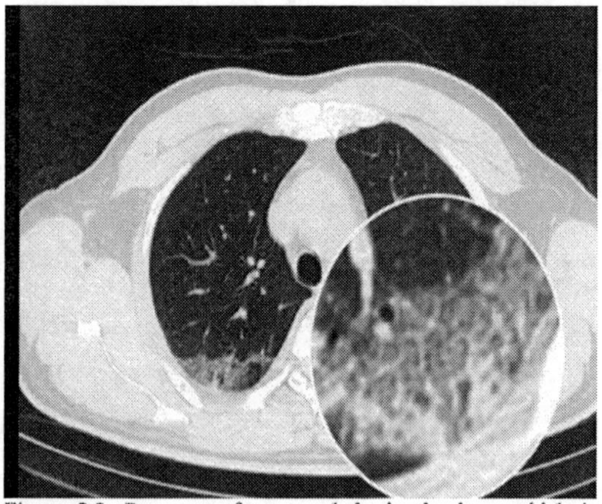

Figure 2.3: Presence of reverse halo in the lung which is enlarged for convenience [4]

3 SYSTEM DESIGN

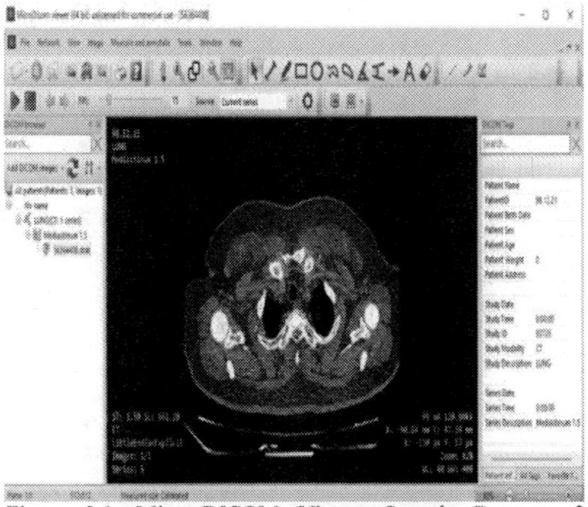

Figure 3.1: Micro-DICOM Viewer Sample Dataset of a patient's lungs diagnosed with COVID-19

In this section, we will give a comprehensive interpretation of how our model will operate and an in-depth reasoning of the technical aspects.
Firstly, we are recommending the application of CT scans as a CT scan administers much more specific images with the acclimatization of tissues, bones and organs very well defined. This will be beneficial in training the model.
The dataset used is LIDC (Lung Image Database Consortium), which consists of tomographic images (in DICOM format) with designated lesions. This is readily accessible on the internet, and is effective in lung segmentation as well as diagnosis. The dataset accommodates CT scans with 20 '. nii' files, each possessing around 180 images. It is a public access consortium founded by NCI and further advanced by FNIH [16]. It incorporates a sum total of 1018 reported cases. However, given that the dataset size aggregates to almost 125GB, an substitute dataset must be used. In this case, there is another dataset that can also be used to similar effects. This dataset contains chest CT scans of over 1000 patients diagnosed with COVID-19. It is an open repository base. It encompasses data of patients from March 2020 to January 2022 till date. The dataset subsists of grayscale images of size 512x512 pixels stored in DICOM form, i.e., with an extension of '.dcm'. Figure 3.1 maintains an instance of an image from the dataset as seen using MicroDICOM Viewer Software. The patient's confidentiality is abetted, while only revealing the enforced particulars compulsory to launch that this is the CT scan of a patient diagnosed with COVID-19. The CT scan evidently demonstrates the presence of GGOs, and reverse halo, signifying that the patient has been detected with Covid-19.

3.1. Image Segmentation Using Otsu's Binary Thresholding

For segmentation, we can employ marker-based Watershed Algorithm using "Otsu's Binary Thresholding". For this methodology, a greyscale input is needed, which implies we shall proselyte the DICOM image into grayscale [4]. Both these methods will develop a threshold automatically which curtails the demand for human interface. These trchniques will assist in retaining the essential lineaments, while cloaking the gratituous minutiae in the image.
Otsu's approach desires to develop the image histogram and segment an image by diminishing the variation on each of the accessible classes. The histogram composed by these images accommodates two clear crests, characterising fluctuating intensity. The essential principle is separating the image histogram into a couple of clusters with a threshold defined as a result of minimization of the weighted variance of these classes.
In Otsu's Binary Thresholding method, the general algorithm is as follows:

- Process Input Image.
- Obtain the pixel distribution in the image, i.e., obtain histogram.
- Evaluate a threshold value.

If saturation is higher than the threshold value, convert the pixels into white regions, else convert the pixels into black regions.

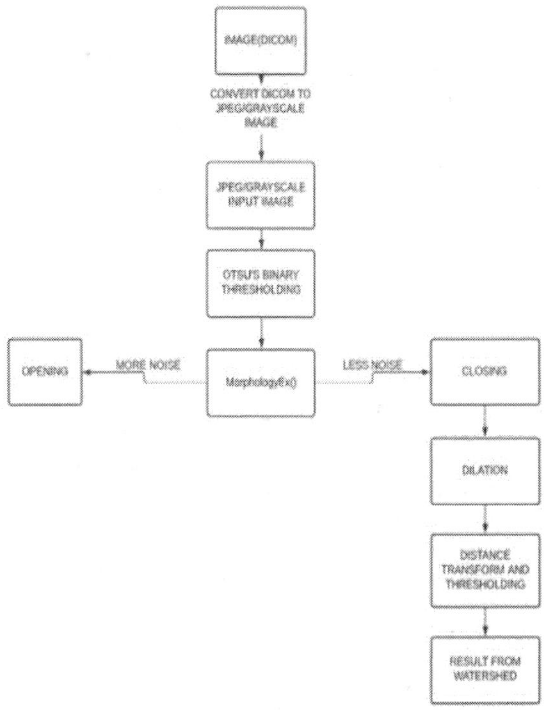

Figure 3.2: Otsu's Thresholding Flow Diagram

3.2. Image Segmentation Using Otsu's Binary Thresholding

https://doi.org/10.1051/itmconf/20224403003

After the segmentation is done, we plan to detect the presence of COVID-19 in the lungs, by using CLAHE equalization. We plan to use this technique as it increases the contrast in images, making the detected locations more prominent, thereby improving the visibility of the same.

As seen from Figure 3.3, we must first analyze the images, i.e., we must obtain the images from the dataset and perform segmentation in order to mask the unnecessary regions in the image. After this, we shall normalize the pixel values to a range within 0 and 1. This is a mandatory step before inputting the image to the deep learning model. Normalizing the pixel intensity will help avoid the effect of high frequency and very low frequency noise. It also ensures that the image is normally distributed. This will improve the convergence of the neural model. This is then subjected to CLAHE Equalization in conjunction with Interpolation. The equalized model is then sent to the ResNet50 deep learning model, where it undergoes MaxPooling, Filtering, Feature extraction, and so on. For the Deep Learning model, the ReLU activation function is the most preferred option as it is easy to train, and it is a high-performance function. The GRAD-CAM algorithm is used to generate a heat-map, which makes it easier to understand the strike points of the COVID-19 virus.

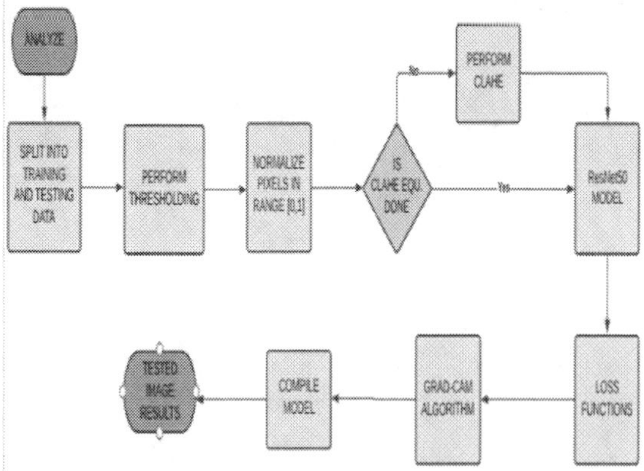

Figure 3.3: COVID-19 Detection Flow

This data is fed onto a ResNet50 model on the UNET base. Usually, a neural network model is trained via a stochastic gradient descent optimization algorithm while the weights are updated using the backpropagation of the perceptron error algorithm. For our model, either MSE or Cross-Entropy method can be employed. After this is done, then boundary loss segmentation is done as a finishing touch to help the model

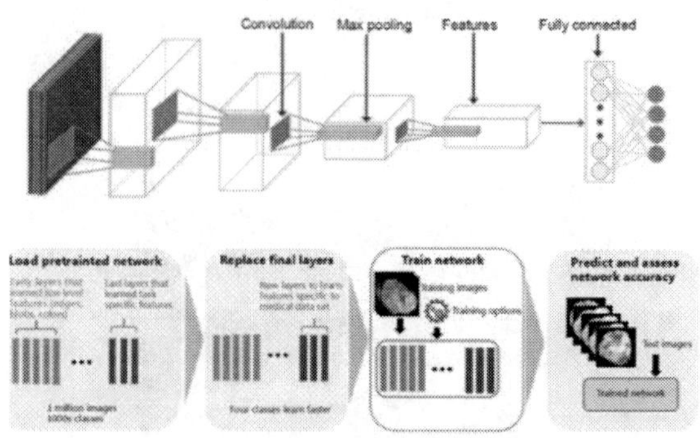

Figure 3.4: CNN Layered Classification [5]

in recognizing the signs of Covid-19 easily. The model can then be compiled, trained and tested, to check if it's working as per expectations.

4 SIMULATION

For the implementation of our project, we will use TensorFlow 2.0, which supports both high-level and low-level APIs, and is written in python language. It runs effortlessly on both CPU and GPU. Our model is tested on an Intel i7-10750H CPU 12GB RAM with a GPU processor NVIDIA GeForce GTX 1650 with 6GB memory to run the huge dataset being employed without any issues.

The Otsu's thresholding is carried out first, and is compared with other thresholding schemes to check which method is more appropriate, and this is seen in Figure 4.1. From this, it can be deduced that the Otsu's method provides a much clearer image with the contrasts clearly visible.

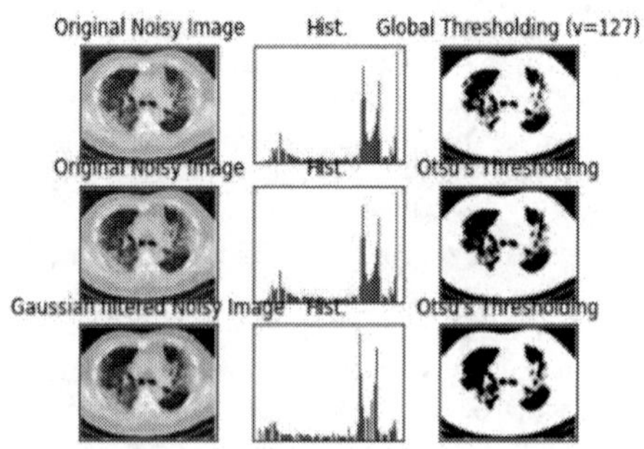

Figure 4.1: Otsu's Thresholding Comparison with Global Thresholding

Figure 4.2 shows the CLAHE equalized image on the right. It can be clearly viewed that the hidden features described in Section 2 can be made accesible using this technique.

https://doi.org/10.1051/itmconf/20224403003

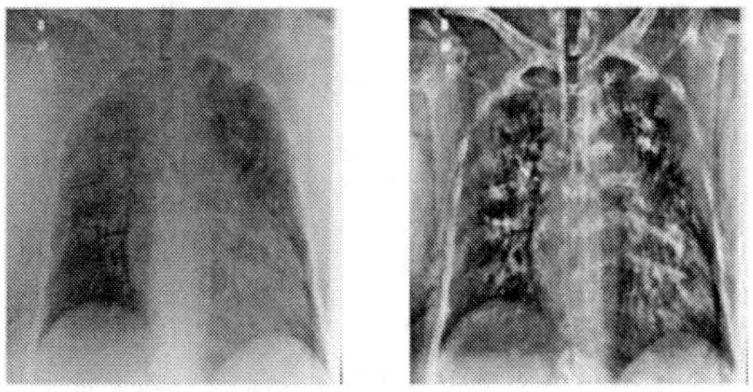

Figure 4.2: Neural Model Layers

Based on the model composition, a sample image is used to test the functionality of the model, as shown in Figure 4.3. The model predicts if the patient is infected or not, based on the CT scan, and also a mathematical analysis of how probable the chances of infection are. The same is also tested by locating the nodes of infection, in Figure 4.4. The colored regions (red, orange, green) represent the regions where the COVID-19 virus is predominantly present. This includes presence of GGOs, Reverse Halo, and so on.

```
The given X-Ray image is of type = Covid

The chances of image being Covid is : 99.99998807907104 percent

The chances of image being Normal is : 2.3658344616706017e-05 percent
```

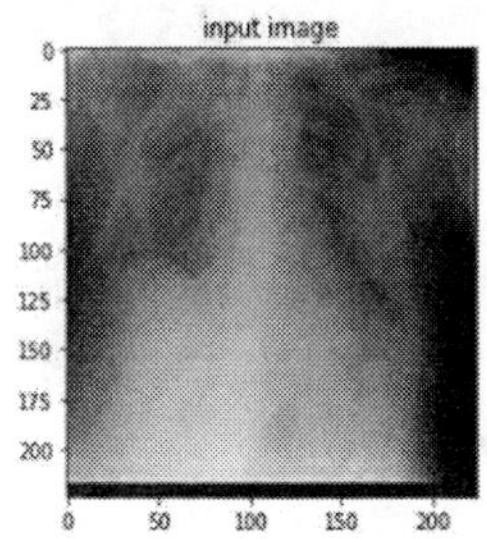

Figure 4.3: Simulation of Model with sample image

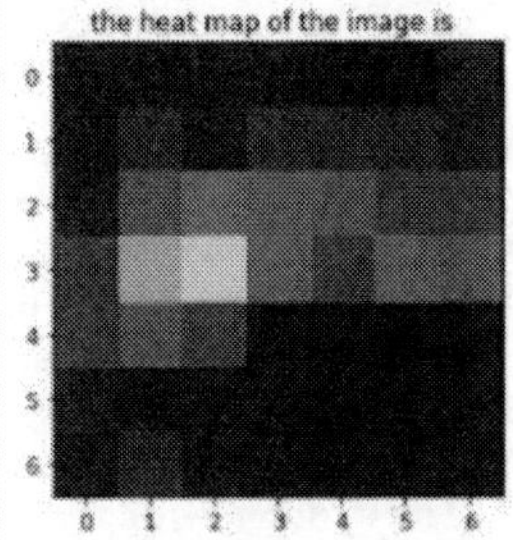


```
The given X-Ray image is of type = Covid

The chances of image being Covid is : 99.99498128890991 %
The chances of image being Normal is : 0.00034904455787909683 %

image with heatmap representing region on interest
```

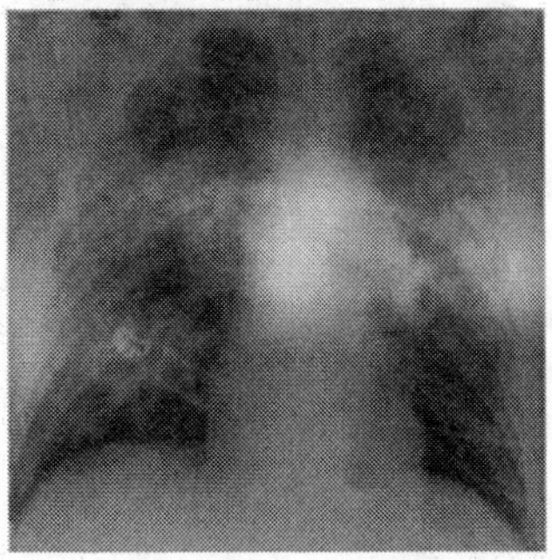

Figure 4.4: Simulation of Model of sample image with Heat-Map

The overall simulation results in an accuracy of 97.5 percentage (Figure 4.5), as opposed to the traditional accuracy of about 84 with global thresholding for image classification.

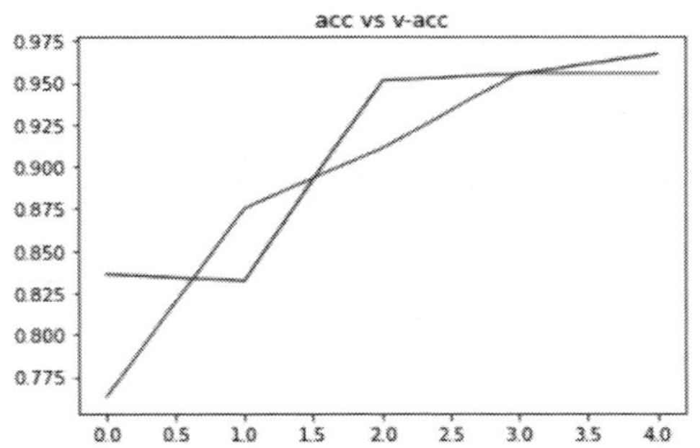

Figure 4.5: Simulation Result

5 CONCLUSION AND FUTURE SCOPE

Radiological imaging provides a unique perspective in addition to the assessments done for the early prognosis and remedy of the disorder. Chest radiography or CT experiment can be subjected to certain irregularities withinside the lung related to disorders in comparison to a healthy patient's lung. Deep studying version or the CNN fashions are well-equipped in detecting COVID-19 lung segmentation and therefore the diagnostic accuracy price is pretty exorbitant. The version is a hit-to-hit upon consolidation areas and nodular opacities, which can be pathogenic findings of patients suffering from COVID-19, on X-ray radiography. In COVID-19, bilateral, decrease lobe, and peripheral involvement is observed, and the proposed version can identify localization of the lesion. These fashions are especially critical in figuring out early presence of COVID-19 victims. Early prognosis of the disorder is critical to offer on-the-spot remedy, and to prevent disorder transmission. The model can also can play a vital role in victims missing early symptoms. The medical and radiological pictures of higher-degree patients are more vulnerable. The function of deep studying models are more distinguished in screening and prognosis while the contamination is in its early stages.

Technological advances have resulted in excellent image quality and reliability, and by turning to universal use in clinical medicine.

Temporary repair improvements can be expected, with short rotation periods of less than 200 milliseconds, 80-millisecond scan times in single-source systems and 40 milliseconds of dual-source systems, as well as additional protection from residual movement using integrated reconstruction methods. the movement of the remaining object in the calculation. The use of repetitive updates will continue to increase significantly as the algorithms become more robust and the rebuilding times become shorter.

If X-ray detectors with the best holes and photon counters are readily available, they are likely to be introduced first for the intended purpose. It can be expected that they will produce significant improvements in the landscape, as well as in the capacity of detectives, especially among high frequencies. Low radiation exposure will be achieved by using these highly efficient detectors and advanced reconstruction algorithms.

CT tools with a special purpose have been developed in recent years, mainly for research purposes but marketing has begun. If this trend continues, there will be an increased variety of CT scans that are still distributed in the clinic. Each of these may lead to new clinical use as other adjustments occur.

6 REFERENCES

1. https://images.app.goo.gl/EC8wmVehweg4BKL56.
2. Christian Szegedy, Alexander Toshev and Dumitru Erhan,"Deep Neural Network For Object Detection" NIPS , 2013.
3. https://microbeonline.com/rt-pcr-principles-applications/
4. Adnan Saood and Iyad Hatem, "COVID-19 lung CT image segmentation using deep learning methods: U-Net versus Segnet",Feb 2021.
5. Zbontar, Ling ,"Barlow Twins: Self-Supervised Learning via Redundancy Reduc- tion",2021
6. Shokouh Shakouri, Mohammad Amin Bakhshali, Parvaneh Layegh, Behzad Kiani, Farid Masoumi, Saeedeh Ataei Nakhaei and Sayyed Mostafa Mostafavi, "COVID19- CT-dataset: an open-access chest CT image repository of 1000+ patients with con- firmed COVID-19 diagnosis", May 2021.
7. Shervin Minaee, Yuri Boykov, Fatih Porikli,et al, "Image Segmentation Using Deep Learning: A Survey", *arXiv*, Nov 2020
8. Nassir H. Salman, "Image Segmentation Based on Watershed and Edge Detection Techniques", DBLP , Jan 2006
9. Samuel G. Armato, Geoffrey McLennan, "The Lung Image Database Consortium (LIDC) and Image Database Resource Initiative (IDRI): A Completed Reference Database of Lung Nodules on CT Scans", ISSN: 38(2): 915–931,Feb 2011.
10. K. He, X. Zhang, S. Ren and J. Sun, "Deep Resifual Learning for Image Recogni- tion", CVPR , 2018
11. Jasjit S. Suri, Sushant Agarwal and Rajesh Pathak, "COVLIAS 1.0: Lung Seg- mentation in COVID-19 Computed Tomography Scans Using Hybrid Deep Learning Artificial Intelligence Models" MDPI Diagnostics ,11,2021,1405 February 2014.

Machine Translation Systems for English Captions to Hindi Language Using Deep Learning

Arvinder Singh , *Ninad* Bhase, *Manav* Jain , *and Tushar* Ghorpade

Department of Computer Engineering
Ramrao Adik Institute of Technology
D.Y. Patil Deemed to be University
Nerul, Navi Mumbai, India.

Abstract. Machine Translation is the process of translating text from one language to another which helps to reduce the conversation gap among people from different cultural backgrounds. The task performed by the Machine Translation System is to automatically translate between pairs of different natural languages, where Neural Machine Translation System stands out from all because it provides fluent translation along with reasonable translation accuracy. The Convolution Neural Network encoder is used to find patterns in the images and encode it into a vector that is passed to the Long Short Term Memory decoder which finds the caption word-by-word to best describe the image. Upon reaching the end-line token, the entire description of the image in English is generated and that is our output for that particular image. Automatically creating the description of an image in English using any natural language sentences and then translating it using Neural Machine Translation to Hindi is a very challenging task. It requires expertise in both image processing as well as natural language processing. In this paper, the aim is to compare the two Machine Translation Systems: Google Translation System and the proposed Neural Machine Translation System to convert the text obtained from an image in English to Hindi language.
Keywords - Image captioning, English language, Hindi language, encoder-decoder framework, Neural network, Machine translation.

1. Introduction

English is the universally used language and Hindi is the national language of India, as per the analysis, English to Hindi machine translation system is very important to enhance the knowledge of Indians without any language barrier. Machine translation [1] is an application of Natural Language Processing (NLP) which aims to fill the communication gap between different parts of society. Machine Translation works with large amounts of source and target languages that are compared and matched against each other by amachine translation engine. The benefit of Machine Translation is that it is possible to translate large amounts of text in a very short time.

Further the paper is divided as follows: - in the section 2 we discuss about the Literature Survey, section 3 includes the problem definition of proposed system, section 4 contains the details of the Proposed System, section 5 shows the Results and Analysis for the proposed system and in the section 6 the paper is concluded and discusses about the future work.

2. Literature Survey

In this section, many research papers were reviewed that are revolving around the topic. From these papers, the main focus was on the methodologies used and their brief descriptions along with their limitations.

The research paper by Lucia Benkova and Lubomir Benko on Neural Machine Translation as a Novel Approach to Machine Translation aims to introduce a novel approach i.e., Neural Machine Translation. The Neural machine translation structure is developed on the coding frame. The encoder converts the sentence in the source language into a continuous space representation with a recurring neural network. The results from this paper reveal that neural machine translation system provides with a high accuracy result but still needs to be assessed in the future.

*Corresponding Author : arvindersingh0912@gmail.com

 https://doi.org/10.1051/itmconf/20224403004

By referring to the research paper written by Ankit Rathi on Deep Learning Approach with Image Captioning in Hindi [2], it is seen that the model used in this paper is trained to predict the image caption using the image feature vector and the previous word. This paper has tested their model on four databases and concluded that image definition quality increases after training the model with a pure database. This study gives an idea of how to generate captions in Hindi using a coding model. It is clear from their experiment that a model with a single definition defined for each image produces high quality captions.

The Literature Survey involves studying about many machine translation systems like Rule-Based Machine Translation (RBMT) [3], Corpus-Based Machine Translation (CBMT) [4], Hybrid-Based Machine Translation (HBMT) [5], Neural Machine Translation (NMT) [6] along with Google Translation System (GTS) [7]. The paper published by Sabine Hunsicker, talks about HBMT being a combination of two or more machine translation techniques where the advantages of the individual techniques are combined to achieve an overall better translation. This paper points out that HBMT has a drawback that it needs extensive editing and human translators are also required sometimes. To learn about CBMT, a research paper published by B. Premjith, M. Anand Kumar and K.P. Soman was very helpful. According to this paper, Statistical Machine Translation (SMT) comes under CBMT and the advantage of SMT is that translation can be done without linguistic language. Still, creation of parallel corpus is one of the drawbacks of SMT. Daniel Torregrosa and Nivranshu Pasricha in their research paper suggest that RBMT can achieve high accuracy within narrow subsets of language. Although RBMT does not require much data, it needs expert human involvement along with many rules in order to improve the quality which results in a very complex system. In order to grasp the concept of Google Translation System, research paper published by Santosh Kumar Mishra and Rijul Dhir was very informative. This paper talks about using a dataset in English Language and then translating it into Hindi Language. Accuracy of GTS is mainly dependent on the source and language pairs. However, Google translation sometimes creates translations which are grammatically incorrect as a result the meaning of these translations are lost.

3. Problem Definition

This proposal aims to define, or rather in appropriate terminology, "formulate" the problem statement as follows: The concept behind machine translation is to generate grammatical rules for the source and target language. Machine Translation works on those set of rules and acts as a kind of translation between languages. The problem that arises is the addition of new content, language pairs because maintaining and extending such a set of rules was too tedious and expensive. Aim is to work on a comparative study between two Machine Translation Systems: - Neural Machine Translation and Google Translation System (GTS).

4. Proposed Methodology

4.1 Proposed Work:

The main motive is to look into generating captions in English and then converting it to Hindi with the help of Machine Translation System Viz. Neural Machine Translation System and Google Translation System. Further on the basis of different BLEU scores analyzing the standard of the existing Machine Translation output.
In addition, natural language should define semantic knowledge i.e., to develop language models in terms of visual comprehension.

4.2 Implementation Details:

For Image Captioning, we host ResNet50 [8], which is a deep neural network with 50 layers. The deeper the neural network, the more difficult it is to train. The ResNet50 architecture facilitates network training and allows them to be more in-depth, leading to increased performance in various tasks.

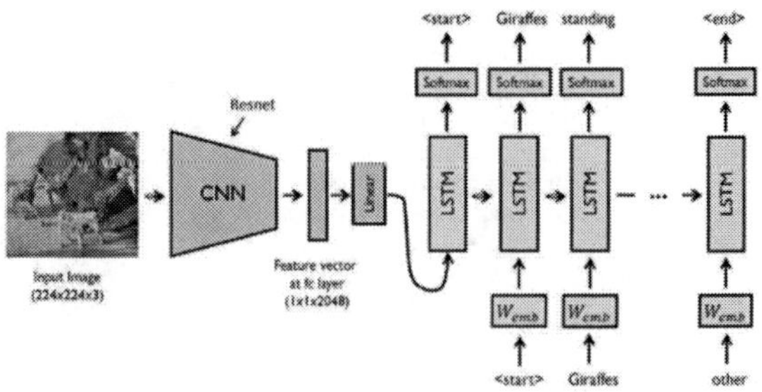

Fig. 1. Image Captioning Architecture [9]

To find the captions for the images, the encoder decoder framework model is used as shown in Fig.1. Here the system uses the ResNet50 model as an encoder to produce a vector containing the image features. After that, LSTM is adopted as a language decoder model to decode the vector into a sentence. Here, image is taken using OpenCV [6]. It gives image in BGR format but ResNet50 only accepts in RGB format. Also, ResNet50 only accepts image of size 224x224. After this pre-processing of image, it is loaded into the ResNet50 model. ResNet50 model produces 3D vector of 2048 which contains all the image features. Typically, a pre-trained CNN removes features from our installation image. The feature vector is converted linearly to the same size as the input LSTM network input. This network is trained as a language model in the vector tool. LSTMs have become an innovative solution for problems related to sequencing and time series, but LSTMs are difficult to train. Even the simplest models require a lot of time and system resources. Due to limitations of SMT, translation with the help of Neural Network was introduced to deal with problems of accuracy and ability to analyse context. Here a Feed-Forward Neural Network is used to calculate the result of phrase pairs by considering fixed size phrases. However, this is not always the case as the length of phrase may change depending on the scenario. It is observed that small phrases perform well when Recurrent Neural Networks (RNN) is used for neural machine translation. For lengthy sentences Statistical Machine Translation is preferred. It is tough to control complicated context-dependent cases by RNN. Hence, RNN combined with LSTM has potential to recollect long-term features. The other aspects which enhance the effectiveness of NMT system are stacked RNNs, test-time decoding using beam search, input feeding using attention mechanism.

The implementation details are as follows:

• Data Analysis Model: It is implemented by the use of the environment provided by Google Colab for data analysis and machine learning, which is a perfect online platform to analyse/train/share data of system with the bindings as per required. The model is implemented in this environment using key Python libraries.

• Model Analysis: It is implemented by the use of the environment provided by Python IDLE- 3.7/Anaconda/Spyder for data analysis and machine learning, which is a perfect platform to analyse/train/share data of system with the models as per required. The model is implemented in this environment using key Python libraries.

• GUI Web Application: It is implemented by the use of the environment provided by Flask for Python HTML5 for web designing and ML based API construction, which is a perfect online platform to present/showcase the system with an interactive interface for the User. The model is implemented in this environment using key aspects of Flask, and an online form is constructed using HTML5 with the form connected to the project.

4.3 System Design

4.3.1 Image Captioning

The image captioning is divided into two logical modules-one image based-model-it is used for feature selection of the images, and two language model-it converts those features of image into natural sentence. CNN is used for image feature extraction and RNN is used for caption generation. ResNet stands for Residual Network. It is the new terminology that is introduced in residual learning. There is sequence of breakthrough Deep Convolution Neural Networks in the field of image captioning. ResNet model has a different model called as ResNet50 having 50 layers which contains 48 Convolution layers, 1 MaxPool and 1 Average Pool layer. Basically, in a deep convolutional neural network, various layers are stacked together and trained to perform a task. The network learns various features at end of its layers. In residual learning, instead of trying to learn certain aspects, it tries to learn some residuals. Long Short Term Memory (LSTM) [4] is an advanced RNN model that was designed to read a long list of sentences that did not occur in normal RNN. The LSTM model will learn a function that creates a sequence of previous recognition as input detection output. Thus, the sequence of views must be converted into multiple examples where LSTM can learn. The NMT system is trained using sentence pairs directed at the English and Hindi source. The parallel corpus contains 40,455 cases, including data sets from a various source.

• This technique mainly uses in-depth reading and deep learning to produce new captions. In this method the image features are first extracted using a CNN model and then it creates captions for the images using the LSTM model.
• Use CNN as an image editing encoder and LSTM as a decoder to generate descriptive sentences.
• This method uses image data sets and their meaning in the native language and requires a word-for-word connection between the description and the visual data.
• The first model aligns pieces of sentences with visual cues, and then creates a single meaning with multiple embedding. This definition is treated as learning data for a second model of a repetitive neural network that has learned to detect captions. The proposed Machine Translation Systems adopted the concept of Open NMT in NMT-1 and GTS in NMT-2 system respectively –

• The system has trained the Neural Machine Translation System 1 & 2 with the help of English to Hindi parallel training corpus to obtain the required models. So, for training purpose of the models, two test cases were taken under consideration.

• Both the systems were then re-trained with the help of the same corpus. Then with increasing the number of iterations and testing the model, it was saved, thus acquiring the predicted results at thirteen and fourteen for machine 1 and 2 respectively.

• The last step was to validate the results obtained from both the systems using the sample images to check the confluence of the training process.

4.3.2 Machine Translation

Machine Translation (MT) is the process of converting a source language into target language without any human entanglement such that the meaning of the source language is preserved in the translated text.
With a large amount of research available in the present-day, several types of NMT are already being looked into and being established in the industry. The Encoder-Decoder model is the most popular kind of the NMT model. The NMT model aims to take an individual sentence as an input and convert this sentence

 https://doi.org/10.1051/itmconf/20224403004

to a different language which is the final output as shown in Fig.2.

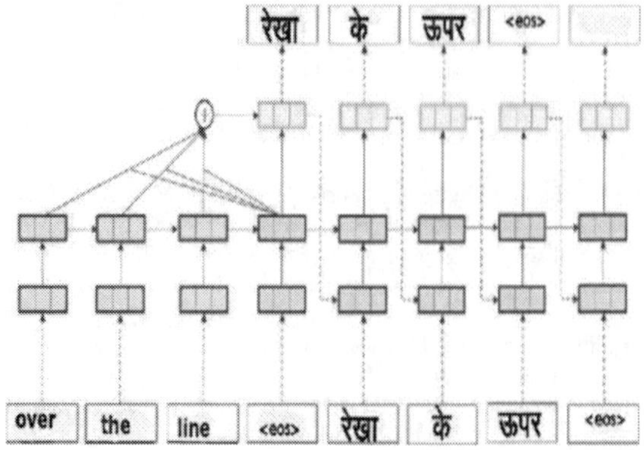

Fig. 2. System Design [3]

The encoder takes a phrase from the input language and develops a vector using this phrase. This vector then holds on to the meaning of the phrase and dispatches it to the decoder which manifests the translation into the target language. The decoder issues predictive sentences of variable sizes, where the decoder continues to predict words till it reaches the <EOS> tag.

Once the <EOS> tag is detected, the work of decoder is done and the sentence is completely translated. The translation is more exact when Attention mechanism concept is used in NMT.

In this concept basically at every step, the decoder creates a memory vector by recalling all the encrypted vectors of the encoder. For prediction of the following word in the translated sentence, it uses the memory vector along with hidden vector in the decoder. In this process, decoder only uses crucial details from the encoder which would otherwise cause damage.

5. Results And Analysis

5.1 Dataset

A cluster of picture-based captions, consisting of more than 8,000 images as shown in Fig 3. each paired with five different captions that provide clear descriptions of important businesses and events.

Images are selected from six different Flickr groups, that usually doesn't contain any celebrities or locations, but are hand-picked to reflect different scenarios.

In this proposed system, the dataset has been prepared by combining the Flickr8k English and Hindi Caption dataset [10]. This dataset consists of three columns: source, english_sentence and hindi_sentence.

```
[ ] images_path = '/content/drive/MyDrive/Images/'
    images = glob(images_path+'*.jpg')
    len(images)

    8091
```

Fig. 3. Total number of images in the dataset

The source column consists of all the image names, english_sentence column contains five different descriptions of the source image in English and the hindi_sentence column consists of the corresponding descriptions of these images in Hindi as shown in Fig.4.

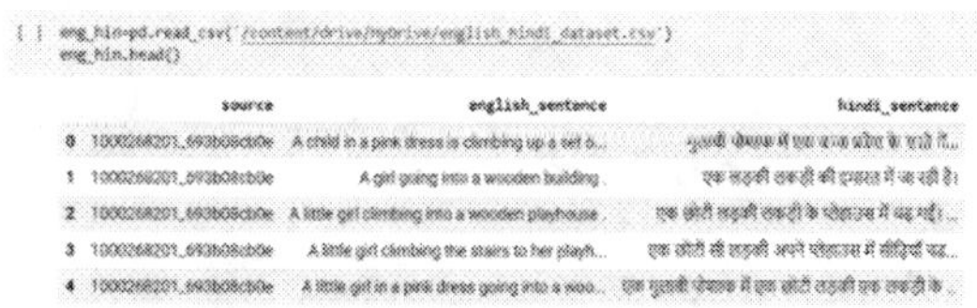

Fig. 4. English-Hindi Dataset

```
eng_hin.dropna(inplace=True)
eng_hin=eng_hin[:50000]
eng_hin.drop(['source'],axis=1,inplace=True)
eng_hin.shape

(40455, 2)
```

Fig. 5. Total number of captions in the dataset

Fig 5. shows that there are a total of 40,455 captions for both English and Hindi languages in the dataset for 8091 images. The accuracy of the model is shown in Fig 6.

 https://doi.org/10.1051/itmconf/20224403004

```
Epoch 47/50
189/189 [==============================] - 33s 173ms/step - loss: 1.0620 - accuracy: 0.7308
Epoch 48/50
189/189 [==============================] - 33s 174ms/step - loss: 1.0243 - accuracy: 0.7409
Epoch 49/50
189/189 [==============================] - 33s 173ms/step - loss: 1.0024 - accuracy: 0.7451
Epoch 50/50
189/189 [==============================] - 33s 173ms/step - loss: 0.9732 - accuracy: 0.7518
<keras.callbacks.History at 0x7f1b5b24c0d0>
```

Fig. 6. Corresponding Accuracy and Loss of the trained model at Epoch value 50

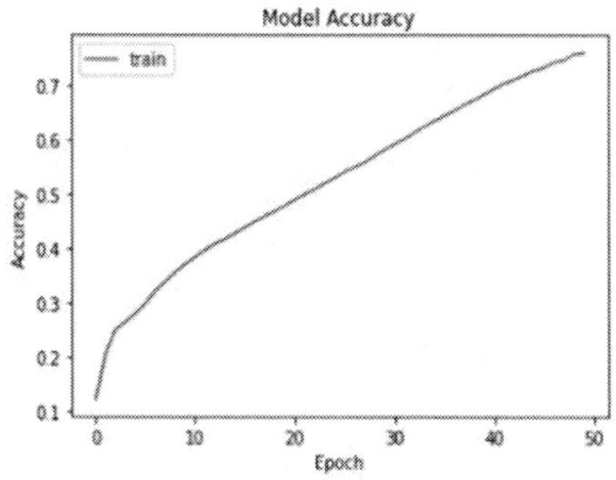

Fig. 7. Epoch vs Accuracy Graph

From Figure 7 it's seen that, as the Epoch value increases, the accuracy of the model increases simultaneously.

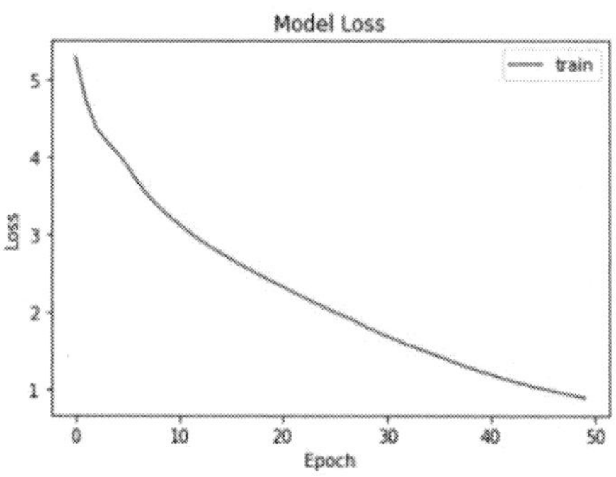

Fig. 8. Epoch vs Loss Graph

From Figure 8 it's seen that, as the Epoch value increases, the loss of the model decreases simultaneously.

5.2 Results on comparison of NMT and GTS

Table 1. Comparison of the proposed NMT Model with GTS

Input Image	Neural Machine Translation (NMT)	Google Translation System (GTS)
	एक पर्वतारोही एक बर्फ से ढंके पहाड़ पर चढ़ते हुए एक पेय लेने के लिए रुकता है।	एक पर्वतारोही एक स्नोसूट गियर पहनता है जो एक पहाड़ के सामने है।
	एक लाल ट्रक एक चट्टानी सतह पर चल रहा है।	एक लाल जीप एक चट्टानी सतह के निशान क्षेत्र पर लटकी हुई है।
	एक व्यक्ति एक चट्टानी पहाड़ी के नीचे एक बाइक की सवारी करता है।	एक आदमी अपनी पथरीली राह पर चलता है।

Table 1 shows a few examples of the input image along with its caption in English and the translated text in Hindi Language.

5.3 About BLEU Score

The result which are being expected, are based on evaluation metrics of BLEU score i.e., Bilingual Evaluation Understudy Score. In BLEU score, it does the comparison between reference caption and generated caption. So, the evaluation of the Machine Translation can be done on the entire dataset, however here evaluation is done on generated captions i.e., sentences, which means that comparing calculations will be done phrase by phrase. To calculate the adjusted accuracy of the test corpus, the total n-gram count is added for all candidates, and it is divided by the n-gram number of test corpus. A BLEU score value of 0 means that the translated output and the reference translation are not overlapping i.e., the translated text is of low quality. A BLEU score value of 1 means that there is perfect overlap with the reference translation i.e., the translated text is of high quality. It is observed that even human translators cannot achieve a perfect score of 1.0.

Table 2 shows the three sentences from our training dataset which we took under consideration for now to compare the BLEU Score of both, the proposed NMT and GTS models.

https://doi.org/10.1051/itmconf/20224403004

Table 2. BLEU Score of the proposed NMT and GTS model

Neural Machine Translation (NMT)	Google Translation System (GTS)	BLEU Score
एक पर्वतारोही एक बर्फ से ढंके पहाड़ पर चढ़ते हुए एक पेय लेने के लिए रुकता है।	एक पर्वतारोही एक स्नोसूट गियर पहनता है जो एक पहाड़ के सामने है।	0.242
एक लाल ट्रक एक चट्टानी सतह पर चल रहा है।	एक लाल जीप एक चट्टानी सतह के निशान क्षेत्र पर लटकी हुई है।	0.306
एक व्यक्ति एक चट्टानी पहाड़ी के नीचे एक बाइक की सवारी करता है।	एक आदमी अपनी पथरीली राह पर चलता है।	0.626

6. Conclusion

The proposed Neural Machine Translation system will be used for English to Hindi translation and this system will be compared with the Google Translation System with respect to the BLEU score. After exploring the results, it can be concluded that Machine Translation based on neural network needed improvements when it comes to recognizing unspecified words, repetition of words and varied translation of the generated phrase.

In addition, it is seen how total loss changes during training. As the value of epoch increases i.e., the model is being trained a greater number of times, it is seen that the total loss of model decreases. As seen from the results, the accuracy of the model increases over the time.

References

1. Benková, Lucia & Benko, Ľubomír. (2020). Neural Machine Translation as a Novel Approach to Machine Translation, (2020).

2. A. Rathi, "Deep learning apporach for image captioning in Hindi language," 2020 International Conference on Computer, Electrical & Communication Engineering (ICCECE), 2020, pp. 1-8, doi: 10.1109/ICCECE48148.2020.9223087, (2020).

3. Aspects of Terminological and Named Entity Knowledge within Rule-Based Machine Translation Models for Under-Resourced Neural Machine Translation Scenarios. (2020)

4. B., Premjith & Kumar, M. & Kp, Soman. (2019). Neural Machine Translation System for English to Indian Language Translation Using MTIL Parallel Corpus: Special Issue on Natural Language Processing. Journal of Intelligent Systems. 28. 10.1515/jisys-2019-2510. (2019).

5. Nair, Jayashree & Krishnan, K & Deetha, R. (2016). An efficient English to Hindi machine translation system using hybrid mechanism. 2109-2113. 10.1109/ICACCI.2016.7732363. (2016).

6. P. Anderson, X. He, C. Buehler et al., "Bottom-up and top-down attention for image captioning," in Proceedings of the IEEE Conference on Computer Vision and Pattern Recognition, Salt Lake City, UT, USA, June (2018).

7. Laskar, Sahinur Rahman et al. "Neural Machine Translation: English to Hindi." 2019 IEEE Conference on Information and Communication Technology (2019): 1-6. (2019).

8. JalFaizy Shaikh. "Automatic Image Captioning using Deep Learning (CNN and LSTM) in PyTorch", (2018).

9. K. Loganathan, R. Sarath Kumar, V. Nagaraj, Tegil J. John, CNN & LSTM using python for automatic image captioning, Materials Today: Proceedings, 2020, ISSN 2214-7853, doi.org/10.1016/j.matpr.2020.10.624. (2020).

10. https://github.com/manavjain179/Machine-Translation. (2022)

College Election System using Blockchain

Aanchal Mani[1, *], *Samarjeet Patil*[2], *Soham Sheth* [3] and *Lakshmi Sudha Kondaka*[4]

[1]Department of Information Technology, SIES Graduate School of Technology, Nerul, Navi Mumbai, India

[2]Department of Information Technology, SIES Graduate School of Technology, Nerul, Navi Mumbai, India

[3]Department of Information Technology, SIES Graduate School of Technology, Nerul, Navi Mumbai, India

[4]Department of Information Technology, SIES Graduate School of Technology, Nerul, Navi Mumbai, India

Abstract. *Blockchain is a distributed system that confirms security and reliability. It has started a new era of a solid and consensus system. Because of blockchain's emphasis on security, many other operations and processes are adopting the same reliable approach. Almost all processes and operations are now encouraged to be computed electronically in the digital Ethereum network that has been presented. In the current situation, we have seen what the conduction of elections has done to the worsening situation of the Covid-19 Pandemic. The proposed system uses an Ethereum network on a blockchain platform to implement a college voting system, making use of the Aadhar API. E-voting will help to support transparency and voters' trust to reduce corruption and unreliability in the voting processes. The use of this system will allow a voter to vote from home while also having the assurity that this process is completely secure and reliable due to the use of blockchain* technology.

1 Introduction

Today, the vast majority of our lives are led over the web or internet innovations. We perform most of our banking transactions and finance via the web/mobile application. This includes transferring money to individuals via payment portals, registering our vehicles on the web and connecting with loved ones on the internet, etc. We hardly see any industry that isn't leveraging technology to improve their service or industry function. Claims of fraud or extortion and outside impact can come up if technology is not used to dispose of election tamperingor altering. There is a need for an electoral platform that streamlines the process of checking cast a ballot. All while giving transparency and straightforwardness to the election outcomes. Blockchain voting systems make use of smart contracts, which permits ticket confirmation and vote counts to be performed in a decentralized way and confirmed by system members. Under the blockchain voting system, the hackers would have to break into an entire system of PCs to gain access to the data, which is highly unlikely. Due to the ongoing pandemic, it has become extremely difficult to step out for everyone, be it teachers or students. This project will provide a platform for college elections to be carried out without anyone having to get out of their houses. Blockchain technology has been used to ensure that the process of vote casting is secure and there is no tampering. Each block on the chain contains the id of the voter, the person they're voting for and their slogan. This information is encrypted and stored as a block on the blockchain in every node. This means that even if someone attempts to change the data on one node, as the change won't be reflected on the other nodes, the data will be secure.

2 Literature Review

Fraij et al. proposed a voting system using blockchain and Ethereum networks to build the trust in the voting system [1]. In this paper they have proposed a distributedsystem, so that authorities cannot or will not interfere in the voting records, so that corruption and cheatings can be omitted. Many new changes can be accommodated byintroducing new systems, where, instead of performing high-level computational tasks for mining, mining can bebased on a predetermined turn for each node to mine a new block in the blockchain, as proposed in [2].

A blockchain-based electronic voting system can utilize smart contracts to enable secure and cost-efficient elections while guaranteeing voters privacy and integrity [3]. Above mentioned system uses the "voter districts" having different smart contracts connected to a boot node that isn't connected to the actual blockchain.

Syeda Sumbul Hossain et al.[4]. A distributed network comprises an enormous range of interconnected nodes , in which each node has their own copy of distributed ledger that contains the history of all the transactions the system has processed. Bitcoin scripting language cannot support the creation of complex smart contracts that contain rich logic but ethereum supports building and running complex applications based on smart contracts on the blockchain, as stated by M. Alharby et al..[5]

[6], Using an Ethereum private blockchain, it is possible to send hundreds of transactions per second onto the blockchain, utilizing every aspect of the smart contract to ease the load on the blockchain. Pali et al.[7] It focused the launch of the AADHAR api on the interface of various e-governance functionalitiesto ensure the optimal utilization of information, communication, and Technology Infrastructure. There

* Corresponding author: aanchal.ganeshmani18@siesgst.ac.in

 https://doi.org/10.1051/itmconf/20224403005

are also benefits and usage of the api followed by the Privacy and Security Issues in the Aadhaar Life Cycle.
Decentralized blockchains are immutable. For bitcoin, this means transactions are permanently recorded and viewable to anyone. Each stores a collection of transactions with a display of a public ledger that records and exhibits any bitcoin transaction that takes place as is shown by Nakamoto, Satoshi.[8] It can remain in a computer and easily stored by a code word. You have better control over your money.
E-voting, as discussed [9] is a potential solution to the lack of interest in voting amongst the young tech savvy population. For the process to become less opaque, open for all, and independently auditable, a potential solution would be to base it on blockchain technology.
S. Aruna et al. [10] mentions that, despite the fact that blockchains are exceptional at securing data that is stored in them, the main problem occurs with the identification of the voter before the voting process.
Although blockchain is a newly emerging technology, there are multiple applications of blockchain outside of cryptocurrency, in various fields if the legacy issues can be cleared as mentioned by Kitsantas et al.[11]
Ali Mansour Al-madani et al.[12] blockchain is a better choice for building a voting application. In the blockchain, all the data is not allocated on a centralserver, but the data allocated via decentralization is called a distributed database. The data distributed across each device connected to the blockchain using a peer-to- peer network of nodes which talk to one another. In this review, a comprehensive view of e-voting on a thematic basis like e-voting with iot and fingerprint, e-voting withblockchain and Aadhar verification, etc is present. An attempt has been made to include every aspect and approach taken by individuals to make a robust systemby Kanika Garg et al.[13]
Rabeya Bosri et al.[14] EVM is introduced in this that also provides the ease of tabulation of ballots into result, that gives a more accurate and faster outcome compared with traditional paper-based voting systems. Data of an individual will be collected, shared, and used in a way so that no entity other than the system or who are directly involved in the process will be able to know or crack the information.

3 Proposed System

The proposed system will make use of react.js to make the front-end of the Election System. Smart Contracts, using Solidity, will be used to ensure the authenticity of the votes along with ensuring that one person is allowed to vote only once. To authenticate a particular user, we will be emulating the Aadhar number verification with the help of firebase's PhoneAuth Api. This system will enable students to vote for their Class Representatives, College Team Members and officials of the Student Council.

The entire blockchain runs on the smart contracts that have been written using the solidity language. The contract defines the basic instances of candidates and elections. As soon as the contract is deployed, the account deploying the election is automatically initialized as the admin for the election. When the website is accessed using that account, the admin panel is automatically visible which enables the admin to authorize elections. The contract specifies how candidates are to be initialized and how a vote is to be registered.

Web3.js is the element which connects the web application to the blockchain. The testing environment Ganache is used for setting up our blockchain and performing basic testing of smart contracts. It is an environment that is free of cost and helps deploy contracts and run tests.

A number of platforms have been used to facilitate the process of creating this system. They are as follows:-

a. Ganache: Ganache is a testing environment which is used for testing the functionality of smart contracts and provides a private blockchain. It is used to test the smart contracts as it provides 10 accounts with fake ether.

b. Truffle: It is a framework used for deploying the contracts, truffle automatically provides the structure for storing the contracts and migrations folders for connecting it with mainnet and testnet

c. Metamask: Metamask is a crypto wallet- it allows users to store and transact Ethereum. It allows users to access their Ethereum wallet through a browser extension or mobile app, which can then be used to interact with decentralized applications.

A. WORKFLOW

The election can start only after the contract has been deployed. The account deploying the contract will be given admin status and thus will be able to access the admin side of the website. This will enable the admin to initialize, start and end elections.

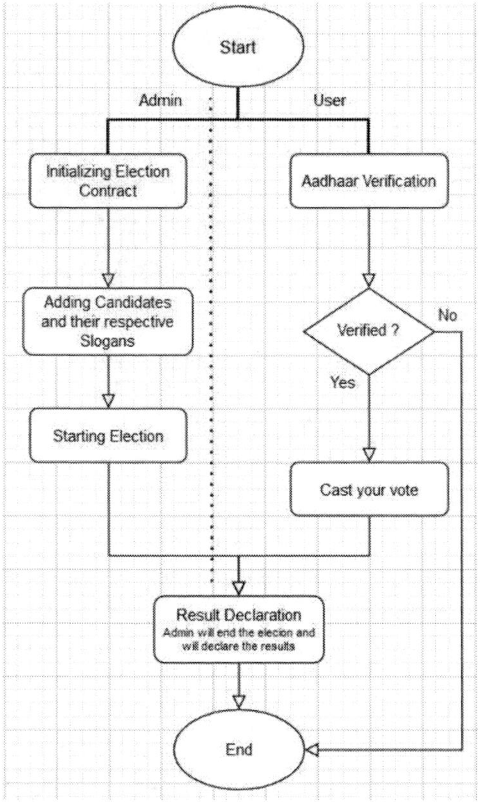

Fig 1 : flow chart of Election System using blockchain

Step 1: Aadhar Verification
To be eligible to vote, all voters will need to verify their Aadhar registered mobile number on the website. Once they enter the otp, they will be directed to the voting portal.

Step 2: Initializing Election
The admin will have to enter their details such as their name and position along with the election details. They will then be able to start the election for a certain amount of gas fee which they will need to approve with their meta mask wallet.

Step 3: Adding Candidates
The admin will be able to add candidates contesting in the election along with their slogans. These are the candidates out of which the winner will be selected after the election ends.

Step 4: Casting a Vote
Once a voter's Aadhar has been verified, they will be eligible to vote. On the voting page they will find a list of all the candidates out of which they will be able to vote for only one candidate that they support.

Step 5: Result declaration
Once the election is over, the admin will have to end the election from their side to declare the winner. After the election has ended, all the voters will be able to view the number of votes for each candidate and the name of the winning candidate.
To start another election, the admin will need to re-deploy the contract.

4 Experimental Results

Aadhar card should be verified, to do so enter the phone number that is linked to it and you will receive an OTP. Enter it in the given dialogue box to proceed further as shown in the figure below.

Fig 2 : Aadhar Verification

After entering, your Aadhar number will be verified, on successful verification you will be directed to the next page. On this page your account number will be displayed, after which you will have to enter your personal details. Next you shall proceed to the election process.

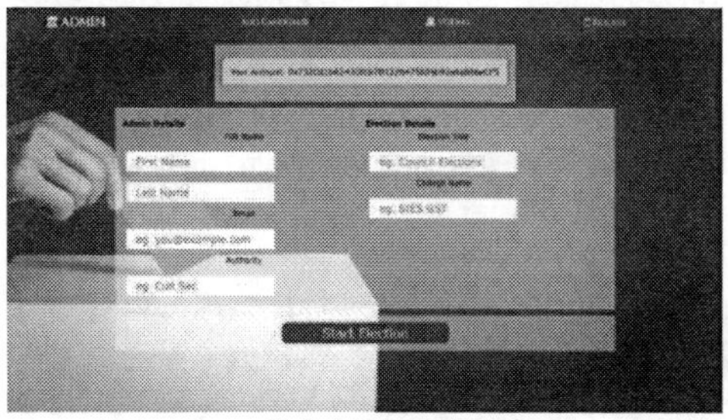

Fig 3 : Initializing Election

Further, in the add candidates page the next step it is required for you to enter your name and also a cheerful and a strong slogan. This slogan might help your party.

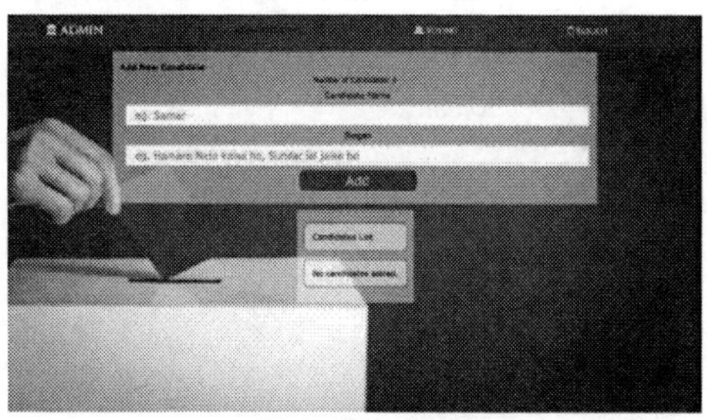

Fig 4 : Adding Candidates

Once after entering the candidates details successfully, name of all the candidates will displayed below in the candidates list.

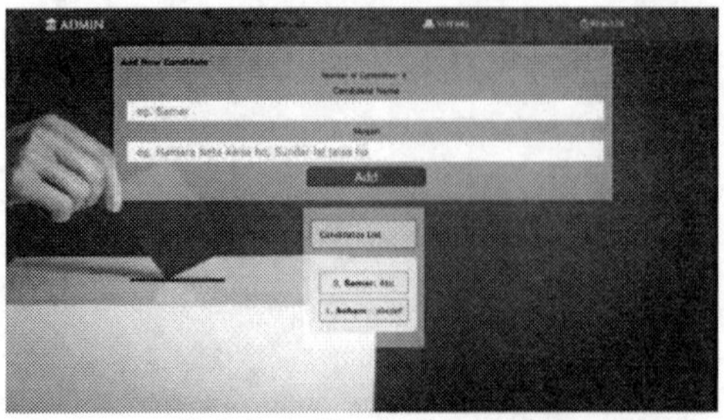

***Fig 5** : Successful registration of candidates*

Furthermore, an admin page will appear displaying the details of the admin which was asked during initializing election (as shown is Fig 3) and show the current status of the election and will also have the right to end the election.

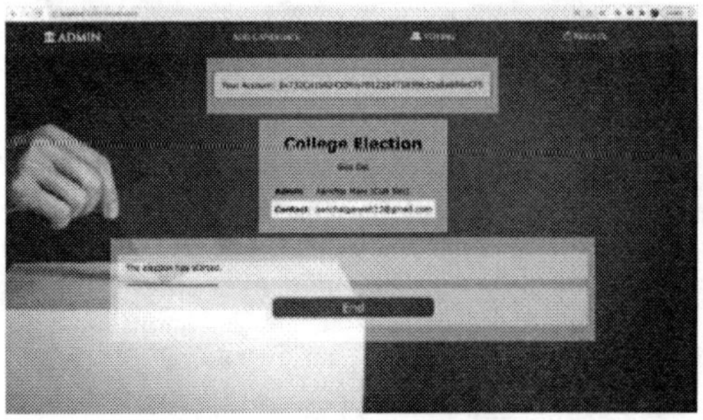

***Fig 6** : Details of Admin*

On homepage, inside the voting page a voting poll will be opened for all the voters where you can choose any candidate of your preference.

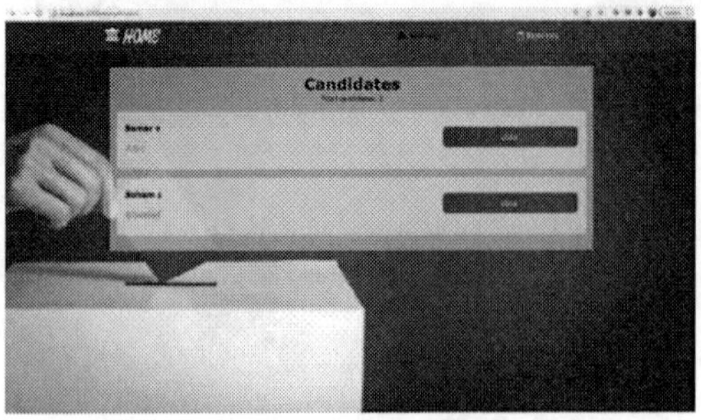

***Fig 7** : Voting for candidates*

In the voting page, to confirm your vote after casting, the user shall receive a dialogue box as shown in the figure below, with their voting details displayed below.

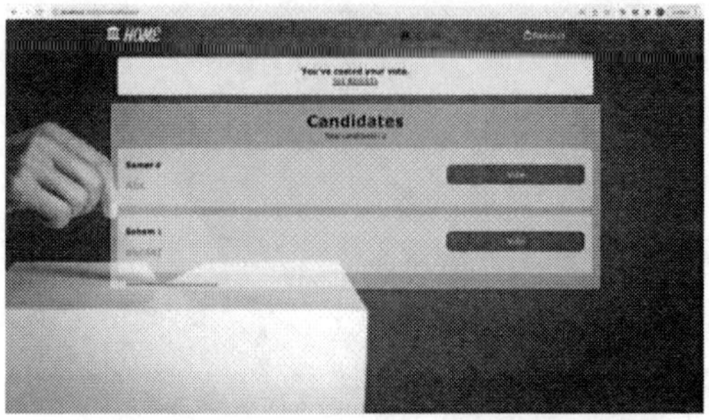

***Fig 8** : Successfully casting of votes*

Each ethereum transaction requires computational resource to execute, hence a gas fee is issued after every transaction is occurred during voting. Gas refers to the fee required to conduct a transaction on ethereum successfully.

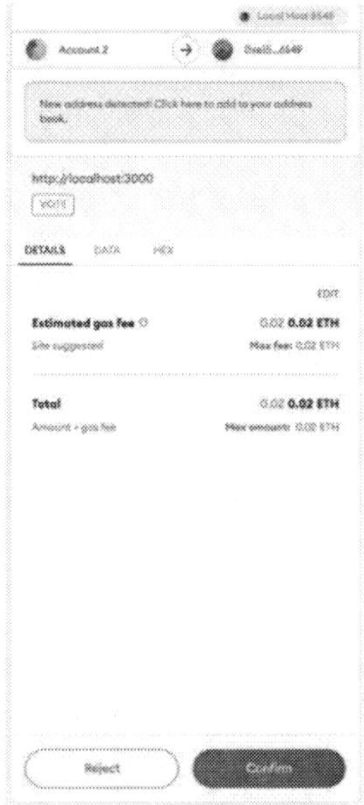

***Fig 9** : Meta mask wallet*

Once admin ends the election. The results will be declared in the results page and one winner from all the candidates will be announced who has received the maximum number of votes during the election process.

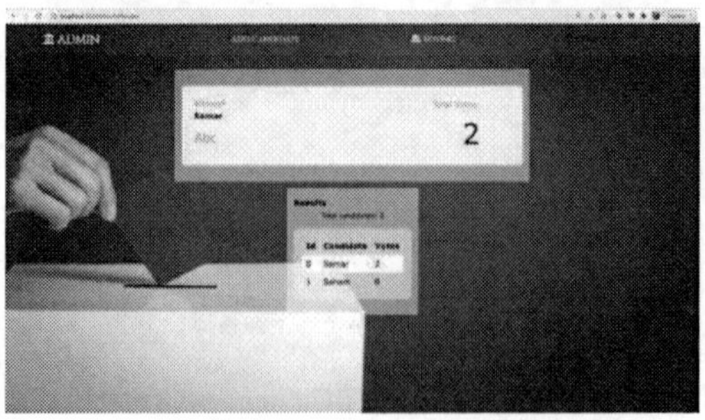

***Fig 10** : Successfully casting of votes*

5 Conclusion

A web application that enables the smooth and secure conduction of a college election has been developed to make the process of voting easier on both the voters and the election committee. The use of Blockchain technology opens a plethora of options for providingconfidentiality throughout the process and Aadhar verification takes care of the integrity of the interaction.

In the future, this project can be worked on to include the actual Aadhar API, providing a higher level of authenticity. A feature can be added which will show the voters different candidates based on their voting district, to give an example for a college election, showing the students of the IT department candidates only from the IT department.

Acknowledgement

The authors would like to thank their family members and the SIES Management for their support towards the development of the project.

References

[1] Fraij, Jihad & Dabbas, Ashraf & Aburumman, Nemer. (2021). BLOCKCHAIN AS AN E-VOTING TOOL. International Journal of Advanced Research. 8. 858-866. 10.21474/IJAR01/12225.

[2] Rahman M.R., Hossain M.B., Arefin M .S., Khan M.I. (2021) A Secured Electronic Voting System Using Blockchain. In: Vasant P., Zelinka I., Weber GW. (eds) Intelligent Computing and Optimization. ICO 2020. Advances in Intelligent Systems and Computing, vol 1324. Springer, Cham.

[3] F. Þ. Hjálmarsson, G. K. Hreiðarsson, M. Hamdaqa and G. Hjálmtýsson, "Blockchain-Based E-Voting System," 2018 IEEE 11th International Conference on Cloud Computing (CLOUD), 2018, pp. 983-986, doi: 10.1109/CLOUD.2018.00151.

[4] Syeda Sumbul Hossain, Samen Anjum Arani, Md. Tanvir Rahman, Touhid Bhuiyan, Delwar Alam, and Moniruz Zaman. 2019.E-voting system using Blockchain technology. In Proceedings of the 2019 2nd International Conference on Blockchain Technology andApplications (ICBTA 2019). Association for Computing Machinery, New York, NY, USA, 113–117.

[5] M. Alharby, A. Aldweesh and A. v. Moorsel."Blockchain-based Smart Contracts: A Systematic Mapping Study of Academic Research (2018)," 2018 International Conference on Cloud Computing, Big Data and Blockchain (ICCBB), 2018, pp. 1-6, doi: 10.1109/ICCBB.2018.8756390.

[6] G. Wood."Ethereum: A secure decentralised generalised transaction ledger", Ethereum ProjectYellow Paper, vol. 151, 2014.

[7] Pali, Isha & Krishania, Lisa & Chadha, Divya & Kandar, Asmita & Varshney, Gaurav & Shukla, Sneha. (2020). A Comprehensive Survey of Aadhar and Security Issues.

[8] Nakamoto, Satoshi. (2009). Bitcoin: A Peer-to-Peer Electronic Cash System.

[9] F. Sheer Hardwick, A. Gioulis, R. Naeem Akram and K. Markantonakis, "E-Voting With Blockchain: An E-Voting Protocol with Decentralisation and Voter Privacy," 2018 IEEE International Conference on Internet of Things (iThings) and IEEE Green Computing and Communications (GreenCom) and IEEE Cyber, Physical and Social Computing (CPSCom) and IEEESmart Data (SmartData), 2018, pp. 1561-1567, doi: 10.1109/Cybermatics_2018.2018.00262.

[10] S. Aruna et al 2020 IOP Conf. Ser.: Mater. Sci. Eng. 993 012103

[11]Kitsantas, Thomas & Vazakidis, Athanasios &Chytis, Evangelos. (2019). A Review of Blockchain Technology and Its Applications in the Business Environment.

[12] A. M. Al-madani, A. T. Gaikwad, V. Mahale and Z. A. T. Ahmed, "Decentralized E-voting system based on Smart Contract by using Blockchain Technology," 2020 International Conference on Smart Innovations in Design, Environment, Management, Planning and Computing (ICSIDEMPC), 2020, pp. 176-180, doi: 10.1109/ICSIDEMPC49020.2020.9299581.

[13] K. Garg, P. Saraswat, S. Bisht, S. K. Aggarwal, S. K. Kothuri and S. Gupta, "A Comparitive Analysis on E-Voting System Using Blockchain," 2019 4thInternational Conference on Internet of Things: Smart Innovation and Usages (IoT-SIU), 2019, pp. 1-4, doi: 10.1109/IoT-SIU.2019.8777471.

[14] Bosri, Rabeya & Uzzal, Abdur & Omar, Abdullah & Hasan, A S M Touhidul & Bhuiyan, Md. (2019). Towards A Privacy-Preserving Voting System Through Blockchain Technologies. 10.1109/DASC/PiCom/CBDCom/CyberSciTech.2019.00116.

Performance Evaluation of routing protocols with and without malicious nodes

Snehal Mendadkar [1,*], and Poornima *Talwai*[1]

[1]Department of Electronics Engineering, Ramrao Adik Institute of Technology, Nerul, Navi Mumbai, 400 706, India

Abstract. Vehicular adhoc networks (VANETs) are made by applying the standards of mobile adhoc networks (MANETs)-the unconstrained making of a remote network of cell phones to the area of vehicles. VANETs were first referenced and presented in 2001 under "vehicle to-vehicle specially appointed versatile correspondence and systems administration" applications, where networks can be shaped and data can be handed-off among vehicles. It was shown that vehicle-to- vehicle and vehicle-to-road side exchange designs will exist together in VANETs to give street security, route, and other side of the road administrations. VANETs are a vital piece of the intelligent transportation frameworks (ITS) system. At times, VANETs are known as Intelligent Transport Networks. They are perceived as having advanced into a more extensive "Web of vehicles". Which itself is relied upon to eventually advance into a "Web of self-sufficient vehicles". The vehicles send some data about street status and traffic. But sometimes the information send from one vehicle to another can be malicious. This malicious information is added by the intruders to cause problem in vehicular network so that accident can occur. In this report, we will analyse the throughput, packet delivery ratio and end to end delay in sending information from vehicle to vehicle when the network is attacked by some malicious nodes.

1 Introduction

Recently, the streets have seen a huge expansion in the quantity of vehicles, which brought about an increment in auto collisions and traffic clog on the streets. [1-3] Due to these, it got important to give security and comfort to the driver on the street. Thus, the requirement for a network named VANET came into existence. Vanet is an invention which utilizes
Vehicles as mobile nodes to set up a remote association between them without the need of any focal base station or any regulator. It permits the vehicles sending and receiving data between one another and the climate encompassing them.

Vanet can be utilized for security, well-being applications such as improved route, area based administrations like finding the nearest fuel station or cafe, entertainment hubs, access to internet on go, vehicle security.[4] From the last decade, mobile adhoc networks have changed the car business by giving whenever anyplace correspondence between various gadgets. This simplicity of correspondence permits trade of significant data between gadgets simply in a hurry. The
Consistent exchange of data on on-going bases has ended up becoming another worldview in the business. Correspondingly, the advances in the data innovation and correspondence have effortlessly upheld the possibility of correspondence between mobile devices. [6]

2. Vehicular Adhoc Routing Protocols

2.1 *Ad-hoc On Demand Distance Vector Routing Protocol (AODV)*

It is on demand convention in which every hub keeps up the routing information by using steering table which is kept up at every hub of the organization. In routing table, target area, next jump IP address and target progression number is taken care of. Course interest (RREQ), Hello message, Route answer (RREP) and Route Error (RERR) are the four kinds of messages used in AODV framework. Hi message is used to screen associates, each organization in network broadcast the Hello Message, and this invite message will be received by all neighbour hubs. Invalid Links are recognized if hub fails to get any of the control or data information message. Course Request is used when a source S needs to send data to Destination ID, this allure is imparted to neighbours. All of the widely appealing hubs among source and target transmission the requesting in two cases, one they don't get the allure already or second
It's everything except the goal hub. If momentary is the target or it having the course to objective, it will reply to source in jump by hopping. [7-9]

2.2 Dynamic source Routing Protocol (DSR)

It is an On-Demand routing protocol in which the sequence of nodes through which a packet needs to travel is calculated and maintained as an information in packet header. Every mobile node in the network needs to maintain a route cache where it

* Corresponding author: snehalmendadkar69@gmail.com

caches source routes that it has learned. When a packet is sent, the route-cache inside the node is compared with the actual route needs to be covered. If the result is positive, the packet is forwarded otherwise route discovery process is initiated again. [10]

2.3 Destination Sequenced Distance Vector (DSDV)

DSDV is table driven routing protocols using Bellman Ford Algorithm to determine a way. As DSDV uses proactive philosophy all hubs in the organization understands a course to every single Hub in the organization going before correspondence. Hence every hub in the organization broadcast HELLO information at typical range to get information about the neighbour hubs, course available to show up at the neighbour hub and to recognize the real or invalid association. A routing table is kept up at every hub to store every one of this information. Every objective course is connected with a characteristic of progression number. [11]

3. Problem Statement

1. The goal of the work is to think about the exhibition of the three routing protocols dependent On-Demand Behaviour, for example AdHoc On-Demand Distance Vector (AODV), Dynamic Source Routing(DSR) and Destination-Sequenced Distance Vector routing, for remote adhoc networks dependent on the presentation and examination has been made based on their properties like the throughput, packet drop ratio (PDR), Average end-to-end delay and packet data loss regarding various situations one by fluctuating the number of nodes, changing the speed of the vehicles with constant number of nodes in the network.

2. The second significant issue in utilizing the VANET network as ITS innovation and its execution in reality is the security and protection of the application. As a wireless network, VANET is vulnerable to few attacks. For instance, an aggressor can infuse false data into the network by sending traffic data that doesn't exist. Mistaken traffic data can make traffic be redirected starting with one street then onto the
Next. The results can cause traffic sticks surprisingly more terrible on the off chance that it's anything but a mishap. The network will be analysed on what results after the malicious nodes enter the network.

4 Methodologies

In this report, we use network test system NS-2 which is a discrete test system and item situated dependent on both C++ and OTcl. Ns-2 additionally plays out the reproduction on the foundation of two sort of document C++ and TCL. It is utilized for both wired and remote conventions and furthermore support
Client expanded conventions with IP convention. Ns-2 has numerous other reasonable highlights like a channel for remote, routing along different ways etc. The Simulations
were performed utilizing Network Simulator NS2.35. At first, situation and traffic records are created. These documents are utilized as contribution for TCL script. After execution of TCL script two records are made for example NAM document and trace file. Trace file are utilized to break down the conduct of network. Trace file is dissected utilizing AWK
scripts. Following steps are performed to run the simulation
1. Select performance parameters. (Throughput, delay, packet delivery ratio and
packet loss ratio).
2. Generate scenario using NSG2.1 and topology files using cbrgen and setdest commands.
3. Create TCL script with .tcl extension.
4. Execute TCL script (Use ns Command)
5. Generate Trace and NAM file.
6. Execute AWK script to measure performance.

5. Simulation Assumptions

The simulations were performed using Network Simulator 2 (NS-2.35). The traffic sources are Constant Bit Rate (CBR). The source destination pairs are spread randomly over the network. The mobility model uses 'random waypoint model' in a rectangular field of 500m x 500m with 20 nodes to 100 nodes. For the simulation we consider five different scenarios by varying packet size, varying number of nodes and spedd of the nodes.

6. Performance Metrics

To evaluate the evalauation of both proactive and responsive protocols different estimations measurements like: Throughput, start to finish delay, packet delivery ratio, number of dropped packets, number of sent packets, number of received packets has been surveyed.

4.4.1 Packet Delivery Ratio

Packet Delivery Ratio is a crucial factor to evaluate the display of routing protocol in any organization. The show of the protocol depends upon various limits picked for the simulation. The critical limits are parcel size, no of hubs, transmission range and the development of the organization. The Packet delivery ratio to be acquired from the hard and fast number of information packets displayed at objective isolated by the full scale information packets sent from sources. As such Packet Delivery Ratio is the extent of number of packets got at the objective to the amount of parcels sent from the source. The show is better when parcel Delivery Ratio is high. Mathematically it will in general be shown as

$$PDR = \frac{\text{Total packets received by all destination node}}{\text{Total packets send by source nodes}} \quad (1)$$

4.4.2 Average End to End Delay

Normal End-to-end delay is the time taken by a packet to course through the network from a source to its objective. The typical beginning to end delay can be procured enlisting
the mean of beginning to end deferral of all viably passed on messages. Consequently, beginning to end delay midway depends upon the parcel conveyance extent. As the distance among source and objective grows, the probability of parcel drop increases. The typical beginning to end postpone recalls all possible deferrals for the association for instance buffering course revelation torpidity, retransmission delays at the MAC, and creating and transmission delay. Mathematically it will in general be shown as

$$D=\frac{1}{n}\sum_{1}^{n}(Tri - Tsi) * 1000[ms] \text{---}(2)$$

Where D = Average E2E Delay
i = packet identifier
Tri= Reception time
Tsi = Send time
n = Number of packets successfully delivered

4.4.3 Packet loss

Packet Loss is the proportion of the quantity of packets that never arrived at the destination to the quantity of packets started by the source. Numerically it tends to be displayed as follows.

$$PL = \frac{(nSentPackets - nReceivedPackets)}{nSentPackets} \quad \text{-- (3)}$$

where nReceivedPackets = Number of recieved packets
nSentPackets = Number of sent packets

4.4.4 Packet Loss Ratio

Packet Loss Ratio is the proportion of the quantity of packets that never arrived at the destination to the quantity of packets started by the source. Numerically it very well may be displayed as follows

$$PLR = \frac{(nSentPackets - nReceivedPackets)}{nSentPackets} * 100 \quad \text{–(4)}$$

Where nReceivedPackets = Number of received packets
nSentPackets = Number of sent packets

4.4.5 Throughput

The end-to-end network throughput measures the number of packets per second received at the destination. It is considered here as an external measure of the effectiveness of a protocol.

$$Throughput = \frac{(recvdSize)}{(stopTime - startTime)} * \frac{8}{1000}$$

Where recvdSize = Store received packet's size
stopTime = Simulation stop time
startTime = Simulation start time

7 Simulation Results

The simulations were performed using Network Simulator 2 (NS-2.35). The traffic sources are Constant Bit Rate (CBR). The source destination pairs are spread randomly over the network. The mobility model uses 'random waypoint model' in a rectangular field of 500m x 500m with 20 nodes to 100 nodes. For the simulation we consider five different scenarios by varying packet size, varying number of nodes and speed of the nodes.

Scenario1:- The nodes are moving with a constant speed of 20m/s. The simulation time is taken to be 600ms. The number of nodes are varying from 20, 40, 60, 80 and 100 with constant packet size of 512Mb.

PARAMETER	VALUES
PROTOCOLS	AODV, DSDV, DSR
NUMBER OF NODES	20,40,60,80,100
TRAFFIC TYPE	CONSTANT BIT RATE(CBR)
TRANSMISSION RANGE	250M
MOBILITY MODEL	RANDOM WAY POINT
SIMULATION AREA	500*500m
NODE SPEED	20m/s
PACKET SIZE	512MB
RADIO PROPAGATION MODEL	TWO RAY GROUND

Table 1;- Constant packet size of 512Mb with variable number of nodes

No of nodes	AODV	DSDV	DSR
20	0.1685	0.6428	1
40	0.0448	0.0491	0.0403
60	0.056	0.0493	0.0482
80	0.0743	0.0968	0.068
100	0.0746	0.0897	0.0586

Table 2:- Packet loss

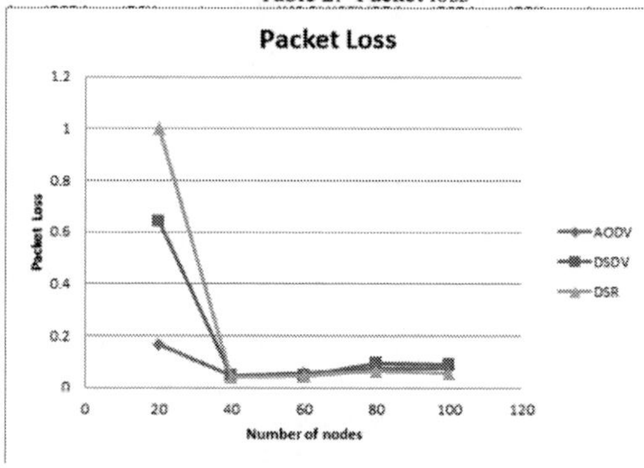

Figure 1:- Packet loss

No of nodes	AODV	DSDV	DSR
20	16.85	64.28	100
40	4.48	4.91	4.03
60	5.6	4.93	4.82
80	7.43	9.68	6.8
100	7.46	8.97	5.86

Table 3:- Packet Loss Ratio

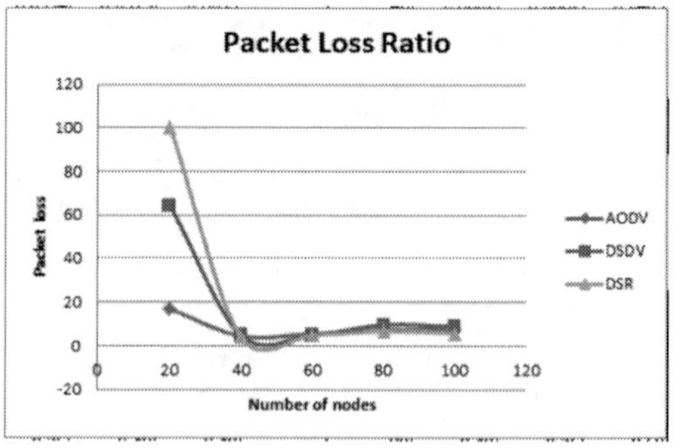

No of nodes	AODV	DSDV	DSR
20	0.14	0	0
40	0.21	0.17	0.36
60	0.17	0.19	0.29
80	0.19	0.19	0.38
100	0.25	0.2	0.39

Table 4:- Average End to end delay

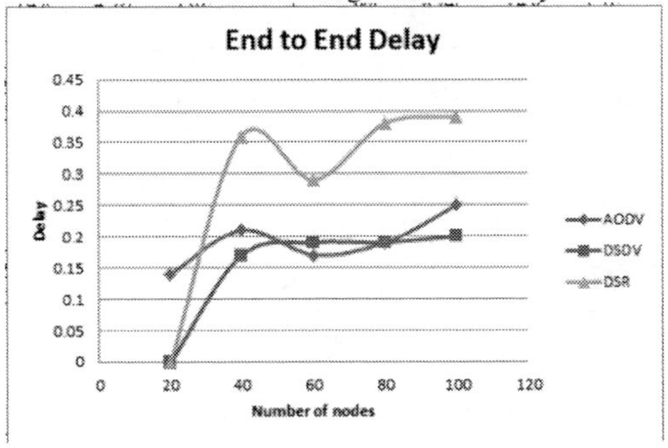

No of nodes	AODV	DSDV	DSR
20	0.83	0.35	0
40	0.95	0.95	0.95
60	0.94	0.95	0.95
80	0.92	0.9	0.93
100	0.92	0.91	0.94

Table5:- Packet delivery Ratio

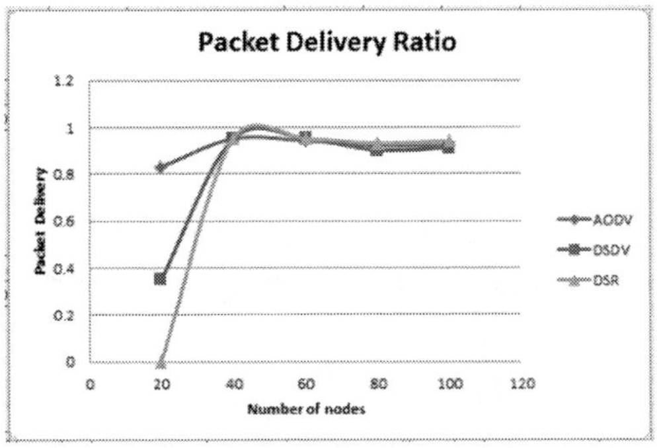

No of nodes	AODV	DSDV	DSR
20	22546	6765	24596
40	27052	6881	24496
60	19328	6709	24600
80	24592	6709	24897
100	17085	6402	20812

Table 6:- Throughput

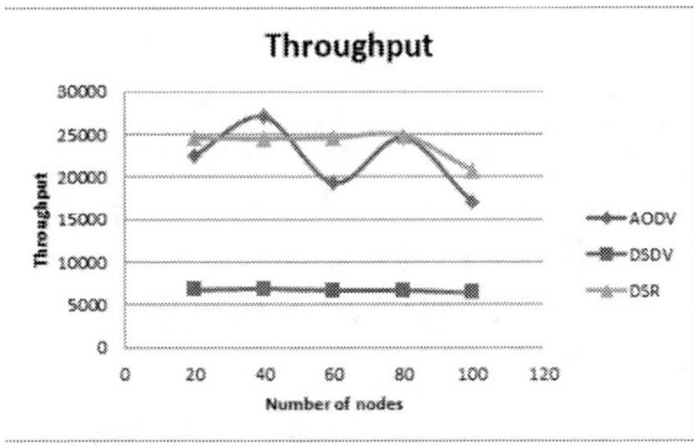

Scenario2:- The nodes are moving with a constant speed of 20m/s. The simulation time is taken to be 600ms. The number of nodes are varying from 20, 40, 60, 80 and 100 with constant packet sie of 1024Mb.

PARAMETER	VALUES
PROTOCOLS	AODV, DSDV, DSR
NUMBER OF NODES	20,40,60,80,100
TRAFFIC TYPE	CONSTANT BIT RATE(CBR)
TRANSMISSION RANGE	250M
MOBILITY MODEL	RANDOM WAY POINT
SIMULATION AREA	500*500m
NODE SPEED	20m/s
PACKET SIZE	1024MB
RADIO PROPAGATION MODEL	TWO RAY GROUND

Table7 :- Constant packet size of 1024 Mb with variable number of nodes

No of nodes	AODV	DSDV	DSR
20	0.2343	0.6428	1
40	0.0594	0.0315	0.0695
60	0.0545	0.0492	0.0707
80	0.0926	0.0795	0.0795
100	0.0782	0.0797	0.0797

Table 8:- Packet loss

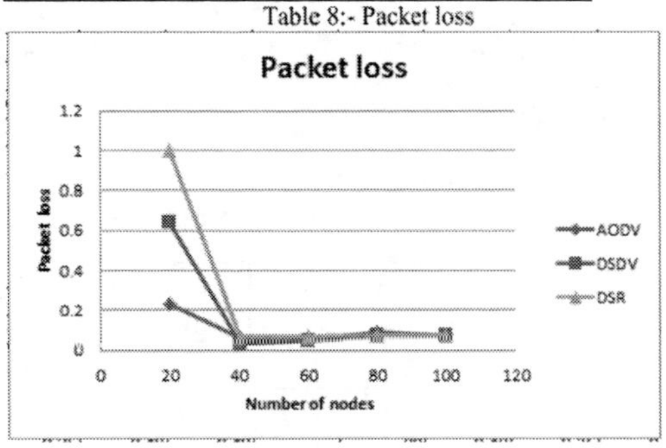

No of nodes	AODV	DSDV	DSR
20	23.43	64.28	100
40	5.94	3.15	6.95
60	5.45	4.92	7.07
80	9.26	7.95	7.95
100	7.82	7.97	7.97

Table 9:- Packet loss ratio

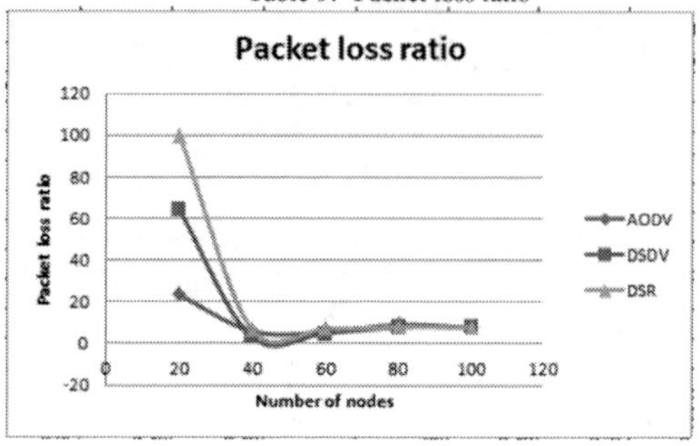

No of nodes	AODV	DSDV	DSR
20	0.21	0	0
40	0.23	0.23	0.46
60	0.31	0.26	0.52
80	0.25	0.49	0.49
100	0.35	0.65	0.65

Table10:- Average end to end Delay

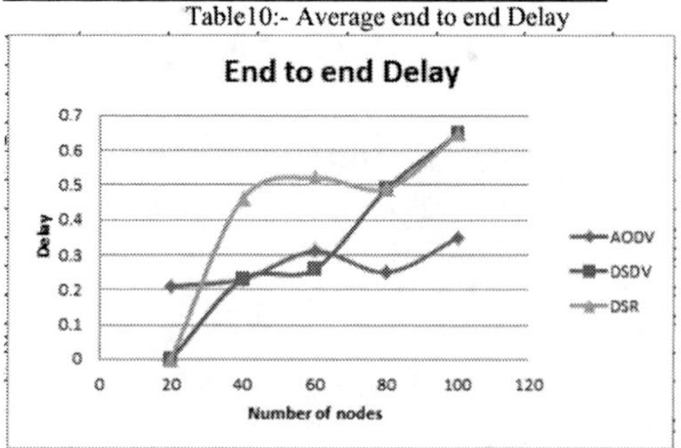

No of nodes	AODV	DSDV	DSR
20	0.76	0.64	0
40	0.94	0.96	0.93
60	0.94	0.95	0.92
80	0.91	0.92	0.92
100	0.92	0.92	0.92

Table11:- Packet Delivery ratio

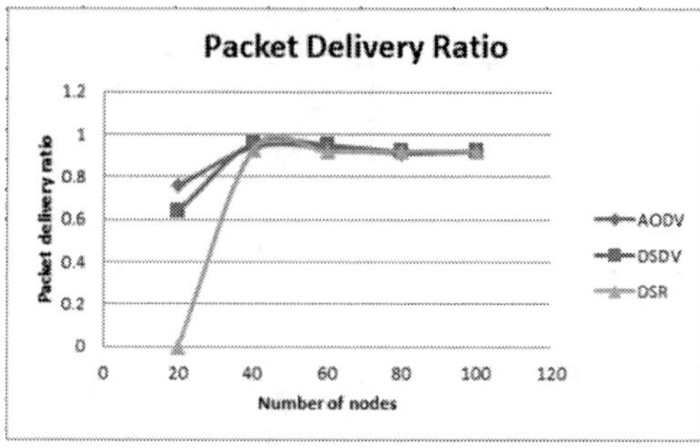

No of nodes	AODV	DSDV	DSR
20	32723	6891	29091
40	33255	6991	28973
60	17547	6891	29091
80	32723	6687	29091
100	22084	5092	26076

Table12:- Throughput

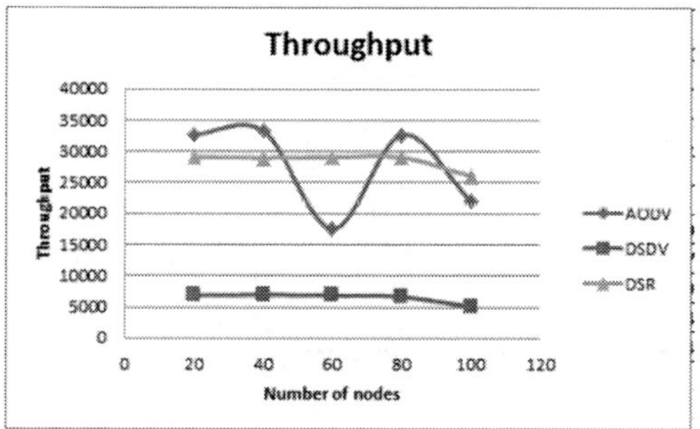

Scenario3:-The nodes are moving with a constant speed of 20m/s. The simulation time is taken to be 600ms. The number of nodes are varying from 20, 40, 60, 80 and 100 with constant packet sie of 2048Mb.

PARAMETER	VALUES
PROTOCOLS	AODV, DSDV, DSR
NUMBER OF NODES	20,40,60,80,100
TRAFFIC TYPE	CONSTANT BIT RATE(CBR)
TRANSMISSION RANGE	250M
MOBILITY MODEL	RANDOM WAY POINT
SIMULATION AREA	500*500m
NODE SPEED	20m/s
PACKET SIZE	2048MB
RADIO PROPAGATION MODEL	TWO RAY GROUND

Table 11:- Constant packet size of 2048 Mb with variable number of nodes

 https://doi.org/10.1051/itmconf/20224403006

No of nodes	AODV	DSDV	DSR
20	0.3444	0.6428	1
40	0.0662	0.0602	0.0403
60	0.0814	0.0545	0.0482
80	0.0942	0.1273	0.068
100	0.0932	0.1378	0.0586

Table 13:- Packet loss

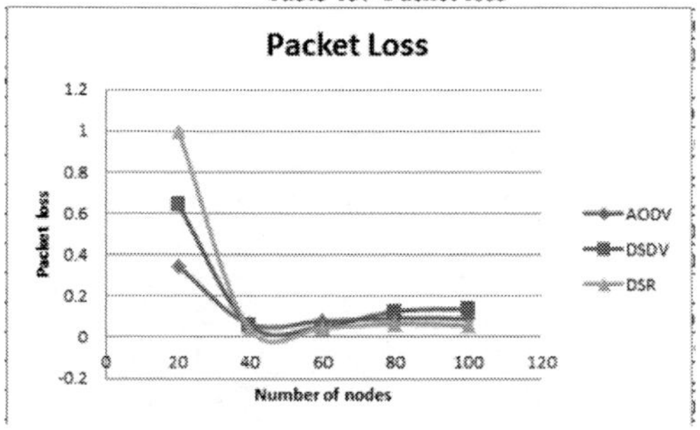

No of nodes	AODV	DSDV	DSR
20	23.43	64.28	100
40	5.94	3.15	6.95
60	5.45	4.92	7.07
80	9.26	7.95	7.95
100	7.82	7.97	7.97

Table 14:- Packet loss ratio

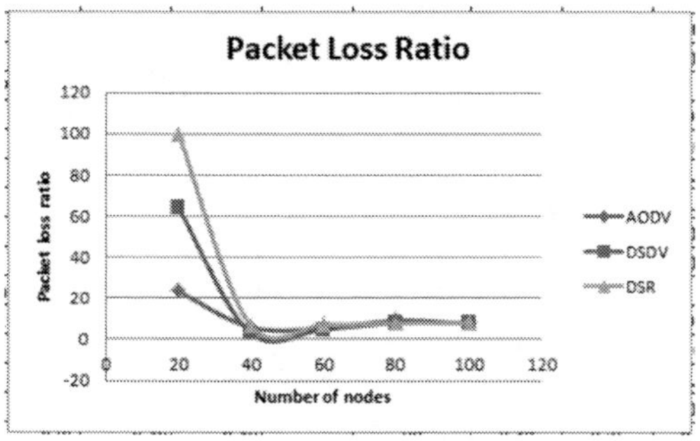

No of nodes	AODV	DSDV	DSR
20	0.21	0	0
40	0.23	0.23	0.46
60	0.31	0.26	0.52
80	0.25	0.49	0.49
100	0.35	0.65	0.65

Table 15:- Average End to End Delay

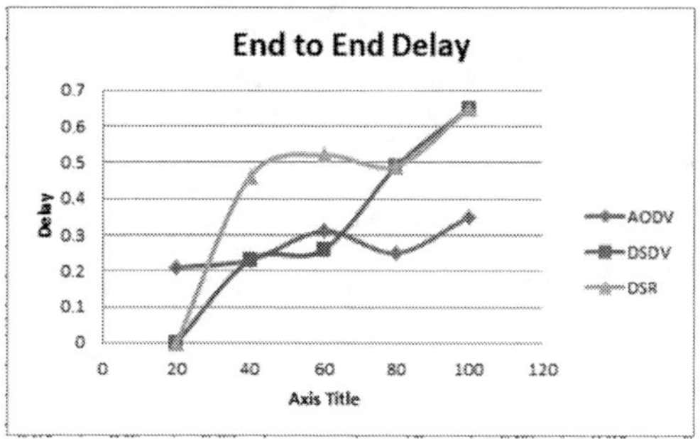

No of nodes	AODV	DSDV	DSR
20	0.76	0.64	0
40	0.94	0.96	0.93
60	0.94	0.95	0.92
80	0.91	0.92	0.92
100	0.92	0.92	0.92

Table16:- Packet Delivery ratio

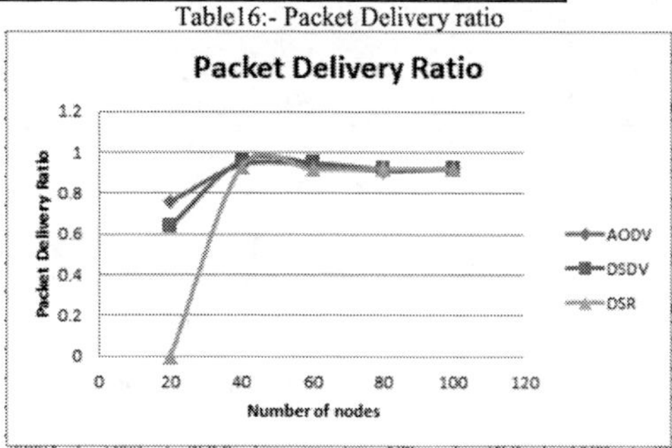

No of nodes	AODV	DSDV	DSR
20	36993	6871	36998
40	38186	7137	36700
60	8631	6776	36998
80	36993	2872	33418
100	7458	6109	30400

Table17:- Throughput

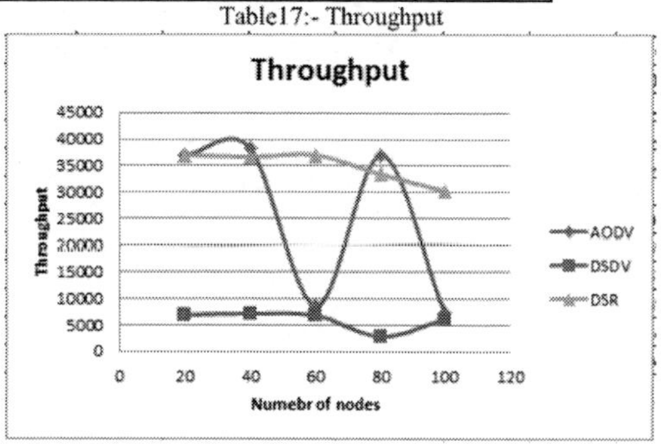

Scenario4:- The nodes are moving with variable speed of 30m/s, 50m/s, 70m/s, 90m/s and 110m/s. The simulation time is taken to be 600ms. The number of nodes are kept constant i.e
100 nodes with a packet size of 2048Mb.

PARAMETER	VALUES
PROTOCOLS	AODV, DSDV, DSR
NUMBER OF NODES	100
TRAFFIC TYPE	CONSTANT BIT RATE(CBR)
TRANSMISSION RANGE	250M
MOBILITY MODEL	RANDOM WAY POINT
SIMULATION AREA	500*500m
NODE SPEED	30m/s, 50m/s, 70m/s, 90m/s and 110m/s
PACKET SIZE	2048MB
RADIO PROPAGATION MODEL	TWO RAY GROUND

Table 18:- Varying speed of 100 nodes

Speed	AODV	DSDV	DSR
30	0.002	0.143	0.115
50	0.001	0.103	0.057
70	0.013	0.103	0.071
90	0.011	0.103	0.066
110	0.001	0.103	0.062

Table 19:- Packet loss

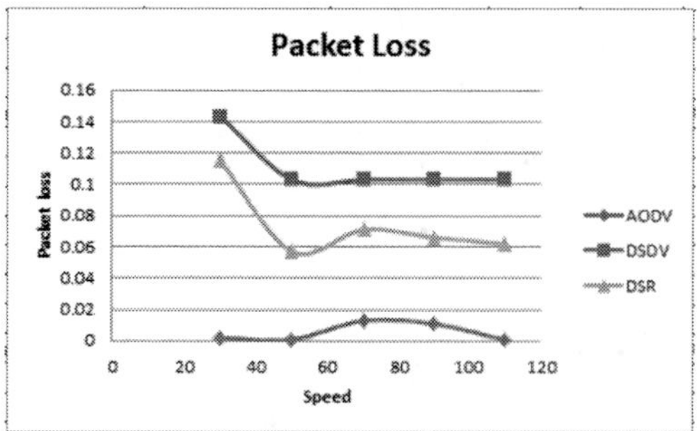

Speed	AODV	DSDV	DSR
30	2	14.3	11.5
50	1	10.3	5.7
70	1.3	10.3	7.1
90	1.1	10.3	6.6
110	0.1	10.3	6.2

Table 20:- Packet loss ratio

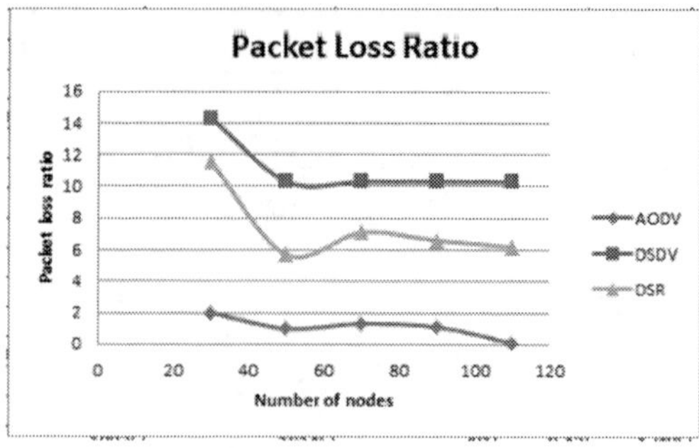

Speed	AODV	DSDV	DSR
30	0.19	0.36	0.87
50	0.21	0.2	0.33
70	0.2	0.2	0.42
90	0.23	0.2	0.45
110	0.21	0.2	0.38

Table 20:- Average End to End Delay

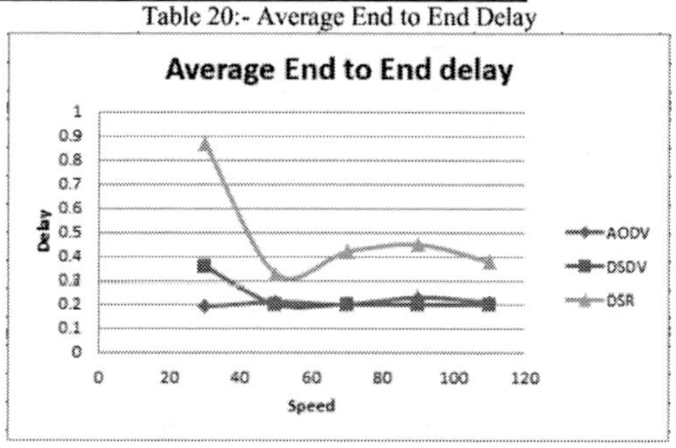

Speed	AODV	DSDV	DSR
30	0.997	0.856	0.884
50	0.995	0.896	0.942
70	0.983	0.896	0.928
90	0.979	0.896	0.933
110	0.997	0.896	0.938

Table 21:- Packet Delivery Ratio

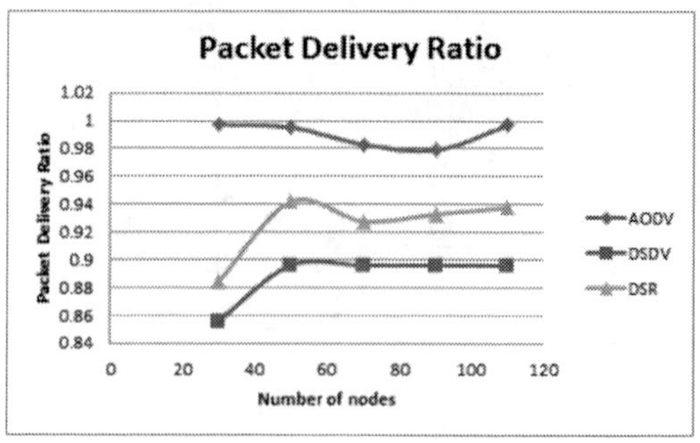

Speed	AODV	DSDV	DSR
30	33082	24217	24669
50	33297	15069	17541
70	30605	15069	17495
90	29861	15069	17218
110	33035	15069	17833

Table 22:- Throughput

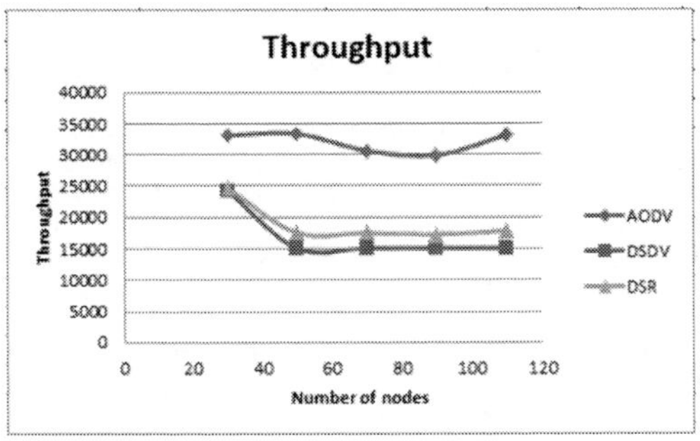

Scenario5:- In this situation, the number of nodes are varied as 10, 20, 30, 40, 50, 60, 70, 80 and 100. The packet size of the data is kept constant to be 2048Mb.

Total number of nodes	Without Malicious node	With Malicious node
10	0.16	0.16
20	0.3444	0.34
30	0.074	0.37
40	0.0662	0.23
50	0.078	1
60	0.0814	1
70	0.087	1
80	0.0942	1
100	0.0932	1

Table 23:- Packet loss

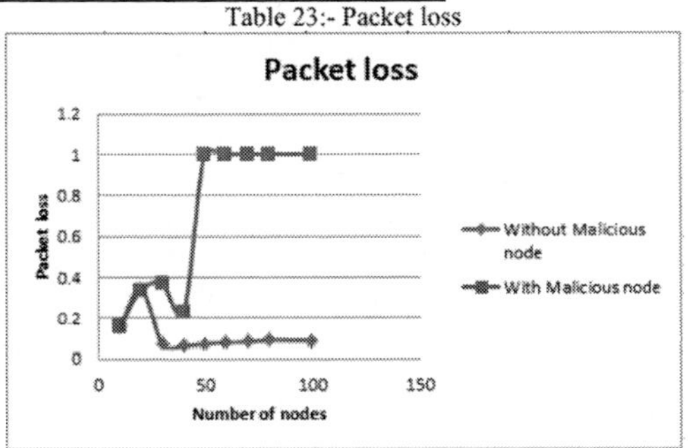

Total number of nodes	Without Malicious node	With Malicious node
10	16	16
20	34.44	34
30	7.4	37
40	6.62	23
50	7.8	100
60	8.14	100
70	8.7	100
80	9.42	100
100	9.32	100

Table 24:- Packet loss ratio

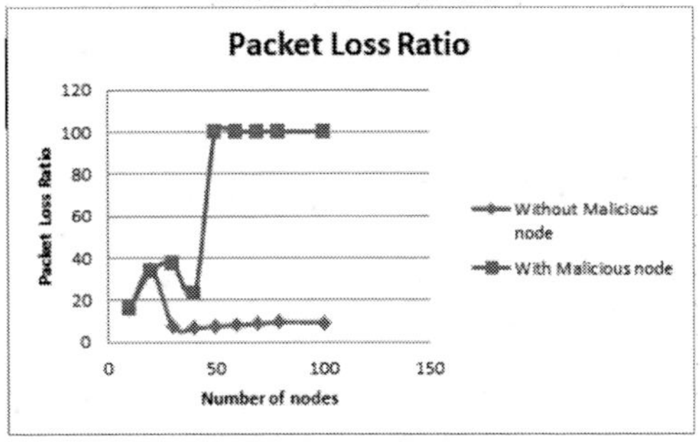

Total number of nodes	Without Malicious node	With Malicious node
10	0.08	0.08
20	0.31	0.31
30	0.33	0.42
40	0.33	0.59
50	0.44	0.65
60	0.48	0.73
70	0.31	0.82
80	0.48	0.93
100	0.39	0.99

Table 25:- Average End to End Delay

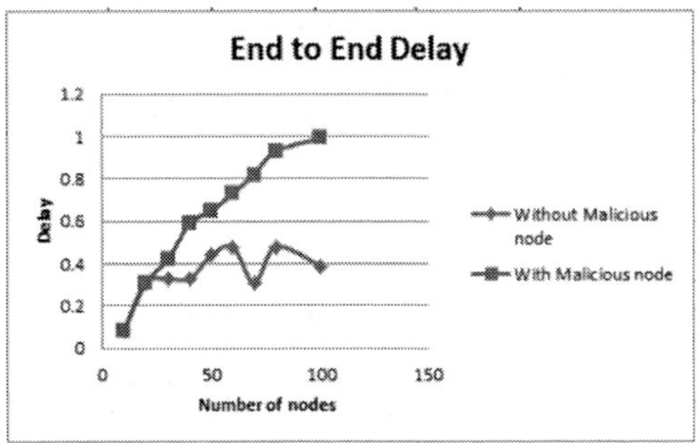

Total number of nodes	Without Malicious node	With Malicious node
10	88539	88539
20	36993	39698
30	28152	5359
40	38186	26120
50	28152	0
60	8631	0
70	24292	0
80	36993	0
100	7458	0

Table 26:- Throughput

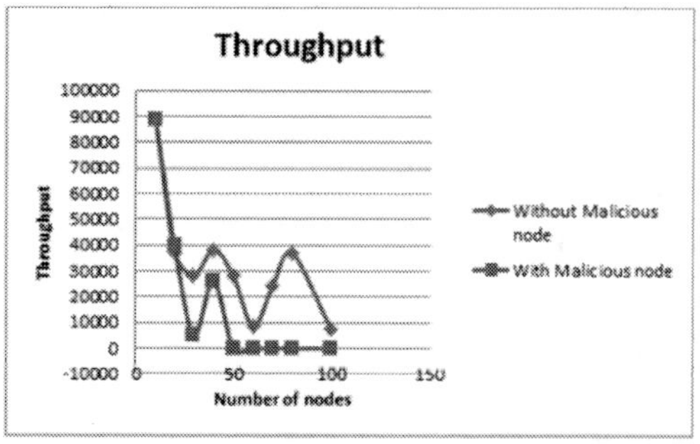

8. Analysis and Results

The simulations were performed using Network Simulator 2 (NS-2.35). The traffic sources are Constant Bit Rate (CBR). The source destination pairs are spread randomly over the network. The mobility model uses 'random waypoint model' in a rectangular field of 500m x 500m with 20 nodes to 100 nodes. For the simulation we consider five different scenarios by varying packet size, varying number of nodes and speed of the nodes.

Scenario1:- In this situation, the number of nodes are varied as 20,40,60,80 and 100. The packet size of the data is kept constant to be 512kB. The packet loss and Packet Loss Ratio is less in AODV as compared to DSDV and DSR. Packet delivery ratio is high in AODV. DSR has high End to End Delay whereas AODV and DSDV has less delay. DSR shows constant line of throughput ghraph whereas DSDV has low Throughput. So we can say that, AODV and DSR show best performance in this scenario.

Scenario2:- In this situation, the number of nodes are varied as 20,40,60,80 and 100. The packet size of the data is kept constant to be 1024Mb. The packet loss and Packet Loss Ratio is less in AODV as compared to DSDV and DSR. Packet delivery ratio is almost same in AODV, DSDV and DSR. DSR has high End to End Delay whereas AODV shows less delay. DSR shows constant line of throughput ghraph whereas DSDV has low Throughput. So we can say that, AODV and DSR show best performance in this scenario.

Scenario3:- In this situation, the number of nodes are varied as 20,40,60,80 and 100. The packet size of the data is kept constant to be 2048Mb. The packet loss and Packet Loss Ratio is less in AODV as compared to DSDV and DSR. Packet delivery ratio is almost same in AODV, DSDV and DSR. DSR has high End to End Delay whereas AODV shows less delay. DSR shows constant line of throughput graph whereas DSDV has low Throughput. So we can say that, AODV and DSR show best performance in this scenario.

Scenario4:- In this situation, the number of nodes is kept constant to be 100. The speed of the vehicles is changed from 30m/s, 50m/s, 70m/s, 90m/s and 110m/s. The packet size of the data is kept constant to be 512kB. As the speed of the vehicles is increasing, the packet loss and Packet Loss Ratio is less in AODV as compared to DSDV and DSR. Packet delivery ratio is high in AODV. DSR has high End to End Delay whereas AODV shows less delay. AODV shows high Throughput whereas DSDV has low Throughput. So we can say that, AODV and DSR show best performance in this scenario.

Scenario5:-In this situation, the number of nodes are varied as 10, 20, 30, 40, 50, 60, 70, 80 and 100. The packet size of the data is kept constant to be 2048Mb. The packet loss and Packet
Loss Ratio is more as the number of malicious nodes are increased in the network. Packet delivery ratio is lowest in AODV when malicious nodes are introduced in the network.
Average End to End Delay increases whereas throughput graph shows decreasing trend. This shows that when the network is infected by malicious nodes, the throughput and packet delivery ratio becomes the least where as the end to end delay and packet loss increases with the increase in malicious nodes.

When the network is moving without any malicious nodes, the performance matrics show different behaviour in each scenario. The packet loss, packet loss ratio is less and packet delivery ratio is more in each case. But when malicious nodes are introduced in the network, they degrade the performance of the network. Packet loss and packet loss ratio increases as the number of malicious nodes increase in the network. As packet loss increases, packet delivery ratio decreases. Packet delivery reduces as the malicious nodes drop the packets in the network. Throughput is dependent on the packets received and the time taken to receive the packets. The malicious nodes do not allow the packets to reach to the destination, hence reducing the throughput and efficiency of the network and increases the end to end delay of the packets reaching the destination node.

9. Conclusion

AODV performs best in the event of packet delivery ratio without malicious nodes in the network. As the number of malicious nodes enter the network, they degrade the network with respect to throughput, packet delivery ratio and End to End delay in the network. Analysis of malicious nodes shows that with malicious node and without malicious node the output of parameters are different. From the graph it is shown that
malicious nodes have less accuracy then malicious-less networks. Therefore security of MANET becomes fragile in presence of malicious nodes.

10. References

[1]Jakubiak, J.,Koucheryavy, Y. (2008, January). State of the art and research challenges
for VANETs. In 2008 5th IEEE Consumer Communications and Networking
Conference (pp. 912-916). IEEE.
[2] Youse_, S., Mousavi, M. S., Fathy, M. (2006, June). Vehicular ad hoc networks
(VANETs): challenges and perspectives. In 2006 6th International Conference on ITS
Telecommunications (pp. 761-766). IEEE.
[3] Taleb, T., Sakhaee, E., Jamalipour, A., Hashimoto, K., Kato, N., Nemoto, Y. (2007).

A stable routing protocol to support ITS services in VANET networks. IEEE Transactions
on Vehicular technology, 56(6), 3337-3347.
[4] Abdullah, A., Ramly, N., Muhammed, A., Derahman, M. N. (2008). Performance comparison
study of routing protocols for mobile grid environment. IJCSNS International
Journal of Computer science and Network security, 8(2), 82-88.
[5] Mahmood, Z., Nawaz, M. A., Iqbal, M., Khan, S., Haq, Z. U. (2015). Varying pause
time e_ect on AODV, DSR and DSDV performance. Int. J. Wirel. Microwave Technol,
1, 21-33.
[6] Ripan, K., Verma, A. K. (2007). Performance Comparison of AODV and DSR Routing
Protocols in MANETs (Doctoral dissertation
[7] F□u_ler, H., Mauve, M., Hartenstein, H., K□asemann, M., Vollmer, D. (2002). A comparison
of routing strategies for vehicular ad hoc networks. Technical reports, 2.
[8] Johnson, D. B., Maltz, D. A. (1996). Dynamic source routing in ad hoc wireless
networks. In Mobile computing (pp. 153-181). Springer, Boston, MA.
[9] Li, F., Wang, Y. (2007). Routing in vehicular ad hoc networks: A survey. IEEE
Vehicular technology magazine, 2(2), 12-22.
[10] Henderson, T. (2011). Radio Propagation Models. Information Sciences Institute,
University of Southern California.[Online] Available: http://www. isi.
edu/nsnam/ns/doc/node216. html.
[11] Siraj, M., Kanrar, S. (2012). Performance of Modeling wireless networks in realistic
environment. arXiv preprint arXiv:1201.0842.
[12] Sto_ers, M., Riley, G. (2012, August). Comparing the ns-3 propagation models. In
2012 IEEE 20th International Symposium on Modeling, Analysis and Simulation of
Computer and Telecommunication Systems (pp. 61-67). IEEE. 28
[13] Abdullah, A., Ramly, N., Muhammed, A., Derahman, M. N. (2008). Performance
comparison study of routing protocols for mobile grid environment. IJCSNS International
Journal of Computer science and Network security, 8(2), 82-88.
[14] Perkins, C. E., Bhagwat, P. (1994). Highly dynamic destination-sequenced distancevector
Routing (DSDV) for mobile computers. ACM SIGCOMM computer communication
Review, 24(4), 234-244.
[15] Ab Rahman, R., Kassim, M., Yahaya, C. K. H. C. K., Ismail, M. (2011, June). Performance
analysis of routing protocol in WiMAX network. In 2011 IEEE International
Conference on System Engineering and Technology (pp. 153-157). IEEE.
[16] Nikam, S., Jadhav, B. T. (2016). Analysis of AODV Protocol against Pause Time
Using NS2. 34.
[17] Broch, J., Maltz, D. A., Johnson, D. B., Hu, Y. C., Jetcheva, J. (1998, October). A
performance comparison of multi-hop wireless ad hoc network routing protocols. In
Proceedings of the 4th annual ACM/IEEE international conference on Mobile computing
and networking (pp. 85-97).
[18]
The Network Simulator ns-2. http://www.isi.edu/nsnam/ns/index.html.

https://doi.org/10.1051/itmconf/20224403007

Multi-class Chest X-ray classification of Pneumonia, Tuberculosis and Normal X-ray images using ConvNets

Rachita Mogaveera[1], *Roshan* Maur[1 *], *Zeba* Qureshi[1] and *Yogita* Mane[2]

[1]Student, Dept. of Information Technology, Universal College of Engineering, Maharashtra, India
[2]Professor, Dept. of Information Technology, Universal College of Engineering, Maharashtra, India

Abstract. Pneumonia and Tuberculosis (TB) are two serious and life-threatening diseases that are caused by a bacterial or viral infection of the lungs and have the potential to result in severe consequences within a short period of time. Therefore, early diagnosis is a significant factor in terms of a successful treatment process. Chest X-Rays which are used to diagnose Pneumonia and/or Tuberculosis need expert radiologists for evaluation. Thus, there is a need for an intelligent and automatic system that has the capability of diagnosing chest X-rays, and to simplify the disease detection process for experts and novices. This study aims to develop a model that will help with the classification of chest X-ray medical images into normal vs Pneumonia or Tuberculosis. Medical organizations take a minimum of one day to classify the diagnosis, while our model could perform the same classification within a few seconds. Also, it will display a prediction probability about the predicted class. The model had an accuracy, precision and recall score over 90% which indicates that the model was able to identify patterns. Users can upload their respective chest X-ray image and the model will classify the uploaded image into normal vs abnormal.

1 Introduction

Pneumonia and Tuberculosis are caused by viruses or bacteria, and less often, other microorganisms[1]. Nowadays, chest X-ray (CXR) imaging is commonly used for health intensive care and analysis of many lung diseases such as Pneumonia and Cancer because of the relatively low costs[2]. Therefore, Chest X-rays are the best tool for diagnosing diseases related to which has played a significant role in clinical care and epidemiological research. However, detecting Pneumonia and Tuberculosis in chest X-rays is a challenging task that relies on the availability of expert radiologists. However, not every doctor has high quality medical tools to diagnose patients . As a result, sometimes, their diagnosis is not very accurate. It is also much harder to judge Pneumonia and Tuberculosis just by looking at chest X-rays images.

With the advancements of information technology, Statistical analysis, Machine Learning, and Deep Learning algorithms have been successfully applied to many healthcare problems and have explained complex relationships and improved clinical predictions[3]. They can solve computer vision problems in the medical domain.. In recent years, researchers have proposed different Machine Learning based solutions for medical problems. The objective of this paper is to create a custom CNN model and detect the diseases from their chest X-rays with higher accuracy, recall, precision and F1-Score. Classification methods are among the most commonly used techniques in medical imaging, where the goal is building classifiers capable of predicting whether X-ray images are normal or abnormal (i.e whether it is Pneumonia or Tuberculosis).

Identifying and analyzing an X-ray image might take a minimum of one day by a medical organization or a doctor and that could be troublesome for patients and therefore, to benefit the doctors as well as decrease the waiting period for patients, this project would truly have an adverse effect on the world. With the help of Deep learning we would be able to build a model and create a website which would return you the result of your Chest X-rays in a few seconds thus, classifying if you are diagnosed with Tuberculosis or Pneumonia or a healthy chest X-ray.

2 Literature survey

Different classic Machine learning algorithms like SVM (Support Vector Machines) or Random Forest Classifiers can achieve a decent F1-score and accuracy score but not as good as a CNN model as discussed in [4].

In [5], the authors used a ResNet-50 model to classify the top ten most common chest diseases through X-ray images but the dataset was highly imbalanced which led to bias and thus, affected the model's performance in inference mode.

Similar approach was adopted in [6], where a DenseNet model was able to classify two pathologies but dataset imbalance affected the model's AUC score.

* Corresponding author:maurroshan17@gmail.com

3 Dataset and methods

3.1 Dataset

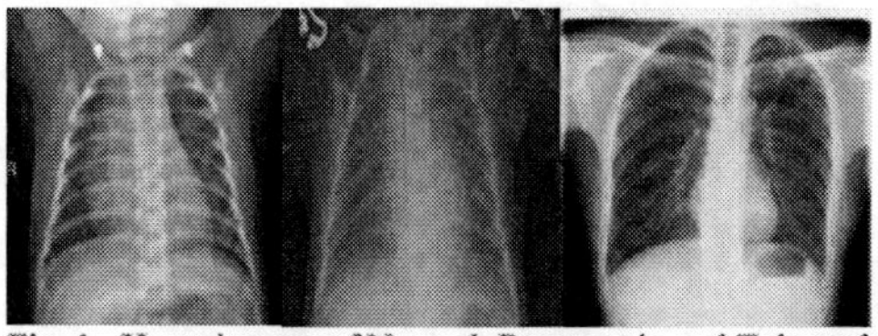

Fig. 1. X-ray images of Normal, Pneumonia and Tuberculosis.

The dataset used for this research is acquired through Kaggle. There were few individual datasets available for each of the three classes and we merged the datasets from all the sources to have an all-in-one dataset for our research. Kaggle is an open source platform for Data Science/Machine learning researchers and practitioners where they can find datasets or participate in Data Science hackathons. The sample images of each class from the dataset is shown in Fig 1.

The dataset contains three sub categories which are Normal, Pneumonia and Tuberculosis. In this dataset there were 9288 images of Normal chest X-rays, 4245 images of Pneumonia chest X-rays and 1888 images of Tuberculosis chest X-rays. All these images are in JPEG format. [7]

3.2 Balancing and data preprocessing

Preparing the data in other words, extracting information from the raw data, taking care of the class imbalance, eliminating duplicate images, preprocessing the data by normalizing the pixel values, resizing the images and adding some data augmentation.

After eliminating the duplicate images, and segregating the images into their respective classes a huge class imbalance occurred between the classes. There were over 9100 images for Normal chest X-rays, 4145 for Pneumonia chest X-rays and 1788 images for Tuberculosis chest X-rays. There are two types of Pneumonia infections: Pneumonia through virus and Pneumonia through bacteria. To avoid bias, a 50-50 ratio for both viral and bacterial infection was maintained in the Pneumonia class. In order to eliminate the class imbalance, we used the downsampling approach. The dataset was balanced by downsampling Pneumonia and Normal chest X-rays images to 1700 thereby maintaining balance between all the classes.

There were 5100 total images in the training set which included 1700 images for each class. There were 150 total images in the validation set which included 50 images for each class. There were 238 images in the test set which included 100 images for Pneumonia and Normal chest X-rays respectively and 38 images for Tuberculosis.

Once the dataset was prepared it was divided into training set, validation set and testing set. A general split ratio of 80:10:10 was used.

All the classes in the train folder have the same number of images. Since the images inside the training set, validation set and test set are of different sizes and ConvNets do not accept images of different sizes as inputs we had to perform data preprocessing.

3.2.1 Data preprocessing

For data preprocessing, we will resize the image, preprocess (normalize/rescale) the images and add some data augmentation techniques.

The images were resized to a certain height x width value. We used 256 x 256 px as the size of the image. The images were also converted into grayscale mode if any of the images were in RGB mode. That results in the image size as (256,256,1). The 1 in the end indicates that the image is in grayscale mode. We used grayscale mode because we are dealing with X-ray images.

Data augmentation techniques can not be applied to validation and test dataset. Data augmentation techniques add some variations to the training set. We used three data augmentation techniques: Random Zoom, Random Contrast and Random Rotation. We have selected a value of 0.3 (30%) for each parameter.

3.3 Creating optimized input pipeline

3.3.1 Prefetching

Rather than passing the whole data i.e the training set, validation set and the test set directly to the model, we created small batches of data. The size of the batches was 32. Normal batches of data will be converted to a prefetch dataset. The process carried out in Fig. 2 is the training part where our model receives the input and learns the patterns from the batch.

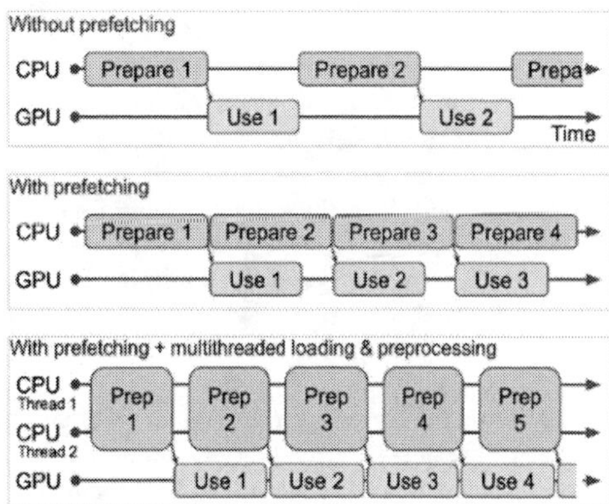

Fig. 2. Prefetch Dataset [8]

https://doi.org/10.1051/itmconf/20224403007

So, the first method is without a prefetch dataset, where the CPU first prepares a dataset and then passes it to the GPU where the model learns the patterns and updates its weights. Until the GPU has not completed training the first batch, the CPU cannot start the preparation for the second batch and thus, CPU and GPU work dependently, meaning they will wait for each other to finish their respective process in order to start working with the new batch.

Prefetch dataset helps us avoid this anomaly, meaning after the CPU passes on the first batch to the GPU for the model to train and learn the patterns, the CPU can start preparing the second batch, the third batch and so on, resulting in an optimized input pipeline and faster training time. Here, both the CPU and GPU are working independently. In the third process, which is prefetch dataset + multithreading processing and loading, is the optimal input pipeline. We have used this optimized input pipeline for our project. This works exactly like the prefetch dataset except it has more than two GPUs, so if the first batch is being used by GPU - 1, the second batch after preparation by the CPU would be used by GPU - 2 and so on. This speeds up the training process by 3x.

3.4 ConvNets

We built a custom model using the functional API. The reason for using the functional API was the flexibility it offers over the Sequential API. Fig. 4. shows the model architecture.

The layers used in this model are Convolutional layers, Max Pooling layers, a Global Max Pooling layer, Dense layers, a Batch Normalization layer and a Dropout layer. Categorical Cross Entropy is used to calculate the loss with a label smoothing of 0.2 (20%) to refrain the model from being too confident on any prediction. Adam optimizer was used as the learning rate optimizer with a default learning rate of 0.001. The three metrics, accuracy, precision and recall were used to calculate the performance of the model along with the loss. The model was trained for 80 epochs and we used two callbacks. The Model checkpoint callback and the Early Stopping callback.

Batch Normalization layer helps to normalize the output from activation layers. It is entirely possible that the output could have a very small value or a very large value and that may affect the weights of the following hidden layers accordingly. Thus, to avoid both the exploding gradient problem and the vanishing gradient problem, the Batch Normalization layer is used to normalize the output from the activation functions.

When defining the dense layer we set the units i.e. the output size of the next layer and the activation function. For our project we used 64 units and the ReLU activation function for our first dense layer. Another dense layer is applied with smaller units i.e. 32 and lastly, another dense layer of units 3 with activation function softmax is applied to the final dense layer from which the output can be decided. Each neuron of the output layer will perform matrix-vector multiplication to decide its value. The output of the neurons is calculated by calculating the dot product of the vector and matrix.

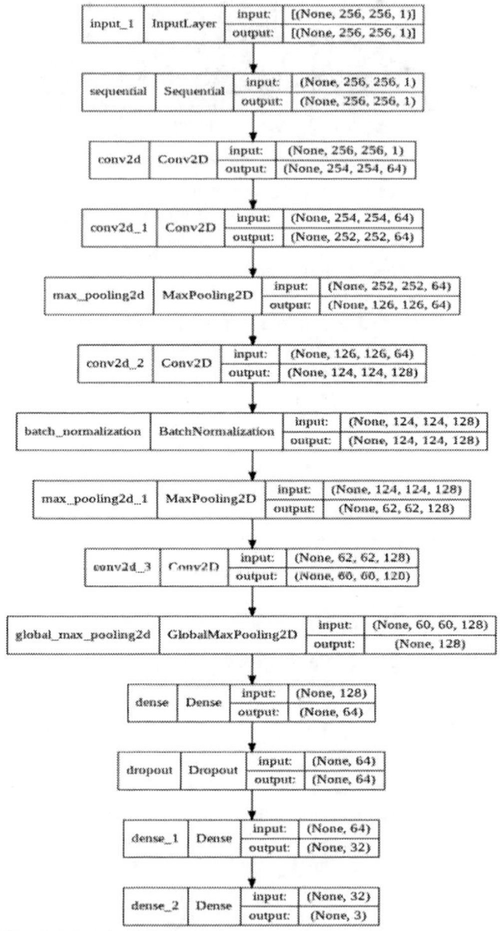

Fig. 3. Model architecture

Model checkpoint callback is a vital callback as it saves the best model after every epoch. By setting the monitor parameter it can monitor that variable for example, we set the monitor to validation loss so every time it had a lowest validation loss recorded, it will save the model in the path defined.

Early Stopping callback similarly monitors the validation loss and it stops training the model after there is no improvement in the model's validation loss. To stop the model from training we have to set the patience parameter meaning how many epochs it should consider before stopping the model from training further. The patience parameter was set to 10.

The Relu activation function formula is

$$f(x) = \max(0, z)$$

 https://doi.org/10.1051/itmconf/20224403007

If a negative value is inputted the output is 0 else it is z. Due to this formula the output value is always a positive value. The purpose of the activation function is to provide non-linearity to our function and to set boundary condition rules.

The last step is to apply the categorical cross entropy function. It is a loss function and it uses the formula.

$$Loss = -\sum_{i=1}^{output\ size} yi\ .\log yi$$

This function is used for multi-level categorical models. The sum of all the values of the neurons to which cross entropy is applied is 1. It is used to give the highest probability to the correct class and lowest to the other class.

4 Streamlit web-app

In this research work, we used Streamlit for deploying the model. Streamlit is one of the best Python frameworks to create a quick web-app with an attractive user-interface. Our Streamlit web-app consists of three pages:

The first page as shown in Fig. 4. as the name suggests is where the user can upload his/her chest X-ray image and the model will classify the image. It will also create a simple bar graph to represent the confidence for each class. The sidebar on the first page will also have the confusion matrix for the universal test dataset and the other metrics of the model.

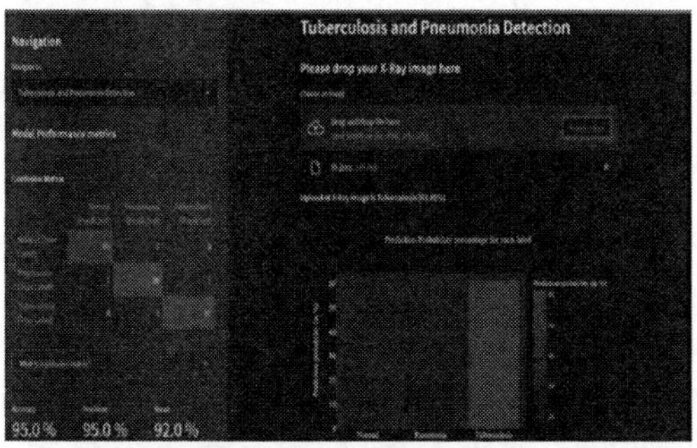

Fig. 4. Tuberculosis and Pneumonia Detection page

The second page will consist of a forum page shown in Fig.5. where users can post their queries, search queries and read about other people's queries. It will help the users to interact with each other, recommend methods that worked for them and even recommend hospitals as to where they received the best treatment for any particular disease.

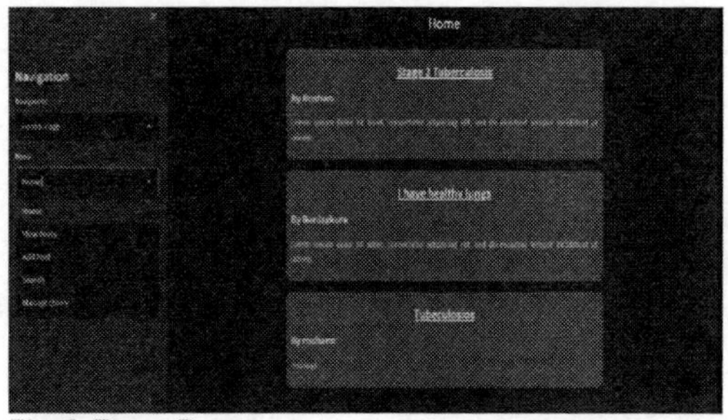

Fig. 5. Forum Page

The third page Fig. 6. is where users can search about different hospitals. The dataset for the following hospitals filtered out by their respective locations was scraped from Wikipedia[9] using Beautiful Soup library. After scraping, it was converted into a Pandas dataframe. There is also a remedies note if you're diagnosed with Pneumonia or Tuberculosis which will automatically get displayed on the screen after you've uploaded your image on the first page.

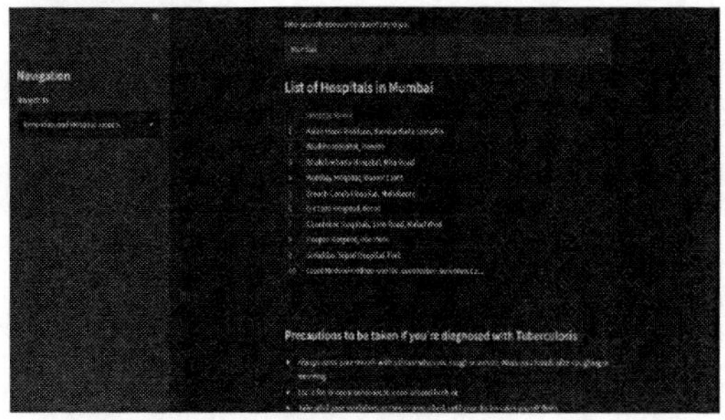

Fig. 6. Hospital search and remedies page

5 Performance Analysis

5.1 Comparision with previous works

There were some research papers in this domain where different diseases were identified from chest X-rays using Deep learning techniques. Few of them built their own custom ConvNets while others used pretrained models like AlexNet or ResNet or VGG networks and thus, leveraging the power of transfer learning. The research works we used for references as shown in Table.1 were some of the classical papers in this field.

Our dataset was created from different open source contributions on Kaggle. Many research papers used one specific dataset but we merged many different datasets to create an all-in-one dataset for Normal, Pneumonia and Tuberculosis. Rather than relying on transfer learning techniques we created our own custom ConvNet model that performed well on two different datasets i.e the balanced dataset and the unbalanced dataset. Also, our model was able to achieve an F1-score over 90% on both the datasets.

Table 1. Comparision with previous works

Name	Metrics used	Dataset imbalance	Pneumonia And Tuberculosis
[4]	Accuracy and F1-score	Chest X-ray images with Pnuemonia were over 4200 whereas total number of Normal Chest X-ray images were around 1200.	Only Pneumonia
[5]	Specificity and Sensitivity	High class imbalance.	10 Chest X-ray images but none included Pnuemonia or Tuberculosis
[6]	AUC score	High class imbalance led to a good AUC score on pulmonary modules but a poor AUC score of 73% on cardiomegaly.	Pathologies like pulmonary modules and cardiomegaly.

5.2 Balanced dataset model vs Unbalanced dataset model

We used the same model architecture to train on the balanced dataset and the unbalanced dataset. To analyze the model's performance metrics on unseen data, we created an universal test dataset. The universal test dataset was created after eliminating duplicate images in the later stages of the data collection process. 100 images from each class were selected at random to create a universal test data to further analyze the model's performance metrics. The model trained on the balanced dataset for 80 epochs and the model trained on the unbalanced dataset for 44/80 epochs. We used the Early Stopping callback for the model trained on the unbalanced dataset to refrain overfitting.

The confusion matrices for the universal test dataset by the model trained on the balanced dataset and the model trained on the unbalanced dataset are in table 2 and table 3 respectively.

Table 2. Confusion matrix of model trained on balanced dataset

True/Pred	Normal (pred)	Pneumonia (pred)	TB (pred)
Normal (True)	85	7	8
Pneumonia (True)	1	98	1
TB (True)	9	.0	91

In table 2 the important takeaway is that the model trained on the balanced dataset was better able to classify both Tuberculosis and Pneumonia compared to Normal. It was able to classify 85 Normal chest X-ray images correctly compared to Pneumonia chest X-rays 98 correctly classified images and Tuberculosis chest X-rays 91 correctly classified images.

Table 3. Confusion matrix of model trained on unbalanced dataset

True/Pred	Normal (pred)	Pneumonia (pred)	TB (pred)
Normal (True)	95	4	1
Pneumonia (True)	9	90	1
TB (True)	14	2	84

In table 3 the important takeaway is that the model trained on the unbalanced dataset had a bias while classifying Tuberculosis. Since the images in the Normal class were 5x that of Tuberculosis, it had a bias factor which is clearly visible in the confusion matrix. 14 chest X-ray images with Tuberculosis were misclassified as Normal which indicates the bias factor that the model trained on the unbalanced dataset.

In the medical image classification domain, a False Negative is as bad as a False Positive. Comparatively, the model trained on the balanced dataset beats the model trained on the unbalanced dataset model slightly.

The performance metrics of the model trained on the balanced dataset and the unbalanced dataset on their respective training set, validation set and test set are in Fig. 7 and Fig. 8 respectively.

In Fig. 7, we have the metrics for the model trained on the balanced dataset, the F1 scores for the training set and test set are above 0.90 which indicates that the model is able to generalize well. Though the loss is over 0.60 for the training and validation set it plummeted to 0.58 which is the lowest recorded loss.

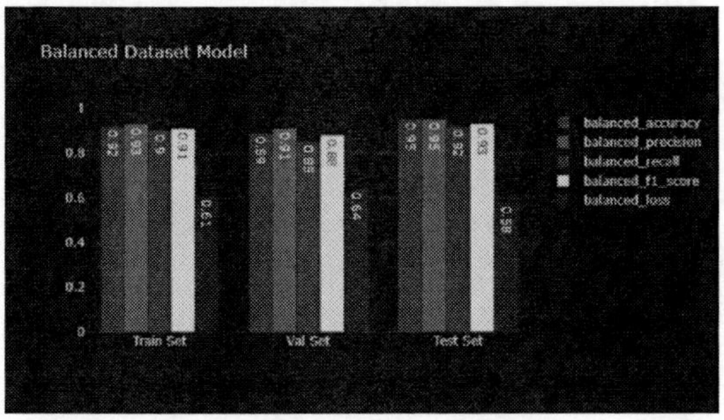

Fig. 7. Balanced dataset model performance metrics

In Fig. 8, we have the metrics for the model trained on the unbalanced dataset, the F1 scores for all the sets is around 0.90 though it was trained for only 44/80 epochs training because of the Early Stopping callback.

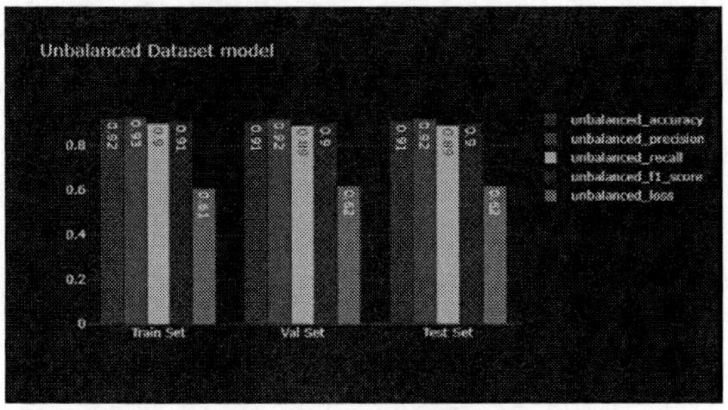

Fig. 8. Unbalanced dataset model performance metrics

6 Future Scope

This research paper has a few limitations for example, the dataset is not diverse enough. Since, the dataset was collected from open source platforms like Kaggle, the model is accustomed to inference bias like the patches of Tuberculosis or Pneumonia will differ for a child to that of an adult, different stages of patches in Tuberculosis and Pneumonia. Another limitation is the hospital feature search, since the list of hospitals were scraped from Wikipedia, there were only a limited number of hospitals added to the dataset, so we would like to add more hospitals from many different locations and add more information about different hospitals in the dataset so the user can feel satisfied. In future, we look forward to work along with medical organizations to create a robust model using a large and diverse dataset that will be able to classify different chest X-ray diseases.

7 Conclusion

This Deep Learning model is not created to replace the jobs of doctors, but rather help them and save more time for the patients. By adding different lung diseases with more images of different classes, medical organizations can identify the type of disease the patient may be diagnosed with as early as possible so as to initiate the medical procedure to cure the disease and on a larger scale, it would benefit both the medical organization as well as the patients. Our proposed model achieved an accuracy, precision and recall score of over 90%.

References

[1] R. T. Sousa, O. Marques, F. A. A. Soares, I. I. Sene Jr, L. L. de Oliveira, and E. S. Spoto, "Comparative performance analysis of machine learning classifiers in detection of childhood pneumonia using chest radiographs," Procedia Computer Science, vol. 18, pp. 2579–2582(2013).

[2] W. H. Organization et al., "Standardization of interpretation of chest radiographs for the diagnosis of pneumonia in children," World Health Organization, Tech. Rep.(2001).

[3] J. De Fauw, J. R. Ledsam, B. Romera-Paredes, S. Nikolov, N. Tomasev, S. Blackwell, H. Askham, X. Glorot, B. O'Donoghue, D. Visentin et al., "Clinically applicable deep learning for diagnosis and referral in retinal disease," Nature medicine, vol. 24, no. 9, pp. 1342–1350,(2018).

[4] R. E. Al Mamlook, S. Chen and H. F. Bzizi, "Investigation of the performance of Machine Learning Classifiers for Pneumonia Detection in Chest X-ray Images," 2020 IEEE International Conference on Electro Information Technology (EIT),pp.098-104,doi:10.1109/EIT48999.2020.9208232(2022).

[5] Y. Dong, Y. Pan, J. Zhang and W. Xu, "Learning to Read Chest X-Ray Images from 16000+ Examples Using CNN," 2017 IEEE/ACM International Conference on Connected Health: Applications, Systems and Engineering Technologies (CHASE), 2017, pp. 51-57, doi: 10.1109/CHASE.2017.59.

[6] M. S. Majdi, K. N. Salman, M. F. Morris, N. C. Merchant and J. J. Rodriguez, "Deep Learning Classification of Chest X-Ray Images," 2020 IEEE Southwest Symposium on Image Analysis and Interpretation (SSIAI), 2020, pp. 116-119, doi: 10.1109/SSIAI49293.2020.9094612.

[7] Roshan Maur, "Imbalanced Tuberculosis and Pneumonia dataset," Kaggle.com, 2022. [Online]. Available:https://www.kaggle.com/roshanmaur/imbalanced-tuberculosis-and-pnuemo nia-dataset. [Accessed: 27-Feb-2022]

[8] Géron, A. Learning with Scikit-Learn, Keras & TensorFlow. 2nd ed. O'Reilly Media, Inc., p.422.(2019)

[9] Wikipedia Contributors, "List of hospitals in India," Wikipedia, 17-Feb-2022. [Online]. Available: https://en.wikipedia.org/wiki/List_of_hospitals_in_India. [Accessed: 27-Feb-2022]

Multiple Disease Prognostication Based On Symptoms Using Machine Learning Techniques

Kajal Patil[1,*], *Sakshee* Pawar[2], *Pramita* Sandhyan[3] and *Jyoti* Kundale[4]

[1,2,3] Ramrao Adik Institute of Technology, Nerul, Navi Mumbai, India

[4] D.Y. Patil Deemed to be University, Ramrao Adik Institute of Technology, Nerul, Navi Mumbai, India

Abstract. Disease Prediction system that uses Machine Learning forecasts the ailments on the basis of the data pertaining to the symptoms entered by the user and provides trustworthy findings based on that data. If the patient isn't in any danger and the user merely wants to know what kind of ailment he or she has had. It is a system that gives the user suggestions and methods on how to keep their health system in good shape, as well as a way to find out if they have a sickness utilizing this forecast. Due to a diversity of diseases and a lower doctor-patient ratio, the use of particular disease prediction technologies as well as concerns about health has risen. We are focusing on offering customers with an instant and accurate disease prognosis based on the symptoms they enter, as well as the severity of the condition projected. It will provide the best algorithm and doctor consultation. Different machine learning algorithms are employed to forecast illnesses, ensuring speedy and reliable predictions.

1 Introduction

Medical science and universal health care are critical components of the economy and human existence. There has been a significant amount of shift between the world we live in now and the world that existed only a few weeks ago. Everything has become dull and erratic.Today's demand for medical industries have a greater requirement for data mining. When certain data mining techniques are applied correctly, important information may be mined from a huge database, which can aid medical practitioners in making early judgments and improving health care. The goal is to aid the physician by using the categorization.

Machine learning has become more popular as technology has advanced, owing to greater computer power and the availability of datasets on open-source sources. Machine learning is used in a number of ways in health care. The health care industry generates a vast quantity of data in the form of photographs, patient data, and other types of information that may be used to spot trends and make predictions. In health care, machine learning is utilize to tackle a variety of issues. Individuals with heart illness have varying degrees of heart disease, and the severity of heart disease varies from person to person.

As a result, creating a machine learning model, training it on the data set, and incorporating unique patient information can aid in prediction. The forecast result will be based on the information provided, and hence will be unique to that person. Type-2 diabetes is a condition that may be avoided by maintaining a healthy weight, lifestyle, and other factors. Corona virus is a condition for which there is no apparent cure. COVID-19 is a corona virus that emerged in China. This Disease Prediction using Machine Learning is totally done with the aid of Machine Learning, and we will forecast the disease using the data set that was previously provided by the hospitals. Numerous scientific methods and methodologies are being used by doctors for the identification and diagnosis of not only common ailments, but also many deadly disorders.

The proper and precise diagnosis is always credited with a successful therapy. Doctors may occasionally make incorrect judgments while diagnosing a patient's sickness; as a result, disease prediction systems that employ machine learning algorithms aid in obtaining accurate findings in such circumstances. It was created to combat general disease at an earlier stage, as we all know, in the competitive environment of economic development, mankind has become so engrossed that he or she is no longer concerned about health. It can detect people who are at risk of sickness or other health problems.

Clinicians can then take the necessary precautions to avoid or reduce the risk, so improving the quality of care and avoiding unnecessary hospitalizations. Disease risk prediction may now take advantage of vast volumes of semantic data, such as demographics, clinical diagnosis and measures, health behavior, test findings, prescriptions, and care mutilation, thanks to recent advancements in data analytic tools and methodologies. Electronic health data might be a good option for constructing illness prediction models in this case. Over time, a great number of illness prediction models based on large-scale electronic health databases have been proposed in the literature. Stakeholders such as the government and health insurance firms may gain from disease prediction. It can detect people who are at risk of sickness or other health problems.

* Corresponding author: 2906kajal@gmail.com

1.1 Purpose and Problem Definition

If a person has been observing a few symptoms and is unsure of the sickness he or she is dealing with, this will lead to a variety of diseases in the future. To avoid this and to learn about the condition in the early phases of the symptoms, this disease prediction will be extremely beneficial to a wide range of individuals, including children, teens, adults, and elderly citizens. Preventive measures can be taken by the individual or can seek for professional medical advice in order to get the proper treatment required to treat the ailment.

The scope of a disease prediction system is specifically vast, conferring how the world is evolving and along with the advancements in the technology comes with a lot of cons such as the various adulterates of food items, lack of proper nutrient supply to the body, unhealthy lifestyles involving improper food consumption as well as problems like obesity or unhealthy weight. All this accompanies with a variety of diseases.

Unfortunately as people are too engrossed in their daily activities they neglect their health. Children and senior citizens can also ignore or fail to recognition the important symptoms which can result in greater issues later on. It is wise to get treated before the disease develops and grows more detrimental. A prediction system can help such people to detect and act upon their health issues at an early stage and take preventive care. This also helps in getting primary health care in remote areas.

2 Literature Survey

Numerous research works have been carried out for the prediction of the diseases based on the symptoms shown by an individual using machine learning algorithms. Mont et al. [1] designed a statistical model to predict whether a patient had inuenza or not. They included 3744 unvaccinated adults and adolescent patients of inuenza who had fever and atleast 2 other symptoms of inuenza. Out of 3744, 2470 were confirmed to have inuenza by the laboratory. Sreevalli et al. [2] used the random forest machine-learning algorithm to predict the disease based on the symptoms.

The system resulted in low time consumption and minimal cost for the prediction of diseases, algorithm resulted in an accuracy of 84.2%. [4] considered medical records of 4920 patients and identified 132 symptoms corresponding to 41 diseases and implemented this dataset by using Naïve Bayes, Decision Tree and Random Forest algorithms. In this system, Naïve Bayes showed the highest security comparatively of 93.61%. The researchers in [5] used Convolutional Neural Networks and K-Nearest Neighbour algorithms are implemented on Disease dataset from UCI machine repository.

The results found through comparative analysis is that the CNN performed better than KNN in accuracy and time comparison. [6] used UCI heart disease repository and implemented Support Vector Machine to classify attributes into different heart disease classes. In this, SVM achieved an accuracy of 89%. In [7], the associated diseases with the symptoms using Random Forest where UCI disease repository is used. In the result, Random Forest achieved an accuracy of 85%.

3 Proposal

3.1. Module Description

Methods of machine learning have been available for a long time, and they've been compared and used in a number of ways for data science analysis. The major purpose of this research project was to examine the strategies for selecting features, preparing data, and processing data in machine learning training models. The challenge we have today with first-hand models and libraries is data, which, in addition to their amount and our cooked models, has a wider variety in the accuracy we see throughout training, testing, and real validation.

Furthermore, because the goal of machine learning is to produce an appropriate computer-based system and decision support that may help in the early diagnosis of diseases, we constructed a model in this project that uses machine learning algorithms to classify whether a patient will have any ailment. As a result, early illness prediction can assist in making decisions about lifestyle adjustments in high risk patients, reducing consequences, which can be a significant milestone in the area of medicine.

In order to look into the verified diseases and their corresponding diseases, an authentic data set was collected which had the list of the diseases and their symptoms listed in the database. An automated technique was used to create a knowledge database of disease-symptom connections based on information from textual discharge summaries of patients hospital to New York Presbyterian Hospital in 2004.

The disease is listed first, followed by the number of discharge summaries with a positive and current mention of the condition, as well as the accompanying symptom. Relevant algorithms were studied and compared with the help of the pre-exist literature and the results were observed. Accordingly, three algorithms were found to be suitable and then compared with the results generated by them individually. The selected algorithms were Naive Bayes, Decision tree and Random Forest.

Given the list of symptoms a patient is currently showing, our proposed system shall predict the likely diseases using an appropriate machine learning algorithms. Based on this data, we shall apply a set of machine learning algorithms to comparatively evaluate which algorithm performs better among them. The algorithms used are Multinational Naive Bayes Algorithm, , Decision Tree Algorithm, Random Forest Algorithm and which will help us in getting accurate predictions.The data set will be derived from many open source assets available on the research groups like John Hop kin university and Columbia University.

We can detect the chronic kinds of disease in a certain location and population using structured analysis. With the aid of algorithms and methodologies, we choose the characteristics automatically in unstructured analysis. Based on the user's symptoms and prior assets, this algorithm forecasts the condition. It also assists in the constant examination of viral illnesses, heart rate, blood pressure, sugar level, and other information stored in the system, and it forecasts the suitable and correct disease based on those symptoms as well as other exterior symptoms.The system captures the user's symptoms and uses these, as well as prior assets, to forecast the condition. It also aids in the constant examination of viral illnesses, heart rate, blood pressure, sugar level, and other information stored in the system, and it forecasts the proper and correct disease based on those symptoms as well as other exterior symptoms.

Step 1:

Data collection and data set preparation The act of gathering, cleaning, and combining data into a single file or data table, particularly for use in analysis, is known as data preparation.

Step 2:

Developing a probabilistic modeling and machine learning approach Probabilistic Models in Machine Learning is the use of statistical coding to data analysis. The use of probabilistic models to define the world is portrayed as a common idiom.

Step 3:

Training and experimentation on database The first data needed to train machine learning models is known as training data (or a training data set). Machine learning algorithms are taught how to make predictions or perform a task using training assets.

Step 4:

Deployment and analysis Deployment is the process of integrating a machine learning model into an existing production environment in order to make data-driven business decisions. It's one of the last steps in the machine learning process, and it's also one of the most time-consuming.

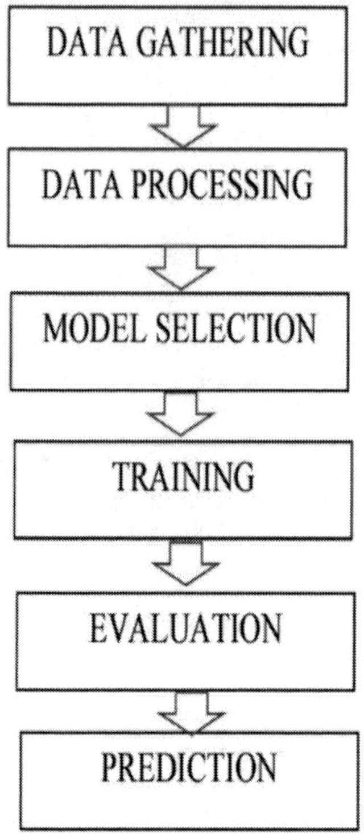

Fig. 1. Contraflow diagram

3.2 Hardware and Software Requirements

Hardware Requirement:

- 4 GB RAM.
- 256 GB HDD.
- Intel 2.8 GHz i3 Processor

Software Requirement:

- Windows 7/8/10
- Python
- Jupyter notebook.

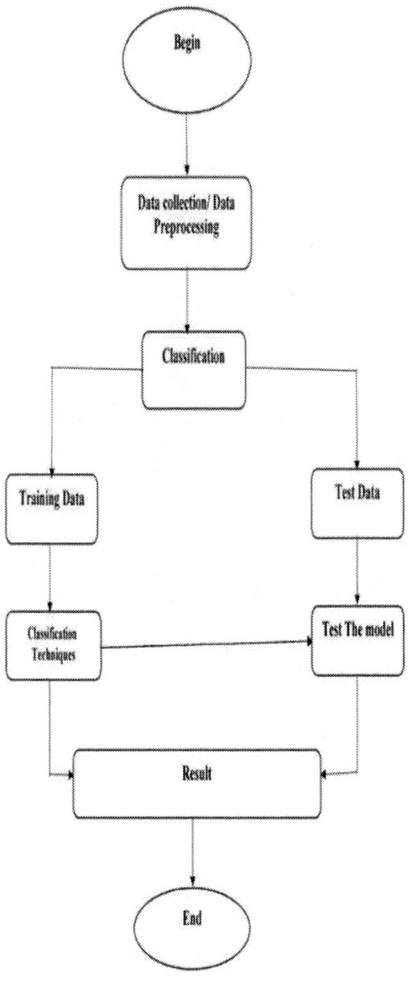

Fig. 2. Work flow of the proposed system

The initial step is to collect the relevant data and process it using the advanced data cleansing techniques. Later, the data set is divided into two divisions, training data and testing data so as to prepare the machine learning model effectively. After the ML model is ready, it is tested using a variety of combinations of symptoms and the result is verified.

4 Experimental Results

4.1 Result and Analysis

Multinomial Naive Bayes

It's a tribe of algorithms and not one algorithm. All naive socio-economic classifiers in Bayes assume that, given the class variable, the value of a certain feature is independent of the value for each other. The Thomas Bayes classifier might be the straight probabilistic classifier that supports the use of theorems with strong naive assumptions of independence (from Bayesian statistics). The Bayes theorem with associated grades of independence between the predictors was supported by classification method. Simply put, the presence of a special feature in an extremely class is assumed by a Naive Thomas Bayes category to be distinct from that of the other.

For example, if this is red, spherical, and around a few inches' diameter, the fruit may also be considered as an association grade apple. Even though those characteristics are interdependent and dependent on other characteristics, a naive classifier in Bayes would regard all these attributes separately to contribute to the likelihood that this fruit is an apple. The class label for each training data set is predicted by these learners. The class label which most of the models predict is voted using the majority voting procedure and the class label is agreed on in the training data set. Regulations are produced from the ensemble models.

The Multinomial Naive Bayes method is a probabilistic learning technique commonly used in Natural Language Processing (NLP). Using the Bayes principle, the algorithm estimates the tag of a text such as an email or a newspaper article. It assesses the likelihood of each tag for each sample and outputs the tag with the highest likelihood.

The Naive Bayes classifier is made up of several algorithms that all have one thing in common: each feature being classed is unrelated to any other feature. The presence or absence of one trait has no bearing on the bother's existence or absence. Accuracy achieved on our dataset: 0.92

Decision Tree

Building classification models using decision tree induction is a non-parametric technique. The challenge of finding the best decision tree is NP-complete. Decision tree construction techniques have been designed to be computationally cheap, allowing models to be built even when the training set size is quite high.

Decision trees, particularly those of a smaller size, are quite simple to read. For learning discrete-valued functions, decision trees give an expressive representation. The presence of noise is not a problem for decision tree algorithms, especially when approaches for diminishing over-fitting are used.

Decision tree induction is a non-parametric approach for creating classification models. Although constructing an optimal decision tree is an NP-complete problem, computationally affordable ways for creating decision trees have been identified, allowing models to be generated even when the training set size is relatively large.

Decision trees, especially ones of a smaller size, are easy to understand. Decision trees provide an expressive model for learning discrete-valued functions. Noise is not a concern for decision tree algorithms, especially when over-fitting minimization techniques are applied. On our data set, we attained an accuracy of 0.97.

Logistic Regression

Logistic Regression is a Machine Learning technique for resolving categorization problems. It's a type of predictive analytic approach based on the probability concept. To forecast the probability of a categorical dependent variable, the classification algorithm Logistic Regression is utilized. In logistic regression, the dependent variable is a binary variable with data coded as 1 (yes, True, normal, success, etc.) or 0 (no, False, abnormal, failure, etc).

The purpose of Logistic Regression is to find a relationship between a set of attributes and the probability of a particular event. A Logistic Regression model is similar to a Linear Regression model, except that instead of using a linear function, it uses a more complicated cost function known as the "Sigmoid function." The accuracy achieved on our data set: 0.89

4.2 Dataset

It is a set of disease-symptom connections developed by an integrated strategy using data from written clinical notes of admitted patients to New York Presbyterian Hospital in 2004. Depending on these notes, the most common illnesses were calculated, and the symptoms were graded based on the strength of relationship. Every sickness has a clinical manifestations that can be used to forecast the condition.

Table 1. Sample dataset with symptoms with the diseases

Symptoms	Disease
Runny nose, sore throat, cough, congestion, body aches, headaches, sneezing, fever	Common cold
Fever, profuse sweating, headache, nausea, vomitting, diarrhoea, anemia, muscle pain, convulsions	Malaria
Poor appetite, abdominal pain, headaches, generalized aches, pains, fever, lethargy, intestinal bleeding or perforation, constipation	Typhoid

In the same manner as shown in the above table, the symptoms are related with the associated disease. That is, the similar set of symptoms are grouped together and linked with a particular disease to indicate that the individual must be suffering from that disease if the symptoms experienced are matching with the dataset. By this method, the disease is predicted based on the symptoms which are taken as the input from the patient. The model is equipped to predict a plethora of diseases from the combination of symptoms, the diseases that can be predicted are fungal infection, allergy, peptic ulcer disease, diabetes, gastroenteritis, bronchial asthma, hypertension, jaundice, malaria, chickenpox, typhoid, dengue, tuberculosis, common cold, pneumonia are some of them.

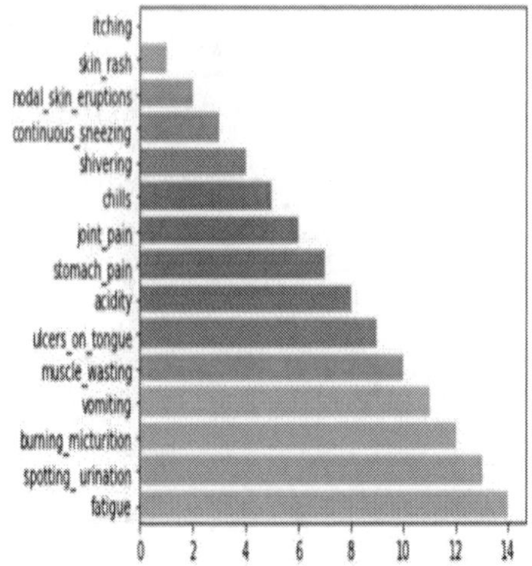

Fig. 3. Occurrence of the symptoms in the test data set

The above figure demonstrates the frequency of the symptoms occuring in the data set and shows which symptom is associated with the maximum of the diseases. The data set contains various diseases and the symptoms associated with them. The model predicts the disease which matches with the symptoms in the data set with the set of symptoms entered by the user. With this methodology, we can predict the disease which is most likely associated with the set of symptoms already present in the data set.

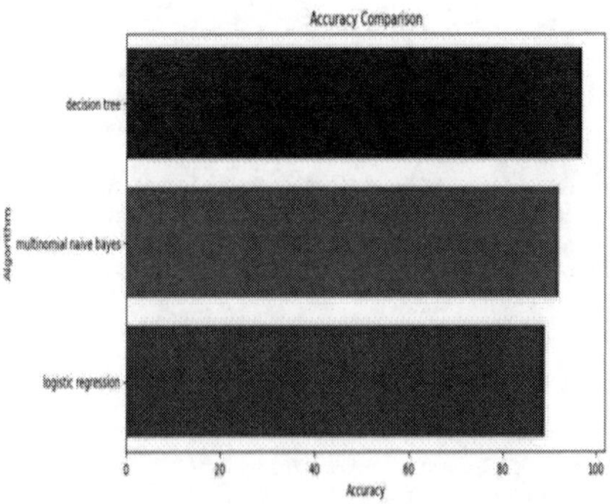

Fig. 4. Accuracy comparison of the algorithms

The accuracy comparison shown in the Fig. 4. portray the comparison between the three algorithms - Decision

Tree, Multinomial Naive Bayes and Logistic Regression plotted against a scale of accuracy percentage. The data set which was collected from an authorized source was then made suitable for use of the machine learning algorithms by data cleaning. The model was developed for Multinomial Naive Bayes algorithm, Decision Tree algorithm and Logistic Regression algorithm. The accuracy of Multinomial Naive Bayes on the chosen data set was 92%. As for Logistic Regression, it was merely 89%.The highest accuracy among all these algorithms was demonstrated by Decision Tree which was 97%.

5 Conclusion

Because medical data is growing at an exponential rate, it is necessary to process existing data in order to predict precise disease based on symptoms. Several general disease prediction systems based on machine learning algorithms such as Decision tree, Logistic Regression, and Multinomial Naive Bayes have been proposed to classify patient data. Sickness and risk prediction may be done in a short amount of time and at a minimal cost using this method. To summarise, people who are always concerned about their health will benefit from our method. Our objective is to develop this system so that people are more aware of their health issues and may lead a healthier lifestyle. Furthermore, depending on the user's symptoms, our system will provide more accurate sickness forecasts, as well as inspiring thoughts and graphics. As a consequence, we can say that our system has no limitations because it can be used by everyone.

References

1. A.S. Monto, S. Gravenstein, M. Elliott, M. Colopy, J. Schweinle, Clinical signs and symptoms predicting inuenza infection, Archives of internal medicine 160(21), 3243 (2000)
2. B. Qian, X. Wang, N. Cao, H. Li, and Y.-G. Jiang, "A relative similarity based method for interactive patient risk prediction," Springer Data Mining Knowl. Discovery, vol. 29, no. 4, pp. 1070– 1093, 2015.
3. D.R. Langbehn, R.R. Brinkman, D. Falush, J.S. Paulsen,M. Hayden, an International Huntington's Disease Collaborative Group, A new model for prediction of the age of onset and penetrance for huntington's disease based on cag length, Clinical genetics 65(4), 267 (2004)
4. S. Grampurohit and C. Sagarnal, "Disease Prediction using Machine Learning Algorithms,"2020 International Conference for Emerging Technology(INCET),2020, pp. 17, doi: 10.1109/INCET49848.2020.9154130.
5. D. Dahiwade, G. Patle and E. Meshram, "Designing Disease Prediction Model Using Machine Learning Approach," 2019 3rd International Con- ference on Computing Methodologies and Communication (ICCMC), 2019, pp. 1211-1215, doi: 10.1109/ICCMC.2019.8819782.
6. Automatic Heart Disease Prediction Using Feature Selection And Data Mining Technique Le Ming Hung,a, Tran Ding,Journal of Computer Science and Cybernetics, V.34, N.1 (2018), 3347 DOI: 10.15625/1813- 9663/34/1/12665
7. International Journal of Scientific Research in Computer Science, E., & IJSRCSEIT, I. T. (2019). Generic Disease Prediction using Symptoms with Supervised Machine Learning. International Journal of Scientific Research in Computer Science, Engineering and Information Technology. https://doi.org/10.32628/CSEIT1952297.
8. Pingale, Kedar, et al. "Disease Prediction using Machine Learning." (2019).Mr. Chala Beyene, Prof. Pooja Kamat, "Survey on Prediction and Analysis the Occurrence of Heart Disease Using Data Mining Techniques", International Journal of Pure and Applied Mathematics, 2018.
9. S. Patel and H. Patel, "Survey of data mining techniques used in healthcare domain," Int. J. of Inform. Sci. and Tech., Vol. 6, pp. 53-60, March, 2016.
10. Balasubramanian, Satyabhama, and Balaji Subramanian. "Symptom based disease prediction in medical system by using Kmeans algorithm." International Journal of Advances in Computer Science and Technology 3.
11. Dhenakaran, K. Rajalakshmi Dr SS. "Analysis of Data mining Prediction Techniques in Healthcare Management System." International Journal of Advanced Research in Computer Science and Software Engineering 5.4 (2015).
12. T. Vivekanandan and N. C. S. N. Iyengar, ''Optimal feature selection using a modified differential evolution algorithm and its effectiveness for prediction of heart disease,'' Comput. Biol. Med., vol. 90, pp. 125–136, Nov. 2017.
13. A. Gavhane, G. Kokkula, I. Pandya, and K. Devadkar, ''Prediction of heart disease using machine learning,'' in Proc. 2nd Int. Conf. Electron., Commun. Aerosp. Technol. (ICECA), Mar. 2018, pp. 1275–127

Secure asset transactlon using blockchain

Saarabi Parkar[1], *Pavitra* Venkatraman[2], *Vikrant* Pawar[3], and *Mrinal* Khadse[4]

[1]*SIES Graduate School of Technology, Information Technology Department, Navi Mumbai, India*
[2]*SIES Graduate School of Technology, Information Technology Department, Navi Mumbai, India*
[3]*SIES Graduate School of Technology, Information Technology Department, Navi Mumbai, India*
[4]*SIES Graduate School of Technology, Information Technology Department, Navi Mumbai, India*

Abstract -- The current property exchange methodology includes a lot of susceptibilities and people use it to delude the common people and the government. Using blockchain, this project utilises majority consensus to implement secure asset transactions. The main objective behind the implementation of asset transaction using blockchain is to secure property transactions from vulnerabilities and to easily maintain records. The proposed solution aims at avoiding scenarios related to the same property being sold multiple times to different clients. Since very little documentation and record-keeping effort are required, the time required for performing a transaction is also less. The asset transaction will consist of n nodes which will be responsible for verifying a transaction and adding new blocks to the blockchain. The proposed implementation of the secure asset transaction using blockchain thus offers a reduction in manual effort spent in record keeping.

Index Terms— Asset transaction, Blockchain, IPFS, Smart Contract.

1. Introduction

Asset transaction in India as well as in many parts of the world is a very slow and tedious process[1]. There are also many intermediaries involved in the process of property transactions. Our aim with this project is to implement blockchain to develop a system that not only speeds up the process of property transaction, but also make it easier for various parties involved in an agreement to transfer the asset ownership from seller to a new buyer, by creating a distributed system that stores all the details of the transactions made during the process of property exchange.

BLOCKCHAIN TECHNOLOGY: Blockchain technology offers a shared record-keeping which is designed to be difficult to tamper with. Blockchain technology functions through decentralised peer-to-peer platforms, building a firewall against the spread of corrupted information and boosting resistance to fraud[2]. Blockchain technology has impacted the real estate industry in a variety of ways, by offering a new avenue for buyers and sellers to interact with one another. Through the use of blockchain technology, real estate transactions could be made cheaper by eliminating middlemen.

Removing the MiddleMan with Machine Consensus: A peer-to-peer network of computers running the blockchain protocol validate the transactions by consensus instead of a single trusted third party validating them through their servers with authority[3]. By formalising consensus rules for validating transactions on the P2P network, the blockchain protocol facilitates management and auto enforcement of transactions between participants.

Smart Contracts: Smart contracts run when certain conditions are met within a blockchain. Contracts run automatically when certain conditions are met. They appear as a contractual agreement but should not be confused with legal contracts. On a blockchain, smart contracts are simply programs that are triggered by preset conditions. Therefore by making use of smart contracts, various conditions can be put into place and upon whose satisfaction only the transaction can move ahead. Also, the use of smart contracts speeds up the procedure of reflecting the new ownership by updating the record automatically. Simple economic transactions like sending money from A to B can be implemented using smart contracts. In addition to registering ownership and property rights, smart access control can also be used for managing sharing economy services.

2. Literature survey

Krishnapriya S et al [4] proposed a seamless, easy to use and hassle-free platform which can be used for making the land registration easy. There are many problems such as involvement of brokers or middlemen, time delays, etc. This platform will eliminate the problems associated with land registration in India as well as in many parts of the world. Blockchain technology is emerging very rapidly due to the secure features it offers. Hence using blockchain to save the land record transaction is the way to create the immutable records.

[5] Dealt with different possible applications of blockchain like health care, IoT, supply chain etc. along with their security and privacy challenges.

[6] Defined the protocol of Ethereum. A node can be implemented by the reader on the ethereum network using this protocol. By implementing this protocol, readers can join a decentralised, secure, and social operating system that is built on the Ethereum network.

[7] The consensus algorithms that are used in Blockchain provides the valuable computation thus allowing the transaction by means and verification and validation. These algorithms might be required to compute more which requires a huge amount of computing power. These algorithms work the best when the blocks in the network are limited in number

[8] With smart contracts, Blockchain can be used to provide a decentralised/ distributed environment where multiple parties can interact with each other without any trusted third party.

IPFS [9] is a distributed storage system that allows users to access files, websites, applications, and data. All the file logs collected from the IPFS are embedded in the blockchain as transactions.

3. Proposed system

The aim behind this project is to make the process of asset transactions efficient, transparent and also decrease the cases of fraud and malpractices in the process. This system simulates the process of buying and selling a property, using blockchain and smart contract technology. A smart contract is defined which includes identifying the roles that are involved in our asset transaction and the different components and transactions in the agreement process.

4. Objectives

a. To transform core CRE (Commercial Real Estate) operations of property transactions like purchase, sale and management transactions from the traditional approach to the blockchain technology.

b. Accelerating the Process: This platform can be used to create a distributed database where anybody can record and access information without having to rely on a central authority.

c. Reducing Fraud Cases:
In a blockchain asset transaction platform, you can upload the title documents onto the blockchain network and other parties can verify them as needed. Blockchain can prove the ownership of the property and prevent forgery of documents by keeping an immutable record of transactions.

d. Bringing Efficiency with Smart Contracts:
Smart contracts improve efficiency by automating verified transactions. As an owner or buyer, one can create a decentralized, digital ID using this blockchain platform. Using the blockchain platform, ownership transfers would be more seamless and quicker than using traditional methods.

e. To avoid the same property being sold twice by a user.

5. Modelling

Smart Contracts are the fundamental building blocks on which the system is based. All the traditional processes pertaining to asset transactions have been replicated using smart contracts. These contracts trigger all events related to asset transactions. Smart contracts will be used to automate the transaction. Each block will consist of a block number`, hash of previous block, Property details, cost. These contracts, written in Solidity, have been tested and compiled on Remix IDE.

Web3.js acts as the connecting link between the web application and blockchain. Ganache, which is a testing environment, is used to set up a private blockchain which is used to test smart contracts where one can develop applications, deploy contracts, run tests and perform other tasks without any cost.

Following is the list of platforms used throughout the process -

a. Truffle: It is a framework used for deploying the contracts, truffle automatically provides the structure for storing the contracts and migrations folders for connecting it with mainnet and testnet

b. Metamask: Metamask is a crypto wallet- it allows users to store and transact Ethereum. It allows users to access their Ethereum wallet through a browser extension or mobile app, which can then be used to interact with decentralized applications.

c. Remix IDE: Remix IDE (Integrated Development Environment) is a web application that can be used to write, debug, and deploy Ethereum Smart Contracts[10].

d. Ganache: Ganache is a testing environment which is used for testing the functionality of smart contracts and provides a private blockchain. It is used to test the smart contracts as it provides 10 accounts with fake ether.

6. Result

The owner will upload all the details about the property along with its document. The listed property will be verified by an Admin. Once the documents are verified, the buyers will be able to see the displayed properties. The buyer can request to access the documents and after approval by the owner a transaction will happen.

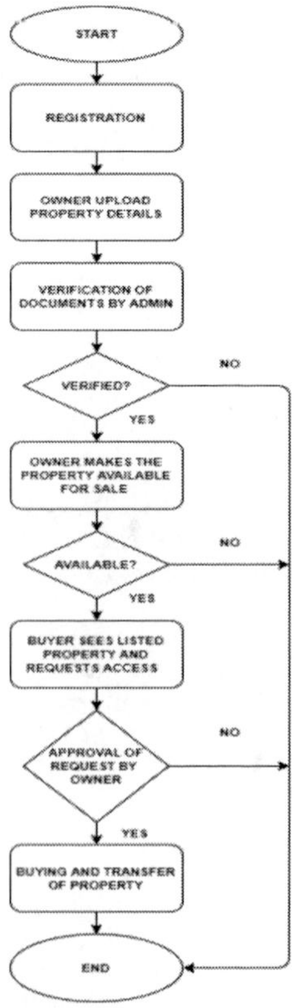

Fig 1: flow chart of asset transaction using blockchain

The skeleton of the project in detail will be along these lines:

Step 1: User registration
Buyer and owner both have to register on the web app. Users will register to the platform using their credentials. Once registered, the user will be redirected to the login page where they'll login using their registered Metamask account address. A user, say Pavitra, registers themselves into the webapp using a unique Metamask account address. The Metamask account consists of digital currency which is used for transactions and gas fees.

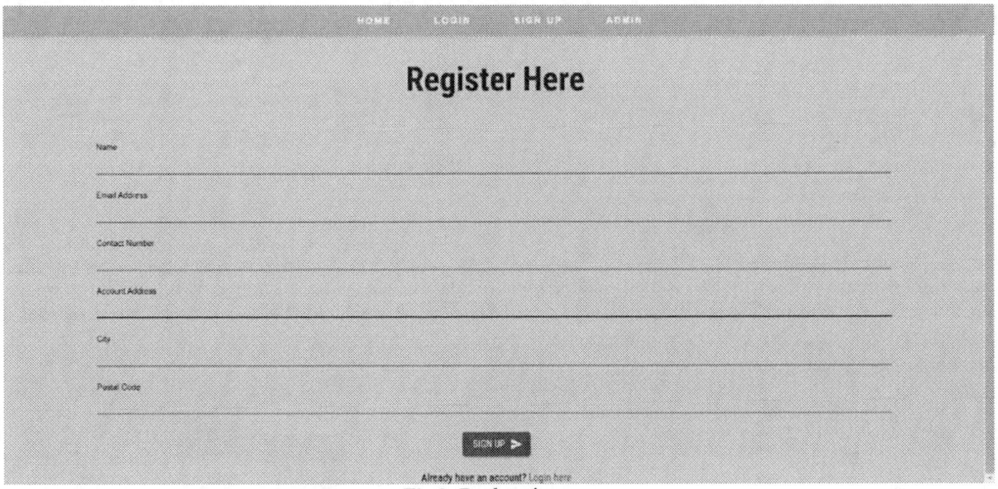

Fig 2: Registration page

Step 2: Property listing
After the user gets registered, they have the option to either list properties or purchase properties. Once the property gets listed, every consequent action along with the property detail will get recorded on the Blockchain nodes. If an owner logs in, they'll be asked to list out all the details pertaining to their property like name, description, price in ether, PId etc. Also, documents and images of the property have to be uploaded for verification. All these details will be stored in the Blockchain ledger.

Fig 3: Listing of property

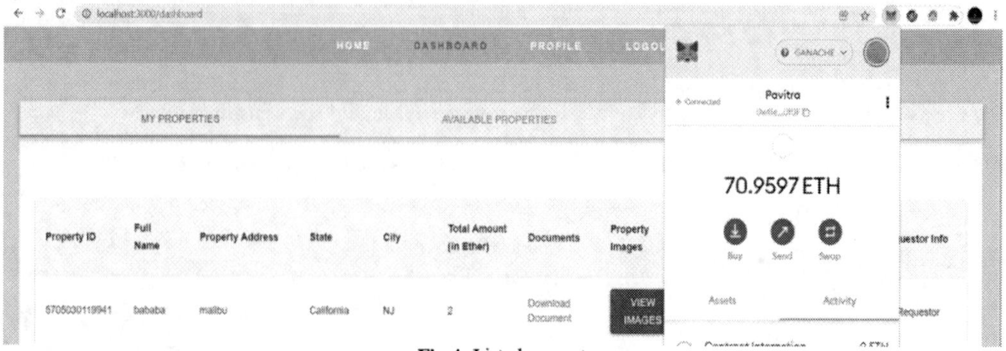

Fig 4: Listed property

Step 3: Admin Verification

An admin account will verify the property details/documents. If there is no discrepancy, the admin gives a go ahead to the owner to list the property for sale. If the property gets rejected by the admin for some reason, the property will not be visible in the 'Available property' section. The listed property gets loaded onto the blockchain irrespective of its status.

Fig 5: Before admin approval

Fig 6: After Admin approval

Step 4: Request to purchase

Once verified by the Admin, the owner makes the property available for sale. The logged in buyer will be able to see all the listed properties and accordingly make a request.

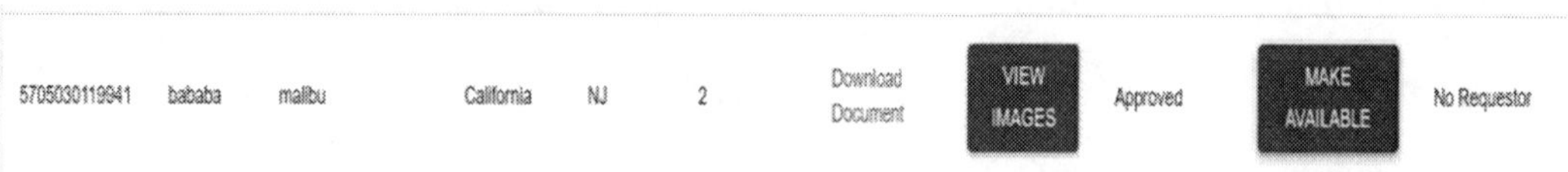

Fig 7: Making property visible for buyers

 https://doi.org/10.1051/itmconf/20224403009

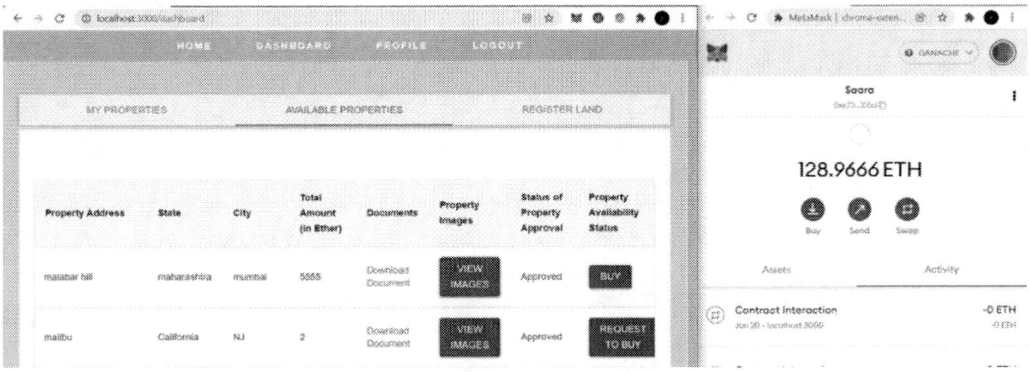

Fig 8:` Property is visible in the 'Available Properties' section of the dashboard.

Step 5: Ownership transfer

The owner has the choice whether to accept a request or not. If the request goes ahead, the buyer will send the required ether to the owner and subsequently the ownership will get transferred. As shown in fig 6.10, the property is now transferred to the buyers 'My Properties' tab from the owners 'My Properties' tab.

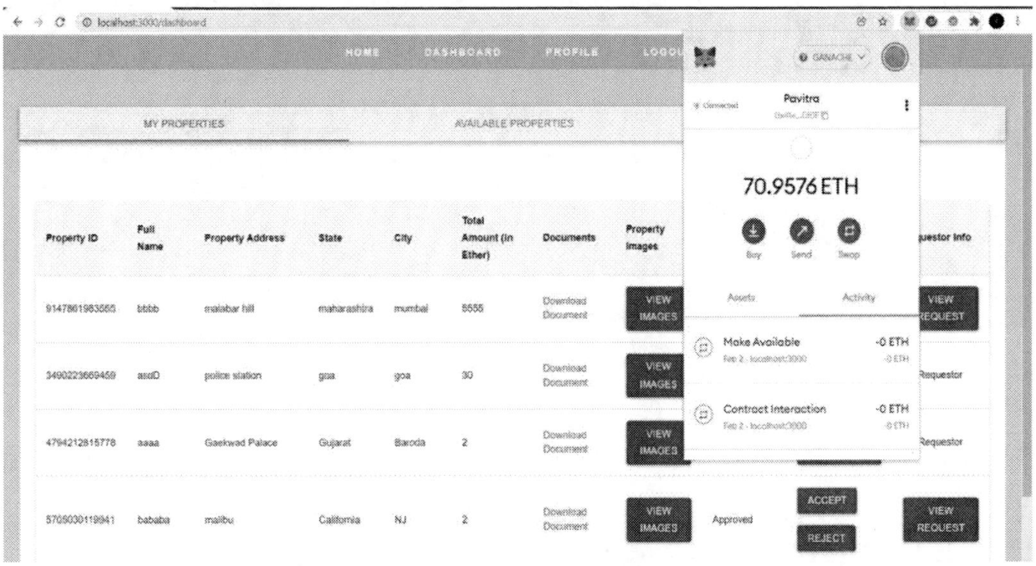

Fig 9: Acceptance/Rejection of request by owner

Fig 10: Payment

Fig 11: Ownership transfer

7. Conclusion

The traditional system is prone to various types of tampering at every stage and indirectly affects the costing in the form of paper resources, storage requirements of vast record keeping, and security issues of these records. The system is also time-consuming and requires much time in the verification and updating process that give rise to bribery and chances of selling the same piece of property simultaneously to more than one buyer increases.

Blockchain technology has the potential to counter all these issues. The proposed system is very economical, as it requires very less human resources and is more reliable. If this system is upgraded further and integrated with useful APIs then it will lead to faster transactions and eventually ease the entire process, thus making the system more hassle free and convenient.

References

[1] Oprunenco A, Akmeemana C. Using blockchain to make land registry more reliable in India. LSE Business

[2] Eric Piscini (Deloitte U.S.), David Dalton (Deloitte Ireland) and Lory Kehoe (Deloitte Ireland), "Blockchain & Cyber Security"

[3] Cong T. Nguyen; Dinh Thai Hoang; Diep N. Nguyen; Dusit Niyato, Huynh Tuong Nguyen; Eryk Dutkiewicz "Proof-of-Stake Consensus Mechanisms for Future Blockchain Networks - Fundamentals, Applications and Opportunities," 2019, accepted June 17, 2019, date of publication June 26, 2019, date of current version July 16, 2019. Digital Object Identifier 10.1109/ACCESS.2019.2925010

[4] S, Krishnapriya & Sarath, Greeshma. (2020). Securing Land Registration using Blockchain. Procedia Computer Science. 171. 1708-1715. 10.1016/j.procs.2020.04.183.

[5] Mohanta, Bhabendu & Jena, Debasish & Panda, Soumyashree & Sobhanayak, Srichandan. (2019). Blockchain Technology: A Survey on Applications and Security Privacy Challenges. 8. 100107. 10.1016/j.iot.2019.100107.

[6] Dr. Gavin Wood: ETHEREUM- A secure decentralized transaction generalized ledger

[7] N. Chaudhry and M. M. Yousaf, "Consensus Algorithms in Blockchain: Comparative Analysis, Challenges and Opportunities," 2018 12th International Conference on Open Source Systems and Technologies (ICOSST), 2018, pp. 54-63, doi: 10.1109/ICOSST.2018.8632190.

[8] B. K. Mohanta, S. S. Panda and D. Jena, "An Overview of Smart Contract and Use Cases in Blockchain Technology," 2018 9th International Conference on Computing, Communication and Networking Technologies (ICCCNT), 2018, pp. 1-4, doi: 10.1109/ICCCNT.2018.8494045.

[9] Rajalakshmi, A. & Lakshmy, K.V. & Sindhu, M. & Amritha, P.P.. (2018). A blockchain and IPFS based framework for secure Research record keeping. International Journal of Pure and Applied Mathematics. 119. 1437-1442.

[10] Remix - Solidity IDE. (2018). Remix.ethereum.org. [Online]. Available: https://remix.ethereum.org/, Dated: 15/01/2022.

[11] B. Kothari, S. Mudaliar, S. Nadar and M. Khadse, "Securing IoT with Blockchain," 2020 International Conference on Inventive Computation Technologies (ICICT), 2020, pp. 785-788, doi: 10.1109/ICICT48043.2020.9112526.

Natural disaster detection in social media and satellite imagery

Stuti Ahuja[1], *Melvina Michael*[1,*], *Malvika Selvan*[1] and *Vaishnavi Mantri*[1]

[1]Department of Information Technology, SIES Graduate School of Technology, Navi Mumbai, India

Abstract. Natural disasters caused by natural processes may lead to significant losses in terms of property and human lives. The timely collection of information about the damage caused by natural disasters is very important and can help reduce losses and speed recovery. Social media has become an important source of information for communication and dissemination of information in emergencies. Under such circumstances, inferring disaster events through the information available in social media will be very useful Satellite data has also been widely used to analyze the impact of natural disasters on the surface of the earth. In this paper, a detailed analysis of how social media and satellite imagery can be used to detect natural disasters is discussed.

1 Introduction

Natural disasters are fatal events caused by natural or natural processes in the world. The severity of disasters is measured in terms of lost lives, financial losses, and thus the ability of the population to rebuild. All natural disasters cause damage and loss to the environment and the people who live there. Natural disasters include floods, earthquakes, tsunamis, landslides, volcanic eruptions, and hurricanes. An example of a flood-affected area is shown in fig.1.

The degree and extent of damage can be further divided into three categories.
Minor disasters: Minor disasters are disasters that extend over 50km or up to 100 km. Fires can be counted as minor disasters.

1 Corresponding author: pushpammit118@gst.sies.edu.in

Fig. 1. Flood-affected area

Medium-scale disasters: Medium-scale disasters range from 100km to 500km. These do more damage than a minor catastrophe. Erosion, landslides, tornados, etc can be considered medium-scale disasters.

Catastrophes: These disasters cover an area of more than 1000 km. These cause the most serious environmental damage. Moreover, at a high level, these disasters can even involve the country. Earthquakes, floods, tsunamis, etc can be considered a large-scale disasters.

In such disaster events, satellite imagery and social media emerged as key sources of information-gathering tools. Most of the times researchers and disaster relief agencies focus on these sources for information. Satellite images serve as a practical solution to analyze the impact created by disaster over the surface of the Earth. Remote sensing data also has some challenges since sometimes it has the low temporal frequency of satellite images, and oftentimes it only gives a bird's-eye view of an event. For social media we divide the detection into two parts -i) Text ii) Images.

For both the above detection we use texts and images from Twitter. But the relevance and credibility of content shared via social media are one of the biggest challenges involved. For the above detection, we use three modules mainly data extraction, sorting, and analysis.

2 Literature Survey

We examined and read the research papers listed below to gain additional information and ideas for our project's execution.

Naina Said, Kashif Ahmad, Michael Riegler, Konstantin Pogorelov et al. [1] proposed natural disaster data extraction from social media platform(Twitter) and satellite imagery; Disaster Detection in Satellite Imagery CNN-based framework for the detection of natural disasters in satellite imagery, MediaEval for flood detection, RF classifier is used for the classification of the satellite image patches into flooded and non-flooded regions and VGGNet model. The next one is Data Extraction from Twitter using Geo-tagged filter, K-Nearest Neighbour (K-NN) algorithm, Natural Language Processing (NLP) Bayesian approach for the identification and classification of disaster-related tweets, Support Vector Machines (SVMs), Ashktorab et al proposed Tweed, a text-mining tool for the extraction of useful information from tweets during natural disasters. Imran et al. proposed a platform, called AIRD. The last method Disaster Detection in images from Social Media, User tags, geolocation has been proved effective, HSV low-level color feature, SPCPE, Textual features are extracted through word frequency and CNN models on IMAGENET are used as the feature descriptor.

Sheharyar Ahmad1, Kashif Ahmad, Nasir Ahmad, Nicola Conci et al. [2] proposed A system called "JORD" has been introduced to automatically collect information from various social media platforms and combine it with remote sensing data to provide a more detailed idea of a disaster and they have also discussed CNN, Imagenet.

Olga Ostroukhova, Pal, Halvorsen, Nicola Conci, Rozenn Dahyot et al [3] proposed an active learning framework intending to collect, filter, and analyze social media contents for natural disasters, For data collection, a publicly available system, namely AIDR, has been used to crawl social media platforms a crowd-sourcing activity for data annotation.

Angela Maria Vinod Dharathi Venkatesh Dishti Kundra Jayapandian N [4] focused on using cognitive computing for industrial purposes and this study examined the proposed predictive models, specifically using ANN (Artificial Neural Networks), sentiment models, and smart disaster prediction application (SDPA) to forecast the flash flood.
Jigar Doshi Saikat Basu Guan Pang et al [5] proposed a framework using CNN to detect which areas were most affected by disaster, and they also used the Disaster Impact Index to quantify the impact of two natural disasters.
Siti Nor Khuzaimah Binti Amit Soma Shiraishi Tetsuo Inoshita et al [6] proposed an automatic disaster detection system by implementing one of the advance deep learning techniques, convolutional neural network (CNN), to analyze satellite images. They created their own disaster detection training data patches. The results reveal an accuracy of 80%~90% for disaster detection.
The results which were obtained presented may help in facilitating detecting natural disasters efficiently by developing an automatic disaster detection system.

3 Existing System

The existing system categories disaster detection of available content into three groups: I disaster detection in text, (ii) analysis of disaster-related content from social media images. (iii) Detection of disasters using satellite imagery

In the traditional system, the relevance and authenticity of content shared via social media is always a massive obstacle for applications aimed at disaster detection and analysis of disaster-related data available in social media. It lacks large-scale annotated datasets for training and evaluating machine learning techniques for disaster analysis in tweets, images from social media, and satellite imagery. Due to the low temporal frequency of satellite imagery, satellite imagery is not frequently available. Furthermore, the quality of the satellite imagery may be compromised.

4 Proposed System

We have proposed a model wherein three different factors are considered for the analysis of how data from satellite and social media can play an important role in disaster situations Communication and dissemination of messages. The three approaches taken into consideration are i)Detection from Satellite imagery ii) Detection from Social media text iii) Detection from Social media images.

Dataset:

Three different datasets are used for the detection of natural disasters.

i) Satellite Imagery: Here we have a set of images consisting of pair of satellite images wherein the first image is taken before the disaster event and the next is taken after the incident.

ii) Social Media Text: Here we have text and location related to disaster events.

iii) Social Media Images: We have a set of 5 different disaster-related data, which are flood, earthquake, volcanic eruption, fire, and some random images.

For all three groups, the data is divided into training and testing data.

The proposed system have the following modules listed below: data extraction and analysis. The data extraction/collection is responsible for data collection and pre-processing (cleaning the retrieved data). The next module analyzes the data to extract useful information about the disaster scope, distribution, Geo-tagging, occurrence frequency, and sentiment rating. The approach proposed relies on the use of matching keywords and hashtags for the identification of relevant messages from the social media stream using Natural language processing (NLP) trained on a diverse collection of linguistic characteristics including raw and lexical textual features. For extracting images from social media, we have used the CNN algorithm, wherein we have used the datasets, in which we have samples of 5 categories they are: earthquake, flood, fire, volcanic eruption, and some random images. We are training our model in such a way that it predicts the disaster from the image provided to test. For satellite imagery, the main aim is to find the dissimilarity between a pair of satellite images of the same region i.e. flooded and non- flooded. If the dissimilarity between the provided images is high then it means, that particular region is flooded else the region is non-flooded.

Let's look at the above three approaches:

A. Satellite Imagery

The motive here is to find the dissimilarity between a pair of satellite images of the same particular region which are taken before and after the incident. The input taken here is as a pair of images, in which the first image is taken before the incident occurred and the other image is taken after the incident. Now both the images are compared using the Siamese network as shown in fig. 2. Siamese network is a class of neural networks containing two or more sub-network which are identical. Siamese network is trained using PyTorch, Considering the pair of images, the first image is passed through the network and then the second image is passed and then the loss is calculated considering the output coming from the first and second image and the gradient is calculated. To test the model the images are passed through the network and the distance is calculated from the output and then the sigmoid is calculated. Then dissimilarity is calculated, if it's high then the result is declared as the region is flooded and if the dissimilarity is low then the region is declared as not flooded.

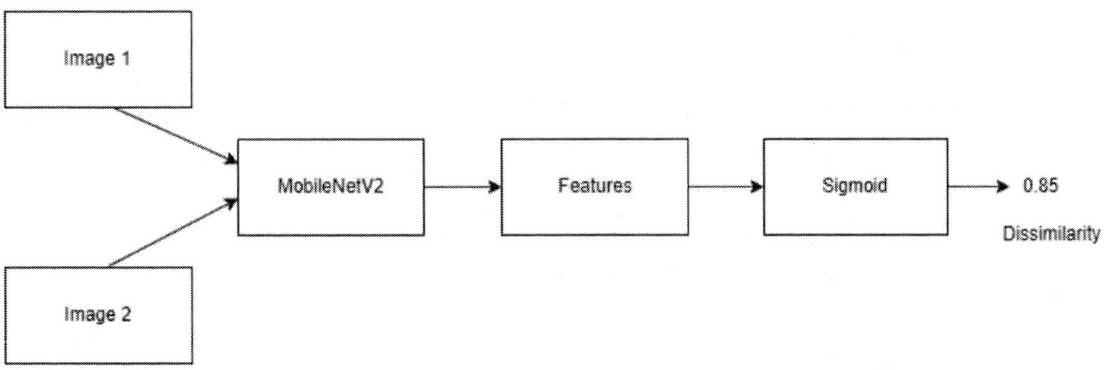

Fig. 2. Flowchart for Disaster detection using Satellite Imagery

A. Social Media Images

For social media images, we have used the CNN algorithm and Keras call-backs a Python interface for artificial neural networks. CNN is a deep learning algorithm that consists of convolution layers for extracting features from the given network as shown in fig. 3. The approach is that we have taken as a set of 5 datasets wherein there are disaster-related images and they are as follows flood, earthquake, fire, volcanic eruption, and random images. Firstly we get a sample batch of 25 images and train the images in epochs. We have used batch normalization and dropouts to avoid overfitting. The Keras callbacks control the learning rate of the model. After the model gets trained to the dataset, we provide the image to the model and it predicts whether it is a disaster image or not along with the name and accuracy. The feature we have used here is the early stopping method wherein we can set the number of epochs and we can see that whether the learning rate is improved or not then we can stop the model at that particular stage. The learning rate is reduced for the next epoch of training, once that level of accuracy is achieved then call-back switches. The confusion matrix is used over here to evaluate the performance of the training model.

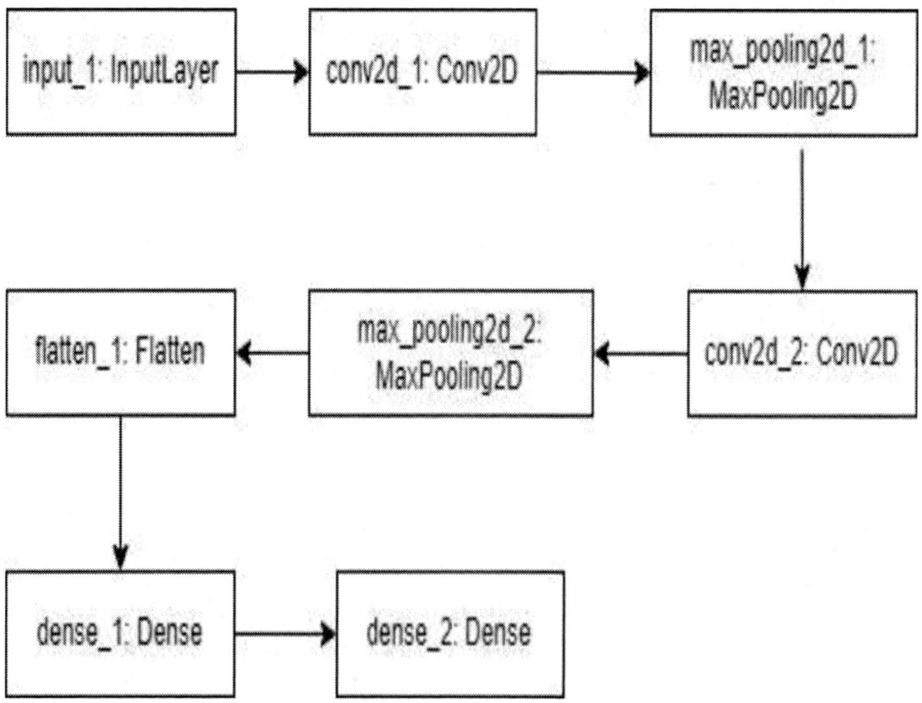

Fig. 3. CNN Architecture

B. Social Media Text

In Twitter Text analysis we are analyzing the text and then trying to differentiate between disaster twitter text and non- disaster twitter text. The input taken here consists of 4 attributes the first attribute is the keyword of the Twitter text the 2nd attribute is a location from where the twitter text is written and the next attribute is the text apart from these attributes, we have the target attribute. The target attribute has values as 0 and 1. Then we are creating a separate column that tells us whether the text is disaster-related or non-disaster According to the text values. We classify the value 0 as Non-Disaster text and value 1 as Non-Disaster related text. We have plotted attributes in the form of a pie chart to classify between Disaster text and non-disaster text. We have also tried to find the most common keywords in the disaster data. The classification is also done based on the length of the tweets and location values. As the Twitter text might contain certain emojis and unnecessary punctuations. For that, we have used pattern matching we have first done data cleaning. Then with the help of the Natural Language Toolkit (NLTK), It is a set of libraries for Natural Language Processing. We have also used Stop words. We have also used Snowball Stemmer. With the help of Word cloud, we have displayed the Words that have been used most in Disaster Tweets and Non-Disaster Tweets. Here we have also tried to find out to what extent do the keywords overlap between the data sets and for that, we have used the basic Venn diagram structure. The final result is stored in an Excel sheet differentiating both types of tweets. For testing purposes, we have used 2 algorithms Logistic Regression and Random Forest Classifier. Logistic regression is a statistical model. The Accuracy of the model has also been calculated for Logistic Regression it is 90% and for Random Forest Classifier it is 98% . So Random Forest Classifier shows better accuracy in consideration with Logistic Regression.

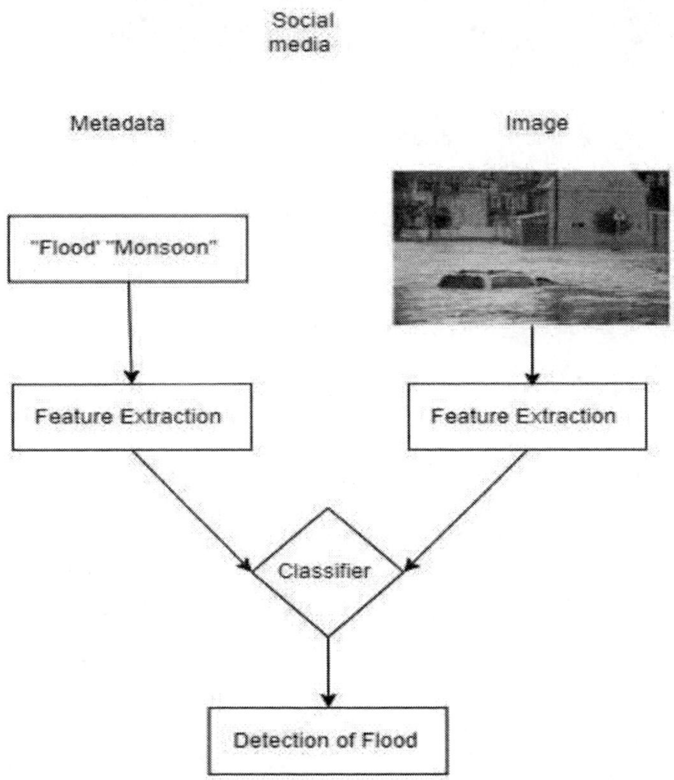

Fig. 4. Flowchart for Disaster detection using Social Media

5 Results

The implementation of our project has been shown below:

1) Satellite Imagery: Fig. 5 shows the result of disaster detection using satellite imagery. The dissimilarity, If the dissimilarity between the provided images is high then it means, that a particular region is flooded else the region is non-flooded.

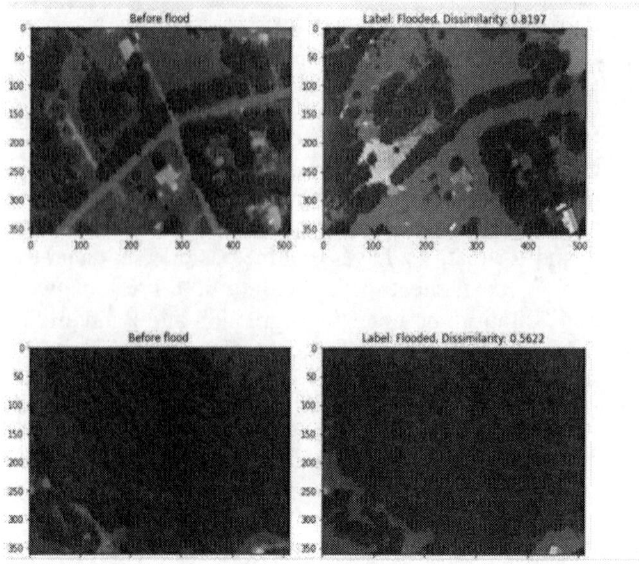

Fig. 5. Disaster detection using Satellite Imagery

2) Social Media Text: Table 1 shows the result of disaster detection using social media text.

Table 1. Disaster detection using Social Media Text

	id	keyword	location	text	target	class	length
0	1	NaN	NaN	Our Deeds are the Reason of this #earthquake M...	1	Disaster	13
1	4	NaN	NaN	Forest fire near La Ronge Sask. Canada	1	Disaster	7
2	5	NaN	NaN	All residents asked to 'shelter in place' are ...	1	Disaster	22
3	6	NaN	NaN	13,000 people receive #wildfires evacuation or...	1	Disaster	8
4	7	NaN	NaN	Just got sent this photo from Ruby #Alaska as ...	1	Disaster	16
5	8	NaN	NaN	#RockyFire Update => California Hwy. 20 closed...	1	Disaster	18
6	10	NaN	NaN	#flood #disaster Heavy rain causes flash flood...	1	Disaster	14
7	13	NaN	NaN	I'm on top of the hill and I can see a fire in...	1	Disaster	15
8	14	NaN	NaN	There's an emergency evacuation happening now ...	1	Disaster	12
9	15	NaN	NaN	I'm afraid that the tornado is coming to our a...	1	Disaster	10
10	16	NaN	NaN	Three people died from the heat wave so far	1	Disaster	9
11	17	NaN	NaN	Haha South Tampa is getting flooded hah- WAIT ...	1	Disaster	27
12	18	NaN	NaN	#raining #flooding #Florida #TampaBay #Tampa 1...	1	Disaster	12

3) Social Media Images: Fig. 6 shows the result of disaster detection using social media images. Here it predicts the image is of which class with its probability.

Fig. 6. Disaster detection using Social Media Image

6 Conclusion

This research proposes a method for detecting natural disasters using social media and satellite imagery. Different approaches which are used for detecting disasters in satellite images, social media text, and social media images are discussed in detail. Results of implementation of all three techniques are shown and these results clearly show the effectiveness of the proposed method for disaster detection. The above techniques is yet to be combined. The above models do not retrieve real-time data; this will be worked upon hereafter.

References

1. Naina Said, Kashif Ahmad, Michael Riegler, Konstanin Pogorelov et al. *Natural Disasters Detection in Social Media and Satellite imagery* (2019).
2. Sheharryar Ahmad1, Kashif Ahmad, Nasir Ahmad, Nicola Conci et al. *Convolutional Neural Networks for Disaster Images Retrieval* (2017).
3. Olga Ostroukhova Pal, Holvorsen, Nicola Conci, Rozenn Dahvot et al. *Automatic detection of passable roads after floods in remote sensed and social media data* (2019).
4. Angela Maria Vinod, Dharathi Venkatesh, Dishti Kundra, Jayapandian N et al. *Natural Disaster Prediction by Using Image Based Deep Learning and Machine Learning* .
5. Jigar Doshi, Saikat Basu, Guan Pang et al *From Satellite Imagery to Disaster Insights* " .

6. Siti Nor Khuzaimah Binti Amit Soma Shiraishi Tetsuo Inoshita Yoshimitsu Aoki *Analysis of satellite images for disaster detection*.
7. Meedinti Gowri Namratha, Sanklap Chauhan, Swarnalatha P *Social Media use During Crisis Management, Disaster Response and Recovery Phases* (2022)
8. Lida Huang, Panpan Shi, Haichao Zhu2, Tao Chen1 *Early detection of emergency events from social media: a new text clustering approach* (2022)
9. P Jedsada, S Tejal , Nipun Balan Thekkummal et al *Use of Social Media Data in Disaster Management: A Survey* (2021)
10. S. Ahmad, K. Ahmad, N. Ahmad, and N. Conci. *Convolutional neural networks for disaster images retrieval* (2017).
11. K. Avgerinakis, A. Moumtzidou, S. Andreadis, E. Michail, I. Gialampoukidis, S. Vrochidis, and I. Kompatsiaris. *Visual and textual analysis of social media and satellite images for flood detection* @ multimedia satellite task mediaeval. *(2017)*
12. B. Bischke, P. Bhardwaj, A. Gautam, P. Helber, D. Borth, and A. Dengel. *Detection of flooding events in social multimedia and satellite imagery using deep neural networks*. In Working Notes Proceedings MediaEval Workshop, page 2 (2017)
13. C. Castillo. Big crisis data: *Social media in disasters and time-critical situations*. Cambridge University Press, 2016.
14. J. B. Houston, J. Hawthorne, M. F. Perreault, E. H. Park, M. Goldstein Hode, M. R. Halliwell, S. E. Turner McGowen, R. Davis, S. Vaid, J. A. McElderry, et al. *Social media and disasters: a functional framework for social media use in disaster planning, response, and research. Disasters* (2015).

https://doi.org/10.1051/itmconf/20224403011

Resume Classification using various Machine Learning Algorithms

Riya Pal[1,*], *Shahrukh* Shaikh[2], *Swaraj* Satpute[3] and *Sumedha* Bhagwat[4]

[1,2,3] Ramrao Adik Institute of Technology, Nerul, Navi Mumbai, India
[4] D.Y. Patil Deemed to be University, Ramrao Adik Institute of Technology, Nerul, Navi Mumbai, India

Abstract. With the onset of the epidemic, everything has gone online, and individuals have been compelled to work from home. There is a need to automate the hiring process in order to enhance efficiency and decrease manual labour that may be done electronically. If resume categorization were done online, it would significantly save paperwork and human error. The recruiting process has several steps, but the first is resume categorization and verification. Automating the first stage would greatly assist the interview process in terms of speedy applicant selection. Classification of resumes will be performed using Machine Learning Algorithms such as Nave Bayes, Random Forest, and SVM, which will aid in the extraction of skills and show diverse capabilities under appropriate job profile classes. While the abilities are being extracted, an appropriate job profile may be retrieved from the categorised and pre-processed data and shown on the interviewer's screen. During video interviews, this will aid the interviewer in the selection of candidates.

1 Introduction

Interviews are becoming time-consuming affairs. Employees are required to travel to locations and conduct interviews, and it is difficult to manually remember each and every aspect of a candidate or the interview process. In many instances, the artificial intelligence system assists us in simplifying things.

Using the conventional method of recruitment, an organization's HR department invites individuals based on their resumes to an interview for a specific position. This HR department manually evaluates a candidate's skills based on their résumé to determine if he or she is qualified for the position or not. HR's conduct interviews, and the panel plays a significant role in determining who is the best applicant for the post. They examine not just the candidate's talents, but also his or her personality.

Maintaining resumes and profiles of all candidates becomes a very tedious job when it comes to big mass recruitment companies because they provide employment in bulk, and thus maintaining or storing data physically is not possible.

Machine Learning enables the path through which a computer can be trained to follow specific instructions again and again to make human life easy. The most common usage of machine learning is for the classification of objects. In machine learning, iteration is important because models are exposed to new data and adapt accordingly. Machine learning models learn from previous results and computation to produce correct and reliable decisions. In statistics, classification is a supervised learning concept in which segregation of data into similar groups is done depending upon various factors and observations.

A resume is one of the most important necessities when it comes to the selection of a candidate for any job. When the company's hiring team receives the resume of any candidate, the skills, if extracted using automation, will save a tremendous amount of time for the hiring team as they no longer need to sit and read through each and every word. But for this, first dataset needs to be scraped or made, then preprocessing has to be done on that dataset. Once the preprocessing of words is done, then using various machine learning algorithms, the data needs to be classified into different classes of job profile in accordance to the skillset...

2 Literature Survey

Finding the need for resume classification: As humans, everyone is bound to make mistakes. Storing and sharing physical copies safely is very inconvenient and inefficient. For this purpose of overcoming the drawbacks, need for resume classification significantly increases for which Artificial intelligence tools and Machine Learning algorithms [1] have been used widely.

Categorization of similar data together: With the advancement of computer and information technologies, a large number of research papers have been published both online and offline, and as new study topics continue to emerge, users are having a difficult time discovering and classifying their relevant research articles. A classification method that can cluster research articles into

* Corresponding author: palriya292@gmail.com

a meaningful class in which publications are extremely likely to have similar subjects is needed to overcome the restrictions. The suggested method uses K-means Clustering [2] and the Latent Dirichlet allocation (LDA) scheme [2] to extract representative keywords from the abstracts of each publication and subjects.

Importance of Automation: the issue regarding lack of automated systems for medical institutions and hospitals have caused a splurge for development of automation for hospital system [3]. Natural Language Processing (NLP)[11] benefits society as a whole as text-based information handling is very difficult manually.

Preprocessing of data using TF-IDF Vectorization: A well-developed categorization system that can group research papers into relevant classes based on their subjects [4] helps in finding out relevant data in the most time efficient manner. The suggested approach retrieves representative keywords from each paper's and topic's abstract. Then, using the Term frequency-inverse document frequency (TF-IDF) values [4] of each article, the K-means clustering method [4] is used to categorize the entire set of papers into research papers with comparable themes.

Modelling of semi-structured documents to fetch job postings from resume: Matching semi-structured resumes with positions in a big size real-world collection is a tough challenge. Experiments reveal that the SRM technique [5] yielded encouraging results and outperformed traditional unstructured relevance models in the first try. W. Bruce Croft, Xing Yi, and James Allan [5] Furthermore, we compared the suggested system's efficiency and efficacy to those of state-of-the-art online recruiting methods. A system that utilizes machine learning to match job postings and resumes for huge data sets; in our work, it is less difficult than previous papers; and using text mining, the extracted data from the resume is matched with the keyword recorded in the database and categorized for each job category.

3 Methodology

This section will describe the methodology and concepts that facilitate the building of classification model capable for resume classification and displaying the output with a suitable job profile for the candidate. The system works in the following phases as given below.

3.1 Data Gathering

Data Gathering includes collection of datasets from various websites like kaggle.com, glassdoor.com and indeed.com. The datasets are not classified and are unstructured datasets in which the data will be cleaned, classified, and stored in "25_cleaned_job_descriptions.csv" including some parts from Kaggle and some from glassdoor and indeed.com. 70% of the data is being used for training data and the remaining 30% will be used for test data.

3.2 Data Cleanup

The dataset containing a huge number of records is still very rough and unclassified. Data cleaning will be done by removing any blank spaces from the data, then changing all the text to lowercase to avoid confusion and removing stop words from the data. Stop words are those words which don't play an important role in sentence formation, such as "are", "we", "is", etc. Cleaned data is stored in a separate dataset containing 10,000 entries with two main classes of "query" and "description".

3.3 Tokenization

In this step, each entry in the corpus i.e., each entry in the document will be broken down into a set of words. To begin the tokenization process, we look for concepts or words that make up a character sequence. This is significant because we will be able to deduce meaning from the original text sequence using these terms. Tokenization is the process of separating large chunks of text into smaller pieces known as tokens. This is accomplished by deleting or isolating characters like as whitespace and punctuation. Tokens are phrases that are divided into individual words after being tokenized out of paragraphs. We may obtain information such as the number of words in a text, the frequency of a specific term in the text, and much more by doing Tokenization. Tokenization can be done in a variety of methods, such utilising the Natural Language Toolkit [NLTK], the spaCy library, and so on. Tokenization is a required step for subsequent text processing such as stop word removal, stemming, and lemmatization.

3.4 Stemming and Lemmatization

It is common to see a single English word employed in a variety of different ways in different phrases based on its grammatical rules. "Describe", "describing" and "described", for example, are all various tenses of the same verb. This condition necessitates the reduction of all changed or derived versions of a word to its primary stem or base, so that these derivationally related terms with comparable meanings are not deemed distinct from one another. Both stemming and lemmatization have the same goal but take different approaches to achieve it.

"Stemming is the mechanism of reducing inflected or derived words to their word root, or stem. It is a crude heuristic process that involves chopping off the ends of words to achieve this objective, and often includes the removal of derivational affixes [7]" These are rule-based algorithms that analyse a certain word under a variety of scenarios and then decide how to shorten it based on a list of recognized suffixes. It is worth noting that the root generated after stemming may not be the same as the word's morphological root. Stemming is prone to under and over-stemming. Porter-Stemmer, Snowball stemmer, and Lancaster stemmer are some common stemming algorithms. Lemmatization, on the other hand, is the process of accurately reducing words to root words using a language dictionary. Lemmatization, as opposed to Stemming, which merely chops out tokens by basic pattern matching, is a more sophisticated technique that employs language vocabulary and morphological study of words to provide linguistically proper lemmas. This implies that lemmatization makes use of context information and may thus distinguish between words with various meanings based on parts of speech. For the English language, our system uses the NLTK python package's WordNet Lemmatizer (based on the WordNew Database).

The concept of stemming and lemmatization can be understood by a simple example. The word "loving" will turn into "lov" after stemming which has no meaning while if lemmatized, "loving" will turn into "love" which now has a proper meaning and use case.

3.5 Parts of Speech (POS) Tagging

It is the process of associating grammatical information with a word depending on its context and relationship to other words in the sentence [8]. According to its usage in the phrase, the part-of-speech tag identifies whether the word is a noun, pronoun, verb, adjective, or other. These tags must be assigned in order to grasp the right meaning of a phrase and to create knowledge bases for character recognition. This procedure is not as straightforward as mapping a word to its appropriate part of speech tags. This is because a word may have a distinct part of speech depending on the context in which it is spoken.

For example, "writing" is a Verb in the statement "I am writing an essay," yet "building" is a Noun in the line "I stay in the tallest building in the entire town." It is a supervised learning approach that analyses information such as the preceding word, following word, initial letter capitalised or not, and so on to label the words after tokenization. It is also known as grammatical tagging.

3.6 TF-IDF Vectorization

TF-IDF stands for Term Frequency – Inverse Document Frequency and it tells us how important a word is from the set of words in the dataset and assigns a tfidf value indicating the importance of the word as per frequency.

The formula for term frequency is the division of number of occurrences of a word in a sentence by the total number of words in the sentence (1). Inverse document frequency is calculated by the log of total sentences in the document divided by the sentences that actually contain the word (2). The multiplication of TF formula and IDF formula will give us a value in the form of a vector with a graph of importance on one hand and set of words on the other hand (3). Fig. 1 depicts a graph that shows TF-IDF vectorization which explains that the word with higher importance is placed according to ascending order of its importance and hence an inverse growth curve.

$$TF = \frac{No. of\ occurence\ in\ sentence}{Total\ no. of\ words\ in\ sentence} \quad (1)$$

$$IDF = \log \left[\frac{total\ sentences\ in\ document}{sentences\ which\ actually\ contain\ the\ word}\right] \quad (2)$$

$$\text{TF-IDF Vectorization} = TF \times IDF \quad (3)$$

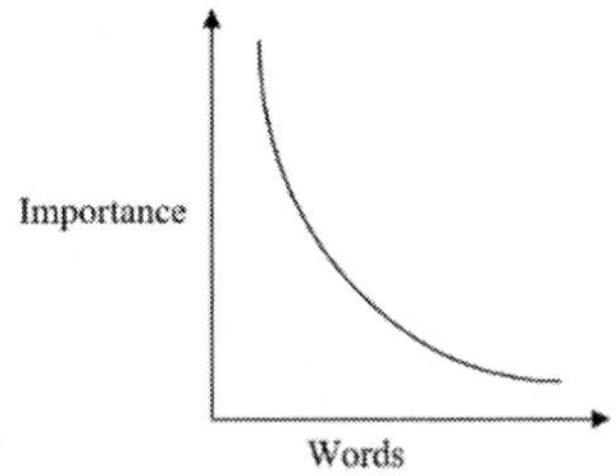

Fig. 1. Shows the graph of TF-IDF Vectorization

This will mark all the important sets of words related to jobs and skill sets with the highest term frequency value which can be then used to fetch important words and use them to train data on various algorithms.

3.7 Applying Classification Algorithm To Dataset

The following classification algorithms have been used for classification and model training.

Naïve bayes is a classification algorithm that works on probabilistic output as in whether the event is going to occur or not provided with a set of conditions. On the NB Classifier, the training data is fitted. Then, in the validation dataset labels are being predicted. To get accuracy, use accuracy_score function. This classification model gave an accuracy of around 45% and did not get the expected results.

Support Vector Machine (SVM) algorithm classifies data by drawing a hyperplane between two or more items. The Hyperplane which best classifies the items is considered as ideal output. The working flow of SVM is similar to that of Naïve Bayes. This classification model gave an accuracy of 60% which proved to be better than naïve bayes model.

Random Forest is a classification algorithm that works on the principle of decision trees. It takes in input of many decision trees and gives the best majority output from all the inputted decision trees. On the RF Classifier the training data is fitted. Then, in the validation dataset labels are being predicted. To get accuracy, use accuracy_score function. Random forest model gave an accuracy of 70% and gave the most correct predictions as per the comparison.

The comparison between the naïve bayes model, the SVM model and the random forest model with their accuracy, precision, recall and F1 score is given in TABLE 1.

4 Methodology Flowchart

The methodology is important to understand, and hence it becomes necessary to make a flowchart for easy understanding of the flow of system. Fig. 2. Shows the flowchart of the system.

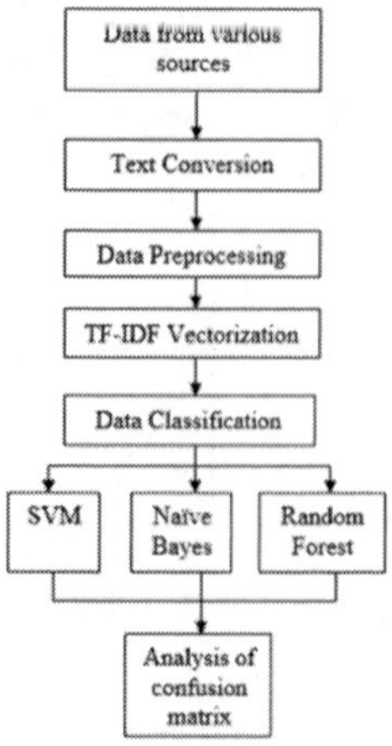

Fig. 2. Proposed Flowchart for the system

First the dataset needs to be scraped which can be done using various websites like indeed.com, glassdoor.com and kaggle.com. Once the dataset has been scrapped, preprocessing of data is to be done by doing proper stemming and lemmatization, removing stop words and filler words store the relevant words in a separate column for further process. After the preprocessing of data is done, making sure that the relevant data have the words which have occurred the highest amount of times need to be put in a term-frequency document using TF-IDF Vectorization. Once the previous steps are done, the cleaned data is now ready to test and form classification models using machine learning algorithms. Analysing the model by their accuracy and the forming confusion matrix for the same is required for better understanding of the results.

5 Results

The dataset has been taken from various websites like indeed.com, glassdoor.com and kaggle.com. These datasets are unstructured and thus preprocessing has been done which includes, tokenization, removing stop words, stemming and lemmatization with POS Tagging. The output for preprocessing data is shown in Fig. 3. Which consists of 3 main columns in its corpus having 10,000 rows. "Query" displaying all the job profiles (Fig. 4), "Description" containing raw data of words (Fig. 5), and "text_final" displaying the main set of preprocessed words (Fig. 6). The zoomed version of Fig 3. Has been represented in Fig. 4, Fig. 5, and Fig. 6.

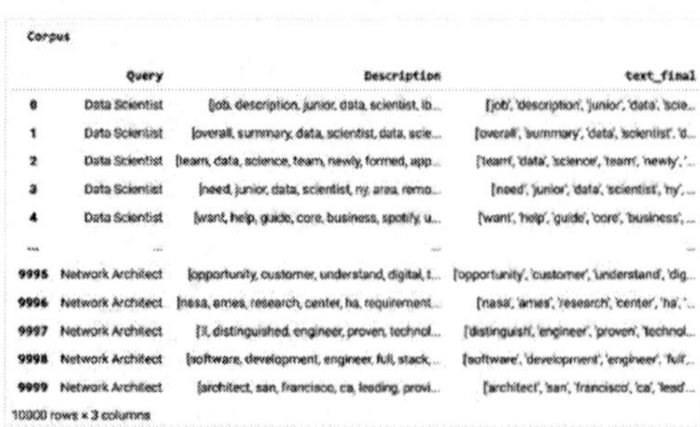

Corpus

	Query	Description	text_final
0	Data Scientist	[job, description, junior, data, scientist, ib...	['job', 'description', 'junior', 'data', 'scie...
1	Data Scientist	[overall, summary, data, scientist, data, scie...	['overall', 'summary', 'data', 'scientist', 'd...
2	Data Scientist	[team, data, science, team, newly, formed, app...	['team', 'data', 'science', 'team', 'newly', '...
3	Data Scientist	[need, junior, data, scientist, ny, area, remo...	['need', 'junior', 'data', 'scientist', 'ny', ...
4	Data Scientist	[want, help, guide, core, business, spotify, u...	['want', 'help', 'guide', 'core', 'business', ...
...	...	...	...
9995	Network Architect	[opportunity, customer, understand, digital, t...	['opportunity', 'customer', 'understand', 'dig...
9996	Network Architect	[nasa, ames, research, center, ha, requirement...	['nasa', 'ames', 'research', 'center', 'ha', '...
9997	Network Architect	[ll, distinguished, engineer, proven, technol...	['distinguish', 'engineer', 'proven', 'technol...
9998	Network Architect	[software, development, engineer, full, stack,...	['software', 'development', 'engineer', 'full'...
9999	Network Architect	[architect, san, francisco, ca, leading, provi...	['architect', 'san', 'francisco', 'ca', 'lead'...

10000 rows × 3 columns

Fig. 3. Output of Preprocessing of Data

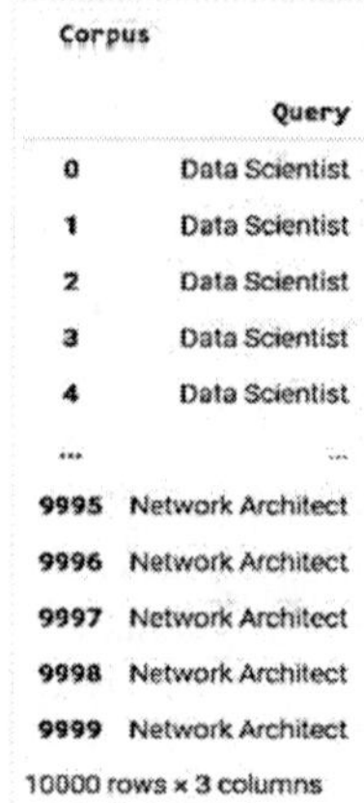

Corpus

	Query
0	Data Scientist
1	Data Scientist
2	Data Scientist
3	Data Scientist
4	Data Scientist
...	...
9995	Network Architect
9996	Network Architect
9997	Network Architect
9998	Network Architect
9999	Network Architect

10000 rows × 3 columns

Fig. 4. Query Column from Fig. 3.

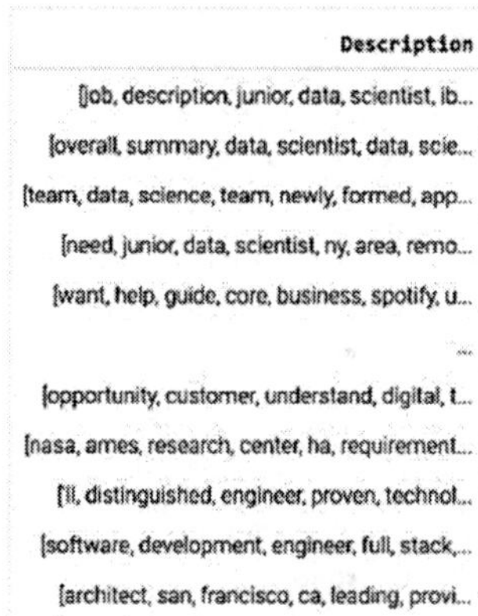

Description
[job, description, junior, data, scientist, ib...
[overall, summary, data, scientist, data, scie...
[team, data, science, team, newly, formed, app...
[need, junior, data, scientist, ny, area, remo...
[want, help, guide, core, business, spotify, u...
...
[opportunity, customer, understand, digital, t...
[nasa, ames, research, center, ha, requirement...
[ll, distinguished, engineer, proven, technol...
[software, development, engineer, full, stack,...
[architect, san, francisco, ca, leading, provi...

Fig. 5. Description Column from Fig. 3.

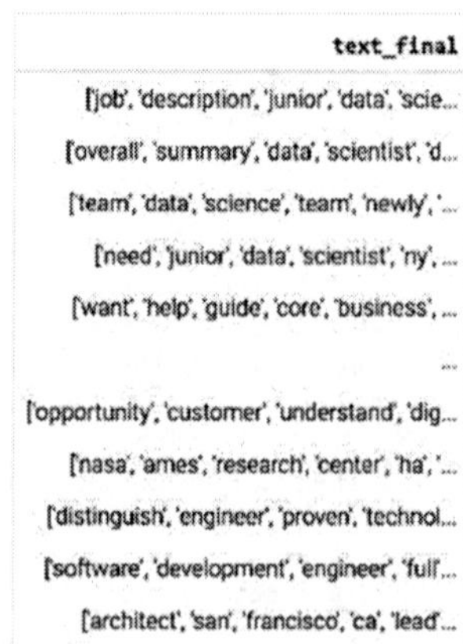

text_final
['job', 'description', 'junior', 'data', 'scie...
['overall', 'summary', 'data', 'scientist', 'd...
['team', 'data', 'science', 'team', 'newly', '...
['need', 'junior', 'data', 'scientist', 'ny', ...
['want', 'help', 'guide', 'core', 'business', ...
...
['opportunity', 'customer', 'understand', 'dig...
['nasa', 'ames', 'research', 'center', 'ha', '...
['distinguish', 'engineer', 'proven', 'technol...
['software', 'development', 'engineer', 'full'...
['architect', 'san', 'francisco', 'ca', 'lead'...

Fig. 6. text_final Column from Fig. 3.

After preprocessing of data, the dataset containing all the words are given TF-IDF values in order to arrange the word matrix in ascending order of their importance. Result of TF-IDF vectorization shows columns with their frequency number which have term frequency of more than 5000 is shown in Fig. 7. With output like: 'job': 2456, 'description': 1221, 'junior': 2480, 'data': 1134, 'scientist': 3984, 'ibm': 2175, 'work': 4959, 'part': 3216, 'team': 4448, 'solve': 4182, 'problem': 3498, 'use': 4763, 'technique': 4455, 'apply': 266, 'scientific': 3983, 'method': 2816, 'business': 578, 'scenario': 3971, 'clean': 734, 'prepare': 3445, 'statistical': 4286, 'machine': 2667, 'learn': 2556, 'model': 2877, 'variety': 4798, 'analytical': 215, 'algorithm': 165, 'build': 570,...

```
print(Tfidf_vect.vocabulary_)

{'job': 2456, 'description': 1221, 'junior': 2480, 'data': 1134, 'scientist': 3984,

print(Tfidf_vect.vocabulary_)

'scientist': 3984, 'ibm': 2175, 'work': 4959, 'part': 3216, 'team': 4448, 'solve': 4182,
```

Fig. 7. Output of TD-IDF Vectorization

Data after being cleaned into a new dataset containing 10,000 entries with two main categories "query" and "description". Fig. 8. Depicts the main 25 job classes that are being evaluated which are: 'Artificial Intelligence', 'Big Data Engineer', 'Business Analyst', 'Business Intelligence Analyst', 'Cloud Architect', 'Cloud Services Developer', 'Data Analyst', 'Data Architect', 'Data Engineer', 'Data Quality Manager', 'Data Scientist', 'Data Visualization Expert', 'Data Warehousing', 'Data and Analytics Manager', 'Database Administrator', 'Deep Learning', 'Full Stack Developer', 'IT Consultant', 'IT Systems Administrator', 'Information Security Analyst', 'Machine Learning', 'Network Architect', 'Statistics', 'Technical Operations', 'Technology Integration'.

```
array(['Artificial Intelligence', 'Big Data Engineer', 'Business Analyst',
       'Business Intelligence Analyst', 'Cloud Architect',
       'Cloud Services Developer', 'Data Analyst', 'Data Architect',
       'Data Engineer', 'Data Quality Manager', 'Data Scientist',
       'Data Visualization Expert', 'Data Warehousing',
       'Data and Analytics Manager', 'Database Administrator',
       'Deep Learning', 'Full Stack Developer', 'IT Consultant',
       'IT Systems Administrator', 'Information Security Analyst',
       'Machine Learning', 'Network Architect', 'Statistics',
       'Technical Operations', 'Technology Integration'], dtype=object)
```

Fig. 8. Classification of jobs

After the data is cleaned and arranged, classification of data into various job profiles and skill sets can be achieved using various machine learning classification algorithms. The developed systems have excellent true predictions when confusion matrix is taken into consideration. The following diagrams below depict the confusion matrix of various classifications, where true positive values are those values where the model predicted the correct job profiles as expected value. Fig. 9 shows the confusion matrix of all the job classes.

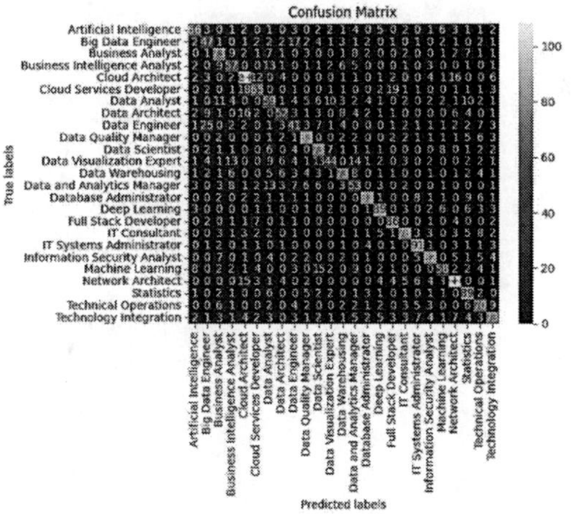

Fig. 9. Confusion matrix of job classes

The confusion matrix for naïve bayes classification model is shown in Fig. 10 depicting the true positive, true negative, false positive and false negative values that have been detected. A diagonal line of dark colored boxes can be observed which shows the true positive values for each label using Naïve Bayes Classification.

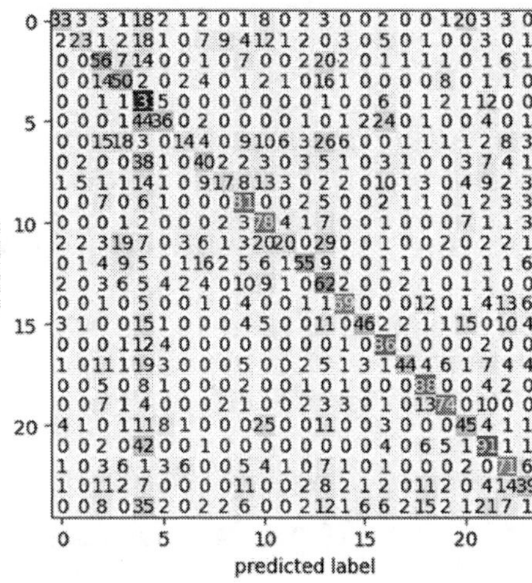

Fig. 10. Confusion matrix for naïve bayes classification

The confusion matrix for SVM classification model is shown in Fig. 11 depicting the true positive, true negative, false positive and false negative values that have been detected. A diagonal line of dark colored boxes can be observed which shows the true positive values for each label using SVM classification.

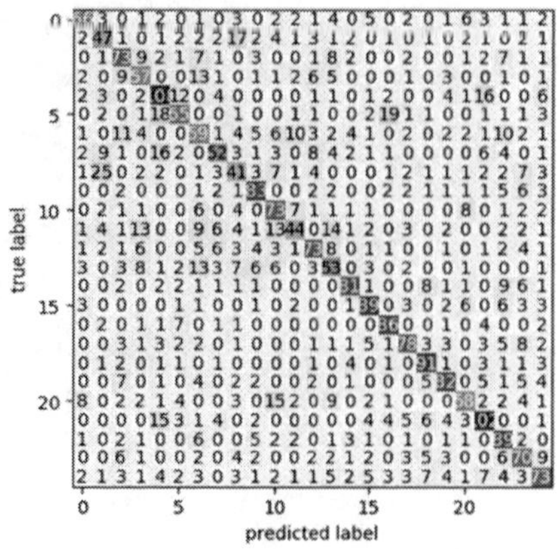

Fig. 11. Confusion matrix for SVM Classification

The confusion matrix for Random Forest is shown in Fig.12 depicting the true positive, true negative, false positive and false negative values have detected. A diagonal line of dark colored boxes can be observed which shows the true positive values for each label using random forest classification.

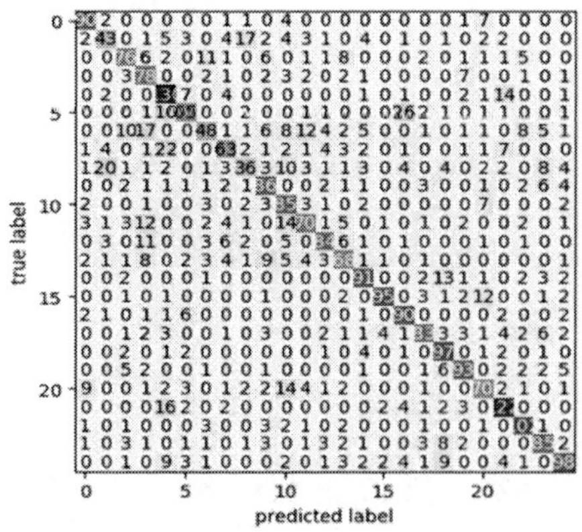

Fig. 12. Confusion matrix for random forest classification

The end result of the confusion matrix and classification models shows that Random Forest is best suited for classification of resume data with 2500 decision trees taken into consideration. TABLE 1. shows the result and comparison of various classification algorithms.

Table 1. Comparison of various classification algorithms

Algorithm	Accuracy	Precision	Recall	F1 Score
Naïve Bayes	45	0.521	0.452	0.448
SVM	60	0.598	0.597	0.594
Random Forest	70	0.687	0.683	0.678

Despite of average accuracy being 70%, the confusion matrix shows excellent results. Also, accuracy is not the only measure that needs to be taken into consideration while building a model but the confusion matrix plays an equally important part for accurate results for how many true positive values have been returned. As per the comparison in TABLE 1, Random forest has the highest accuracy and has the best confusion matrix.

The random forest accuracy can be improved by increasing the number of trees or by increasing the size of data. In our case, we took 2,500 trees and increasing the number would increase the accuracy of the model but at the same time the learning time would get slower. Another technique that can be applied for improvement is to use hyper parameter tuning for random forest classifier.

Conclusion and Future Work

The concept of classification is grasped by Resume Classification, and classification models have been built using numerous techniques. This resume categorization platform will make the e-recruitment process more efficient and user-friendly. This approach will assist businesses and save time throughout the recruitment process.

Future work includes combining these classification models with frontend resume portal where the user will be able to upload their resume and resume classification can work in the backend. From the results and discussions, we can conclude that various algorithms have given different accuracy score percentages with Random Forest having the highest accuracy among all three of them. Also, the confusion matrix of random forest is well suited for all the observed classes.

Thereby, resume classification is achieved using preprocessing of data, cleaning of data and by applying various classification algorithms.

This research paper can also be extended with the combination of other video interviewing features such as facial recognition, voice to text generation and voice analysis. Building a plugin for people to include resume classification models into their project is also one of the future goals of this research paper and project work.

References

1. B. Balci, D. Saadati, D. Shiferaw, Handwritten Text Recognition using Deep Learning, STAN. U. (2017).
2. SW. Kim, JM. Gil, Research paper classification systems based on TF-IDF and LDA schemes. Hum. Cent. Comput. Inf. Sci. **9**, 30 (2019).
3. C. Friedman, G. Hripcsak, Natural language processing and its future in medicine. Acad Med. August (1999)
4. J. Ramos, Using TF-IDF to Determine Word Relevance in Document Queries, Rutgers University.
5. E. Xing Yi, James Allan and W. Bruce Croft Center for Intelligent Information Retrieval, Department of

Computer Science 140 Governor"s Drive, Matching Resumes and Jobs Based on Relevance Models, SIGIR, (2017)

6. F. Aseel B. Kmail, Mohammed Maree, Mohammed Belkhatir, Saadat M. Alhashmi -An Automatic Online Recruitment System based on Exploiting Multiple Semantic Resources and Concept-relatedness Measures, IEEE 27th International Conference on Tools with artificial intelligence (2015)
7. Jivani, A.G., A comparative study of stemming algorithms. Int. J. Comp. Tech. Appl, (2011)
8. Gelbukh, A., Computational Linguistics And Intelligent Text Processing. Berlin, Heidelberg: Springer Berlin Heidelberg (2014)
9. Laumer, S. and Eckhardt, A., May. Help to find the needle in a haystack: integrating recommender systems in an IT supported staff recruitment system. In Proceedings of the special interest group on management information system's 47th annual conference on Computer personnel research, (2009).
10. A. Zaroor, M. Maree, MuathSabhaJRC, A job post and resume classification system for online recruitment, International Conference on tools with Artifical Intelligence, (2017)
11. Bird, S., Klein, E. and Loper, E, Natural Language Processing With Python. Bejing: O'Reilly, (2009).
12. Kim S. B., Rim H. C., Yook D. S. and Lim H. S., "Effective Methods for Improving Naive Bayes Text Classifiers", (2002)
13. F. Dernoncourt and J. Y. Lee, Pubmed 200k rct, a dataset for sequential sentence classification in medical abstracts (2017).
14. D. E. Johnson, F. J. Oles, T. Zhang, T. Goetz, A decision-tree-based symbolic rule induction system for text categorization, IBM Systems Journal, September (2002).
15. S. Sanyal, N. Ghosh, S. Hazra, S. Adhikary, Resume Parser with Natural language Processing, IJESC (2007).
16. X.Chen,X. Lin, Big Data Deep Learning: Challenges and Perspectives, IEEE, (2014)
17. Bao Y. and Ishii N., Combining Multiple kNN Classifiers for Text Categorization by Reducts.
18. Kolluri, Johnson and Razia, Shaik and Nayak, Soumya Ranjan, Text Classification Using Machine Learning and Deep Learning Models. International Conference on Artificial Intelligence in Manufacturing & Renewable Energy (ICAIMRE) (2019)
19. S. Brindha, K. Prabha and S. Sukumaran, A survey on classification techniques for text mining, 3rd International Conference on Advanced Computing and Communication Systems (ICACCS) (2016)

Analysis of hyperparameters in Sentiment Analysis of Movie Reviews using Bi-LSTM

Amankumar Singh[1,*], *Riya* Thapliyal[1,**], *Ritika* Vanave[1,***], *Rajashree* Shedge[1,****], and *Snehal* Mumbaikar[1]

[1]Computer Engineering Department, Ramrao Adik Institute of Technology, Navi Mumbai, India.

Abstract. Movie reviews are an important factor in determining a film's success because instead of depending solely on the number of views as a parameter for the success of the movie, movie reviews are used to acquire additional insights into the movie. Existing systems use LSTM for sentiment analysis but there is no study available how various hyperparameters affect the performance of the model. Bi-LSTM along with dropout layers provide good accuracy in sentiment analysis. The suggested method outperforms CNN and Natural Language Toolkit in terms of accuracy.The proposed model is tested using different hyper parameters including dropout rate,number of Bi-LSTM layers and Bi-LSTM nodes. 64 LSTM nodes, 2 Bi-directional Layers, and a 0.2 Dropout rate should be used for optimal accuracy. Effect of different text vectorization algorithms and activation functions was also studied. The combination of Tf-idf text vectorization and the ReLU activation function yields the best results.

Keywords: Sentiment Analysis, Text-vectorization, Bi-LSTM.

1 Introduction

Sentiment Analysis on movie reviews can help readers make more informed decisions by providing a more comprehensive description of the film[1]. Sentiment analysis is used to get the performance of the movie based on user reviews. Any sequence of words that is usually associated with a specific opinion is referred to as a sentiment. Sentiment analysis not only focuses on polarity but also on feelings, emotions, and intentions. This type of analysis can be extremely beneficial in marketing, customer feedback, and customer service. In this scenario, using machine learning to learn and predict whether a movie review is positive or negative, the analysis is the act of looking at data and generating judgments. Two main strategies for sentiment analysis are rule-based and machine learning-based. The machine learning model identifies sentiment based on words and their order, whereas the rule-based technique employs sentiment scores. News and social media datasets can also be used for sentiment analysis.

Long Short-Term Memory (LSTM) networks are a type of recurrent neural network that may learn order dependence in sequence prediction challenges[8]. LSTM preserves historical information by establishing a memory unit which can substantially reduce the long-distance reliance problem of Recurrent Neural Networks. The LSTM is made up of three sections, each of which serves a distinct purpose. The first portion determines whether the previous timestamp's information should be remembered or is irrelevant and can be ignored. The cell attempts to learn new information from the input in the second section. Finally, the cell passes updated information from the current timestamp to the next timestamp in the third component. The gates are the three components of an LSTM cell[8]. The Forget gate is the first part, the Input gate is the second, and the Output gate is the final part. Future information is ignored by LSTM. BiLSTM[3] adds another LSTM layer to the mix, reversing the flow of data. At a nutshell, it means that in the additional LSTM layer, the input sequence flows backward. The outputs from both LSTM layers are then combined in a variety of methods, including average, sum, multiplication, and concatenation. BiLSTM can provide a more relevant output by merging LSTM layers from both directions, because every component of an input sequence carries information from both the past and the present[3].

Section II of the paper gives a brief description about the literature survey done on the existing systems and comparative analysis of the survey.The problem definition and the need for such a system is covered in section III.Section IV discusses the proposed system and architecture of the system.The result of the system is covered in Section V .Section VI draws the conclusion of the paper.

2 Related Work

To train machines to perform sentiment analysis, a variety of techniques and complicated algorithms are used. Each one has their benefits and drawbacks. However, when used in tandem, they can provide extraordinary outcomes. In recent years, there has been a lot of research on sentiment analysis of text based on movie reviews. For sentiment analysis several methodologies were used.

*e-mail: aks001235@gmail.com
**e-mail: riyathapliyal2018@gmail.com
***e-mail: ritikavanave@gmail.com
****e-mail: rajashree.shedge@rait.ac.in

https://doi.org/10.1051/itmconf/20224403012

An ML-based model is proposed by Saeed Mian Qaisar[1], in which the imported dataset is first partitioned, then vectorized, and then submitted to the LSTM model. Three layers are used here: a 50-node first layer, a 101-node second LSTM layer, and a final layer with outputs. This paper uses a 50k movie review dataset from IMDb for its research. To identify IMDb movie reviews, Aswathi Sajeevan and Lakshmi K S [2] utilize two separate deep learning models: the first is a hybrid LSTM-CNN, while the second is a hybrid CNN-LSTM. This study uses the IMDb movie reviews dataset, which includes 1000 positive and 1000 negative reviews. Guixian Xu, Yueting Meng, Xiaoyu Qiu, Ziheng Yu, and Xu Wu[3] use the TensorFlow library, as well as Dropout and TF-IDF, to develop a BiLSTM model. Comment text vectors are created from the textual data. ReLu is the activation function employed. The paper is based on a crawl of 15000 hotel comment texts from the Ctrip website.

Anwar Ur Rehman, Ahmad Kamran Malik, Basit Raza, and Waqar Ali offer a model with recurrent and convolutional neural networks in their paper[4]. It starts with a corpus and then preprocesses it. Sentiment Analysis was then performed using a word embedding layer in conjunction with CNN and LSTM models. The categorization layer is then applied at the end. The IMDb movie reviews dataset and the Amazon movie reviews dataset were used. The Nave Bayes algorithm, which is a supervised learning approach as well as a statistical method for classification, is used in this study by G. P Saradhi Varma, A.Govardhan, and I.Hemalatha[5]. It also employs the maximum entropy algorithm, which is a machine learning technique for classification and prediction that may be used in a variety of situations.

The text classification approach is utilized in this study by Madhav Singh Solanki[6], which is a supervised learning approach to recognise a specific type of writing, such as a blog, book, web page, news item, or tweet. Another rule based model is used that employs lexicon to recognise certain terms in the text. Rule-based techniques typically define a set of rules in a scripting language that denotes subjectivity, polarity, or a point of view. Machine learning techniques, POS tagging, dependency parsing, and other techniques are utilized in this article by Amlan Chakrabarti[7] to identify the aspects and user opinions linked to the aspect.The input layer, which consists of word embedding characteristics for each word in the phrase, and two convolution layers make up a seven-layer deep CNN architecture.

Table 2.1. Comparative Analysis of the existing systems

Techniques	Advantages	Limitations
Sentiment Analysis of IMDb Movie Reviews Using Long Short Term Memory[1]	1. The model i very simple to build and understand. 2. The accuracy achieved in this paper is 89.9	1. A better precision can be achieved by further cleaning data and employing the ensemble classification approaches.
An enhanced approach for movie review analysis using deep learning techniques[2]	1. The framework utilizes this advantage of both CNN and LSTM. 2. High performance.	1. Accuracy of CNN-LSTM is low at 51.75%.
Sentiment Analysis of Comment Text Based on BiLSTM[3]	1. F1 score - 92 2. Both forward and reverse sequences are considered.	1. The sentiment analysis using BiLSTM takes a long time for training
A Hybrid CNN-LSTM Model for Improving Accuracy of Movie Reviews Sentiment Analysis[4]	1. 1. Local features extraction. 2. Capture long-term dependencies . 3. Improved the execution time.	
Sentiment Analysis Using Machine Learning Algorithms[5]	1. Machine learning algorithms can achieve high accuracy for classifying sentiment when using this method. 2. A leading advantage of maximum entropy (ME) models is their flexibility.	1. ME models requires considerable computational resources. 2. Estimation of such models also leads to sensitivity to round-off errors.

https://doi.org/10.1051/itmconf/20224403012

Sentiment Analysis of Text using Rule based and Natural Language Toolkit[6]	1. Sentiment Analysis of Text using Rule Based algorithm is fast and has high accuracy. 2. To improve analytical speed and reliability, the map reduction environment can be used.	1. One of the limitations of using lexicon resorting is that the way people express their emotions varies enormously.
A Mixed approach of Deep Learning method and Rule-Based method to improve Aspect Level Sentiment Analysis[7]	1. In comparison with the existing methods, the proposed technique (CNN and Rule-Based) brings better classification accuracy for both the positive and negative classes.	1. The model developed using both the rule based and Machine Learning method is very complex and needs to be tweaked on various aspects for best performance.

For sentiment analysis, the majority of studies have employed either traditional deep learning algorithms like LSTM and CNN. Models employing CNN or machine learning[6] have a long training time and require large computing power.

3 Proposed System

Gauging the performance of the movie based solely on the views of the movie can be mis-leading and understanding the performance of the movie based on user comments will be more realistic. Therefore, a system is proposed that could perform sentiment analysis on the comments of the users and provide users with the rating of the movie based on the comments. The problem is best framed as single-label classification, which generates a numerical measure of positivity and negativity. The comment given by the user will be mapped to a scale of 1 to 5,1 being the most negative and 5 the most positive. Real-life scenarios like emojis, slang language and spelling mistakes will be taken into consideration while doing the classification. The use of LSTM model to solve sentiment analysis is proposed.

A. System Architecture

The proposed system architecture has 5 phases beginning from text vectorization,then model building and evaluation followed by input sanitization. Based on the input, the model calculates a sentiment score.

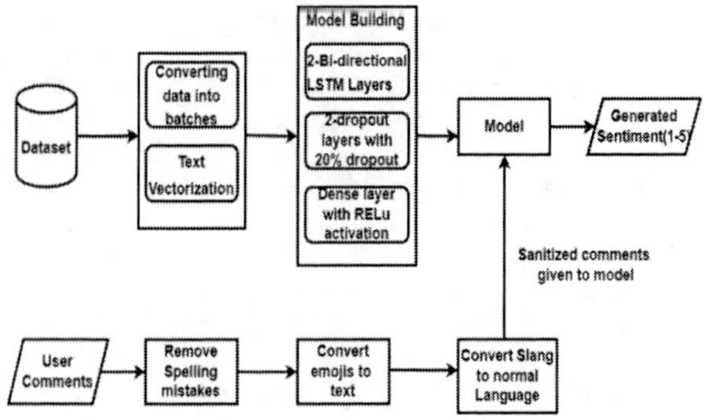

Figure 3.1. Proposed Framework

B. Proposed Methodology

1. Stage 1(Data preparation) : Convert data into batches for better training and perform text vectorization on data.
2. Stage 2(Model Building): 2 Bi-directional LSTM layers along with two dropout layers with a 20% dropout rate are used. At the output layer, use the Dense layer with the ReLu Activation function.
3. Stage 3(Model Evaluation): Model is evaluated on Binary cross entropy loss and Adam Optimizer. Metric used is accuracy.
4. Stage 4(Input Sanitization): Spelling errors are removed and emojis are converted into text.
5. Stage 5(Feeding data to model and Output generation): The entire sanitized input is then sent to the model and it generates the output.

C. Bi-LSTM

The proposed system uses Bi-LSTM. LSTM networks are designed to overcome the long-term dependency problem faced by recurrent neural networks. LSTMs can process entire sequences of without having to handle each point in the sequence separately, instead, preserving useful information about previous data in the sequence to help with the processing of new data points. LSTMs employ a number of 'gates', which regulate how data in a sequence enters, is stored in, and exits the network. A typical LSTM has three gates: a forget gate, an input gate, and an output gate. These gates are each their own neural network and can be thought of as filters. A Bi-LSTM is a sequence processing model made up of two LSTMs, one of which takes the input in one way and the other in the opposite direction. BiLSTMs effectively improve the quantity of data available to the network, providing the algorithm with more context.

4 Results and Discussion

Dataset:

- Using the imdb reviews dataset from tfds(TensorFlow datasets) which contains 50,000.
- Convert data into a batch size of 64 for batch training.
- Perform text vectorization on the data to convert text data into vectors of numbers.

There are 8 layers in model architecture. The first layer is the text vectorization layer followed by an embedding layer. Then there are two pairs of Bidirectional and Dropout layers. Finally, there is a dense layers as the output layer.

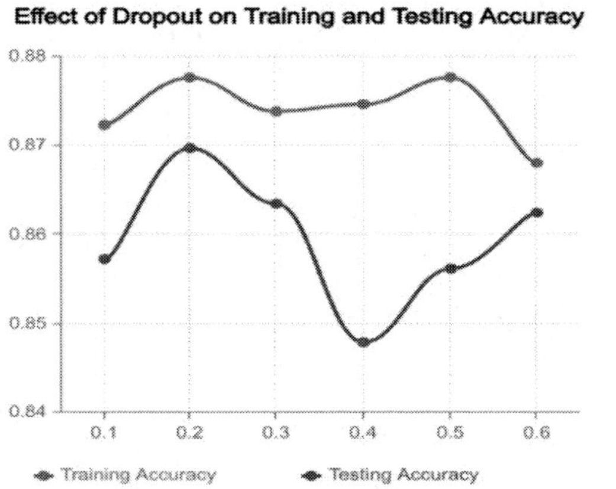

Figure 4.1. Effect of Dropout on Training and Testing Accuracy

In Figure 4.1, the X-axis represents the dropout rate ranging from 0.1 to 0.6, and the Y-axis represents the accuracy. The accuracy becomes optimum at 0.2. Further training accuracy increases, but testing accuracy decreases due to over fitting. So 0.2 dropout rate is optimum.

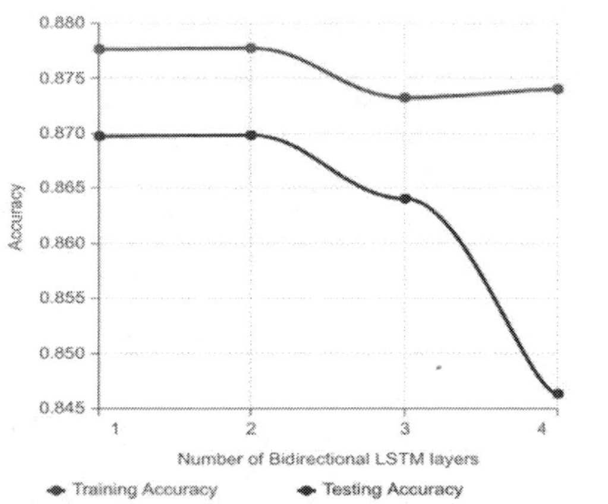

Figure 4.2. Effect of Number of Bidirectional LSTM layers on Training and Testing Accuracy

In Figure 4.2, the X-axis represents the number of Bidirectional LSTM layers ranging from 1 to 4, and the Y-axis represents the accuracy. In this case the accuracy becomes optimum at 2 layers. On further increasing number of layers both training and testing accuracy decreases.

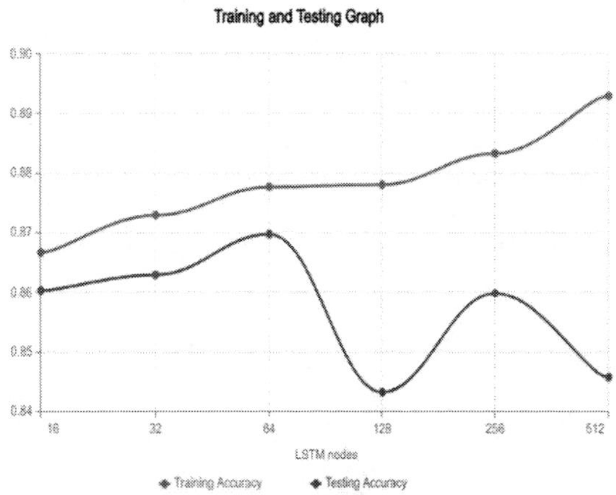

Figure 4.3. Effect of Number LSTM nodes on Training and Testing Accuracy

In Figure 4.3, the X-axis represents the number of LSTM nodes ranging from 16 to 512, and the Y-axis represents the accuracy. The accuracy becomes optimum at 64 nodes.

Table 4.1. Effect of Dataset on model Accuracy

Dataset	Training Accuracy	Testing Accuracy
IMDb Dataset	87.42%	86.56%
Rotten Tomatoes Dataset	82.53%	82.86%

As is evident from Table 4.1, it can be inferred that using the proposed model the training and testing accuracy on both the datasets are within the same range. Both the datasets are of different sizes, therefore the accuracy varies slightly. If two datasets of similar sizes are taken then one could get near the same accuracy.

Table 4.2. Effect of Text Vectorization on model Accuracy

Text Vectorization	Training Accuracy	Testing Accuracy
tf-idf	87.42 %	86.56%
count Vectorizer	53.99 %	52.29 %
int	55 %	56.29%

As is evident from Table 4.2, tf-idf vectorizer gives the optimum training and testing accuracy.

Table 4.3. Effect of Activation Function on modal Accuracy

Activation Function	Training Accuracy	Testing Accuracy
Relu	87.42 %	86.56%
Sigmoid	87.42 %	85 %
Softmax	83.1 %	81.94%
Selu	88 %	86%
Elu	87.78 %	84.85%

The activation function specifies a neuron's or node's output in response to a single or multiple inputs. Selection of activation function is important for the performance of the model. As shown in Table 4.3, Softmax has slightly poor performance, otherwise all other activation function has similar performance, relu having the best training and testing accuracy.

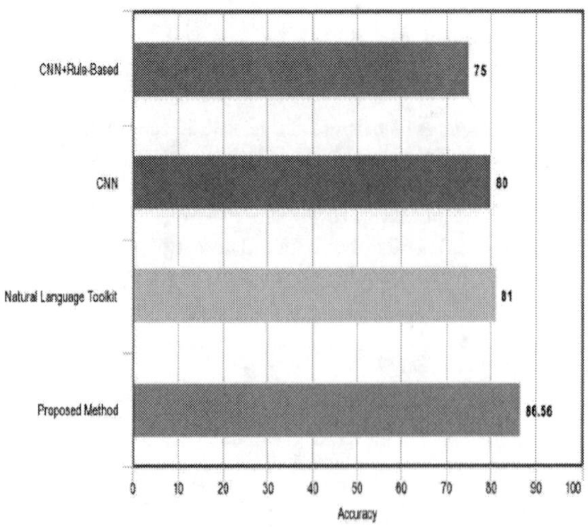

Figure 4.4. Effects of different systems on Accuracy

Upon testing on multiple sets of hyperparameters ,it was found that 64 LSTM nodes,2 Bi-directional Layers and a 0.2 Dropout rate provides optimum accuracy. Tf-idf text vectorization and ReLU activation function provides best accuracy. Using these parameters the proposed model has the testing accuracy of 86.56% while existing systems using CNN provides 80% accuracy. Existing systems using Natural Language Toolkit achieves an accuracy of 81%.

5 Conclusion

The proposed binary classification model could rate comments as positive or negative using Bi-LSTM. For optimal accuracy 64 LSTM nodes,2 Bi-directional Layers and a 0.2 Dropout rate should be used. Tf-idf text vectorization along with ReLU activation function provides best results.Accuracy of the proposed system is better than of CNN and Natural Language Toolkit. The model was tested on two different datasets and achieved comparable results.

References

[1] S.M.Qaisar, "Sentiment Analysis of IMDb Movie Reviews Using Long Sort Term Memory", 2020, *College of Engineering, Effat University, 21478, Jeddah, Saudi Arabia, IEEE.*

[2] A.Sajeevan & Lakshmi.K.S, "An enhanced approach for movie review analysis using deep learning techniques", Proceedings of the Fourth International Conference on Communication and Electronics Systems(ICCES 2019), pp. 1788-1794.

[3] Guixian Xu, Yueting Meng, Xiaoyu Qiu, Ziheng Yu & Xu Wu, "Sentiment Analysis of Comment Text Based on BiLSTM", *IEEE Access*, vol. 7, Mar 31. 2019, pp. 51522-51532.

[4] A.U.Rehman, Ahmad K.Malik, B.Raza & W.Ali, "A Hybrid CNN-LSTM Model for Improving Accuracy of Movie Reviews Sentiment Analysis", 2019, *Springer.*

[5] G.P.S.Varma, A.Govardhan & I.Hemalatha, "Sentiment Analysis Tool using Machine Learning Algorithms", *Elixir Comp. Sci. & Engg. 58, Elixir International Journal*, Jul. 2013, pp. 14791-14794.

[6] M.S.Solanki, "Sentiment Analysis of Text using Rule Based and Natural language Toolkit", *International Journal of Innovative Technology and Exploring Engineering (IJITEE), Blue Eyes Intelligence Engineering & Sciences Publication*, vol. 8, Oct. 2019, pp. 164-168.

[7] A.Chakrabarti, "A Mixed approach of Deep Learning method and Rule-Based method to improve Aspect Level Sentiment Analysis" *Applied Computing and Informatics, Emerald Publishing Limited*, Feb. 2019, pp. 163-178.

[8] P.Srivastava, Essentials of Deep Learning: Introduction to Long Short Term Memory, Dec 10, 2017. Accessed on: Feb 20, 2021. [Online].
Available: https://www.analyticsvidhya.com/blog/2017/12/fundamentals-of-deep-learning-introduction-to-lstm/

[9] R.Wolff, "Sentiment Analysis & Machine Learning", April 20, 2020. Accessed on: Feb 24, 2021. [Online].

Available: https://monkeylearn.com/blog/ sentiment-analysis-machine-learning/

[10] C.Olah, "Understanding LSTM Networks", Aug 27, 2015. Accessed on: Mar 15, 2021. [Online]. Available:https://colah.github.io/posts/2015-08-Understanding-LSTMs/

[11] Shiao-li Green, "Sentiment Analysis — A how-to guide with movie reviews", Towards Data Science, Dec 20, 2019. Accessed on: Mar 20, 2021. [Online]. Available: https://towardsdatascience.com/sentiment-analysis-a-how-to-guide-with-movie-reviews-9ae335e6bcb2

Algovis – An Enhanced Way to Visualize Algorithms

Rahul Lanjewar[1], *Nidhish* Mehra[1], *Aashka* Doshi[1,*] and *Yogita* Mane[2]

[1]Bachelor's of Engineering (Information Technology), Universal College of Engineering, Kaman, Vasai, India
[2]Head of Department (Information Technology), Universal College of Engineering, Kaman, Vasai, India

Abstract. It has been observed that learning using visual aids is less complicated than learning using traditional methods. As they say, "Words usually cannot do a picture's justice". Keeping the above fact in mind, we have come up with 'Algovis'. The aim of this project is to build a website that enables a user to visualize the steps and iterations involved in Data Structures and Algorithms using custom input and various animations. The program is menu driven to fit to the convenience of the user, depending on how many inputs he wants to enter, and implement the algorithm of his/her choice. The website also aims to become a convenient pedagogical tool. Algovis helps in visualizing various Data structure algorithms using graphical animations and with 100% accuracy.

1 Introduction

It is broadly seen that algorithm visualizations can give a strong option in contrast to static composed presentations (from course books), or verbal portrayals upheld by illustrations (from lectures). Algorithm Visualizations can improve the understanding of fundamental Data Structures and Algorithms.

Certainly, algorithm visualizations and animations have a long history in the field of computer science. Even so, most of these visualizers have been proven to have a bad user experience. Secondly, most of them have been least useful when it came to teaching, which means they give little to no information of how the algorithm being visualized works and so, they have no value in classrooms.

The purpose of this project will be to tackle these issues by extensive research, understanding and studying of various algorithms and developing a website using MERN stack that provide a superior client experience alongside a wide scope of elements. The venture will likewise plan to end up being a superior instrument for the teaching of Data Structures and Algorithms by utilizing custom visualizations and animations.

1.1 Motivation

We have observed and experienced as well that students struggle to understand the working and implementation of algorithms while learning them in a class. Even if they are successful in applying and running the algorithm practically, they might not necessarily understand what really goes on in the processing part when they are trying to run that algorithm. The dry runs can only be effective to a limit. Also, it is not possible to make dry runs for each iteration of every algorithm. Adding to it, due to the covid pandemic, almost the whole world shifted to online learning mode. Drawing and explaining the algorithms using dry run method was very difficult. Hence, to counter these problems to make learning easier and to encourage digital learning we have come up with this idea.

1.2 Problem statement

Currently, students learning Data Structures and Algorithms in various languages learn mostly through traditional methods like books, lectures, and innumerable dry runs. Through these methods they do understand what the algorithm does, but most of the students have problem imagining the implementation and working of these algorithms.

Secondly, even the traditional 'chalk and board' classrooms are being replaced by modern classrooms that use technological pedagogical aids. To add to the teacher's convenience, there should be a tool that explains by visualization of the processes that go on "behind the scenes" while implementing an algorithm. Though we realize that animations in the form of videos are available on various websites like YouTube, we can't have custom inputs and various use cases tested over there.

In an era where it's possible to experience a whole new virtual world through technology, there should be a way for students to learn with and teachers to teach with the use of these advanced technologies instead of using pen and paper.

2 Review of Literature

In paper [1], the authors have created a wiki of over 350 algorithm visualizers. They have analysed the quality, coverage, and effectiveness of these visualizers. The theoretical establishments for making compelling visualizations give off an impression of being consistently getting to the next level. More papers are seeming regarding successful utilization of visualization in courses, and a collection of work has started to shape with respect to how to foster compelling visualizations in any case. Collectively, the community is learning how to improve. *The disadvantage of this paper is that it is seen that most existing algorithm visualizations are of inferior quality, and the substance inclusion is slanted vigorously toward more straightforward points. There are no*

*Corresponding author: aashkadoshi7@gmail.com

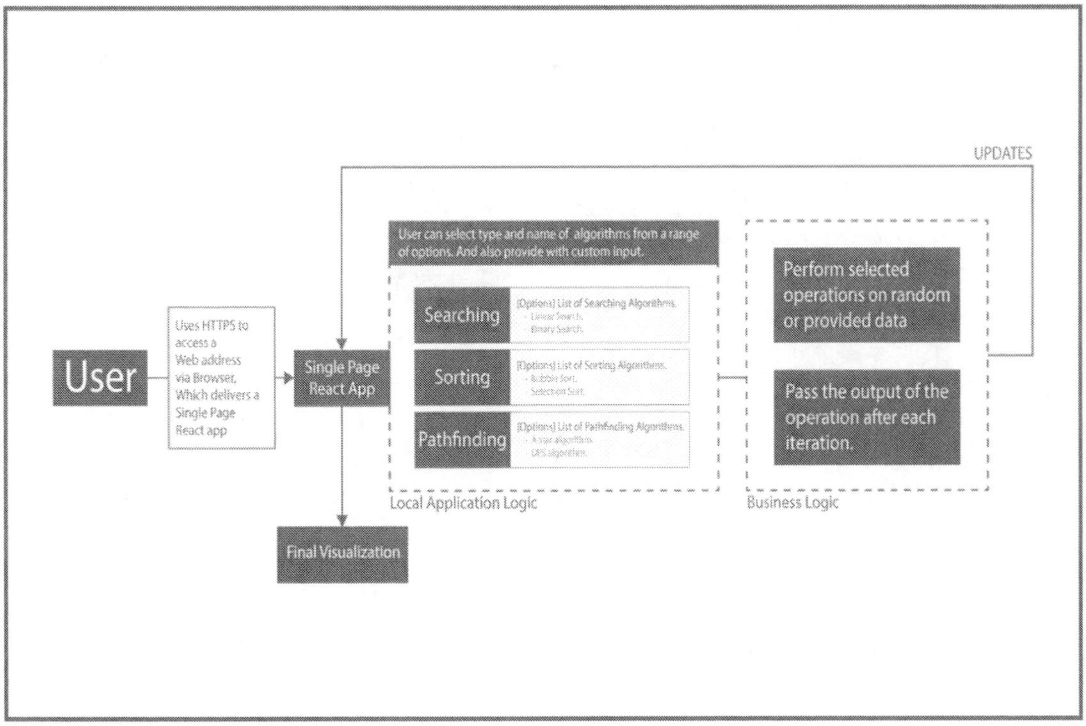

Fig.1. Algovis: A System Architecture

powerful repositories or coordinated assortments of algorithm visualizations as of now accessible.

Algorithm visualization uses computer graphics or interactive media to show the steps of an algorithm. In this research, the algorithm visualization is utilized to assist understudies with understanding the idea of selection sort algorithm and creating coding to picture it. The objective of the paper [2] is to create visualization of Selection Sorting Algorithm. The visualization shows how all data move to the proper position to be sorted. It can be easily understood by students how the algorithm should be implemented while writing actual code. Students easily understand the concept of selection sort by looking at the visualization.

Learning becomes easier when graphic, animation, or video means are used by student. Adobe Flash is a timeline-based authoring and object-oriented programming tools that can be utilized to foster a logical visualization. Adobe Flash with ActionScript programming tools can be utilized to make logic and mathematics operation. *The disadvantage here is that there are no dynamic input options. The visualization does not show which step in the algorithm is being implemented. The design is not very interactive.*

The authors in this paper [3] have analysed different software visualizers and noted down why the development of software visualizations is difficult and what features are lacking in current technologies. The researchers have worked towards the development of generalized algorithm visualizers. *This paper does not show searching, sorting and path finding algorithms.*

Algovis visualizes searching, sorting and pathfinding algorithms with 100% accuracy and the animations are easy to understand. It is an open source webapp whose repository is available on GitHub "https://github.com/RahulLanjewar93/AlgoVisualizer" [4]. It enables the user to give dynamic input and shows which exact step is being implemented currently.

3 System Architecture

There are three modules in our system namely searching, sorting and path finding which can be seen in Fig. 1. The user can navigate between these modules through the navbar at the top.

The user can access our web app which is developed using react. With the help of react it is possible to update the DOM (Document object model) effectively. The user can select the algorithm they want to perform on the landing page as well as a navigation bar is provided for ease of access. After selecting the algorithm, the user will be redirected to a dedicated algorithm page which will display the steps and visualization of the algorithm.

In sorting, we have provided options on which type of sorting the user wants to perform. The user can use the random array generated by the app or enter a custom array. After clicking on the start button, the user will be able to see the steps of the algorithm and which step is currently being performed in real-time and the animated graph where user will be able to see how the sorting works. A new array will be formed after each iteration unless it's sorted.

In searching, the user either uses randomly generated array or input a custom array. Then chooses which type of searching must be performed. Then the user enters the number which must be searched and press the start button. Then the user will be able to see the steps of the algorithm and which step is currently being performed in real-time and the animated graph where user will be able to see how the searching works. The index number on which the entered number is present will be given as a result. If the number is not there, then "Number not found" message will be shown.

In pathfinding, we show a grid having walls and nodes. There are various algorithms from which the user can choose. After selecting the algorithm, the user will be able to see the animation through which he will understand how the algorithm works.

4 Requirements

- Personal Computer/ Laptop.
- Internet Connection.
- Any browser.

5 Performance Analysis

5.1 Bubble Sort

Considering the size of the array and varying speed of users, the entire dry run might take on an average 5 minutes to complete. Apart from that, this method might not be 100% accurate if we keep human errors in mind.

The system that we have developed (Fig. 2) visualizes the bubble sorting of an array input of [41, 6, 43, 50, 23, 23, 57, 47, 32, 52, 22, 1, 20, 35, 15, 27, 36, 16, 44, 61] along with constantly highlighting the step which is currently being performed in approximately a minute. Along with that, the accuracy of the visualization is 100%. [5, 6]

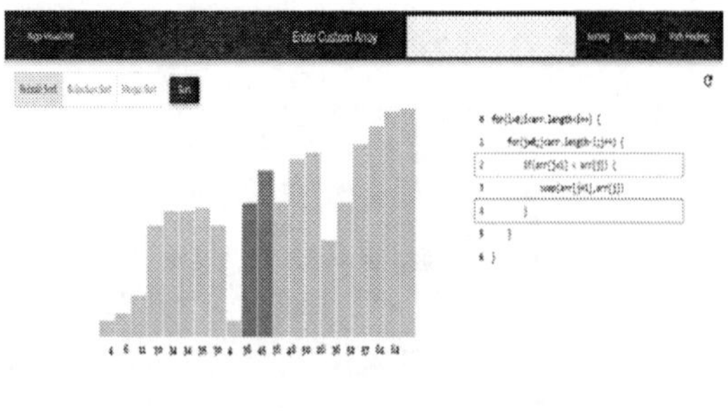

Fig.2. Bubble Sort Animation

Note the total time taken by the algorithm is directly dependent upon the animation speed + $O(n^2)$.

5.2 Selection Sort

Considering the size of the array and varying speed of users, the entire dry run might take on an average 10 minutes to complete. Apart from that, this method might not be 100% accurate if we keep human errors in mind.

The system that we have developed (Fig 3) visualizes the selection sorting of an array input of [1, 3, 6, 63, 40, 55, 38, 25, 21, 19, 8, 52, 36, 35, 8, 13, 37, 34, 62, 7] along with constantly highlighting the step which is currently being performed in approximately two minutes. Along with that, the accuracy of the visualization is 100%. [5, 6]

Note the total time taken by the algorithm is directly dependent upon the animation speed + $O(n^2)$.

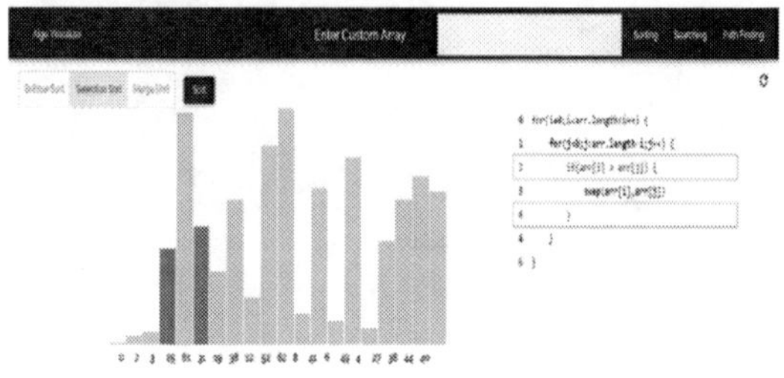

Fig.3. Selection Sort Animation

5.3 Merge Sort

Considering the size of the array and varying speed of users, the entire dry run might take on an average 12 minutes to complete. Apart from that, this method might not be 100% accurate if we keep human errors in mind.

The system that we have developed (Fig. 4) visualizes the merge sorting of an array input of [46, 44, 8, 32, 45, 25, 35, 62, 15, 2, 36, 33, 26,47, 20, 34, 27, 38, 33, 56] along with constantly highlighting the step which is currently being performed in approximately two minutes. Along with that, the accuracy of the visualization is 100%. [5, 6]

Note the total time taken by the algorithm is directly dependent upon the animation speed + O(nlogn).

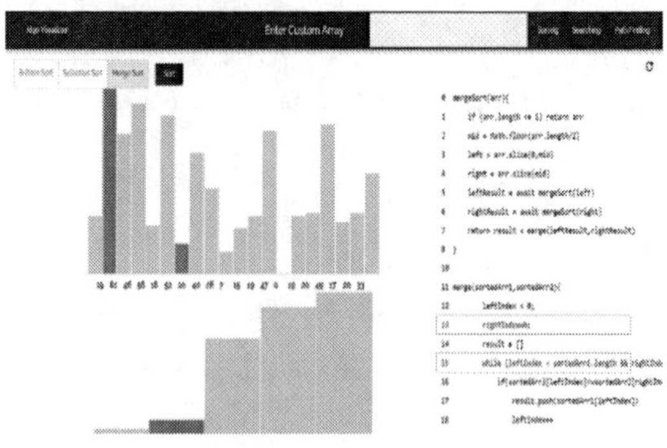

Fig.4. Merge Sort Animation

5.4 Linear Search

Considering the size of the array and varying speed of users, the entire dry run might take on an average 5 minutes to complete. Apart from that, this method might not be 100% accurate if we keep human errors in mind.

The system that we have developed (Fig. 5) visualizes the linear searching of an array input of [16, 35, 55, 37, 25, 27 ,46, 53, 36, 62, 37, 3, 57, 54, 51, 60, 54, 46, 47, 32] along with constantly highlighting the step which is currently being performed in approximately under a minute. Along with that, the accuracy of the visualization is 100%. [5, 6]

Note the total time taken by the algorithm is directly dependent upon the animation speed + O(n).

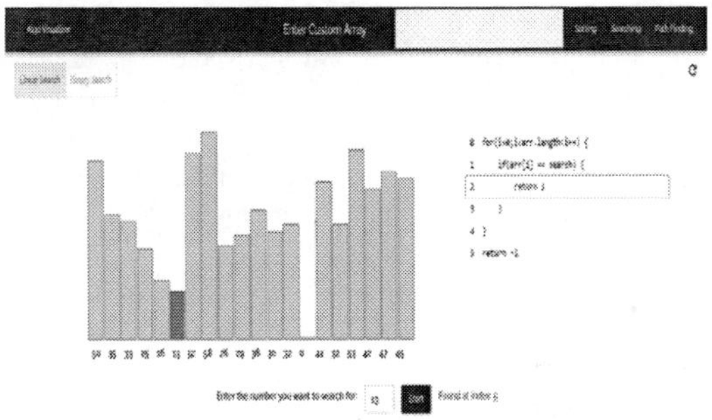

Fig.5. Linear Search Animation

5.5 Binary Search

Considering the size of the array and varying speed of users, the entire dry run might take on an average 10 minutes to complete. Apart from that, this method might not be 100% accurate if we keep human errors in mind.

The system that we have developed (Fig. 6) visualizes the binary searching of an array input of [0, 7, 8, 13, 15, 15, 16, 18, 18, 24, 29, 36, 37, 37, 47, 50, 51, 58, 58, 63] along with constantly highlighting the step which is currently being performed in approximately under a minute. Along with that, the accuracy of the visualization is 100%. [5, 6]

Note the total time taken by the algorithm is directly dependent upon the animation speed + O(logn).

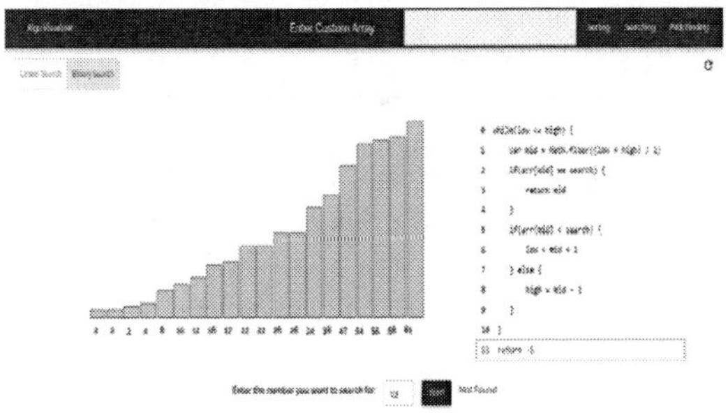

Fig.6. Binary Search Animation

5.6 A star path finding algorithm

Considering the number of nodes and varying speed of users, the entire dry run might take on an average 15 minutes to complete. Apart from that, this method might not be 100% accurate if we keep human errors in mind.

The system that we have developed (Fig. 7) visualizes the A star path finding while constantly highlighting the step which is currently being performed in approximately under a minute. Along with that, the accuracy of the visualization is 100%. [5, 6]

Note the total time taken by the algorithm is directly dependent upon the animation speed.

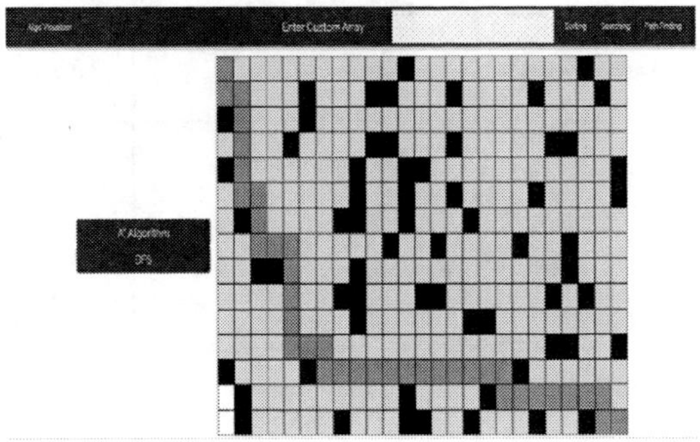

Fig.7. A star path finding Algorithm Animation

5.7 Depth first search algorithm

Considering the number of nodes and varying speed of users, the entire dry run might take on an average 12 minutes to complete. Apart from that, this method might not be 100% accurate if we keep human errors in mind.

The system that we have developed (Fig. 8) visualizes the Depth First Search algorithm while constantly highlighting the step which is currently being performed in approximately under a minute. Along with that, the accuracy of the visualization is 100%. [5, 6]

Note the total time taken by the algorithm is directly dependent upon the animation speed.

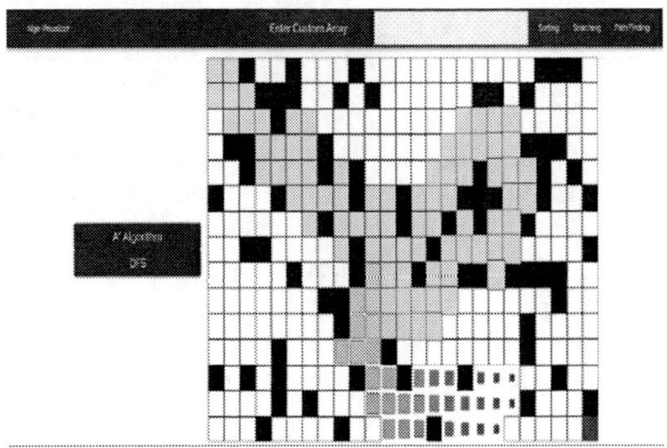

Fig.8. Depth First Search Algorithm Animation

Table 1 shows performance analysis of Algovis.

Algorithm	Time	Accuracy
Bubble Sort	Animation speed + $O(n^2)$	100%
Selection Sort	Animation speed + $O(n^2)$	100%
Merge Sort	Animation speed + O(nlogn)	100%
Linear Search	Animation speed + O(n)	100%
Binary Search	Animation speed + O(logn)	100%
A* Algorithm	Animation speed + $O(b^d)$	100%
DFS	Animation speed + O(V)	100%

Table 1. Performance Analysis

6 Comparative Analysis

	algorithm - visualizer [7]	Visualgo [8]	UFSCA [9]	Proposed work
Good UI	No	No	Yes	Yes
Custom Input	No	No	No	Yes
Real-time algorithm steps	Yes	Yes	No	Yes

Table 2. Comparative Analysis

Table 2 gives brief idea about comparisons of different tools available online.

7 Conclusion and Future Plans

Our future plan is to add more data structures' visualizations. Along with that we aim to improve the UI/UX of the website.

Apart from that we will add various features that will ease the work of students as well as teachers. Also, we aim to remove as many bugs as possible.

In conclusion, Algovis is a free, open-source pedagogical tool that enables the teachers to teach data structures conveniently and the learners to learn data structures easily while promoting remote and digital learning.

References

1. C. Shaffer, M. Cooper, S. Edwards,Algorithm visualization: a report on the state of the field, SIGCSE Bull. 39, 1 (March 2007), 150-154 (2007)
2. H. Sutopo, selection sorting algorithm visualization using flash, Int. J. of Multimedia & it's Applications, vol. 3, no. 1, (February 2011)
3. V. Karavirta, Effortless creation of Algorithm Visualization (2002)
4. "Source Code", Accessed: 12th October 2021 [Online], Available: https://github.com/RahulLanjewar93/AlgoVisualize
5. "Algorithms", Accessed: 18th October 2021 [Online], Available: https://www.geeksforgeeks.org/fundamentals-of algorithms/?ref=shm
6. "Algorithms", Accessed: 18th October 2021 [Online], Available: https://www.tutorialspoint.com/data_structures_algorithms/index.htm
7. "Algorithm-Visualizer", Accessed: 27th February 2022 [Online], Available: https://algorithm visualizer.org/
8. "Algorithm-Visualizer", Accessed: 27th February 2022 [Online], Available: https://algorithm visualizer.org/
9. "Algorithm-Visualizer", Accessed: 27th February 2022 [Online], Available: https://algorithm visualizer.org/
10. "USFCA", Accessed: 27th February 2022 [Online], Available: https://www.cs.usfca.edu/~galles/visualization/Algorithms.htm
11. "DSA", Accessed: 27th February 2022 [Online], Available: https://www.programiz.com/dsa/
12. "Learn DSA using Animations", Accessed: 27th February 2022 [Online], Available: https://www.youtube.com/playlist?list=PLmYemBEXGmVjAlx4K9JkyVTiFFayzDNwi

ITM Web of Conferences **44**, 03014 (2022) https://doi.org/10.1051/itmconf/20224403014
ICACC-2022

Smart QR-based Restaurant Dine-in System with Sales Analysis

Riteshkumar Singh[1], *Rupal* Sonje[1], *Soham* Salkar[1], and *Ashish* Jadhav[1,*]

[1]Ramrao Adik Institute of Technology, DY Patil Deemed to be University, Nerul, Navi Mumbai, India

Abstract. The modern-day customer has new priorities when it comes to dine-out and the top of them is safety. Most of the restaurant's meal orders rely on the interaction with waiters to place orders into the kitchen. Due to the pandemic situation, there happens to be discomfort interacting with waiters and other staff members of the restaurant. To solve such problems this system is developed. This system covers the whole order process of a restaurant including the interaction between the customer, the waiter, the kitchen, and the cashier through a web application. Additionally, restaurant owners will be able to enhance sales by using this system to analyze data gathered. This system will attempt to replace the traditional manual ordering process.A better user experience that includes food recommendations can indirectly boost customer loyalty to the restaurant. It is a complete product for managing restaurant services with minimal human interaction while providing maximum contactless service.

1 Introduction

Almost every industry is embracing the road to digital transformation and the restaurant sector is no exception. Despite it being classified as an essential business and continuing to reach out to its customers in one way or the other, it incurred heavy losses. The traditional paper based ordering method is being used in most of the restaurants worldwide. In this system the whole process of checking the menu, ordering food to bill receipt is done using paper. Here, the waiter takes the order from the customer and forwards it to the kitchen. There are a lot of problems associated with this system. The most common stumble is that waiters may make mistakes with customers' orders. At times, a waiter may forget to add a specific item ordered by the customers and make changes and forget to give the updated order to the kitchen. It is extremely important for the waiters to rely on the chefs to inform them whether the food is ready or not. The restaurant business has been adversely affected because of the covid pandemic. Hygiene and safety is the prime concern of consumers when they think of moving out to dine. People are afraid of catching the virus in public places and thus, order everything online. This has immensely affected the footfall that a restaurant receives and eventually their sales. Due to the covid norms imposed by the government, the restaurants have to reduce their opening hours and capacity. Many restaurants are using printed laminated paper menus. Since printed menus are one of the most touch surfaces in any restaurant as they go through different hands. They require constant sanitation and also such menus are harmful to the environment. The designed system will help the restaurants to overcome such consequences by means of having a digital menu which is an essential part of contactless dining. This system aims to automate the entire traditional paper-based menu ordering to a digital and smart ordering system using the QR code. The digital ordering system enables the user to register and log in followed by ordering food and viewing the status of their order. Digital menus can be equipped with recommended and top-selling dishes based on past orders and restaurant suggestions. They can easily add items to their order and pay online. Order details are also forwarded to their registered WhatsApp number. They can also access their previous orders in the restaurant. The manager can view all the orders and the table number from where the order is placed with the customer's details. Any new incoming order will be notified to the chef where he can accept the order and start with his preparation. The manager can also view the payment status. Manager can also see the real-time data of sales and orders. The sales data is graphically visualized to give better insight to the restaurant owner. Also during rush hours the load on the waiters and the chefs will be reduced, and restaurants will perform better than usual and also the human error that occurs when performing tasks manually is also minimized and the presence of queues in the system to accept orders and assign tasks to chefs can reduce congestion in the kitchen. This can be used for better service and planning. This system thereby reduces the manpower and eases the systematic operation of the restaurant.

*e-mail: ashish.jadhav@rait.ac.in

2 Motivation

The motivation for the project comes from the fact that, although there are multiple types of management systems available, they do not follow contact-less dining practices, which is a necessity today. Restaurant dining is unsafe unless we follow contactless practices. In light of the current social conditions, having in-contact dining practices becomes a concern and a threat. To provide a healthier and safer digital restaurant environment in the upcoming days of a better world, our project is taking measures. Developed under a highly efficient digital platform with a visually understandable user interface and application that captures accurate data and generates efficient reports for restaurant management, the solutions combine all the modules necessary to manage a restaurant under one roof.

3 Literature Survey

As technology advanced, a variety of new systems were developed to automate the ordering process. A familiar system which was implemented in various restaurants was a Bluetooth based ordering system where the peripheral interface controller consists of a keypad which acts as a remote control and is placed on the customers table connected to a monitor placed in the kitchen using a Bluetooth module [6]. The Microcontroller will transmit the data via Bluetooth transmitter which will be processed and displayed on the monitor [11]. This system was improved by replacing the PIC board with a touch screen panel and transmitting the data through a Wifi module [7]. Speech recognition technology were also added to place the order using speech commands [15]. Since touch screen devices are currently in vogue, it did not take long for tablet-based menus to make an appearance on the market. Each table will be equipped with a touch screen tablet with an android application installed [10]. A smart menu will be available, and customers can order food by selecting the dish they want and placing the order [9]. A chef's interface will display the placed order, and the order will be marked as completed once it has been prepared [4]. It is possible for the manager to view orders and make changes to the menu at any time [1]. Customers can make their payments in cash or by swiping a credit or debit card after the order process [12]. In addition to checking and verifying the current status of orders, the admin can also check if the payments have been received [8]. The system allows customers to provide real-time feedback at the end of the meal [5]. The recommendation system were also added which focuses on customer reviews and location to provide personalized recommendations to customers [14].Based on previous orders, the Recommendation algorithm suggests dishes to the customers so that they can build his or her order and easily view the most popular dishes [13]. An android application was made available on the Play Store in order to reduce device costs [2]. In order to further improve the system, offers on food and bill payment have been added [3] .

4 Problem Statement

It was recognized that the management system of a restaurant is still being manually operated by waiters and managers and is not automated yet. Due to the covid situation, interacting with restaurant personnel has created discomfort.It is found that transmission of viruses is higher when people interact in a closed environment. Also, most staff and customers wear face masks which muffle conversation, the chance of miscommunication, misordering, or misinterpretation when receiving orders is greater. It is also a problem that customers have to wait until the waiter responds to them. There are constraints on updating available dishes due to printed menu cards. Customers need to ask the waiter if a particular dish is available or not. Customer has to toil for improvising the dish according to his taste in a crowded quarter. Poor customer service on holidays and weekends when the number of customers increases. This project aims to provide a full-featured system that includes order and payment integration, rather than a menu-only solution.

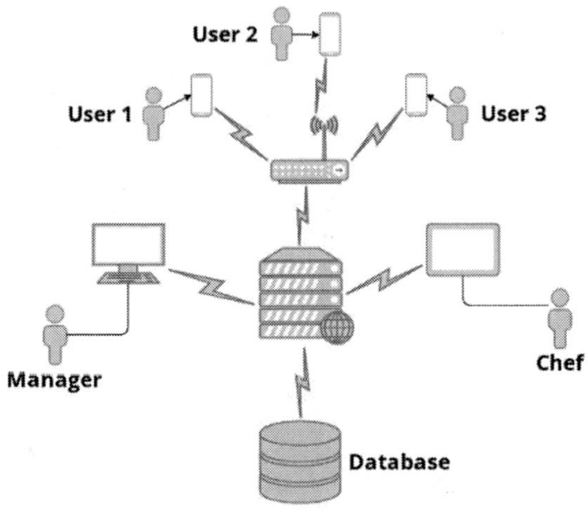

Figure 1. Proposed system architecture

5 Proposed system

The proposed system automates all aspects of the dining experience. The restaurant follows a contact-less dining norm by reducing the human contact in terms of waiters at service to a great extent with the customers. The QR Code on each table redirects the user to the UI of the system with the specific table number. The graphical interface allows registered users to order food and pay bills using their smartphones. The user can also view their past orders, their feedback and ratings, and favorite food items. A confirmation of the order is sent to the customer's registered Whatsapp number after the order is placed. Chefs are notified when new orders arrive. System access is restricted to each department. Data about sales can be analyzed for purposes such as tracking the most popular products, new customer counts, and revenue stats for daily or monthly periods. Restaurant managers can edit the menu items of their restaurant as required and can manage users as well. A daily and monthly report of sales data and user data can be generated and downloaded in CSV and PDF formats. Proposed system includes following modules:

1. Customer Modules

https://doi.org/10.1051/itmconf/20224403014

2. Manager Modules
3. Chef Modules

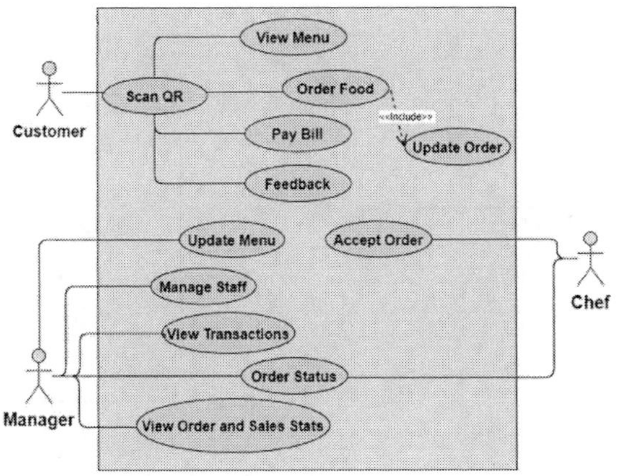

Figure 2. Use Case Diagram

6 Detailed Description of System

Customer modules

The application provides all the modules the customer needs when they visit the restaurant, such as viewing the digital menu, placing an order, and paying online.

Customer modules contains the following:

- **Scanning module:** This is the first module when a customer is starting with the process. When the customer arrives at the restaurant, he scans the QR code on the table which redirects him to the restaurant's website to continue. The QR code contains a unique table ID on top of a URL that will be stored for later process.
- **Login/ Registration module:** In this module, customers are required to log in to the website if they are already registered, using their unique username. New customers must complete the registration process before logging in. The customer must select the number of people with him and continue with the process after logging in.
- **Menu display module:** Customers can view the current menu of the restaurant using this module. In addition to the different categories of dishes, the top selling and recommended dishes are also displayed so they can see what is popular.The product image is included, along with the product name, price, and some description.
- **Cart module:** The customer's cart gets automatically updated as soon as an item is added, showing the subtotal as well as the order quantity. Customers can edit their cart at any time before they place an order.To prevent data loss, items in the cart are also stored in the session. When an order is placed, customers can add additional items and update their orders before making a payment.
- **View previous orders module:** This module allows customers to view past orders in detail, including total prices, payment IDs, and the date of each order.
- **Payment module:** In this module, as soon as a customer clicks on the 'PAY' button, he can view his final bill and proceed with payment. The payment module is integrated with the testmojo API, which provides a secure payment gateway. After completing the payment process, the customer can either order again or end his session followed by a feedback.
- **Feedback Module:** In this module, customers can evaluate each dish, comment on the ordering process, and suggest improvements to the restaurant's hospitality.
- **Notification module:** A notification will be sent by WhatsApp with order details upon placing an order. After successful completion of payment, customers will receive a notification with the amount received for an order ID.Payment receipts are also sent by testmojo API to their registered email address.

Manager modules

This includes all the features a manager needs to manage a restaurant. This includes analysis of orders, product management, order and sales management.

Various modules included here are:

- **Analysis module:** This module provides a variety of data visualizations to the restaurant manager including bar graphs, line graphs, and pie graphs. The module also gives customer and product statistics as well as order information and daily and monthly revenue estimates.
- **Product management module:** Managers can perform CRUD operations on products, such as adding categories, adding new products based on categories, updating the current products, and deleting them. Furthermore, they can manage the current date's menu, based on its availability.
- **Sales module:** Managers can get all the details of a product's sales with this module. It contains the details such as product details, customer ID, quantity ordered, order ID and date of order. Incoming orders are notified by beep sound to the manager and a record of the seen/unseen status of each order is maintained.
- **Bill history module:** This module allows the manager to view all billing information, including past orders, and delete them. It provides additional information such as total price, payment status, and payment ID for each order.
- **Report module:** This module helps the manager to download the report of sales, orders and customers in various formats like PDF and CSV.

Chef Modules

It includes modules which help the chef to view the incoming orders in a detailed manner and update the status of each order.

Modules included here are:

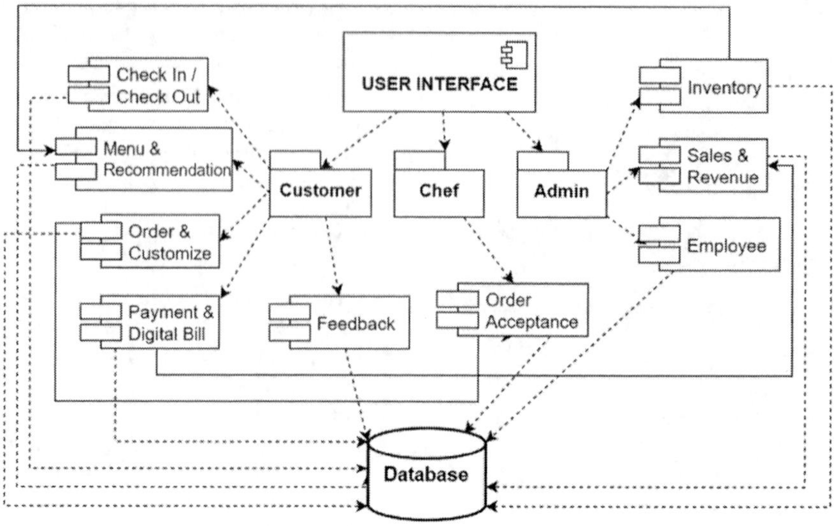

Figure 3. Detailed design of system

- **View orders module:** It allows the chef to be notified with a beep sound of when an order is coming in. The chef can then view a detailed list of what is in each order along with the quantity.
- **Update Status module:** This module allows the chef to update each order's status to 'Accepted', 'In Progress', or 'Rejected'.

7 Outcome and Results

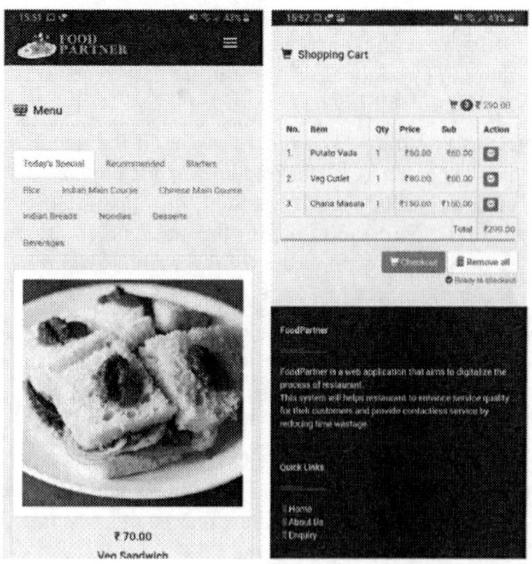

Figure 4. Smart Menu and cart view

The customer lands on the index page as shown in Fig. 4, after scanning the QR code placed on the table where he can see the menu, including today's special, recommended dishes, and various dishes sorted by category. On this page, the customer can view each dish details and put them in the cart.

The order is sent to the chef for confirmation, and the customer is notified after confirmation. The customer then reviews the final bill and makes payment. Upon completing the payment part, the customer receives a notification on their registered WhatsApp number, after which they can then submit their feedback.

In the Manager dashboard, the manager can see various analysis data, such as the number of customers present and revenues today. They are notified of any incoming orders, and they can view all sales data including past orders. They can also download these data as PDFs or CSVs.This dashboard is built using technologies like PHP and various data visualization libraries, including CanvaJs and HighCharts Js.

Figure 5. Daily Revenue and order count

Manager dashboard includes various graph based on sales data like date-wise/monthly revenue is shown with a bar chart, the orders count is shown with a line graph, the new-customers ratio and top-selling products are shown with a pie chart, and the current date order details are shown with a bubble graph. This dashboard is built using technologies like PHP and various data visualization libraries, including CanvaJs and HighCharts Js.

An example of this is seen in Fig. 5, where the bar graph shows the revenue generated on a daily basis and the line graph shows the order count on that day. This information allows the manager to analyze revenue generation and the frequency of orders being placed in the restaurant.

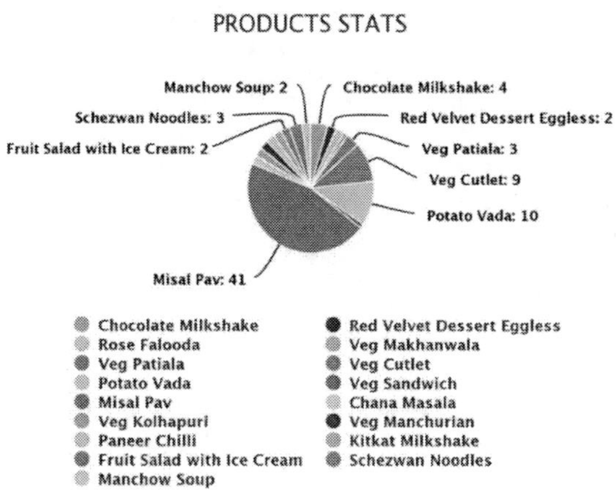

Figure 6. Product Stats

The pie chart in Fig. 6 shows the sales of dishes in the restaurant in real time. This information helps the restaurant manager to identify the most popular dishes for customers to order. Manager can also view this in full screen mode and even download it in form of report.

Manager can also the see the count of new customer coming to restaurant on that specific day. Also the manager can view the order details of specific day.

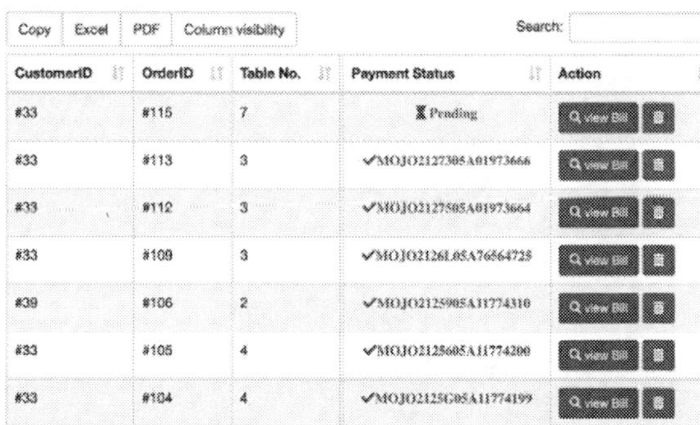

Copy | Excel | PDF | Column visibility   Search:

CustomerID	OrderID	Table No.	Payment Status	Action
#33	#115	7	Pending	view Bill
#33	#113	3	✓MOJO2127305A01973666	view Bill
#33	#112	3	✓MOJO2127505A01973664	view Bill
#33	#109	3	✓MOJO2126L05A76564725	view Bill
#39	#106	2	✓MOJO2125905A11774310	view Bill
#33	#105	4	✓MOJO2125605A11774200	view Bill
#33	#104	4	✓MOJO2125G05A11774199	view Bill

Figure 7. Bill History view

Fig. 7 represents the bill history where all the order details are present along with its payment status. Manager can view each detail order, can also delete them if they want and can also download these data as reports in PDF and CSV format.

8 Conclusion

The proposed system provides an easy, convenient, and low-cost method for ordering meals with a contactless dining experience. By encouraging no-contact interaction and personal hygiene, restaurants can lower their risk of transmitting viruses. As a result, diners who find that hygiene and safety standards are satisfactory are more likely to become repeat diners. Such repeat diners promote restaurants for their value to customers, ultimately improving the goodwill of the restaurant. Furthermore, contactless dining helps improve order accuracy, preventing miscommunication between staff and diners. This allows diners to focus on food, ambiance, and service. Machine learning can be applied to gathered data to make food recommendations and boost restaurant revenue. In the restaurant business, data analysis is essential for increasing income and sales. During rush hours, the restaurant load would be balanced effectively, and customer service would be better than usual. By including reservation and take-away features, restaurants can improve even further. Thus, investing in a contactless dining solution will not only provide a safer dining experience in the short term but also position the restaurant to meet changing consumer behavior in the future.

References

[1] A. Patil, R. Kalani, B. Patil, S. Shinde, Prof S.M. Shedole, IJTRA, Smart restaurant system using android, **5**, Issue 3 pp. 78-80 (2017)

[2] Kavitha S. , Shailaja K. , Suma T.N. , V. Shreenidhi, Asst Prof. Hari H. , IJTRE, Paperless restaurant system, **6**, Issue 9 pp. 5554-5557 (2019)

[3] S. Khedikar, S. Baghele, V. Muley, P. Shah, A. Bagade, P. Wagh, S. Pusdekar, IRE, Digital food ordering system for restaurants, **3**, Issue 10 pp. 182-185 (2020)

[4] S. Sarkar, R. Shinde, P. Thakare, N. Dhomne, K. Bhakare, IJTRA, Integration of touch technology in restaurants using android, **5**, Issue 3 pg. 721-728 (2017)

[5] S. Dimbar, A. Kumbhakarna, P. Shend, IJEET, Survey of digital food ordering system based on android system for restaurant, **5**, Issue 1 (2017)

[6] N.M.Z. Hashim, N.A. Ali, A.S. Jaafar, N.R. Mohamad, L. Salahuddin , N.A. Ishak, IJCTT, Smart ordering system via bluetooth, **4**, Issue 7 pg. 2253-2256 (2013)

[7] S. Thakre, P. Sakharkar, S. Bhalerao, S. Meshram, Prof. V. Dhumal, IRJET, Review on implementation of touchscreen digital menu card for hotel, **5**, Issue 2 pg. 932-934 (2018)

[8] M.P. Chorage, S.S. Bhande, S.S. Lale, M.A. Powar, K.V. Mane, IJARIIE, Digital restaurant system, **7**, Issue 1 pg. 756-760 (2021)

[9] Prof. N.M. Yawale, Prof. N.V. Pardakhe, Prof. M.A. Deshmukh, Prof. N.A. Deshmukh, IAETSD-JARAS, Online restaurant management system, **4**, Issue 7 pg. 92-96 (2017)

[10] K. Bhandge, T. Shinde, D. Ingale, N. Solanki, R. Totare, IJARCST, A proposed system for touchpad based food ordering system using android application, **3**, Issue 1 pg. 70-72 (2015)

[11] Sadiq B.G., Shanthi A., Madumitha P., IJIRSET, Wireless menu ordering system for restaurant, **4**, Issue 12 pp 12284-12291 (2015)

[12] V. Gupta, N. Gaddam, L. Narang, Y. Gite, IRJET, Digital restaurant, **7**, Issue 4 pg. 5340-5344 (2020)

[13] V. Ranjan, N. Masiwal, N. Verma, Dr. V.A. Bharadi, IJACSA-ICWAC, e-Restaurant: online restaurant management system for android, **3**, Issue 1 pg. 48-54 (2013)

[14] N. Kadale, P. Bansod, R. Pillai, S. Sane, S. Pratape, S. Pawar, IJLTEMAS, Wireless customizable food recommendation system using apriori And K-Means algorithm, **IV**, Issue X pg. 97-100 (2015)

[15] M.F.A. Khan, V. Swapna, IJERT, Design and implementation of ordering system for restaurants, **1**, Issue 10 pg. 01-05 (2012)

Detection of Counterfeit Products using Blockchain

Kunal Wasnik[1,*], *Isha Sondawle*[1], *Rushikesh Wani*[1], and *Namita Pulgam*[1]

[1]Ramrao Adik Institute of Technology D Y Patil Deemed to be University Navi Mumbai

Abstract. Supply chain management frequently faced issues such as service redundancy, poor coordination between several departments, and lack of standardization as a result of the lack of transparency. Product counterfeiting is something which is very common now-a-days and it's almost impossible to detect a counterfeit product just by looking at it. Counterfeiters cause significant challenges for legitimate firms, yet far too many people have no idea of the entire amount of counterfeit items' influence on brands. There are several methods devised in the past to get away with this problem of product counterfeiting. The most popular methods are using RFID tags, Artificial Intelligence, QR code based systems, etc. But each of them had few disadvantages such as the QR code can be copied from a genuine product and placed on a fake product, artificial intelligence uses CNN and machine learning which needs heavy computational power and so on. The idea of this project is to improve detection of fake products by tracking its supply chain history. This is achieved with Blockchain technology which ensures the identification and traceability of real products throughout the supply chain. Blockchain based system, makes everything decentralized that may be accessed by several parties at the same time. One of its main advantages is that the recorded data is difficult to change without the consent of all parties concerned which makes the data extremely secure and protect from all vulnerabilities. This paper presents system designed using blockchain technology for detection of counterfeit products.

1 Introduction

Product counterfeiting happens when a product is sold pre- tending to be another product. It is consumer fraud and commonly defined as deceptive business practices that cause consumers to suffer financial or other losses. According to the Authentication Solution Providers' Association reports it costs the Indian economy INR 1 trillion every year. Counterfeit incident are increasing by 20% average in between 2018-20.[1]. Counterfeit goods include counterfeit handbags, clothing, cosmetics, and electronics. It not only has negative effects on the economy, but on citizens too. For example poor cosmetics can affect skin and cause skin diseases and rashes, counterfeit electronic components can cause malfunction in gadgets and can lead to unfavorable situations and mishaps. Poor quality clothes, shoes when worn can cause discomfort. Hence this issue necessitates finding some solution for the sale of counterfeit products.

Another consequence of counterfeiting is that a company's reputation suffers. Because many customers are clueless that the object they are holding is a knock-off, they will accuse the genuine company if the knock-off product fails to perform properly, comes apart rapidly, or fails to satisfy their expectations. Customer demand recompense, either in the form of a refund or a new product, and they seek it out directly from the legitimate company. A lot of affected businesses may find themselves in a scenario where they are dealing with an unhappy customer who is complaining about the bad quality of the item, and the customer care representative is unaware that the item in question is a counterfeit. Companies are caught between a tough situations, attempting to avoid wasting time and effort dealing with poor imitations of their goods while yet trying to keep their customers pleased. The harm caused by counterfeiters extends beyond customer relationships. Because of the behaviors of counterfeiters, distributors, retailers, and other business partners frequently lose faith in legitimate enterprises.

The most successful mitigation measures for overcoming misleading counterfeit risk in global supply chains include network transparency, cost control and pre-supply evaluation approaches, and supplier relationship management. Hence the objective of this paper is to present the system designed for anti-counterfeit using Blockchain technology and to give end user and supplier power to track supply chain of product in a secured environment. In an overview of proposed system, it is aimed to solve the problem of brand counterfeiting and provide the chance to the customer, vendors and suppliers to check the integrity of the product.

The paper is organized as follows: The detail explanation of Blockchain with its working and features is mentioned in section 2. The section 3 offers a comprehensive review of the literature. The proposed System in Section 4 includes the system model and the flow of the system. The simulation results for the proposed method are presented in Section 5. Section 6 is where the paper comes to a close with the conclusion.

*e-mail: wasnik.kunal4@gmail.com

https://doi.org/10.1051/itmconf/20224403015

2 Blockchain

Blockchain is collection blocks that are linked together which stores information. Each block has a timestamp, transaction data and hash of its own and hash of previous block, so it is difficult to tamper with data. Blockchain is a decentralized system. It ensures that every new block added to the blockchain is the one and only true version that is agreed upon by all nodes in the Blockchain. It refers to the collective maintenance of a technical solution that maintains a continuous record file as a reliable database through decentralization[2][3][4].

2.1 Working of Blockchain

When a new transaction is entered, it is then transmitted in a network of peer-to peer computers scattered across the world. The network of computers then solve the equations to confirm the validity of the transaction. They are called miners. Once confirmed to be legitimate transactions, they are clustered together into blocks. The miner then receives an award as a proof of work. These blocks are then chained together creating a long history of all transactions that are permanent. The transaction is complete.Whole procedure is done as shown in figure 1.[5][6]

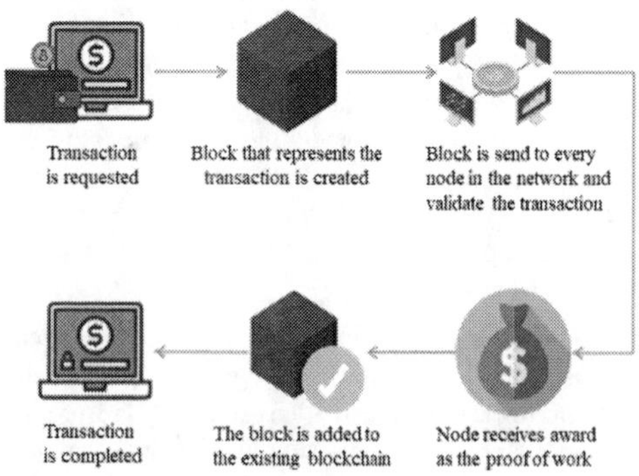

Figure 1: Working of Blockchain [6]

2.2 Blockchain Features

Blockchain can add data records to its database which does not depend on any centralized authority as a arbitrator, instead it works on its own consensus algorithms. Blockchain is openly available database and is highly reliable. The features of Blockchain technology are described in detail below. The features of Blockchain are shown in figure 2.

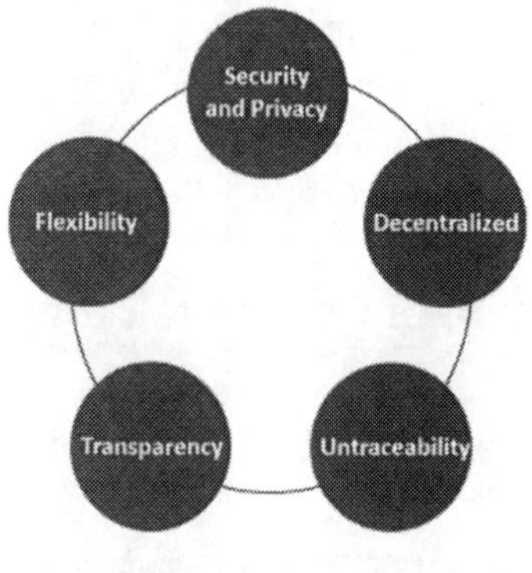

Figure 2: Features of Blockchain [7].

1. Security and privacy : Blockchain uses cryptography to secure its data. Private key is used to sign the data, using public key we can verify whether the data has been tampered or not and check its genuineness. A user should protect its private key similar to bank OTP and passwords and prevent it from leaking to ensure the security of its data on blockchain [5][7][8].

2. Decentralized : In a decentralized blockchain network, no one has to know or trust anyone else. Each member in the network has a copy of the exact same data in the form of a distributed ledger. If a member's ledger is altered or corrupted in any way, it will be rejected by the majority of the members in the network [5][7].

3. Untraceability : Once the block is entered into the blockchain, it cannot be tampered. Due to this if the block in the Blockchain is altered and is immediately rejected or deleted.

4. Transparency The data in Blockchain is completely public and can be viewed by the participants.

5. Flexibility : Being open source is one of biggest advantages of blockchain. Various public and private blockchains are available to the users, which can be used based on type of application which has to be created [5][7].

2.3 Importance of Blockchain

Blockchain increases trust as we don't have to depend on any third party. The smart contracts which are basically programs on blockchain are run only when certain conditions are met. Since all the blocks stores its data along with hash of previous block it becomes difficult to modify the blockchain with false information. If attacker changes information of block , it's hash also change but the hash of next block remains same. To alter chain would require the consensus of more than half of the participants which is unlikely, since lots of resources and financial amount is required. Also other members would come to know of this drastic change [5].

3 Literature Survey

The survey focused on understanding the sources of counterfeits, impact on the society.There exist various systems of fake product detection,which use Artificial Intelligence, QR codes , Machine Learning and Blockchain.

The methods discussed by Shaik included the use of providing product with public and private keys as QR code, the app used to scan the QR should have cryptographic functionality to decrypt the QR code. The manufacturer is also supposed to run server to accept request and match the buyers name, and items code. The scanning app should have cryptographic functionality to decrypt ciphertext of the item code encoded in the QR code [9]. Benatia and Baudry et.al explains traceability-CPS based architecture for supply chain management consists of several layers that interact to form a traceability-CPS. Also, the proposed architecture allows supply chain monitoring and data analytic to enhance product. Safety and quality. The proposed algorithm con- sist on computing the most frequent item sets in the prod- uct transaction database. This item sets are then used as genuine product trajectories and can serve in detecting ab- normal product behavior[10].

Khalil and Doss et.al comes up with the solution of using RFID based system to reduce counterfeiting. This system allows consumers to query in-store the tag attached to an item to verify its legitimacy. RFID-based anti-counterfeiting and anti-theft schemes are suitable for large scale implementation in retail environments. The proposed scheme is lightweight and suitable for implementation using low-cost passive RFID tags. Tran and Hong's anti- counterfeiting protocol are used. This system is immune to DOS attacks [11]. Habib and Sardar et.al gives explanation on SCM trends. They are examined in their work process that executives' difficulties and transaction issues are problems featured in the SCM. Hence proposed a solution, SCM by considering the blockchain as a technological feature for solving them. Primary method for structuring new models should find the transaction process at a plan level [12].

Daoud and Vu et.al focuses on the architecture of AI Application. It has three main parts: the data set, detection models, and trained model. Anti- counterfeiting machine learning-based solution to detect fake products. Training models step and detecting logo step are the two steps required. Faster R-CNN achieves high accuracy and low training speed [13]. Chen and Shi et.al explains SCQI. Framework for blockchain based SCQI provides a theoretical basis to intelligent quality management of supply chains based on blockchain technology. RFID technology is used to record quality information, trans- action information. Smart contracts are used to execute quality control and improve the efficiency of the supply chain [14]. Toyoda, Kentaroh and Mathiopoulos, P Takis et.al Proposed system to detect fake product with the help of QR code. End users can scan the QR code assigned to product to get the product details and transaction history, the steps involved Product enrolment, ship product to distributor, and ship product to retailer, end user gets details about the produc [15].

In a Blockchain based system the data is stored on each node, then the nodes exchange information with each other over the network. Each node maintains all Blockchain data. The node verifies the received transactions and include them in the new block based on its own Blockchain data, and try to obtain the rights of the new block. Ethereum as the back-end Blockchain operating system. Store relevant information on product sales in Blockchain which is accessible to everyone. It is cost efficient [7]. In this blockchain technology for information sharing is proposed. Is this the information is in the control of the owner so third party interference is difficult. Users are always aware of the data that is being collected about them and how it is used. The blockchain block contains sender, amount, receiver, transaction id, product id and metadata [16]. Ethereum is a open-source Blockchain.Ethereum is a technology that's home to digital money,global payments and applications.The process is simple as to get into the portal,pick a wallet that lets you connect to Ehereum and manage your funds , Get the ETH, use applications powered by Ethereum, start building [17].

Abhijeet and Adrew et. al.[18] discusses various findings on counterfeiting in global supply chain environments based on various papers and online surveys of professionals targeted at a national purchasing body and affiliated UK purchasing groups. It was found that counterfeiting is widely increasing in areas of low-cost spare parts and sectors like drug market. Strategies used by industries to tackle this problem include avoidance, prevention based on previous experience, destruction. The counterfeit products were difficult to identify for customers due to availability of forged certificates.

The limitations in the existing systems are that brands used QR codes on products to prove the validity of the product.But the QR code can be copied and used to label counterfeit products[9]. In the RFID based system that low Cost RFID tags can be used for auto identification of products, but due to cloning of RFID tags, this method is not suitable [14]. In AI and machine learning application, CNN takes more time and memory.It needs training and testing phase before its actual deployment.Artificial Intelligence fails to detect tag reapplication attacks, wherein a counterfeiter removes a legitimate tag from a genuine product and reapplies it to a counterfeit or expired product [13]. There is no power for the customer,suppliers and retailers to check the integrity of product.

4 Proposed system

Counterfeit has spread worldwide and has huge effects on organizations, manufacturers, and consumers. It affects the influence of the organization and the wellbeing of the consumers. India is not excluded. The proposed system is aimed at consumer products, and it helps track the products by maintaining the product and the supply chain integrity by using Blockchain. This gives the customers the power to track the history of the entire product from manufacturer to customer using blockchain and QR code.

4.1 System Model

The proposed system will be a decentralized application (Dapp) which will be implemented using the Ethereum Network as the main blockchain for keeping all the records and managing the transactions regarding the products of the companies listed on Dapp. The basic system architecture is shown in figure 3.

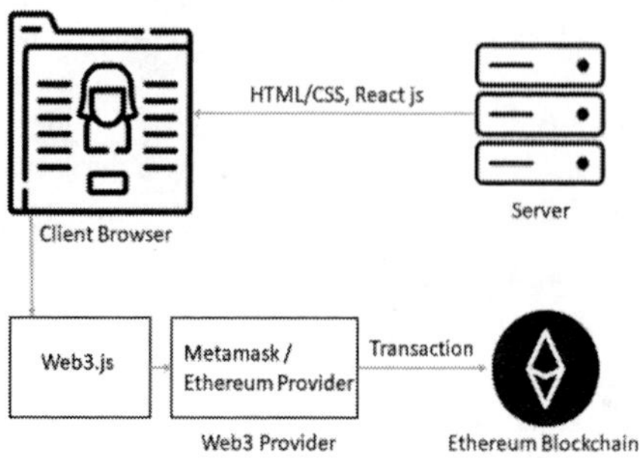

Figure 3: System Architecture [19].

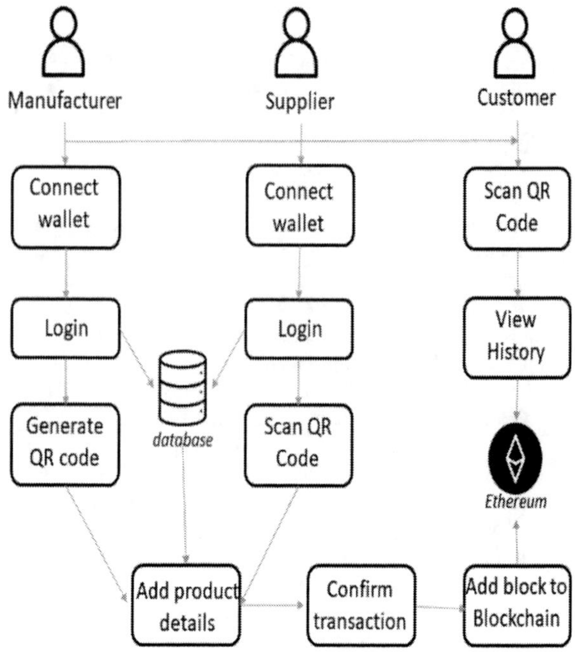

Figure 4: System Flow.

Ethereum
It is a decentralized blockchain which uses a proof-of-work consensus mechanism. Proof-of-work is adding block to the blockchain by solving the mathematical expressions. Solving the puzzle "proves" that nodes have done the "work" by using computational resources. It confirms that the block is added and recorded in the blockchain. This process is known as mining. Mining is typically brute force trial and error, but successfully adding a block is rewarded in ethereum (ETH) [17][19].

Smart contract
Smart contracts are programs that are stored inside Blocks. Smart contracts replaces the involvement of third-party members. These are basically protocols that execute when the conditions are satisfied. They never change, that means no one can tamper with the contract [19].

4.2 Flow of Proposed System

The main aim of this proposed system is to maintain the Genuity of the product by helping the customer track the supply chain history of the product. System give customers the power to track the history of an entire product from manufacturer to customer using blockchain. This product anti-counterfeiting system based on Blockchain is composed of three roles, the Manufacturer role, the Seller role, and the Consumer role, as discussed and shown in figure 4.

Manufacturer:
Manufacturer logs into the manufacturer account and generates QR Code for Product and adds other required details of the product and by using his ethereum wallet, the manufacturer adds a block to Ethereum blockchain. The userid of our local database and the wallet address of the entity will be mapped together, if both the things are there, that is a manufacturer logs in from his own account and uses his own wallet then only the block will be added to the digital ledger.

Supplier:
Supplier logs into supplier account and scans the QR code on the product. The seller can access information about his products that the manufacturer has entered. It adds its own details of the product like shop destination and pushes it into the Blockchain. Those details can be viewed by the buyer.

Customer:
Customers can check the integrity of the product by scanning QR code which will list the history of transactions and thus verifying the genuinity of the product. At the time of customer purchasing the product after the QR scan in supply chain history, if the last location is not matched with the purchase location, the customer will know that the product is not genuine. It concludes that the QR code was copied and the customer becomes aware of counterfeiting.

https://doi.org/10.1051/itmconf/20224403015

The process of detecting a counterfeit product by the customer while purchasing is shown in figure 5.

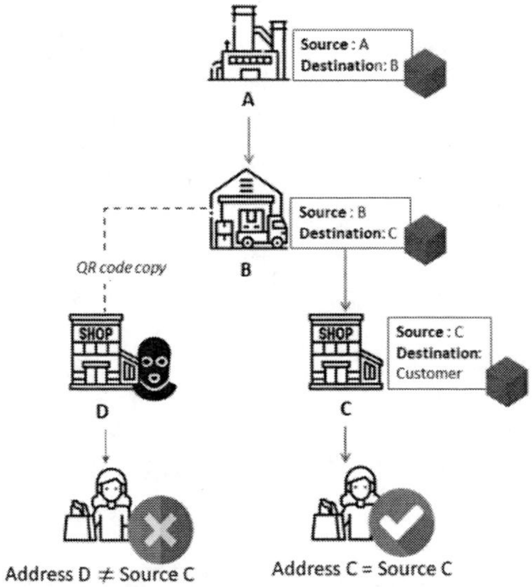

Figure 5: Dealing with Counterfeit Product.

5 Result and Discussion

The proposed system allows both manufacturers and suppliers to interact with the system to add their respective block containing the transaction details to blockchain without modifying other's block. The contracts for the manufacturer and supplier block are written using solidity. Since the code is running on local network ganache have been used for local testing. The host"127.0.0.1", and port 7545 is configured in truffle-config.js file. The contracts are then compiled and deployed using truffle. Migrations files are created for deployment. Migrations are files that help us to deploy contracts on an ethereum blockchain network.

The interface is created using React. To allow interacting with ethereum blockchain Web3.js library is used which is used to perform actions like sending ether, confirming transactions reading and writing data from smart contracts. Metamask is installed on a browser which is a wallet to interact with ethereum blockchain, to allow accessing ethereum wallet through a browser. Accounts from ganache are imported into the metamask. To add supplier and manufacturer blocks they have to confirm the transactions using their account using metamask wallet which is connected using Web3.js.The end-user can then check the supply chain by scanning the QR code to check the product integrity.

Accounts can be categorized into manufacturer ,supplier and customer as shown in Figure 6. The manufacturer connects to his Ethereum account using the Metamask Wallet as shown in Figure 7.

Figure 6: Landing page.

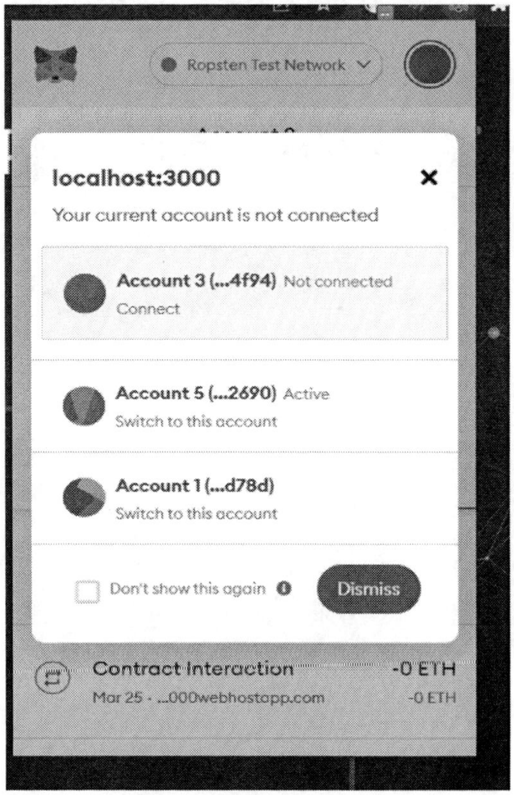

Figure 7: Connecting to Ethereum using Metamask wallet

The Manufacturer logs into his account using his credentials like username and password as seen in Figure 8.

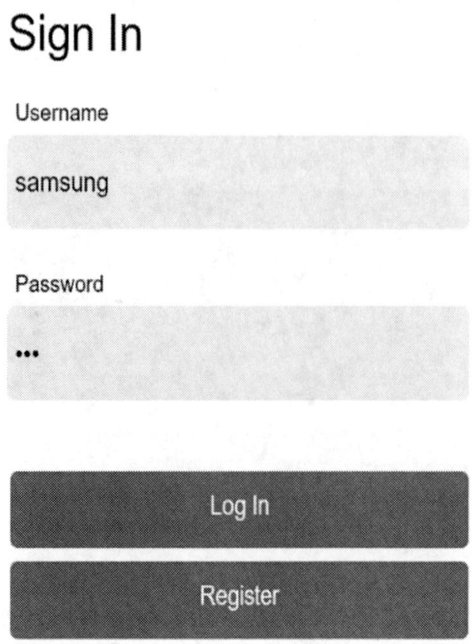

Figure 8: Manufacturer Login page

The credentials are stored in SQL server as shown in Figure 9. SQL database is used for storing manufacturer and supplier login details and their address.

id	name	uid	addr	category	password
10	samsung	26513	Samsung Electronics Huizhou Co., Ltd., Industrial ...	Electronics	202cb962ac
59	louis_vuitton	65464	116 Greene St, New York, NY 10012, United States	Fashion	df2a3e9e50

Figure 9: SQL database.

As shown in Figure 10, after logging into his account the manufacturer assigns unique serial number for the product and generates its QR Code. This QR code is placed on the product when it is transported to other places.Along with this the manufacturer fills other details of product like its name , current address that is the source and destination where it is currently headed.Once all the details are filled the manufacturer clicks add block button which is used to add all the filled details to blockchain.

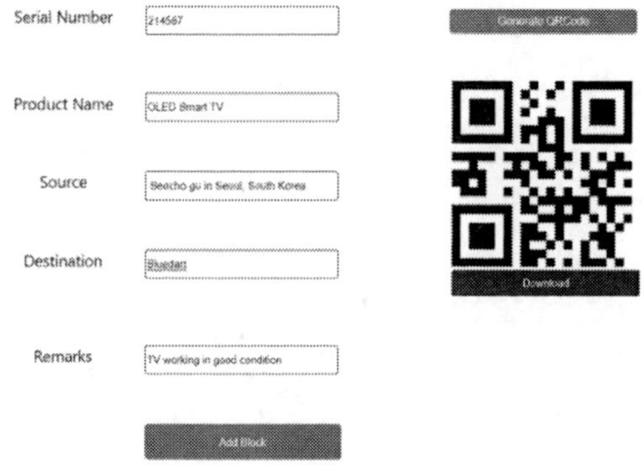

Figure 10: Manufacturer adding the details of the product.

A metamask confirmation popup is displayed which asks for the confirmation as in Figure 11.

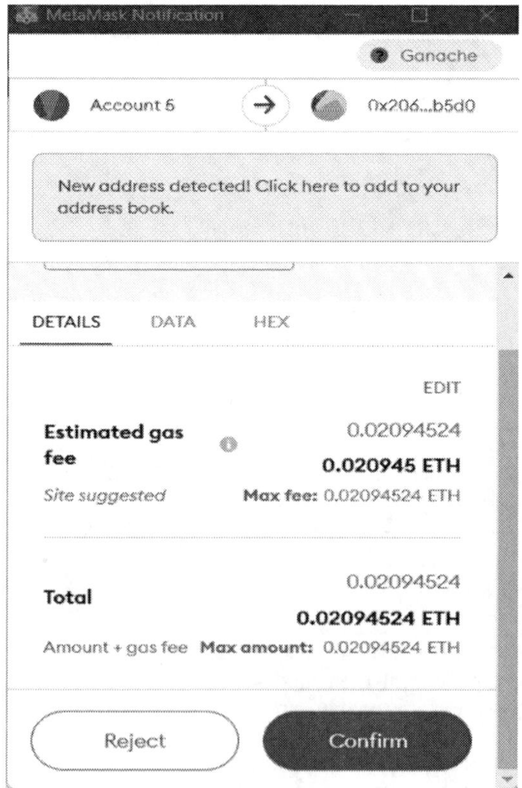

Figure 11: Completing the transaction with metamask wallet

Once confirmed the block containing all details are added to blockchain and success page is displayed as shown in Figure 12.

https://doi.org/10.1051/itmconf/20224403015

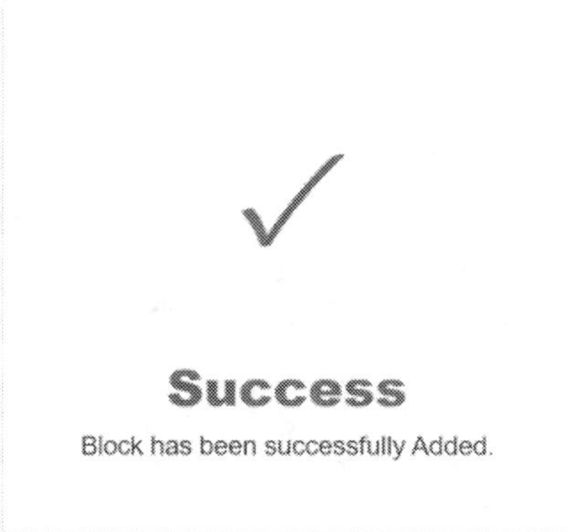

Figure 12: Message after addition of block to blockchain.

Once the product reaches the supplier destination , the supplied needs to login into his account shown in Figure 13 and connect his metamask wallet.

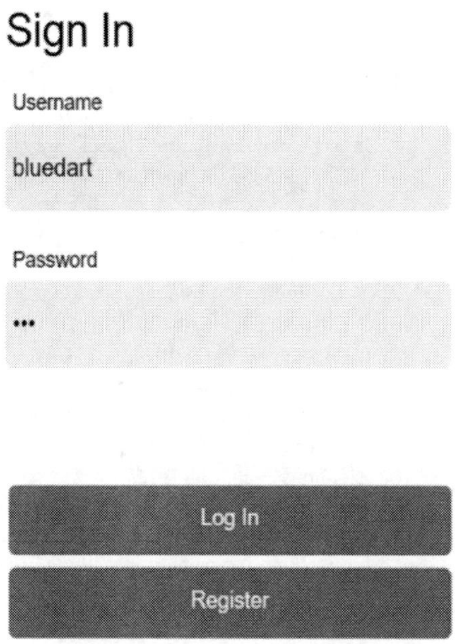

Figure 13: Supplier Login page.

Once done the supplies is presented with screen to fill in the required details of product.This time the supplier doesn't generate the QR code ,the supplier clicks on scan qr and scans the qr code.The supplier enters the required details for the product and clicks on add block as shown in Figure 14.The block gets added after the supplier confirms his transaction through the metamask wallet and logs out.

Figure 14: Supplier adding details for the product.

Other supplier involved similarly log into their account and adds their respective blocks to the blockchain. After the product reaches the customer he can visit the customer page , scan the QR code as shown in Figure 15 and check the complete supply chain history of product.

Figure 15: Customer page to check Supply chain history of the product.

The supply chain history as shown in Figure 16 and Figure 17 shows various information of like the the product id , its name , source , destination address related to the entities involved , their ethereum account address , timestamp of when the block was added ,and any additional remarks if added.

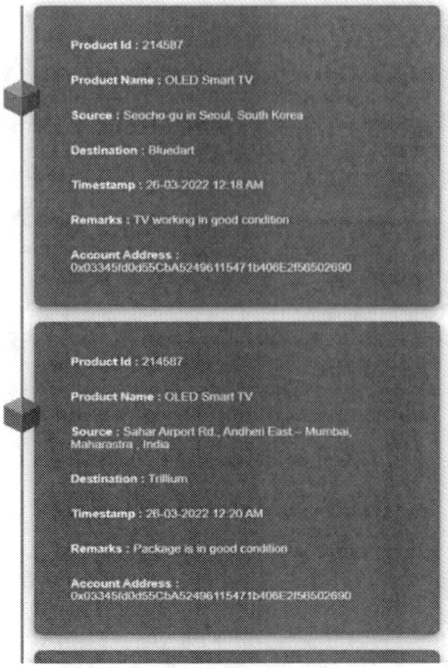

Figure 16: Supply chain history - 1.

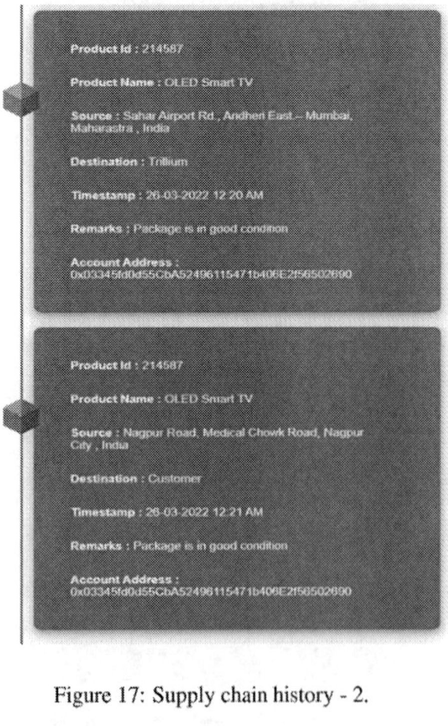

Figure 17: Supply chain history - 2.

At the time of customer purchasing the product after the QR scan in supply chain history, if the last location is not matched with the purchase location, the customer will know that the product is not genuine. It concludes that the QR code was copied and the customer becomes aware of counterfeiting.

6 Conclusion

Blockchain is a decentralized system, therefore the local suppliers cannot interfere with the checking or counterfeiting of the product in the proposed system. Manufacturers and Suppliers can use the system to store product details in Blockchain which offers certain properties such as tamper-resistance, data consistency and confidentiality that assure the security and privacy of the data on the network. The customer views the product supply chain history and verifies if the product is genuine. Customers can be sure about the integrity of goods they purchase. The proposed system can effectively lower the rate of counterfeiting of branded goods and provide the companies with an easier approach to provide consumers with the confidence that they will not purchase counterfeit goods. This system will help to build trust and good bonding between manufacturer and customer and in deed it will help in improving economy and reducing corruption. Further system can be extended to avoid frauds done in banking, healthcare, voting system, online shopping and so on.

References

[1] ASPA, *The state of counterfeiting in india 2021*, https://www.aspaglobal.com/pre_upload/nation/1623216858-4730baa0efdb83aba174859af0a3a6a5-Report%20The%20State%20of%20Counterfeiting%20in%20India%202021.pdf (2021)
[2] Y. Lu, Journal of Management Analytics **5**, 1 (2018)
[3] F. Casino, T.K. Dasaklis, C. Patsakis, Telematics Informatics **36**, 55 (2019)
[4] M. Peck, IEEE Spectrum **54**, 26 (2017)
[5] S. Idrees, M. Nowostawski, R. Jameel, A. Mourya, Electronics **10**, 951 (2021)
[6] Zignuts Technolab, *How blockchain architecture works? basic understanding of blockchain and its architecture.*, https://www.zignuts.com/blogs/how-blockchain-architecture-works-basic-understanding-of-blockchain-and-its-architecture/ (2022)
[7] J. Ma, S.Y. Lin, X. Chen, H.M. Sun, Y.C. Chen, H. Wang, IEEE Access **8**, 77642 (2020)
[8] M.J.L.I.N.M. J.M. Bohli, N. Gruschka, IEEE **10**, 9 (2013)
[9] C. Shaik, Computer Science & Engineering: An International Journal (CSEIJ) **11** (2021)
[10] M.A. Benatia, D. Baudry, A. Louis, Journal of Ambient Intelligence and Humanized Computing pp. 1–10 (2020)
[11] G. Khalil, R. Doss, M. Chowdhury, IEEE Access **8**, 47952 (2020)
[12] M.A. Habib, M.B. Sardar, S. Jabbar, C.N. Faisal, N. Mahmood, M. Ahmad, *Blockchain-based supply chain for the automation of transaction process:*

Case study based validation, in *2020 International Conference on Engineering and Emerging Technologies (ICEET)* (IEEE, 2020), pp. 1–7

[13] E. Daoud, D. Vu, H. Nguyen, M. Gaedke, *Improving Fake Product Detection Using Ai-Based Technology*, in *18th International Conference e-Society* (2020)

[14] S. Chen, R. Shi, Z. Ren, J. Yan, Y. Shi, J. Zhang, *A blockchain-based supply chain quality management framework*, in *2017 IEEE 14th International Conference on e-Business Engineering (ICEBE)* (IEEE, 2017), pp. 172–176

[15] K. Toyoda, P.T. Mathiopoulos, I. Sasase, T. Ohtsuki, IEEE access **5**, 17465 (2017)

[16] M. Nakasumi, *Information sharing for supply chain management based on block chain technology*, in *2017 IEEE 19th conference on business informatics (CBI)* (IEEE, 2017), Vol. 1, pp. 140–149

[17] G. Wood et al., Ethereum project yellow paper **151**, 1 (2014)

[18] A. Ghadge, A. Duck, M. Er, N. Caldwell, Supply Chain Forum: An International Journal **22**, 87 (2021), `https://doi.org/10.1080/16258312.2021.1908844`

[19] I. Singhal, International Journal for Research in Applied Science and Engineering Technology **9**, 291 (2021)

 https://doi.org/10.1051/itmconf/20224403016

Music Genre Classification Using Neural Network

Tarannum Shaikh[1*], *Ashish* Jadhav[1]

[1]Ramrao Adik Institute of Technology, D. Y. Patil Deemed to be University, Nerul, Navi Mumbai, India

Abstract. Music Genre classification on Neural Network is presented in this article. The research work uses spectrogram images generated from the songs timeslices and given as input to NN to do classification of songs to their respective musical genre. The research work focuses on analyzing the parameters of the model. Using two different datasets and implementing NN technique we have achieved an optimized result. The Convolutional Neural Network model presented in this article classifies 10 classes of Music Genres with the improved accuracy.

* Tarannum Shaikh: tanu99@gmail.com

1 Introduction

1.1 Music Genre Classification

On Internet web you could be connected to world of Music and experiencing it inclusive of live concerts, Remixes, Old Classic, etc. like never before [1]. Depending upon listener what he wants to listen, he simply downloads the music from songs database and listens without storing it in the device, saving the memory space. The music can be liked or formed a playlist so as to listen again whenever mood swings.

Humans are very intelligent to listen to the short samples of songs and differentiate the singer, beat, lyrics, the title, album, as well the genre of the song. Studying this intelligence, new technology is been applied on several NN (Neural Network) approaches, showing successful achievements [2].

1.2 Machine Learning and Neural Network

Today machine learning is very popular. There are many ML (machine learning) techniques for numerous applications. ML is divided into following techniques, those are supervised learning, unsupervised learning, semisupervised learning, and reinforcement learning [3].

Supervised Machine Learning algorithm is a technique which utilizes labelled training data to attain desired results. It helps the model to learn quickly. In unsupervised learning it uses unlabeled dataset, this technique extracts useful features and performs the complex processes. Semi supervised learning algorithm uses a dataset containing small number of labelled and large number of unlabeled data. Reinforcement learning technique uses a feedback mechanism. In reinforcement learning, it is about to take correct action to maximize reward. For instance, reinforcement learning is usually used in online entertainment when correct prediction is rewarded.

A NN is a technology of machine learning used explicitly to extract the essential features from complicated database and designing a framework that demonstrate the features [4]. The Neural Network uses the training dataset initially to train the designed model. Afterwards the NN is applied to new or test data and observed that the classification of the data is done correctly assumed by the trained model.

1.3 Neural Network

A standard Music Genre Classification usually includes two major elements, one is feature extractor and another is classifier model [5]. From both of these elements, feature extractor is very important parameter that gives the performance of MGC precisely.

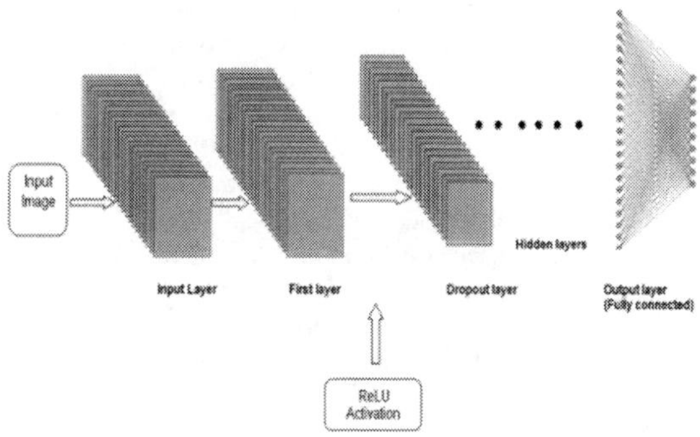

Fig. 1. Proposed CNN Architechture

Above figure 1 shows a proposed architechture of a Convolution Neural Network (CNN), CNN is a class of neural network which is specialized to perform certain process on multidimensional vectors having a grid structure for eg: images [4]. In order to attain optimized result, a CNN is applicable for binary classification as well as for multi classification. They are mostly similar but they vary in the output classes only. For instance, a data collection of floral photos may be used to train a model. The input applied to CNN is a vector of pixel values of an image processed from the dataset which is associated with the right label of vector defined as output class (eg. If floral images then output class is rose, tulip, lily, etc…). Once an input image is applied to CNN model for training purpose, the output will be mapped to particular class after classifying the images. While performing classification, every training dataset image is matched with the predicted output class label. During training the accuracy of the CNN is increased by back propagation technique in which the weights throughout the network are updated iteratively. Number of iterations helps to shape the CNN structure. This structured model does feature extraction of training datasets improving its capability to perform accurate classification of images.

The main contributions made in this literature is focusing both feature extraction and genre classification

The optimized model depicted here has good scope and growth in numerous technologies of classification.

Organization of topics in the paper is as follows. Related work section gives an overview of the current technology for MGC (Music Genre Classification). The Proposed architechture explains the designed model step by step. The next section is Implementation work and results in which complete implementation of the research work using various techniques and logic is explained in detail along with their metric results. At the last, the conclusion section comments over the research work done and mention future scope.

2 Related Works

In 2020 Nikki Pelchat and Craig M. Gelowitz presented a research work on music genre

https://doi.org/10.1051/itmconf/20224403016

classification which classified the songs into their respective musical genres. He used two different music datasets consisting of 1880 different songs, these songs were splitted into 2.56 seconds section which eventually gathered with 132000 spectrograms snippets. This increased the number of training spectrogram slices per genre, typically 128X128 pixel slices were used for training the Neural Network. Activation functions used were softmax and Rectified Linear Unit (ReLU). Also used Hamming windows functions to convert mp3 files to spectrograms and increased the number of CNN layers to achieve 85% accuracy [6].

In 2021 Jaime Ramirez Castillo and M. Julia Flores presented a web based application in which retrieval of YouTube songs was done for music genres classification. Chunks of 10 seconds songs were classified into one or more musical genres. The models are made training several machine learning techniques, namely, Support Vector Machines (SVM), Naive Bayes classifiers, Feed forward deep neural networks and Recurrent neural networks and embedded those in a web application. These models let the user to envisage how each model "percept" the music in terms of music genre, at particular instant of song [7].
In 2020 Wing W. Y. NG Music genre consists of various features that are heterogeneous with abstractions in numerous levels. To eliminate same level feature extraction (FE), he proposed a combination of Convolutional Neural Network (CNN) with NetVLAD and self-attention to extract the information across different levels for learning long-term dependencies. Finally a meta classifier is utilized to make Music Genre Classifier.

His work shows that the proposed approach provides higher accuracies than other state-of-the-art models on GTZAN, ISMIR2004, and Extended Ballroom dataset [8].

Binary Pattern, which is the best result on this dataset In 2016 Costa et.al worked on Music genre recognition considering different datasets and their spectrograms. He trained the classifiers with various techniques like Gabor filters, Local Binary Patterns, Phase Quantization, to extract spectrograms and attained state-of-the-art outcome on numerous audio database. He has done the comparison of results produced by Convolutional Neural Network (CNN) with the outcomes procured by handcrafted features. His research was done on three different audio datasets with definite attributes, specifically a western music dataset (ISMIR 2004 Database), a Latin American music (LMD dataset), and African music dataset.

His research proved that the CNN is best alternative as it is favourable to all classifiers for music genre recognition. For the African database, the CNN excelled the handcrafted representations and showed the state-of-the-art by a margin. Another dataset i.e LMD database attained the recognition rate with 92% by combining CNN and Robust Local till then. On the ISMIR 2004 dataset, the CNN though did not improved the state of the art, but it gave comparatively better performance than the classifiers centered on other kind of attribute [3].

3 Proposed architecture of MGC

3.1 Overall Depiction of presented model

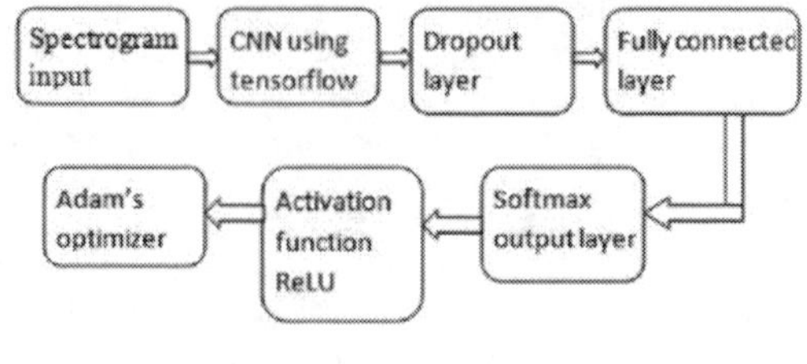

Fig. 2. Architechtural model

In the proposed architechture CNN model, a dense layer is used with 512 features, immediate after each dense layer a dropout layer is added by the factor of 0.2.

Figure 2 depicts the architechtural design of music Genre classification model, build using Keras and Tensorflow framework. The model consists of four hidden layers along with input layer.

These all layers are using the ReLU activation function. The output layer is fully connected network layer which uses the Softmax activation function. Evaluation metric considered here is loss and accuracy. Sparse categorical cross entropy function calculates the loss of the presented model. Adam's optimizer is used for updating the weights in neural network while training the model. The epoch chosen for training the proposed model is 600, for each epoch the model takes nearby 20ms (milli seconds).

Dropout is used to reduce the problem of overfitting by making all the parameters to zero. i.e making the neurons inactive. A dense layer along with the ReLu activation function is used to predict the data for classification. Evaluation metric can be improved more by adding more epochs, but there are certain limits to do so, hence the value for epochs must be considerate.

3.2 Music Database

In this article two different databases are utilized, a western database "GTZAN", and a Music Database "MusicNet", with different genres. [9]

GTZAN dataset consists of total 1,000 music excerpts each of 30 seconds long. The downloaded zip file consists of three different folders.

- First one is Genres Original: which consists of 1000 audio formats of .wav file of 30 seconds each equally divided in 10 genres. Therefore, each genre containing 100 songs.
- Second is Images original: which consist of graphical form or spectrograms of each 30 sec audio file. This is used as input image to the model.

- Third folder is CSV files having multiple features of audio files of above folders.

MusicNet is second dataset used in proposed research which has compilation of 330 freely-licensed classical music files [11]. This accumulation is compiled at three top-level files:

- musicnet.tar.gz - This file has the MusicNet dataset as it is, composed of PCM-encoded audio wave files (.wav) and related CSV-ciphered note labelled files (.csv). The data is arranged according to the train/test split required for training and validation.
- musicnet_metadata.csv – This metadata file is composed of music track-level details of recordings contained in MusicNet. The name of data and label files are done with MusicNet ids, hence it is used to cross-index the data and labels with this file.
- musicnet_midis.tar.gz - This file consists of reference Musical Instrument Digital Interface (MIDI) files which are useful to construct the MusicNet labels.

3.2.1 Data Analysis

Using GTZAN and Musicnet datasets, which are freely available on internet [10-11]. The audio files are all sampled with 22050Hz Mono 16-bit audio files in .wav format. The core of dataset is feature analysis for developers, including genres and images. The size of GTZAN dataset is 9990 by 60 where each row represents a single audio piece and each column signifying one feature.

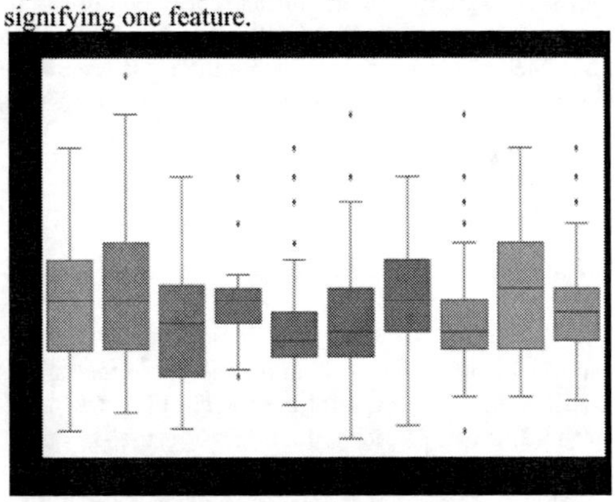

Fig. 3. Boxplot of Music Genre

Above figure illustrates Boxplot for genre and beats per minute (BPM) of the dataset, which shows detail analysis and distribution of the data in the dataset.

Using both the dataset the accuracy and loss is calculated. This shows the presented model can be used on any dataset for classification purpose which yields good results.

Principal Component Analysis (PCA) is a unsupervised technology which is used to reduce dimensionality of database. It is an algorithm applied to compress the high dimensional data into lower dimensional feature without any class label information. PCA reduces the data in two dimensions as shown in figure below (Principal Component 1(PC1), Principal Component 2(PC2)) or in three dimensions (PC1, PC2, PC3).

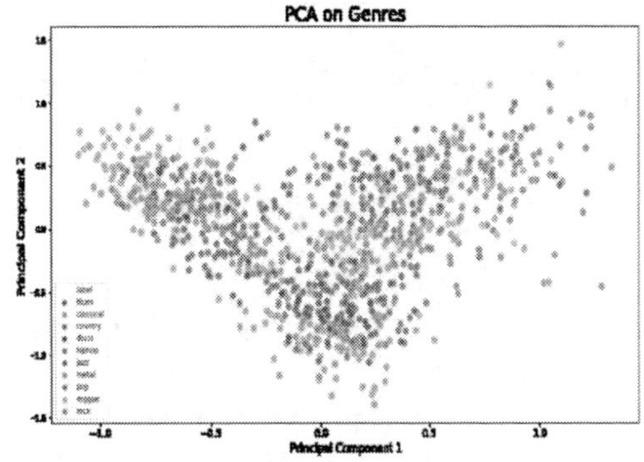

Fig. 4. Principal Component Analysis

Figure 4. Shows Principal Components Analysis (PCA) of the music dataset, indicating the principal components of the data, which reduces the dimensionality of dataset. (i.e) dataset can be reduced to 2 dimensions without incurring any loss in data. We see here in the plot that the data is scattered more along PC1 than PC2 that means any classifier can separate or classify the classes or genres perfectly.

3.3 Preprocessing

Preprocessing of data is an important step done on the dataset before applying the data to the input of CNN for machine learning models. It assures to improve the performance and efficiency of the network also saves time during training the model. Data preprocessing actually involves pre-emphasis, framing and widowing techniques on the given data doing so useful information can be retrieved from the data.

3.3.1 Scaling the features

To standardize the audio features Standard scalar is best preprocessing technique which removes the mean and scaling of audio signal to unit variance.

Consider sample x, which is calculated as

$$z = (x - u) / s$$

Where, u is training samples mean value

s stands for training samples standard deviation

Standardization of a dataset is a widely recommended for machine learning or Neural Network classification else the dataset response will be unfavourable.

3.3.2 Visualizing Audio Files

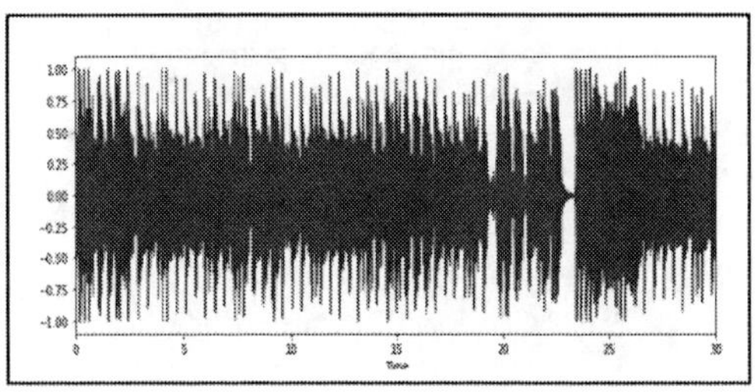

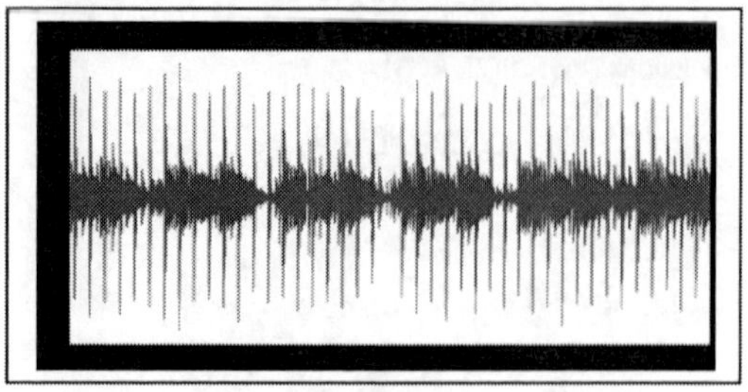

(b)

Fig. 5. Raw wave plot of audio file (a)hiphop 14, (b) reggae 36

We have used librosa to plot raw wave files. Figure 5 illustrates a waveform that is a visual representation of sound showing amplitude Vs time where the Y axis is amplitude and X axis is time. Such graphical images enable to process the audio data instantly and relate the similarity index.

3.4 Feature Extraction

Feature Extraction is done to analyse and find relation between different things. Using feature extraction techniques like PCA, Independent Components Analysis (ICA), and Linear Discriminant Analysis (LDA) dataset dimensionality can be reduced efficiently.

The dataset to be understood by models directly is converted to understandable format. FE is essential analysis for classification, prediction and recommendation algorithm.

We brief the features extracted in the research work [12].

3.4.1 Zero-crossing rate

Zero crossing rate (ZCR) [13] is the rate in which the audio signal passes number of times from upper half cycle to lower half cycle or from lower half cycle to upper half cycle through zero. Application of this feature is in audio or speech classification, recognition and music information retrieval.

$$\text{ZCR} = \frac{\sum_{n=1}^{N} |sign\,(An) - sign\,(An-1)|}{2N} \quad (1)$$

Where, N is the number of samples in the frame and
An is the amplitude of the n-th sample.
sign $(\cdot)$ represents the sign function.

3.4.2 Spectral Centroid

Spectral Centroid [13] feature gives essential facts about the frequency and energy distribution of the audio signal, we evaluate the center of gravity of the spectrum as

$$\text{Centroid} = \frac{\sum_{k=1}^{N} P(fk)fk}{\sum_{k=1}^{N} P(fk)} \quad (2)$$

Where, fk is the k-th frequency,
N is the number of frequency bins.
$P\ (fk)$ is the spectral amplitude on the k-th frequency [8].

3.4.3 Chroma Feature

Chroma Feature is a robust tool for semantic analysis of the musical audio features. It's a 12 element feature representing spectral energy level in the signal. It's a powerful way for finding out similarity index of audio samples.

Short Time Fourier Transform (STFT) are used to extract the Chroma feature from audio signal. Better the extraction quality, best are the results.

3.4.4 Spectral Roll-Off

Spectral roll-off [14] is the specified amount of frequency lying below the total spectral frequency.
The spectral rolloff point is calculated as described in rolloffPoint = i

$$\text{Such that } \sum_{k=b1}^{i} sk = k \sum_{k=b1}^{b2} sk \quad (3)$$

Where, s_k is the spectral value at bin k.
b_1 and b_2 are the band edges, in bins, over which to calculate the spectral spread.
k is the percentage of total energy contained between b_1 and i.
k can be set using Threshold.

3.4.5 Mel-Spectrogram

Mel-spectrogram is a graphical representation of the frequencies spectrum of an audio signal with respect to time. Mel spectrogram is the magnitude spectrogram of the input audio data on the log scaled frequencies on the Y-axis. The expression of mel-frequency is as follows [13]

$$\text{Mel}(f) = 2595\log 10\ \ 1 + f/700 \quad (4)$$

3.4.6 Mel-Frequency Cepstral Coefficient (MFCC)

Mel-Frequency Cepstral Coefficient (MFCC) [12] feature is broadly used in automatic audio and speech classification. The cepstral coefficient obtained on the mel-frequency is called (MFCC) mel-frequency cepstrum coefficient.

Above features in the form of vectors are used to build the input feature vector.

4 Implementation

The proposed Music Genre Classification model is implemented in Python Programming and the corresponding output were achieved. For training and testing purpose, we have split the dataset into 67% and 33% respectively. All the images are preprocessed using Librosa and then given as the input to the proposed CNN model.

Using Deep Convolutional network the NN model was implemented applying Tensorflow framework. The batch size given here is 128. The first four layers are convolutional layers and dropout layers with 0.2

probabilities, which avoid a curb overfitting and is implemented after each successive layer. The activation function used here at input and hidden layers is ReLU and Softmax for the output layer. The last layer is fully connected layer to which each and every output of the previous layer is given. This process gives a lengthy array. A softmax layer is applied to regulate the 10 outputs, these 10 outputs is nothing but 10 Genre.

Adopting "sparse categorical cross entropy" with the standard initial learning rate the loss function is measured. The Adam's optimizer is used to converge towards the minima fast. It is computationally efficient and is used for huge data or maximum parameters.

To see that the designed neural network has the same accuracy throughout different datasets, the same NN architechture was used to train another data set. The two datasets used in this research has different genre, but number of songs or music piece in each genre of both dataset is equal. This is done intentionally for consistency and comparing test accuracy.

Table 1. Table of GTZAN[10].

Genre			
Name	**No. of files**	**Name**	**No. of files**
Blues	100	Classical	100
Country	100	Disco	100
Hiphop	100	Jazz	100
Metal	100	Pop	100
reggae	100	rock	100

[a] Genre details, (files are 30 sec long .wav format).

Table 1 shows respective number of files of each Genre. The other dataset had different genre with same number of files in each genre. Equalizing number of songs per genre in two different dataset, a test accuracy attained was 92.65%. Hence using different dataset on the same model is possible that too with higher accuracy.

Following are the comparison results of accuracy and loss for both the datasets. Table 2 gives the metric evaluation of the results obtained by the proposed MGC model.

Table **2.** Accuracy And loss result.

Dataset	Accuracy	Loss
GTZAN	92.65%	57.37%
Musicnet	91.70%	25.46%

- GTZAN accuracy

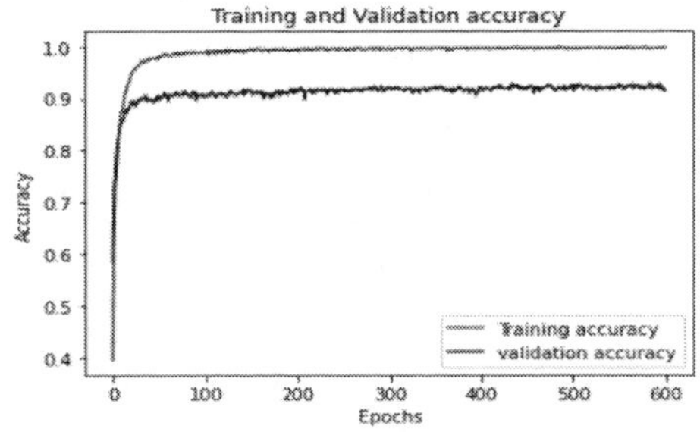

- Musicnet accuracy

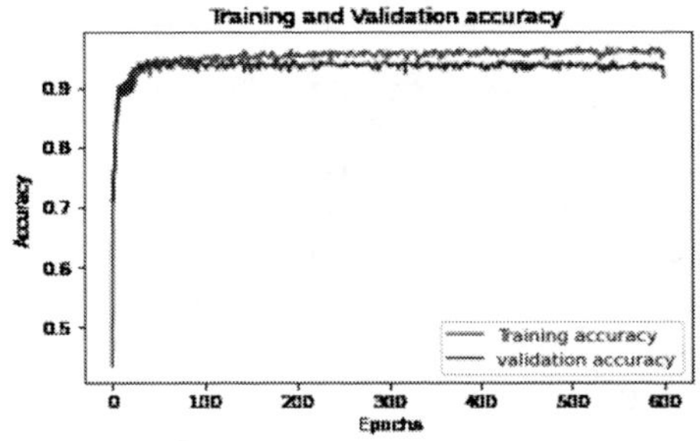

- GTZAN loss

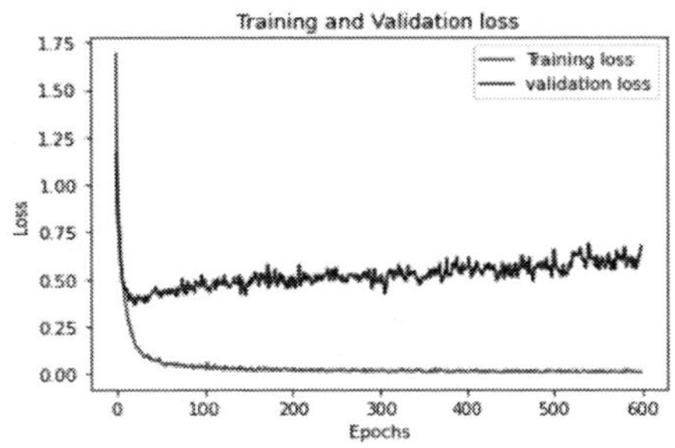

- Musicnet loss

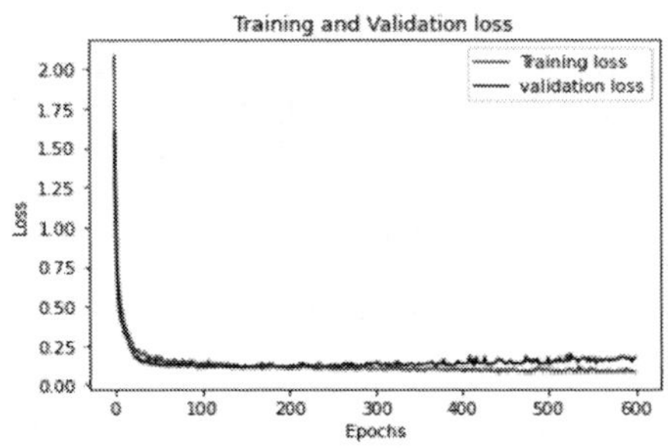

Fig. 6. Training accuracy and training loss of two datasets.

Figure 6 illustrates graphical representation of table 2. The plot shows both the training and validation data acuuracy and loss curves.

Table 3. Comparison of Pre-Trained Models [1]

Authors/Models	Accuracy
Stochastic Gradient Descent	65%
Pelchat, Gelowitz	85%
Naïve Bayes	51.9%
Costa *et.al.*	67%
Lopes *et al.*	60%
Despois	90%

Table 3 shows an overview of accuracy obtained by pre trained models in recent years, which indicates that to improve the performance only increasing the number of layers is not sufficient. On the contrary we can increase the accuracy by increasing the model depth.

5 Conclusion

This research work proposes an approach of implementation of a Music Genre classification model. The work involved of preprocessing and extracting the features of the music. This extracted feature like spectrograms are provided as an input to the CNN. The NN comprises of four layers including a input layer with ReLU, and finally a fully connected output layer with 10 classes including Softmax activation function predict the probability of each genre. In addition we have used a song library of MIDI data for contemplation of datasets sequential features. Concluding further, we consider that Neural Networks are most efficient in machine learning technology. Framework Tensorflow is very important in building the CNN model and to execute it effectively.

Future work may have some modification in initialization of weights and including full song instead of 30s slice classification. Implement pooling, for individual output to nullify false positives is a future task. Also contemplating binary classifiers to find out the user may like a predicted song, found on their personal music library.

References

[1] M. Goto and R. B. Dannenberg, "Music Interfaces Based on Automatic Music Signal Analysis: New Ways to Create and Listen to Music," in *IEEE Signal Processing Magazine*, vol. **36**, no. 1, pp. 74-81, Jan. 2019, doi: 10.1109/MSP.2018.2874360.

[2] Y. M.G. Costa, L. S. Oliveira, C. N. Silla, "An evaluation of Convolutional Neural Networks for music classification using spectrograms", in *Applied Soft Computing*, Volume **52**, 2017, Pages 28-38, ISSN 1568-4946, https://doi.org/10.1016/j.asoc.2016.12.024.

[3] S. Gollapudi, *Practial Machine Learning*. Birmingham, U.K.: Packt, 2016.

[4] Y. LeCun, Y. Bengio, and G. Hinton, "Deep learning.(Report)," *Nature*, vol. **521**, no. 7553, p. 436, May 2015, 2015.

[5] T. Kim, J. Lee, J. Nam, Comparison and analysis of sampleCNN architectures for audio classification, *IEEE* J. Sel. Top. Sign. Proces. 13 (2) (2019) 285–297.

[6] N. Pelchat and C. M. Gelowitz, "Neural Network Music Genre Classification," in *Canadian Journal of Electrical and Computer Engineering*, vol. **43**, no. 3, pp. 170-173, Summer 2020, doi: 10.1109/CJECE.2020.2970144.

[7] J. R. Castillo and M. J. Flores, "Web-Based Music Genre Classification for Timeline Song Visualization and Analysis," in IEEE Access, vol. **9**, pp. 18801-18816, 2021, doi: 10.1109/ACCESS.2021.3053864.

[8] W. W. Y. Ng, W. Zeng and T. Wang, "Multi-Level Local Feature Coding Fusion for Music Genre Recognition," in *IEEE Access*, vol. **8**, pp. 152713-152727, 2020, doi: 10.1109/ACCESS.2020.3017661.

[9] G. Peeters, 2021, 15 August 2021 [https://ismir.net/resources/datasets/]

[10] G. Tzanetakis, 2015, 15 August 2021, [http://marsyas.info/downloads/datasets.html]

[11] J. Thickstun, Z. Harchaoui, S. M. Kakade, November 30, 2016 ,15 August 2021 [https://zenodo.org/record/5120004#.YbnBRjNBzIV]

[12] D. Yu, H. Duan, J. Fang and B. Zeng, "Predominant Instrument Recognition Based on Deep Neural Network With Auxiliary Classification," in *IEEE/ACM Transactions on Audio, Speech, and Language Processing*, vol. **28**, pp. 852-861, 2020, doi: 10.1109/TASLP.2020.2971419.

[13] J. D. Deng, C. Simmermacher, and S. Cranefield, "A study on feature analysis for musical instrument classification," *IEEE Trans. Syst., Man, Cybern., Part B (Cybern.)*, vol. **38**, no. 2, pp. 429–438, Apr. 2008.

[14] Scheirer, E., and M. Slaney, "Construction and Evaluation of a Robust Multifeature Speech/Music Discriminator," *IEEE International Conference on Acoustics, Speech, and Signal Processing*. Volume **2**, 1997, pp. 1221–1224.

 https://doi.org/10.1051/itmconf/20224403017

Ensemble approach for fake news classification using machine learning

Gopi Pogul[1,*], *Sanket* Rohokhale[2,**], *Priya* More[3,***], and *Pallavi* Chavan[3,****]

[1]Department of Information Technology
[2]Ramrao Adik Institute of Technology
[3]Navi Mumbai India

Abstract. During the covid 19 outbreak, fake news has grown highly, affecting people's mental and physical health. There is a wide range of solutions for fake news classification which are machine learning-based proposed models. Research shows that the existing proposed models have less accuracy, and they are only text-based models. In our research paper, we are focused on different algorithms, and we are comparing these algorithms in our proposed model in this research paper. We are considering the title author and text in the proposed model. Based on our experiments, Logistic Regression has high accuracy, recall, and precision score values. This research paper suggests using a logistic regression model to classify fake news.

1 Introduction

In today's world the use of Internet and Social Media has increased a lot.This results in increase in accessibility of any news and information through text, image and multimedia.Most of the users are tend to use Digital form of news than traditional form of hard copy of newspaper.The people who reads on this online form of news get so many false information also.This can also affect and negative impact on students. Most of the news we read on internet is not that much trustworthy. This will result in misunderstanding and confusion in people and it is very easy for anyone to post any news on social media . There is so much information available on the internet that separating the truth from the fake has become impossible. As a result, Fake news classification is needed. With the rise in popularity of social media, more people are turning to it for news instead of newspapers.This caused social media used to distribute false information, which has bad impact for every users. The subject of false news was investigated in this paper by analysing existing literature in two phases: characterization and identification. We looked at existing false news detection algorithms from a data mining standpoint during the detection phase. The main reason for this shift is because of following reasons: 1)It is easy , convenient and less expensive to read news on social media than hard copy of newspaper.2)It is easy to share, comment and discuss the news with friends or other readers on social media. It is cheap to provide news online and much faster and easier confuse people through social media.

*e-mail: gopipogul47@gmail.com
**e-mail: sanketrohokale0401@gmail.com
***e-mail: priyapmore2000@gmail.com
****e-mail: pallavi.chavan@rait.ac.in

2 Related Work

Mykhailo Granik et. al.[1] in their paper shows simple approach for fake news classification.This way of classification is tested against a Facebook news posts.They are collected from various pages on Facebook which is most popular.Also they collected news from popular news websites like CNN,BBC news. They successfully achieved 74this model.The accuracy of this model is not that great.Becaue of the unstructured data which was in fromat of text,images and multimedia,has been used in datasets. Himank Gupta et. al. [2] given a classification model that supported completely different machine learning approach which deals with varied issues as well as accuracy of the model.They used (BotMaker) high time interval to handle thousands of news in one sec. Firstly, they got collected 450000 news from Hspam dataset. Then they divided these this news into 150000 spam news and 250000 non- spam news. They additionally developed a model that can improve model by using Bagof words model.They able to 1 achieve accuracy of 91improvent. Marco L. Della Vedova et. al. [3] In this model they used a completely different technique which was of cubic centimetre pretend news detection.They combined news and social media content which was a success giving accuracy of 81their model among a Facebook traveller Chabot and validate it with a real-world use.Getting a accuracy of eight one.Thier approach was to classify a point as particular point from where they can sure of news is real or fake.They 1st collected the datasets they used for their model,then they divided the content primarily based approach.Also the approach they used is to mix it with a social-media based approach out there. The ensuing dataset consists of 150,500 posts,out of which 14 are conspiracy pages and eighteen scientific pages They have likes of about 30,000 likes by 900,000+ users. 8,923 (57.6 posts square measure hoaxes and half dozen. Cody Buntain et. al. [4] develops a model for automatically pretending fake news detection on Twitter by learning to predict accuracy assessments in 2 credibleness targeted Twitter datasets.One is of CREDBANK, a dataset of accuracy checking for events in Twitter,and PHEME, a dataset of rumours in Twitter.The media assessments of their accuracy is done by this model.They used BuzzFeeds pretend news dataset for using their model. This model identifies options that square measure most prognosticative for accuracy calculation, results of that square measure in line with previous work. They think about characteristic most viewed news and use the options of those threads to classify stories, limiting this work's pertinence solely to the set of standard news.Since the bulk of news square measure seldom retweeted,this technique thus is just helpful on a particular speech threads. In his paper, Shivam B. Parikh et. al. [5] .Aimed to characterize the reports of story of trendy news article and he combined the unstructured content with variety of news article.Then afterward, we tend to dive into existing pretend news detection approaches that square measure heavily supported NLP which text based approach,and additionally use different news datasets. We tend to summarize the paper by four characteristics.The challenges which will guide future analysis. it's a theoretical Approach which supplies Illustrations of false news detection by working the psychological way.

3 Proposed system

The News industry has been witnessing technological advancement in it and industries surrounding it. The use of online news and social media have risen exponentially especially during Pandemic Periods. This resulted in increase in fake news on internet and social media. There are already many existing systems to do classification on the fake news data but there the use of them is restricted and it is not user friendly. The proposed system has a user-friendly dashboard for showing the classification of news with accuracy. The news shared by online sites and social media are generally in unstructured format. The system is divided into pre-processing of the data and training the system and testing the algorithms. The pre-processed data is spit into two datasets one is train data and second is test data. Out of which 80is training data and 20% data is testing data. The first training the model using the train data to get used to the data and then using the remaining test data to perform actual classification of fake news.

1. Fetching the data: In this phase we will fetch data through system.
2. Cleaning and transforming data: removing anomalies and filter out Unwanted data
3. Training the Model: Train the data using algorithms.
4. Detecting Fake News: Comparing original label value with predicted value.
5. Show result on Dashboard: Showing accuracy of the news being real or fake.

In Proposed System we used Python for classification of fake news. We used Tf-Idf Vectorizer for Vectorizing

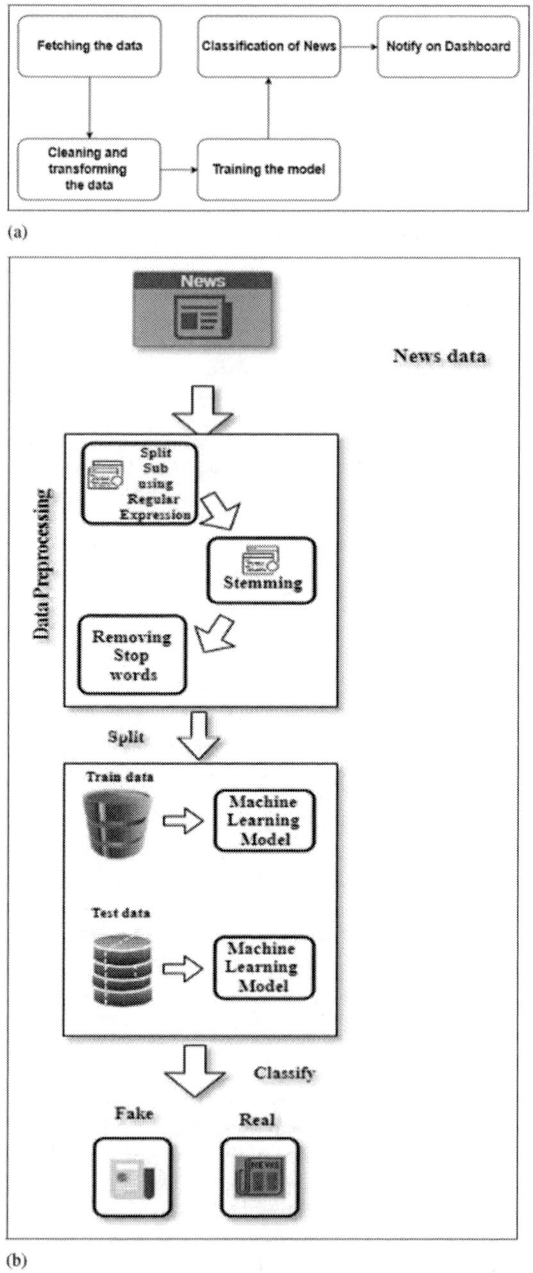

Figure 1. Proposed Model of Ensemble approach for Fake News Classification

of data with sklearn

Working – We are using three approaches for solving the problem.

1) Use natural language processing to verify the accuracy of a news story.

2) If a person has a question about the legitimacy of a searching of news, he or she may come on our platform and receive a accuracy of classification using our own

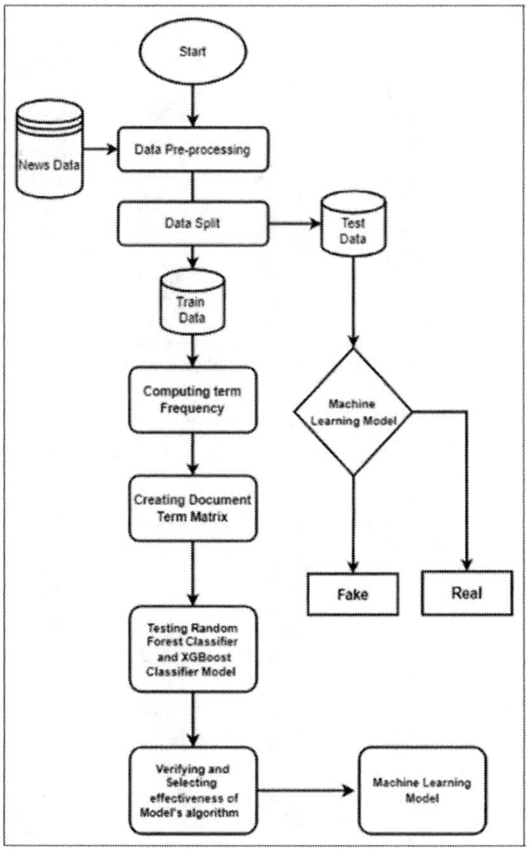

Figure 2. Work flow of Proposed System

algorithm.
3) The news source credibility is verified.

The Regular Expression has been used in the system for split and sub function. The regular expression is the search pattern expression having sequence of characters. In Split it returns a list where the string has been split at every match case. In sub it replaces one or many matches with a string .

Stop words are removed because these words are not important in sentences, so we remove them from our database which helps in focusing on only important words. These words don't contain any valuable information. These words are a,an,but,after,when,where etc. The suffix from the word is removed to return the root words. In Stemming process, the adverb , adjective ,verb ,noun, etc. are transformed into its root word. For example, the words playful , played , playing is stemmed to the word play. In some cases, the root word is not meaningful, so we are checking the root words with English Dictionary to remove these words having no meaning. The pre-processed data is divided into two data sets. First is training data containing 80% of data and second is the test data containing 20% of data.

Confusion Martrix		**Actual Classes**	
		Positive Class	**Negative Class**
Predicted Class	**Positive Class**	**True Positive (TP)**	**False Positive (FP)**
	Negative Class	**False Negative (FN)**	**True Negative (TN)**

Table 1. Confusion matrix for fake news.

In Vectorizing of data the Tf-Idf checks the frequency of the word repeats in a document.Then it compares its frequency across data of database. Tf-Idf weight calculate the importance of the word in the data of database. In Tf-Idf the Tf stands for Term frequency. Term Frequency calculates the frequency or repetition of word in a document. The weight of Tf-idf various from length of document as the term will have more repetition in a long document than a short document. The Idf stands for Inverse Document Frequency. The Idf calculates the unique terms of the document excluding the stop words.The count of each word is kept in Document Matrix as the Tf-idf is used in body text. The Document matrix is used for further calculation.

3.1 Formulas

These are the formulas we used in the proposed system.
Document Term Matrix:-

$$\begin{pmatrix} D_1 \\ D_2 \\ \vdots \\ D_n \end{pmatrix} \begin{pmatrix} W_{11} & W_{12} & W_{13} & \cdots & W_{1n} \\ W_{21} & W_{22} & W_{23} & \cdots & W_{2n} \\ \vdots & & & & \vdots \\ W_{m1} & W_{m2} & W_{m3} & \cdots & W_{mn} \end{pmatrix} \tag{1}$$

The Equation displays the equation for the Term Frequency.

$$TermFrequency : TF = \frac{Nij}{Di} \tag{2}$$

The accuracy is number of correct classification of algorithm for all prediction.

$$Accuracy = \frac{TP + TN}{TP + FP + TN + FN} \tag{3}$$

The percentage of correct classification that were classified is called precision.

$$Precision = \frac{TP}{TP + FP} \tag{4}$$

Recall is correct positive outcomes divided by the addition of all classification(it includes classifications samples that were meant to be positive).

$$Recall = \frac{TP}{TP + FN} \tag{5}$$

F1-Score is calculated by harmonic mean.(The precision and recall values.)

$$F1 - Score = \frac{2 \times P \times R}{P + R} \tag{6}$$

 https://doi.org/10.1051/itmconf/20224403017

Models	XGBoost	LSTM	Random Forest Classifier	Logistic Regression
Accuracy	0.7902	0.8501	0.9489	0.9812
Precision	0.7685	0.8106	0.9612	0.9726
F1 Score	0.7895	0.8690	0.9449	0.9012
MCC	0.5849	0.7298	0.8786	0.8961
Recall	0.7994	0.9289	0.9455	0.9823
Specificity	0.7759	0.7789	0.9579	0.8923
G-mean	0.7902	0.8497	0.9348	0.9487

Table 2. Comparing results of different models with proposed models

MCC calculates correlation coefficient among the actual values of sophistication and expected values of the class label.

$$MCC = \frac{TP+TN-FP+FN}{\sqrt{(TP+FP)\times(TP+FN)\times(TN+FP)\times(TN+FN)}} \quad (7)$$

Specificity is calculation of fraction of negative values classified correctly.

$$Specificity = \frac{TN}{TP+FN} \quad (8)$$

G-mean is nothing but Geometric mean.It is calculated as square root of the multiplication of recall and Specificity.

$$G-mean = \sqrt{TP \times TN} \quad (9)$$

4 Results

We used random search method for better performance of different parameters. The batch size and Kernel size are the parameters which are important for model's overall performance.To get the right Batch length and kernel size we've got finished experiments with different batch size and kernel size. For different batch sizes we considered each of the kernel sizes. Following Graphs shows the experimental results.

From this results we have been observed following findings :

- The accuracy of Logistic Regression is very good than other Classification Models.The accuracy of XGBoost and Random Forest is 96 and 92 respectively.Even with different batch it gives better accuracy.
- Precision of Logistic Regression with kernel size is better than other classification Models.Even with different batch size it gives better precision.Precision Score of Random Forest , MultinomialNB and XgBoost is 96,96,99 respectively.
- The Recall of classification model is different for all models.But the Logistic Regression Model has highest recall value.

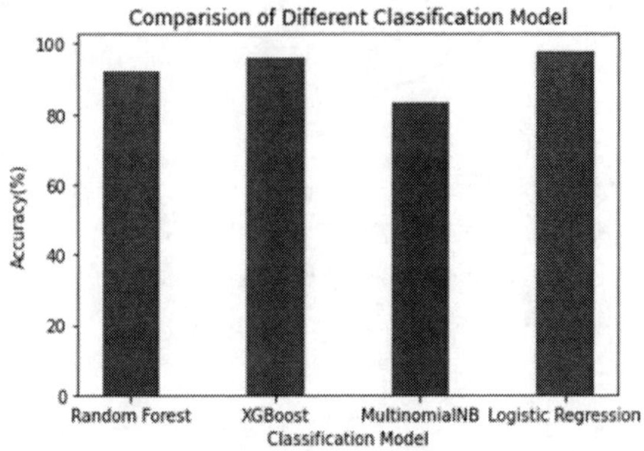

Figure 3. Accuracy

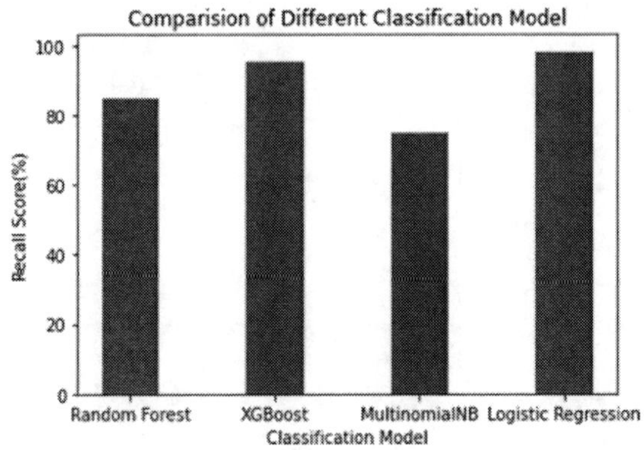

Figure 4. Recall

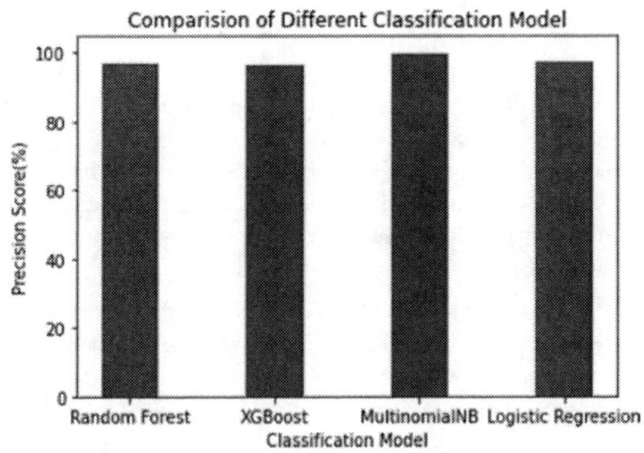

Figure 5. Precision

- Based on accuracy,recall score,precision score the Classification Model of Logistic Regression

Thus by Comparison with other model the performance of Logistic Regression have been selected in this work.

https://doi.org/10.1051/itmconf/20224403017

5 Conclusion

In this paper we proposed the Fake news classification model using Logistic Regression Classification. We have gone through with four algorithms as Random Forest, XG-Boost , LSTM(Long Short-Term Memory) and Logistic Regression. Because these three algorithms have different Approach for Classification of Fake News. The Logistic Regression is the most optimized algorithm among them as it gives more accuracy than other algorithms. The presented model is using three parameters which are title , author and text for its credibility.The model is tested on four datasets.From the experimental results,the model effectively classifies real news and fake news.This model is adoptable and it supports multiscale feature learning from the datasets of news article.

In future , the model could be used for multi-model learning. In future We could use images with news text for fake news classification.

References

[1] Mykhailo Granik, Volodymyr Mesyura, "Fake News Detection Using Naive Bayes Classifier", 2017 IEEE First Ukraine Conference on Electrical and Computer Engineering (UKRCON).

[2] H. Gupta, M. S. Jamal and M. S. Desarkar, "A framework for real-time spam detection in Twitter," 2018 10th International Conference on Communication Systems Networks (COMSNETS), Bengaluru, 2018, pp. 380-383

[3] Marco L. Della Vedova, Eugenio Tacchini, Gabriele Ballarin, Luca de Alfaro,"Automatic Online Fake News Detection Combining Content and Social Signals", ISSN 2305-7254,2017

[4] Cody Buntain, Jennifer Golbeck "Automatically Identifying Fake News in Popu- larTwitter Threads", 2017 IEEE International Conference on Smart Cloud.

[5] Shloka Gilda,"Evaluating Machine Learning Algorithms for Fake News Detection" ,2017 IEEE 15th Student Conference on Research and Development (SCOReD)

[6] Kai Shu, Amy Sliva, Suhang Wang, Jiliang Tang, and Huan Liu, —Fake News Detection on Social Media:A Data Mining Perspective|| arXiv:1708.01967v3 [cs.SI], 3 Sep 2017

[7] Kushal Agarwalla, Shubham Nandan, Varun Anil Nair, D. Deva Hema, —Fake News Detection using Machine Learning and Natural Language Processing,|| International Journal of Recent Technology and Engineering (IJRTE) ISSN: 2277-3878, Volume-7,Issue-6, March 2019

[8] Wang, F., Hussels, P. and Liu, P. (2009) 'Securely and flexibly sharing a biomedical data management system', SPIE.

[9] Rada Mihalcea , Carlo Strapparava, The lie detector: explorations in the automatic recognition of deceptive language, Proceedings of the ACL-IJCNLP

[10] Deligiannis, N.; Huu, T. D.; Nguyen, D. M.; and Luo, X. 2018. Deep Learning for Geolocating Social Media Users and Detecting Fake News.

[11] Gross, N., and Simmons, S. 2006. Americans' views of political bias in the academy and academic freedom. In annual meeting of the American Association of University Professors.

[12] Ma, J.; Saul, L. K.; Savage, S.; and Voelker, G.M. 2009. Beyond blacklists: learning to detect malicious web sites from suspicious urls. In Proceedings of the 15th ACM SIGKDD international conference on Knowledge discovery and data mining, 1245–1254. ACM.

[13] Long, Y.; Lu, Q.; Xiang, R.; Li, M.; and Huang, C.-R. 2017. Fake News Detection Through Multi-Perspective Speaker Profiles. *In Proceedings of the Eighth International Joint Conference on Natural Language Processing (Volume 2: Short Papers)*, 252–256. Taipei, Taiwan: Asian Federation of Natural Language Processing.

[14] Main, T. J. 2018. *The Rise of the Alt-Right.* Brookings Institution Press..

[15] Katsaros, D.; Stavropoulos, G.; and Papakostas, D. 2019. Which machine learning paradigm for fake news detection? In *2019 IEEE/WIC/ACM International Conference on Web Intelligence (WI)*, 383–387

Hand gesture based X-ray image controlling using Convolutional Neural Network

Rutika Mhatre[1], *Bhakti* Dhage[2], *Vishesh* Kwatra[3] ,and *Pallavi* Chavan[4]

[1]Department of Information Technology,Ramrao Adik Institute of Technology,Navi Mumbai,India rutikamhatre21@gmail.com

[2]Department of Information Technology,Ramrao Adik Institute of Technology,Navi Mumbai,India dhagebhakti0123@gmail.com

[3]Department of Information Technology,Ramrao Adik Institute of Technology,Navi Mumbai,India kwatra.vishesh@gmail.com

[4]Department of Information Technology,RAIT,D.Y. Patil Deemed to be University,Navi Mumbai,India pallavi.chavan@rait.ac.in

Abstract.This paper proposes a novel computer vision based system that allows doctors, surgeons and other physicians to control X-Ray images just by using simple gestures thus eliminating the need of traditional devices like mouse and keyboard. This will help reduce the risk of contamination in sterile environments like those found in the hospitals and it will also help in preventing the spread of covid by not allowing contact with contaminated surfaces.It is implemented using CNN model. CNN is specially used for image recognition as well as processing.The system detects gestures through in-built webcam and converts it into corresponding computer commands to perform its associated tasks.
Keywords Computer Vision, X-ray, Keyboards, Contamination.

1. Introduction

Computer information technology is continuously growing in the Hospital domain. It is necessary to handle these technologies in a safe manner to avoid serious mistakes, because this may lead to possible fatal incidents. Real-time imaging review is important in surgery, particularly during the operation. Currently, Doctors and Nurses use keyboards and mouse to handle the images but these are not the appropriate devices for this task. This may lead to the spreading of disease causing pathogens from one user to another.

Many studies have tried to solve this problem by using various devices like leap motion devices as well as the use of Microsoft Kinect Technology to enable contactless control of the computers allowing medical images to be controlled without the risk of bacterial contamination.

However the main issue with all of these approaches is that all of them require some sort of expensive hardware device which may or may not be available in some regions. Also, not all hospitals and clinics can afford to buy expensive hardware devices like kinect or the leap motion sensor. Often small clinics have limited budgets and they have to spend most of their budget in buying the essential things that are very much crucial to keep the clinic up and running. So our project is aiming to resolve this issue by enabling the identification of gestures just by using a common webcam that is found in-built in most computers and this will allow our system to be accessible to even the smallest of clinics.

Hand gestures are a kind of body language that can be conveyed through different actions, for example one can use their palm to move the cursor, use their index finger for clicking, etc. These gestures are again divided into two types i.e. static and dynamic gestures. Static hand gestures means stable hand gestures whereas the dynamic gesture consists of hand movements such as waving. One of the main advantages of using static hand gestures is that they are read more accurately and require less complex algorithms for their recognition making the system fast and resource efficient. That is why we have opted to use static gestures instead of dynamic gestures for faster gesture recognition and more accuracy.

We are using hand gestures for manipulation of x-ray images. The major advantage of sterility as gesture is basic form of communication and it is often used when people speak to each other . It can perform actions like moving images, zooming images,etc. Related work i.e. various literature surveys appear in section 2 and proposed system of the project specified in section 3 which include initial planning,training a CNN,mapping of gesture,debugging,etc.Section 4 contain the result of the project.Final conclusion of the project are provided in section 5.

2. Related Work

Reviewing real-time x-ray images during surgery is very important, especially during operations. Traditionally devices like mouse, keyboard and touchpads are used. But these devices are reliant on physical contact which can lead to the spreading of infection from one user to another. So we have developed a solution that will allow doctors to manipulate and view x-ray images without the usage of devices that require physical contact like keyboard and mice.

[1] In this paper, tests have been done to check for the presence of pathogenic microorganisms on keyboards and mice. A total of 35 samples were collected from different computer labs of an institute in Libya. The samples collected were cultured for further analysis. The results demonstrated that all 35 samples were contaminated with five pathogenic bacteria (E.coli supp. Salmonella spp. Klebsiella spp. Pseudomonas spp. and Staphylococcus spp.). All of the samples in this study were contaminated. This shows how common electronic devices like keyboard and mouse can act as a carrier for pathogenic microscopic organisms. Regular Cleaning of these devices minimizes the bacterial level necessary to diminish the bacterial level.

[2] In this paper, 100 keyboards were tested for bacterial contamination and in the results it was found out that ninety five of them were contaminated with some sort of bacteria. Streptococcus, Clostridium perfringens, Enterococcus (including one vancomycin-resistant Enterococcus), Staphylococcus aureus, fungi, and gram-negative organisms were solded. This really shows the importance of keeping computer equipment clean so that it does not lead to the spreading of disease causing pathogens from one user to another.

[3] During surgery, the surgeon usually is not the person who directly controls the computer. Instead the doctor instructs his/her assistant to perform operations on the computer. This type of communication is not efficient at all and it can be a source of frustration in the operating room if there is misunderstanding between the surgeon and the assistant. This paper describes the design of a joystick like device that can be used to control the computer without needing verbal communication with the assistant. The surgeon controls the computer by placing the probe into the bese and manipulating it like a joystick. The base provides force feedback and permits intuitive clicking with the probe.

[4] This paper proposes a touchless visualization system for computer aided surgery (CAS) which can be used to manipulate a patient's 3D anatomy model through the use of Microsoft kinect. Such anatomy models play a really important role during surgery as they help surgeons to get a good idea on how to go about the procedure and where the problem actually is within the patient. Microsoft Kinect was released alongside the Xbox 360 game console as an interactive way to play games where your body acts as a controller to provide inputs to the game. The motion tracking provided by the Kinect was very accurate and responsive. The depth and skeleton information from Kinect is used effectively to recognise the various gestures made by the user and translate them into actions to control various types of medical images.

[5] In this paper, a gestural interface based on the Leap Motion sensor device is developed and the gesture operations are integrated into PetaVision3D, and in-house 3D PACS software for clinical use. An optional toot pedal was also implemented so that the user could make gestures using just one hand. The main goal was to translate gestures into actions to control medical images. The Leap Motion Controller is an optical hand tracking module that captures the movement of your hands and allows you to interact directly with the digital content on Windows PCs. The Leap Motion sensor itself consists of two cameras and some infrared LEDs. It also has a wide angle lens to increase the interaction space so that gestures could be detected over a larger area.

[6] This paper is more of a general overview of the work that has been done in the field of hand gesture recognition and the comparison of various techniques that have been used for the same. The hand gesture recognition techniques have been compared on a variety of criterias like performance, algorithms used, drawbacks, type and number of gestures, the dataset used and so on and so forth. This paper also discusses the applications of hand gesture recognition in a variety of fields giving us a very clear iden on how useful hand gesture recognition can be in the future.

[7] This research paper aims at offering new possibilities for people to interact with machines using intuitive gestures that are then translated into computer commands. Their system is based on computer vision and it allows their system to recognise a number of static and dynamic gestures. Their system can recognise six static and eight dynamic hand gestures. There are three main steps involved when it comes to gesture recognition in their system : Recognising the shape of the hand, tracing the detected hand only if it is a dynamic gesture and then converting the acquired data into a computer command to perform an action. Their system shows an accuracy of around 93.09 %.

[8] This research paper cites that with the ever growing inclusion of computers into our society, the existing modes of interactions with computers (mouse and keyboard) can become a bottleneck hindering the information flow between the machines and humans. In order to avoid this bottleneck in the future and allow for effective exchange of information between humans and computers, Vision based Gesture recognition can be a powerful tool for achieving the naturalness

required for Human Computer Interaction (HCI). This paper also highlights the work that has been done in the field of computer-vision based analysis by many researchers and it emphasizes the importance of having such a system in place for the future.

[9] This paper highlights that the attractiveness of computer games can be enhanced by vision-based user inputs. A system like this should have a very quick response time (less than a video frame) and should not be costly so that it can be produced for the masses. These constraints were met by using special algorithms that were tailored to a particular hardware. A chip called the artificial retina chip was developed which allowed for fast image processing. The algorithms were then developed to use the capabilities of the chip to provide a very interactive response to player's hand or body movements at 10 msec frame time and at a low cost. These interactions were then demonstrated in several games.

[10] In this paper, research has been carried out for the recognition of dynamic hand gestures. The gestures that have been selected are a sequence of distinct hand poses which can undergo motion and different changes. A recognition engine was developed that was able to accurately detect these gestures despite individual variations. The engine also has additional capabilities like detecting the starting and ending of gesture sequences in an automated fashion. One of the main advantages of this system is that it is able to stay accurate despite the background clutter that might be present and it uses skin color for tracking and recognition. This was implemented on standard hardware which allowed for the recognition of dynamic hand gestures in real time.

[11] This paper uses a camera and two colored markers colored red and green that are worn on finger tips that are used to generate the desired hand gestures. For detecting the markers and tracking they have used template matching with Kalman Filter. When the gestures are detected the commands are issued to the system to perform the desired action. Their system can perform a variety of system actions like movement of the cursor, mouse button clicks and zooming. This system is feasible on devices where touch-screen is not an option like large screens or projected screens and it even works on desktops.

[12] This paper cites the importance of human computer interaction and how traditional mouse and keyboard can be a huge barrier and bottleneck when it comes to HCI. In this paper, a robust and marker-less hand gesture recognition has been proposed that allows their system to effectively track both static and dynamic gestures and perform various system tasks like opening websites, applications, moving through slides on microsoft powerpoint. Their overall system consists of two parts which includes the front end and the back end. They have used three hardware modules which include the detection module, the camera module and the interface module.

[13] This paper makes the use of sensible shape based features like COM (Center of mass), orientation, fingers that are folded, fingers that are raised and their position in the image. A web cam is used to get the image frames which are then pre-processed to remove any background noise. Then with the usage of K-means clustering, the hand is separated from the rest of the background in order to calculate the shape based features. A voice processor is also used that plays a voice through the speaker whenever a gesture is recognised that allows even blind users to use their system.

3. Proposed System

Hand gestures are a form of nonverbal communication that can be used in several fields such as communication between deaf - mute people, robot control, human-Computer interaction (HCI), home automation and medical applications. The system can operate and perform specific user tasks efficiently. A system that can accurately and quickly recognise gestures and perform tasks makes the interaction between a computer and a human very smooth and hassle free.

3.1 - Planning of our Project

Our proposed idea was executed in 5 major phases:

Phase 1 : Planning our Proposed System
Phase 2 : Determining the libraries and software required
Phase 3 : Using a dataset to train a CNN model
Phase 4 : Using the trained model to recognise gestures and mapping them to keys
Phase 5 : Cleaning the code and adding the final touches.

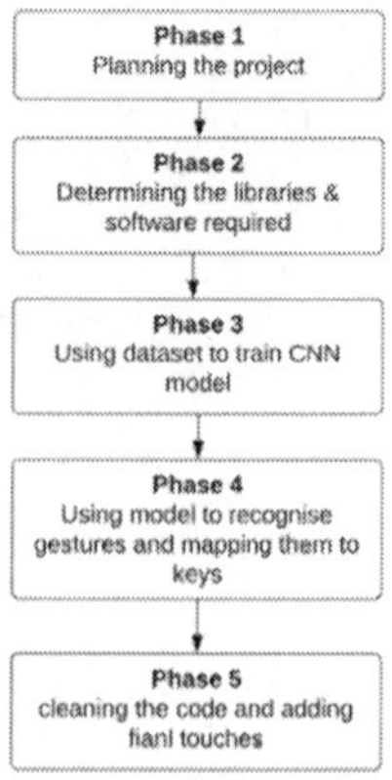

Figure 1 - Planning of our project

3.1.1 - Initial Planning

In the first phase, we planned our project in order to achieve our desired goal. This phase mostly involved reconnaissance and surveying to determine the extent to which our project can help and make the lives of medical professionals easier. We also did literature surveys and studied a number of research papers to determine the amount of work that has been done in this field and the technologies that already exist and used by medical professionals.

3.1.2 - Libraries

In the next phase we determined the libraries that we wish to use and the environment that we want to use for coding our project. To accomplish this, we studied and selected the libraries that are available for us to use for free of cost to make sure our system is as cost effective as possible. Once we selected the libraries that we require, we researched and analyzed these libraries so as to import only the packages which are essential for our use case so as to not slow down our system by importing any unnecessary packages. For our proposed system we have used a number of libraries, some of which are:

- Numpy
- Keras
- Tensorflow
- OpenCV
- PyAutoGUI

NumPy is a library for the Python programming language that adds support for large, multi-dimensional arrays and matrices, along with a number of comprehensive mathematical functions and linear algebra routines to operate on these arrays. NumPy arrays are way faster than Python lists for performing large numerical operations as NumPy array functions are evaluated internally in C++ which makes their runtime way faster. This library is open-source and its high level syntax makes it very easy to understand for programmers of any skill level.

Keras is an open-source software library that uses Python and provides a way to interact with the TensorFlow library to create artificial neural networks and it is also used for other machine learning and artificial intelligence applications. It has a number of commonly used neural-network building blocks (layers, activation functions, optimizers) and a large number of tools that allow for the easy processing of a variety of data types and makes the coding process for writing a deep neural network easier. Keras is used by a large number of individuals and scientific organizations around the world including CERN, NASA and NIH

TensorFlow is an open-source software library developed by Google used for machine learning and artificial intelligence purposes. It can be used for a variety of purposes but it mainly focuses on the creation and training of deep neural networks. The Keras library runs on top of TensorFlow. It is written in C++, Python and CUDA. Having a GPU (Graphical Processing Unit) greatly speeds up the deep learning processing rather than just relying on the CPU for training. It accepts data in the form of arrays with multiple dimensions called tensors.

OpenCV is an open-source software library that allows for real-time computer vision. It can be used in conjecture with other libraries like Keras and TensorFlow to build complex neural networks and other machine learning applications. It is written in C++ but it also has a Python API. By using this library, it is possible to process images and videos for a variety of applications such as street view image stitching, medical image analysis, interactive art installations etc. It also supports GPU acceleration that allows it to use the CUDA cores for faster processing of data.

PyAutoGUI is a python library that allows for control of mouse and keyboard strokes through python scripts. It can be used for automating interactions with applications, taking screenshots, displaying messages, interacting with application windows (moving, resizing, minimizing, maximizing), moving the mouse cursor to a specific position on the screen etc. It can also be used to press multiple keys at once which are called macros. This can be very useful especially in photo and video editing applications where there is a macro for every task.

3.1.3 - Convolutional Neural Network

Deep learning is a subset of machine learning and it basically tries to imitate the structure and functioning of a human brain and allows computers to think and perform tasks like humans and even allows them to "learn" things without being explicitly coded. A Convolutional Neural Network is a deep learning model that can be used to work with data that is far more complicated, unstructured and varied such as audios, images or text files. This kind of data cannot be used with traditional machine learning algorithms. A neural network has the following components :-

- Input Layer
- Hidden Layer
- Output Layer

The **input layer** is the layer from which the data is fed as input to the neural network. The data can be anything ranging from text files to audio or images.

The **hidden layer** is responsible for the processing of data and feature extraction by performing complex computations. This layer/layers has weights and biases that are continuously updated as a part of the training process. After the computation is complete, the output is passed on to the output layer.

The **output layer** is responsible for predicting the values by using suitable activation functions. The output can be either numerical or categorical in nature.

Keras offers two ways to build neural networks :-

1. Sequential
2. Functional

The **Sequential Model** is the easiest way to build a model in Keras. It allows the user to build a model layer by layer and the data flows from one layer to another until it reaches the final layer. But the drawback of this model is that the sharing or branching of layers is not allowed. Also, it is not possible to have multiple inputs and outputs.

The **Functional Model** is more flexible than the sequential model as it allows for branching and sharing of layers and enables you to create more complex models. In fact, you can connect layers to any other layer. This model also allows for multiple inputs and outputs.

3.1.4 - Training a CNN

In the next phase of our project, we trained a sequential CNN model to recognise various gestures that the user makes in front of the webcam. In order to train our model, we have made use of various datasets available to us from Kaggle. These datasets had a large number of images for each gesture which we have used to train our model. Before training our model these images had to be pre-processed so as to make the training process as seamless and as accurate as possible. These datasets are imported and the CNN model is trained until a good accuracy rate is achieved. Our trained CNN model is capable of recognising 5 static hand gestures :-

1. Closed Palm
2. Open Palm
3. Point Left
4. Point Right
5. Point Up

3.1.5 - Mapping of Gestures

In the next phase, we mapped the hand gestures to specific key(s) or keyboard strokes which are used to control the X-Ray images within the software. In order to accomplish this we mainly made use of the PyAutoGUI library. The user needs to have a webcam for inputting the gestures. The webcam is used to gather video frames and each frame is then run through our CNN model to determine what gesture is the user doing and our system accordingly performs the tasks within the software to control the X-Ray images.

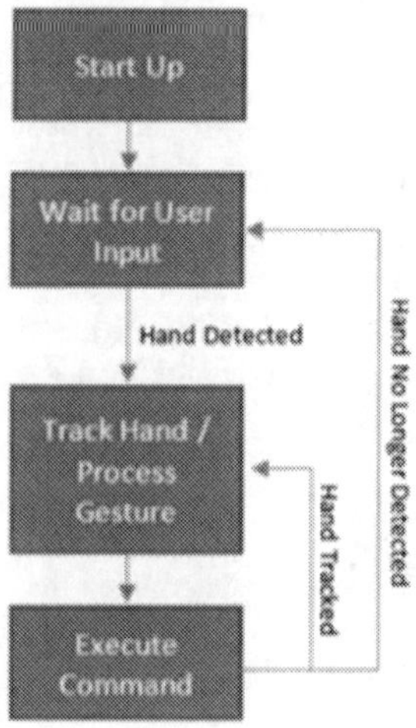

Figure 2 - Block Diagram of our System

3.1.6 - Debugging

The final stage of our project involved adding final touches to the code and documenting it so that it is easy to understand whenever any user tries to modify the code for their own usage. We also removed any unwanted lines of code and any unused libraries in order to further optimize our code.

3.2 - X-Ray Imaging

X-Ray Imaging is one of the most widely used radiography techniques allowing the doctor to get a look inside the patient's body without performing any incision which helps them to diagnose, monitor and treat many medical conditions. As the name suggests, it uses X-rays which is a type of electromagnetic radiation which is not harmful to human tissue in any way. These X-ray beams are able to pass though your body and depending on the material they pass through, these rays are absorbed in different amounts. For example, dense materials such as bones show up as white in X-ray images while fat and muscle appear as shades of gray. It is a very quick and painless procedure and usually requires little to no special preparation on the part of the patient unlike MRIs which may require fasting or the ingestion of special fluids by the patient. This X-Ray technology has also been employed in other types of diagnostic procedures such as fluoroscopy, CT scans and arteriograms.

3.3 - DICOM Files

DICOM stands for "Digital Imaging and Communications in Medicine" and it is the industry standard that was created by the National Electrical Manufacturers Association (NEMA) to aid the distribution and viewing of medical images such as X-Ray Images, CT Scans, Ultrasound, MRI etc. A

https://doi.org/10.1051/itmconf/20224403018

DICOM is both a file format and a communication protocol which means it can store medical information and patient's information all in one file ensuring that all the data stays together.

A single DICOM file contains a header (which stores information like the name of the patient, type of scan, dimensions of the image etc) and image data which can also be in three dimensions. This is better than the older Analyze format which stored this information in two different files (image data in a .img file and header data in a .hdr file). DICOM files also support compression. Since this file format is now the industry standard when working with medical images and is widely adopted by the hospitals, doctors and researchers can now easily exchange medical images with each other without any compatibility issues.

The DICOM files can be viewed using a DICOM image viewer. Most of the softwares that are actually used by hospitals for the analysis of medical images are closed source and are not sold to normal consumers. However, there are plenty of open-source DICOM image viewers which can be downloaded for free by people who want to view their medical images and analyze them at the comfort of their home. MicroDicom is one such open-source software that can be downloaded for free and that's why we have decided to use it and integrate it with our system to perform various tasks related to X-Ray images.Here is a flowchart to show the functioning of our system.

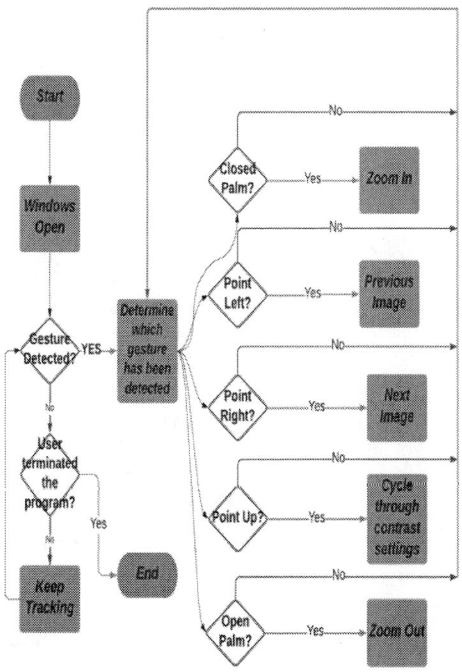

Figure 3 - Functioning of system

4. Result and Discussion

When our code is run, our trained CNN model is loaded and a message is printed in the command prompt to give the user an indication that the execution is in progress and the program is working correctly. After this, a couple of small windows pop up giving the user an idea as to what the webcam is capturing and how it is processing the frames to recognise the gestures. The name of the windows are as follows :-

- Trackbar
- Frame
- Processed
- Cam

The **Trackbar** window can be used by the user to adjust various settings using sliders so as to tweak the system to the lighting conditions around them for better and more accurate gesture recognition.

The **Frame** window shows the user what the webcam is capturing and gives the user an idea on how to position their hand for the best possible experience.

The **Processed** window gives the user an insight into how the system is processing the frames captured using the webcam. This is mostly used for debugging purposes and is not essential for the user experience.

The **Cam** window is the most important window out of the bunch. This is the main window that the user will be interacting with to input their gestures and perform tasks. It consists of a small green rectangular box where the user is supposed to perform the gesture they want to. Only the hand gestures inside the rectangular frame are tracked and any movement outside the rectangular box is ignored. This ensures that the system does not accidentally register commands and perform tasks that the user does not intend to. This window also has a number indicator at the top left. In our system, each gesture is given a number as follows :-

S.No.	Hand Gesture	Number Assigned
1.	Closed Palm	0
2.	Point Left	2
3.	Point Right	3
4.	Point Up	4
5.	Open Palm	6

Table 1 : Table showing what number is each gesture assigned

Whenever the system detects a gesture inside the rectangular frame, it updates the number indicator accordingly to tell the user as to what gesture is the system recognising. It also gives the user an idea if the system is detecting the gestures accurately or not. If not, the user can adjust the sliders in the Trackbar window for better gesture detection and accuracy. The number indicator displays 5 if no gesture is detected within the rectangular frame. This is the NONE state of our program. In this state, no actions are performed and the system stays as is.

Each gesture is mapped to a particular action that allows the user to control the X-Ray images within the MicroDicom software. Below is a table that shows which action is performed by which gesture :-

Hand Gesture	Action Performed
	Zoom In
	Previous Image
	Next Image
	Cycle through contrast settings
	Zoom Out

Table 2 : Table showing which gesture performs which action

As soon as a gesture from the above table is detected by the system, it performs the action corresponding to it inside the MicroDicom software. In order to make the actions more controllable and less eccentric, an artificial delay has been added between every consequent action so that the user gets the time to precisely control what they want to do with the X-Ray images.

In order to determine how well our system is performing, we did testing and documented the results. In our tests, we performed each gesture 100 times under ideal lighting conditions with the Trackbar settings set to their default values and we noted the number of times each gesture was recognised successfully to perform the assigned action. The success rate and failure rate for each gesture was then calculated. The results have been documented below with the formulas :-

$$Success\ Rate\ \% = \frac{Action\ performed\ successfully \times 100}{Total\ No.\ of\ times\ the\ gesture\ was\ performed}$$

Failure Rate % = 100 - Success Rate %

Hand Gesture	Action Performed	Success Rate
	Zoom In	94 %
	Previous Image	95 %
	Next Image	91 %
	Cycle through contrast settings	90 %
	Zoom Out	96 %

Table 3 : Success Rate of each gesture

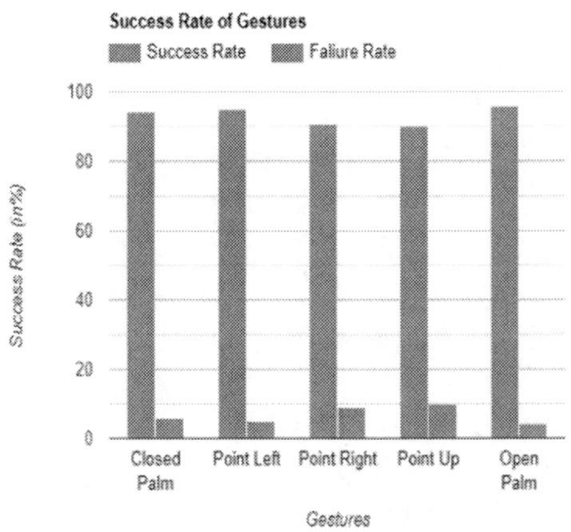

Figure 4 - Success rate of gestures

5. Conclusion and Future Work

The authors have built a gesture based system for controlling X-Ray images. We hope that our project helps a lot of doctors to perform their tasks more effectively while keeping them free from the risk of contamination especially in the light of the current Covid situation.

In the future, more hand gestures can be added and applied as input to the computer. These hand gestures can then be converted into computer commands to perform tasks in real time. Since our code will be flexible, it can be adapted so that it can be used in any application and it is not just limited to X-Ray image viewing. The gestures can be mapped to various keyboard commands and these commands can then be used to implement various operations within different apps.

6. References

[1] Hamida Amer, Ahmed Atia, Khaled Tawil : *Bacterial Contamination of Computer Keyboards and Mice in University and Hospital Settings*

[2] Maureen Schultz, Janet Gill, Sabiha Zubairi, Ruth Huber, Fred Gordin : *Bacterial Contamination of Computer Keyboards in a Teaching Hospital* **24** , 4 (2003)

[3] Dumitru Onceanu, A. James Stewart : *Direct Surgeon Control of the Computer in the Operating Room*

[4] Jiaqing Liu, Ryoma Fuji, Tomoko Tateyama, Yutaro Iwamoto, Yenwei Chen: *Kinect-Based Gesture Recognition for Touchless Visualization of Medical Images* **9** , 2 (2017)

[5] Ben Joonyeon Park, Taekjin Jang, Jong Woo Choi, Namkung Kim : *Gesture Controlled Interface for Contactless Control of Various Computer Programs with a Hooking Based Keyboard and Mouse-Mapping Technique in the Operating Room*

[6] Munir Oudah, Ali Al-Naji, Javaan Chahl : *Hand Gesture Recognition based on Computer Vision : A Review of Techniques* (2020)

[7] Soeb Hussain, Rupal Saxena, Xie Han, Jameel Ahmed Khan : *Hand gesture recognition using deep learning* (2017)

[8] G.R.S. Murthy & R.S. Jadon : *A Review of Vision Based Hand Gestures Recognition* **2** , 2 (2009)

[9] W.T. Freeman, K. Tanaka, J. Ohta, K. Kyuma : *Computer Vision for Computer Games , IEEE 2nd Intl. Conf. on Automatic Face and Gesture Recognition, Killington, VT, October,* (1996)

[10] Aditya Ramamoorthy A, Namrata Vaswani A, Santanu Chaudhury A, Subhashis Banerjee B : *Recognition of Dynamic Hand Gestures* , (2002)

[11] Sayem Mohammad Siam, Jahidul Adnan Sakel, and Md. Hasanul Kabir : *Human Computer Interaction Using Marker Based Hand Gesture Recognition* , (2016)

[12] Aashni Hariaa, Archanasri Subramaniana, Nivedhitha Asokkumara, Shristi Poddara, Jyothi S Nayaka : *Hand Gesture Recognition for Human Computer Interaction , 7th International Conference on Advances in Computing & Communications, ICACC-2017, 22-24 August 2017, Cochin, India* (2017)

[13] Dr R Suguna, Neethu P S : *Hand Gesture Recognition using Shape Features* **117** , 8 (2017)

Loan Prediction System Using Machine Learning

Anant Shinde[1], *Yash* Patil[2], *Ishan* Kotian[3], *Abhinav* Shinde[4] *and Reshma* Gulwani[5]

[1,2,3,4]Department of Information Technology, RAIT, Nerul, India
[5]D.Y. Patil Deemed to be University, Ramrao Adik Institute of Technology, Nerul, Navi Mumbai, India

Abstract. As the needs of people are increasing, the demand for loans in banks is also frequently getting higher every day. Banks typically process an applicant's loan after screening and verifying the applicant's eligibility, which is a difficult and time-consuming process. In some cases, some applicants default and banks lose capital. The machine learning approach is ideal for reducing human effort and effective decision making in the loan approval process by implementing machine learning tools that use classification algorithms to predict eligible loan applicants.

1 Introduction

Machine Literacy is a subset of artificial intelligence that allows computer programs to automatically learn from former tasks. It works by analysing the data, relating patterns, and incorporating minimum mortal intervention. Nearly any work that can be done using a data description pattern or set of rules can be done using a machine learning machine. This allows companies to modify processes that preliminarily only humans could make hypotheticals for client service calls, accounts, and reviews.

Loan distribution is the middle enterprise of virtually every bank. Loan distribution is the middle enterprise of nearly every bank. Utmost of a bank's means come directly from the gains it makes from the loans it makes. The main thing of the banking terrain is to put wealth in a safe place. Currently, many banks / financial companies approve loans after a verification and validation redemption process, but it is not yet clear if the selected applicant is correct among all applicants. Through this system, it is possible to predict whether a particular applicant is safe and the entire process of verifying the characteristics will be automated by machine learning technology. Credit forecasting is very useful for both bank employees and applicants.

The goal of this system is to provide a quick, immediate and easy way to select good applicants. It can offer banks special benefits. The credit forecasting system can automatically calculate the weights for each feature that participates in credit processing, and the new test data will process the same features for the assigned weights. The model can set a deadline to see if the applicant can approve the loan. Credit analysis allows to jump to specific applications and check according to priority. This system is exclusively for bank / financial company management authorities, the entire forecasting process is carried out privately and no stakeholders can change the process. The results of a particular credit ID can be sent to various departments of the bank so that they can take appropriate action on demand. This helps all other departments handle other paperwork.

2 Literature Survey

According to the authors, the forecasting process begins with data clean-up and processing, missing value substitution, data set experimental analysis, and modelling, and continues to model evaluation and test data testing. A logistic regression model has been executed. The highest accuracy obtained with the original dataset is 0.811. Models are compared based on performance measurements such as sensitivity and specificity. As a result of analysing, the following conclusions were drawn. However, other characteristics of customers that play a very important role in lending decisions and forecasting defaulters should also be evaluated. Some other traits, such as gender and marriage history, do not seem to be considered by the company [1]. A credit credibility soothsaying system that helps companies make the right opinions to authorize or reject the credit claims of guests. This helps the banking assiduity to open effective distribution channels. This means that if the customer has a minimum repayment capacity, their system can avoid future risks. Including other techniques (using the Weka tool) that are better than the general data mining model has been implemented and tested for domains [2]. The author suggests that, a credit status model for predicting loan applicants as valid or standard customers. The proposed model shows a score of 75.08 when classifying loan aspirants using R-Package. Lenders can use this interpretation to make mortgage choices for mortgage operations. In addition, comparative studies were conducted at different iterative levels. The replication position is a 30- grounded ANN model that offers a more advanced delicacy than other situations. This model can be used to avoid large losses in marketable banks [3]. Six machine learning classification models were used to predict Android applications. The model is available in open-source software R. This application works well and meets the requirements of all banks. The downside of this model is that it gives each element a different weight, but in reality, it may be possible to approve a loan only based on a single powerful element, which is not possible with this system. This component can be easily connected to many other systems. There are cases of computer failure, and the most important weights of content errors and features are fixed by the automatic prediction system, and soon, so-called software may be safer, more reliable and more [4]. Risk assessment and forecasting is an important task in the banking industry in determining whether a good and lazy loan applicant is applicable. To improve the accuracy of risk, risk assessments are conducted in primary and secondary education. Customer data is extracted and related attributes are selected using information gain theory. Rule forecasting is performed for each credit type based on predefined criteria. Approved and rejected applicants are considered "Applicable" and evaluated as "Not Applicable". Corresponding experimental results have shown that the method proposed predicts better accuracy and takes less time than existing methods [5]. The main purpose of this design is to prognosticate which customers will be repaid with a loan because the lender needs to anticipate the problem that the borrower won't be suitable to repay the threat. Studies of three models show that logistic regression with a rating is superior to other models, random forests, and decision trees. Poor credit seekers aren't accepted, presumably because they have the option of not paying. In utmost cases, high-value appliers may be eligible for a reduction that may repay the loan. Certain sexual orientations and marriage status appear to be out of the reach of the company [6].

3 Feature Engineering

Predicated on the field knowledge, this system can develop new features that can affect the target variables. Created three new functions:

3.1 Total Income – In Fig 1, as explained in the bivariate analysis, combined the income of the applicant with the income of the co-applicant. The higher the total income, the more likely there are to get loan approval.

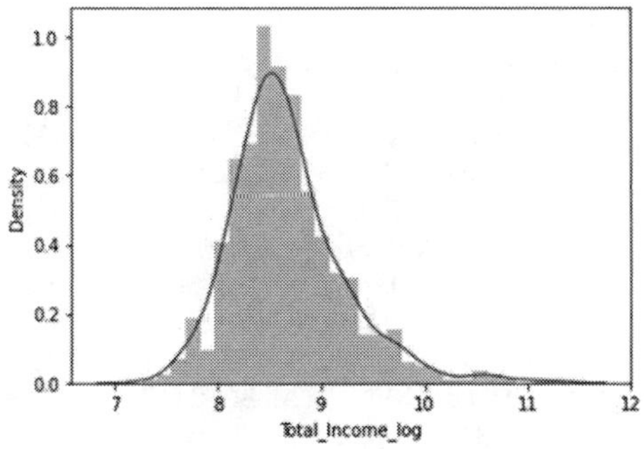

Fig. 1. Density vs Total Income Log

3.2 EMI – In Fig 2, EMI is the yearly volume that the seeker must pay to reimburse the loan. The model behind this variable is that people with lofty. EMI may possess challenges with the prepayment of their loans. EMI can exist figured by taking the rate of the loan volume to the majority the loan volume rate of the loan volume to the majority of the loan volume.

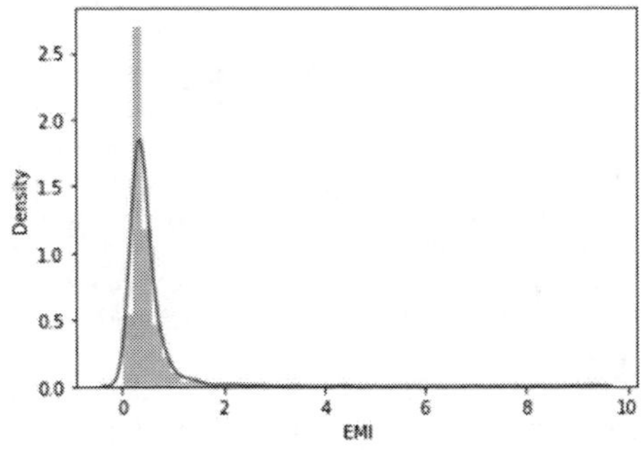

Fig. 2. Density vs EMI

3.3 Balance Income – In Fig 3, this is the return deserted over after compensating the EMI. The model behind creating this variable is that the advanced the valuation, additionally probable a person is to reimburse the loan and thus additionally probable it's to authorize the loan.

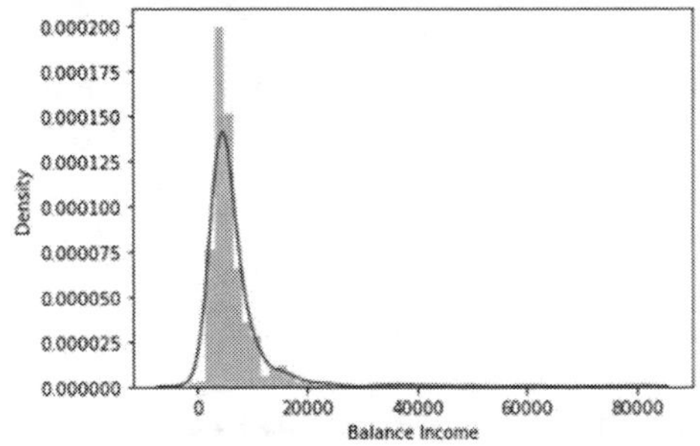

Fig. 3. Density vs Balance Income

4 Proposed Framework

4.1 Business understanding:

In the early stages, the base is on deriving the design from a custom outlook and rephrasing that lore into data mining challenge delineations and primary designs.

4.2 Data convention:

The data convention aspect focuses on the original data library, data command, relating data rate outcomes, and a subset of stake for undertaking retired data.

4.3 Data processing:

https://doi.org/10.1051/itmconf/20224403019

The data processing aspect includes all the conditioning to produce the concluding dataset.

4.4 Modelling:

The algorithm which will be used for data modelling is Logistic Regression using stratified k-folds cross-validation and Random Forest.

Logistic Regression using stratified k-folds cross-validation: This system uses validation to see how robust the model is against hidden data. This is an approach for booking distinct exemplifications of reports that don't train the model. Latterly, the system tests the model in this illustration and also finalize it. Some of the generally applied confirmation styles are the confirmation incubate path, k-fold cross-validation, Leave one outcross-validation (LOOCV), and stratified k-fold cross-validation. In Table 1, analyzed the mean validation and f1-score of Logistic Regression with the k-folds model.

Table 1. Accuracy for Logistic Regression model

The mean validation accuracy for this model turns out to be	0.7214
The mean validation f1 score for this model turns out to be	0.8279

In Fig 4, Visualized the roc curve:

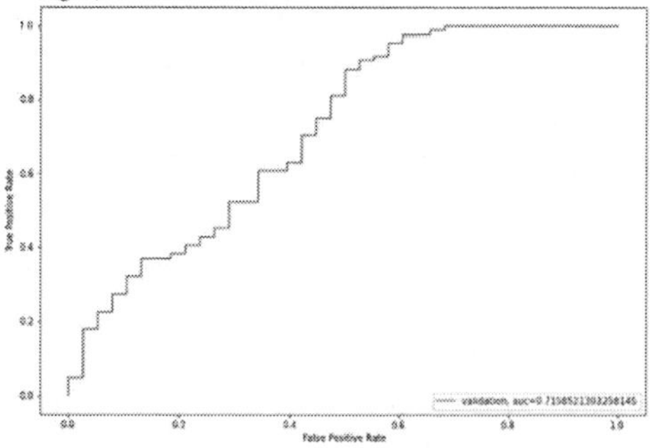

Fig. 4. AUC value of 0.5626

Random Forest: This system was tested to reduce the exactness by conforming to the hyperparameters of this model. The model used grid search to master optimized valuations for hyperparameters. Grid - search is a way to elect the stylish one from the family of hyperparameters parameterized by the parameter grid. adapted the max_depth and n_estimators' parameters. max_depth determines the maximum depth of the tree and n_estimators determine the number of trees used in the random forest model. In Tabel 2, generated the mean validation accuracy for the hyperparameters.

Table 2. Mean Validation Accuracy of Hyperparameters

Mean Validation Accuracy	0.7947

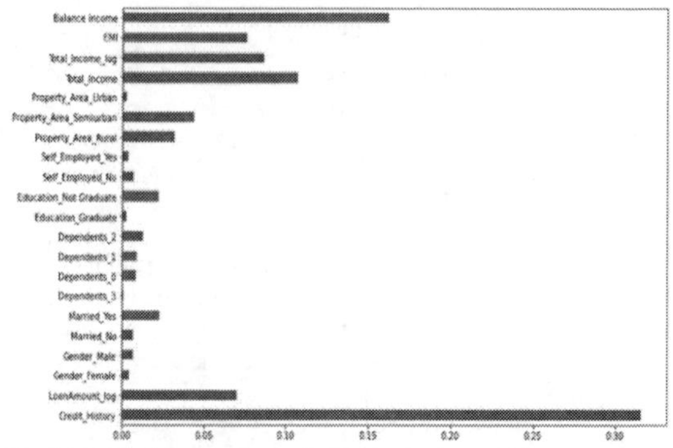

Fig 5. Feature importance predicting the target variable

In Fig 5, Credit_History is the most major point succeeded by Balance Income, Total Income, EMI. Accordingly, feature engineering assisted the model in forecasting the target variable.

5 Conclusion

Borrowers use a loan application to qualify for a mortgage. The above research employs a logistic regression algorithm-based prediction model. To create a logistic classification model that predicts loan status, over 600 sample data were collected and evaluated. The algorithm can obtain a maximum accuracy of about 82 percent and regression models are used to obtain such precision. The model can anticipate outcomes and is quickly adaptable to a wide range of inputs. Also, this strategy saves the banking industry and its staff a significant amount of time.

References

[1] M. Sheikh, A. Goel, T. Kumar, "An Approach for Prediction of Loan Approval using Machine Learning Algorithm," International Conference on Electronics and Sustainable Communication Systems (ICESC), (2020).

[2] S. M S, R. Sunny T, "Loan Credibility Prediction System Based on Decision Tree Algorithm," International Journal of Engineering Research & Technology (IJERT) Vol. 4 Issue 09, (2015).

[3] A. Kumar, I. Garg and S. Kaur, "Loan Approval Prediction based on Machine Learning Approach," IOSR Journal of Computer Engineering, (2016).

[4] Dr K. Kavitha, "Clustering Loan Applicants based on Risk Percentage using K-Means Clustering Techniques," IJARCSSE - Volume 6, Issue 2, (2016).

[5] P. Dutta, "A STUDY ON MACHINE LEARNING ALGORITHM FOR ENHANCEMENT OF LOAN PREDICTION", International Research

Journal of Modernization in Engineering Technology and Science, (2021).

[6] G. Arutjothi, Dr C. Senthamarai, "Prediction of Loan Status in Commercial Bank using Machine Learning Classifier," Proceedings of the International Conference on Intelligent Sustainable Systems, (2017).

[7] P. Supriya, M. Pavani, N. Saisushma, N. Kumari and K. Vikas, "Loan Prediction by using Machine Learning Models," International Journal of Engineering and Techniques, (2019).

[8] R. Salvi, R. Ghule, T. Sanadi, M. Bhajibhakare, "HOME LOAN DATA ANALYSIS AND VISUALIZATION," International Journal of Creative Research Thoughts (IJCRT), (2021).

[9] B. Srinivasan, N. Gnanasambandam, S. Zhao, R. Minhas, "Domain-specific adaptation of a partial least squares regression model for loan defaults prediction," 11th IEEE International Conference on Data Mining Workshops, (2011).

[10] M. V. Reddy, Dr B. Kavitha, "Neural Networks for Prediction of Loan Default Using Attribute Relevance Analysis," International Conference on Signal Acquisition and Processing, (2010).

[11] G. Chornous, I. Nikolskyi, "Business-Oriented Feature Selection for Hybrid Classification Model of Credit Scoring," IEEE Second International Conference on Data Stream Mining & Processing August (2018).

https://doi.org/10.1051/itmconf/20224403020

Intelligent Medical Chatbot System For Women's Healthcare.

Shrushti Polekar[1], *Shivani Wakde*[2], *Mayuresh Pandare*[3], *Priyanka Shingane*[4]

[1] *Department of Information Technology,RAIT,D.Y.Patil Deemed to be University Nerul*

Email: shrushtipolekar@gmail.com

[2]*Department of Information Technology,RAIT,D.Y.Patil Deemed to be University Nerul*

Email: shivaniwakde1211@gmail.com

[3]*Department of Information Technology,RAIT,D.Y.Patil Deemed to be University Nerul*

Email: mayuresh.pandare21@gmail.com

[4] *Department of Information Technology,RAIT,D.Y.Patil Deemed to be University Nerul*

Email: priyanka.shingane@rait.ac.in

Abstract— Even today talking about women health is big taboo in India. The society is unbiased about women especially about their health issues hence they face disparity in the society. Women feel hesitant to talk about or speak about their health issues or problems openly. Thus, the goal of this chatbot is to help women to find information and remedial solutions about their health. Query is processed by the bot and response will be displayed on web application. This chatbot will provide helpful information instantly. It will also provide prediction about the disease that the women might be suffering.So this bot will help to make Right decision and give right advice to women on 24/7 basis. It will act as an helping hand for working women to keep check on their health in their busy routine. At the same time this will also help women in rural areas who are apprehensive to talk about their health issues publicly.

I. INTRODUCTION

The idea of this system is to create a chatbot that can help girls and women find information and medicinal solutions about their health. Women sometimes tend to ignore their health problems due to their busy schedules and hectic lifestyles. Also normally ladies aren't aware of all the treatments regarding the diseases they have. Many of the diseases can be cured if they are treated early and have proper treatments at the early stages itself. This will even decrease the chances of having major health issues later on in life.

On the other hand, sometimes ladies have to visit doctors even for some small problem which is time consuming. Such problems can be solved by using a medical chatbot which can give them proper guidance. Time to time guidance can help girls to remain updated about their health and have a routine check-up of their health. Girls or women can share their health-related problems or symptoms to the bot and the bot will provide proper guidance according to the symptoms they have.

The bot will process the query and will provide a proper response to the user. In this system, we have given women option to chat and also get prediction for the diseases. The chatbot processes the question and answers the women's questions. We have also made use of DialogFlow for connecting women with bot .We have added multiple questions in our Dialogflow ,but which women will get answers to their questions. We have majorly taken two diseases in account ,that are Breast Cancer and PCOS.

These diseases can be cured if they are detected at an early stages and proper medication is taken time to time. For prediction of the Breast Cancer we have used Logistic Regression algorithm. Where the accuracy is 96% which is maximum compared to rest of the algorithms. For prediction of PCOS we have made use of multiple algorithms who accuracies are KNN is 72%,MLPClassifier is 63%,Random Forest I 81% and for Logistic Regression is 89%.

II. LITERATURE SURVEY

The approach hired in Chatbot Utilization for Medical Consultant System [1] The chatbot is carried out in IM utility, in which Line utility is utilized in our study. Then the software program transfers the message to Dialogflow, that is the engine of the chatbot. The message is extracted to achieve the motive. In a few case, to react to the request message, the gadget desires to select out up the statistics from an outside database or outside APIs. To do this, the extra coding is necessary. The structures will generate the actionable statistics that person can apprehend and ship returned to the utility. In closing segment the person will obtain responses in types of video,image, and textual content etc.

The paper Exposes Chatbot for Healthcare System Using Artificial Intelligence [2] .This chatbot is a conversational AI assistants which automates the interactions with the users. Chatbot is powered by AI using ML techniques to interpret natural language. The aim of the paper is to provide the user with minor or major medical and health information. Initially when the user's visits the website , he has to get registered. And later the user can ask the bot their queries. The system uses an expert system to answer the questions. The domain experts also have to register themselves on the platform.The content is stored in form of pattern template. It uses SQL for interacting with database.The durability and performance can be increased by using a Structured Query Language.

The paper Study on Artificial Intelligence in Healthcare[3] emphasizes on how technology can be used in the medical domains ,increase productivity ,Increase efficiency and make diagnostics.as during this ongoing pandemic of Covid-19 there is a need for increase in medical services and technologies . In this paper they have used Support Vector Machine(SVM)to inspect patients suffering from Heart failure.The paper solves various problems by making use of AI ,Machine Learning, Agent based system .Use of ELM algorithm is discussed in this paper to improve the accuracy .

III. PROPOSED SYSTEM

- Authentication Module : The user can get herself registered on the platform by signing up. Firebase has been used for handling authentication.

- Chatbot Module: The chatbot will process all the queries of the user and based on the input it will provide relevant response. The patient or user can ask the questions regarding their symptoms and disease in the chatbot and the chatbot will respond accordingly. We have made use of DialogFlow for connecting women with bot. We have added multiple questions in our Dialogflow ,but which women will get answers to their questions. This chatbot service will be available to the user on 24/7 basis.

- Disease Prediction Module: Whenever the user tells about Symptoms it recognizes the symptoms and analysis by machine learning logic algorithms and predicts the disease and gives response to the user. We have consider two disease ,that are Breast Cancer and PCOS. These diseases can be cured if they are detected at an early stages and proper medication.Breast Cancer is been predicted using LogisticRegression algorithm.Where the accuracy is 96%.For prediction of PCOS we have made use of multiple algorithms whose accuracies are KNN is 72%,MLPClassifier is 63%,Random Forest is 81% and for Logistic Regression we got accuracy as 89%.

Logistic Regression can be used to determine the probability of certain class or event when the data is linearly seperable and outcome is binary in nature.

The equation for Logistic Regression is given as:-

$a=e^{(x0 + x1*b)}/(1+e^{(x0+x1*b)})$

where:

b is the input value

a is the predicted output

x0 is the bias or intercept term

x1 is the coefficient for the single input value(b)

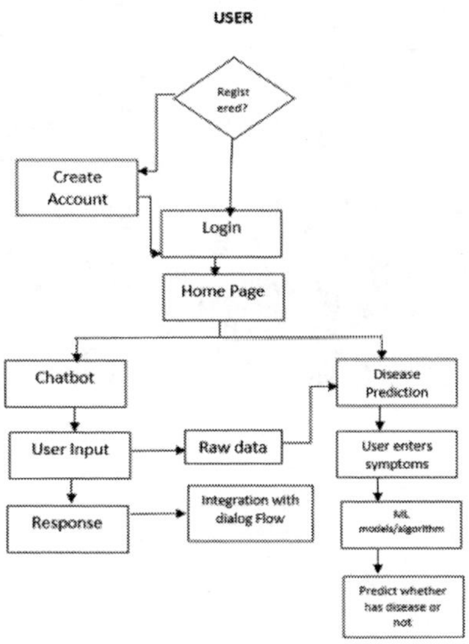

Fig. 1. System Architecture

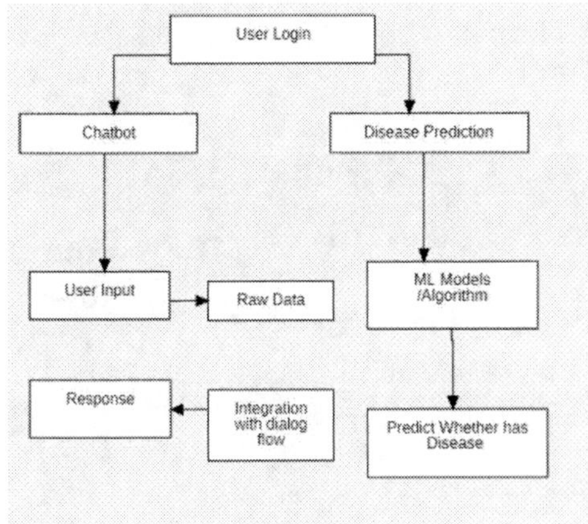

Fig. 2. Training Flow

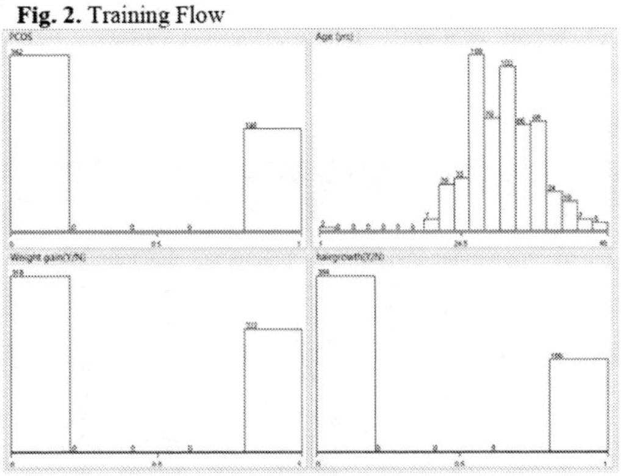

Fig. 3. Methodology of Visualizing data set.(Weka Tool).

In the above figure we have plotted graphs for all the attributed like Age, BMI, Weight loss, Weight gain, Hair loss, Fast food,Skin darkening that we have considered in this application. We have made use of weka tool for visualization.

IV. RESULT

After basic authentication women will get redirected to landing page, she will be getting different options. Figure number 4,5 shows chatbot facility that is been provided to women. In this system, we have given women option to chat and also get prediction for the diseases. The chatbot processes the question and answers the women's questions.

Figure number 7,8,9,10,11 depicts prediction of diseases. We have majorly taken two diseases ,that are Breast Cancer and PCOS. Logistic Regression algorithm is used for prediction of breast cancer .Where the accuracy is 96% which is maximum compared to rest of the algorithms. For prediction of PCOS we have made use of multiple algorithms who accuracies are KNN is 72%,MLPClassifier is 63%,Random Forest I 81% and for Logistic Regression is 89%.

Figure number 14,15,16,17,18 shows Dialogflow. We have made use of Dialogflow for connecting women with bot.We have added questions in our Dialogflow , by which women will get answers to their queries.

Fig. 4. Output 1

Fig. 5. Output 2

Fig.6. Output 3

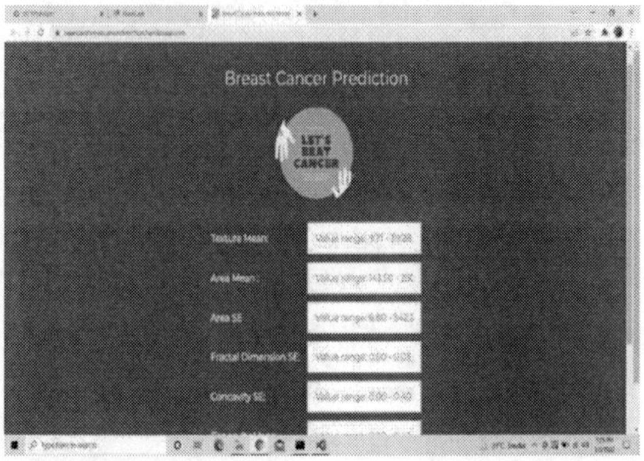

Fig. 7. Output 4

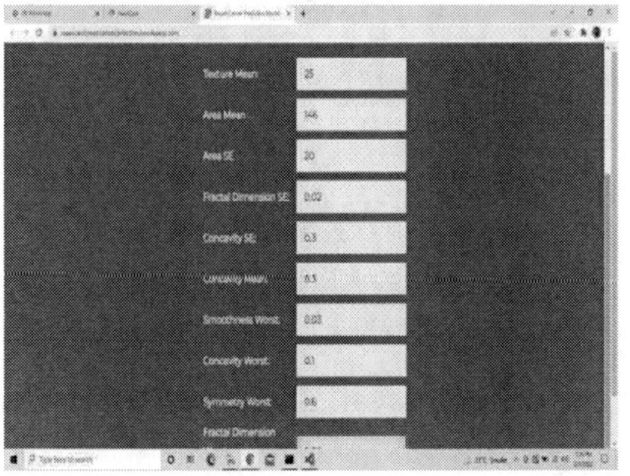

Fig. 8. Output 5

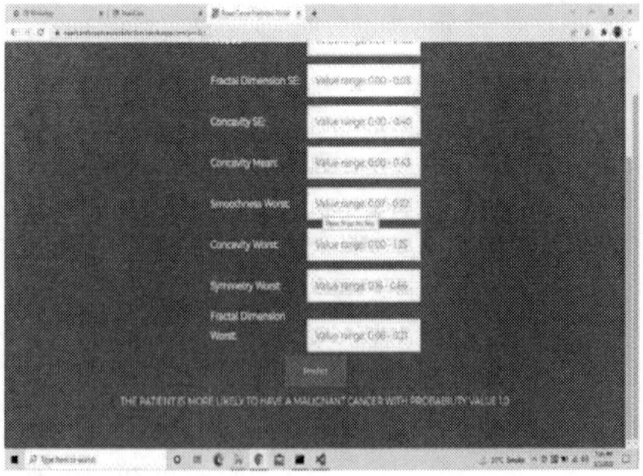

Fig. 9. Output 6

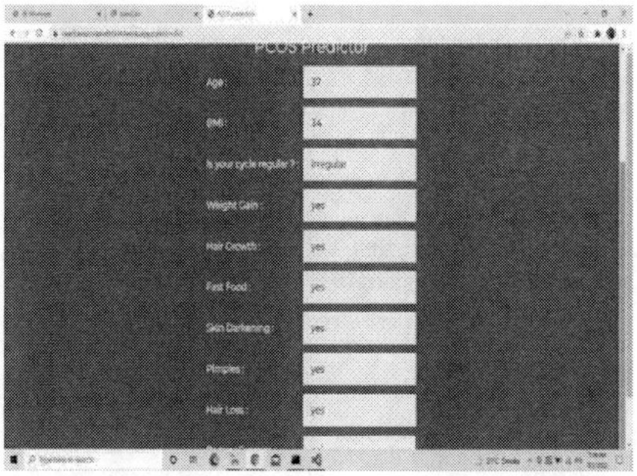

Fig. 10. Output 7

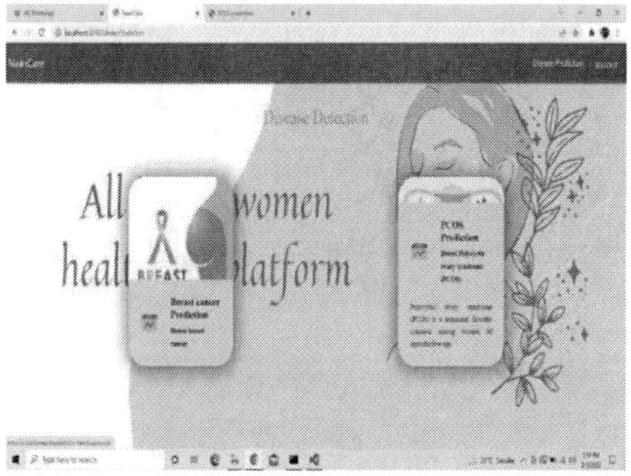

Fig. 13. Output 10

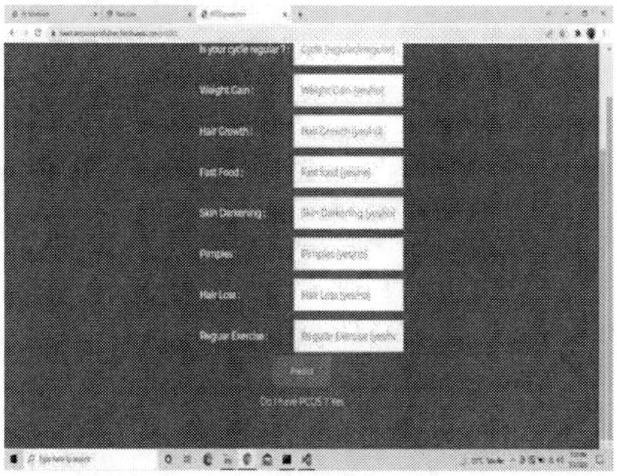

Fig. 11. Output 8

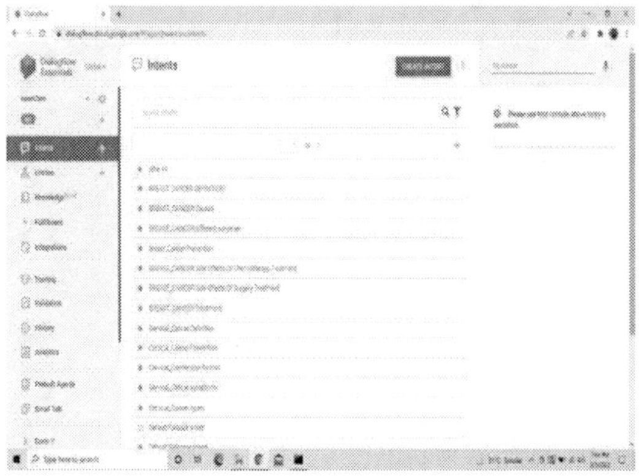

Fig. 14. Output 11

Fig. 12. Output 9

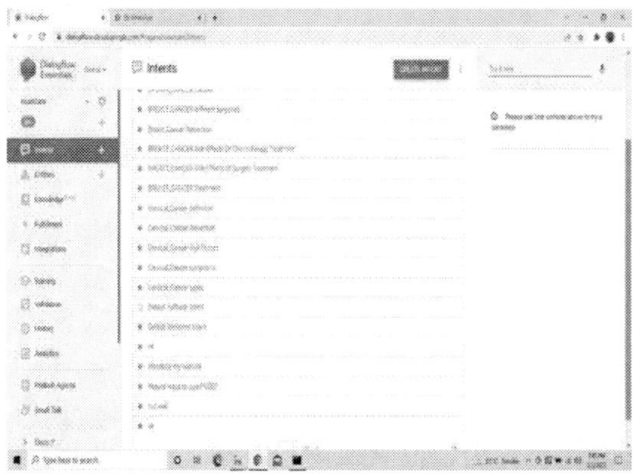

Fig. 15. Output 12

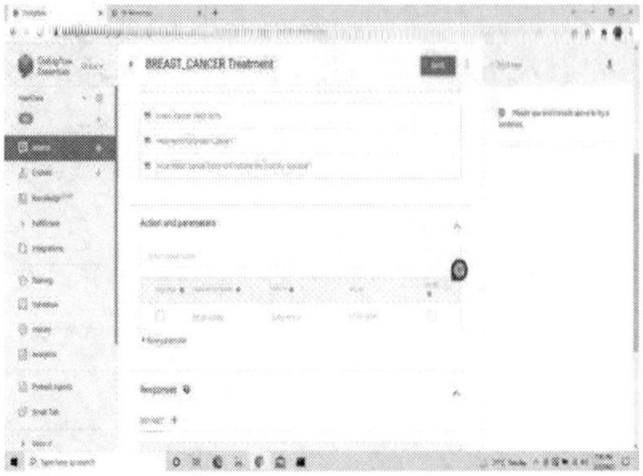

Fig. 16. Output 13

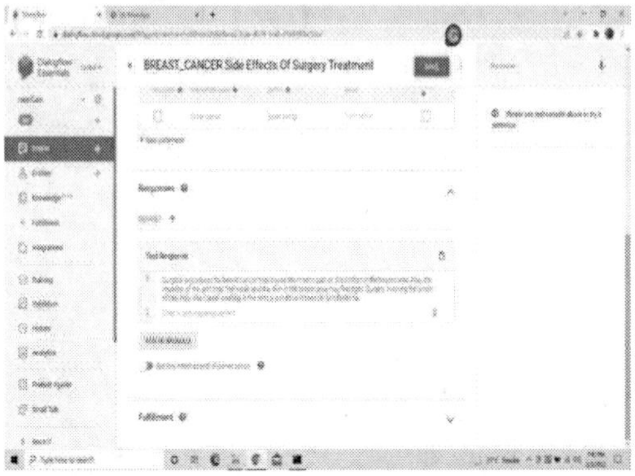

Fig. 17. Output 14

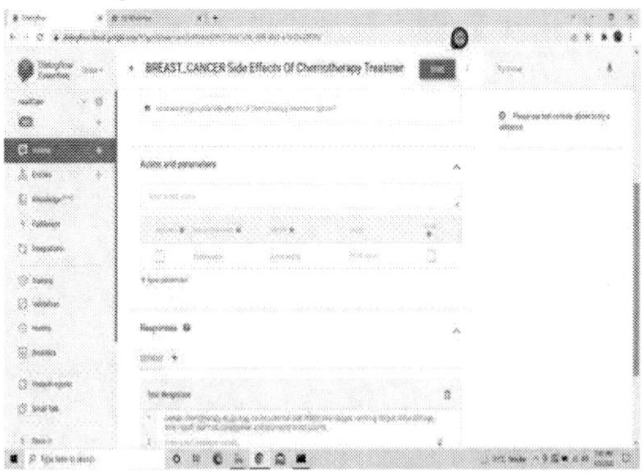

Fig. 18. Output 15

V. Conclusion

This web application does the implementation of women health chatbot where the girl user can login into the web application at any time and ask question about the her symptoms and gets disease prediction. This application can help women to remain updated about their health issues and will prevent them from having any major health issues in their future life.

After each registration of the women on our application, dataset will be formed. By making use of dataset of registered women's we can also make analysis on how many women's are suffering from particular disease and how many of them are healthy. For example if we consider two attributes age and BMI .On basis of which graph can be made and future analysis can be made.

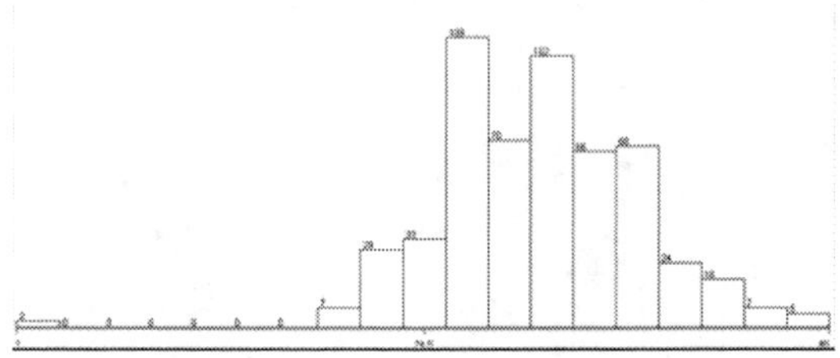

Fig. 19. Visualizing graph of age and BMI(Weka Tool).

VI. Limitations

The women/girl user will not be able to detect whether she has any disease other than the diseases that we have mentioned in our website.

References

[1] Rosruen, Nudtaporn; Samanchuen, Taweesak .Chatbot Utilization for Medical Consultant System.(2018).

[2] Athota, Lekha; Shukla, Vinod Kumar; Pandey, Nitin; Rana, Ajay. Chatbot for Healthcare System Using Artificial Intelligence. (2020).

[3] Gaikwad, S. Study on Artificial Intelligence in Healthcare. (2021).

[4] Nadia Tripp, Kirsten Hainey, Anthony Liu, Alison Poulton, Michael Peek, Jinman Kim, Ralph Nanan, "Women and birth?" Women and Birth, November(2013)

[5] "Natural Language Processing based Human Assistive Health Conversational Agent for Multi-Users" (2021)

[6] Wahidah Husain, Lee Ker Xin, Nur'Aini Abdul Rashid, Neesha Jothi, "Predicting Generalized Anxiety Disorder among women using random forest approach",(2016)

[7] Harshil Kanakia,Preeti Chauhan ,Pooja Patil, Neha Rane,"Comparative Analysis of Machine Learning Algorithms for Prediction of PCOS"(2021)

[8] Salim Nabhan,Dwi Susanto, Akhmad Alimudin,Farah Adila,Moh. Hasbi Assidiqi "Developing English Conversation Chatbot Using Dialogflow"(2020)

[9] Palwinder Kaur Mangat; Kamaljit Singh Saini "Health CARE Prediction using Predictive Analytics"(2021)

[10] Shamsuddin S. Khan; Rohit Barve; Umesh Kulkarni"Proposed model on Prediction and Analysis using application of Health care"(2018)

Crop Guidance and Farmer's Friend – Smart Farming using Machine Learning

Tanvi Daware[1,*], *Pratiksha* Ramteke[1,**], *Uzma* Shaikh[1,***] and *Smita* Bharne[1,****]

[1]Ramrao Adik Institute of Technology, D Y Patil Deemed to be University, Navi Mumbai, India

Abstract: Agriculture is the centre of our country's social and economic development. But the majority of the farmers are still struggling for better agricultural facilities. The government is also helping farmers with their benefits with different government policies. Today's farmers are well aware of using the internet and smartphone for smart farming. There are expanding quantities of instructed individuals in the farming field and they began utilising cell phones in rustic areas. Thus, the main aim of this Crop Guidance and Farmer's Friend application is to assist the farmers with expanding their yield as well as to be aware of the current agriculture-based data used for smart farming. The proposed system is designed to track weather conditions at a specific location, such as temperature, humidity, and changes in the environment. This application diminishes the time and efforts of farmers and assists them with getting the day-to-day market cost for various harvests, fertilisers and vegetables without visiting the market. The data (value, climate, most recent rural technique) will be shipped off to the farmers, through this application. This application helps farmers a great deal and keeps them updated. It offers a livelihood for almost all of the population, contributing to countrywide income and gainful employment.

1 Introduction

Agriculture is the backbone of our Indian economy, which plays an important role in increasing our country's economic development. The main focus of this app is on the agricultural society because the farmers have the highest contribution to our country's GDP.[1] But the farmer still does not get the price for the harvest. This mainly happens due to improper watering or incorrect plant selection, or sometimes the crop yield is lower than expected. By analysing the soil and atmosphere in a given region, the best crop will produce a higher yield and the net yield of the crop can be predicted using Water level, distance depth, and soil ph. In the wake of innovation, numerous farms and agribusinesses are gradually receiving innovation into farm activities as it extensively affects usefulness and productivity.[2] Agribusinesses and other horticultural businesses, both administrative and non-legislative, are going to different programming and equipment computerised cultivating instruments, for example, ranch the board arrangements, Internet of things (IoT), and different types of precision agriculture. Current innovation in farming is something other than the utilisation of versatile applications or an equipment framework to robotize repetitive cycles and lessen reliance on human work.[5] This prediction will help farmers choose the appropriate crops for their farm based on soil type, temperature, humidity, water level, depth of spacing, soil pH, season, fertiliser, and months.[7] Rising technologies are used to improve the productivity of the crops by changing traditional farming to e-farming. About 70 per cent of primary and secondary activities are based on agriculture. Therefore, to improve agriculture, many farmers began to use new technologies and methods.[9] The key issue however to be resolved is cultivating precise crops at precise times. This could be done with the help of machine learning algorithms that are found to be an efficient methodology for predicting the appropriate crop.[11]

2 Literature Survey

In paper [1] a system has been designed based on sensors and IoT systems to monitor the crop field with the help of a sensor. The system is proposed for monitoring weather conditions in a particular place like temperature, humidity, and changes in the environment.

In paper [2] Machine Learning Algorithms are used and designed to make a system easier to compare prices. It achieved an accuracy rate of 95. The algorithm was created to predict modal prices at 14-day intervals across the major mandis in MP.

In paper [3] Time series and Regression algorithms are used, as a means to assess Moisture Content (MC) in their grain. It calculates Moisture Content (MC) with high accuracy, using Relative Humidity (RH) and Temperature (T) time series.

In paper [4] Random Forest Algorithm is used and a broader dataset results in a higher rate of accuracy. Random forest shows to be a superior prediction algorithm. To predict the yield of a particular state that uses some of the various factors which affect the manufacturing of crops.

* Corresponding author: tanvidaware17@gmail.com
** Corresponding author: pratikshasramteke@gmail.com
*** Corresponding author: ukshaikh210595@gmail.com
**** Corresponding author: smita.bharne@rait.ac.in

In the paper [5] authors used Image Processing, Machine Learning, the Internet of things and Artificial Intelligence. Comparative research is conducted between the advanced framework and the present framework. Survey of different methods for Crop selection, sowing of the crop, weed detection, and monitoring of the system thereby yielding productive output.

From the literature survey, we analysed that the existing methods, the process of finding the soil type, identifying the leaf disease, and the suitable fertiliser had been all carried out manually. Even when the framework became digitised, it had certain issues such as predicting numerous fertilisers for a soil kind, plant sickness, or fertiliser. The existing systems lack efficiency due to limited algorithm-based prediction.

3 Proposed Methodology

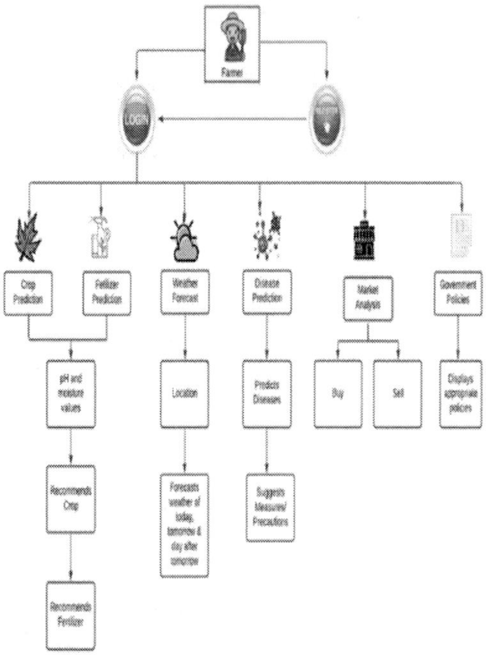

Fig. 1. Proposed System Architecture

Figure 1 shows the detailed system architecture of the proposed system. It portrays all the modules of the system and the functionalities in each module.

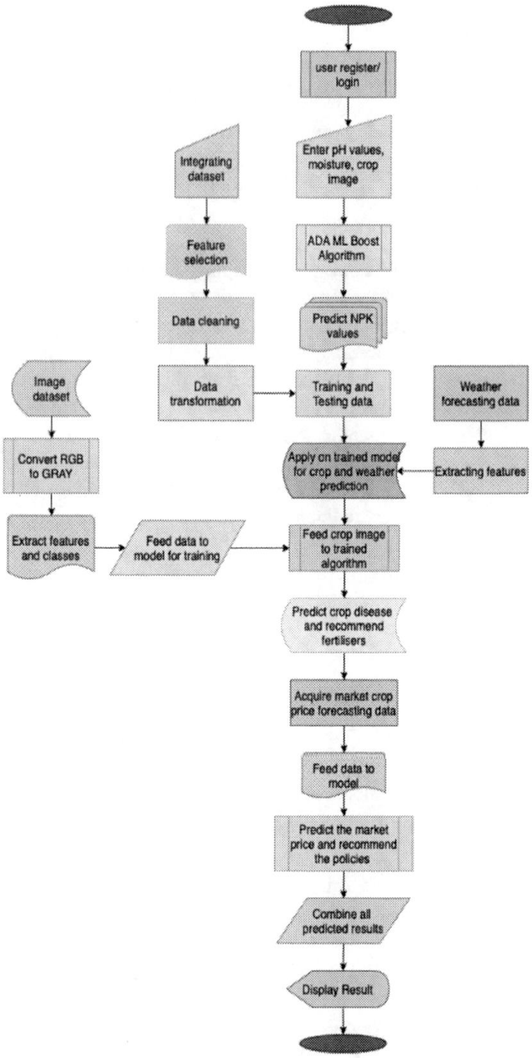

Fig. 2. Detailed flow of the system

Figure 2 shows the detailed flow of the proposed system. It portrays the control flow from user registration/login to displaying the results based on the inputs given by the user.

Composts and fertilisers are added to get the nutrient supplement levels to the proportion needed by the crop the farmer intends to develop, prompting loads of costly base work of developing soil.[3] Be that as it may, our model as of now predicts which harvest develops best given current conditions. Following are the descriptions of the modules included in the proposed system.

3.1 Crop Prediction

The system will use the Ada boost algorithm on moisture and PH values to predict the N-P-K values which are further used in the ensemble method technique which includes algorithms like Naive Bayes, logistic regression, ADA Boost, Random forest, Decision tree, and SVM for crop recommendation.

3.2 Fertiliser Prediction

Enter the nutrient contents of your soil and the crop you need to grow. The set of rules will inform which nutrient the soil has extra or lacks. Accordingly, it's going to supply pointers for getting fertilisers.

3.3 Government Policies

This provides various schemes offered by the Government of India in the field of agriculture. It also describes the eligibility criteria, benefits, procedure to apply, and documents required to take the advantage of a particular scheme.

3.4 Weather Prediction

Anticipating the real weather conditions in current scenarios and then accordingly providing information to the end-user to take the correct measures and also plan the development of the crop.

3.5 Market Price

There will be an analysis of the market to provide the user with the best deal. We will be supplying statistics of the huge gamers of the marketplace including Big Basket, scraping the retail charges from a genuine supply and permitting farmers to make their bargain and ship quotations.

3.6 Disease Prediction

Upload a picture of the leaves of your plant. The set of rules will inform the crop kind and whether or not it's far diseased or healthy. If it's diseased, it'll inform you of the purpose of the sickness and advocate to you the way to prevent the sickness accordingly.

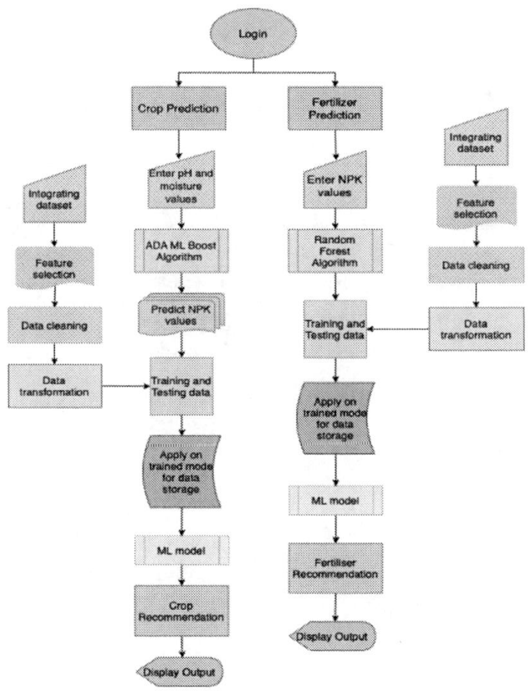

Fig. 3. Flowchart of Crop and Fertiliser Prediction

The given Fig. 3 shows the system module that consists of a portal that will predict the crop and fertiliser suitable for a particular user. When the user enters moisture and pH value for the recommendation of the crop, the ADA Boost Classifier will fit the Decision tree adjusting the incorrect weight such that subsequent classifiers focus on difficult cases. Later it will predict the NPK values and then we will integrate, clean, and transform the dataset which is followed by training and testing the data. The next stage is to apply the trained models which in turn will predict crops based on the ML models and finally display the name of the crop. Similarly, for the fertiliser prediction, our model will take inputs of NPK values. On integrating the dataset, cleaning, and transforming the data, we will train and test the data and apply the trained data to the ML model and display the name of the fertilisers.

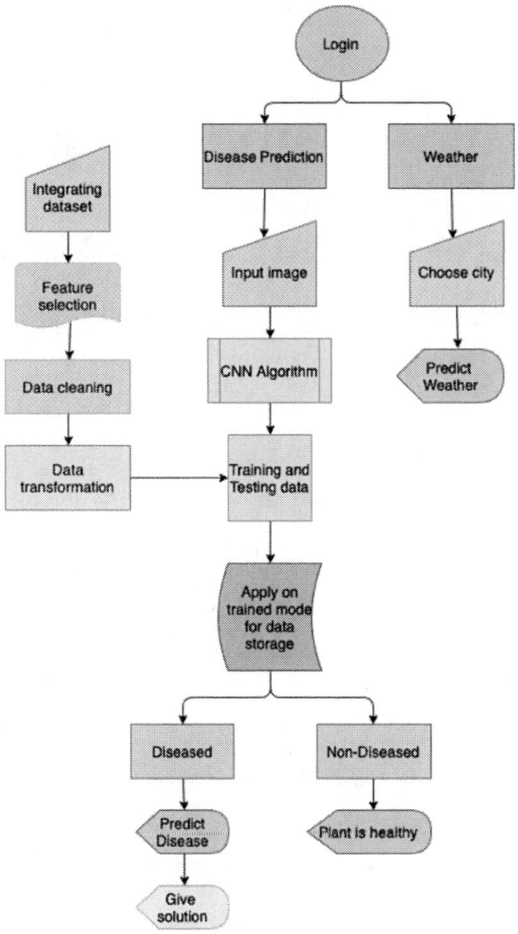

Fig. 4. Flowchart of Weather and Disease Prediction Module.

Figure 4 describes the flowchart for the plant disease detection and weather prediction modules. For the prediction of disease, the user has to give an image as an input for the system, using the CNN algorithm, and then we will integrate, clean, and transform the dataset which is followed by training and testing of data. The next stage is to apply the trained models which in turn will predict whether the plant is diseased or not based on the ML models. For the prediction of weather, the user has to choose a particular city that will finally predict the weather pattern.

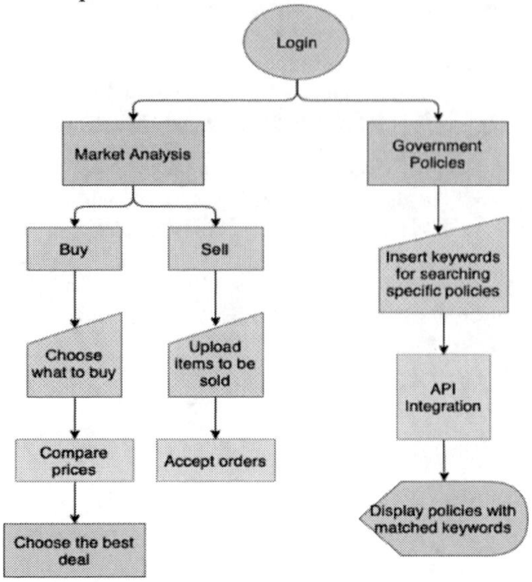

Fig. 5. Flowchart of Market Analysis and Government Policies

Fig.5 describes the flowchart for the market analysis and government policies modules. Our system allows users to buy and sell products and also compare prices to choose the best deals. This provides information regarding various government policies of India related to agriculture.

4 System Details

Machine learning is an important support tool for crop expectations, including support options for plant development and what to do during the plant development phase. The following algorithms are applied to the various features:

4.1 Decision Tree

The decision tree creates characterization or fallback models within the level of a tree structure. It divides the dataset into a moderate subset and gradually evolves a decision tree. It deals with all kinds of downright and mathematical information.

4.2 Naive Bayes

Naive Bayes is used to construct classifiers: models that assign to drawback cases and labels described as vectors of entity values wherever the class labels are drawn from a finite set. It takes into consideration that the worth of a particular feature is insignificant to the real worth of alternative characteristics given class variables.

4.3 Support Vector Machine

SVM makes use of a subset of training factors in the decision function (referred to as guide vectors), so it's also memory efficient.

4.4 Random Forest Algorithm

The Clustered Random Forest (CRF) methodology is used to identify intruders into the network by dividing the entire network into character systems so that glowing comments can be made on communications initiated within the network. It examines every node in the network and creates log files that are given as input for the proposed technique.

$$Gini\ Index = 1 - \sum_{i=1}^{n} (Pi)^2$$

4.5 ADA Boost

It is used as an ensemble method and it adapts a series to differently weighted training data. It starts with predicting the original data set and giving equal weight to each observation.

$$H(x) = sign\ (\alpha_t h_t(x))$$

$h_t(x)$ is the output of weak classifier for input x.
Where α_t = 0.5 * ln ((1 — E)/E) (where E = error rate)

4.6 Neural Network

A neural network is a computational model with network architecture and specific parameters that can be used to modify it to perform specific tasks.

4.7 Dataset

Crop Prediction data is collected from the Kaggle website for the prediction of crops and fertilisers. It has 22 different crops as labels and 7 features - Nitrogen, Phosphorus, Potassium, Temperature, Humidity, Rainfall and Moisture. The input dataset consists of around 3000 data values of crops (rice, maize, orange, moth beans etc) for different PH values and moisture content in the soil.

	N	P	K	ph	Moisture
51	76	60	39	6.766240	51
1656	17	16	14	6.625539	27
245	35	64	78	7.496645	49
357	34	59	18	5.744118	44
267	41	69	82	6.715587	47
...	...	...	...	...	...
643	2	47	15	6.401456	37
131	63	43	19	6.641906	44
1663	26	11	11	7.609348	25
130	87	35	25	6.178056	56
1137	35	18	26	5.279389	35

660 rows × 5 columns

Fig. 6. Dataset for crop prediction module

 https://doi.org/10.1051/itmconf/20224403021

	Crop Type	Nitrogen	Potassium	Phosphorous	Fertilizer Name
0	Maize	37	0	0	Urea
1	Sugarcane	12	0	36	DAP
2	Cotton	7	9	30	14-35-14
3	Tobacco	22	0	20	28-28
4	Paddy	35	0	0	Urea
...	...	...	...	...	...
94	Pulses	24	0	19	28-28
95	Tobacco	4	17	17	10-26-26
96	Wheat	39	0	0	Urea
97	Millets	15	0	41	DAP
98	Sugarcane	12	0	10	20-20

99 rows × 5 columns

Fig. 7. Dataset for fertiliser prediction

Based on the given pH and moisture content, we can predict the NPK values to recommend suitable crops and using NPK value and crop type we can predict the suitable fertilisers.

Fig. 8. Dataset for Disease Prediction

For the disease prediction module, the dataset is collected from the Kaggle website, the system takes the image of the infected leaf as input and predicts the disease remedy. Input dataset consists of 2000 different images of healthy and infected leaves of plants like tomatoes, potatoes, bell pepper etc.

4.8 Mathematical formulae

1. Precision: Precision is the ability to correctly identify the number of positive predictions.
   $$\text{Precision} = \frac{True\ Positive}{True\ Positive + False\ Positive}$$

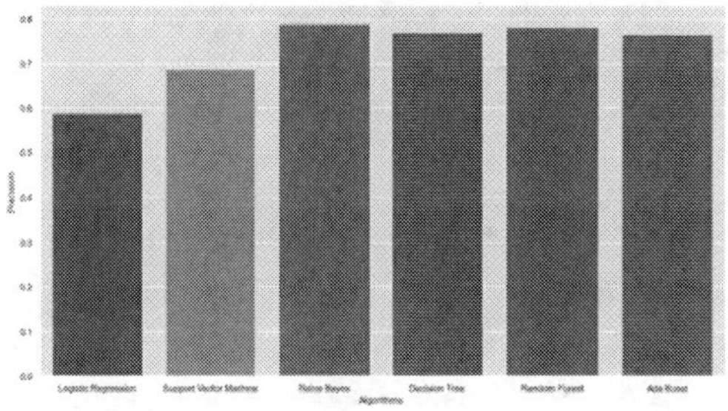

Average Precision = 0.72467

2. Recall: It finds all positive results from the confusion matrix.
   $$\text{Recall} = \frac{True\ Positive}{True\ Positive + False\ Negative}$$

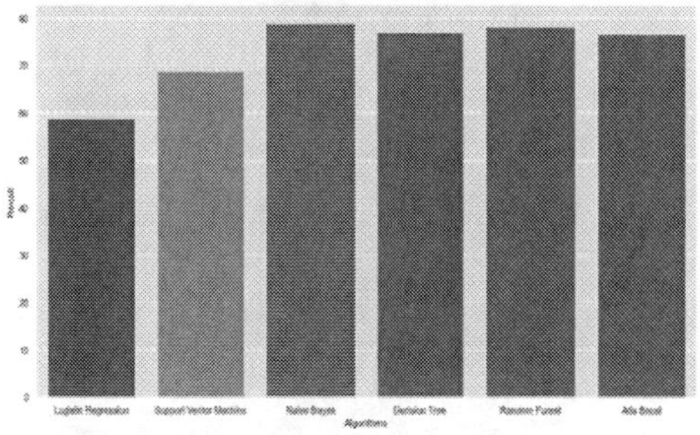

Average Recall = 73.015

3. F1 Score: It finds a balance between precision and recall.
   $$\text{F1 score} = 2 * \frac{Precision * Recall}{Precision + Recall}$$

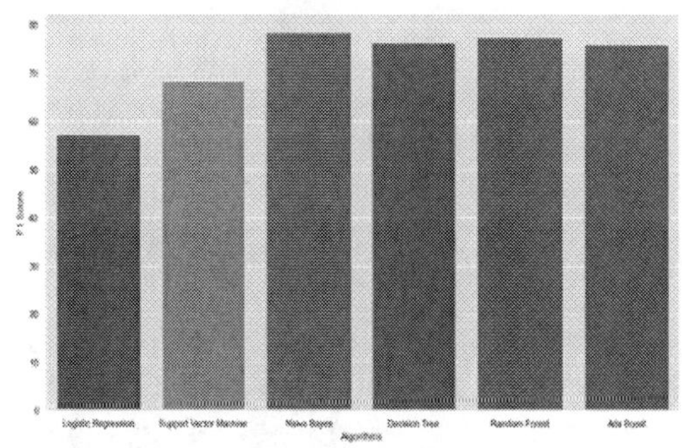

Average F1 score = 72.2867

4. Accuracy: It is calculated by dividing correct predictions by total predictions.
   Accuracy =
   $$\frac{True\ Positive + True\ Negative}{True\ Positive + True\ Negative + False\ Positive + False\ Negative}$$

5 Results

In particular, our contributions are as follows:

• Predicts the suitable crop and fertiliser with maximum accuracy.

• Detects the disease on a plant with the help of an image of its leaves.

• Provides the latest updated government schemes.

• Provides a marketplace so as to compare the prices and sell or buy the required item. Here, we propose a methodology for crop and fertiliser prediction using the ensemble method and the disease will be predicted using the CNN algorithm. The fetching of weather and government schemes will be done by integrating the APIs.

This model is trained using a dataset of around 100 epochs and 3000 values compiled with sparse categorical cross-entropy as the loss function for crop and fertiliser prediction.

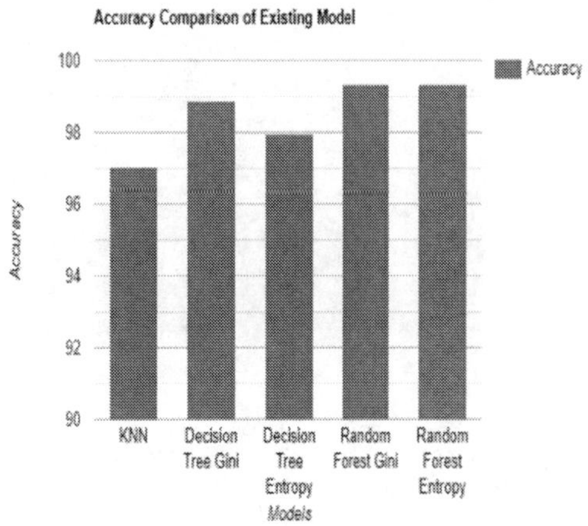

Fig. 9. Accuracy comparison of Existing Model

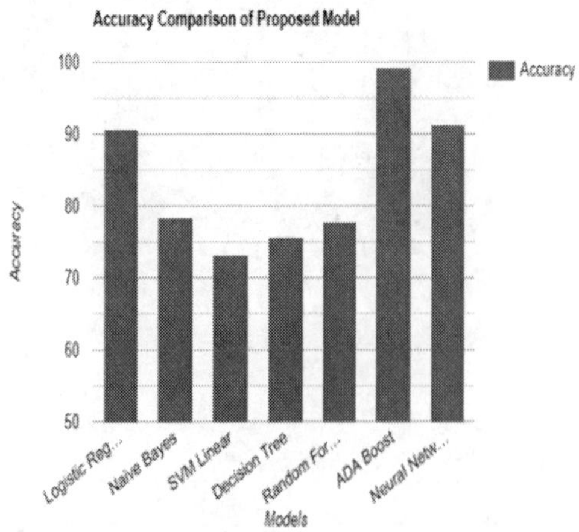

Fig. 10 Accuracy Comparison of Proposed Model for Crop Prediction

Table 1. Accuracy Comparison for Crop Prediction

Sr No.	Machine Learning Model	Accuracy
1	Logistic Regression	60.61%
2	SVM Linear	78.3%
3	ADA Boost	73.18%
4	Decision Tree	75.61%
5	Random Forest	77.73%
6	Naive Bayes	79.09%
7	Neural Network	61.18%

Table 1 summarises the model's accuracy of the crop prediction module. The highest accuracy obtained is 79.09% for the Naive Bayes Model.

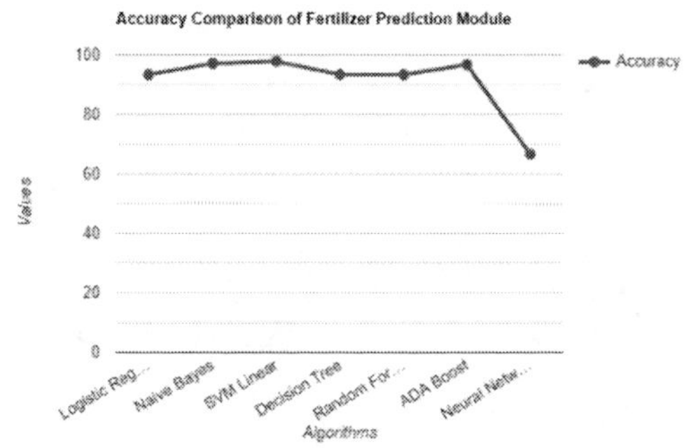

Fig. 11. Accuracy Comparison of Proposed Model for Fertiliser Prediction

Table 2. Accuracy Table for Fertiliser Prediction

Sr No.	Machine Learning Model	Accuracy
1	Logistic Regression	93.33%
2	SVM Linear	97.1%
3	ADA Boost	97.8%
4	Decision Tree	93.33%
5	Random Forest	93.33%
6	Naive Bayes	96.67%
7	Neural Network	66.67%

Table 2 summarises the model's accuracy of the fertiliser prediction module. The highest accuracy obtained is 97.8% for ADA Boost Model.

In the weather prediction module, the user has to enter a city name, which in turn will predict the weather pattern for the subsequent days.

As mentioned earlier, for the disease prediction module, we have taken the images dataset with 2000 images and the system got an average accuracy of 71.20% with the help of convolutional neural networks.

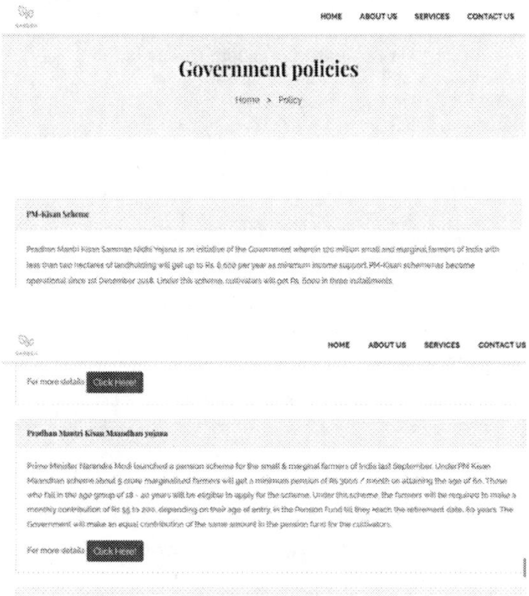

Fig. 12 Government Policies

The government policies will help farmers to access various government policies according to their needs and it will redirect them to the government portal.

6 Conclusion

The comparative study of different machine learning algorithms helps us understand which is best suited for the prediction of crops and fertilisers which will increase the yield. Hence Crop Guidance and Farmers Friend make farming easier by providing proper guidance to the farmers about irrigation, fertilisers, weather conditions, soil quality, and suggesting alternative crops and seeds. Since we are using the ensemble method, we got the highest accuracy of 79.09% for the Naive Bayes Algorithm for crop prediction and 97.8% for ADA Boost in Fertilizer Prediction and 71.2% for disease prediction using CNN.

References

1. V. R. V. S. S. T. Sooraj Krishna, Sreevijay, *Farmer friend*, IEEE, **10**, pp. 9–19 (2021)
2. M. Dr Anitha Govindraj, *Crop price prediction using machine learning*, IEEE, **6**, pp. 14–20 (2020)
3. C. B. W. W. S. Ghionea, *Algorithms to predict the moisture content of the grain*, IEEE, **1**, pp. 1–7 (2020)
4. P. M. R.Vidhya, *Predicting yield of crop using ml algorithm*, IEEE, **9**, pp. 1–13 (2020)
5. S. K. M. G. S. Prince Samuel S., K. Malarvizhi, *Machine learning and internet of things based agriculture*, IEEE, **2**, pp. 3–11 (2020)
6. S.D.M, Subramaniya Raman M.K, *E- Farming: A Breakthrough for Farmers, IJERT,* ***9*** *(2020)*
7. V. J. Monika Jadhav, *Android application for farmers*, IEEE, **6**, pp. 1–3 (2019)
8. P. K. M. B. Karthikeyan S.Sainavinraj M., Praveen Kumar A., *Farmer's friend*, IEEE, **6**, pp. 1–9 (2019)
9. G. A. Anirudha Vachaspati Vempati, *Agricultural problems and technology-based sustainable solutions*, IEEE, **5**, pp. 6–13 (2020)
10. R. C. N.Vanitha, N.Archana, *Agriculture analysis using data mining and machine learning techniques*, IEEE, **3**, pp. 8–15 (2019)
11. K. P. H. Trupti Bhange, Swati Shekapure, *Prediction of crop yield and suitable crop*, IEEE, **8**, pp. 1–5 (2019)
12. W. G. Parul Sharma, Yash Paul Singh Berwal, *Krishimitr (farmer's friend)*, IEEE, **4**, pp. 893–899 (2018)

Farming Assistance for Soil Fertility Improvement and Crop Prediction using XGBoost

Mangesh Deshmukh[1,*], *Amitkumar* Jaiswar[1,**], *Omkar* Joshi[1,***], and *Rajashree* Shedge[1,****].

[1] Department of Computer Engineering, Ramrao Adik Institute of Technology, Nerul, Navi Mumbai, India.

Abstract. In India most vital and widely practiced occupation is Agriculture and it plays a vital role in the development of our country. Soil properties, rainfall, temperature, humidity and soil pH are the factors on which agriculture is depended. In agriculture, the selection of the wrong crop may reduce crop production. Farmers should know which crops can be grown in their area. Machine Learning-based solutions are widely used in the agriculture sector. This proposed work is a recommendation system in which Machine Learning techniques are used to recommend best three crops based on soil and weather parameters. The top three crops are recommended because farmers may not have access to a particular crop if only one crop is recommended. Previous studies in this field have been done by using different Machine Learning algorithms such as Random Forest, KNN, Naïve Bayes, etc. In this proposed system XGBoost Machine Learning algorithm is used which gives better results than other algorithms. In addition, the system provides information about how to improve the soil for growing the desired crop and gives the weather forecast for next five days. As a result, this system will help farmers minimize their financial losses while also increasing crop productivity.

Keywords- *Machine Learning, Crop Recommendation System, XGBoost, Weather Forecast*

1 Introduction

Improving crop production is necessary in agriculture because 1.2 billion people depend on agriculture and country's 60 percent land is used for agriculture [1]. Farmers are unable to select the most suitable crops depending on soil and environmental factors, and the process of manually predicting the right crops has more often than not, resulted in failure. They only choose the crops which are generally grown in their neighborhood or trending crops. Initially studies were conducted to recommend suitable crop for sowing but in many of them only soil parameters were taken into consideration, not taking weather parameters into consideration can affect the result significantly. Some studies have used algorithm which are not generating accurate results. Also the Farmers don't have knowledge about soil nutrients which are present in the land. Applying nutrients in an inadequate amount to the soil and no proper rotation of crop may lead to reduction in soil fertility which will reduce the yield of the crop [2]. Bad weather condition also affects the crop production and farmer has to face huge loss.

The designed system was made taking the above problems into consideration. The proposed system uses Machine Learning which a game changer for agriculture sector. XGBoost machine learning algorithm is used for recommending the best three crops based on the soil and weather parameters. Also, for the farmers who don't have data of their soil ingredients the system recommends best three crops that has the highest production in that season in their region. The designed system also gives recommendation about the soil improvements based on the parameters of soil and the desired crop that farmers want to grow. Also the system forecast weather for next five days which help farmers to take necessary precautions.

Further the paper is structured as follows: section 2 takes a glance at previous work done in crop recommendation system. In section 3 data extraction, pre-processing and model are discussed. Section 4 depicts the model accuracies and result analysis. Finally, In section 5 concluding remarks are mentioned.

2 Related Work

With more and more research been done in the field of agriculture and advanced Machine Learning, many Research and Studies were conducted over the past few years that provides valuable knowledge about the potential of modern day technologies in the agricultural sector.

Kevin Tom Thomas et.al [1] Crop Prediction work done using KNN with cross validation and has taken only soil parameters into consideration. Manhendra N. et.al [2] Crop Prediction is done by using Decision Tree algorithm and have taken both soil and weather parameters into consideration.

*email: man.des.rt18@rait.ac.in
**email: ami.jai.rt18@rait.ac.in
***email: omk.jos.rt18@rait.ac.in
****email: rajashree.shedge@rait.ac.in

Among the weather parameter rainfall is calculated by using previous year data with Support Vector Machine (SVM) algorithm.

Mansi Shinde et.al [3] Crop Recommendation is done using Random Forest algorithm and only soilparameters has been taken into consideration i.e. NPK (Nitrogen, Phosphorous and Potassium) values of the soil.

Kiran Shinde et.al [4] Recommendation system for farmers using Random Forest Algorithm and gives recommendation about fertilizer. Swapneel Chakraborty et.al [5] has implemented Crop Recommendation using Neural Network and have taken soil parameters and Environmental factors into consideration.

Sonal Jain et.al [6] Crop Selection is done based on soil parameters and weather but some important soil parameters such NPK values are not taken into consideration. A. Suresh. et.al [7] Crop yield prediction of major crops of Tamilnadu using modified KNN and K-Mean is done in this study and other crops are not taken into consideration. S.Pudumalar et.al [8] Recommendation of crop is implemented using ensemble approach by only taking only soil type into consideration. Naive Bayes, K-Nearest neighbour and random tree are combined as an ensemble approach.

This research paper [9] enumerates the significance of selecting the correct crop and factors such as government policies, market price and production rate which affects the crop selection. The paper presents a Method for selecting crop, also known as Crop Selection Method (CSM). It also helps in improving the rate of net yield of the crop. Shilpa Pande et.al [10] has implemented Crop Recommendation system using Random Forest Algorithm and taken location and soil properties as input and predict the crop yield for the desired crop. Angu Raj et.al [11] Crop Recommendation using Internet of things (IOT). The soil parameters such as temperature, humidity, soil moisture and PH are collected from the sensor using IOT and then applied Random Forest and Naïve Bayes algorithm on the data.

T. Ragunthar et.al[12] has implemented Crop Selection problem by making use of the Apriori algorithm and Decision tree induction by taking attributes as duration, water needed, budget, soil type, season sowed, profit per hectare and market price into consideration. Sridhar Mhasikar et.al [13] Survey for Predicting Agricultural Cultivation using IOT. The sensor used to get data for the values of moisture, PH and temperature of the soil and then Random Forest and Naïve Bayes algorithm was used but only soil factors are taken into consideration.

Lakshmi N. et.al [14] proposed a crop recommendation system considering factors such as drainage, texture, color, depth, soil erosion, soil pH, permeability and water holding using big data. M.V.R Vivek et.al [15] Based on wind, precipitation, and temperature, I did a complete survey on the use of several machine learning algorithms for crop recommendation, including Naive Bayes, multi-layer perceptron, JRIP, Jf48 and SVM. Hao Zhang et.al [16] Meteorological data, crop data and soil data are used to discuss the implementation and design of crop recommendation and fertilizer recommendation systems.

Following Table 2.1 shows the limitations of reviewed work.

Table 2.1. Limitations of reviewed work

Authors	Technologies Used	Limitations
Kevin Tom Thomas et.al [1]	• KNN with cross validation, KNN, Decision Tree and Naïve Bayes algorithm areused. • Parameters: Nitrogen, Phosphorous, Potassium and soil pH value.	No Environmental factors are used for recommending suitable crop.
Mansi Shinde et.al [3]	• Random Forest, Naïve Bayes and ID3 machine learning algorithmare used. • Parameters: Nitrogen, Phosphorous and Potassium.	Only soil parameters are taken into consideration. Weather parameters arenot taken into consideration.
Sonal Jain et.al [6]	• Random Forest algorithm is used. • Parameters: Temperature, Rainfall, soil type,soil pH, watercapacity and soil fertility.	Soil Ingredients like Nitrogen, Phosphorous and Potassiumare not taken into consideration.
A. Suresh et.al [7]	• KNN algorithm is used. • Parameters: Rainfall, groundwater, soiltype and cultivated area.	Crop recommendation is done only for majorcrops i.e. Rice,Sugarcane, Maize, Raggi and Tapioca. Other crops are not taken into consideration.

3 Proposed System

The backbone of country's economy is agriculture. It is the important source of income. Soil type, PH value, rainfall, environment factors are major factors on which agriculture is dependent. Selection of wrong crop may lead to reduction of crop production and farmer may have to bear huge loss. If it is possible to get the right crop before sowing, it can be very helpful for farmers to increase crop production and to make the right decisions about storage and the business side. Farmer is not able to grow the crop that he desires as the soil may lack fertility or some ingredients required for the desiredcrop. With proper recommendation of improving soil fertility, the farmer can grow the crop that he desires. Also, bad weather condition may affect the crop production on large scale so weather forecast of next five days will help farmer to take necessary precautions.

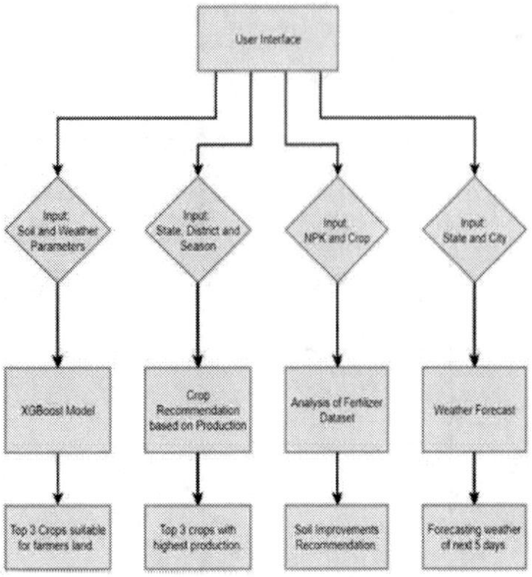

Figure 3.1 Crop Recommendation Model

3.1 Crop Prediction

After the dataset is processed and split into training and testing dataset. Different Machine Learning algorithms were used such as Decision Tree [17], Naïve Bayes[18], SVM [19], Linear Regression [20], Random Forest[21] and XGBoost [22]. From these algorithm XGBoost gives us the best accuracy.

XGBoost:

It is a gradient boosting algorithm which enhances performance and speed of a tree based (decision trees) machine learning model. In the first iteration, XGBoost will have less accuracy as initially the model will be basic. But with increase in iterations the model will optimise the loss function using Gradient Descent technique. This process is repeated until model reaches a threshold i.e. loss can't be optimised further. Hence the model's accuracy will improve with increase in number of iterations.

In the proposed system the train size is kept to 0.8 i.e. 80% of the data in dataset will be used for training purpose. Later the trained model will be tested on 20% of remaining test data. Then the XGBoost model is trained using input parameters like Nitrogen, Phosphorous, Potassium, pH, rainfall, temperature and humidity. This values will be given as input by the farmer. Then using predict_proba() function we find the probability of each of the output class. Then top three values with maximum probability is selected and then we get the corresponding crop for this given probabilities. Finally these three crops are recommended as they are more suitable to grow for given input features.

3.2 Crop Recommendation based on Production

Some Farmers may not have the data of Ingredients present in their soil so for them the system will take input as State, District and Season. After analyzing the input and dataset the system will recommend top three crops which have the highest production in the given State, District and Season.

3.3 Soil Improvements Recommendation

Fertility of the soil depends upon its NPK value present in the soil. The proposed system will take the input as soil properties i.e. Nitrogen, Potassium, Phosphorus and the crop that farmer desire to grow.

The System then analyze the input and dataset and gives recommendation about what can be added in the soil to improve its fertility to grow the desired crop.

3.4 Weather Forecast

Climate forecasting and prediction include gathering and communicating data about future weather conditions based on weather predictions.

The farmer has to give its state and city as input then the system will fetch the weather by using Weather API [23] this will help farmer to take necessary precaution and can avoid huge loss.

4 RESULTS AND DISCUSSION

This section contains implementation and result of all the features used in the proposed system.

To train and test any Machine Learning model, you'll need a dataset. The dataset for the proposed crop recommendation system was obtained from Kaggle [24], a well-known website. The data is collected from soil testing lab, it contains the soil specific attributes. Different crops like maize, rice, coffee, etc are considered in this system. The attributes considered different nutrients like nitrogen, potassium, phosphorus and pH value of the soil in the dataset. It also contains weather factors such as temperature, humidity and rainfall as its parameters. The other dataset that is used in this system is also extracted from Kaggle [25] and has attributes such as Season, State name, Area, District name and Production.

Database collected from open source should be cleaned. Noisy, incomplete and incompatible data cannot be processed by Machine Learning Algorithm so they must be cleaned.

4.1 Crop Prediction

After preprocessing of dataset then the data set is split into 80% training data set and 20% testing data set. Different Machine Learning Algorithms such as Decision Tree, Naïve Bayes, SVM, Logistic Regression, Random Forest and XGBoost were used on the dataset. After calculating the accuracies of the model, the accuracy of Decision Tree is 90%, Naïve Bayes is 99%, SVM is 10%, Logistic Regression is 95.22%, Random Forest is 99% and XGBoost is 99.3%. The same is shown in the following Figure 4.1.

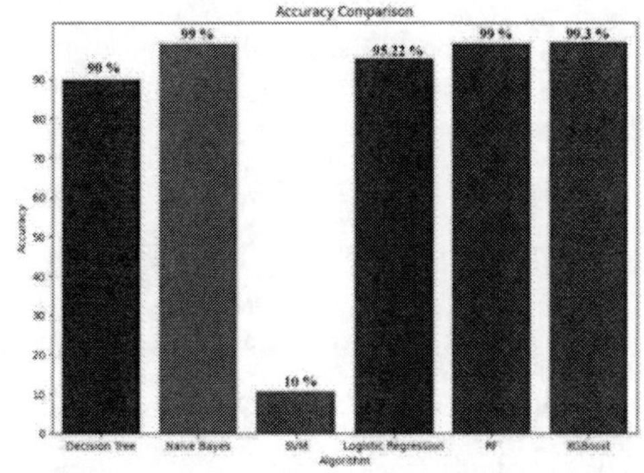

Figure 4.1 Comparing Accuracy of different Models

The proposed system uses the XGBoost algorithm to predict the best top three crops for farmers based on soil and weather parameters as it has the highest accuracy.

Table 4.1 Tabular Comparison of the Classifiers (in %)

Algorithm	Accuracy (%)	Precision	Recall	F1-Score
Decision Tree	90	84.29	88.16	85.36
Naïve Bayes	99	99.43	98.86	99.04
SVM	10	66.06	12.52	13.93
Logistic Regression	95.22	95.40	94.86	95.04
Random Forest	99	99.29	98.90	99.03
XGBoost	99.3	99.34	99.44	99.37

The Columns of Table 4.1 are defined as follows.

- Classifiers: Classifiers are models that are used to train and test data.
- Accuracy: The percentage of times the algorithm successfully classifies a data point.

$$\text{Accuracy} = \frac{TP+TN}{TP+FP+TN+FN} \qquad \text{eq. 4.1}$$

Where,
TP = True Positive
FP = False Positive
TN = True Negative
FN = False Negative

- Precision: The number of accurately predicted positive observations divided by the total number of predicted positive observations.

$$\text{Precision} = \frac{TP}{TP+FP} \qquad \text{eq. 4.2}$$

- Recall: The percentage of accurately anticipated positive observations to the total number of observations in a class is known as recall.

$$\text{Recall} = \frac{TP}{TP+FN} \qquad \text{eq. 4.3}$$

- F1 Score: The F1 Score is the weighted average of Precision and Recall.

$$F1 = \frac{2 * \text{Precision} * \text{Recall}}{\text{Precision} \div \text{Recall}} \qquad \text{eq. 4.4}$$

4.2 Crop Recommendation based on Production

The Proposed system takes inputs as State, District and Season as shown in the following Table 4.2.

Table 4.2 Input for Crop Recommendation based on Production

Sr. No.	State	District	Season
1	Maharashtra	Dhule	Kharif
2	Gujarat	Surat	All Season

After taking the input the system then recommends the best three crops which has the highest crop production in the given season in that region as shown in the Table 4.3.

Table 4.3 Output for Crop Recommendation based on Production

Sr. No.	Crop	Production(t/ha)	Area(ha)
1	Cotton	2386223	1658151
	Sugarcane	2355044	43800
	Bajra	1962191	2193245
2	Sugarcane	126331300	1778100
	Banana	2448600	36300
	Rice	149100	81900

4.3 Soil Fertility Improvement

The Farmer gives the input such as the desired crop he wants to grow and NPK of the soil. After analyzing the dataset the system will tell which ingredient is low or high inthe soil and recommend improvements in soil to make it suitable for the desired crop.

For example, the desired NPK value for Rice are 80-40-40 respectively. If the soil with composition of NPK values is 80-80-40 i.e. the phosphorous value is high, then the system will tell that the phosphorous value of the soil is high and will recommend improvements as follows:

a) Avoid using manure since it includes a lot of important nutrients that your soil needs, but it also contains a lot of phosphorus. Reducing the amount of manure applied will assist to limit the amount of phosphorus added.

b) Usage of phosphorus free fertilizer can help you to control the quantity of phosphorous that is been added to your soil. This will help the soil as plants will absorb the phosphorous that is already present in the soil while still supplying other essential nutrients like Potassium and Nitrogen

c) Nitrogen fixing vegetables like peas and beans can be planted to boost nitrogen without increasing phosphorous.

d) Water the soil - letting your soil soak the water will help in the removal of phosphorous from the soil. This is a last ditch effort.

4.4 Weather Forecast

Farmer give input as State and City then the system will forecast weather for next five days which includes parameters as temperature, minimum temperature, maximum temperature, pressure, sea level, ground level, humidity, weather description, wind speed, wind gust, visibility, date and time. The following is an example of the same.

Table 4.4 Sample result of weather forecast

Parameters	Date (2022-02-10 21:00:00)
Temperature	14.52
Temperature Minimum	11.55
Temperature Maximum	14.52
Pressure	1016
Humidity	76
Weather	Clear
Description	Clear Sky
Wind Speed	3.55
Wind Degree	294
Wind Gust	5.61
Visibility	10000

In the above Table 4.4 we have shown the sample result of weather forecast for one day. Similarly weather forecast of next five days will be done.

5 Conclusion

Farmers can choose the most profitable crop with the help of this system. It helps the farmer to make better decision on the storage and business side if they know the best crop before sowing. In this proposed system most widely used techniques, Random Forest, SVM, Decision Tree, Logistic Regression, Naive Bayes and XGBoost are compared to choose the optimal classifier for recommending best three crops. After the comparison, it can been seen that XGBoost algorithm has the highest accuracy of 99.3%.Also soil improvement recommendation was done by the system so that the soil is suitable to grow the crop that farmer desires. The proposed system provides weather forecast of next five days so that farmer can take necessary precaution. Thus the result shows that there is significant potential to develop the agricultural sector using machine learning based techniques in the most of the situations.

References

1. Kevin Tom Thomas, Varsha S, Merin Mary Saji, Lisha Varghese and Er. Jinu Thoma, “Crop Prediction Using Machine Learning”, International Journal of Future Generation Communication and Networking, Vol. 13, Issue: 3, pp. 1896-1901, 2020
2. Mahendra N , Dhanush Vishwakarma , Nischitha K , Ashwini, and Manjuraju M. R, “Crop Prediction using Machine Learning Approaches”, INTERNATIONAL JOURNALOF ENGINEERING RESEARCH & TECHNOLOGY (IJERT) Volume 09, Issue 08 (August 2020).
3. Mansi Shinde, Kimaya Ekbote, Sonali Ghorpade, Sanket Pawar and Shubhada Mone, "Crop Recommendation and Fertilizer Purchase System", (IJCSIT) International Journal of Computer Science and Information Technologies, Vol. 7 (2), 665-667,2016.
4. Kiran Shinde, Jerrin Andrei and Amey Oke, "Web-Based Recommendation System for Farmers", International Journal on Recent and Innovation Trends in Computing and Communication ISSN: 2321-8169, Volume: 3 Issue: 3 1444 – 1448.
5. P. A, S. Chakraborty, A. Kumar and O. R. Pooniwala, "Intelligent Crop Recommendation System using Machine Learning," 5th International Conference on Computing Methodologies and Communication (ICCMC), , pp. 843-848, doi: 10.1109/ICCMC51019.2021.9418375, 2021.
6. S. Jain and D. Ramesh, "Machine Learning convergence for weather based crop selection," IEEE International Students' Conference on Electrical, Electronics and Computer Science (SCEECS), pp. 1-6, doi: 10.1109/SCEECS48394.2020.75, 2020.
7. Suresh, P. Ganesh Kumar and M. Ramalatha, "Prediction of major crop yields of Tamilnadu using K-means and Modified KNN," 3rd International Conference on Communication and Electronics Systems (ICCES), pp. 88-93, doi: 10.1109/CESYS.2018.8723956, 2018.
8. S. Pudumalar, E. Ramanujam, R. H. Rajashree, C. Kavya, T. Kiruthika and J. Nisha, "Crop recommendation system for precision agriculture," Eighth International conference on Advanced Computing (ICoAC), pp. 32-36, doi: 10.1109/ICoAC.2017.7951740, 2017.
9. R. Kumar, M. P. Singh, P. Kumar and J. P. Singh, "Crop Selection Method to maximize crop yield rate using machine learning technique," International Conference on Smart Technologies and Management for Computing, Communication, Controls, Energy and Materials (ICSTM), pp. 138-145, doi: 10.1109/ICSTM.2015.7225403, 2015.
10. S. M. PANDE, P. K. RAMESH, A. ANMOL, B. R. AISHWARYA, K. ROHILLA and K. SHAURYA, "Crop Recommender System Using Machine Learning Approach," 5th International Conference on Computing Methodologies and Communication (ICCMC), , pp. 1066-1071, doi: 10.1109/ICCMC51019.2021.9418351, 2021
11. Raj, Angu & Balashanmugam, Dr.Thiyaneswaran & Jayanthi, J & Yoganathan, N & Srinivasan, PCrop. . Recommendation on Analyzing Soil Using Machine Learning. Turkish Journal of Computer and Mathematics Education (TURCOMAT). 12. 1784-1791, 2021.
12. T. Ragunthar, S. Selvakumar and G. Ilamurugan, " Counsel System for Effective Farming using Data Mining Algorithm", International Journal of Pure And Applied Mathematics, Volume 117 No. 21, 921-924 ISSN: 1311-8080 (printed version); ISSN: 1314-3395 (online version), 2017.
13. Shridhar Mhaiskar, Chinmay Patil, Piyush Wadhai, Aniket Patil and Vaishali Deshmukh, “A Survey on Predicting Suitable Crops for Cultivation Using IoT ”, International Journal of Innovative Research in Computer and Communication Engineering (An ISO 3297: 2007 Certified Organization), Website: www.ijircce.com, Vol. 5, Issue 1, January 2017.
14. Lakshmi N, Priya M, Mrs. Sahana Shetty and Mr. Manjunath C.R ,"Crop Recommendation System for Precision Agriculture ", International Journal for Research in Applied Science & Engineering Technology (IJRASET), ISSN: 2321-9653; IC Value: 45.98; SJImpact Factor: 6.887, Volume 6

Issue V, May 2018.

15. Lakshmi. N, Priya. M, Mrs. Sahana Shetty and Mr. Manjunath C.R ,"Crop Recommendation System for Precision Agriculture ", International Journal for Research in Applied Science & Engineering Technology (IJRASET), ISSN: 2321-9653; IC Value: 45.98; SJImpact Factor: 6.887, Volume 6 Issue V, May 2018.
16. M.V.R. Vivek, D.V.V.S.S. Sri Harsha and P. Sardar Maran, "A Surveyon Crop Recommendation Using Machine Learning", International Journal of Recent Technology and Engineering (IJRTE), ISSN: 2277-3878, Volume-7, Issue-5C, February 2019.
17. Zhang H. et al. Design and Implementation of Crop Recommendation Fertilization Decision System Based on WEBGIS at Village Scale. In: Li D., Liu Y., Chen Y. (eds) Computer and Computing Technologies in Agriculture IV. CCTA 2010. IFIP Advances in Information and Communication Technology, vol 345. Springer, Berlin, Heidelberg, 2011.
18. Patel, Harsh & Prajapati, Purvi. "Study and Analysis of Decision Tree Based Classification Algorithms". International Journal of Computer Sciences and Engineering. 6. 74-78. 10.26438/ijcse/v6i10.7478, 2018.
19. Kaviani, Pouria & Dhotre, Sunita. "Short Survey on Naive Bayes Algorithm". International Journal of Advance Research in Computer Science and Management. 04, 2017.
20. Evgeniou, Theodoros & Pontil, Massimiliano. Support Vector Machines: Theory and Applications. 2049. 249-257. 10.1007/3-540-44673-7_12, 2001.
21. Rong, Shen & Bao-wen, Zhang. "The research of regression model in machine learning field". MATEC Web of Conferences. 176. 01033. 10.1051/matecconf/201817601033, 2018.
22. Schonlau, Matthias, and Rosie Yuyan Zou. "The Random Forest Algorithm for Statistical Learning." The Stata Journal, vol. 20, no. 1, pp. 3–29, doi:10.1177/1536867X20909688, 2018.
23. Santhanam, Ramraj & Uzir, Nishant & Raman, Sunil & Banerjee, Shatadeep. Experimenting XGBoost Algorithm for Prediction and Classification of Different Datasets, 2017.
24. Weather Forecast API- https://openweathermap.org/api
25. https://www.kaggle.com/atharvaingle/crop-recommendation-dataset
26. https://www.kaggle.com/abhinand05/crop-production-in-india

ITM Web of Conferences **44**, 03023 (2022) https://doi.org/10.1051/itmconf/20224403023
ICACC-2022

EtherRider: A Decentralized Intercity Ride-Sharing Platform using Block-Chain Technology

Chaitali Narkhede [1,*], *Atharva* Hankare [1,**], *Rushi* Redkar and [1,***], *Smita* Bharne[1,****]

[1] Ramrao Adik Institute of Technology, D Y Patil Deemed to be University, Navi Mumbai, India

Abstract: Ridesharing entails the sharing of journeys in order to make optimal use of fuel by allowing people to go along the same route to share rides. It allows regular passengers to share trips with others, having the additive benefit of lowering travel costs and reducing traffic congestion. Most current ride choices rely on a centralized authority to enable the system, leaving it vulnerable to faults at a specific point in the system and raising concerns about privacy disclosure to attackers acting both within and outside. Furthermore, they are vulnerable to external threats and fraud, and the payment made by the current ride-sharing service provider is rather costly. As a result, we have proposed the system named EtherRider, based on the Ethereum blockchain technology. EtherRider enables drivers to provide transportation services without the need for a central system. Both the passenger and the driver will know about sharing ride details, secure their travel details, such as pick-up and drop-off locations, arrival/departure times, and secure payment through the ethereum blockchain.With a distributed ledger, drivers and riders could create a more user-driven,value-oriented marketplace.In the context of car-sharing systems, our work also indicates that the design of such an integrated platform is dependent on striking the correct balance between important design concepts (such as security and privacy, authenticity, traceability and reliability, scalability, and interoperability).

1 Introduction

With the rapid advancement of vehicular technologies, today's crowded world necessitates ultra-fast moving cars. Transportation should be enhanced to be convenient and safe for every entity with a well-planned traffic system. All of these things depend on the vehicle's speed, which is proportional to the amount of traffic on the road. According to statistics, in recent years the amount of automobiles per person has increased around the world, resulting in delayed vehicle traffic because of the limited capacity of the roads. To steer clear of them, the key purpose should be to encourage efficient usage of car capacity, i.e., people who own vehicles should be advised to make use of them as part of mass transit [8-9]. One of the most extensively used and effective modes of public transportation is the mass transit system. Although mass transit can mitigate some of the negative effects of private vehicles, it lacks flexibility and reliability. The majority of ridership is centered on a few routes. It also has the disadvantage of having a lower occupancy per vehicle, with most buses moving empty seated during off-peak hours and regularly becoming overloaded during peak hours. As a result, consumers who seek a comfortable ride rarely use regular public transportation. The Ride-Sharing System, which matches drivers with other riders who want to travel the same or a similar route, is a more effective and efficient technique. It can be a manually matched ride-sharing where drivers who want to create a carpool pick up passengers waiting by the roadside.[10] The Ridesharing technology automatically compares the information to those of other users in the database and suggests possible Rideshare partners. The goal is to plan requests in real-time and reduce user trip time while ensuring service quality. Blockchain is different from the conventional client-server approach. Blockchain is a provable, immutable, and distributed ledger that allows mistrusting entities to deal with one another without depending on a central third party [1]. Smart contracts are self-contained computer programs that operate on a Blockchain network. These computer programs function as smart contracts, having the capacity to execute and enforce without the necessity for a centralized authority [6].

2 Literature survey

In paper [1], "B-Ride: Ride Sharing with Privacy-preservation, Trust and Fair Payment atop Public Blockchain" uses the time-locked deposit protocol in which both the rider and driver make deposits are used along with leveraging smart contract and zero-knowledge proof to make sure the rider and driver have goodwill. But in this system the initial deposit made by the rider directly goes to the driver

* Corresponding author: narkhede.chaitali@gmail.com
** Corresponding author: atharvahankare@gmail.com
*** Corresponding author: rushi.redkar45@gmail.com
**** Corresponding author: smita.bharne@rait.ac.in

upon the driver's arrival, a dishonest driver can just collect this down payment without committing to the offer.

In paper [2] "Privacy-Preserving Authentication Scheme for Connected Electric Vehicles Using Blockchain and Zero-Knowledge Proofs", zero-knowledge proofs to Blockchain are used for enabling privacy-preserving authentication. The author uses a token-based method for anonymous authentication. This system is only used for anonymous registrations of electrical vehicles and not for ride-sharing.

In paper [3] "PEBERS: Practical Ethereum Blockchain-based Efficient Ride-Hailing Service" uses a system in which smart contracts are deployed by the driver. It is profitable for the driver and reduces the expenses of the passenger. This system does not track the completion of rides and a driver can cheat the system by reporting false elapsed distance, therefore, lacking fair payment.

In paper [4] "Boosting Ridesharing Efficiency Through Blockchain: GreenRide Application " uses Truffle, Ganache Metamask for the development and deployment of smart contracts, and Google Firebase for data storage. This system developed GRT(Green Ride Token)which is rewarded to users for every kg of $CO2$ reduced via sharing rides. This application is used only for corporate and universities and not for long trips.

In paper [5] "Implementation of a secure ride-sharing DApp using Smart Contracts on Ethereum Blockchain" uses MongoDB to store driver's documents on profile creation, which are later validated by the legal authorities present on the Blockchain. This system involves a third party to authenticate the users in the system and also perform background checks on the user's personal data.

From the literature survey, it has been observed that the existing system has some limitations. In paper [1] time-locked deposits are used to ensure the trust of rider and driver but the driver can still cheat the system by not committing to the offer.

In paper [2] zero-knowledge proofs are used but the system is only used for the registration of users on the Blockchain and not for developing a secure ridesharing platform.

In paper [3] deployment of smart contracts is done by the driver which is profitable for the driver and rider as it costs fewer GAS fees but the driver can sometimes cheat the system as there is no ride tracking in the system.

In paper [4] the application distributes GRT which encourages the users to carpool but it is not used for long journeys, it is only used for corporate carpooling and in universities.

Considering all the above-mentioned systems, the proposed system has a fair payment system where a driver cannot cheat and it must be suitable for ride tracking for long journeys.

3 Proposed methodology

The flow of the proposed system is depicted in Figure 1. The rider fills out the necessary information for the ride, such as the pickup location, the destination, and the pickup time. Following that, the application will present the rider with a list of drivers who match the information provided by the rider. In order to ride, the Rider must choose one driver based on their preferences. A notification is sent to the Driver when the Rider selects a Driver. The driver will proceed to pick up the Rider from the predetermined pick-up location at the predetermined time if the request is confirmed by the rider.

When the ride begins, the Rider and the Driver will each place a small amount of money into the time-bound deposit. The Driver is required to provide the rider with proof of the distance traveled at each time interval. The Rider must determine whether or not the distance traveled is accurate. As the ride progresses, the rider will deposit money into the time-bound deposit as soon as the distance has been confirmed. The payment is made possible through the use of Metamask[7] which serves as a wallet for transactions conducted over the Blockchain.

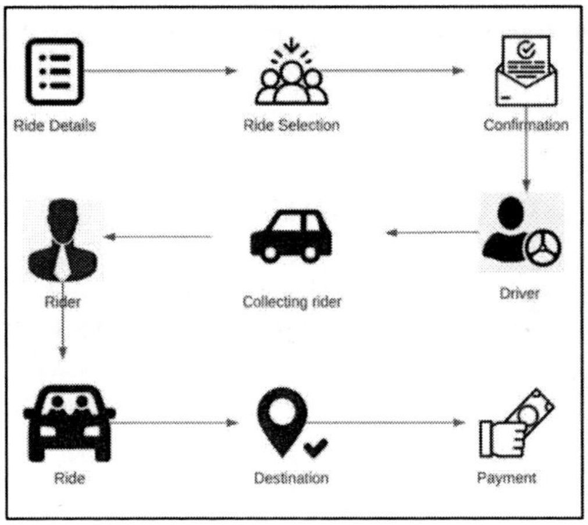

Fig.1. Flowchart for Ride-Booking Application

The following are our primary contributions:

1. A time-bound deposit mechanism and set membership in zero-knowledge are used to build confidence and belief between rider and chosen driver.

2. Create a prove-or-mulct mechanism that works as stated: The rider must produce a smart contract with a time-bound deposit amount, as well as a list of obscure spots on the map.

3. The accepted driver should deposit a sum as a commitment to his/her offer. In addition, verification of the driver's arrival at the pick-up place must be updated on the Blockchain. Finally, the smart contract validates all proofs in a zero-knowledge approach and either rewards or penalizes the driver based on their validity.

4. We also recommend an approach for assuring fair remuneration between the driver and the passenger. To validate an elapsed distance, the passenger must sign it with

 https://doi.org/10.1051/itmconf/20224403023

his private key. The smart contract sends the collected fare to the driver when the rider generates the elapsed distance receipt. The driver is paid while driving in this manner.

The Blockchain also only stores elapsed distances, exposing no other vital data to the public. Based on previous acts, the algorithm calculates a driver's reputation. Unlike current centralized reputation systems, our decentralized profile management system on the Blockchain is self-enforcing if a set of prerequisites are met.

3 System Design

On the registration page, the user will enter his/her email id, Name, Phone Number, and Password. Then the system will check if a user with the same email id exists or not. If the email id already exists. Then the user will be directed to the login page if the email id does not exist, then-new an id will be created and the user will be directed to the Mainpage of the system.

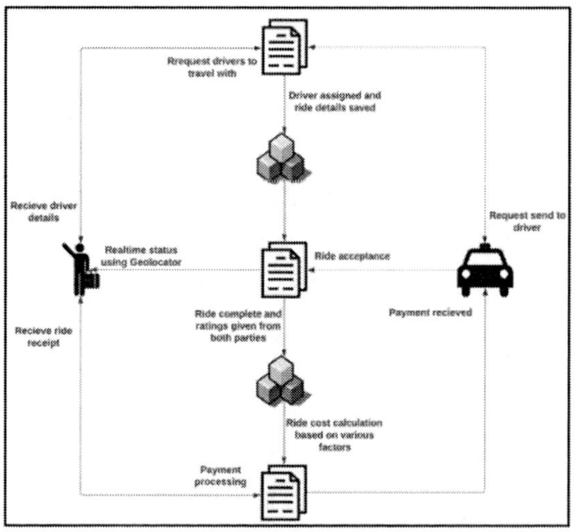

Fig.2 System Architecture

Figure 2 shows the system architecture of the proposed system. The rider will select the location they want to travel to and then will see a list of drivers who are also going the same way. The Rider then needs to select the Driver he/she wants to travel with based on their rating. Then the Driver will meet the rider at the pickup location specified by the Rider.

After the ride starts the ride details will be saved and after the end of the ride, the Ride cost will be calculated based on various factors like distance covered, the type of vehicle. We will be using pseudo-codes to implement smart contracts.

The Rider will be redirected to their metamask account. Using the metamask account the rider will pay the driver the calculated amount. After that, the Rider and Driver can rate each other on the app. This Rating will be displayed in the profile of both the driver and Rider and will help others to select their rides based on it.

4 Results

4.1 Implementation details

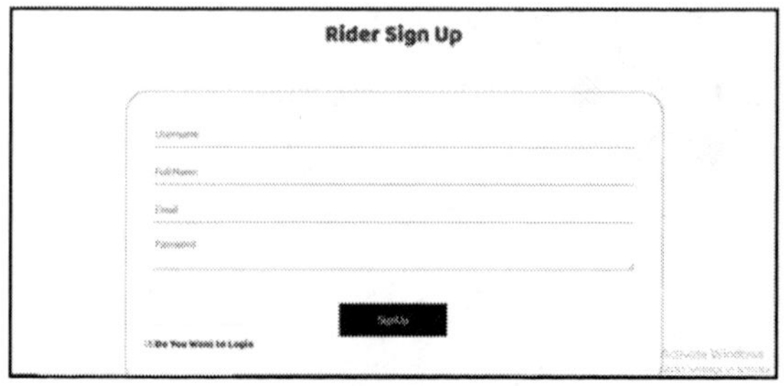

Fig.3 Rider Sign UpPage

From the user, the registration page takes four inputs: Name, Email, Phone, and Password; and the data is stored on the Blockchain. After logging in to the app the user requires to enter the registered email address and password on the login page.

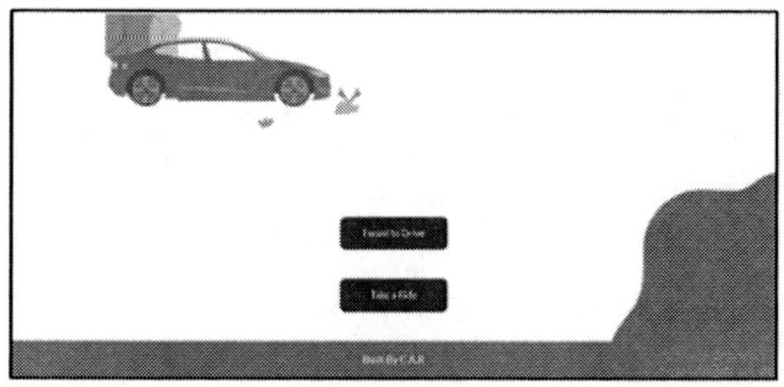

Fig.4 Role Selection Page

Fig.4 shows the role selection page, on this page the user will be able to select if they want to be the rider (Book a Ride) or the driver (Offer a Ride).

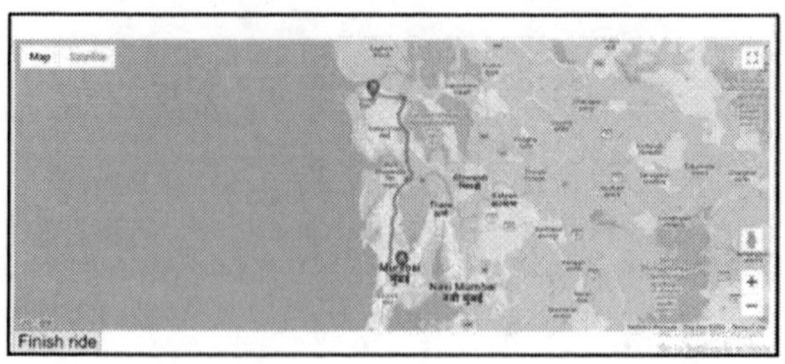

Fig.5 Live Location Tracking

Fig.5 shows the live location tracking page, the user location tracking is an important factor as it is being used to get the periodic live location of both the driver and the rider. The

Point A and B depicts source and destination respectively of the journey.

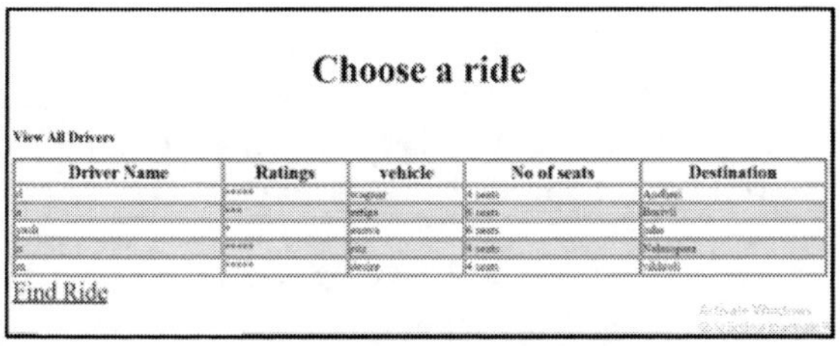

Fig.6 Driver selection

Fig.6, displays the available drivers who happen to be traveling towards the location the rider intends to be; the application provides details of the drivers(name, ratings, type of vehicle, destination) to the rider for selecting a better option to travel with.

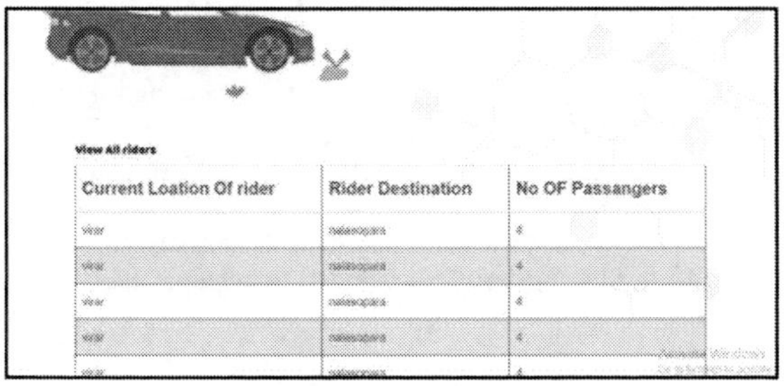

Fig.7 Rider selection

Fig.7 displays the current location of the rider along with the destination and number of passengers each rider has requested from the driver. The driver can select a rider based on this data.

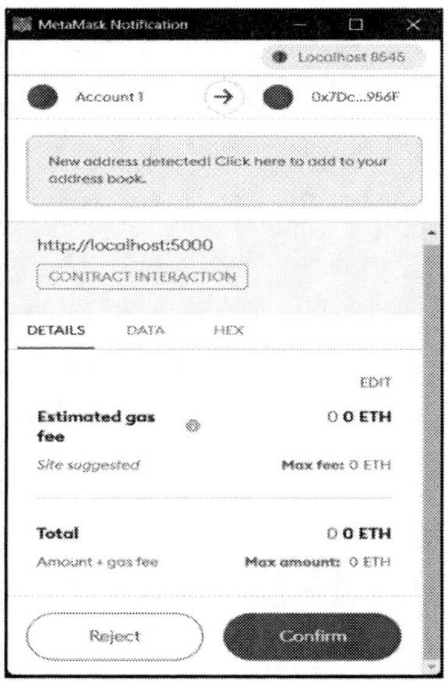

Fig. 8 Payment Request Page on Metamask

Fig.8 Metamask transaction request page, this page displays gas fees the user has to pay.

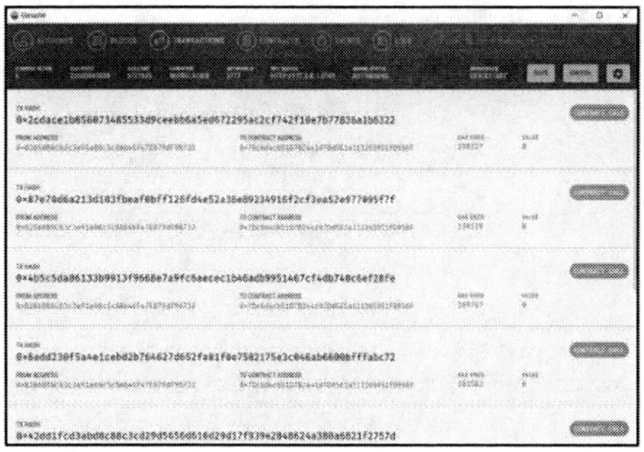

Fig.9 Ganache Transactions screen

Fig.9, Shows the Ganache Transactions screen, this page displays the previous transactions made by the user.

4.2 Performance Analysis

Table 1 describes the comparison between the traditional system and the Blockchain system proposed. Since there are many benefits of using Blockchain in a way that covers the limitations of the existing system and provides benefits of the proposed system.

Table 1: Comparison of Traditional System and Proposed system

Criterion	Traditional / Existing Charity Donations	Proposed System
Methodology	Physical records	Digital ledger
Reliable	Can lead to mishandling official documents.	More reliable.
Cost	Cost for registration and maintenance.	No additional cost.
Safety / Immutable	Physically stored data can be tampered	Impossible to tamper data due to security algorithms.
System Design	Centralized	Decentralized network
Security	Less Secure	More Secure
Speed	Slow	Faster

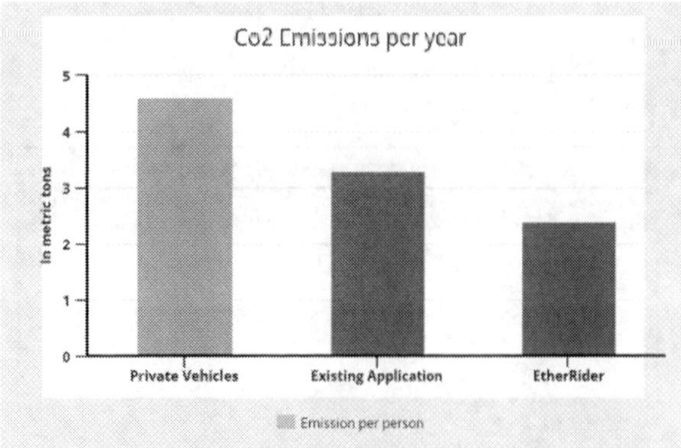

Fig10. CO2 Emissions comparison of private vehicles, existing centralized ride sharing system and proposed system decentralized ride sharing system, EtherRider

Figure 10 depicts the comparison of CO_2 emission values (in metric tons) by Private Vehicles, the Existing ride-sharing system; and our proposed system using blockchain. Private cars produce a burst of pollution every time the engine is started; previous research has found that cold-start air pollution is roughly equal to driving a car 200 miles. Since ride sharing runs more or less continuously, they avoid that.

5 Conclusion

This work has been carried out to provide a more secure and transparent application for ride-sharing to the users. As a result, we have proposed EtherRider -a ride-sharing system with a secure decentralized payment facility through the use of progressive public Blockchain technology. The current trends in the application of Blockchain technology will aid in the development of a system more fraud-proof and more reliable.

References

[1] M. Baza, N. Lasla, M. Mahmoud, G. Srivastava, M. Abdallah, "B-ride: Ridesharing with privacy-preservation, trust and fair payment atop public blockchain," IEEE, vol. **8**, pp. 1214 – 1229, (2019).

[2] D. Gabay, K. Akkaya, M. Cebe, "Privacy-preserving authentication scheme for connected electric vehicles using blockchain and zero-knowledge proofs," IEEE, vol. **20**, pp. 36–58, (2020).

[3] S. Kudva, R. Norderhaug, S. Badsha, S. Sengupta, A. Kayes, "Pebers: Practical Ethereum blockchain-based efficient ride-hailing service," IEEE, vol. **1**, pp. 1–24, (2020).

[4] S. Khanji, S. Assaf, "Boosting ridesharing efficiency through blockchain: GreenRide application," Research Gate, vol. **2**, pp. 12–35, (2019).

[5] B. Banik, S. Ar, "Implementation of a secure Ride-Sharing DApp using Smart Contracts on Ethereum Blockchain," IIETA, vol. **11**, pp. 167–173, (2021).

[6] M. Baza, M. Mahmoud, G. Srivastava, W. Alasmary, M. Younis, "A light blockchain-powered privacy-preserving organization scheme for ride-sharing services," IEEE, vol. **10**, pp. 9–19, (2021).

[7] W. Li, C. Meese, H. Guo, M. Nejad, "Blockchain-enabled identity verification for safe ridesharing leveraging zero-knowledge proof," IEEE, vol. **11**, pp. 41–63, (2020).

[8] P. Pal, S. Ruj, "BlockV: A blockchain-enabled peer-peer ride-sharing service," IEEE, vol. **13**, pp. 142–167,(2019).

[9] M. Li, L. Zhu, and X. Lin, "Efficient and Privacy-Preserving Carpooling Using Blockchain-Assisted Vehicular Fog Computing," IEEE, vol. **6**, pp. 4573 - 4584, (2019)

[10] R. Gupta, R. Gupta, S. Shanbhag "A Survey of Peer-to-Peer Ride Sharing Services using Blockchain," IJERT, vol. **10**, (2021)

Deepfake Video Detection using Neural Networks

Nimitt Patel[1], *Niket* Jethwa[2], *Chirag* Mali[3], *Jyoti* Deone[4]

[1,2,3,4] Department of Information Technology Engineering

[1,2,3,4] Ramrao Adik Institute of Technology University, DY Patil Deemed to be University, Nerul, Maharashtra, India

Abstract—In today's era, software tools based on deep learning have made the people work easier to make credible faces exchanges in video with little signs of manipulation, nicknamed "DeepFake" videos. Manipulation in digital media has been performed for decades through the appropriate use of visual effects; nevertheless, current breakthroughs occurred in deep learning have resulted in a significant rise to gain reality of fake material or contents using the simple ways. This are Artifical Intelligence-generated media (known as DF). Using tools of artificial intelligence to create the DF is an easy task. However, detecting these DF poses a significant barrier. Because it is difficult to teach the algorithm to detect the DF. Using Convolutional Neural Networks and Recurrent Neural Networks, we have made progress in detecting the DF. The system employs a Convolutional Neural network (CNN) on frame level to extract features. These observations are noted and this can train a Recurrent Neural Network (RNN), which has the ability to learn and classify whether or not a video has been tampered with and identified the temporal irregularities in the frame introduced by DF tools. We demonstrate how utilizing a simple architecture, our system may get competitive outcomes in this job.

I. Introduction

The growing era of mobile technology and integration of cameras, as well as the expanding reach of social media and sharing media portals, has made the creation and dissemination of digital video easier than before. Lacking in the advanced tools and high demand to expertise the time-consuming steps which are difficult and steps involved to limit the false videos and degree of realism until recently. However, the required time to create and manipulate videos has been reduced in the past years, this all is possible because of large amounts of training data and computing power, majorly the advancements in computer vision techniques and machine learning that replaces the requirement of manual editing. [1]

Tools like the Adobe Photoshop are used for video editing, but editing videos by replacing the faces is tedious task for this software, like if we want to process 20 second video with 25 frames per second, then it will edit about 500 images. So, software like this cannot edit this large number of images. [2]

Nowadays, any small video of any person or identity of a person can be forged very easily by replacing the facial image. [4] A fully deepfake audio-video of any person can be created by the techniques developed by Suwajanakorn and othes. [4]

A lot of attention has been attracted recently by the new vein of fake video generation using AI-based technology for its generation. It takes an input video of a particular individual and provides an output video with the individual's face replaced with another person's and the result is provided. [6]

Deep neural networks developed and trained on face images to automatically map and detect expressions of facials from the source to target which act as a backbone for DeepFake video generation. A high level of realism is achieved with effective post-processing. [1]

The importance of DF detection in such a situation cannot be overstated. As a result, we present a novel deep learning-based strategy for distinguishing false videos generated by AI technology from actual(real) videos. It's critical to have technologies that can detect fake videos so that they can be tracked down and avoided from getting viral over the internet. An example of deepfake is show in Figure 1.

Fig 1. A deepfake manipulate images examples [14]

It is critical to comprehend how the Generative Adversarial Network (GAN) generates the DF in order to detect it. GAN takes a video and extracts an image of a person (target) as input and provides a video with the face of target being replace with another person's face (source). Deep learning alongside neural networks being trained on the face cropped photos and target videos provides the backbone of DF, which automatically transfers the source's faces and facial emotions to the target. [1]

The produced movies can achieve a high level of realism with suitable post-processing. The GAN performs the function of breaking the videos down into frames and replacing each frame with input image. It goes on to rebuild the video.

Autoencoders are commonly used to do this. We provide a new deep learning-based strategy for distinguishing DF videos from actual real-world videos. The solution is based on the same mechanism as GAN's DF creation. [7] The approach is based on DF video attributes; because of production time constraints and computational resources, the DF algorithm only synthesizes face pictures of limited size and must undergo the step of affinal warping to fit and save the source's face configuration. Due to the inconsistency in resolution between the surrounding context warped face area, this warping leaves some noticeable artifacts in the output deep fake video. [1]

By splitting the video into frames and comparing the created face areas and their surrounding regions, our approach detects such artifacts. Using a LSTM along with RNN to capture the inconsistencies between frames produced by GAN during the process of DF reconstruction which is temporal, and getting the features with a ResNext Convolutional Neural Network. [8]

The process of training the ResNext CNN model is simplified by making models of inconsistency in affine face wrapping. The GAN consist of a discriminator which is simply a classifier. It is used to differentiate between the real data from data generated. [15-17]

A simple GAN architecture diagram is shown in Figure 2, which gives a clear idea about how it works, and processes the image as fake or real.

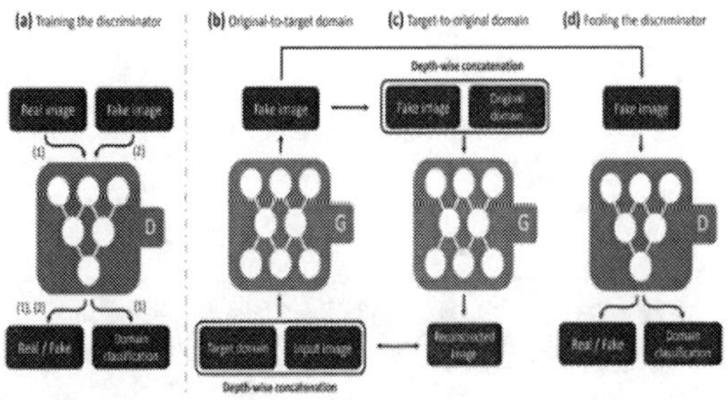

Fig. 2. GAN's General Architecture

II. LITERATURE SURVEY

Deep fake video's fast development and illicit use pose a serious danger to democracy, justice, and public trust. As a result, the demand for fraudulent video detection, analysis, and intervention has increased. Following are some of the relevant words in deep fake detection:

The method employed in ExposingDF Videos by Detecting Face Warping Artifacts[1] was to compare the surrounding regions of the face and the relative face area with artifact detection using Convolutional Neural Network. This work consists of two Face Artifacts.

Such an idea was based on the observation that images with limited resolutions must be altered further to match the source face in the approach of the present algorithm being applied over the video.

The paper Exposing AI-Created False Videos by Detecting Eye Blinking[2] explains a unique method of exposing deepfake videos which are created using deep neural network models. This approach depends on the identification of blinking of eyes in the video as it's a physiological signal which is difficult to present in bogus videos.

The method works on the eye blinking datasets and provides promising results when it comes to detecting videos created with the Deep Neural Network-based program DF.

Their strategy relies solely on the lack of blinking as a detection clue. However, additional factors such as teeth enchantment, wrinkles on the face, and so on must be considered when detecting a thorough fake. All these parameters are considered in our project.

Detecting forged images and videos [3] with capsule networks is a method that employs a capsule network to detect forged, modified photos and videos in a variety of circumstances, such as computer-generated video detection and replay attack detection. Figure 3 shows the total number of papers published from the year 2016-2021.

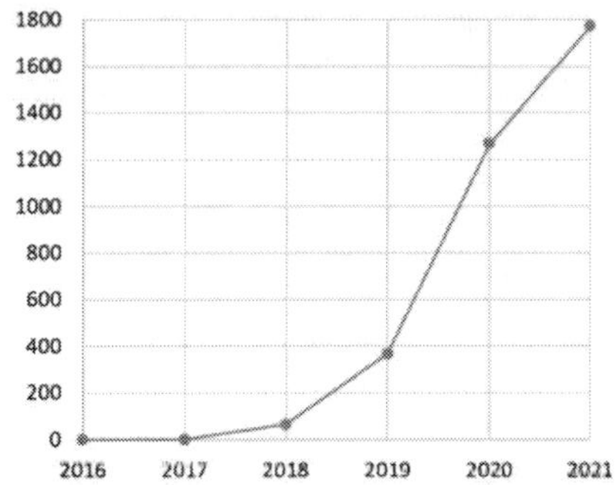

Fig 3. Graph showing the count of papers published from 2016-2021 based on deepfake project [18-20]

Our approach is designed to be trained on datasets that are both noiseless and real-time. Comparison between the different papers is shown in Table 1 based on the key parameters that distinguish each project.

Table 1. Comparison Between Different Projects

Parameters	**Exposing DF Videos by Face Warping Artifacts [1]**	**DF Detecting using Eye Blinking [2]**	**DF detection by capsule networks [3]**
Key Identification	Face Wrapping artifacts	Involuntary eye blinking movements	Inbuild capsule networks
Technology used	Neural Networks	Neural Networks	Neural Networks
Accuracy	This model is tested for moderate to low accuracy	This model is tested for medium accuracy	This has the highest accuracy among all
Consumption of resources	High consumption of computing resources	Moderate to High consumption of computing resources	Highest consumption of resources accounting a proprietary capsules.

The Biological Signals Approach for Detecting Synthetic Portrait Videos extracts biological signals while performing authentic and fraudulent portrait video pairings, such biological signals can be gained from facial areas. Train a probabilistic SVM and a CNN using temporal consistency, capture signal characteristics in feature sets, convert to compute spatial coherence and PPG, and acquire the signal properties in advancement sets. Checks whether the video is real or not. [10]

Cele-DF is another project created to detect the deepfake video but it predicts based on the low resolution by improving the synthesized face to 256 × 256 pixels, check the colour mismatch, and by reducing the temporal flicking of the fake videos. [11]

In Effective and Fast Deepfake detection method based on Haarwavelet Transform, this project finds the deepfake by haar wavelet transform. It works by retrieving sharpness from blur pictures, edges of the images and also the synthesized surrounding area by using the haar function. They have used the UADFV dataset which basically have about 49 fake and 49 real videos. The accuracy proposed by these was 90.5%. It also works on the videos frames and each frame is inspected and face surroundings are extracted. [12]

III. Proposed system

Many ways are present to make DeepFake videos but to detect them there are only a few possible ways. The technique used here to detect DF will secure the internet from the spread of DeepFake videos. The project provides a Django application that enables the users to upload the videos and justify whether it's real or fake. The project provides a web-based platform via the browser plugins made available for the DF detections.

The project can be incorporated in various applications such as WhatsApp and Facebook for identifying the DF videos before transferring them to a connected group or individual. The significant goal that can be achieved is in its performance and reliability in terms of usability, accuracy, reliability, and security.

The technique used here specializes in identifying various types of DeepFake like retrenchment DeepFake, replacement DeepFake.. Figure 4 represents the system architecture.

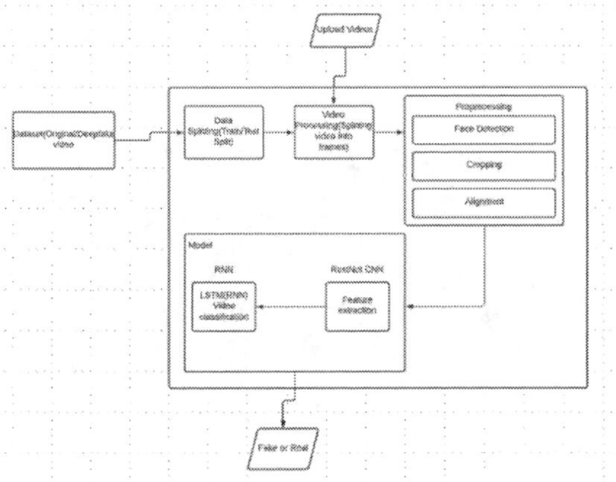

Fig. 4. System Architecture

A. Dataset Used:

The dataset used here is mixed with videos of equal numbers from various sources like Face Forensics++ followed by YouTube, various challenges datasets. By considering all these videos a new dataset is made which has fifty percent of the first video and fifty percent of tampered DF videos. Finally, the dataset is made by an appropriate division of thirty percent for test and seventy percent for the train.

Table 2 shows the total dataset created for deepfake till date.

Table 2. Basic details of the version dataset based on DeepFake [11]

Dataset	# Real		# DeepFake		Release Date
	Video	Frame	Video	Frame	
UADFV	49	17.3k	49	17.3k	2018.11
DF-TIMIT-LQ DF-TIMIT-HQ	320*	34.0k	320 320	34.0k 34.0k	2018.12
FF-DF	1,000	509.9k	1,000	509.9k	2019.01
DFD	363	315.4k	3,068	2,242.7k	2019.09
DFDC	1,131	488.4k	4,113	1,783.3k	2019.10
Celeb-DF	590	225.4k	**5,639**	2,116.8k	2019.11

B. Preprocessing Part

The pre-processing done on the dataset includes various steps like the splitting video into number of frames, detecting the face in frame and then cropping only the face part from it. In order to maintain consistency within the total count of frames, mean is been calculated of the dataset video, and also a new dataset is created which be having the cropped face. This dataset will have frames which will be equal to the mean calculated earlier.

The pre-processing part ignores some frames like which don't have face in it. As we know that if we process a 10 second video at frame count of 30 per seconds, i.e., about 300 frames, then a huge ton of computational power is been required. So for our hardware requirement match we have trained the model with only the first 100 frames.

C. Model Used

There are many parameters to consider a single layer of LSTM with resnext50 in the model creation. The Data Loader loads the preprocessed face-cropped videos and divides the videos into a test set and train set. Frames gained after preprocessing of the videos are provided to model for testing and training. Table 3 show error rate.

Table 3. Error rate of various models [13]

model	top-1 err.	top-5 err.
VGG-16 [41]	28.07	9.33
GoogLeNet [44]	-	9.15
PReLU-net [13]	24.27	7.38
plain-34	28.54	10.02
ResNet-34 A	25.03	7.76
ResNet-34 B	24.52	7.46
ResNet-34 C	24.19	7.40
ResNet-50	22.85	6.71
ResNet-101	21.75	6.05
ResNet-152	**21.43**	**5.71**

D. Feature Extraction using ResNext CNN

The use of ResNext CNN in our project is to extract the features and accurately getting the frame level features also. Our network is finely tuned by addition of extra layers and then choosing the best rate and precisely converge the model gradient descent. Once the last pooling layer is completed, feature vector of size 2048-dimensional feature is created and it is used as the input of LSTM.

E. Sequence Processing using LSTM

Now, if we take for an example, 2 nodes of neural node and sequence of feature vectors of ResNext CNN of the frames as the input with the probability of sequence which are of the part of untampered or deepfake video. Then main moto of ours is to address the design of the model to continuously process the sequence in the proper sequential order. Figure 5 represents the LSTM for sequence processing.

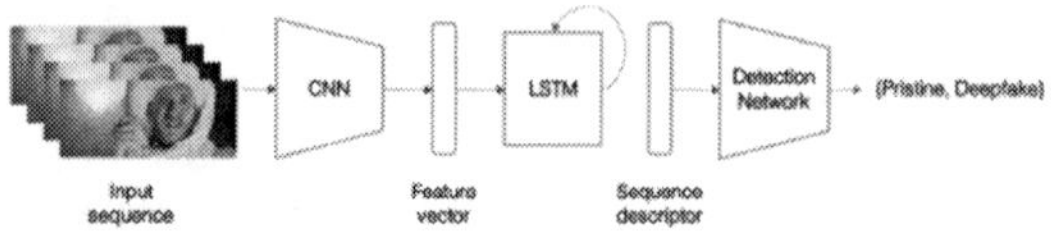

Fig 5. LSTM for Sequence Processing [9]

For achieving this, we are using the 2048 LSTM unit with dropout of 0.4 chance. To process the frames in sequential manner we make use of LSTM so that we do the temporal analysis of the video, by doing a comparison between the frames present at 't' and 't-n' second. Here 'n' represents the total number of frames present before t.

F. Model Accuracy based on frame count

Table 4. Model Accuracy

No of videos	No of frames	Accuracy
6000	10	84.21461
6000	20	87.79160
6000	40	89.34681
6000	60	90.59097
6000	80	91.49818

G. Prediction

To do the prediction of the video, an input video must be passed to trained model. The trained model mainly takes a pre-processed video as input and hence it must be in the same format. Further this video gets split into frames and then face cropping part is done.

Instead of using the local storage and occupying the memory by saving the video, there is a better way of directly passing the cropped frames to trained model. The output given by the model will be the video confidence related to the deepfake part along with the details of video like whether is real or fake one. Figure 6 represent Training flow.

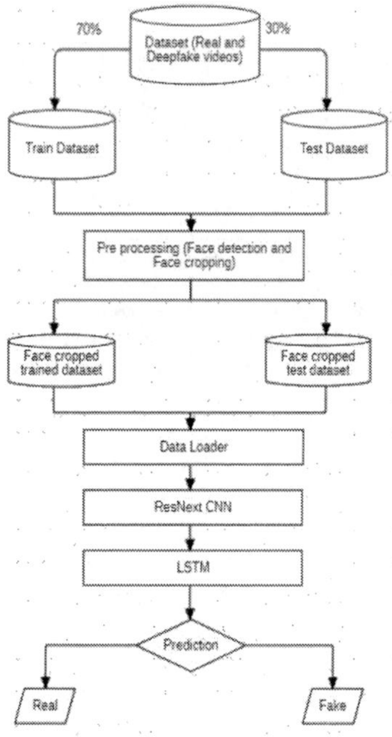

Fig. 6. Training Flow

IV. Result

A GUI will be provided for the user to upload the video and also, they can set the size of frame like 20,40,60,80. Uploading the corrupted video, long length videos, images other will give an error.

The final output will be the confidence level of model along with video detection as fake or real. Figures shown below depicts such occurrence. Outputs are shown in Figure 7, Figure 8, Figure 9. Prediction flow diagram is shown in Figure 10. Figure 11 shows the confidence of the video.

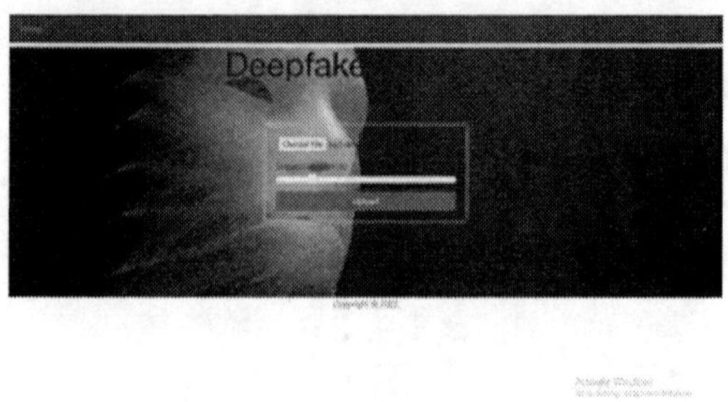

Fig. 7. GUI for video uploading

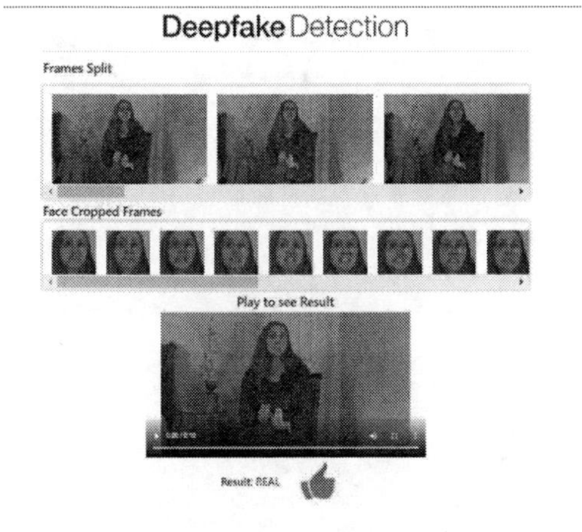

Fig. 8. Detection of Real Video

Fig. 9. Detection of Fake video along with confidence

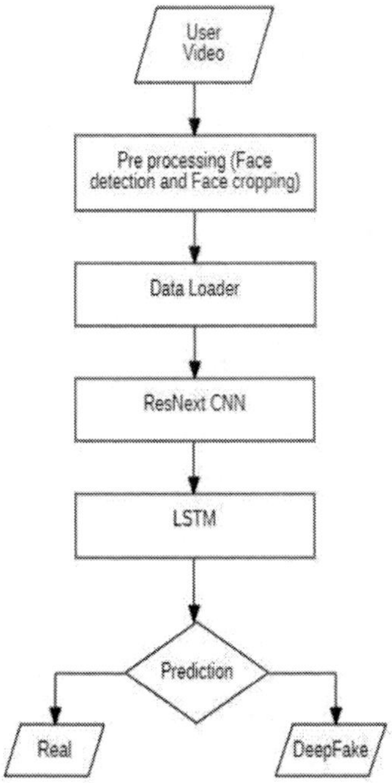

Fig. 10. Prediction Flow

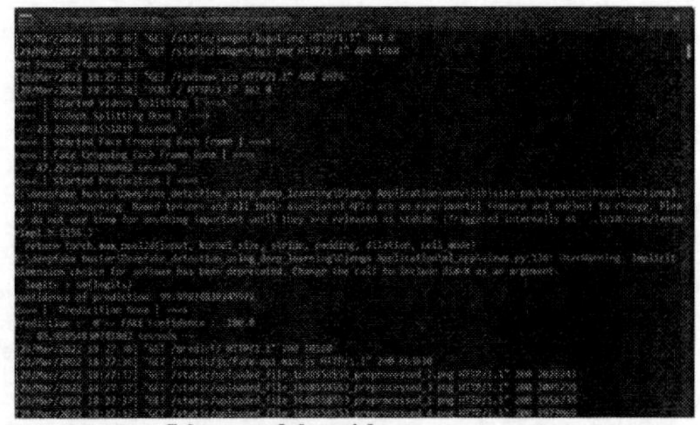

Fig. 11. Confidence of the video

V. Conclusion

We have come up with neural network-based approach for classifying the videos into the real or deepfake and showing the proposed model confidence. The proposed method is been created by keeping in mind the different ways of creating deepfakes using GANs and autoencoders. Our method make use of ResNext CNN for the frame level detection and RNN for the video classification and the LSTM also. The main goal of our project was to detect the accuracy of the video based on the alteration done and then classifying it as real or fake one based on certain parameters mentions in the paper. A high accuracy will be provided once the real time data will be used.

VI. Limitations

Our method has only considered the video part and not the audio. So, for that reason our method will not work for the audio deepfake. Hence, in future we can try to detect the audio deepfake done in the videos.

References

[1] Yuezun Li, Siwei Lyu, "ExposingDF Videos By Detecting Face Warping Artifacts," in arXiv conference May, 2019

[2] Yuezun Li, Ming-Ching Chang and Siwei Lyu "Exposing AI Created Fake Videos by Detecting Eye Blinking" in IEEE conference 2018

[3] Huy H. Nguyen , Junichi Yamagishi, and Isao Echizen " Using capsule networks to detect forged images and videos ",IEEE conference, 2018.

[4] Justus Thies, Michael Zollhofer, Marc Stamminger, Christian Theobalt, and Matthias Nießner, "Face2Face: Real-time face capture and reenactment of RGB videos," in CVPR. IEEE, 2019.

[5] Hyeongwoo Kim, Pablo Garrido, Ayush Tewari and Weipeng Xu "Deep Video Portraits" in arXiv conference, May 2020

[6] MikaWesterlund, The Emergence of Deepfake Technology:, Technology Innovation Management, Version 9, November 2019.

[7] Thanh Thi Nguyena, Quoc Viet Hung Nguyenb, Dung Tien Nguyena, Duc Thanh Nguyena, Thien Huynh-ThecSaeid Nahavandid, Thanh Tam Nguyene, Quoc-Viet Phamf, Cuong M. Nguyen, arXiv conference, Febuary 2022.

[8] Rushikesh Potdar, Ajay Gidd, Shreya Kulkarni, Rohit Chavan, Prof. Nikam, International Research Journal of Modernization in Engineering Technology and Science, Volume:03/Issue:07/July-2021.

[9] David G¨uera and Edward J Delp. Deepfake video detection using recurrent neural networks. In AVSS conference, 2018.

[10] Umur Aybars Ciftci, ˙Ilke Demir, Lijun Yin "Detection of Synthetic Portrait Videos using Biological Signals", IEEE Transactions On Pattern Analysis And Machine Intelligence, Vol. X, No. X, July 2020

[11] Yuezun Li, Xin Yang, Pu Sun, Honggang Qi and Siwei Lyu, arXiv conference, arXiv:1909.12962v4, March 2020.

[12] Karthik P C, Sanjana S, M P Adithya Vijayan, Thushara P, International Journal of Engineering Research & Technology (IJERT), ISSN: 2278-0181, Vol. 10 Issue May-2021

[13] Kaiming He, Xiangyu Zhang, Shaoqing Ren, Jian Sun, Deep Residual Learning for Image Recognition, arXiv conference, arXiv:1512.03385v1, Dec 2018.

[14] Luisa Verdoliva, Media Forensics and DeepFakes: an overview , arXiv:2001.06564v1, IEEE, January 2020.

[15] R. Raghavendra, Kiran B. Raja, Sushma Venkatesh, and Christoph Busch, "Transferable deep-CNN features for detecting digital and print-scanned morphed face images," in CVPRW. IEEE, 2019.

[16] Tiago de Freitas Pereira, Andr´e Anjos, Jos´e Mario De Martino, and S´ebastien Marcel, "Can face anti spoofing countermeasures work in a real world scenario?,"in ICB. IEEE, 2020.

[17] Nicolas Rahmouni, Vincent Nozick, Junichi Yamagishi, and Isao Echizen, "Distinguishing computer graphics from natural images using convolution neural networks," in WIFS. IEEE, 2018

[18] Thanh Thi Nguyena, Quoc Viet Hung Nguyenb, Dung Tien Nguyena, Duc Thanh Nguyena, Thien Huynh-ThecSaeid Nahavandid, Thanh Tam Nguyene, Quoc-Viet Phamf, Cuong M. Nguyen, arXiv conference, Febuary 2022.

[19] A. Rossler, D. Cozzolino, L. Verdoliva, C. Riess, J. Thies, and M. Niessner, "Faceforensics++: Learning to detect manipulated facial images,"in The IEEE International Conference on Computer Vision (ICCV), October 2019.October 2019.

[20] Ayush Tewari, Michael Zollhoefer, Florian Bernard, Pablo,Garrido, Hyeongwoo Kim, Patrick Perez, and Christian Theobalt. High-fidelity monocular face reconstruction based on an unsupervised model-based face autoencoder. IEEE Transactions on Pattern Analysis and Machine Intelligence, 42(2):357–370, 2018.

https://doi.org/10.1051/itmconf/20224403025

Supervisory framework for threat detection with multilayer processing in CNN

Omkar Mule[1,*], *Abhijeet* Pandit[1,**], *Sujoy* Karmakar[1,***], *Atharva* Kavale[1,****], and *Mr. Uttam* Waghmode[1,†]

[1]Ramrao Adik Institute Of Technology, Electronics and Telecommunication, Nerul, Navi Mumbai

Abstract. Face recognition has received huge acknowledgement due to its various uses in Internet communication, security, access control, surveillance, PC entertainment and law enforcement. Conventional methods of recognition based on the ownerships of identity-cards or full knowledge such as a security number or password are not totally solid. Physical ID cards can be lost or forged, passwords can be hacked or forgotten but a face is undoubtedly connected to its owner. It cannot be stolen, borrowed or easily forged. Our current system has a lot of weaknesses wherever it is simply taken and merged. The focus of this paper is to help users for development of the security by utilizing face identification and recognition. The proposed framework principally comprises of subsystems specifically picture capture, face identification and detection, email alerts and metal detection. Furthermore, the improvement in Computer Vision through Deep Learning algorithms has been an impressive achievement, especially with the Convolutional Neural Network algorithm. A convolutional neural network is a feed-forward neural network that is by and large used to break down visual pictures by using grid-like topology. It is also called as ConvNet. The objects in a picture are distinguished and arranged using convolutional neural network. CNN detects various simple complex patterns in images and data in its different layers of Convolution Layer, Max Polling Layer and Fully Connected Layer .This field intends to allow and configure machines to see the world as people do, and utilize the information for doing tasks (such as Image Analysis, Image Recognition and Classification, etc).

1 Introduction

In the 21st century, the main problem is security of human life from attacker. In the event that we can keep an eye on people's movement from a live cameras, so that we will be aware of the security in places where human movement is restricted like jails, airports, border area, VIP buildings etc. These days, the utilization of CCTV has expanded to protect the unattended areas with the low installation and storage cost. A lot of fear and crimes makes the access restricted to any area, a main issue for many circumstances. Ordinary techniques such as secret key and savvy card are spurious and questionable. Equivalently, face recognition is a faithful and a smart bio-metric identification as well as object detection technique. The recognition systems is extremely functional to be utilized in various applications, for example security systems, criminal recognition, explosives detection and identity verification access. Because of the all around created innovations connected with Deep Learning, we can acquire extensively great and fulfilling results of face recognition. The extracted features through images, are examined and analysed with existing face in the dataset. This system is extremely productive in detecting danger. Convolutional Neural Network also known as ConvNet or CNN, is a type of artificial neural network used primarily for image recognition and processing, due to its ability to recognize patterns in images. It is used for applications like face recognition, neural language processing, etc. The convolution steps are shown below from https://thinkingneuron.com.

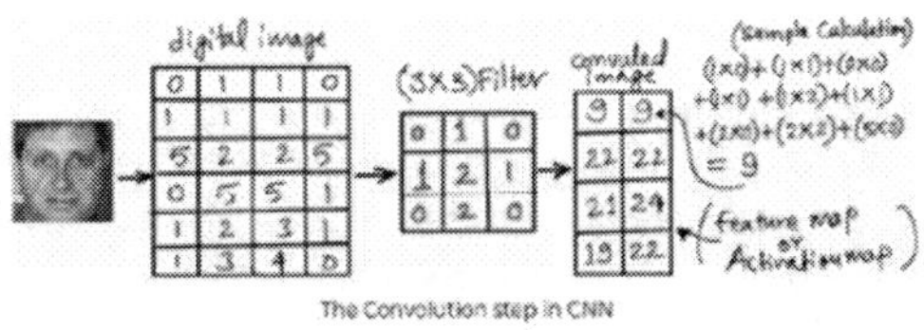

Figure 1. CNN features

A proposed solution for this matter is a face recognition security framework, that can recognize tresspassers in highly sensitive areas, and also help in limiting human error. This security system has two sides: hardware side and software side. The hardware side includes a camera and a metal detector system while the software side comprises of face and object detection and identification/classification programming. In this paper, face recognition is begun by starting the framework. Of course, the camera will click many photos of the individuals in frame. The face examined will be checked in the current data set. If there is any occurrence of unknown face, the unknown face will be stored and sent to the admin with an alert email. In any

*e-mail: muleomkar231@gmail.com
**e-mail: abhijeetpandit7@gmail.com
***e-mail: mstsujoykarmakar@gmail.com
****e-mail: atharvakavale21@gmail.com
†e-mail: uttam.waghmode@rait.ac.in

case, in the event of known face, real face is coordinated with faces put in the database. Moreover, the system also scan for guns through camera as well as a metal detector system which is integrated to the original system.

2 Literature Survey

Mohammad Hoque et al. [1] demonstrates human face recognizing or faces identification strategies such as Camshift, AdaBoost, Haar-Cascade, Hausdorff distance, viola jones, and so on. Among this, they utilized Haar classifier Cascade calculation for location of face in this project. In future, their task will be stretched out to other face recognition algorithm; with face identification from a devoted information base for analyzing their performance in similar conditions.

Apoorva.P [2] et al. represents a technique for face detection in progressively environment. They utilized Haar classifiers in order to follow faces on OpenCV. The primary benefit is that they utilized citizenship data set which already exists.

Kushal M et al. [3] proposes that to work with object recognition in a school environment, the proposed work distinguishes an individual utilizing tensor stream object discovery API, identifies and perceives the face utilizing haar course technique for OpenCV. They presumes that Model testing and preparing were finished utilizing Tensor Flow Object Detection API with the assistance of elements accommodated ID card location, Harr cascade classifier for face discovery and LBPH strategy for face acknowledgment was executed.

Faizan Ahmad et al. [4] shows the real benefits of face based identification over other biometrics. They want to assess different face detection and recognition strategies, give total answer for picture based face detection and acknowledgment with higher precision, better response rate as an underlying advance for video observation. In their work, they fostered the framework to assess the face discovery and acknowledgment strategies which are viewed as a bench mark.

Hong Zhao et al.[5] presents the particular face recognition technology which depends on installed platform and advances an answer, which weights on face identification calculation, face acknowledgment calculation, and application improvement. They expresses that working vigorously depends on acknowledgment by the back-end and is restricted by the transfer speed and soundness of information transmission organization. This system utilizes the benefit of PCA algorithm on feature extraction and the benefits, (for example, quick recognition speed and fast discovery pace) of AdaBoost calculation on the basis of Haar.

E. Omer Akay et al. [6] exhibits two different face recognition algorithms, in particular Histogram of Oriented Gradients and Haar-Cascade algorithms, that are applied and their performances are measured. Deep learning in view of convolutional neural networks (CNNs) is utilized for the identification of the students in the classroom. As a future work, it is expected to incorporate the developed framework into the Student Data Management System of the college/school, that will give programmed generation of the data set for every class.

Ayman Ben Thabet et al. [7] expects to replace expensive picture handling sheets using Raspberry pi board with ARMv7 Cortex-A7 as the main for Opencv library. This project is basically made on image handling by connecting the Opencv library to the Raspberry Pi board. They audit the related work in the sector of home automation and presents the framework plan, programming algorithm, execution and results. In this paper, face acknowledgment framework has been created to read up the likely application for home computerization entryway security with constant reaction and better acknowledgment rate.

Aashna R. Bhatia et al. [8] centers around the advantages and the disadvantages of the Viola-Jones and why it is the most important face detection algorithm and how it can be improvised to address the issues of the present time. The essential goal is to speed up the calculation which might be accomplished through parallel execution utilizing CUDA and OpenCV on Graphics Processing Unit (GPU) and then give analysis of the better computational outcomes between the serial and parallel executions.

Nandhini R et al. [9] states that in their face recognition project, a PC framework will actually identify and perceive human faces quick and definitively in pictures and recordings captured by a surveillance camera. Various algorithms and techniques are created for improvising the face recognition results, yet the idea to be carried out is Deep Learning. The point of this paper is to capture the video of students, convert it into frames, compare them with the data set in order to guarantee their presence or absence and mark the following to keep up with the record.

Danish Ali Chowdhry et al. [10] designs and implements a smart security framework for restricted places where access is restricted to individuals whose faces are stored in the data set. Simple methods are taken on to depict the possibility of a smart system which is a demand of numerous organizations from investigation, intelligence to military.

Dr. V Suresh et al. [11] mentions that the main purpose for their system is to construct a face recognition based attendance monitoring framework for academic institutes for improvising and redesigning the current attendance system into more proficient and powerful when compared with previous. They states that the face datasets will be created for pumping data into the recognizer algorithm. Hence, the facial recognition inserted in the attendance monitoring system can not just guarantee attendence to be taken precisely but furthermore eliminates the defects in the past framework.

Samridhi Dev et al. [12] states that the improvement of their system is meant to achieve digitization of the conventional system for taking attendence by calling names and keeping up with pen and paper records. The proposed framework utilizes Haar classifiers, Gabor filters, Generative adversarial networks, SVM (Support Vector Machine), CNN and KNN (k-nearest neighbors). To recognize face, three algorithms have been used which are K-closest neighbour, convolutional neural networks, and support vector machine, among these, the KNN algorithm demonstrated the most elevated precision of 99.27%.

Md Khaled Hasan et al. [13] presents a broad study on face recognition techniques, classifying them essentially into feature based and image based methodologies. They additionally explores different types of the accessible face detection algorithms in five stages, including history, working method, benefits, limitations, and use in different sectors related to face detection. Their paper gives point by point comparisons among the algorithms in order to have a comprehensive viewpoint.

Smitha et al. [14] proposes a framework to check the participation by means of face Id. This framework comprises of four stages database creation, face detection, face recognition, attendance updation. Database is made with the pictures of the children in classroom. Face recognition is done by utilizing Local Binary Pattern Histogram algorithm and Haar-Cascade classifier. Faces are identified from live real-time video of the study hall. Attendance will be sent to the individual staff toward the finish of the meeting.

Shivam Singh et al. [15] proposed a automated face recognition system. They utilized KLT Algorithm, Viola-Jones Algorithm face identification that identify human face utilizing Haar course classifier, but camera is consistently recognizing the face each frame, PCA algorithm for feature selection. In real- time situations, PCA beats other algorithms.. The future work is algorithm recognition.

Michel Owayjan et al. [16] proposed a Face Recognition System, that can recognize intruders to limited regions, and help in limiting human erros. This system is comprised of face identification and recognition algorithms. At the point when an individual enters to the zone in question, the camera captures the images and they are sent to the software for analysis and comparision with current data set of trusted people.

Kruti Goyal et al. [17] introduced a system for tracking as well as detecting faces in videos that can be utilized for many tasks. Different algorithms like Adaboost and Haar cascades were utilized. They presumed that the algorithms such as camshift algorithm or haar cascade are relaible and give more confirmed result than detection by means of movement. Yet to the extent time is considered, camshift algorithm and identification through motion algorithm is a better choice.

Xiao Han et al. [18] gave different research areas of interest on face recognition based on depth in the field of biometrics combined with the significant theory and techniques for depth learning, face recognition technology. Their research concluded that Deep learning has a critical upper hand over machine learning for other face recognition methods, deep learning can not just figure out how to get more valuable data, but also fabricate a more exact model.

H. Zhang et al.[19] proposes a face recognition model based on LBP feature and CNN. In this, the LBP feature map is used as the input of CNN to train and identify CNN. After comparing the experimental results, they concluded that the face recognition is easier and better by CNN.

Tejashree Dhawle et al. [20] states that this report contains the manners by which deep learning is a significant piece of engineering field can be utilized to decide the face involving a few libraries in OpenCV alongside python. The systems help in recognizing the human face progressively. This execution can be utilized at different platforms in machines and cell phones, and a few programming applications. The utilization of python and OpenCV creates it a simpler and helpful tool that can be made by anybody anyone according to their needs.

Kushsairy Kadir et al. [21] assesses two techniques for face detection, their highlights and Local Binary Pattern features in view of detection accuracy and speed. The algorithms were tried on Microsoft Visual C++ 2010 Express with OpenCV library. Related research was conducted between these two algorithms by involving three face datasets as tests which are Taarlab database, MIT CBCL database and Color FERET database. Every one of the three data sets furnish single face images with different kinds of faces. Results have been assessed in light of speed detection for each picture and recognition's accuracy which prompts precision. The experiment's outcomes displays that Local Binary Pattern highlights are generally effective and solid for the execution of a real-time face detection framework.

X. Peng et al. [22] analyzes the effects of different activation on face recognition rate. Majorly they focused on experimenting with small samples of CNN. They concluded that the ReLU activation function performs better in the siamese neural network than the MFM activation function.

Patrick Laytner et al. [23] states that the most crucial issue is the insufficient contrast between facial features. To beat this test, another face recognition approach, that comprises of Adaboost learning techniques, Haar transformation and Histogram analysis is proposed. The extended Yale Face Database B is utilized to analyze the execution of the suggested strategy and compared against regularly utilized OpenCV's Haar detection algorithm. The experiment's outcomes with 9,883 positive images and 10,349 negative images concluded an impressive enhancement in face accuracy without a critical change in bogus acceptance rates. Experimental results from the proposed strategy end up being a viable approach to managing this issue with an impressive expansion in hit rate without seriously affecting the bogus acceptance rate.

Fan-Fan Peng et al. [24] shows a technique that has insuperable defect, particularly when the lightning changes, the detection effect quickly decreases and is not sufficient for the actual system. The motivation behind this study is to take care standard issue of face detection under various lights, and foster a precise and effective human face detection algorithm using Visual Studio 2015 and OpenCV technology.

Amritha Nag et al. [25] states that their research is to help individuals for development of the entryway security in delicate areas by utilizing detection and identification methods. The suggested framework principally comprises of subsystems in particular taking image, detecting and recognizing face, sending notification on email and

https://doi.org/10.1051/itmconf/20224403025

programming entryway access. OpenCV supported Face Recognition is put forward because it utilizes Eigen faces and minimizes the size of face images without losing crucial elements, pictures of many people can be put in the databset. The entryway lock can be accessed remotely from any region of the planet by utilizing Telegram app. For security issues, the admin is notified with all the images captured by the camera by an email.

Madan Lalet al. [26] , exhibit most endeavoring face highlights like posture invariance, illuminations, maturing and partial occlusion. They are viewed as basic elements in face recognition system when acknowledged over face pictures. After indepth study, they found that when dimension of feature is more for original image, PCA is best method while in frontal face recognition, the eigen face image feature works better. However, point of this research is to give a detailed analysis over face recognition alongside its uses.

Nourman S. Irjanto et al. [27] exhibit CNN Alexnet facial recognition system which is executed in an entryway security system, information is gathered by 1048 facial information with the help of a framework that is trained using the images where the outcomes are very exact. This research incorporates enhancing the facial data enhancement procedure by utilizing a database, good camera, and the most recent Raspberry Pi model to further develop computational capacities.

Omar Abdul Rhman Salim et al. [28] proposes a technique for fostering a detailed embedded class attendance framework and also control the entryway with the use of facial recognition access. The framework runs on Raspbian (Linux) which is the Operating System of Raspberry Pi. By confronting the camera, it captures the picture then, at that point, passes it to the Raspberry Pi that is programmed to deal with the face recognition by executing the Local Binary Patterns algorithm LBPs. When the student's face matches the trained database images, the model entryway opens utilizing Servo Motor, then, MySQL data set stores the attendance. The data set is associated with Attendance Management System (AMS) web server, that allows to access the attendance from any device.

Mayank Srivastava et al. [29] states that their product will work with the automatic attendance system and allow teachers to access student's information simply by keeping data for IN and OUT time by utilizing OpenCV, Face acknowledgment, Authentication and Haar Cascades Algorithm. This can be utilized by overall population for enquiring the image of any individual caught by any camera by the assistance of appropriate approval by dataset.

Prayag Bhatia et al. [30] proposed Local Binary Patter to identify a person from the database. Their system is proposed to be used for home security with the help of Rasberry-pi and camera.

Table 1. Table of Literature Survey

References	Methods	Advantages	Disadvantages
[1]	Haar classifier Cascade algorithm	Success and error rates using three lighting conditions were compared.	This system covers up the requirement of only daylight instead of darkness.
[2]	Different XML files based on Haar cascade classifier	Tracking and identifying faces with different angles.	Only captures timestamp on detecting an unknown
[3]	Tensor flow object detection API	LBPH (Local Binary Pattern Histogram) for face recognition and pattern recognition method in computer vision describes the texture and structure of an image.	Problem while detecting moving targets and object of similar dimensions.
[6]	Deep learning based on convolutional neural networks	Histogram of Oriented Gradients and Haar Cascade algorithms, are applied and compared.	The performance of this method drastically changes with the pose of the head and it is less robust to occlusion.
[9]	Deep Learning and Convolution Neural Network(CNN)	Helps increase the efficiency and speed eventually achieving the high-accuracy real-time monitoring to meet the requirements for automatic attendance system.	Poor lighting condition of the classroom may affect image quality which indirectly degrades system performance

[20]	OpenCV along with python.	Fast processing, automated recognition, accurate results, high security, real time face recognition	Costing is very high, poor image quality may limit the effectiveness of this system, large storage is needed for this system for effective working.
[21]	Haar features and Local Binary Patterns (LBP)	Used three face databases as samples which are Color FERET (Facial Recognition Technology) database, MIT CBCL database and Taarlab database.	Effectiveness of both algorithms are almost similar when database of sequential static images are used to train faces.
[10]	PCA kernel (principal component analysis),linear discriminant analysis (LDA) methods	Motion detection, tracking, multi-camera surveillance	It is designed for the authorized personals and is is used in a confined area
[11]	OpenCV and Numpy through Python.	Faster processing, saves 10 - 30 reference images for every person registered in the system, a hard drive that can compensate with the large volume of images taken	Does not alert on detecting unknown person
[12]	Haar classifiers, KNN, CNN, SVM	digitization of the conventional system by recording head movements by Different camera positions	Personal record of students not available, based on overall group analysis

3 Methodology

The focus of this algorithm is to identify human faces and objects by means of a camera like webcam or cctv. Convolutional Neural Network, otherwise called ConvNet or CNN, is a notable technique in computer vision applications. It is a class of DNN that are utilized for analysing the images. This sort of architecture is prevailing to recognize objects from an image or video. It is utilized in applications like picture or video recognition, neural language processing, etc.

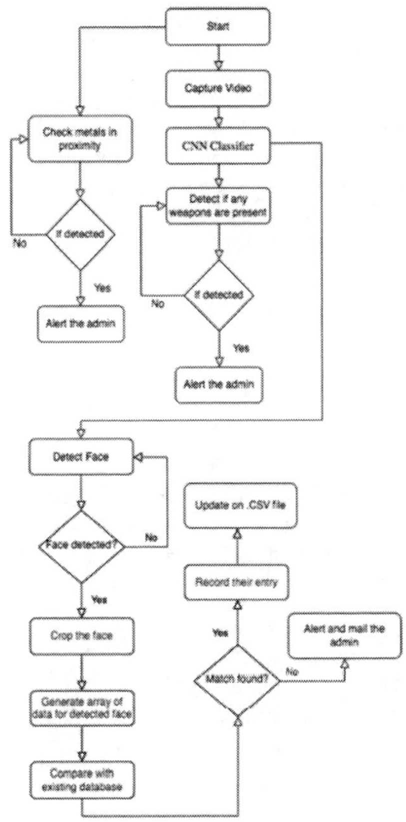

Figure 2. Flowchart of the Security System

The initial segment includes face recognition utilizing CNN, then, at that point, look for similarity in the data set. A flowchart of security system is given below.

For executing the framework, we made use of Convolutional Neural Network's layers function, a camera reads the input and communication is established serially to connect to the Arduino Uno. The metal detector is made of an insulated copper coil of radius 10 cm. The results of metal detector were transmitted to the GUI (graphical user interface) using serial ports of the Arduino.

3.1 Metal Detector:

When some current passes through the coil, it generates a magnetic field around it. Moreover, the modification in the magnetic field induces an electric field. According to Faraday's law, a voltage develops across the coil because of this Electric field, which resists the modification in magnetic field. That is how the coil generates the inductance, which signifies that the generated voltage resists the gain in the current. When any metal comes close to the coil, the coil varies its inductance. This modification in inductance relies upon the metal class. It reduces for non-magnetic metal and boosts for ferromagnetic materials. Arduino generates the block wave or pulse provided to the high pass filter. Due to this, the coil will induce momentary spikes in every transition. The pulse altitude of the developed spikes is proportionate to the coil's inductance. The rising pulse or spike charges a capacitor. Moreover, it needs a few pulses to charge the capacitor to the point where an Arduino analog pin can read its voltage. After reading the voltage, the capacitor quickly discharged. After obtaining the result, we transfer the results via serial port to the application interface and buzzer to detect metal's presence.

3.2 Architecture of CNN:

In comparison to other object detection techniques, the best technique is CNN layers. The CNN structure consists of different layers, like CNN layers, max-pooling layers, and full connected layers. A typical structure has several CNN layers and a pooling layer, followed by 1 or more full connected layers. The convolution and pooling operations can be performed for 2D-CNN as well as three-dimensional (3D)-CNN. The architecture takes a gray scale image with shape (100, 100, 1)

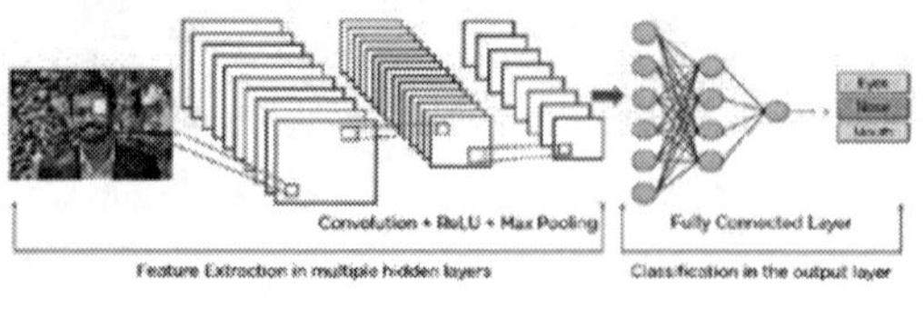

Figure 3. Layers in CNN

A. Convolution layer:
The convolution layer is a key part of the CNN architecture and it extracts features, that normally comprises of a mix of convolution operation and activation function. The results of a convolution operation, for example, convolution are then gone through a activation function that are representations of a biological neuron behaviour, the most well-known nonlinear activation function utilized as of now is the rectified linear unit (ReLU), which computes the function:

$$f(x) = max(0, x) \tag{1}$$

ReLU is short form for rectified linear unit. After extracting the feature maps, they are shifted to a ReLU layer. ReLU processes each element and sets all less than 0 pixels to 0. The graph of a ReLU function is shown in fig. 4.
The spatial size (o/p volume) depends on i/p volume size W and filter size k x k of the CNN layer neurons, the stride size of 0 padding P on edges. The neuron size that present in a given volume is then:

$$ConvolvedImageSize = (\frac{W + 2P - k}{Stride} + 1), (\frac{W + 2P - k}{Stride} + 1) \tag{2}$$

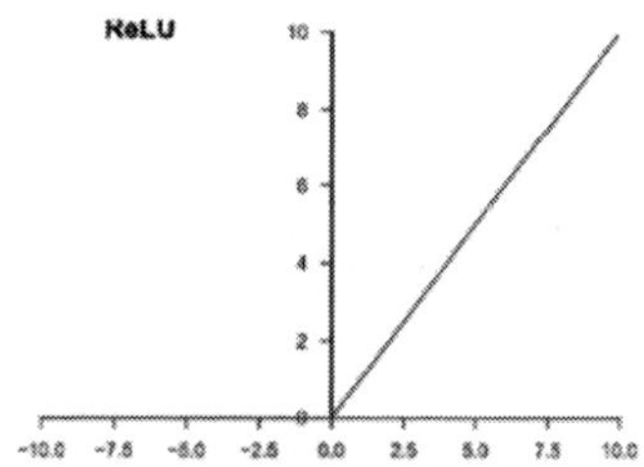

Figure 4. Rectified linear unit (ReLU) [32]

B. Pooling layer:
A pooling layer produces decreased sampling process, this decreases in any of pooling layers include filter size, pdding, stride(hyper parameter) just like a convolution process.
The pooling layer utilizes different filters to distinguish various pieces of the picture like edges, corners, body, quills, eyes, and snout. The most famous type of pooling activity is max pooling, which takes patches from the input feature maps, outputs the maximum value in every patch, and deletes all other values. A max pooling with a filter of size 2 × 2 with a stride of 2 is commonly used in practice. This downsamples the in-plane dimension of feature maps by a factor of 2.

$$f_{X,Y}(S) = \max_{a,b=0}^{1} S_{2X+a,2Y+b} \tag{3}$$

C. Fully connected layer:
The input of the Fully Connected Layer is taken from the output of the last Pooling Layers. High-level features are represented in the output of the convolutional layers. While that output could be flattened and connected to the output layer, adding a fully-connected layer is a (usually) cheap way of learning non-linear combinations

of these features. The CNN layers are providing a low-dimensional, meaningful invariant feature space, and the fully-connected layer is learning a function in that space. Fully Connected Layer are mapping/embedding from the feature maps (attention to the spatial correlation and feature description) to an N-dimensional vector space, where a given input x, represented by the set of its feature maps, is mapped to. Ultimately each image, if appropriately mapped, represents a signal point in the higher dimensional vector space that can make up clusters, surfaces, manifolds.

$$\begin{bmatrix} 1 & 2 & 3 \\ 4 & 5 & 6 \\ 7 & 8 & 9 \end{bmatrix} \xrightarrow{\text{Flatten}} \begin{bmatrix} 1 \\ 2 \\ 3 \\ 4 \\ 5 \\ 6 \\ 7 \\ 8 \\ 9 \end{bmatrix} \quad (4)$$

To accomplish the described working, we divided face recognition procedure into four parts: Detection, Extraction, Recognition and Alerting.

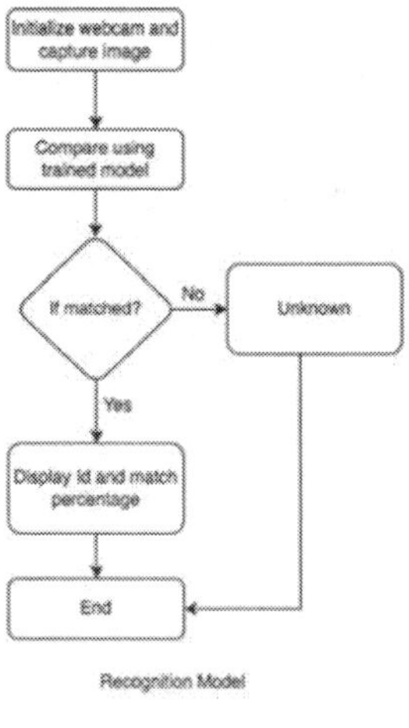

Figure 5. Flowchart of recognition system

A. Detection
The main goal of this segment is to search the images to decide if structures appear, and pinpoint their location and prepare for cropping The subsequent result of this stage is the selection from the each image. In order to increase the success rate of the system, alignment and scaling are used in the input picture. Recognition is also used for locating interest areas, video sorting, retargeting pictures, and many more. By using CNN, the framework could perceive the presence or the shortfall of a human face.

B. Extraction
Subsequent to identifying the face or object in picture, their selction are extracted from pictures. To keep away from natural deficiencies such as lightening, face expressions, impediment and mess, feature extractions are executed to extract data from the picture to reduce size, noticable extraction, and noise reduction. After this process, the selection is extracted and converted into flattened layers.

C. Recognition
After setting up the training file and interpreting the structure of face, the following stage applies the comparison with the database and input image. In this stage, when the system shows input picture, the algorithm will precisely point out the attributes of a face or object, after which feature extraction process only the selected face and distinguish the attributes extracted to the dataset. In this stage, two major features are set up: the first being identification and subsequent being verification.

D. Alerting
After identification, the system makes an note of the time. If in any case it detects an unknown face or dangerous object, an alarm is raised on the screen and an alerting email along with the picture of it is automatically sent to the admin.

4 Output

The realtime outputs of the security framework system are shown below.

Figure 6. Recognition of faces

Figure 7. Gun detection

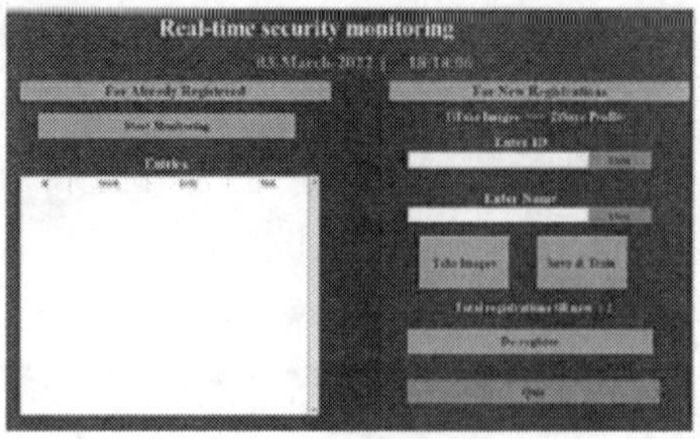

Figure 8. GUI of the system

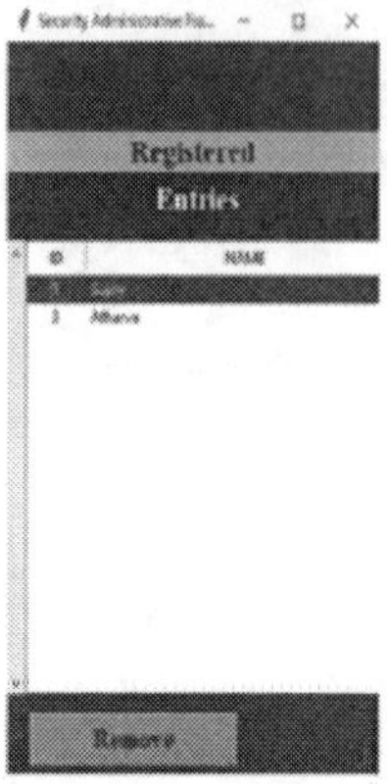

Figure 9. De-registration window

Figure 10. Database

Figure 11. Alert to e-mail

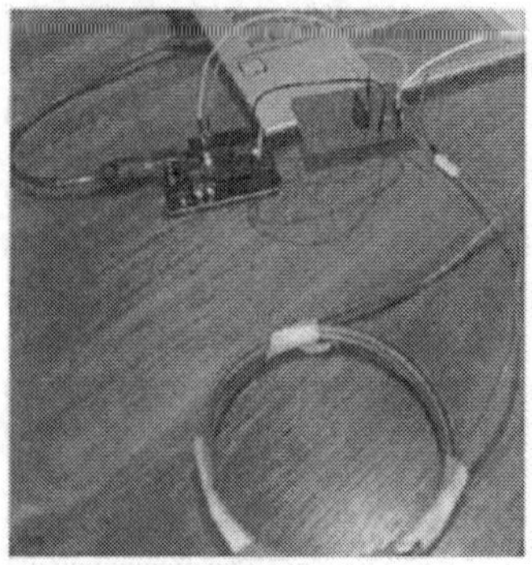

Figure 12. Metal detector module

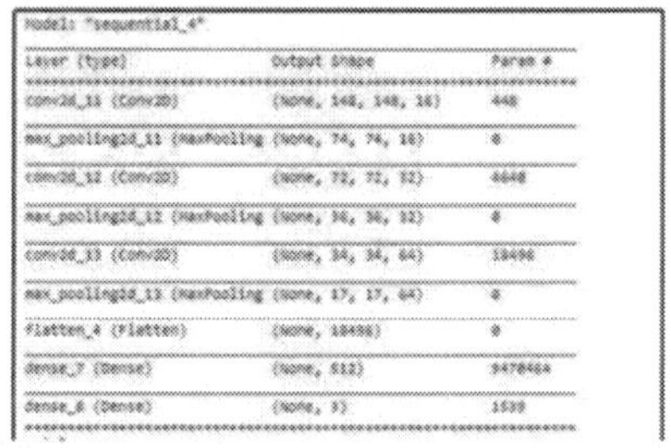

Model: "sequential_4"

Layer (type)	Output Shape	Param #
conv2d_11 (Conv2D)	(None, 148, 148, 16)	448
max_pooling2d_11 (MaxPooling	(None, 74, 74, 16)	0
conv2d_12 (Conv2D)	(None, 72, 72, 32)	4640
max_pooling2d_12 (MaxPooling	(None, 36, 36, 32)	0
conv2d_13 (Conv2D)	(None, 34, 34, 64)	18496
max_pooling2d_13 (MaxPooling	(None, 17, 17, 64)	0
flatten_4 (Flatten)	(None, 18496)	0
dense_7 (Dense)	(None, 512)	9470464
dense_8 (Dense)	(None, 3)	1539

Figure 13. CNN layers

Table 2. Test Table

Sr No.	Condition	Output		Success rate
		yes	No	
1	Low Brightness	41	9	82
2	Medium brightness	42	8	84
3	High brightness	47	3	94
4	Low Distance	47	3	94
5	Medium Distance	42	8	84
6	High distance	41	9	82
		Success rate is		86.667

$$Success\ rate = \frac{Positive\ output}{Total\ output} \times 100 \tag{5}$$

5 Conclusion and Future scope

Thus, in the proposed framework, the field of CNN allows to an extent to configure the machine, and enable them to capture the world as we do and use the data for Image Analysis, Image Recognition and Classification. The proposed threat detection system using CNN presents the particular face identification framework which depends on speed of embedded system and provides a solution, which emphasizes mainly on face detection algorithm, face recognition algorithm, classification and application improvement. This system utilizes a hybrid method using software-hardware utilization to enhance the security of social gathering places like banks, airports etc. This work has been tested to evaluate the success rate of the algorithms in different conditions. It is found that, this could be low cost, easy to use for normal security persons, good as compared to alert triggering systems presently available.

The integration of CNN with Arduino UNO will be helpful for researchers to develop prototype for surveillance systems with metal detector. The hardware (Arduino) is basically used in serial communication mode which could be replaced in future with Raspberry Pi for increasing the efficiency, speed of processing, and to develop multicamera system with one CPU. This system shows potential for future developments in various fields of Internet communication, security, access control, surveillance, PC entertainment and law enforcement.

References

[1] M. A. Hoque, T. Islam, T. Ahmed and A. Amin, "Autonomous Face Detection System from Real-time Video Streaming for Ensuring the Intelligence Security System," 2020 6th International Conference on Advanced Computing and Communication Systems (ICACCS), 2020, pp. 261-265, doi: 10.1109/ICACCS48705.2020.9074260.

[2] P. Apoorva., H. C. Impana., S. L. Siri., M. R. Varshitha. and B. Ramesh., "Automated Criminal Identification by Face Recognition using Open Computer Vision Classifiers," 2019 3rd International Conference on Computing Methodologies and Communication (ICCMC), 2019, pp. 775-778, doi: 10.1109/ICCMC.2019.8819850.

[3] M. Kushal, K. Kumar B.V., C. Kumar M.J. and M. Pappa, "ID Card Detection with Facial Recognition using Tensorflow and OpenCV," 2020 Second International Conference on Inventive Research in Computing Applications (ICIRCA), 2020, pp. 742-746, doi: 10.1109/ICIRCA48905.2020.9183342.

[4] Ahmad, Faizan Najam, Aaima Ahmed, Zeeshan. (2013). Image-based Face Detection and Recognition: "State of the Art". IJCSI International Journal of Computer Science Issues. 9.

[5] Hong Zhao, Xi-Jun Liang, Peng Yang, "Research on Face Recognition Based on Embedded System", Mathematical Problems in Engineering, vol. 2013, Article ID 519074, 6 pages, 2013.

[6] E. O. Akay, K. O. Canbek and Y. Oniz, "Automated Student Attendance System Using Face Recognition," 2020 4th International Symposium on Multidisciplinary Studies and Innovative Technologies (ISMSIT), 2020, pp. 1-5, doi: 10.1109/ISMSIT50672.2020.9255052.

[7] A. Ben Thabet and N. Ben Amor, "Enhanced smart doorbell system based on face recognition," 2015 16th International Conference on Sciences and Techniques of Automatic Control and Computer Engineering (STA), 2015, pp. 373-377, doi: 10.1109/STA.2015.7505106.

[8] Aashna R. Bhatia Parallel Implementation of Face Detection Algorithm on GPU 2016 2nd International Conference on Next Generation Computing Technologies (NGCT-2016) Dehradun, India 14-16 October 2016.

[9] Neha Kumari Dubey , Pooja M. R. , K Vishal , Dhanush Gowda H. L, Keertiraj B. R, 2020, Face Recognition based Attendance System, INTERNATIONAL JOURNAL OF ENGINEERING RESEARCH TECHNOLOGY (IJERT) Volume 09, Issue 06 (June 2020).

[10] D. A. Chowdhry, A. Hussain, M. Z. Ur Rehman, F. Ahmad, A. Ahmad and M. Pervaiz, "Smart security system for sensitive area using face recognition," 2013 IEEE Conference on Sustainable Utilization and Development in Engineering and Technology (CSUDET), 2013, pp. 11-14, doi: 10.1109/CSUDET.2013.6670976.

[11] Dr. V Suresh, Facial Recognition Attendance System Using Python and OpenCv, Quest Journals, Journal of Software Engineering and Simulation Volume 5 Issue 2 (2019) pp: 18-29.

[12] S. Dev and T. Patnaik, "Student Attendance System using Face Recognition," 2020 International Conference on Smart Electronics and Communication (ICOSEC), 2020, pp. 90-96, doi: 10.1109/ICOSEC49089.2020.9215441. [13] Md Khaled Hasan Human Face Detection Techniques: A Comprehensive Review and Future Research Directions.

[13] Hasan, M.K., Ahsan, M.S., Abdullah-Al-Mamun, Newaz, S.H., Lee, G.M. (2021). Human Face Detection Techniques: A Comprehensive Review and Future Research Directions. Electronics.

[14] Smitha , Pavithra S Hegde , Afshin, 2020, Face Recognition based Attendance Management System, INTERNATIONAL JOURNAL OF ENGINEERING RESEARCH TECHNOLOGY (IJERT) Volume 09, Issue 05 (May 2020).

[15] Shivam Singh , Prof. S. Graceline Jasmine, 2019, Face Recognition System, INTERNATIONAL JOURNAL OF ENGINEERING RESEARCH TECHNOLOGY (IJERT) Volume 08, Issue 05 (May 2019).

[16] Owayjan, Michel Dergham, Amer Haber, Gerges Fakih, Nidal Hamoush, Ahmad Abdo, Elie. (2013). Face Recognition Security System.

[17] K. Goyal, K. Agarwal and R. Kumar, "Face detection and tracking: Using OpenCV," 2017 International conference of Electronics, Communication and Aerospace Technology (ICECA), 2017, pp. 474-478, doi: 10.1109/ICECA.2017.8203730.

[18] X. Han and Q. Du, "Research on face recognition based on deep learning," 2018 Sixth International Conference on Digital Information, Networking, and Wireless Communications (DINWC), 2018, pp. 53-58, doi: 10.1109/DINWC.2018.8356995.

[19] H. Zhang, Z. Qu, L. Yuan and G. Li, "A face recognition method based on LBP feature

for CNN," 2017 IEEE 2nd Advanced Information Technology, Electronic and Automation Control Conference (IAEAC), 2017, pp. 544-547, doi: 10.1109/IAEAC.2017.8054074.

[20] Tejashree Dhawle et al., Face Detection and Recognition using OpenCV and Python, IRJET, vol 07, issue 10.

[21] Kadir, Kushsairy Kamaruddin, Mohd Nasir, Haidawati Safie, Sairul Bakti, Zulkifli. (2014). A comparative study between LBP and Haar-like features for Face Detection using OpenCV. 335-339. 10.1109/ICE2T.2014.7006273.

[22] X. Peng, J. Ma, Y. Liu, J. He, W. Wang and Y. Wang, "Research on Face Recognition Based on Small Samples of CNN," 2019 IEEE International Conference on Electron Devices and Solid-State Circuits (EDSSC), 2019, pp. 1-3, doi: 10.1109/EDSSC.2019.8754183.

[23] P. Laytner, C. Ling and Q. Xiao, "Robust face detection from still images," 2014 IEEE Symposium on Computational Intelligence in Biometrics and Identity Management (CIBIM), 2014, pp. 76-80, doi: 10.1109/CIBIM.2014.7015446.

[24] Lee, Hai-Wu Peng, Fan-Fan Lee, Xiu-Yun Dai, Hong-Nian Zhu, Ying. (2018). Research on face detection under different lighting. 1145-1148. 10.1109/ICASI.2018.8394486.

[25] A. Nag, J. N. Nikhilendra and M. Kalmath, "IOT Based Door Access Control Using Face Recognition," 2018 3rd International Conference for Convergence in Technology (I2CT), 2018, pp. 1-3, doi: 10.1109/I2CT.2018.8529749.

[26] Lal, Madan Kumar, Kamlesh Arain, Rafaqat Maitlo, Abdullah Ruk, Sadaquat Shaikh, Hidayatullah. (2018). Study of Face Recognition Techniques: A Survey. International Journal of Advanced Computer Science and Applications. 9. 10.14569/IJACSA.2018.090606.

[27] Irjanto, Nourman Surantha, Nico. (2020). Home Security System with Face Recognition based on Convolutional Neural Network. International Journal of Advanced Computer Science and Applications. 11. 10.14569/IJACSA.2020.0111152. [28] Gomes, Clyde Chanchal, Sagar Desai, Tanmay Jadhav, Dipti. (2020). Class Attendance Management System using Facial Recognition. ITM Web of Conferences. 32. 02001. 10.1051/itmconf/20203202001.

[28] M. Srivastava, A. Kumar, A. Dixit and A. Kumar, "Real Time Attendance System Using Face Recognition Technique," 2020 International Conference on Power Electronics IoT Applications in Renewable Energy and its Control (PARC), 2020, pp. 370-373, doi: 10.1109/PARC49193.2020.236628.

[29] A. Singh, D. Gupta and N. Mittal, "Enhancing Home security systems Using IOT," 2019 3rd International conference on Electronics, Communication and Aerospace Technology (ICECA), 2019, pp. 133-137, doi: 10.1109/ICECA.2019.8821833.

[30] A. Singh, D. Gupta and N. Mittal, "Enhancing Home security systems Using IOT," 2019 3rd International conference on Electronics, Communication and Aerospace Technology (ICECA), 2019, pp. 133-137, doi: 10.1109/ICECA.2019.8821833.

[31] U. Waghmode, S. Tandale, "Real time Unattended Object Identification", International Conference on Computing, Communication, Control and Automation (ICCUBEA), 2017, doi: 10.1109/ICCUBEA.2018.8477233.

[32] Zhukovyts'kyy, Ihor Pakhomova, Victoria Domanskay, Halyna Nechaiev, Andrew. (2019). Distribution of information flows in the advanced network of MPLS of railway transport by means of a neural model. MATEC Web of Conferences. 294. 04007. 10.1051/matecconf/201929404007.

[33] Krizhevsky, Alex; Sutskever, Ilya; Hinton, Geoffrey E. (2017-05-24). "ImageNet classification with deep convolutional neural networks" (PDF). Communications of the ACM. 60 (6): 84–90. doi:10.1145/3065386

https://doi.org/10.1051/itmconf/20224403026

Music Feature Extraction And Recommendation Using CNN Algorithm

Aniket Jha[1,*], *Sagar* Gupta[**], *Priyanshu* Dubey[***], *Aditi* Chhabria[****]

[1]Dept. of Computer Engineering Ramrao Adik Institute of Technology, Nerul, India.

Abstract. In this age of technological advancements, it has become considerably easier for an individual to access a variety of music from a significant number of sources. Today, there are a multitude of songs of varying diversity available to users. Therefore, it becomes difficult for users to manually discover new music that may suit their liking. Thus arises the need for a system that will help the music streaming applications to recommend new music to their users that will befit their music taste, based on some predetermined criteria. With the ever-expanding user and song database, the system must also be dynamic and its recommendations must be up-to-date and accurate. Therefore, there is a strong demand for a well-qualified music recommendation system. The proposed system focuses of technical features of audio. The main purpose of this systems is to classify songs in different genre using Deep Learning Algorithm. There are two main approaches for implementing these system, viz, Feature Extraction and Content Based Filtering.

Keywords: Machine Learning, Logistic Regression, KNN, CNN, Decision Tree, Librosa Library.

1 Introduction

Music Recommendation and Classification System focuses on technical features of audio. This extracted features are used to classify songs in different genre using Deep Learning Algorithm. Further we apply Cosine Similarity on lyrics of songs of same genre to find similarity between them. The top 5 songs with most similarity in lyrics compared to current song played by user is recommend. In this way, we are trying to depict the mood of user based on current song that he is playing and then recommending songs according to his mood.

The objective of this project can be summarized as follows:

1. Use Content-Based Filtering.
2. Categorize songs into different Genre and Emotion.
3. Use lyrics as a parameter to find similarity between songs.
4. Identify the mood of user based on current song that he is playing and recommend songs accordingly.

The growth in the amount of information is drastic in the 21st century .In today's time, it is not rare for situations where users have to interact with large sets of items, to occur. In such cases, the Recommendation system comes into picture.

*e-mail: aditi.chhabria@rait.ac.in
**e-mail: ani.jha.rt18@rait.ac.in
***e-mail: sag.gup.rt18@rait.ac.in
****e-mail: pri.dub.rt18@rait.ac.in

In such scenarios, the system recommends the user items that the system thinks the user may be interested in based on contents of data (Content Based Filtering). Therefore, the user gets to see items that he/she may like without having to search for the same.

In this work, we will use technical features of song such chroma_vector, frequency, amplitude, energy, loudness, valence, liveness, mfcc, etc. This parameters will be used for classification of songs into different genre and emotion using CNN algorithm. The songs of same genre will be further filtered based on lyrics using Cosine Similarity. So basically songs recommended will have same genre and they will have most similar lyrics. People of all age love listening to music may be their music taste differ. With so many songs around people look for best recommendations. This Application will recommend songs based on current song user is playing. So, if user is listening to devotional song then recommended songs will also be devotional in nature.

2 Literature Survey

Jiosaavn, Spotify, Gaana are examples of application which recommends songs to user. They recommend songs to user on basis of their song preferences, songs they have played frequently, by comparing likings with other users. Most often popular or trending songs are recommended [1].

Following are some recommendation parameters used by existing systems:

Collaborative filtering – In Collaborative Filtering

recommendation are done by comparing liking and disliking between users of particular system [2].
For e.g. If user1 and user2 listens to similar songs then now a new song played by user1 will also be recommended to user2.

Popular and Trending songs - Here recommended songs are either most popular songs or songs which are listen more often these days. Songs are sorted by how frequently they are played in recent time and accordingly most played or most liked songs are recommended. This method doesn't take users preference into account.

Most frequent songs played by user - Here songs are recommended from user's previous listen history. These may be songs which has listened most recently or most frequently.

Genre based classification - In this method of recommendation songs belong to some genre i.e. they may be type classical, pop, jazz, etc. For e.g. If user is currently listening a classical song then recommended songs will also be classical. This is one of the most common music recommendation method.

Outcome of Literature Survey

1. Songs played by user some other day may not depict his current mood.
2. Songs of same genre may have totally different lyrics.
3. Popular and Trending songs may not be of same genre which user is looking for.
4. Filtering based on technical features is a better measure as compared to Collaborative filtering for music recommendations [3].

3 Proposed Methodology

We have used Content Based Recommendation technique i.e. we are recommending using technical features of song. We have extracted technical features of song using librosa library. Further songs are classified into different genre and emotion category using CNN algorithm. Also, we have used cosine similarity to find similarity in song's lyrics.

Content-Based Filtering

Content-based Filtering is an algorithm that uses features to make recommendations. If two objects are similar in feature then if user chooses one the other one can be recommended [4]. Refer Fig 1. Below.

	Feature 1	Feature 2	Feature 3	Feature 4
Product 1	1		1	2
Product 2		1	4	
Product 3	3			1
User Data	2		1	1

Fig 1. Content-Based Filtering

Feature Extraction using LIBROSA

Zero-Crossing Rate – It implies the rate at which signal moves from positive side to negative side or vice-versa. It gives the rate of amplitude variation of song.

Spectral Centroid - It gives the center of mass for a sound for the song. For e.g. Let say a classical song has same frequency throughout its length then its center of mass lies in center. If a rock song has more frequency at its end then its center of mass lies at end.

Spectral Rolloff - It helps to visualize shape of song represented as signal. It gives frequency in relation to total spectral energy, e.g., 85%, lies.

Mel-Frequency Ceptral Coefficients – The MFCC of a song is a small set of features that describe the shape of spectral envelope. It gives the human voice characteristics in song i.e. pitch at which song is sung.

Chrome Frequencies – It represents the 12 distinct chroma of the octave. This is a very powerful representation of audio [5, 6, 7].

Convolutional Neural Network :

CNN is a deep learning algorithm which is used for classification of data indifferent category. It applies various filter matrix on given input features [1,9]. Fig 2. Represents various layers of CNN.

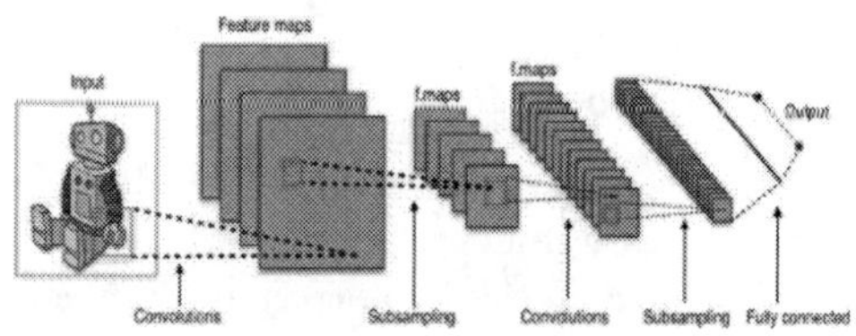

Fig 2. CNN Model (Wikipedia)

CNN for Genre Classification

1. In our CNN model, we have used the Adam optimizer for reducing losses. We have trained our model on 20 epochs.
2. The output layer is using softmax function. All other layers are using RELU as activation function. Sparse categorical crossentropy function is used for evaluating.
3. Adam optimizer gave us the best results as compared to other optimizers.
4. After a certain number of epochs the accuracy of the model achieves threshold i.e. it doesn't increase. We achieved accuracy of 98% for train set and 87% for the test.

Here is a summary of the final architecture:
Layer-1: Input Features (55 Columns)
Layer-2 : 256 Neurons
Layer-3 : 128 Neurons
Layer-4 : 64 Neurons
Output layer: 10 neurons representing 10 different genres

.

Cosine Similarity for lyrics

Cosine Similarity is used to find similarity between two objects represented as vectors. The similarity can be found by cosine angle between two vectors. The greater the angle the more dissimilar the objects. While if the angle is less, then the objects are similar in nature. For e.g. If angle is 0 (degree) then the objects are similar while If the angle is 90 (degree) then the objects are dissimilar [10, 11, 12]. Refer Fig 3.

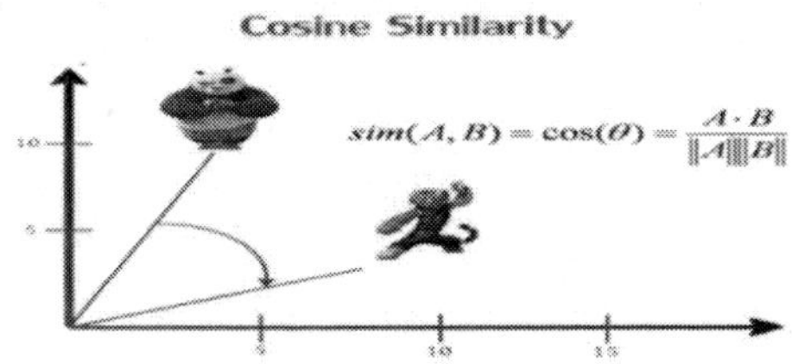

Fig 3. Cosine Similarity

Songs of same genre may have completely different lyrics. For e.g. Songs "Lag ja Gale" and "Lakshya" belongs to the same genre but completely different lyrics and meaning. To remove this inefficiency in recommendation, we found the similarity between songs based on their lyrics. Cosine Similarity will find the similarities in lyrics of two songs by comparing the frequencies of presence of certain words in song lyrics. In "Teri Mitti" song nation-related words are used more frequently. Hence it will be more similar in lyrics with other patriotic songs like "ae watan", "sandhese aate hain", etc. Hence recommended song will have similarity in words and meaning.

Song Genre Classification

We have used Jio Saavn Api to fetch song details such as song url, song name, image, artist name, etc. If the song is played for the first time then its features and lyrics are extracted and stored in a csv. Based on its features we are able to classify the song in a particular genre. Lyrics of the song is used to find top 5 similar songs of same genre. Finally we are recommending songs to the user [13]. Refer Fig. 4.

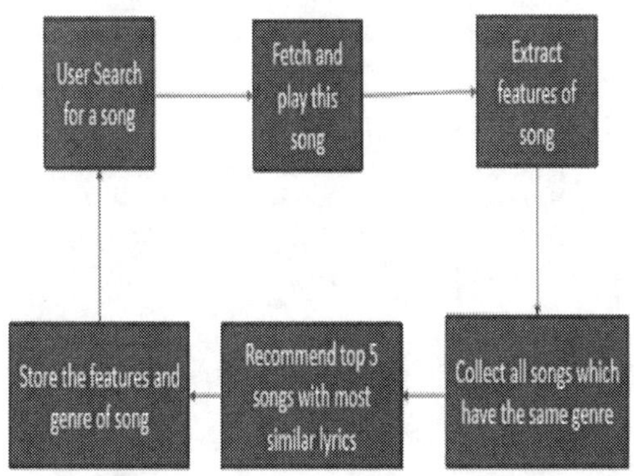

Fig 4. System flow diagram for Song Recommendation

Song Emotion Classification

We have used python library openCV and haarcascade_frontalface_default XML document to detect face. Then we have used CNN algorithm to classify facial expression into emotion. We also classified songs into different emotion category based on technical features such as energy, beats, chroma_vector, mfcc, tempo, valence, etc. Now we recommend songs with emotion similar to that of user. [14]. Refer Fig. 5.

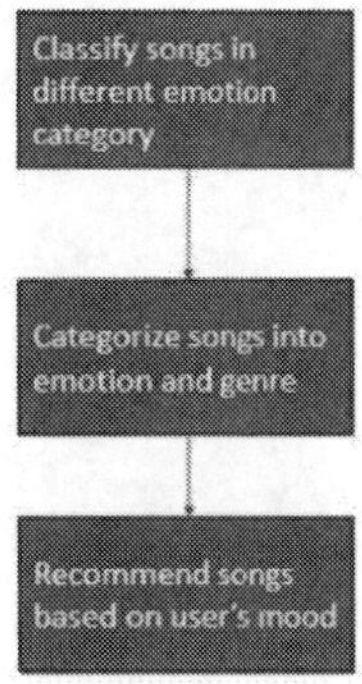

Fig 5. System flow for Song Emotion Classification

4 Results and Simulation

Sr no	Model	Train Accuracy	Test Accuracy
1	CNN	98%	87%
2	KNN	86.04%	82.73%
3	Random Forest	99.81%	82.21%
4	Logistic Regression	73.40%	72.53%

Fig 6. Comparison of different ML Models for Song Genre Classification

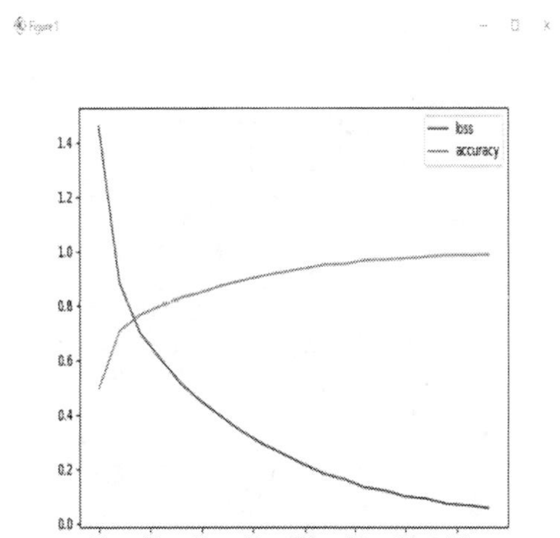

Fig 7. Loss and Accuracy graph for Genre Classification using CNN Model

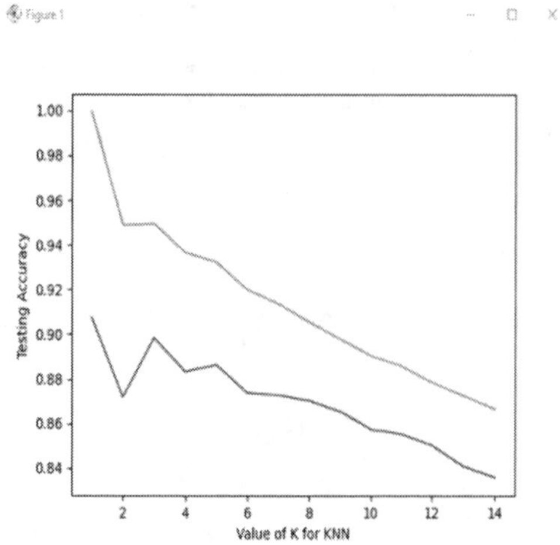

Fig 8. Accuracy Graph for KNN Model

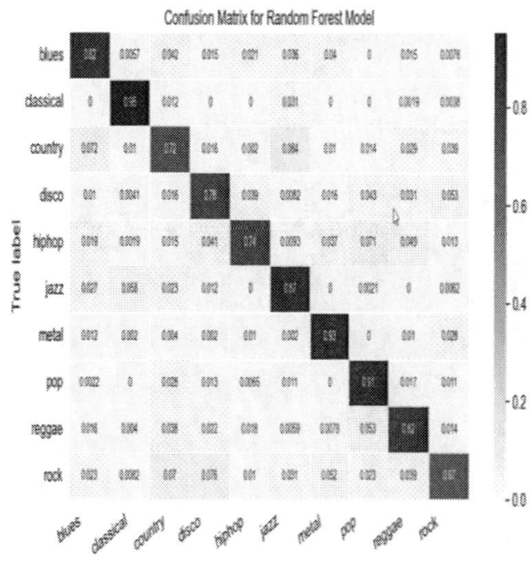

Fig 9. Confusion Matrix for Random Forest

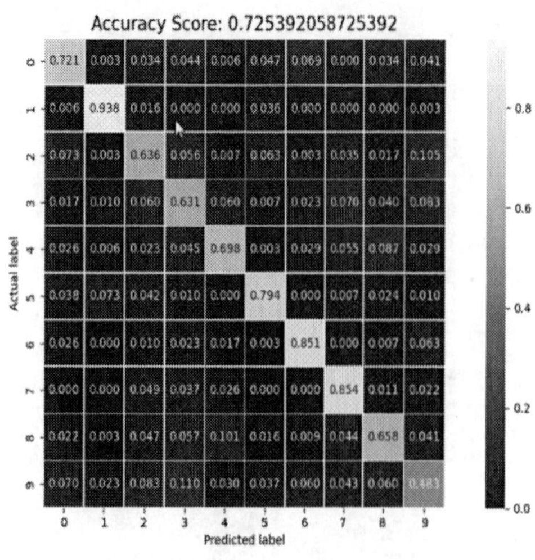

Fig 10. Confusion Matrix for Logistic Regression

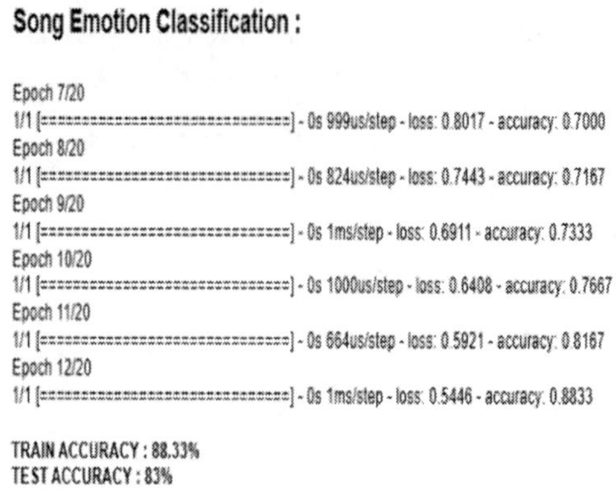

Fig 11. Result for Song Emotion Classification

5 Conclusion

In this work, we have used feature based recommendation scheme which provides a way to classify song in different genre. Lyrics of the song is our second parameter to find similarity between songs. We have trained our model on GTZAN dataset with 10000 songs and achieved an accuracy 98% in training phase and 87% in test phase. We have classified songs in 10 genre these are reggae, pop, classical, blue, country, disco, hiphop, jazz, metal, rock. Further we have classified songs into different emotion using CNN algorithm and achieved an accuracy of 88% in training phase and 83% in test phase.

6 References

1. T-RECSYS: A Novel Music Recommendation System Using Deep Learning. 2019 IEEE International Conference.
2. Badrul Sarwar, George Karypis, Joseph Konstan, and John Ried. Item-Based Collaborative Filtering Recommendation Algorithms (2004).
3. A Survey of Music Recommendation Systems and Future Perspectives, Conference: The 9th International Symposium on Computer Music Modeling and Retrieval (CMMR) At: London, UK (June, 2012).
4. Application of Content-Based Approach in Research Paper Recommendation System for a Digital Library. (October, 2014) International Journal of Advanced Computer Science and Applications 5(10).
5. librosa: Audio and Music Signal Analysis in Python. Conference: Python in Science Conference (January, 2015)
6. https://librosa.org/doc/latest/index.html.
7. Music Feature Extraction. Sanket Doshi (Dec, 2018)

8. Saad Albawi; Tareq Abed Mohammed; Saad Al-Zawi. Understanding of a convolutional neural network. 2017 IEEE International Conference.

9. Keiron Teilo (O'Shea Aberystwyth University). An Introduction to Convolutional Neural Networks. (November, 2015)

10. Alfirna Rizqi Lahitani; Adhistya Erna Permanasari; Noor Akhmad Setiawan. Cosine similarity to determine similarity measure. 2016 4th IEEE International Conference.

11. Dani Gunawan (University of Sumatera Utara), Mohammad Andri Budiman (University of Sumatera Utara). The Implementation of Cosine Similarity to Calculate Text Relevance. March 2018 Journal of Physics Conference Series 978(1):012120

12. https://towardsdatascience.com/understanding-cosine-similarity-and-its-application-fd42f585296a

13. C. de la Bandera, A.M. Barbancho, L.J. Tard´on, Simone Sammartino, and Isabel Barbancho. Humming Method for Content-Based Music Information Retrieval. ISMIR 2011, (Ismir):49–54, 2011.

14. Cristóbal Veas. Predicting the Music Mood of a Song with Deep Learning. Published in Towards Data Science (Aug 15, 2020).

https://doi.org/10.1051/itmconf/20224403027

Diabetic Retinopathy Detection From Fundus Images Using Multi-Tasking Model With EfficientNet B5

Yash Bhawarkar[1,*] ,*Kaustubh* Bhure[1,**] , *Vinayak* Chaudhary[1,***] and *Bhavana* Alte[1,†]

[1]Dept. of Computer Engineering Ramrao Adik Institute of Technology, Nerul, India.

Abstract: Diabetic Retinopathy (DR) is a common eye disease that affects over 3 million people annually. People with diabetes are more prone to suffer from Diabetic Retinopathy. This condition can cause blurring of vision and blindness. Early detection and treatment are the most effective ways to manage Diabetic Retinopathy. Due to the huge number of diabetic patients and the need for more accurate and automatic diagnosis, the development of deep neural networks has been acknowledged. One of the issues with deep learning classification tasks, particularly in the medical field, is a lack of labelled training data. Transfer learning enables the deep learning model to be trained with a minimal Training Dataset. In Deep convolutional networks, transfer learning can be utilized to solve the problem of insufficient training data. Previous studies on deep neural networks have been promising. Through a color fundus photography study, we have developed a model to identify the different stages of Diabetic Retinopathy using Deep learning. Our proposed model has achieved 87% of accuracy with the help of EfficientNet model. The main aim of this work is to develop a robust system for detecting DR automatically.

Key words: CNN, Focal Loss, ResNet, EfficientNet

1. INTRODUCTION

A diabetic eye is a chronic illness that damages the blood vessels in the eye. It is considered the leading cause of vision impairment in the world. A comprehensive eye examination is performed to diagnose diabetes and other vision-related problems. It involves analyzing the central vision and the retina using various tests and procedures. Some of these include: Vision acuity measurements, visual complexity, and the use of a dilated pupil. Currently, the traditional method of detecting diabetic retinopathy involves manual examinations. Doing so requires a trained specialist to thoroughly examine the photos of the fundus images. Before, manual feature extraction was mainly used for detecting diabetic retinopathy. However, it can be very challenging to extract effective options from large-scale datasets due to the complexity of the task. Also, hand-woven features are sensitive to certain conditions, such as noise and photography equipment. Furthermore, as for earlier classification techniques, their generalization is limited and cannot be utilized for detecting Diabetic Retinopathy.

Our proposed model is based on Convolution Neural Network with multi-tasking model. As CNN can make predictions directly from images at the pixel level, thus they require minimal pre-processing. Using deep learning we will take the human factor of the equation. This makes the predictions reliable and superfast.

Diabetic Retinopathy affects nearly 80% of diabetics who have had the disease for more than 20 years. According to studies, 90% of such cases can be avoided if the right remedy is used and the patients' eyes are tracked. Diagnosis of Diabetic Retinopathy calls for professional understanding and we are able to locate the usage of deep studying strategies however it calls for large dataset which it lacks in healthcare and takes a lot time which we are able to dispose of in transfer learning.

2. RELATED WORK

In the past decades, many tasks related to automatic DR diagnosis have been performed. This paper presents a brief summary of these works, which are mainly focused on image classification techniques.

Supriya Mishra[1] proposed comparison based result of vgg16 and DenseNet121 models in Diabetic Retinopathy Detection using Deep Learning IEEE 2020.In preprocessing cropping, resizing and removing of black images is done and converted to NumPy. In proposed vgg16 training data is used without ImageNet dataset and in denseNet121 pretrained convolutional neural network

*e-mail: yas.bha.rt18@rait.ac.in
**e-mail: kau.bhu.rt18@rait.ac.in
***e-mail: vin.cha.rt18@rait.ac.in
†e-mail: bhavana.alte@gmail.com

(CNN) using transfer learning. Quadratic kappa score 0.893 generated using matrix of observed score, the matrix of expected scores based on chance agreement and the weight matrix. Accuracy of vgg16 without ImageNet dataset is 73.26 which is less than DenseNet mode with ImageNet dataset. *ebin pm and p. ranjana*[2] proposed comparison based result of vgg16 and Inception V3 models in An Approach Using Transfer Learning to Disclose Diabetic Retinopathy in Early Stage IEEE Explore 2020. Proposed model pretrained on convolutional neural network (CNN) using transfer learning. Data preparation is done by using RGB Channel (Matrices). Four images are captured. First with all RGB channel. Second with only G and B channel. Third with only R and B channel. Fourth with only R and G channel. Final result generated is classified in three classes only (0,1,2). Sensitivity of proposed models for vgg16 is 0.082 and for inception V3 is 0.293. *Md. Ahsan Habib Raj, Md. Al Mamunand Md. Farukuzzaman Faruk*[3] did preprocess on fundus images as conversion to grey scale and then applied filter for noise reduction. Then proposed model masked the image for blood vessel and Micro aneurysm Detection and used VGG19 for classification. The preprocessing of the dataset as well as the feature extraction technique have a big impact on the performance of a classification model. *Nishi Doshi, Urvi Oza and Pankaj Kumar*[4] did preprocessing on fundus images as cropping, Down sampling by 8x used Bilinear and Learned Image Down sampling LID. Then they applied padding, Inception V3 network is trained to classify images. Multi-Channel CNN are used to train multiple networks in parallel. The last layers of the Inception V3 network were concatenated. The proposed model obtained accuracy of 85.2% and it was stated that eep CNN can increase the performance. *Jayant Yadav, Manish Sharma and Vikas Saxena* [5] proposed model focuses to detect DR using Computer vision and neural network with a fully automated approach and implemented using opensource tools. Model highly focuses on classification of dot hemorrhages and exudates. The accuracy achieved by the proposed model was 75%. *Savita Choudhary*[6] discussed a deep learning ConvNet model that is trained on different data sets throughout the experiment. The experiment consisted of various factors such as batch size, epoch, preprocessing and dataset training. The hyper parameters played a key role in the performance of a model and deriving concrete methods to find the optimal hyper-parameters quickly. The observation made under the experiment was the constantly increasing with tweaking of hyper parameter. *Sagar Suresh Karki and Pradnya Kulkarni*[7] firstly did Intense augmentations for better generalization, they also tried different protocols (median blur, various parameters) for augmentation but obtained results here similar. Including methods to segment only retinal blood vessels did not improve performance. Training was done on EyePacs dataset using EfficientNet. In training they freeze the classification head for 1 epoch, then unfreeze model and trained for various EfficientNet architectures using different learning rate and Optimizers. Their worked concluded stating EfficientNet B4 and B6 using their approach overfitted. Proper results can be achieved with selective training data and proper loss function. *ZHENTAO GAO* proposed system that can give better severity ratings are more practical for clinical application[8]. The model proposed a new labeling scheme for a large dataset. A pre-processing pipeline was proposed to enable to transform fundus images into uniform formats. The testing of this pipeline's efficiency was done against several mainstream CNN models and achieved a accuracy of 88.72%. *Sumit Thorat and Akshay Chavan*[9] proposed convolutional Neural newtwork deep learning model Diabetic Retinopathy Detection by means of Deep Learning IEEE 2020.Imabalnce Dataset of EyePacs consisting of 35126 images was balanced using data augmentation of only images of class 1,2,3,4 i.e. DR leveled images. In preprocessing black images are dropout which result in better precision and recall. Neurons from convolutional neural network consisting of convolution, pooling, fully connected and dropout layer is finally passed to classification layer. Precision and recall in Severe DR i.e., class 3(0.256) and 4(0.447) is less compared to mild and no DR class.

3. METHODOLOGY

The proposed Diabetic Retinopathy (DR) detection from Fundus Images using Multi-Tasking Model with EfficientNet B5 model focuses on estimating the severity of the disease on the basis of their retinal fundus images. Model will help the medical staff to increase the efficiency and accuracy. The result produced by the Model will be from the five class. According to severity and intensity of diabetic retinopathy, one of the following class scale results is predicted: scale will be in range of 0 to 4.

- No DR (scale 0)
- Mild DR (scale 1)
- Moderate DR (scale 2)
- Severe DR (scale 3)
- Extreme Proliferative DR (scale 4).

Project will be having 4 subtasks:

1. EDA (Exploratory Data Analysis): Data will be collected, analyze in the form of fundus images and appropriate class of DR.
2. Image Processing and Augmentation:
   2.1 Image Augmentation: Images will be rotated, flip, padded and added to dataset to increase robustness and strength of dataset.
   2.2 Image processing: Images will be passed with Gaussian filter and converted to sharp grey images.
   2.3 Images are cropped in circular shape and Vector representation is used for images
3. Training of model: A Multi Task Learning model which will have backbone of feature extraction as ResNet50/ EfficientNetB5 and further will contain two sections, namely Classification, Ordinal regression section followed by a Single neuron with linear activation function.

4. Model Evaluation.

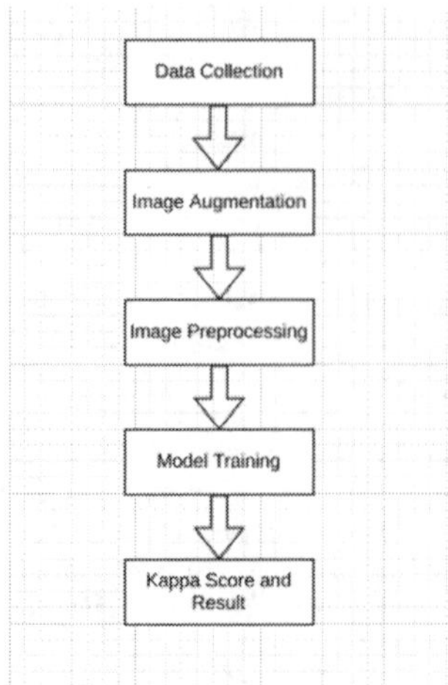

Figure 1: Workflow of System

3.1 Data Collection

We have used the dataset of Fundus Images. Fundus images of retinal eye part will be used. Reflected light will be used to generate 2D image from 3D representation of semitransparent retinal images of eyes.
Specialized fundus camera which contains microscope and flash is used to capture magnified and flashed images. Intensity of image in photo signifies light reflected while capturing image [10]. Collected Data is in form of fundus images and appropriate class of DR of image.

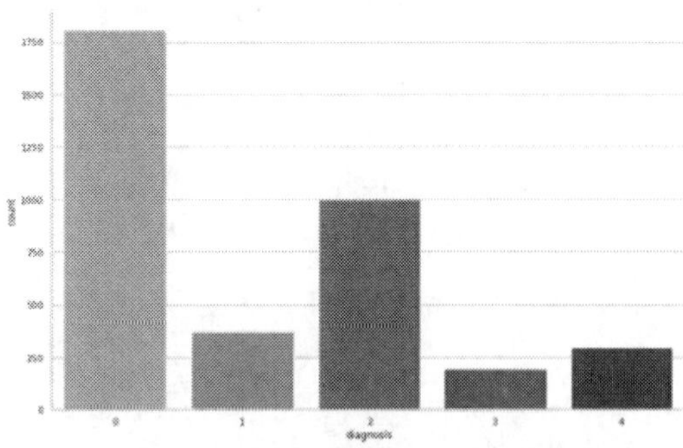

Figure 2: Distribution of dataset over classes

3.2 Image Augmentations

Image Augmentation is the method which is used to make dataset stronger and robust. Additional images are added in dataset using previously existed images. This can be achieved by using various techniques like flipping of images, resizing images, cropping images, rotation of images, padding and shift in images [12]. Above techniques combined or alone can be used to generate more images which helps to make dataset stronger. Mirror image is generated using image augmentation.

We have created and applied following image Augmentation techniques using keras image Data Generator class. We have kept height and width of images as 320 and 320 resp.
1)Rescale = 1/255,
2)Rotation = 360 and 60
3)Brightness_range = [0.5, 1.5]
4)Zoom_range = [1, 1.2]
5)Zca_whitening = True
6)Horizontal_flip = True
7)Vertical_flip = True

3.3 Image Preprocessing

In Diabetic Retinopathy, Higher Hemorrhages and exudates means chances of blindness is more. In our scale range 3-4 is considered as Highly affective DR. So, our main aim is to spot these Hemorrhages and exudates in given dataset of image. Images in dataset are not clear and ready to use in project [14]. Some of the images in dataset are out of focused, under focused, over focused. Some have low brightness while some have high brightness to overcome above problem, we had used low pass filter to avoid unnecessary details. We had used OpenCV library and Gaussian function which is low pass filter used to remove high frequency component from image. This is known as Gaussian Blur technique [13]. To sum up we have converted RGB fundus image to sharp grey image. We have also crop images in circular form as eye shape is circular, so we can extract more accurate features from circular image.
To carry out circular crop, we have set center at height/2 and width/2 and radius is set to min(height,width) with thickness 1.

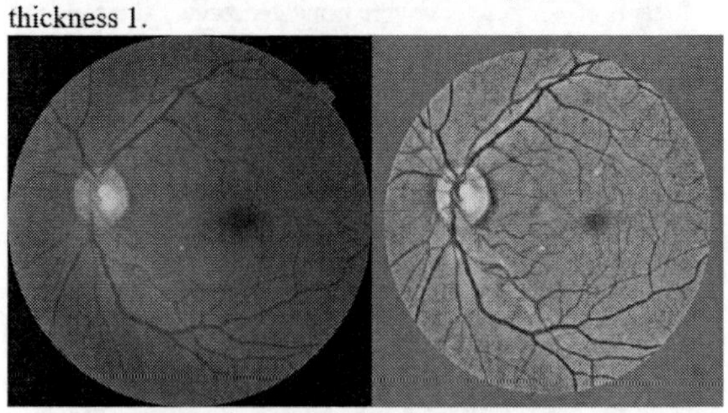

Figure 3: Fundus image on left and preprocessed image on right i.e., after gaussian blur and circular crop

3.4 Metric and Losses

The quadratic weighted kappa (QWK) is agreement or similarity between two predictions. The QWK is between the range of 0 to 1 i.e., from no agreement to complete agreement. There are five steps in calculating QWK

$$\kappa = 1 - \frac{\sum_{i,j} \mathrm{w}_{i,j} O_{i,j}}{\sum_{i,j} \mathrm{w}_{i,j} E_{i,j}}$$

where w is the weighted matrix, O is the histogram matrix and E being the expected matrix. 0.61 to 0.8 QWK score means Substantial agreement and 0.81 to 1 QWK score means almost perfect agreement.

4. SYSTEM DESGIN

The EfficientNet is considered as dynamic than the ResNet, DesNet, ResneXt, Xception and some other CNN models. While creating EfficientNet the developers did scaling uniformly and gradually i.e., they scaled depth, resolution and width. Stem has input layer, rescaling, normalization, zero padding, Conv2D, batch norm., activation and then final layer block. The five modules and stem make all layers of EfficientNet.

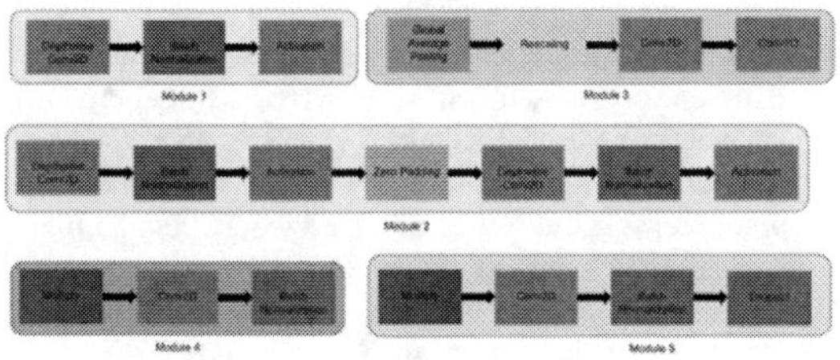

Figure 4: Five Modules of EfficientNet

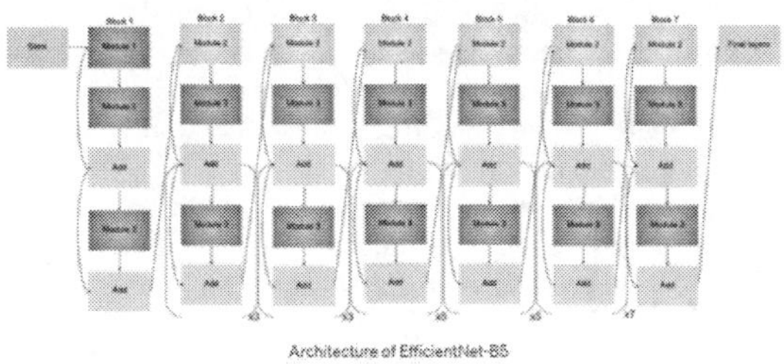

Figure 5: Layers of EfficientNet-B5

We have proposed a multi-task learning model which will have backbone of feature extraction as EfficientNet B5. The pretrained model in which include_top layer will be set to false which will not load the Fully Connected layer for giving output.
The weights of EfficientNet on ImageNet dataset are been initialized but the last layer weights are not taken since we want to extract only feature and we don't want to predict classes of the ImageNet Dataset.

Then the extracted features are further given to Global Average Pooling layer and Global Max Pooling layer. Later the concatenation of the output is performed and then passed it to dense layer for the reduction in the length of the vector. Then dropout is implemented and the output of dropout is given to two sections namely:

1. Classification section
2. Ordinal - Regression section

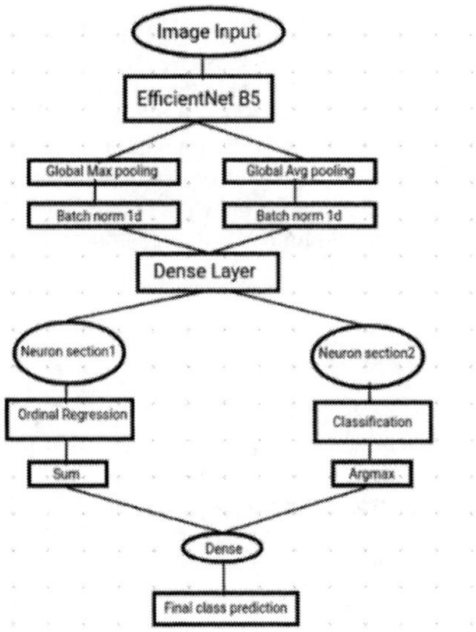

Figure 6: Flowchart of Proposed Model

5. MODEL TRAINING

Classification section is dense layer with 5 neurons with activation function of SoftMax, we have 5 neurons because we have to predict severity and severity scale has 0,1,2,3,4 as classes to predict. Ordinal regression section is dense layer with 5 neurons with activation function of SoftMax. Ordinal Regression section is different from Classification section with respect to loss function i.e., classification has categorical cross entropy and ordinal regression has binary cross entropy.
Further, we train the model for 20 epoch which is done for pretraining for model. Later, we change the loss function and freeze the feature extraction layer and train the model to carry out the warm up for main training.
While training the model we have changed the loss function to focal loss from cross entropy. The focal loss deals better with the data imbalance in one staged detector. The data imbalance cause ineffectiveness in training which causes degeneration of model.
The loss function reshapes the cross-entropy_loss which reduces weights assigned to properly detected/classified examples. The scaling factor reduces to zero as the classification class prediction becomes confident. It focuses on hard negative examples. 'γ' is called the focusing parameter.

$$\mathrm{FL}(p_t) = -\alpha_t(1 - p_t)^{\gamma}\log(p_t)$$

**α_t is a weighted term
**p (model's estimated probability for the class with label y = 1)
Then we unfreeze all layers and train our model for around 40 to 50 epochs. The main training is done after this step. Then we take the output of classification section and apply

argmax function. The output of ordinal regression section is given to sum function.
Then processed output is given for post training to a single neuron with linear activation function.
The single neuron with be trained on output for 50-100 epochs for reducing the error in the prediction.
This neuron will be giving us the predicted value and we will set threshold for determining our severity-on-severity scale from 0 to 4.

6. RESULTS

We have proposed a model which have image augmentation, dataset size reduction, multi-task learning model with feature extraction backbone as EfficientNet B5 with two sections as Classification and Ordinal Regression section. This is further lead by a single dense neuron with linear activation function which will predict the severity of diabetic retinopathy on severity scale 0 to 4. Hence the problem of Prediction of Diabetic retinopathy severity using Fundus image will be solved by using the proposed model. Using our approach, we got Train Cohen Kappa score as 0.870, Train Accuracy score as 0.878, Test Cohen Kappa score as 0.856 and Test Accuracy score as 0.877.

7. DISCUSSION AND ANALYSIS

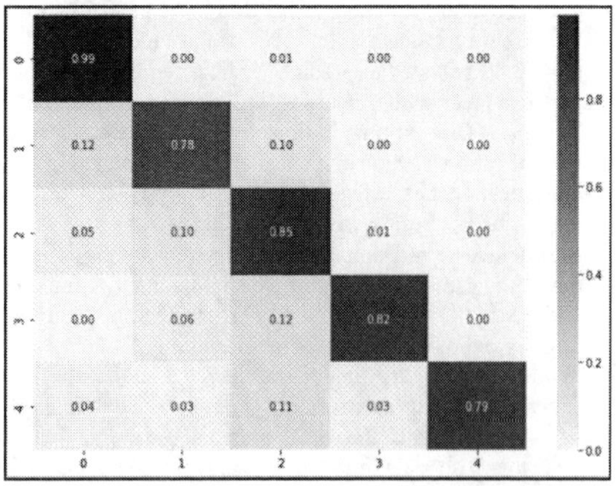

	ResNet50	EfficientNet B5	EfficientNet B5 (Classification + Ordinal Regression)
Train Cohen Kappa score	0.832	0.841	0.870
Train Accuracy score	0.747	0.806	0.878
Test Cohen Kappa score	0.803	0.822	0.856
Test Accuracy score	0.775	0.821	0.877

We have compared three models on basis of kappa score and accuracy. Using Resnet50 model we got train Cohen kappa score as 0.832 and Train Accuracy as 0.747. changing CNN model to EfficientNet B5 we received Train Cohen kappa score as 0.841 and Train Accuracy as 0.806 which is slightly greater than ResNet50. By dropping regression section with EfficientNet B5 with only two section (classification, ordinal regression) we got Train Cohen kappa score as 0.870 and Train Accuracy as 0.878.
Test Cohen kappa score for Resnet Model is 0.803, for EfficientNet B5 is 0.822 and for EfficientNet with two section is 0.856. Test Accuracy score for Resnet Model is 0.775, for EfficientNet B5 is 0.821 and for EfficientNet with two section is 0.877.
As above observation, we found more accurate result for EfficientNet B5 with two sections (Classification, Ordinal variable). We tired different protocols such as median blur and changes in the parameter in augmentation and preprocessing but results obtained where very similar. We also tried to segment retinal blood vessel and pass that along with hemorrhages and hard exudates but no improvement in performance was found. We also found multi-tasking models performs better than a uni-tasking model in terms of F1 score, recall and accuracy.
Due to the multiple sections used in our model, the model becomes more robust. This happens when the two sections produce different output results fed to the last neuron to predict the correct cumulative result based on previous prediction losses.
The key benefit of this approach is that it decreases variance by integrating many sections of networks that have been pretrained on a vast dataset and finely tuned on the target dataset.

8. CONCLUSION

We have compared ResNet50 with three sections (Classification, Regression and Ordinal Regression), EfficientNet B5 with three sections (Classification, Regression and Ordinal Regression), and EfficientNet with classification and ordinal regression section. Using EfficientNet B5 with classification and ordinal regression section, we obtained Train Cohen Kappa score as 0.870, Train Accuracy score as 0.878, Test Cohen Kappa score as 0.856 and Test Accuracy score as 0.877. We found that the preprocessing of data had a major impact on the results also the model performance is decreased if data imbalance is not considered and taken into account. The results of EfficientNet B5 with two sections also displayed that regression section does not perform quite well with classification and ordinal regression.
The research's future goals include implementing unsupervised learning and training on a unified dataset of all open-source datasets like Missidor and IDRid, as well as making optimal use of computational resources and combining them with supervised learning methods and models. Furthermore, we may conduct trials using pretrained encoders on various tasks related to eye disorders. This technique might also be used to examine meta-learning, although this would need separate additional research.

9. REFERENCES

1. P.M. Ebin, P.Ranjana, "An Approach Using Transfer Learning to Disclose Diabetic Retinopathy in Early Stage ", Presented at 2020 International Conference on Futuristic Technologies in Control Systems & Renewable Energy (ICFCR),Malappuram, India, (Sept 2020).
2. S. Mishra, S. Hanchate, Z. Saquib, "Diabetic Retinopathy Detection using Deep Learning", Presented at 2020 International Conference on Smart Technologies in Computing, Electrical and Electronics (ICSTCEE), Bengaluru, India, (Oct 2020).
3. A. Habib Raj, A. Mamun, F. Faruk, "CNN Based Diabetic Retinopathy Status Prediction Using Fundus Images", Presented at IEEE Region 10 Symposium (TENSYMP), Dhaka, Bangladesh, (June 2020).
4. N. Doshi, U. Oza and P. Kumar, "Diabetic Retinopathy Classification using Downscaling Algorithms and Deep Learning", Presented at 7th International Conference on Signal Processing and Integrated Networks (SPIN), Noida, India, (Feb. 2020).
5. J. Yadav, M. Sharma and V. Saxena," Diabetic Retinopathy Detection using feedforward Neural Network", Presented at Tenth International Conference on Contemporary Computing (IC3), 9August 2017).
6. Y. Kanungo, B. Srinivasan and S. Choudhary," Detecting Diabetic Retinopathy using Deep Learning", Presented at IEEE International Conference on Recent Trends in Electronics Information & Communication Technology (RTEICT), (May 2017), India.
7. S. Karki and P. Kulkarni," Diabetic Retinopathy Classification using a Combination of EfficientNets", Presented at International Conference on Emerging Smart Computing and Informatics (ESCI), Pune, India, (March 2021).
8. Z. Gao, J. Li, J. Guo, Y. Chen, Z. Yi and J. Zhong," Diagnosis of Diabetic Retinopathy Using Deep Neural Networks".
9. S. Thorat, A. Chavan, P. Sawant, S. Kulkarni, N. Sisodiya and A. Kolapkar," Diabetic Retinopathy Detection by means of Deep Learning", Presented at 2021 5th International Conference on Intelligent Computing and Control Systems (ICICCS)Madurai, India, (May 2021).
10. S. M. Saiful and M. M. Hasan, "Deep Learning based Early Detection and Grading of Diabetic Retinopathy Using Retinal Fundus Images," arXiv preprint arXiv:1812.10595. (Dec 2018).
11. F. DS, A. L, G. TW, K. GL, B. G, C. JD, K. R, and A. D. Association, "Diabetic retinopathy," Diabetes care, vol. 26, no. 1, pp. 226–229, (2003).
12. D. Silver, A. Huang, C. J. Maddison, A. Guez, L. Sifre, G. Van Den Driessche, J. Schrittwieser, I. Antonoglou, V. Panneershelvam, M. Lanctot et al., "Mastering the game of go with deep neural networks and tree search," nature, vol. 529, no. 7587, p. 484, (2016).
13. A. D. Hoover, V. Kouznetsova, and M. Goldbaum, "Locating blood vessels in retinal images by piecewise threshold probing of a matched filter response," IEEE Transactions on Medical Imaging, vol. 19, no. 3, pp. 203– 210, (March 2000).
14. T. Walter, P. Massin, A. Erginay, R. Ordonez, C. Jeulin, and J.-C. Klein, "Automatic detection of microaneurysms in color fundus images," Medical Image Analysis, vol. 11, no. 6, pp. 555–566, (2007).

Face recognition Attendance system using HOG and CNN algorithm

Aditya Kapse[1], *Tejas* Kamble[1], *Ashutosh* Lohar[1], *Shubham* Chaudhari[1], and *Digambar* Puri[2]

[1]Student, Department of Electronics and Telecommunication Engineering, Ramrao Adik Institute of Technology,Nerul
[2]Assistant professor, Department of Electronics and Telecommunication Engineering, Ramrao Adik Institute of Technology,Nerul

Abstract. Recognition of faces is one of the most useful applications and has a critical role in the technological field. Recognizing the face is a lively concern for authentication,specifically in the context of taking attendance. Attendance system using face recognition is a process of recognizing the profile of the person by using facial features supported by various computing technology and monitoring. The evolution of this process is focused on achieving the digitizing of the orthodox system of taking manual attendance. Current approaches for taking attendance are monotonous and inefficient. Manual records of attendance can be easily manipulated. The orthodox process of checking attendance such as current fingerprint or card scanning systems are susceptible to proxies. To tackle these issues, this paper has been proposed. The proposed system makes the utilization of various algorithms such as histogram of oriented Gradient (HOG), convolutional neural network (CNN) and support vector machine (SVM). After the face is recognized, the reports of attendance are going to be created, maintained and stored in excel format. The system is examined in various situations like illumination, head movements, and the variation of distance between the face and cameras. The proposed system was found to be efficient and reliable for marking attendance during a classroom with negligible time consumption and no manual work. This system is inexpensive as less installation is required.

1 Introduction

Attendance is very important for administration purposes, but usually it can become a tedious activity, with lots of inaccuracies [1]. The orthodox procedures for maintaining attendance have many limitations because it can be extremely difficult to take roll calls and maintain records when the total count of students is high [2]. Every organization has developed their own steps to take attendance. Some organizations follow a document based approach and others are using digital methods such as fingerprinting techniques and card swapping techniques. However, these methods also have some limitations because students will have to stand in long queues. If someone fails to bring his identity card, then he won't be marked present. The digital way of taking attendance is usually carried out with the help of biometric features [3]. Recognizing facial features is one such biometric way to improve digital systems for taking attendance. This method was found to be very efficient in terms of checking attendance. The usual face recognition methodologies do not handle challenges like posing, variation in light, face movements, and obstructions. The method proposed is meant to unravel the downside of present systems. Face recognition has been improved a lot till now, but the necessary stages are face detection, feature extraction, and face recognition [4]. First of all, multiple cameras depend upon the necessity and the size of the room has to be installed on the walls of the class so that it covers the entire class. Video captured from these cameras will be used by the system to get the names of the students who are present. There may be a possibility of getting the image blurred as all the students will be moving. So an enhanced image will be passed to the system for face detection [5, 6]. The system will now first detect the faces in the image , then it will extract the facial features, and at the end it will recognize the students names using those features. Face recognition is accomplished using the CNN (Convolutional neural networks), and SVM (Support Vector Machine) algorithms. After completion of all the steps, the system gives the names of the students who are present in the class. Then attendance of the student is marked in the excel format with respective name and time of arrival. Hardware requirements of this system are less and hence it is a cost-friendly system.

Keywords

HOG (Histogram of oriented Gradient), FLE (Face landmark estimation), CNN (convolutional neural network), SVM (support vector machine)

2 Literature survey

Thida Nyein, et al. [1] proposed a Face recognition attendance system using Facenet and support vector machine this system was divided into three parts first one was preprocessing of raw data in which face alignment is done on the raw data then that data is converted in training dataset in which the images are trained with model and classifier where they used facenet for feature extraction and support vector machine for classification. The flow of the proposed system is machine starts from taking image as an input and then face detection is done by using opencv and then feature extraction and embedding is done by using Facenet and feature matching is done by using support vector machine by matching with trained dataset. After that, it records the attendance by faces and then generates an excel file for attendance records. The accuracy they were getting was 80%. This system was foundation of idea for the face recognition attendance system but the limitations of the process was that with Facenet if person is wearing any acessories then it becomes difficult for the algorithm to detect faces and the accuracy was also low.

Ketan Mahajan , et al. [2] author proposed two phase attendance system scheme were making use of Probability based face mask Pre-Filtering (PFMPF) and Pixel based hierarchical Feature Ad boosting (PBHFA) . The author were making use of this algorithms to solve the problem in Harr cascade method . The system was divided into two parts first part is training phase which has two steps first is face detection using Viola Jones algorithm and next is feature extraction which is done by using PCA algorithm. The testing part was divided into 2 section training dataset and testing dataset limitations with the system was training time was very slow and it also has high False detection rate and the system was effective when the face is in frontal view . They have not used any face centralization method.

Vidya Patil, et al. [3] proposed an automatic students attendance marking system using Kth nearest neighbour (KNN). Here, they were acquiring images from the camera, then preprocessing of the image is done by using Histogram equilization method and the face is detected using HAAR cascade algorithm and then features are extracted using the LDA algorithm and then face recognition is done by using three algorithm namely LDA, SVM and KNN .The limitations with system was KNN dont work properly when the dataset is too large and also it is sensitive to noisy and missing data. proposed system has covered all the drawbacks of this system.

In Ali Elmahmudi, et al. [4] proposed a deep face recognition system for partial faces. In this system, the methodology adopted by the author uses a VGGF model which is pre-trained and will be used for feature extraction. This system uses cosine similarity and SVM for classification. In this system, when the partial faces were present in the images, the accuracy of SVM starts dropping, but on the other hand, Cosine similarity keeps holding the accuracy.

Yohie Kawaguchi , et al. [6] has proposed a model which is based on face recognition and constant surveillance . The author has proposed a system which uses active student detecting (ASD) which has two cameras on install inside on the walls of the classroom where one was a sensing camera and other camera were used for the face detection. He has proposed a plan in which estimation of seating area is done by using ASD after that capturing student using capturing camera.The present student has been estimated using background frame subtraction and inter frame subtraction. He has resolve the issue of linear sum assignment by giving correlation of seats and students.

3 Proposed System

3.1 Architecture

The given system is extremely effortless and manageable with coherent functioning. It has data on pupil's face images and their name. Multiple cameras are installed on the walls of the classroom such that the entire area is covered. Video of the lecture will be obtained by using these cameras.

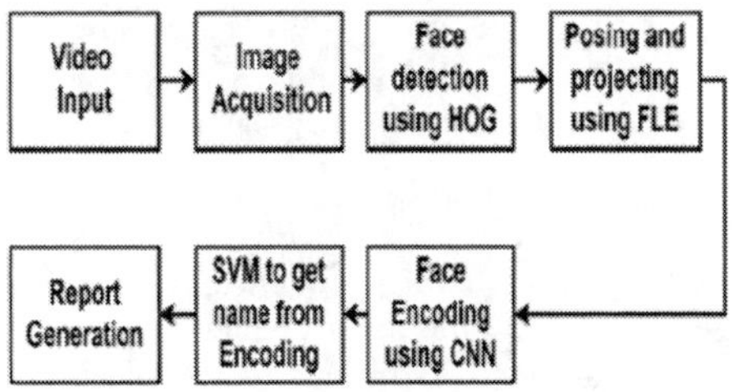

Fig 1. System flowchart

This will increase the effectiveness of our system because if one camera is not able to cover certain students then other cameras will get them. There are numerous poses which a pupil can perform. If in any case the system fails to detect faces due to the pupil's movements, then those faces can be detected in the next iteration of image acquisition. Now, face detection has to be performed to get the faces from the acquired image. As the system detects the face it will further process it to extract facial features and then it will compare those with the facial features present in the database. Now, the name of the student whose image is matched will be updated in the excel sheet. As there will be cases where a pupil is already marked present but his face is again detected in other acquired images. So this system has a provision to make sure that each pupil will be marked as present only once during a lecture. The proposed system flowchart is shown in figure 1.

3.2 System Design

In our system, there will be cameras installed on ceiling of the classroom which will take video input and pass it to the system where system will perform various operations and algorithms to generate attendance report. This system is composed of the following things:

- Image acquisition
- Face detection
- Posing and Projecting using FLE
- Face encoding Using CNN
- Comparison Using SVM
- Report generation

3.3 Image Acquisition

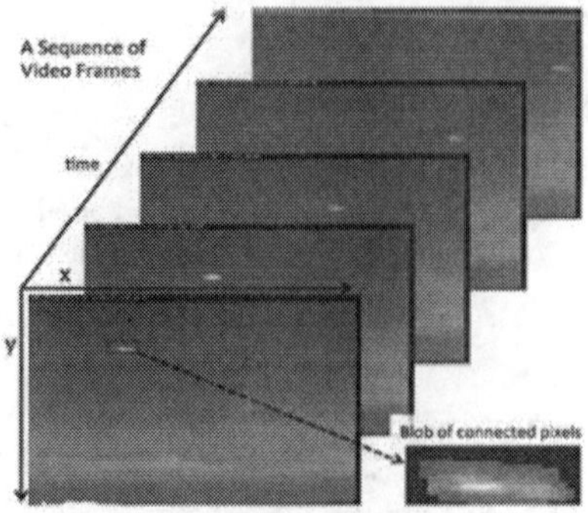

Fig 2. Image Acquisition

We all know that a video is a collection set of multiple images captured in a fraction of seconds. So we will take each frame from the video obtained by the cameras and try to find faces in it. If we found face then we will go ahead for centralization of face else we will go for the next frame of the video [7]-[8]. This process of image acquisition is shown in figure 2.

3.4 Face Detection using HOG

Face detection is one of the most demanding technologies in the field of machine learning. we have done face detection with the help of HOG. Here HOG features are extracted from multiple face images to be used as part of the recognition mechanism. The HOG (Histogram of Oriented Gradient) is a feature descriptor used in machine vision for processing of digital images for the purpose of detecting objects. HOG is extensively known for their use in the detection of moving objects [10]-[11].

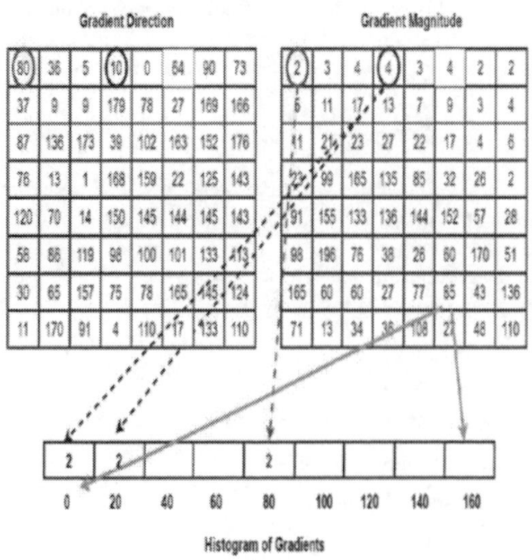

Fig 3. Feature Extraction using HOG

In Figure 3 you can see that HOG is doing feature extraction using two things first one is Gradient direction and other is Gradient magnitude. The formulas for the calculation of Gradient and Magnitude are given in figure 4.

Magnitude : $$||\nabla f|| = \sqrt{\left(\frac{\partial f}{\partial x}\right)^2 + \left(\frac{\partial f}{\partial y}\right)^2}$$

Direction : $$\theta = \tan^{-1}\left(\frac{\partial f}{\partial y} / \frac{\partial f}{\partial x}\right)$$

Fig 4. Formulas for Calculation of HOG

After successful calculation of Gradient direction and Gradient Magnitude for each pixel, features can be extracted as shown in figure 5.

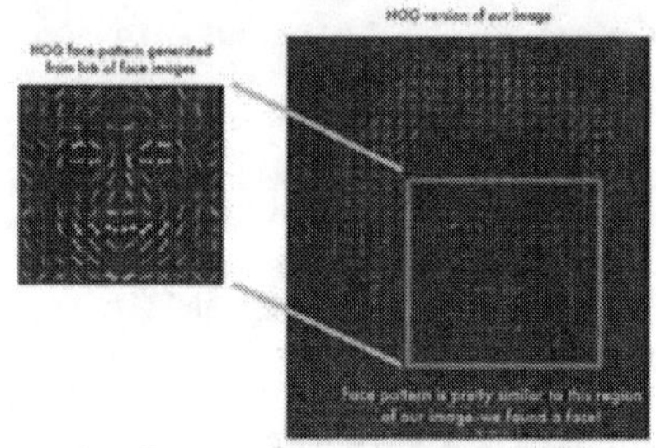

Fig 5. Image Detection using HOG

3.5 Posing and projecting using FLE

Head movements of students in different directions are possible during the lecture. These movements can affect the accuracy of our system. Suppose a same person is looking in different direction; then it is a totally different image for the computer as shown in the figure 6.

Fig 6. Image of Person Looking in Different Direction

So to overcome this issue, we will use Face Landmark Estimation Algorithm (FLE) [12]. The idea of FLE is it will compute 68 landmark points on all over the the faces like edge of both eyes, top of the chin, edge both eyebrows and then it will compute the distance between eyes, nose and the chin. It is shown in the figure 7. After putting 68 landmark points on the face, the final centralized image will be obtained as shown in figure 8. So this image will be used by CNN to generate 128 encoding out of it.

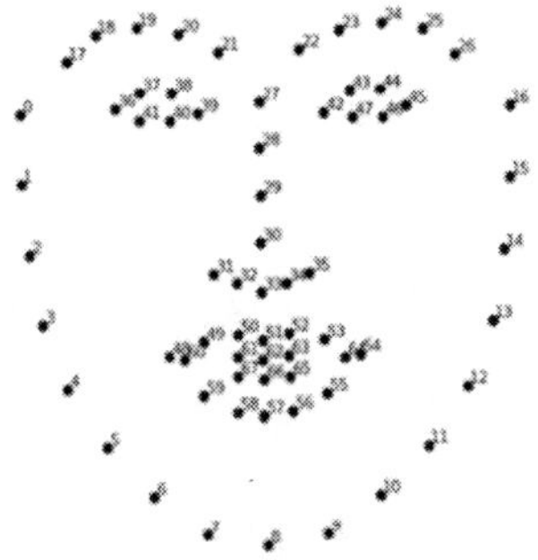

Fig 7. 68 landmark points on Face

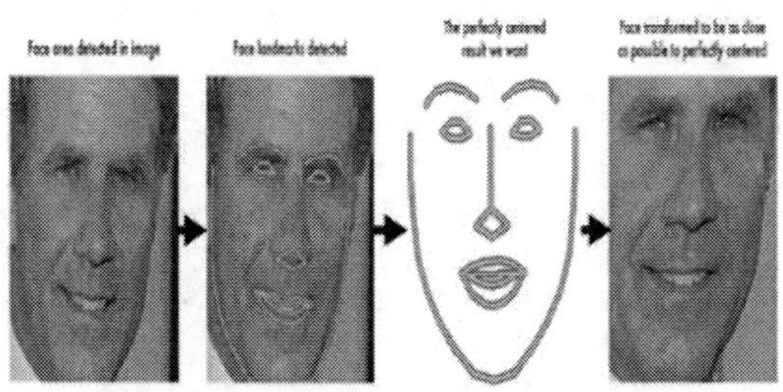

Fig 8. Centralize image using FLE

3.6 Face Encoding using CNN

The data set training steps are shown in figure 9 . First we will take image of a person and generate 128 encoding out of it. Afterwords we will take test image of another person and also generate encoding out of it and same with the another image; then we will compare the result and tweak the neural network slightly according to the comparison. CNN stands for Convolutional Neural Network [13]-[15]. It is a deep learning technique which is a Mask-based approach. CNN is composed of many layers of neuron; the first layer is used to extract basic features, namely horizontal and vertical edges; the next layer is used to detect more complex features such as corners; as we again move to downwards, we get more complex things such as objects, faces . We will use convolutional neural network model to extract few basic measurements from each face. We will train the model in order to generate 128 measurements for each face as shown in figure 10.

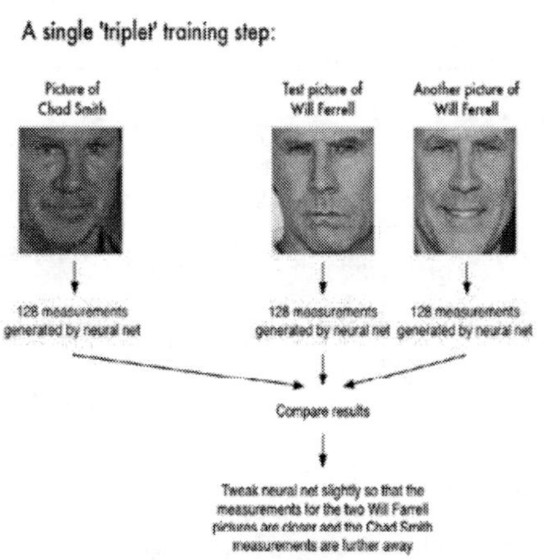

Fig 9. Training steps of Neural Network

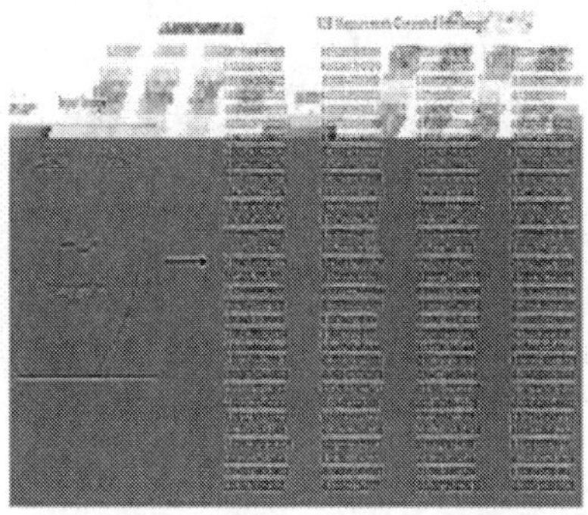

Fig 10. 128 Encoding generated by CNN

3.7 Comparison using SVM

Support Vector Machine (SVM) is a linear classifier. Our system will use this classifier in such a way that it can take all the measurements of unrecognized faces and it will compare them with the measurements of all the images present in our database [16]-[18]. After comparing, it will give us the name of that person whose measurements were closest match. This classifier takes milliseconds to execute. Basically, it introduces the imaginary plane between the two encoding and tries to find the closest match.

3.8 Report Generation

After the face recognition process is done, the attendance report will be updated and the name of the recognized student is inserted into the report along with the time at which the face was recognized. The attendance of all the students is updated only once. It means there will be only 1 occurrence of each student in the report.

3.9 Implementation Details

To implement the proposed system there are some hardware and software requirements.

3.9.1 Hardware

This system needs:

- A high quality camera of resolution 1600x1200 or more to capture the videos [19].
- A device having operating system(Windows, Linux, MacOS), operating voltage of 19V, input voltage of 20-40V, 1 USB port, i7 11th gen or equivalent processor, minimum 8GB RAM, clock speed of 2.4GHz and storage of atleast 100GB.

 https://doi.org/10.1051/itmconf/20224403028

3.9.2 software

To run this system we need :

- An IDE such as pycharm, vscode, etc to make the development process convenient.
- Python version 3.0 and higher.
- Some modules such as OpenCV (computer vision library), numpy, CSV, datetime, os and time.

3.9.3 Training images

- STEP1 : create a folder /train-images/ inside the openface folder.
  mkdir train-images.
- STEP 2: create a subfolder for every single person you want to recognize. For example,
  mkdir ./train-images/Aditya/
  mkdir ./train-images/Tejas/
  mkdir ./train-images/Ashutosh/
- STEP 3: make sure all images of each person are moved into the correct sub-folders. Also make sure only one face appears in each image. The images don't need to be cropped. It is done by OpenFace automatically.
- STEP 4: From inside the openFace root directory run the openFace scripts:
  - First, do pose detection and alignment:
    ./util/align-dlib.py ./train-images/ align outer Eyes And Nose ./aligned-images/ –size 96
    This will create a new ./aligned-images/ subfolder with a cropped and aligned version of each of your test images.
  - Second, generate the representations from the aligned images:
    ./batch-represent/main.lua -outDir ./generated-embeddings/ -data ./aligned-images/
    After you run this, the ./generated-embeddings/ sub-folder will contain a csv file with the embeddings for each image.
  - Third, train your face detection model:
    ./demos/classifier.py train ./generated-embeddings/
    This will generate a new file called ./generated-embeddings/classifier.pkl.
- STEP 5: Recognize faces Get a new picture with an unknown face. Pass it to the classifier script like this:
  ./demos/classifier.py infer ./generated-embeddings / classifier.pkl your_test_image.jpg
  You should get a prediction that looks like this:
  === /test-images/Aditya-1.jpg === Predict Aditya with 0.73 confidence.

3.9.4 Implementing system

To implement the system, a camera of recommended specifications should be connected to the computer where all the videos will be stored and all the computations will be done. The computer should have one of the specified operating system with an installed IDE, a specified python interpreter and all the required modules [20].
Once all the requirements are fulfilled then we will first build a program to create an excel file for each lecture on a particular day. Name of this excel file will be subject_date.csv so that every day for the same subject new file will be created. This file will be stored in the corresponding teacher's folder.
Now, we have to create a program which stores the captured video and updates the names of the recognized faces into the newly created excel file. This program will first acquire images from the video then it will detect all the faces in it using HOG. If any faces are found then it will use FLE so that the faces are centralised. Now, we will pass these faces one by one in the CNN model to get encoding for each of them. Now, it will compare these encoding with the encoding of images of all the students using SVM. This will give the face distance for each comparison and then it will record the attendance of the student whose face distance was least amongst the all. It will keep repeating the whole process during the lecture and then for next lecture the previous program will generate new excel file.

4 Result Discussion

Table 1. Comparison of different algorithms

Algorithm	SVM [14]	KNN [3]	Proposed System
Overall Accuracy	80.15 %	97%	99.2 %
Overall Time Complexity (S-seconds)	480 S	124 S	120 S
Precision	0.78	0.96	0.99
Recall	0.75	0.97	0.98
F1 Score	0.764	0.962	0.992

In Table 1, we have compared different algorithms and found that SVM has the lowest overall accuracy which is 80.15% followed by KNN which has 97% accuracy while CNN has the highest accuracy among them which is 99.2%. The overall time complexity for SVM was 480 seconds which was slowest amongst all followed by KNN which was able to give time complexity of 124 seconds while CNN was fastest amongst all which was 120 seconds. precision score for svm was 0.78 out of 1 while for KNN is 0.96 and for CNN its 0.99 which is highest amongst the all. From table 1, we have observed that our proposed system is better in terms of accuracy, overall time complexity, precision and F1 score.

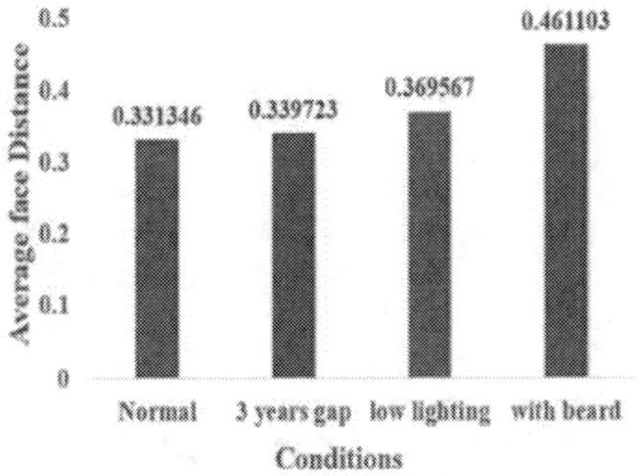

Fig 11. Variation of face distance under various conditions

The graph presented in figure 11 describes the changes in face distance in different conditions. Face distance is basically difference between the facial features of two faces. So face distance will increase, if there is some difference in the facial features of the two faces. From the graph in figure 11, we can see that there is slight increase in face distance if the lighting condition is bad. Also there is a significant increase in face distance if two faces of same person but one face with beard and other without beard is compared.

Table 2. Classification report

Measure	value
Accuracy	0.9920
Sensitivity	0.9930
Specificity	0.9908
Precision	0.9918
Negative Predictive Value	0.9921
False Positive Rate	0.0092
False Discovery Rate	0.0082
False Negative Rate	0.0070
F1 Score	0.9924
Mathews Correlation Coefficient	0.9838

The classification report of our proposed system is shown in the Table 2. Here we get the sensitivity score 0.9930 and the specificity score is 0.9908. Our system was able to get the F1 score of 0.9924.

5 Conclusion

The proposed system is able to achieve high precision and more accuracy for recognizing faces and marking attendance in less computational complexity. The economical and physical interference of the system is also less. We have successfully used HOG and CNN algorithms. Using these advanced algorithms, we were able to achieve accuracy of 99.20% and F1 score of 0.9924. We have also observed that our systems accuracy varies with various conditions such as lightning but still we were able to achieve accuracy of 99.20% . We have observed that our proposed system is efficient than any other system using different algorithms such as KNN and SVM. Our system can be further improvised and can be used for different purposes such as for tracking person by including features such as instead of using a single camera, we can use multiple cameras at different locations and interconnect them to form a network which can track a person. This system can also be configured to be used in ATM machines to keep track of the people who are using the ATM so that at the time of fraud or robbery the bank will be able to take immediate action. Also, at the time of elections this system can be used to identify the voter by recognizing his face.

References

[1] Nyein, Thida and Aung Nway Oo. "University Classroom Attendance System Using FaceNet and Support Vector Machine." *2019 International Conference on Advanced Information Technologies (ICAIT)* (2019): 171-176.

[2] Ketan N. Mahajan and Nagaraj V. Dharwadkar, "Classroom Attendance System Using Surveillance Camera",*Int. j. innov. eng. res. technol.*, pp. 1–4, Mar. 2021.

[3] V. Patil, A. Narayan, V. Ausekar and A. Dinesh, "Automatic Students Attendance Marking System Using Image Processing And Machine Learning," *2020 International Conference on Smart Electronics and Communication (ICOSEC)*, 2020, pp. 542-546, doi: 10.1109/ICOSEC49089.2020.9215305.

[4] A. Elmahmudi and H. Ugail, "Experiments on Deep Face Recognition Using Partial Faces,"*2018 International Conference on Cyberworlds (CW)*, 2018, pp. 357-362, doi: 10.1109/CW.2018.00071.

[5] S. Dev and T. Patnaik, "Student Attendance System using Face Recognition," *2020 International Conference on Smart Electronics and Communication (ICOSEC)*, 2020, pp. 90-96, doi: 10.1109/ICOSEC49089.2020.9215441.

[6] Yohei Kawaguchi, Tetsuo Shoji, "Face Recognition-based Lecture Attendance System", 2005.

[7] Hubballi, Kirti Bhat, Vandana, "Attendance System for Face Recognition using GSM module", *International Journal of Computer Sciences and Engineering*, (2018), 6. 1180-1184. 10.26438/ijcse/v6i7.11801184.

[8] S. Bhattacharya, G. S. Nainala, P. Das and A. Routray, "Smart Attendance Monitoring System (SAMS): A Face Recognition Based Attendance System for Classroom Environment," *2018 IEEE 18th International Conference on Advanced Learning Technologies (ICALT)*, 2018, pp. 358-360, doi: 10.1109/ICALT.2018.00090.

[9] E. Varadharajan, R. Dharani, S. Jeevitha, B. Kavinmathi and S. Hemalatha, "Automatic attendance management system using face detection," *2016 Online International Conference on Green Engineering and Technologies (IC-GET)*, 2016, pp. 1-3, doi: 10.1109/GET.2016.7916753.

[10] J. -M. Guo, C. -C. Lin, M. -F. Wu, C. -H. Chang and H. Lee, "Complexity Reduced Face Detection Using Probability-Based Face Mask Prefiltering and Pixel-Based Hierarchical-Feature Adaboosting," in *IEEE Sig-*

nal Processing Letters, vol. 18, no. 8, pp. 447-450, Aug. 2011, doi: 10.1109/LSP.2011.2146772.

[11] S. Z. Li, R. Chu, S. Liao and L. Zhang, "Illumination Invariant Face Recognition Using Near-Infrared Images," in *IEEE Transactions on Pattern Analysis and Machine Intelligence*, vol. 29, no. 4, pp. 627-639, April 2007, doi: 10.1109/TPAMI.2007.1014.

[12] Samal, Ashok and Prasana A. Iyengar. "Automatic recognition and analysis of human faces and facial expressions: a survey." *Pattern Recognition*. 25 (1992): 65-77.

[13] R. Brunelli and T. Poggio, "Face recognition: features versus templates," in *IEEE Transactions on Pattern Analysis and Machine Intelligence*, vol. 15, no. 10, pp. 1042-1052, Oct. 1993, doi: 10.1109/34.254061.

[14] Guodong Guo, S. Z. Li and Kapluk Chan, "Face recognition by support vector machines," *Proceedings Fourth IEEE International Conference on Automatic Face and Gesture Recognition (Cat. No. PR00580)*, 2000, pp. 196-201, doi: 10.1109/AFGR.2000.840634.

[15] Puri, D., Ingle, R., Kachare, P., Awale, R., "Wavelet Packet Sub-band Based Classification of Alcoholic and Controlled State EEG Signals", *International Conference on Communication and Signal Processing (ICCASP)*, Atlantis Press, 2016, pp. 562-567, DOI:10.2991/iccasp- 16.2017.82.

[16] Kehri V., Puri D., Awale R.N., "Entropy-Based Facial Movements Recognition Using CPVM", *Applied Computer Vision and Image Processing. Advances in Intelligent Systems and Computing*, vol 1155. 2020. Springer. Singapore. https://doi.org/10.1007/97898115402952.

[17] Puri D., Nalbalwar S., Nandgaonkar A., "EEG Based Diagnosis of Alzheimer's Disease Using Kolmogorov Complexity", *Applied Information Processing Systems. Advances in Intelligent Systems and Computing*, vol 1354. Springer, Singapore. https://doi.org/10.1007/978-981-16-2008-9-15.

[18] M. Swapna, Y. K. Sharma, and B. Prasad, "A survey on face recognition using convolutional neural network" in *Data Engineering and Communication Technology*. Springer, 2020, pp. 649–661.

[19] A. El-Sawy, E.-B. Hazem, and M. Loey, "Cnn for handwritten arabic digits recognition based on lenet-5" in *International conference on advanced intelligent systems and informatics*. Springer, 2016, pp. 566– 575.

[20] Z. Zhiyao and B. M. Di, "Multiplet selection: a practical study of multi- faces recognition for student attendance system" in *Proceedings of the 2nd International Conference on Image and Graphics Processing*, 2019, pp. 42–46.

Sentiment Analysis of Images using Machine Learning Techniques

Yash Gherkar[1], *Parth* Gujar[1], *Amaan* Gaziyani[1], and *Siddhi* Kadu[2]

[1]Student, Department of Computer Engineering, Ramrao Adik Institute of Technology, DY Patil Deemed to be University, Nerul, Navi Mumbai

[2]Assistant Professor, Department of Computer Engineering, Ramrao Adik Institute of Technology, DY Patil Deemed to be University, Nerul, Navi Mumbai

Abstract. Sentiment analysis is the process of identifying the idea of a text. People share the comments on social media stating their knowledge of the event and would like to know if most other people had a good or bad experience at the same event. This distinction can be made through Emotional-Analysis. Sentiment analysis captures informal text comments, posts and images from all comments shared by different users and classifies comments into different categories as neutral, negative or positive. This is also called as polarity separation. Various different types of ML and in-depth learning methods may be utilised in Sentiment Analysis like Support Vector Machines, NB, Haar Cascade, LBPH, CNN, etc. Emerging rise in popularity in Social Media has established a trend of posting images in restaurants to express their opinion on the food, ambience, etc which can be a useful resource to obtain opinion and feedback from the Customers. In this paper, the implementation of Sentiment Analysis on images containing users along with their faces from the restaurants review revealing it more efficacious in classifying and identifying sentiments of review-images.

Keyword :

Sentiment Analysis, Product Reviews,
Image Sentiment Analysis, LBPH, Haar Cascade

1 Introduction

Online updates become one of the main ways to explore a wide range of options. The user might check out Yelp reviews when choosing a restaurant. Review's usefulness is found in their role in accurately capturing prior customer information. In reality, one important piece of information that needs to be obtained get from the reviews is the feeling that the consumer is expressing, in the end whether you have had a good or a bad experience. Understanding emotions is an important and important function of analysing reviews, as emotions may express users' preferences, furthermore weaknesses and strengths. Information similar to these are crucial for designing the product, their recommendation, advertising, etc.

There has been extensive work done on detecting sentiments from Text reviews on any product but not many focus on the review images people post online. Most of the research paper and projects have implemented either Text based Sentiment Analysis on reviews such Amazon products review, etc or have implemented facial sentiment analysis without any real word application.With such massive growth in social media, it has provided a huge platform for influencers, critiques, bloggers, who post each and every experience they have with a business or a product. Even when a casual social media user visits a restaurant, it is observed that they take multiple Image to share in online with their peers, followers, etc. Yet even with such a massive food trend online, there has been a very few projects that have researched in extracting opinions from these images. The images that were found on Yelp indicate that a good majority of people are willing to upload their own pictures while dining. These images which contain facial expression can be a great resource for extracting opinions, it can also be used by business to understand their user's feedback better.

The main contribution of this study is that it provides the users a platform to evaluate business with the aid of Image reviews posted by others users. It also enables the businesses to acquire a more in-depth feedback which involves sentiments expressed by the users utilizing facial expressions. By the information gathered from such different platforms, a model was developed that would help users find the restaurant that serves the type of food they are looking for. Also, it would help the restaurant owners in improving the quality of food based on the reviews submitted by their customers.

The paper starts with the Literature Survey, that includes various papers that were critically analysed and were used for the research purpose. The next part contains the Proposed Methodology, consisting of all the techniques used in the project and it also includes the System Design of the project. Result Analysis, the next part, contains the analysis of the project, its accuracy calculated through Confusion matrix and the various results, the project produced in the testing stage. Then, the next part was the Conclusion of the project and its future scopes. The last part,

 https://doi.org/10.1051/itmconf/20224403029

References, includes the work that were used to produce the paper.

2 Literature Survey

There are various scholarly publications such as journal articles, research work etc. that are done in the field related to Sentiment Analysis. Many such articles and research reports were referred before making the project report. The reports/articles that have influenced the project, the most are :

Jin Ye, et al.[1] proposed a project focusing on product reviews. They have used a multi-modal approach to detect sentiments in an e-commerce reviews. Pytorch method is used to implement this method. A pre-trained model called as ImageNet is used to fine tune the dataset and for textual models, words are inserted using 'nn.embedding' in Pytorch. Both of these are then used in an CNN or GRU layer. A pre-trained model called as ImageNet is used to fine tune the dataset and for textual modes, words are embededed using "nn-embedding" in Pytorch. Both of these are then used in an CNN or GRU layer.The accuracy for only Image based analysis was low and text was used to reinforce the accuracy.Anthony Hu, et al.[2] proposed the visual models, a pre-trained model called "Inception" model. This model is trained to recognize image features with a deep architecture of 22 layers. Different features such as colors and shapes of an image are extracted. As for the textual model, Natural Language Processing (NLP) is used which implements methods such as Word embedding, Sequence input, etc to extract sentiments from the text. The input image, Inception network The test start network appears in a high-resolution area that exceeds the LSTM layer. The two methods are then combined and fed by a dense layer. Also no consumer driven application was developed.The proposed method also provide an important tool for the increasing research of images both-memes and photographs-in the social-network.

Quoc-Tuan Truong, et al.[3] proposes a method that is particularly focused on the sentiment that are a review image is displayed. It's not a multi-modal approach and only focus on Image Sentiment Analysis. They realized that the feeling expressed in the review-image, probably is governed by 3 factors: the image aspect (the emotions coded in image); user feature (emotions that were indicated by the reviewer in the images); & the factor of the user (feelings associated with image due to something). The paper utilizes two different model i.e., User-oreiented Model: uVS-CNN and then item-oriented CNN. User Oriented model has a slightly lower accuracy. It focuses on Review Images.The experiments proved much successful in recognizing review image's sentiments and it was followed on the images that were gathered from the reviews from restaurant.Pawel Tarnowski, et al.[4] proposed a model that was capable of recognising 7 states of emotions i.e. fear, anger, disgust, sadness, joy, neutral and surprise based on expression on their face. The classification was done with the help of six subject were performed using a kinect device , and the classifications was performed using MLP neural network and k-NN classifier.The emotions were calculated using 2 methods-> subject-independent(all users together) and subject-dependent(separate for each user). They found out after analysing that the user-independent method was more accurate than user-dependent method in recognising the emotions from facial expressions.

Santosh Kumar, et al.[5] proposed a Image sentiment analysis model which was used on to detect sentiments from Facial Features using Eigen faces, Fisherfaces, LBP, Speeded Up Robust Features(SURF) and FLANN. It focuses on extracting sentiments from Facial Expressions only. Image Sentiment Analysis is used on to detect sentiments from facial features using Eigen faces, Fisherfaces, LBP , Speeded Up Robust Features(SURF). It focuses on extracting sentiments from facial sentiments from Facial Expressions only therefore No real-world application. Paul Viola, et al.[6] proposed a 38 layered classifier which was trained to detect frontal upright faces and set of faces and non-faces images were utilized in classifier's training. The classifier they presented had high detection rate with minimum computation time required. The classifier was also trained to classify and detect faces in certain difficult circumstances.

Xing Fang, et al.[7] proposed a project in which he collected the data in form of product reviews from amazon.com in which the product mainly belongs to 4 major category namely beauty, electronic, book, home.Then the reviews are categorised into negative, positive or neutral by max-entropy POS tagger. After that, multiple ML-Algorithms like the SVM ,Random Forest, NB are used compared with each other to implement Text Based Sentiment Analysis. DataSet containing Sentiment Tokens/Values is used. Each word to be evaluated is given a polarity according to the words from the data set. The aggregate Sentiment score of these words are considered to determine the sentiment of provided text.

Akriti Jaiswal, et al.[8] suggested a model that is based on Deep-learning architecture and worked on Convolution Neural Network(CNN) to find the emotions of the images. The working performance of this model was calculated with the help of two datasets namely JAFFE and FERC-2013. The accuracy of the model produced was 98.65 for JAFFE dataset and it was 70.14 for the FERC-2013 dataset. Stuti Jindal, et al.[9] suggested a model where the prediction of the sentiment was done with the help of Convolution Neural Networks or CNN. Some exclusive experiments were conduction on the Flickr image dataset that were already labeled. They reached on the conclusion that their proposed CNN training achieved better performance than the other competitive image sentiment analysis models.

3 Proposed Methodology

3.1 Dataset

The Dataset used was majorly gathered from the GitHub repository that contained various different facial expressions along with the respective files that were sorted according to their respective sentiments expressed. A CSV file containing the names of the image files with their Sentiment score is used to sort the images. Furthermore, the dataset was also gathered by the process of Web-Scraping using a python library 'BeautifulSoup' from the popular restaurant review website- Yelp.com and then the data that contained the reviews with the facial data was sorted and used. The rest of images for the dataset was gathered by downloading the images from the dataset used in [3].

3.2 System Design

Fig.1 shows the entire system flow. The data set is fetched from a GitHub repository that has Facial data set. Using a list of predetermined images classify the images into different folders. Train the classifier using the classified images. Using Haar-Cascade and LBPH the model is trained. The Images are then scraped off the URL of the restaurant page on yelp provided by the user, then the review images from that web page are fetched and then processed to detect the sentiments in those Images. The overall Sentiment along with processed Images are then presented to the user.

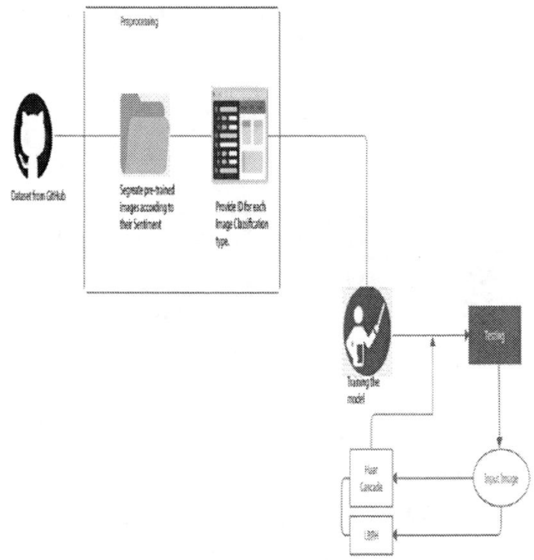

Fig.1. System design

3.3 Implementation Details

1.For this project, a data set which includes about 12,000 text and images from GitHub repository along with the images scraped of Yelp reviews that contained facial data. In that repository there is a separate CSV file which gives us the information of the images sentiments whether it was positive, neutral or negative.
2.The repository has separate folders based on the sentiments of the images. In that folder all the images with sentiment matching to that of folder's name has been included.
3.Then, the pre-processing of the data is carried out by removing the background of the images which have nothing to do with the sentiment analysis.
4.Then, a training file is created, which is made by going through each of the images in our dataset. This training file will contain the information of all the expression their emotions from their faces in images.
5.After that, the model is provided with images which are obtained by scraping off the reviews in order to carry out testing.

3.4 Techniques Used

3.4.1 Haar-Cascade

Haar Cascade classifiers are a very effectual and productive and classifier way for object detection. Haar-Cascade is a ML built method where lots of negative positive images are being used in training the classifier. The main goal is to calculate the summation of all the pixels that are on dark-side of Haar-features & the summation of every pixels on the light-side of the Haar-features. Haar-Cascade separators is an effective way to find an object. This method was suggested by M. Jones P. Viola in [6] and the working of the same is shown in Fig.2.

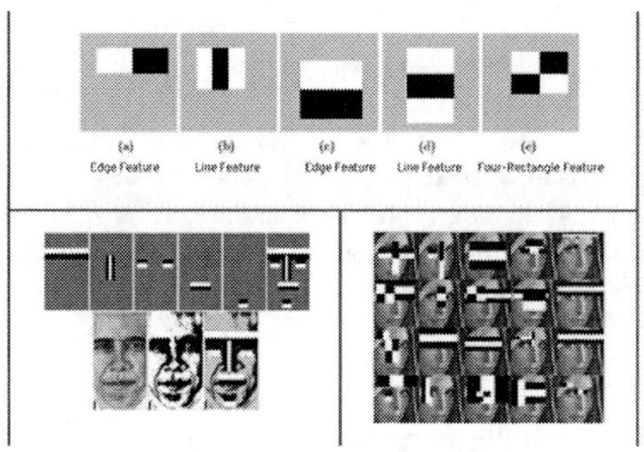

Fig.2. Haar Cascade Working[8]

Making a Haar Cascade Classifier
This algorithm is explained in 4 stages:
-Calculation of the Haar-Features
-Creation of Integral Images
-Use of Ada-boost
-Implementation of the Cascading Classifiers
It is of utmost importance remembering this algorithm needs lots of good facial images and negative non-facial images in training the classifier, similar to other ML models.

3.4.1.1 Calculation of the Haar Features

The primary goal is the collection of Haar-Features. A Haar feature is mainly the computation which is done performed in nearby rectangular areas somewhere in the detection-window. The computation includes adding up the intensities of the pixel in particular area or region and calculation of the difference between both the summation. Some of the examples of Haar Features are shown in Fig.3.

https://doi.org/10.1051/itmconf/20224403029

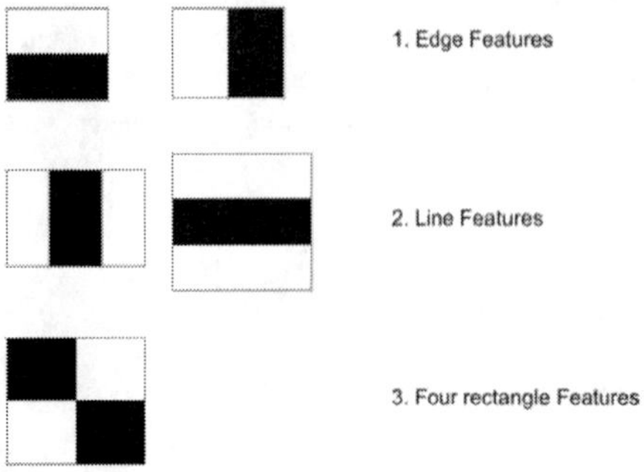

Fig.3. Haar Features[8]

These factors may be a little tricky in finding out for a big and huge picture. That's where the integral-pictures start to play as the no. of tasks is gets reduced by the use integral picture.

There are several advantages of Haar-Cascade that are it has low level of complexity, it has high level of computing. It captures the face more accurately and it has faster rate of recognition.

Some of the disadvantages of Haar-Cascade is that it is genearlly complex and hence slow, it's not suitable for black faces and turns out to be less accurate in this case. Haar- Cascade also requires long training time and if the lighteneing conditions are not favourable it have losts of false positive and false negatives. It is also less robust in case of occlusion.

3.4.2 LBPH

LBPH - Local Binary Pattern (LBPH) is an uncomplicated yet excessively effective texture-operator that highlights image's pixels with the help of thresh-holding the area of each of the pixel & then considering the output to be a binary number. Because of its computer simplicity & discriminatory power , the texture-operator based on LBPH have became one of the most popular feature in various programs. It can be viewed as way of integrating different mathematical models with a structure and analysis of texture. Probably, the majorly necessary attribute of LBPH-Operator in actual life programs is that it is robust in terms of changes in the monotonic gray-scale caused by, for ex- light changes. Another utmost-important asset is that it is simple, computation wise, which makes possible for analyzing the pictures in real time challenging situations.
The LBPH algorithm utilizes four parameter:
1.Neighbors: No. of sample points that are used in building the circular local binary pattern.
2.Radius: radius that is used to create a round binary area pattern design and constitutes the radius throughout the pixel in the center. It is normally adjusted as 1.
3.Grid X: Grid X represents summation of all the cells that are located in the horizontal-direction. The greater, the number of cells ,the finer or superior the grid will be, the higher the dimensional of the emerging feature-vector. It's normally adjusted to 8.
4.Grid Y: Grid Y represents the summation of all the cells that are present in the vertical-direction. The greater the no. of cells, the finer or superior the grid will be, the higher would be the dimensional of results in feature-vector. It's generally adjusted to 8.

3.4.2.1 Training the Algorithm

The foremost step forward is the algorithm's training. To do that, a database that contains the faces of people is used. There is also a need to set the ID for every images, so that the algorithm will use the information in identifying the inserted image will give the output. Photos of same individual must contain the unique identity. The LBPH calculating steps are.

Applying LBPH-Operation: The foremost part is to generate an intermediary image that grant the information of the actual image in a much superior way, by underlining the facial features. In order to do this, the algorithm utilizes the theory of a sliding-window, which is formulated on the radius, neighbors & parameters.

The figure(Fig.4.) shown beneath explains that method:

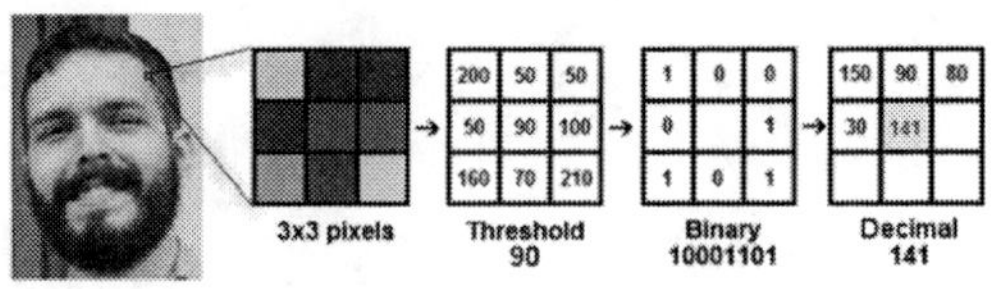

Fig.4. LBPH Working[9]

Based on the picture that was used in the above space, following steps are taken in the algorithm:

-Suppose,the face-image is present in gray-scale.
-Then the part of the picture is obtained in a 3x3 window.
-It maybe shown like a 3x3-matrix that contains intensity or magnitude of every pixel(0 255).
-After that, the value in the middle of the matrix is taken as thresh-hold.
-This thresh-hold value would be utilized in setting the values from 8 neighbours.
-For every adjacent of central value that is the thresh-hold, a fresh binary-value will be set. 1 will be set for values that are equal than or greater than the thresh-hold and 0 will be set for values that are lower than the thresh-hold.
-The matrix must have only binary value, then combination of each and every binary value from the matrix is taken and adjusted line-by-line accordingly to make a new binary-value. For example 10001101 .
-Transform this obtained binary-value into the corresponding decimal value and set this decimal value at the center of the matrix, which actually is one of the pixel from the original picture.

-Finally, in the end of the LBP-Procedure, a latest image which have will have superior properties than the actual picture will be obtained as shown in Fig.5.

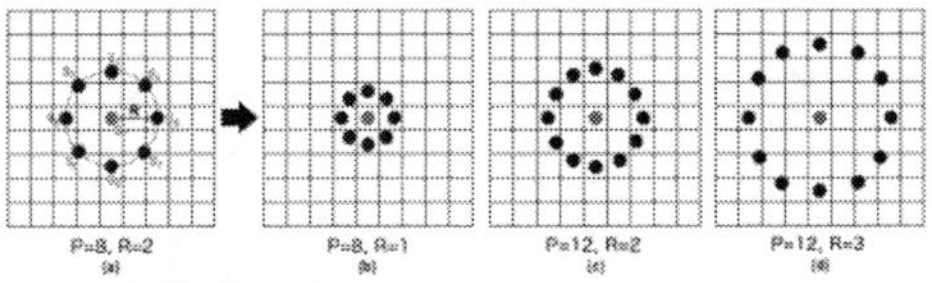

Fig.5. LBPH Bilinear Interpolation[7]

4 Result Analysis

```
id = 0
names = ['Positive', 'Neutral', 'Negative', 'None', 'None', 'None']
cam = cv2.VideoCapture(0)
cam.set(3, 640)
cam.set(4, 480)
minW = 0.1*cam.get(3)
minH = 0.1*cam.get(4)
img = cv2.imread("img1.jpg")
gray = cv2.cvtColor(img,cv2.COLOR_BGR2GRAY)
faces = faceCascade.detectMultiScale(
    gray,
    scaleFactor = 1.2,
    minNeighbors = 5,
    minSize = (int(minW), int(minH)),
    )
for(x,y,w,h) in faces:
    cv2.rectangle(img, (x,y), (x+w,y+h), (0,255,0), 2)
    id, confidence = recognizer.predict(gray[y:y+h,x:x+w])
    # Check if confidence is less them 100 ==> "0" is perfect match
    if (confidence < 100):
        id = names[id]
        confidence = "  {0}%".format(round(100 - confidence))
    else:
        id = "unknown"
        confidence = "  {0}%".format(round(100 - confidence))
    cv2.putText(img, str(id), (x+5,y-5), font, 1, (255,255,255), 2)
    # cv2.putText(img, str(confidence), (x+5,y+h-5), font, 1, (255,255,0), 1)
cv2.imwrite("imgFinal.jpg",img)
print("\n [INFO] Done detecting and Image is saved")
```

Fig.6. Code to categorize images as positive,negative,neutral

The figure above(Fig.6.) represents the code to categorize sentiment of images (categories are positive, negative, neutral).Fig.7. shows the classification report based on our testing data. It consists of factors like false negative rate, accuracy, false discovery rate, etc. Fig.8. represents the confusion matrix for our testing data.Our model gives out an accuracy of 80.46 .

Measure	Value
Sensitivity	0.7963
Specificity	0.8182
Precision	0.8776
Negative Predictive Value	0.7105
False Positive Rate	0.1818
False Discovery Rate	0.1224
False Negative Rate	0.2037
Accuracy	0.8046
F1 Score	0.8350
Matthews Correlation Coefficient	0.6011

Fig.7. Classification Report

Fig.7. represents the various different parameters including Sensitivity, Specificity, Precision, F1 score, Accuracy etc that were calculated with the help of the number of images tested with the model. Similarly, the Confusion Matrix shown in Fig.8. is obtained with the help of number of true-positive,false-positive,true-negative and false-negative results obtained when the model was already trained with the Dataset and is now being tested. The accuracy of our model came out to be 80.46% that was calculated by the summation of True Negatives and True Positives.

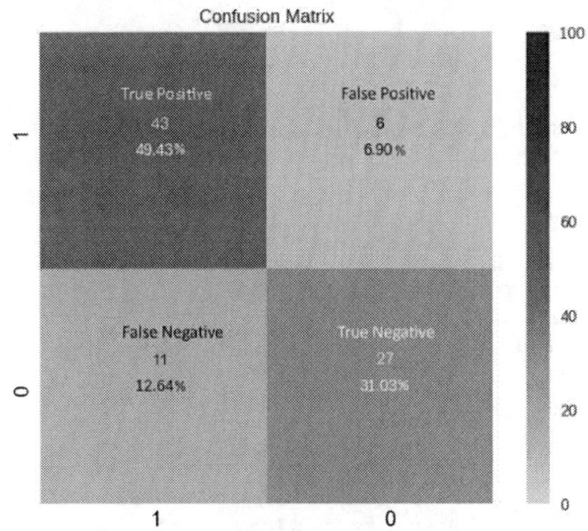

Fig.8. Confusion Matrix

The figures below[Fig.9., Fig.10., Fig.11.] represents images classified as either positive or neutral or negative based on the sentiments derived by the model from these images.

Fig.9. Image identified as a positive image

 https://doi.org/10.1051/itmconf/20224403029

Fig.10. Image identified as a neutral image

Fig.11. Image identified as a negative image

5 Conclusion

In the proposed system, the primary aim was to construct an accurate and dynamic sentiment analysis model that would help the restaurant owners to get accurate feedback from their customers. Our system uses Haar-Cascade to calculate the summation of every pixels that lies under the dark-area of the Haar-feature and summation of every pixels that lies under the light-area of the Haar-feature and LBP (Local Binary Pattern) that labels photo pixels by thresh-holding neighbor-hood of every pixel. Objectives like face recognition and emotion recognition were the primary objectives as they were the main point of concentration in the entire project. Other objectives like data collection, treating false positives were also taken care of to the best.

In future, detecting sentiments from food review images that don't contain any user using SIFT feature extraction method will be the primary goal along with using text shared by the reviewer in order to make the model more efficient and accurate.Also, the restaurant owner will be given a detailed report of the food items reviewed by their customers that would help them make changes accordingly and serve their customers with the type of food they crave for.

References

[1] J. Yin, X. Peng, Y. Qiao, H. Xing, J. Li, R. Ji, *Visual-textual sentiment analysis in product reviews, IEEE ICIP,* ,pp. 869-873, doi: 10.1109/ICIP.2019.8802992 (2019)

[2] A. Hu, S. Flaxman, *Multimodal Sentiment Analysis to explore the structure of emotions, ACM SIGKDD*, International conference on Knowledge Discovery and Data Mining, pp. 350-358, doi: 10.1145/3219819.3219853 (2018)

[3] Q.Tuan Truong, HW. Lauw, *Visual Sentiment Analysis for review images with item-oriented and user-oriented CNN, ACM International Conference on Multimedia* , pp. 1274-1282 , doi: 10.1145/3123266.3123374 (2017)

[4] P. Tarnowski, M. Kołodziej, A. Majkowski, RJ. Rak, *Emotion recognition using facial expressions, ICCS* , pp.1175-1184 , doi : 10.1016/j.procs.2017.05.025 (2017)

[5] S. Kumar, S. Jaiswal, R. Kumar, and SK. Singh , *Emotion Recognition Using Facial Expression, IGI Publications, USA* , pp. 1-19 , doi: 10.4018/978-1-4666-8723-3.ch013 (2016)

[6] P. Viola, M. Jones. *Rapid object detection using adaboosted cascade of simple features, IEEE* Proceedings of the 2001 IEEE computer society conference on computer vision and pattern recognition. CVPR 2001. Vol, pp. 1-1 , doi : 10.1109/CVPR.2001.990517 (2001)

[7] X. Fang, J. Zhan, *Sentiment analysis using product review data, Journal of Big Data 2.1* , article no. 5, doi : 10.1186/s40537-015-0015-2 (2015)

[8] A. Jaiswal, AK Raju, S. Deb,*Facial emotion detection using deep learning, IEEE INCET*, pp. 1-5, doi : 10.1109/INCET49848.2020.9154121 (2020)

[9] S. Jindal, Sanjay Singh, *Image Sentiment Analysis using Deep Convolutional Neural Networks with Domain Specific Fine Tuning, International Conference on Information Processing IEEE* , pp. 447-451 , doi : 10.1109/INFOP.2015.7489424 (2015)

[10] S. Ramalingam, F. Garzia ,*Facial expression recognition using transfer learning, IEEE ICCST* , International Carnahan Conference on Security Technology, pp. 1-5 , doi : 10.1109/CCST.2018.8585504 (2018)

[11] AM. Jagtap, V. Kangale, K. Unune and P. Gosavi, *A Study of LBPH, Eigenface, Fisherface and Haar-like features for Face recognition using OpenCV, IEEE ICISS* , International Conference on Intelligent Sustainable System , pp. 219-224, doi: 10.1109/ISS1.2019.8907965 (2019)

[12] N. Mehendale, *Facial Emotion Recognition using Convolutional Neural Networks (FERC).* SN Applied Sciences 2, 446 , pp.1-8 , doi : 10.1007/s42452-020-2234-1 (2020)

 https://doi.org/10.1051/itmconf/20224403030

Drowsiness Detection using EEG signals and Machine Learning Algorithms

Adinath Joshi[1], *Atharva* Kamble, *Akanksha* Parate, *Siddhesh* Parkar, *Digambar* Puri, and *Chandrakant* Gaikwad

[1]Ramrao Adik Institute of Technology, Electronics and Telecommunication Department, Navi Mumbai, India.

Abstract. Drowsiness is described as a state of reduced consciousness and vigilance accompanied by a desire or want to sleep. Driver tiredness is frequently detected using wearable sensors that track vehicle movement and camera-based systems that track driver behavior. Many alternative EEG-based drowsiness detection systems are developed due to the potential of electroencephalogram (EEG) signals to observe human mood and the ease with which they may be obtained. This paper applies Deep learning architecture like Convolutional Neural networks (CNN) and algorithms for the classification of EEG data for Drowsiness Detection. The key measures of video-based approaches include the detection of physical features; nevertheless, problems such as brightness limitations and practical challenges such as driver attention limits its usefulness. The main measure of video-based methods is the degree of closure of the eyelids; however, its success is limited by constraints like as brightness restrictions and practical challenges such as driver distraction. We have extracted statistical features and trained using various classifiers like Logistic Regression, Naïve Bayes, SVM, and K Nearest Neighbours and compared the accuracy using a deep learning CNN model. Results demonstrate that CNN achieved an accuracy of 94.75% by delegating feature extraction on itself. Upon comparing existing state–of–the–art drowsiness detection systems, the testing results reveal a higher detection capability. The results show that the the suggested method can be used to develop a reliable EEG-based driving drowsiness detection system.

1 Introduction

Drowsiness is a leading cause of vehicular accidents where drivers may experience loss in sense of vision, and overall vigilance. Driving when drowsy would result in longer reaction time, especially at high speeds, and reduced performance on attention-demanding events [1]. Many contributions have been made to explore drowsiness detection and can be categorised into two major classes - one that uses physical features viz. Steering Wheel Movement (SWM), Eye Blinking Based Technique, Yawn Detection, and Head Nodding detection while the other uses physiological features viz. electroencephalogram (EEG), electrocardiogram (ECG) [2]. We consider EEG in this paper due to its high temporal resolution, inexpensiveness, and practicality [3]. Comparing with existing systems, we see that although EEG can provide an accurate assessment of alertness levels, contamination from persuading EEG features such as eye activity, muscle noise, and acoustic anomalies such as line noise, electronic interference, and so on is one of the many technical challenges that must be overcome when implementing EEG-based monitoring systems. To overcome this, outlier removal or dimensionality reduction algorithms on raw data can help remove unusual readings that can be excluded to save computational energy. According to Jung et al. [3] and Lin et al., EEG correlates of fatigue and drowsiness appear to be variable amongst different subjects, which suggests the need for individually customizable systems. In this paper, we present a detailed comparison of two EEG drowsiness detection approaches where one uses machine learning (ML) algorithms namely K-Nearest Neighbors (KNN), Support Vector Machine (SVM), Logistic Regression (LR) and Naive Bayes while the other uses Convolutional Neural Network (CNN) - a deep learning (DL) algorithm. Conventionally CNN is considered a notable methodology for solving problems in image classification. The architecture for CNN seems suitable for other data as well since there is no need for manual feature extraction [4]. The following is how this paper is structured. Section 2 discusses the dataset used for analysis as well as feature extraction for machine learning algorithms. Section 3 describes every algorithm we have used along with their respective results.

2 Materials and Methods

2.1 Datasets and Inputs

We have used the "EEG Eye State Dataset" from UCI Machine Learning Repository as the common dataset for all computations [5]. This data is acquired from an EEG neuroheadset (Emotiv EEG Neuroheadset) for a duration of 117 seconds and collects 14304 samples. Along with 14 electrode values recorded at time t the dataset also provides whether or not the subject had their eye state opened (EO) or closed (EC) at time t. This data is then split into two sections randomly at each run - 80% of rows are set aside for training while 20% is set aside for testing. The acquisition for the 14 electrode measurements, originally labelled TF7, O2, F3, P8, T8, F4, FC6, AF3, FC5, T7, F8, P7, O1, and AF4 are shown in Fig 1.

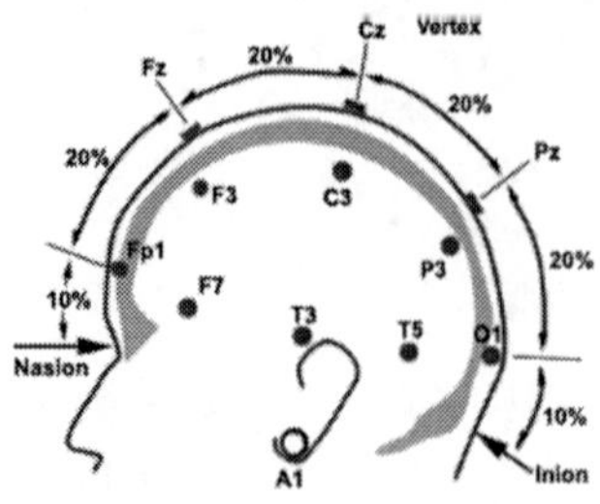

Figure 1. EEG acquisition sensor placement.

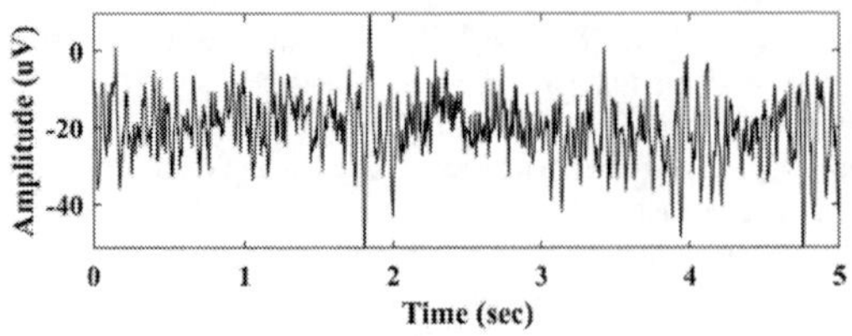

Figure 2. Amplitude values for Awake state.

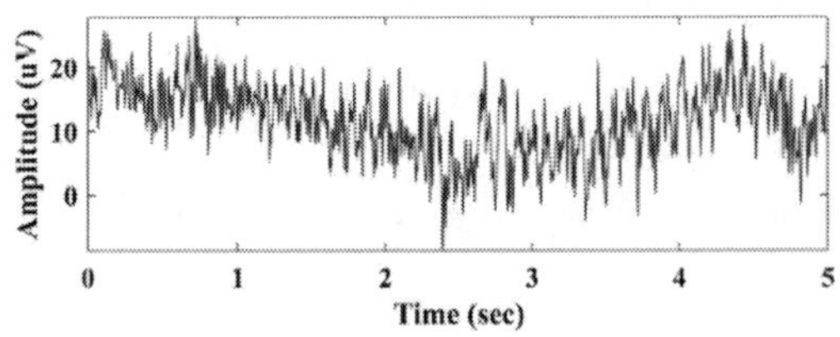

Figure 3. Amplitude values for Drowsy state.

2.2 Feature Extraction and Selection

For ML algorithms, we select the following features[6]: Output (OP), Statistical features viz. mean (μ), standard deviation (σ), kurtosis (β_1), skewness (β_2) and discrete cosine Transform (DCT) [7]. Fig 2 and Fig 3 show that the amplitude values are higher in the Awake state (Mean value = 4316.6) than in the Drowsy state (Mean value = 4302.8). These observations assist us in developing intuitions for extracting statistical features from this time series. The Discrete Cosine Transform (DCT) can significantly reduce the dimensionality of time series, improving the accuracy of time series classification and clustering. Table 1 shows the rows obtained by feature extraction on the original dataset. The extracted feature data frame is divided into X and Y where-

$$X_i(\mu, \sigma, \beta 1, \beta 2, DCT) = [\mu, \sigma, \beta_{1i}, \beta_{2i}, DCT_i] \quad (1)$$

$$Y_i(OP) = [OP_i] \quad (2)$$

Table 1. Visualisation of Mean and Standard Deviation for Drowsy and Awake states.

	Drowsy		Awake	
Channel	**Mean (μ)**	**Standard Deviation (σ)**	**Mean (μ)**	**Standard Deviation (σ)**
TF7	4350.8	3719.03	4322.56	71.735
O2	4005.7	33.03	4018.46	54.012
F3	4265.95	37.71	4282.56	49.177
P8	4121.45	26.527	4148.46	7026.93
T8	4341.75	22.61	4344.61	42.12
F4	4618.6	18.99	4603.33	3939.91
FC6	4073.55	34.214	4087.43	6197.8
AF3	4616.85	18.533	4634.87	35.727
FC5	4241.58	3188.565	4213.84	48.386
T7	4233.33	19.74	4233.07	48.00
F8	4205.22	39.77	4200.25	35.948
P7	4281.45	30.89	4281.28	48.454
O1	4632.85	1801.62	4627.18	77.44
AF4	4367.2	35.723	4384.11	7936.029

3 Algorithms

3.1 Support Vector Machine

SVM classification training aims to find the optimum hyperplane that maximises the difference between classes using inputs X and outputs Y [8].The accuracies obtained after tuning various parameters for SVM are shown in Table 3.

3.2 K-Nearest Neighbors

The K Nearest Neighbors algorithm assumes similarities between new cases / data and available cases and assigns new cases to the category closest to the available categories[9]. We have observed that maximum accuracy is achieved when the number of neighbors is set at 5 as depicted in Fig 5.

3.3 Logistic Regression

Logistic regression (LR) is a popular machine learning technique for performing predictive analysis on data that is categorical or has binary classes.The accuracies obtained using various solvers are as shown in Table 4. The maximum accuracy is observed when the 'lbfgs' solver is used.

3.4 Naive Bayes Classifier

The Naive Bayes technique is a series of supervised learning algorithms based on applying Bayes' theorem to each pair of class variable value features under the "naive" assumption of conditional independence. The accuracy obtained is 58.58%.

3.5 Convolutional Neural Network

A schematic diagram of the CNN used in our drowsiness detection system is depicted in Fig 4. The EEG signal propagates through two convolutional layers, followed by three fully connected layers. First, two fully connected layers are equipped with the Rectifier Linear Unit (ReLU) activation function. Whereas the last fully connected layer is equipped with the activation function of the sigmoid. Moreover, dropout processing [10] is used to minimize the risk of over-fitting.

https://doi.org/10.1051/itmconf/20224403030

3.5.1 Convolutional layers

These layers mainly focus on filter application and feature extraction [11] based on input EEG signals. The convolution operation is represented by following equation:

$$Y_i = b_i + \sum_n W_{in} * X_n \quad (3)$$

where $*$ is the convolution operation, Y_i presents feature map, b_i bias term, W_{in} sub-kernel of the channel, X_n input signal. We have used 2 convolutional layers.

3.5.2 Dense layers

The role of the dense layer is to describe the connection with the intermediate and next layers of neurons. In our architecture, we used 3 fully connected layers. For the best classification results, in the first dense layer of our model, we used a hidden layer of 10 neurons. In the second dense layer, we used a hidden layer of 100 neurons for improved accuracy. In the third dense layer, the value of the last neuron is equal to 1. A single neuron is sufficient to designate class "1" or "0", hence, binary classification is applied in this work.

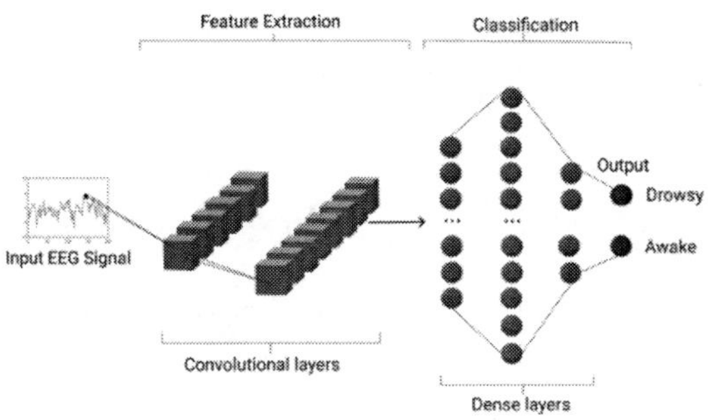

Figure 4. Flow diagram of CNN model.

4 Evaluation Metrics

For evaluating the performance of ML and DL techniques mentioned in this paper, we used metrics like Accuracy, Precision, Recall, and F1-score.

Table 2 shows the Formula and Description of the Evaluation Metrics that we have used to measure the quality of the machine learning models.

Table 2. Evaluation metrics with Description

Metric	Formula and Description
Accuracy	$(TP+TN)/(TP+TN+FP+FN)$
Precision	$TP/(TP+FP)$
Recall	$TP/(TP+FN)$
F1-Score	$2TP/(2TP+FP+FN)$

Where, $TP \Rightarrow$ True Positive, $FP \Rightarrow$ False Positive, $TN \Rightarrow$ True Negative, $FN \Rightarrow$ False Negative.

5 Results

In this paper, we have compared the performance of ML algorithms to DL algorithms. In the algorithms like SVM, Logistic regression, Naive Bayes, and KNN we have used extracted statistical features as input, whereas CNN does not require feature extraction. CNN achieves an accuracy of 94.75% and 97.5% for one dense layer and two dense layers respectively with 11.5% aggregate error loss. Table 5 depicts the performances of various models in decreasing order of their accuracy. Fig 6 shows the the evaluation metrics of the models, we can observe that CNN shows the highest accuracy among all other classifiers.

Table 3. Accuracies after tuning various parameters for SVM.

C	Kernel	Shrinking	Gamma	Accuracy (%)
1.0	RBF	True	Auto	82.98
2.0	linear	False	Auto	91.86
4.0	poly	True	Auto	84.14

Table 4. Accuracies after tuning solver for Logistic Regression.

Solver	Accuracy (%)
lbfgs	79.31
sag	73.4
liblinear	78.44
saga	74.8

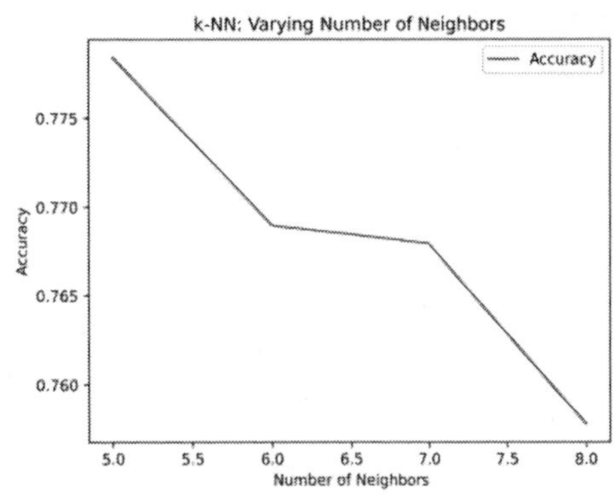

Figure 5. Graphs of accuracies vs. the number of neighbors.

6 Conclusion and Discussion

In this work, a comparison of ML approaches to DL approaches for drowsiness detection using EEG is presented. Out of all techniques mentioned in this paper, CNN is found to be highly efficient as it uses convolution of signals and filters for generating invariant features that are passed to the next layer. Multiple pooling operations aggregate feature maps obtained from convolution layers to an extreme spatial scale, which is a key factor for increasing the overall efficiency of CNN architecture. Fig 7 demonstrates

Table 5. Performance metrics and ranking of classifiers.

Sr. No.	Models	Accuracy (%)	Precision (%)	Recall (%)	F1 score (%)
1	CNN	94.75	93.6	95.7	94.6
2	KNN	78.6	81.1	78.4	79.7
3	Naïve Bayes	59.4	65.7	51.2	57.6
4	LR	77.3	79.4	65.0	62.1
5	SVM	83.7	83.7	100	89.9

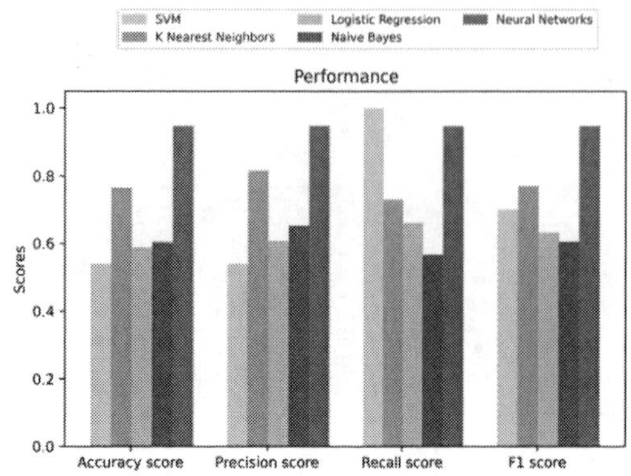

Figure 6. Comparison of model performances.

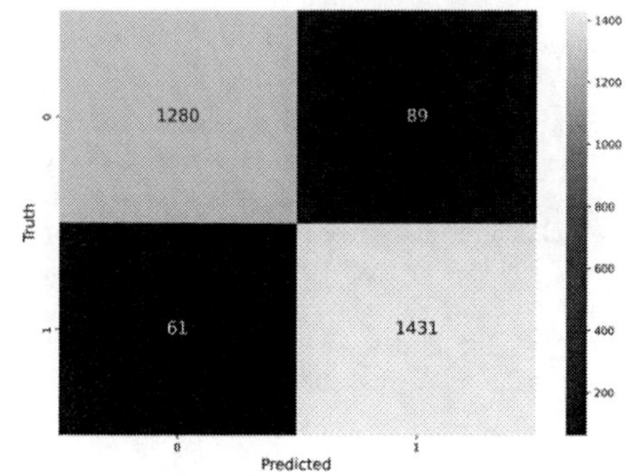

Figure 8. Confusion matrix for the performance of CNN.

~~ICACC-2022~~

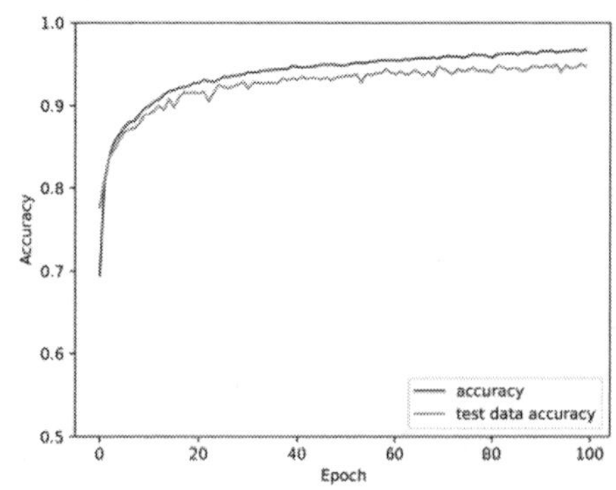

Figure 7. Graph of accuracy calculated from train data vs test data for CNN.

an accuracy graph with 80% training data and 20% testing data varying the number of epochs. When all 14 electrode values and two classes (0 and 1) are used, the confusion matrix obtained by CNN is shown in Fig 8, where the x-axis depicts the true class labels while the y-axis depicts CNN class predictions. The correct classifications are shown along the first diagonal, while all of the other entries show misclassifications.

References

[1] Chin-Teng Lin, Ruei-Cheng Wu, Sheng-Fu Liang, Wen-Hung Chao, Yu-Jie Chen and Tzyy Ping Jung, "EEG-based drowsiness estimation for safety driving using independent component analysis," *in IEEE Transactions on Circuits and Systems I: Regular Papers*, vol. 52, no. 12, pp. 2726-2738, Dec. 2005, doi: 10.1109/TCSI.2005.857555.

[2] Saini, Vandna, and Rekha Saini. "Driver drowsiness detection system and techniques: a review." International Journal of Computer Science and Information Technologies.

[3] Jung TP, Makeig S, Stensmo M, Sejnowski TJ. Estimating alertness from the EEG power spectrum. IEEE Trans Biomed Eng. 1997 Jan;44(1):60-9. doi: 10.1109/10.553713. PMID: 9214784.

[4] Puri D., Nalbalwar S., Nandgaonkar A., "EEG Based Diagnosis of Alzheimer's Disease Using Kolmogorov Complexity", *Applied Information Processing Systems. Advances in Intelligent Systems and Computing*, vol 1354. Springer, Singapore. https://doi.org/10.1007/978-981-16- 2008-9_15.

[5] Kehri V., Puri D., Awale R.N., "Entropy-Based Facial Movements Recognition Using CPVM", *Applied Computer Vision and Image Processing. Advances in Intelligent Systems and Computing*, vol 1155, 2020, Springer, Singapore. https://doi.org/10.1007/97898115402952.

[6] Digambar Puri, Sanjay Nalbalwar, Anil Nandgaonkar, Abhay Wagh, "Alzheimer's disease detection from optimal electroencephalogram channels and tunable Q-wavelet transform", *Indonesian Journal of Electrical Engineering and Computer Science*, Vol. 25, No. 3, pp. 1420-1428, March 2022, ISSN: 2502-4752, DOI:

10.11591/ijeecs.v25.i3.pp1420-1428.

[7] Puri, D., Ingle, R., Kachare, P., Awale, R., "Wavelet Packet Sub-band Based Classification of Alcoholic and Controlled State EEG Signals", International Conference on Communication and Signal Processing (ICCASP), Atlantis Press, 2016, pp. 562-567, DOI:10.2991/iccasp.16.2017.82.
CHAO-YING JOANNE PENG,KUK LIDA LEE,GARY M. INGERSOLL "An Introduction to Logistic Regression Analysis and Reporting" (2002)

[8] Lin, CT., Wu, RC., Jung, TP. et al. "Estimating Driving Performance Based on EEG Spectrum Analysis". EURASIP J. Adv. Signal Process. 2005, 521368 (2005).

[9] Hussain, Mahbub, Jordan J. Bird, and Diego R. Faria. "A study on CNN transfer learning for image classification." *UK Workshop on Computational Intelligence. Springer*, Cham, (2018).

[10] Srivastava, N.; Hinton, G.; Krizhevsky, A.; Sutskever, I.; Salakhutdinov, R. "Dropout: A Simple Way to Prevent Neural Networks from Overfitting". J. Mach. Learn. Res. (2014).

[11] D. Puri, R. Chudiwal, J. Rajput, S. Nalbalwar, A. Nandgaonkar and A. Wagh, "Detection of Alcoholism from EEG signals using Spectral and Tsallis Entropy with SVM," *International Conference on Communication information and Computing Technology (ICCICT)*, 2021, pp. 1-5, doi: 10.1109/ICCICT50803.2021.9510071.

Automated Digital Presentation Control using Hand Gesture Technique

Salonee Powar[1], Shweta Kadam[2], Sonali Malage[3], Priyanka Shingane[4]

[1] *Department of Information Technology, RAIT, D.Y. Patil Deemed to be University Nerul Email:* saloneesp16@gmail.com
[2] *Department of Information Technology, RAIT, D.Y. Patil Deemed to be University Nerul Email:* shwetakadam235@gmail.com
[3] *Department of Information Technology, RAIT, D.Y. Patil Deemed to be University Nerul Email:* sonalimalage3@gmail.com
[4] *Department of Information Technology, RAIT, D.Y. Patil Deemed to be University Nerul Email:* priyanka.shingane@rait.ac.in

Abstract- In today's digital world Presentation using a slideshow is an effective and attractive way that helps speakers to convey information and convince the audience. There are ways to control slides with devices like mouse, keyboard, or laser pointer, etc. The drawback is one should have previous knowledge about the devices in order to manage them. Gesture recognition has acquired importance a couple of years prior and are utilized to control applications like media players, robot control, gaming. The hand gesture recognition system builds the use of gloves, markers and so on However, the utilization of such gloves or markers expands the expense of the system. In this proposed system, Artificial intelligence-based hand gesture detection methodology is proposed. Users will be able to change the slides of the presentation in both forward and backward directions by just doing hand gestures. Use of hand gestures cause connection simple, helpful, and doesn't need any additional gadget. The suggested method is to help speakers for a productive presentation with natural improved communication with the computer. Specifically, the proposed system is more viable than utilizing a laser pointer since the hand is more apparent and thus can better grab the attention of the audience.

Keywords- OpenCV, Hand Gesture Recognition, Human Computer Interaction, Machine learning, Artificial Intelligence, Presentation Slides, Media Control, Virtual Pointer.

1. INTRODUCTION

In recent decades hand gesture recognition is considered as a new technique of Human-Computer Interaction because of its automatic, natural and easiness without requiring input from devices like keyboard and mouse. For example, detection of the language spoken can be done by analysing lip movements; gaming is also using hand gestures [1]. Now a day's though there are number of techniques for hand gesture recognition are in existence like wearable devices like a ring, armband, glove, leap motion, and controllers-based motion recognition such as Wii-mote, and ordinary web-camera, stereo camera, and even using radar but still needs improvement [2]. So, the gesture recognition has gained tons of importance and wont to control various applications. Presentation software is one among the various which will be controlled by hand gestures.

Machine captures gesture and recognizes it to perform the task. Machine will capture the hand gesture through camera and recognize it. First it will remove the background from captured image and filters-out foreground. This recognized gesture is then used for verifying the sign of the gestures. Aim of this proposed work is to implement AI in hand gesture recognition system and to use it to control the digital presentations by just using hand gestures.

2. LITERATURE SURVEY

As per the analysis of many other techniques referred by the researchers, the main aim is to help speakers for an effective presentation with natural improved interaction with computer.

Damiete O. Lawrence and Dr. Melanie J. Ashleigh author presented "Impact of Human-Computer Interaction (HCI) on Users in Higher Educational System: Southampton University as A Case Study". In this paper, Human-Computer Interaction (HCI) perception and impact in the University of Southampton UK, an advanced literacy terrain was measured. The impact HCI in Southampton University has been positive and it's shown that getting familiar with HCI generalities ameliorate a stoner's commerce and effectiveness. In conclusion, it can be said that HCI has impacted the literacy terrain as it has impacted other corresponding surroundings [4]. Sebastian Raschka, Joshua Patterson and Corey Nolet author presented "Machine Learning in Python: Main Developments and Technology Trends in Data Science, Machine Learning, and Artificial Intelligence". They covered extensively used libraries and generalities, collected together for holistic comparison, with the thing of educating the anthology and driving the field of Python machine learning forward [9]. Xuesong

Zhai, Xiaoyan Chu,Ching Sing Chai,Morris Siu Yung Jong, Andreja Istenic, Michael Spector, Jia-Bao Liu, Jing Yuan, and Yan Li author presented a Review of Artificial Intelligence (AI) in Education from 2010. This study handed a content analysis of studies aiming to expose how artificial intelligence (AI) has been applied to the education sector and explore the implicit exploration trends and challenges of AI in education [10].

Jadhav & Lobo proposed the idea were both static and dynamic gestures are used together to control power point presentation. To capture and recognize images Segmentation methodology is used. It also introduces the feature of motion detection to change slide [1]. Zhou Ren, Junsong Yuan, Jingjing Meng, and Zhengyou Zhang author presented "Robust Part-Based Hand Gesture Recognition Using Kinect Sensor". They presented a robust part- grounded hand gesture recognition system using the Kinect detector. A new distance metric, Finger-Earth Mover's Distance (FEMD), is used for diversity measure, which represents the hand shape as a hand with each cutlet part as a cluster and penalizes the empty cutlet- holes. More specifically, our FEMD grounded hand gesture recognition system achieves 93.2 mean delicacy and runs in 0.0750s per frame when using the thresholding corruption cutlet discovery system [6]. Harika et al. authors proposed and approached a method by using vision-based gesture recognition for computer-assisted slide presentation. Techniques like Kalman filter, HSL colour model, Skin colour sampling are used. If we consider accuracy of this proposed model, Skin colour detection has overall success about 72.4%, Single fingertip detection has accuracy of 74.0%, success rate in moving slides is 77% and Success rate in controlling the finger pointing is 80% [2]. Wahid et al. proposed and approached a method to Recognize Hand Gestures Using Machine-Learning Algorithms. If we consider accuracy of this proposed model, The SVM algorithm yielded the highest classification accuracies using both original EMG features (97.56%) and normalized EMG features (98.73%) among NB, RF, KNN and DA [3].

Ajay Talele, Aseem Patil, Bhushan Barse author presented "Detection of Real Time Objects Using TensorFlow and OpenCV". This paper introduced a cutting-edge computer imaginative and prescient-based totally impediment detection technique for cellular generation and its packages. Each character image pixel is classified as belonging either to an impediment based totally on its look. This paper presented a brand new approach for impediment detection with a single webcam digital camera [7]. Ahmed Kadem Hamed AlSaedi, Abbas H. Hassin Al Asadi author presented "A new hand gestures recognition system". They introduced a low cost system to fete the hand gesture in real- time. Generally, the system divided into five way, one to image accession, alternate to pre -processing the image, third for discovery and segmentation of hand region, four to features birth and five to count the figures of fritters and gestures recognition. The paper answered the challenge of gyration, exposure, scaling, and got the same results when used right or left hand. The system that handed used only bare hand and webcam of Laptop so it's veritably flexible for the stoner. The results of system shown that the rate of recognition was 96.6% and this result is considered veritably good compared with other exploration papers [8]. Dhall et al Discovered Hand gesture technology along with Convolutional neural network to build a hand gesture recognition application in "Automated Hand Gesture Recognition using a Deep Convolutional Neural Network model" paper. In this paper author used a CNN which has specific layers such as input layer, output layer and in-between them there are some hidden layers. First hidden layer is convolutional layer which detects and extracts features from images, then Max pooling layer used for dimensionality reduction [5].

3. METHODOLOGY

3.1 Image Pre-Processing

The point of pre-processing is to improve the standard of the image all together that we will examine it in an exceptionally better manner. By pre-processing we will smother undesired distortions and upgrade a few elements which are essential for the real application we are working for. Those features might vary for different applications.

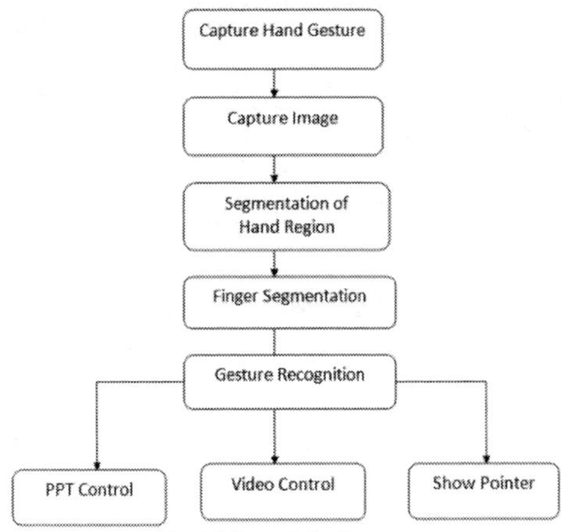

Fig 1. Flow of the System

Steps for Image pre-processing:

- Select a boundary of the input image within which we'll scan for the presence of a person's hand.
- Produce a mask by opting only pixels that match a specified colour range.
- Blur the mask image so that missing data points can be filled.
- Draw a contour of the hand and use Open CV to identify the fingers.

 https://doi.org/10.1051/itmconf/20224403031

3.2 Anaconda Framework

Anaconda is used for scientific computing like data science, AI applications, huge scope information handling, predictive analysis, and so forth), that expects to improve on package executives and management.

- PyCharm

PyCharm is a devoted Python Integrated Development Environment (IDE) executing an enormous scope of required devices for Python engineers, firmly incorporated to establish helping climate for useful Python, web, and data science improvement.

3.3 Libraries

1) OpenCV

OpenCV is used in computer vision, image processing, and machine learning applications. OpenCV guide a large diversity of programming languages like Python, C++, Java, etc. The identification of objects, faces, or even the handwriting of a human can be done by the action of images and videos. It's an open-source library that used to accomplished task like face detection, object tracking, landmark detection and many also.
It is used to capture the video and to perform hand detection process:

```
cap = cv2.VideoCapture(0)
```

2) PyAutogui

Mouse and keyboard can be controlled to do different things. It is a cross-stage GUI automation Python module for human reality. This third-party library can be installed by using the command pip install pyautogui.
It is used to press HotKeys:

```
pg.hotkey('escape')
```

3) MediaPipe

MediaPipe is a framework which is used for working in a machine learning pipeline. It is a high-fidelity hand and finger hunting solution. It is just a part of single frame that from employs machine learning (ML) to derive 3D marker of a hand.
It is used in hand detection process:

```
self.mpHands = mp.solutions.hands
self.hands = self.mpHands.Hands(self.mode, self.maxHands, self.detectionCon, self.trackCon)
self.mpDraw = mp.solutions.drawing_utils
```

4) Keyboard

It records the keyboard action and assist to enter keys thus can also block the keys until a stated key is entered and affect the keys. It captures onscreen keyboard events as it takes all the keys. This module provides groups hotkeys.

It is used to press particular keys.

```
keyboard.press_and_release('ctrl + 1')
```

5) Numpy

NumPy may be used to carry out an extensive variety of mathematical operations on arrays. It provides effective information systems to Python to assure treasured computations with arrays and matrices, and it offers a huge library of high-stage mathematical features to perform on those data structures like arrays and matrices.
It is used for Creating 3*3 kernel:

```
kernel = np.ones((3,3),np.uint8)
```

To define range of skin color in HSV:

```
lower_skin = np.array([0,20,70], dtype=np.uint8)
upper_skin = np.array([20,255,255], dtype=np.uint8)
```

To Convert the coordinates:

```
x3 = np.interp(x1, (frameR, wCam-frameR), (0, wScr))
y3 = np.interp(y1, (frameR, hCam-frameR), (0, hScr))
```

6) Time

The Python time module helps in representing time in code; representation can be in the form of objects, strings, and numbers. By using this module other functionality can be implemented like representing time and measuring the efficiency of your code.
It is used to set frame rate:

```
cTime = time.time()
fps = 1/(cTime-pTime)
pTime = cTime
```

7) Math

The Python Math Library implements some math features and constants in Python. These features can be used in code to perform plenty complicated mathematical computations. The library does not require any installation as it is an essential Python module.

It is used to find length of all sides of triangle:

```
a = math.sqrt((end[0] - start[0])**2 + (end[1] - start[1])**2)
b = math.sqrt((far[0] - start[0])**2 + (far[1] - start[1])**2)
c = math.sqrt((end[0] - far[0])**2 + (end[1] - far[1])**2)
s = (a+b+c)/2
ar = math.sqrt(s*(s-a)*(s-b)*(s-c))
```

To apply cosine rule:

```
angle = math.acos((b**2 + c**2 - a**2)/(2*b*c)) * 57
```

3.4 Hardware Requirements

- Web cam: Real time streaming of image or video to or through network can be done by using video camera. In this proposed system first, we will capture the images of hand gesture that user made in parallel to webcam.
- Processor: intel Pentium 4 or more

 https://doi.org/10.1051/itmconf/20224403031

- RAM: 1GB or more
- Hard disk: min 20GB

4. RESULT

4.1 Gestures

1) Ok (Thumbs Up): By doing Ok gesture presentation will start in presentation mode (Ref. Fig.2).

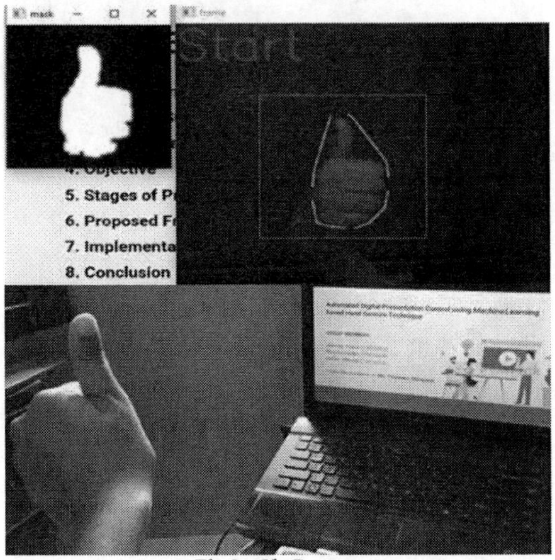

Fig 2. Ok Gesture

2) Two Fingers (Victory): By doing two finger gesture video will be played or paused in presentation slide (Ref. Fig.3).

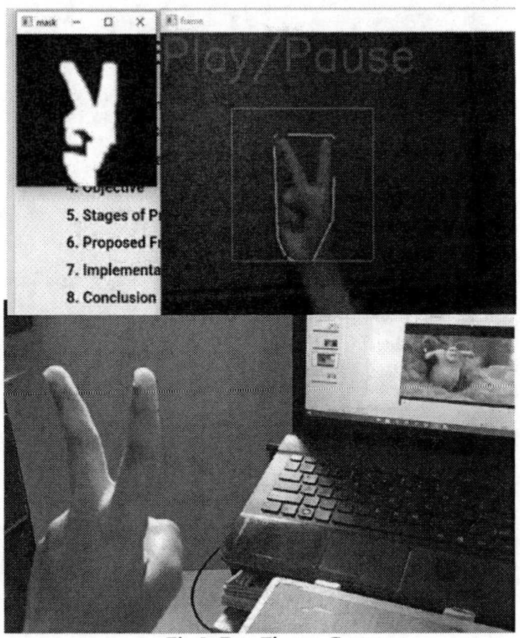

Fig 3. Two Fingers Gesture

3) Good: By doing Good gesture presentation will show previous slide. User can change slide backward (Ref. Fig.4).

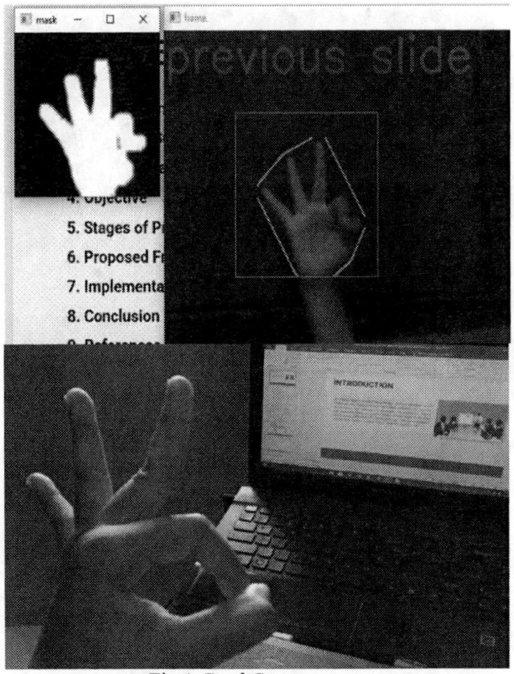

Fig 4. Good Gesture

4) Three Fingers: By doing Three Fingers gesture Next slide will be presented (Ref. Fig.5).

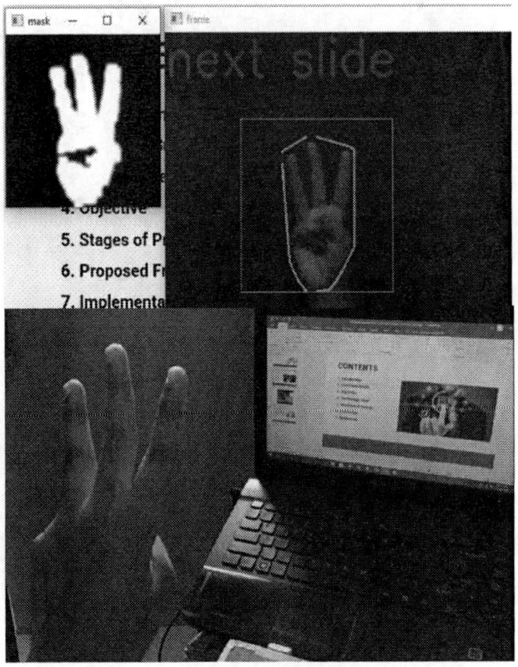

Fig 5. Three Fingers Gesture

 https://doi.org/10.1051/itmconf/20224403031

5) Four Fingers: By doing Four fingers gesture video's volume will be decreased (Ref. Fig.6).

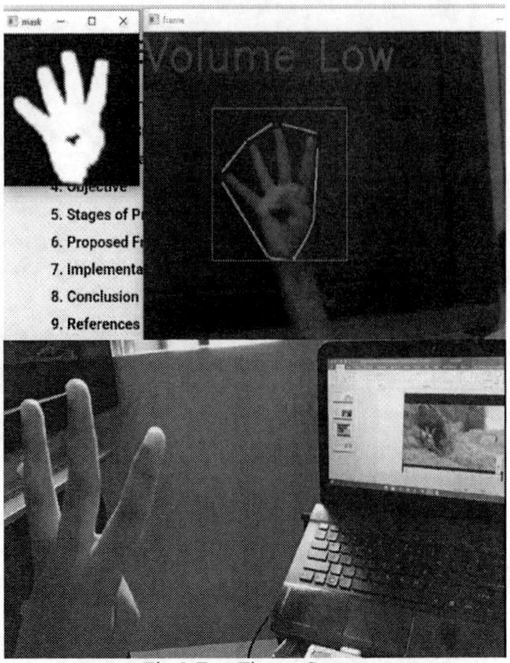

Fig 6. Four Fingers Gesture

6) Five Fingers (Palm): By doing Five fingers gesture video's volume will be increased (Ref. Fig.7).

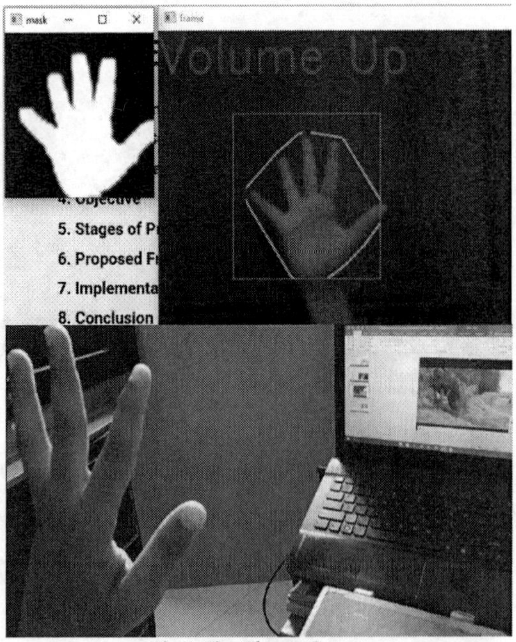

Fig 7. Five Fingers Gesture

7) One Finger: By doing One finger gesture user can point some Information using Pointer (Ref. Fig.8).

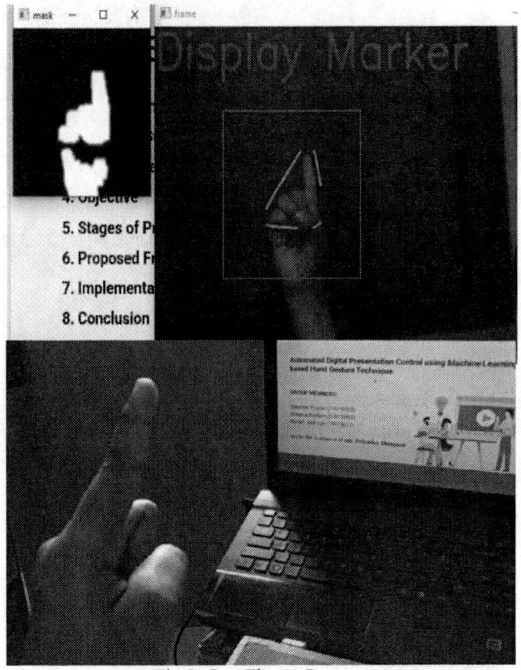

Fig 8. One Finger Gesture

8) Fist: By doing Fist gesture presentation will exit from presentation mode (Ref. Fig.9).

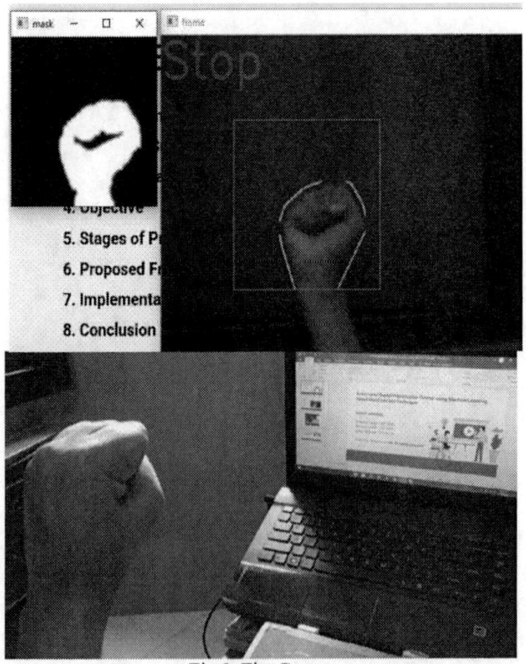

Fig 9. Fist Gesture

5. DISCUSSION

Hand gestures are more natural in interaction as they are important part of body language, compared with other devices. Use of hand gestures does not require any extra device and makes interaction easy. In this proposed system, AI-based hand gesture detection methodology is proposed. Using hand gestures would make it easy for speaker to present more easily. The aim of this proposed system is to develop software that will help presenter to control the slides of presentation by using different hand gestures. With this software there will be no need for any device like keyboard, mouse or even remote for changing slides.

6. CONCLUSION

This proposed system, "Automated Digital Presentation Control Using Hand Gesture Technique" makes presentation easy. Presenter will be able to change slides without using any external device. This will be useful in corporate or institutions where presentation is part of work.

ACKNOWLEDGMENT

I would wish to offer my thanks to every individual who upheld me all through this proposed framework. I'm appreciative for their trying direction, pricelessly developmental audit and cordial exhortation during the proposed framework. I will always be thankful to you during this regard.

REFERENCES

[1] D. Jadhav, Prof. L.M.R.J. Lobo, *Hand Gesture Recognition System To Control Slide Show Navigation* IJAIEM, Vol. **3**, No. 4 (2014)

[2] M. Harika, A. Setijadi P, H. Hindersah, *Finger-Pointing Gesture Analysis for Slide Presentation*, Bong-Kee Sin Journal Of Korea Multimedia Society, Vol. **19**, No. 8, August (2016)

[3] Md.F. Wahid, R. Tafreshi, M. Al-Sowaidi, R. Langari, *An Efficient Approach to Recognize Hand Gestures Using Machine-Learning Algorithms*, IEEE 4th MECBME, (2018)

[4] D.O. Lawrence, and Dr. M.J. Ashleigh, *Impact Of Human-Computer Interaction (Hci) on Users in Higher Educational System: Southampton University As A Case Study*, Vol.**6**, No 3, pp. 1-12, September (2019)

[5] I. Dhall, S. Vashisth, G. Aggarwal, *Automated Hand Gesture Recognition using a Deep Convolutional Neural Network,* 10th International Conference on Cloud Computing, Data Science & Engineering (Confluence), (2020)

[6] Ren, Zhou, et al. *Robust part-based hand gesture recognition using kinect sensor*, IEEE transactions on multimedia 15.5, pp.1110-1120, (2013)

[7] Ajay Talele, Aseem Patil, Bhushan Barse on the *Detection of Real Time Objects Using TensorFlow and OpenCV*, Asian Journal of Convergence in Technology, Vol **5**, (2019)

[8] Ahmed Kadem Hamed AlSaedi, Abbas H. Hassin Al Asadi, *A New Hand Gestures Recognition System*, Indonesian Journal of Electrical Engineering and Computer Science, Vol **18**, (2020)

[9] Sebastian Raschka, Joshua Patterson and Corey Nolet, *Machine Learning in Python: Main Developments and Technology Trends in Data Science, Machine Learning, and Artificial Intelligence*, (2020)

[10] Xuesong Zhai, Xiaoyan Chu, Ching Sing chai, Morris Siu Yung Jong, Andreja Istenic, Michael Spector, Jia-Bao Liu, Jing Yuan, Yan Li, *A Review of Artificial Intelligence (AI) in Education from 2010 to 2020*, (2021)

A Decision-Making Tool for Creating and Identifying Face Sketches.

Manish Bhoir[1,*],*Chandan Gosavi*[1,**], Prathamesh Gade[1,***], *Bhavana Alte*[1,****]

[1]Dept. of Computer Engineering Ramrao Adik Institute of Technology , D Y Patil Deemed to be University, Navi Mumbai ,Nerul,India.

Abstract: A criminal can be quickly identified and prosecuted using a face sketch based on an eyewitness description . Several applications for converting hand-drawn face drawings and using them to automatically identify and recognize the suspect from the police database have been proposed in the past, but the existing system dealt with some drawbacks. It featured a lot of flaws, including as a limited facial features kit and a cartoonish feel to the constructed suspect face, which made it much more difficult to use these apps and acquire the results and efficiency that were required. In this paper, we present a stand- alone tool that allows users to create composite face sketches of suspects without the need for forensic artists. The application offers a drag-and-drop feature that can match the produced composite facial sketch with the criminal database in real time. This can be done considerably more rapidly and efficiently using deep learning and cloud infrastructure.

Keywords: Face Sketch, Face Sketch Construction, Face Recognition, Criminal Identification, Deep Learning.

I. Introduction

A sketch is a design that is created fast and can be made with limited information. Sketches are widely used by artists as a preliminary step in the creation of more complex paintings or drawings. In law enforcement, sketching techniques are frequently employed to determine suspects from a memory by a witness. A face sketch drawn based on the description provided by an eyewitness can easily identify and bring a criminal to justice; however, in today's world of modernization, the traditional method of hand drawing a sketch is not found to be as effective or time saving. They are primarily manual, painstaking, and may not lead to the correct identification of the guilty party.

When the sketch is complete, it is compared to images in the department's possession in order to identify the culprit. Only if a person has been convicted at least once, then their photo is made public. This recognition was previously done manually, which was a time-consuming process. Then came the automatic recognition techniques. Many automatic recognition algorithms have been established to date, and the work listed has attempted to improve these techniques.

*e-mail:manishbhoir50@gmail.com

**e-mail:chandangosavi14@gmail.com

***e-mail: prathameshg.2210@gmail.com

****e-mail:bhavana.alte@rait.ac.in

Uhl et al. [11] began work on automatic face sketch recognition system in 1994, proposing a method based on Principle Component Analysis (PCA).After that Tang and Wang[16] did the majority of the work in this field. An automatic method for recognizing photos from a database using a sketch had been introduced, in which a sketch was not directly matched with a photo due to the significant differences in shape and texture between a sketch and a real photo. As a result, the database was converted from photos to sketches using the Eigenface method, which used the Karhunen-Loeve Transform. It also generated feature vectors for the database sketches and the test image. The sketch and face feature vectors were then matched. Dataset used divided the face sketch into patches giving unsatisfactory results.

Patil and Shubhangi et al[6]created a matching system using a Geometrical Face Model. The AdaBoost algorithm was used to identify the face region, and then the geometrical structure of the face was used to mark the main facial components such as the eyes, nose, and mouth. Then, using Weber Local Descriptor, a texture feature was extracted from each facial component (WLD). For classification, an artificial neural network (ANN) was used. This system was incredibly complex to develop and costes a lot of money to run. Liu, Bae et al.[9] developed an automatic face- sketch recognition system that used joint dictionary learning to match photos from two different modalities (photos and sketches). Dalal, Vishwakarma et al[2] proposed a feature- based matching technique to accomplish this. A feature vector containing the features of interest is used in feature-based matching. The facial image had the characteristics of a histogram of oriented gradients (HoG) and a grey level co- occurrence matrix (GLCM) (whether it is a drawing or a photo). The use of hand-drawn facial sketches was a limitation, as was the time commitment.

The foregoing applications and needs prompted us to consider developing an application that would allow users to upload hand-drawn individual features to the platform, which would then be converted into the application's component set, rather than simply selecting a set of individual features such as eyes, ears, mouth, and so on to create a face sketch. As a result, the produced sketch would resemble a hand-drawn sketch much more closely, making it much easier for law enforcement authorities.

Our platform would also allow the law enforcement team to upload a previous hand-drawn sketch in order to leverage the platform's considerably more efficient deep learning algorithm and cloud infrastructure to identify and recognize the suspect.

The machine learning system would learn from the sketches and the database to recommend to the user all of the comparable facial traits that may be employed with a single selected feature, reducing the platform's timeframe and increasing its efficiency.

The remainder of this paper is organized as follows: Literature survey with reference to our topic is given in section II. Section III provides proposed work. Result & analysis are described in section IV. Section V gives idea of future work.

II. Literature Survey

Face sketch construction and recognition have been studied extensively using a variety of methods. Yasmeen Bashir, Kamran Nawaz, and Anna, along with Dr. Charlie Frowd, Petkovic[8] created a stand-alone programme for building models. The initial system was found to be time consuming and confusing, so a new approach was adopted in which the victim was given a list of faces to choose from and was instructed to select similar faces that resembled the suspect, with the system then combining all of the selected faces and attempting to predict the criminal's facial composite automatically. The results were promising, with 10 out of 12 composite faces correctly named, with 21.3 percent when the witness was assisted by a department employee in constructing the faces and 17.1 percent when the witness attempted to construct the faces on their own. The facial created had limited scope of increasing accuracy and the platform was time consuming as well as confusing.

Fang, Yuke& Deng et al [1] first developed an Identity-Aware CycleGAN (IACycleGAN) model that supervises the image generating network using a new perceptual loss. CycleGAN's photo-sketch synthesis was improved by paying closer attention to the synthesis of crucial facial parts like the eyes and nose, both of which are vital for identifying oneself. They also developed a mutual optimization approach between the synthesis model and the optimization model. IACycle- GAN iteratively synthesises better images, and the triplet loss of the produced images increases the recognition model. With real-life examples Using the widely used CUFS and CUFSF formats, extensive tests were conducted on both photo-to-sketch and sketch-to-photo tasks.This method was extremely difficult to implement and had a significant computational cost.

SahilDalal, Vishwakarma et al[2] introduced a feature-based matching mechanism. A feature vector containing features of interest is used in feature-based matching. In terms of histogram of oriented gradients (HoG) features and grey level co-occurrence matrix (GLCM) features, the facial picture (whether it is a drawing or a photo) The initial step was to compute the characteristics, which enhances the likelihood of correct matches. It can be depicted in a variety of ways. The results of the experiment show that the dataset used to divide the face sketch into patches led to a decline in accuracy.

C.Galea&Farrugia et al [3] used prominent and state-of-the-art algorithms to thoroughly evaluate using publicly available datasets, acting as a standard for future algorithms. The proposed framework had been proved to minimise costs when compared to a leading technique. The mistake rate for seen sketches is reduced by 80.7 percent, and the mean is lower. For real-world forensic sketches, retrieval rank increased by 32.5 percent.The use of hand-drawn facial sketches was a limitation, as was the time commitment.

Wan ,Lee ,Jong et al [4] The original training images and sketches were concatenated with high-pass filtered image patches of the training sketches to create a joint training model. The usage of hand-drawn facial sketches and the length of time required were both limitations.

Sharma & Bhatt et al [5] employed an investigation to automatically detect the person's face. PCA was used to extract crucial information from the input images. Later, linear discriminant analysis, multilayer perceptron, naive Bayes, and support vector machine were used to test it. This method was extremely difficult to implement and came at a high cost in terms of computation.

Another method proposed by Anil K Jain and Brendan Klare [10] was sketch to picture matching, which used SIFT Descriptor and provided results based on the measured SIFT Descriptor distance between the face photos in the database and the sketches. The algorithm first changes the face photos using a linear transformation based on Tang and Wang's proposed model, and then uses the sketch to measure the SIFT descriptor distance compared to the face photo, as well as distance between images in databases for higher accuracy. The experimental results reveal that the datasets used in the experiment were quite comparable to those used by Tang, and that the inclusion in the algorithm was the measurement of the descriptor, which provided a superior result and accuracy than the model offered by Tang and Wang.

Zhang et al.[17] In 2019, presented an automatic and effective photo to sketch synthesising method based on Dualtransfer, which includes both interdomain and intradomain transfer. This method is extremely difficult to implement and has a high computational cost.

The common issue with all of the proposed algorithms was that they compared the face sketches with human faces that were usually front facing, making it easier to map both in drawn sketch and human face photograph. However, when a photograph or sketch was collected with their faces in different directions, the algorithms were less likely to map and match with a front facing face from the database.

There have even been systems proposed for composite face construction, but most systems used facial features that were taken from photographs, then selected by the operator as described by the witness, and finally compiled to form a single human face, making it much more difficult for humans as well as any algorithm to match it with a criminal face because each facial feature was taken from a separate face photograph with various dissimilarities, making it much more complicated for humans as well as any algorithm to match it with a criminal face.

III.Proposed work

In this paper's proposed work, we design and develop our platform in two stages:

https://doi.org/10.1051/itmconf/20224403032

(a) Face Sketch Construction
(b) Face Sketch Recognition

A. Face Sketch Construction:

In this module, an accurate composite face sketch can be created utilising predefined facial feature sets as tools, which can be resized and rearranged as needed/described by the eye-witness.The human face is divided into numerous facial parts such as the head, eyes, eyebrows, lips, nose, ears, and so on are available for usage (as shown in Figure 2).

The flowchart in Figure 1 depicts the users' journey through the platform in order for the platform to construct an accurate face sketch based on the description. The dashboard is designed to be simple in order to encourage no professional training before using the platform, thus saving time and resources for the Department.

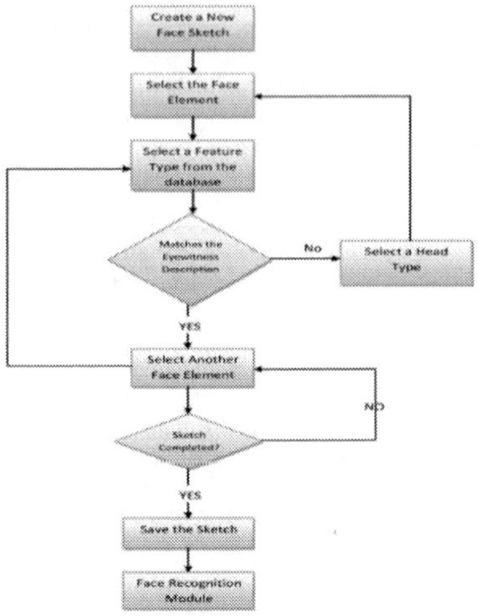

Figure 1: Flow Chart for Creating a sketch in the Application

The dashboard is made up of five primary modules. The first and most essential is the Canvas, which is located in the centre of the screen and houses the face sketch components and features that aid in the creation of the face sketch.

Creating a face sketch would be difficult if all of the face elements were given all at once and in an unorganised fashion, making the process difficult for the user and making it difficult to generate an accurate face, which would be contrary to the proposed system's objective. To solve this problem, we decided to order the face elements according to the face category they belong to, such as head, nose, hair, eyes, and so on, making it much easier for the user to interact with the platform and build the face sketch as shown in Figure 2. This is featured in the left-hand column on Canvas on the dashboard, where clicking on a face category gives the user access to a variety of other face structures.

When it comes to the various face elements in a particular face category, we could have multiple and n number of elements for a single category, so to solve this, our platform will use machine learning in the future to predict similar face elements or predict and suggest the elements to be selected in the face sketch, but this will only work once we have appropriate data to train the model on this algorithm and work to improve the platform.

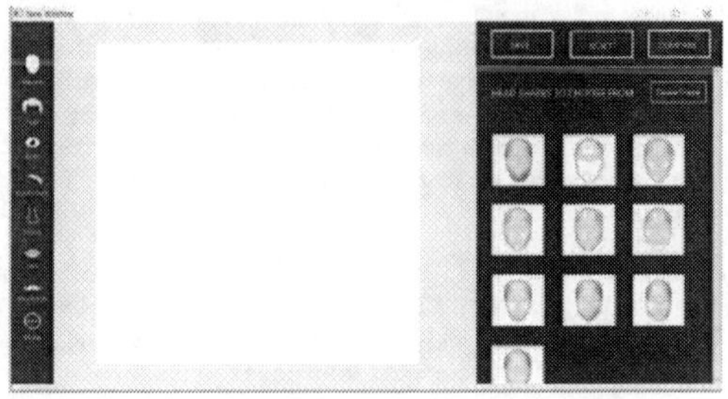

Figure 2: User Interface of Application (blank canvas)

When a user selects a face category, a new module appears to the right of the canvas, allowing the user to choose an element from a list of face elements to build a face sketch. Based on the eye witness's description, this option can be chosen.

When the elements are selected, they appear on the canvas and can be moved and placed according to the eye witness's description to create a more accurate sketch(as shown in Figure 3).However, the elements have a fixed location and order to be placed on the canvas, such as the eye elements being placed above the head element regardless of the order they were selected. The same goes for each facial element.

The final module has options to improve the dashboard's usability. For example, if a user selects an element that should not be picked, this may be corrected by utilising the option to erase that element, which is visible by selecting the face category from the left panel. The most essential buttons are in the right-hand panel, which also features a button to entirely delete anything on the dashboard's canvas, leaving it blank.

Then there's a button to save the completed face drawing as a PNG file for easier access in the future. Depending on the Law Enforcement Department, this could be any position on the host computer or on the server.

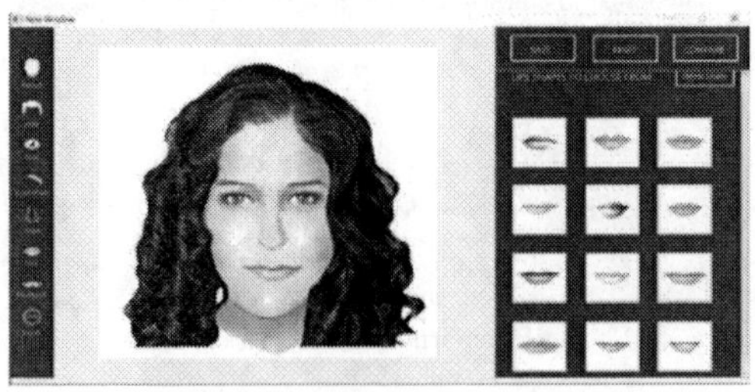

Figure 3: User Interface of Application (with facial features)

B. Face Sketch Recognition:

Basically try to recognise the face if it exists in the database, and if it does, we display it together with its meta data.

The flowchart shown in Figure 4 depicts the user flow as it is followed by the platform to provide an accurate face sketch based on the description. The dashboard is designed to be simple in order to encourage no professional training to be completed prior to using this platform, thus saving time and resources for the Department.

 https://doi.org/10.1051/itmconf/20224403032

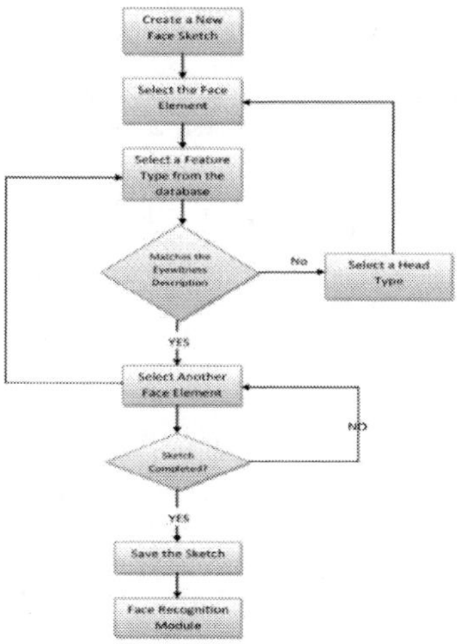

Figure 4:Flow Chart for Recognising a sketch in the Application

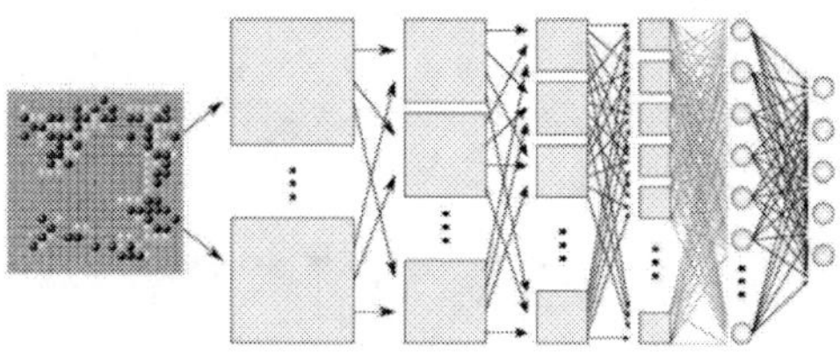

Figure 5: Figure extraction by the platform

The above Figure 5 shows that the first step before using the platform to detect faces is to train and make the platforms algorithm recognise and assign ids to the face photos of the users in the existing data with the law enforcement department. The platform's algorithms link to the records and break down each face photo into smaller features, assigning an ID to each of the numerous features created for a single face shot.

Now, the Module is being executedwhere the user first opens either the hand drawn sketch or the face sketch constructed on our platform saved in the host machine, after which the opened face sketch is being uploaded to the Law enforcements server housing the recognition module so that the process or the data of the record are not tampered with and are secure and accurate.

Once the sketch is sent to the server, the algorithm traces the picture to learn the features in the sketch and maps them as indicated in the Figure 6 below in order to match them with the features of the face photos in the database.

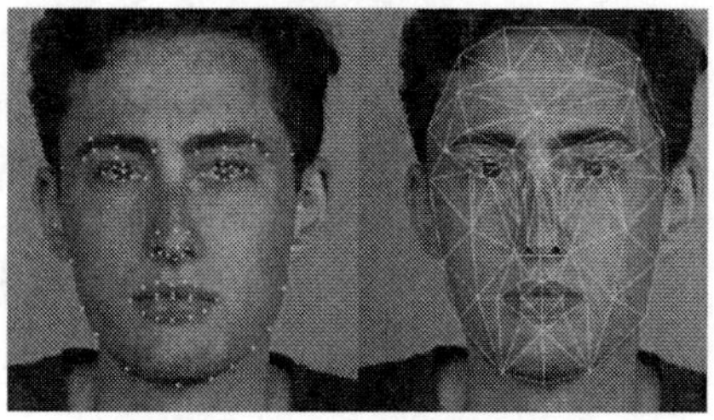

Figure 6: Face sketch been mapped on the platform

The platform presents the matched face along with the similarity percentage and other details of the person (shown in Figure 7)from the records after mapping the sketch and matching the face sketch with the records and finding a match. The platform that displays all of this, as well as the matched person, is depicted in the diagram below.

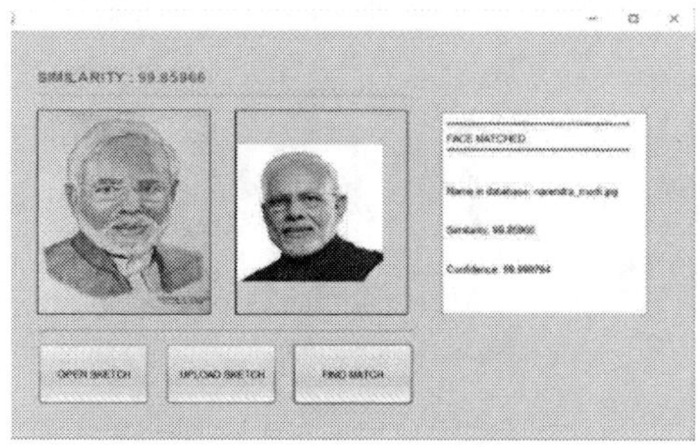

Figure 7: Face sketch mapped to database record

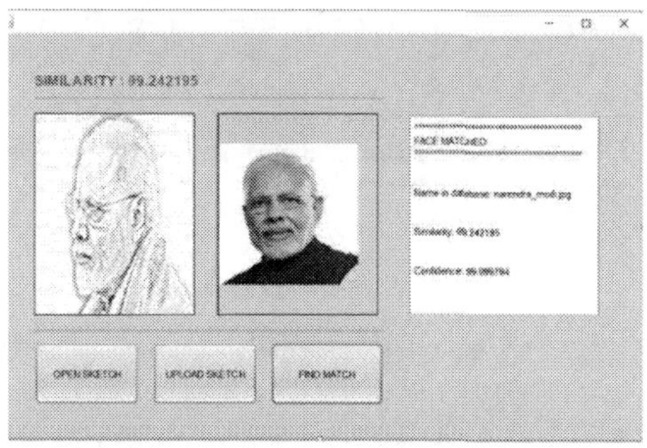

Figure 8: Face sketch mapped to database record (with side facing sketch)

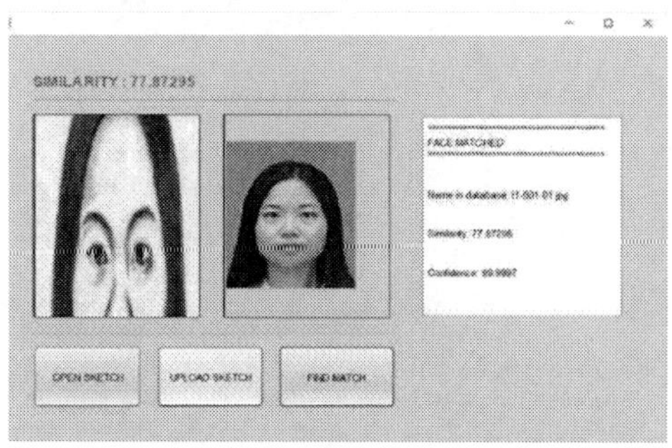

Figure 9: Face sketch mapped to database record (with only single feature)

IV. Results and Conclusion

From the first splash screen to the final screen to harvest data from the records, the project 'Face Sketch Construction and Recognition' will be created, built, and tested with real-world scenarios in mind. Security and accuracy will be paramount in every scenario. When compared to previous research in this sector, the platform even contains elements that are different and unique, boosting overall security and accuracy by standing out among other related studies and proposed systems in this field.

Table 1: Comparison of accuracy with other methods

Method	Accuracy (%)
Proposed Method	94.6
HOG + GLCM Descriptor[2]	94.4
PCA/LDA + Geometric Information[6]	94
Generation using CNN[7]	93.8

Figure 10: Viewed sketch and its corresponding photograph

The experiment was conducted using sketches that were previously seen. The dataset is downloaded and tested from an open store (available for free on the web) where the sketches are viewed. The observed sketches were acquired as a cluster of sketchphotographic sets from the CUHK face sketch database, which contained 188 sets. As a result, there are 188 sets of viewed sketches (as shown in Figure 7) available for testing. For image matching, the sketches are accompanied by photographic sets of the corresponding sketches.

When tested with various test cases, test scenarios, and data sets,(as shown in Figure 7,8,9) the platform showed good accuracy and speed during the face sketch construction and recognition process, providing an average similarity of more than 90%, with system giving accuracy of 94.6% and a confidence level of 100%, which is a very good rate according to related studies in this field.

V. Future Work

The 'Face Sketch Construction and Recognition' project is currently meant to work on a limited number of scenarios, such as face sketches and matching those sketches to face images in law enforcement records.

The platform can be greatly extended in the future to work with a variety of technologies and scenarios, allowing it to investigate numerous media and surveillance mediums and produce a far wider range of outputs. Using 3D mapping and imaging techniques, the platform can be adjusted to match the Face sketch with human faces from video feeds, and the same may be applied to CCTV surveillance to conduct face recognition on live CCTV footage using the Face Sketch.

The platform can also be connected to social media, as social media platforms are a rich source of data in today's world. Connecting this platform to social media would improve the platform's ability to find a much more accurate match for the face sketch, making the process much more accurate and speeding up the process.

When compared to relevant research in this sector, the platform could include features that are different and unique, as well as easy to upgrade, boosting overall security and accuracy by standing out among other related studies and proposed systems in this field.

References

[1]Fang, Yuke& Deng, Weihong& Du, Junping& Hu, Jiani. (2020). Identity-Aware CycleGAN for Face Photo- Sketch Synthesis and Recognition. Pattern Recognition. 102. 107249. 10.1016/j.patcog.2020.107249.

[2] SahilDalal, Virendra P. Vishwakarma, Sanchit Kumar, Feature-based Sketch-Photo Matching for Face Recognition, Procedia Computer Science, Volume 167, 2020, Pages 562-570, ISSN 1877-0509.

[3] C. Galea and R. A. Farrugia, "Forensic Face Photo- Sketch Recognition Using a Deep Learning-Based Architecture," in *IEEE Signal Processing Letters*, vol. 24, no, 11, pp.1586-1590, Nov. 2017, doi:10.1109/LSP.2017.2749266

[4]Wan, Weiguo& Lee, Hyo Jong. (2019). A Joint Training Model for Face Sketch Synthesis. Applied Sciences. 9. 1731. 10.3390/app9091731.

[5]S. Sharma, M. Bhatt and P. Sharma, "Face Recognition System Using Machine Learning Algorithm," *2020 5th International Conference on Communication and Electronics Systems(ICCES)*, 2020, pp. 1162-1168, doi: 10.1109/ICCES48766.2020.9137850.

[6]Patil S, Shubhangi DC (2017) Forensic sketch based face recognition using geometrical face model. In: 2017 2nd InternationalConference for Convergence in Technology (I2CT). pp 450–456

[7]Karasolak, Mustafa &Choupani, Roya. (2019). Face Photograph Recognition via Generation from Sketches using Convolutional Neural Networks. International Journal of Multimedia and Image Processing. 9. 459-465. 10.20533/ijmip.2042.4647.2019.0057

[8]Charlie Frowd, Anna Petkovic, Kamran Nawaz and Yasmeen Bashir, "Automating the Processes Involved in Facial Composite Production and Identification" Symposium on Bio-inspired Learning and Intelligent Systems for Security, 2009.

[9]Liu J, Bae S, Park H, et al (2015) Face photo-sketch recognition based on joint dictionary learning. In: 2015 14th IAPR International Conference on Machine Vision Applications (MVA). pp 77–80

[10]B. Klare and A. Jain, "Sketch to photo matching: a featurebased approach", SPIE Conference on Biometric Technology for Human Identification, 2010.

[11]Uhl RG, da Vitoria Lobo N, Kwon YH (1994) Recognizing a facial image from a police sketch. In: Proceedings of 1994 IEEE Workshop on Applications of Computer Vision. pp 129–137.

[12] D. Liu, C. Peng, N. Wang, J. Li and X. Gao, "Composite face sketch recognition based on components," *2019 8th International Conference on Wireless Communications & Signal Processing (WCSP)*, 2019, pp. 1-5, doi: 10.1109/WCSP.2016.7752734.

[13]Bin Sheng, Ping Li, ChenhaoGao, Kwan-Liu Ma, "Deep Neural Representation Guided Face Sketch Synthesis", IEEE Trans. Vis. Comput. Graph., vol. 25, no.12, pp. 3216-3230, Dec.2019.

[14]N. Wang, X. Gao, L. Sun, and J. Li, "Bayesian face sketch synthesis," IEEE Trans. Image Process., vol. 26, no. 3, pp. 1264–1274, Mar.2017.

[15]HamedKianiGaloogahi and Terence Sim, "Face Sketch Recognition By Local Radon Binary Pattern: LRBP", 19th IEEE International Conference on ImageProcessing, 2012.

[16]Wang X, Tang X (2009) Face photo-sketch synthesis and recognition. IEEE Trans Pattern Anal Mach Intell 31:1955–1967.

[17]Zhang M, Wang R, Gao X, et al (2019) Dual-transfer face sketch--photo synthesis. IEEE Trans Image Process 28:642–657

Study and Analysis of Stock Market Prediction Techniques

Siddhesh Kokare[1], *Anvit* Kamble[2], *Shubham* Kurade[3] *and Deepali* Patil[4]

[1]Department of Information Technology, Ramrao Adik Institute of Technology, Navi Mumbai, India siddheshkoare512@gmil.com

[2]Department of Information Technology, Ramrao Adik Institute of Technology, Navi Mumbai, India anvitkamble@gmail.com

[3]Department of Information Technology, Ramrao Adik Institute of Technology, Navi Mumbai, India shubhamkurade21@gmail.com

[4]Department of Information Technology, RAIT,D.Y. Patil Deemed to be University, Navi Mumbai, India deepali.patil@rait.ac.in

Abstract. Stock marketplace is a complicated and demanding system in which people make more money or lose their entire savings. The stock market prediction having high accuracy yields more profit for stock investors. Stock market data is generated in a very large amount and it varies quickly every second. The decision making in stock marketplace is a very challenging and strenuous task of financial stock market. The development of efficient models for prediction decisions is very difficult because of the convolution of stock market financial data and should have high accuracy. This study attempts to compare existing models for the stock market. Various Machine learning methods like Long Short Term Memory (LSTM), Convolution Neural Networks (CNN) and Convolution Neural Networks – Long Term Short Memory (CNN-LSTM) have been used for the comparison. The models are estimated using conventional strategic measure: MAE (Mean Absolute Error). The measured low values indicates that the models are effective in predicting stock prices.
Keywords: Stock Market Prediction, LSTM, CNN, CNN-LSTM.

1. Introduction

Stock market forecasts are very crucial as it is utilized by many business people and general public. People will both benefit cash or lose their whole existence financial savings in stock market. Building precise models is tough as it relies upon more than one element along with news, social media information, fundamentals, manufacturing of the company, authorities bonds and country' s economics. Prediction model which consists only one factor might not be precise. There are several theories regarding stock market that have been conceptualizing over the years. Those theories try to expalin whether the market can be beaten or try to explain stock market nature. The market price of the stock integrates all the information about that stock in that particular time frame. The changing tendency of the stock prices has been regularly recognized as a total hassle inside the financial field [1]. Stock prices are suffering from diverse inner and outside factors, which includes home and overseas financial environment, global situation, enterprise prospect, economic information of indexed companies, and stock market operation [2, 3].
The conventional analysis is primarily based on finance and economics which uses basic and technical analytical methods. Firstly, the fundamental analysis focuses on the inherent stock values and qualitatively analyses the external factors like interest rates, exchanged rates, inflation, industrial policy, listed companies finances and international relations, etc. which affects the stocks. Secondly, the technical analysis mostly focuses on the stock price direction, trading volumes and psychological expectation of investors that focuses mostly on stock market by using tools like k-line chart or by analyzing individual stock directions in the stock indexes. The above methods are stilled the most generally used methods for many companies and investors [4,5].
Traditional fundamental analysis accuracy is tough to be convincing because the prediction results are highly dependent on the professional quality of the analytics and the influential factors are in a long term cycle. The stock data have the traits of random walk in a financial time series. The accuracy of usage of only time series model is questioned due to unknown and high noise characteristics of the financial time series [6].
There are certain limitations on clearly predicting stock price trends when using linear time series forecast version or the neural network version. Currently, by combining the benefits of different methods to enhance the hybrid approach is now an economic improvement trend for time series deep learning [7].
Therefore, in order to make the better use of the time series, thorough investigation of the characteristics of the records and accuracy improvement of the stock price forecasting can be done. This paper compares stock prices forecasting approaches based primarily on CNN, LSTM and CNN-LSTM.

2. Literature Survey

The financial marketplace is noisy, non-parametric dynamic and there are mainly two types of forecasting techniques: Technical analysis technique and machine learning techniques [8]. The conventional econometric techniques or equations with parameters aren't appropriate for studying complicated large dimensional and noisy financial data. In the paper proposed by Aparna et al. [9], consideration of various parameters of various datasets as done, it was observed that Decision Boosted Tree was performing better when compared to SVM and logistic regression. The paper proposed by Vijh et al. [10], dataset of five companies from 2009-2019 having new parameters for better prediction, such as High - Low , Open-Close, 7 day average stock price, 14 day average stock price, 21 days average stock price , last 7 days standard deviation was used. Comparative analysis based on RMSE, MAPE and MBE results clearly shows that ANN provides better stock prediction when compared to RF. The paper proposed by Hyeong et al.[11], the model performance was validated on both different time periods with several metrics like MSE, MAE and RMSE. By analyzing the testing results it was observed that Arima-Lstm hybid performs far better when compared to other financial models. The paper proposed by Pushpendu et al.[12] it mainly focuses on application of Random Forest and LSTM to predict stock prices directional movements. It was observed that the LSTM outperforms random forests. The paper proposed by Mehtabhorn et al. [13] it basically compares the various types of machine learning techniques and algorithm which is used in finance and stock market prediction. The paper proposed by Wenjie Lu et al. [14], The CNN-LSTM model is used to predict the closing price of a stock price the next day. Experimental results show that CNN-LSTM have highest accuracy and best performance compared to CNN, RNN, LSTM< MLP and CNN_RNN. The paper proposed by Nusrat Rouf et al. [15] comparisons of ANN, SVM, NB and DNN was carried out. SVM was the most popular technique used for SMP. It was observed that ANN and DNN performs more accurate and provides faster prediction. The paper proposed by Jingyi Shen et al. [16] used a comprehensive deep learning system. Prediction was carried on the datasets of Chinese stock market using the LSTM models. It was observed that the LSTM model achieved high prediction accuracy and outperformed the major models. The paper proposed by D. Wei et al. [17], the prediction was performed on the datasets using various LSTM models. It was observed that Vanilla LSTM, Stacked LSTM and Bidirectional LSTM are the commonly used LSTM models. BI-LSTM was having greater accuracy and low error when compared to other models. The Paper proposed by Sheng Chen and Hongxiang He et al. [18], CNN model was used for making Stock prediction which was perform using conv 1d function to process 1d data in convolution layer. It was observed that if source data is sequential then the model is efficient and can even be used to make predictions. The paper proposed by Wu et al. [19] with leading indicators prediction was performed on dataset using hybrid CNN-LSTM model. It was observed that CNN-LSTM model was achieving greater accuracy when compared with CNN and LSTM models. The paper proposed by Xuan Ji et al. [20], MAE, RMSE and R-square values are calculated to evaluate the performance of various prediction models. It was observed that CNN-LSTM model out performs well when applied on various stock prices. The paper proposed by Vanukuru, Kranthi et al. [21], the SVM model was used for predicting the stock index movements. It was observed that model generates higher profit as compared to selected benchmarks. The paper proposed by A M Pranav et al. [22], the sentimental analysis was performed on stock prices to forecast stock price variations. It was observed that machine learning models were performing well on various datasets

3. Comparisons of Existing Machine Learning Models

3.1 Convolution Neural Network (CNN):

Sheng Chen [16], proposed a CNN model for creating stock prediction that use the conv1d function to process the 1D data in the convolution layer. CNN is a feedforward neural network that performs very well in image processing and natural language processing. If implemented correctly, it can even predict forecasting of the time series. The local perception and weight distribution of the CNN can significantly reduce parameter range thereby improving the performance of model learning.

The CNN as shown in Figure 1, particularly consists of a convolution layer as well as the pooling layer. Each convolution layer consist of various convolution kernel and its formula is shown in equation (1).

$$o_v = \tanh(v_i * k_w + v_b) \qquad (1)$$

where o_v is the output value after convolution, tanh is the activation function, v_i is the input vector, k_w is the convolution kernel weight, and v_b is the convolution kernel bias.

The CNN model extracts the features map with varying details across convolution layers of stock data. The stock data includes stock market performance of assets over the period of IPO (initial public offering - private companies offers its share to public in new stock issuance) introduction to current date. This is inherently temporally interdependent data which has been discovered in the EDA (Exploratory Data Analysis – the process of inspection of the dataset to find patterns, irregularity and structure hypothesis based on the comprehension of the stock dataset) phase. This temporal dependency is extracted as a 2D feature map by CNN which in turn is passed through dense layers to generate single continuous output that is target variable which is open price of the stock.

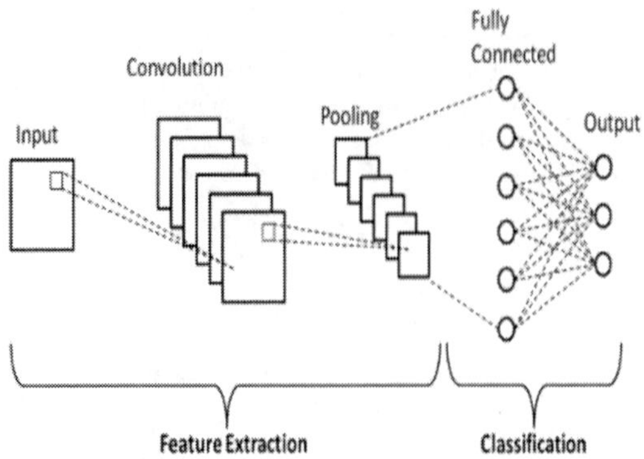

Figure 1. Convolution Neural Network (CNN)

3.2 Long Short term Memory (LSTM):

Xuan Ji [20], developed a new stock price forecasting model based on deep learning technology that uses Doc2Vec, SAE, wavelet transform, and LSTM mode. It mainly focuses on feature selection of stock financial features and text features (through social media like investors' comments and news published regarding stocks published by the companies).

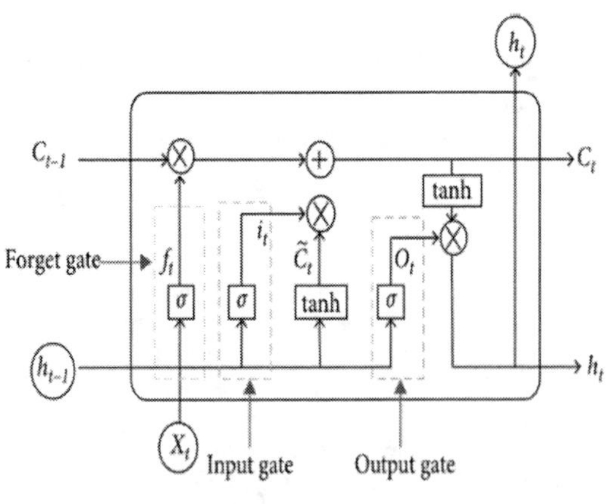

Figure 2. Long Short term Memory (LSTM) [14]

As shown in Figure 2, the LSTM memory cells consists of three parts: a forget gate, an input gate, and output gate. LSTM is used in analytical approach because LSTMs can store important information in the past and forget about other information. The input gate adds cell state information, the forget gate removes the information that is not needed in the model and the current gate selects the information that is displayed as an output. LSTMs combines the result of previous gate and the current gate and further, predicts the next state using the gain correlation.

LSTM is suited to extract long term dependencies in sequential data with temporal dependencies as it eliminates vanishing and exploding gradient in the backpropagation phase. The temporal dependency of stock market data is modelled by unfolding the data in time and passing through LSTM which predicts the output at next time step. The last price, volume and date provides input to the model and the open price is produced as the output of the target variable.

3.3 Convolution – Long Short term Memory Hybrid (CNN-LSTM):

The paper proposed by Wenjie Lu [14], states that a CNN-LSTM model is used to predict the closing price of the stocks of the next day. This method takes opening price, highest price, lowest price, closing price, volume, turnover, ups and downs, and changes in stock data as inputs, uses CNN to characterize the input data, and uses LSTM output to extract. It then learns the characteristic data and predict the closing price of the stock price the next day.

In CNN, it has the notions of listening to the maximum apparent features within side the line of sight, so its miles is extensively utilized in feature engineering. LSTM has the property of increasing overtime and is widely used in time series. The model structure diagram is shown in Figure 3. The CNN captures the spatial dependencies in the stock data inherent to the images. LSTM resolves the issue of vanishing or exploding gradient associated to the long term temporally dependent stock data. The combination of CNN and LSTM is tested in the predictive model. The CNN layers are used as initial layers which extracts the features in the sequential stock data and LSTM is then cascaded to incorporate long term dependency preservation in the features extracted by the CNN layers. At last, fully connected layers have been added to give the single continuous result.

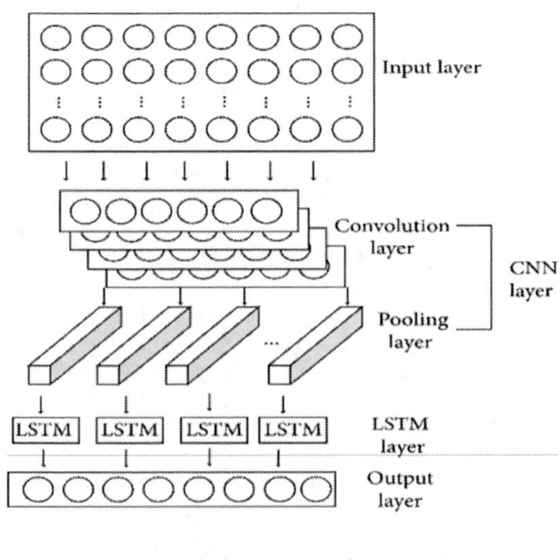

Figure 3. Convolution – Long Short term Memory Hybrid (CNN-LSTM) [14]

The models analyses the dataset in order to remove null values in the columns or replace them with mean or median values and also to detect the relationship between the parameters to determining the important parameters that affects the stock prices. The models handles the categorical and date type data by identifying the columns/attributes datatypes. The date data type is handle by diffusing it into three components. This generates the report regarding the dataset which is returned to the models which is used for further processing. The dataset is divided into training dataset (80% of original dataset) and testing dataset (20% of original dataset). The training of the models is done on the training datasets whose performances (models predicted values on testing dataset) are compared to the testing datasets actual values. This is done by the models which returns the performance metric calculated on the testing dataset. The models then gives the output of the performance report of all the three models.

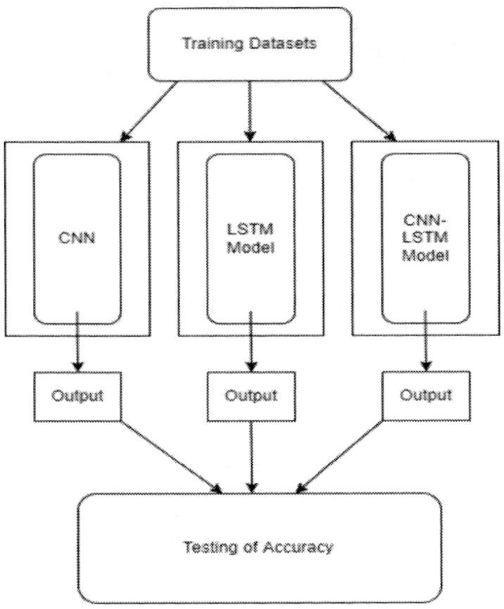

Figure 4. Comparison of CNN, LSTM, CNN- LSTM models

4. Results and Discussion

In the proposed method diverse recognized datasets are used which are TATASTEEL, TATAMOTORS, VEDL, BHARTIAIRTEL and ITC etc. These companies dataset are decided from various diverse sectors such as oil, telecommunication and from many others. The datasets are accumulated from 2000-2021 for evaluation.
The models are evaluated for TATASTEEL, TATAMOTORS, VEDL, BHARTIARTL and ITC dataset stocks.

The parameters used in the stock dataset are date, previous closed, high, low, last, close, volume and turnover. The date is when a stock is purchased/sell or when a stock transaction takes place. The previous closed is the previous closed price of the stock. The open price is the stock price when opened in a stock market on a particular date. The high is the highest stock price on a particular date. The low is the lowest stock price on a particular date. The last is the last price occurred for the last trade of a day. The close is the closing price of a stock is a price at which the stock closes at end of the trading hours in the stock market. The volume indicates how many are sold and bought in a given time duration.

The parameters settings for CNN model are convolution layer filters are 128, convolution layer kernel size is 3, convolution layer padding is same, learning rate is 0.001, loss function is mean absolute error, epochs is 25 and activation function is relu. The parameters settings for LSTM model are activation function is relu, learning rate is 0.001, loss function is mean absolute error, epochs is 15 and dropout is 0.2. The parameters settings for CNN-LSTM are convolution filters are 64, convolution layer kernel size is 3, convolution layer padding is same, batch size is 4, activation function relu, learning rate is 0.001, optimizer is adam, epochs is 15 and loss function is mean absolute error.

The dataset is partitioned into 80% training dataset and 20% testing dataset. The mean absolute error (MAE) is calculated for the evaluation of the forecasting effect on CNN, LSTM and CNN-LSTM models. The MAE equation (2) is given below,

$$MAE = \frac{\sum_{i=1}^{n} |p_v - t_v|}{n} \quad (2)$$

in which t_v is true value and p_v is predictive value. Smaller the MAE value, the higher the accuracy of the models. The ML models in this work predict the open price of the several stocks on the nth day, in which n is the pre-decided value. The analysis of the models shows an overall average error (MAE) of 0.17586 for CNN-LSTM, 0.21204 for LSTM, and 0.22982 for CNN.

The graphs shown in Fig.5, Fig.7, and Fig.9 depict the MAE result for open price of TATASTEEL of CNN-LSTM, LSTM and CNN respectively. Similarly, the graphs shown in Fig. 6, Fig. 8 and Fig. 10 depict the loss for open price of TATASTEEL of CNN-LSTM, LSTM and CNN respectively.

https://doi.org/10.1051/itmconf/20224403033

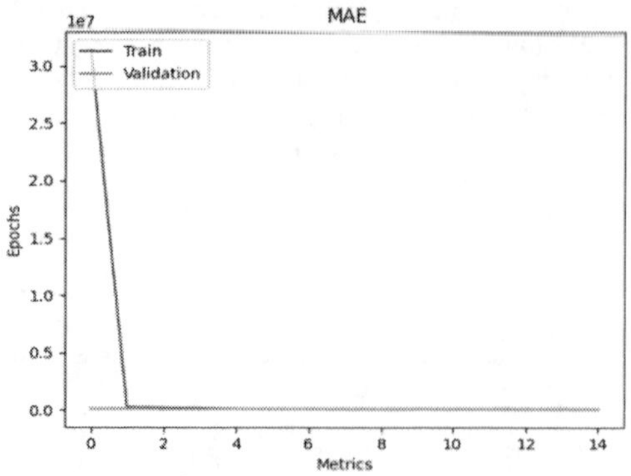

Figure 5. CNN-LSTM MAE

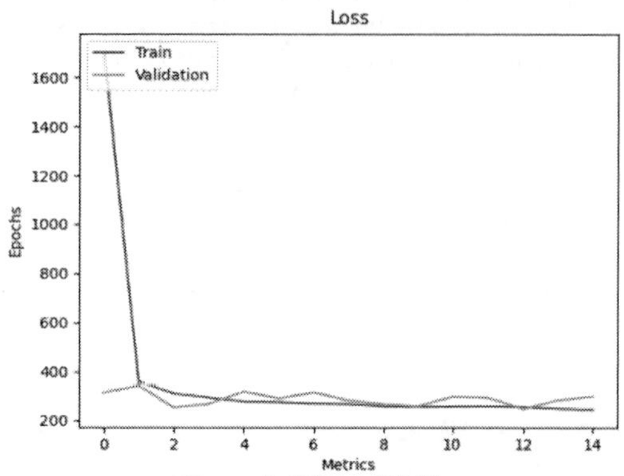

Figure 6. CNN-LSTM Loss

The forecasting of the open price of TATASTEEL using CNN-LSTM model shows that its results are the best among the three models.

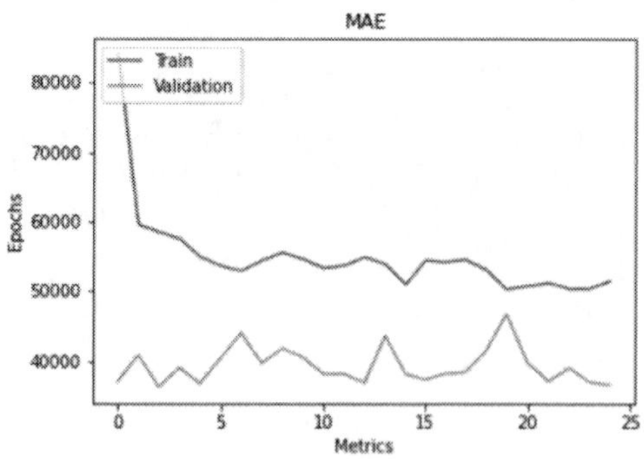

Figure 7. LSTM MAE

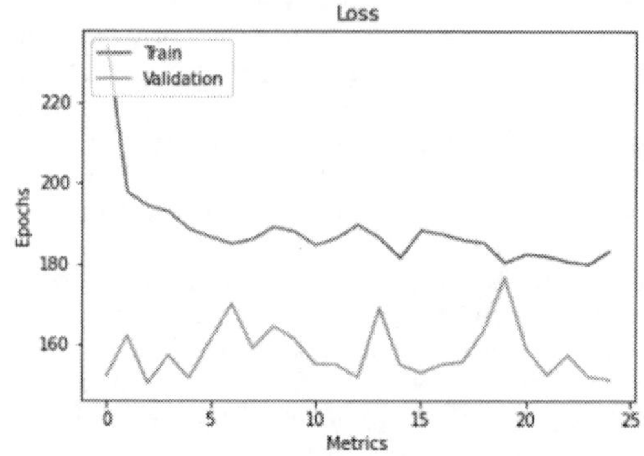

Figure 8. LSTM Loss

The forecasting of the open price of TATASTEEL using LSTM models shows that its results are better than CNN but lags behind.

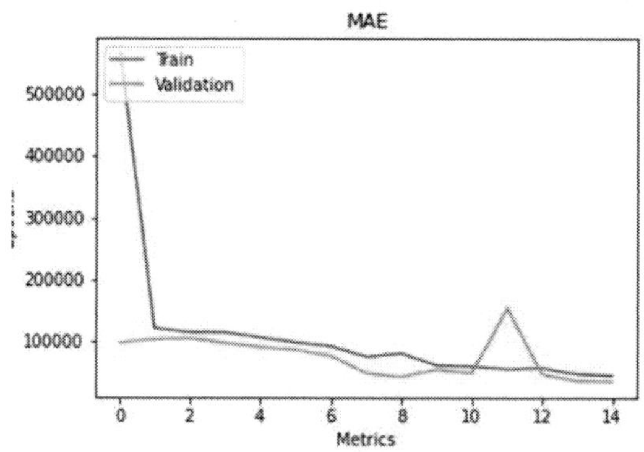

Figure 9. CNN MAE

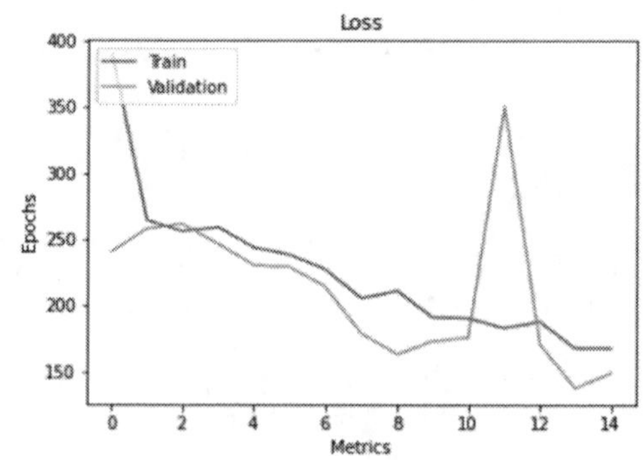

Figure 10. CNN Loss

The forecasting of the open price of TATASTEEL using CNN model shows that its lags behind the CNN-LSTM and LSTM models.

https://doi.org/10.1051/itmconf/20224403033

From Figures 5-10, the MAE and the loss graphs are generated using CNN-LSTM, LSTM and CNN respectively for the stocks datasets. The CNN-LSTM shows overall least variation for the datasets and hence CNN-LSTM has the highest forecasting accuracy in terms of forecasting performance.

Table 1. Comparison of Algorithms

MAE (Mean Absolute Error Values)				
Sr No.	Stocks	CNN	LSTM	CNN-LSTM
1	TATASTEEL	0.2838	0.2373	0.2051
2	TATAMOTORS	0.1878	0.1707	0.1351
3	VEDL	0.0154	0.0151	0.0125
4	BHARTIARTL	0.3092	0.2892	0.211
5	ITC	0.3529	0.3479	0.3156
		0.22982	0.21204	0.17586

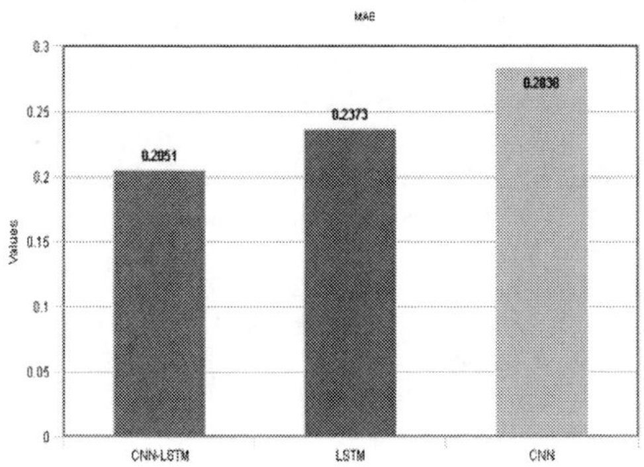

Figure 11. MAE comparison on TATSTEEL

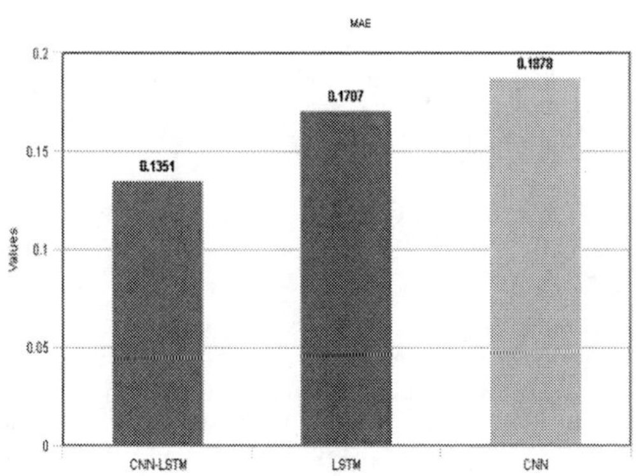

Figure 12. MAE comparison on TATAMOTORS

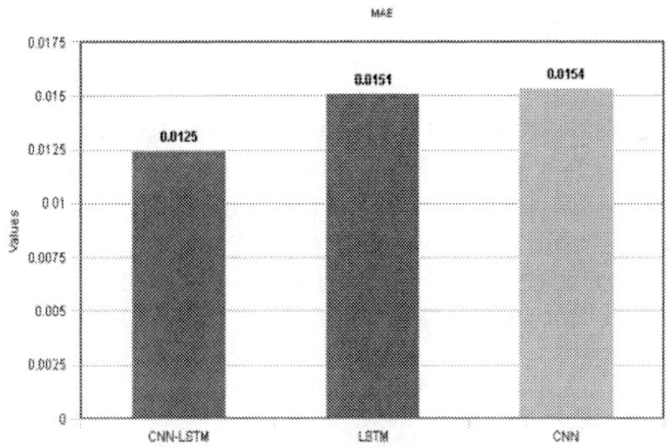

Figure 13. MAE comparison on VEDL

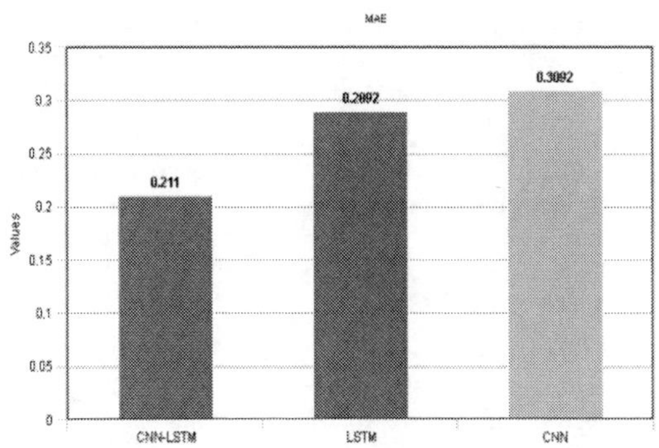

Figure 14. MAE comparison on BHARTIARTL

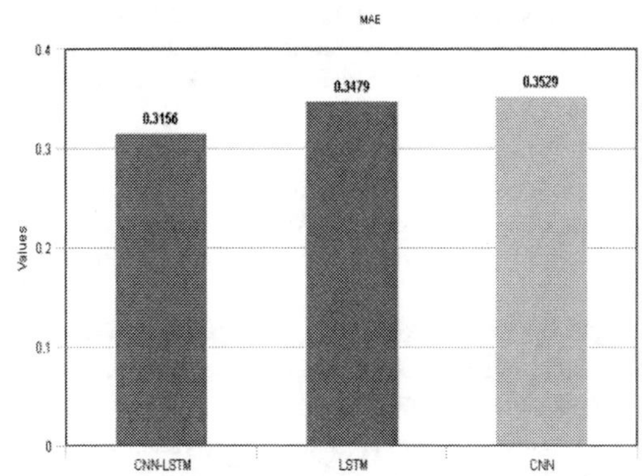

Figure 15. MAE comparison on ITC

In the Figures 10-15, the MAE values of CNN-LSTM, LSTM and CNN are calculated and their bar graphs have been plotted of the stock datasets. It is observed that CNN has the highest MAE values while CNN-LSTM has the lowest MAE values. Thus, this concludes that the CNN-LSTM Hybrid Model has the most smaller MAE values among the three models.

5. Conclusion

This report proposes a CNN, LSTM and CNN-LSTM based models to forecast the stock prices data according to the sequential traits of the stock prices data. The models uses stock datasets parameters such as close price, open price, high price, low price, previous closed price, turnover and volume. These parameters are used as the input to the models for training and testing for stock price prediction. The experimental results shows that in most of the cases, the CNN-LSTM performs very much better than CNN and LSTM models.

References

1. R. Vanaga and B. Sloka, "Financial and capital market commission financing: aspects and challenges," Journal of Logistics, Informatics and Service Science, vol. 7, no. 1, pp. 17–30, 2020.
2. L. Zhang and H. Kim, "-e influence of financial service characteristics on use intention through customer satisfaction with mobile fintech," Journal of System and Management Sciences, vol. 10, no. 2, pp. 82–94, 2020.
3. L. Badea, V. Ionescu, and A.-A. Guzun, "What is the causal relationship between stoxx europe 600 sectors? But between large firms and small firms?" Economic Computation And Economic Cybernetics Studies And Research, vol. 53, no. 3, pp. 5–20, 2019.
4. J. Sousa, J. Montevechi, and R. Miranda, "Economic lot-size using machine learning, parallelism, metaheuristic and simulation," Journal of Logistics, Informatics and Service Science, vol. 18, no. 2, pp. 205–216, 2019.
5. A. Coser, M. M. Maer-Matei, and C. Albu, "Predictive models for loan default risk assessment," Economic Computation And Economic Cybernetics Studies And Research, vol. 53, no. 2, pp. 149–165, 2019.
6. Q. Yang and C. Wang, "A study on forecast of global stock indices based on deep LSTM neural network," Statistical Research, vol. 36, no. 6, pp. 65–77, 2019.
7. K.-S. Moon and H. Kim, "Performance of deep learning in prediction of stock market volatility," Economic Computation And Economic Cybernetics Studies And Research, vol. 53, no. 2, pp. 77–92, 2019.
8. J. Li, S. Pan, L. Huang, and X. Zhu, "A machine learning based method for customer behavior prediction," Tehnicki Vjesnik-Technical Gazette, vol. 26, no. 6, pp. 1670–1676, 2019.
9. Aparna Nayak, M. M. Manohara Pai and Radhika M. Pai, "Prediction Models for Indian Stock Market", (IMCIP) (2016).
10. Vijh, Mehar & Chandola, Deeksha & Tikkiwal, Vinay & Kumar, Arun. (2020). Stock Closing Price Prediction using Machine Learning Techniques. Procedia Computer Science. 167. 599-606. 10.1016/j.procs.2020.03,326
11. Hyeong Kyu Choi, "Stock Price Correlation Coefficient Prediction with ARIMA-LSTM Hybrid Model", arXiv:1808.01560v5 [cs.CE],(2018).
12. Pushpendu Ghosha, Ariel Neufeldb, Jajati Keshari Sahoo, "Forecasting directional movements of stock prices for intraday trading using LSTM and random forests", (2021).
13. Mehtabhorn Obthonga, Nongnuch Tantisantiwong, Watthanasak Jeamwatthanacha c and Gary Willsd, "A Survey on Machine Learning for Stock Price Prediction: Algorithms and Techniques", (2020).
14. Wenjie Lu, 1, 2 Jiazheng Li, 3 Yifan Li, 3 Aijun Sun, and Jingyang Wang, "A CNN-LSTM-Based Model to Forecast Stock Prices", (2020).
15. Nusrat Rouf, Majid Bashir Malik, Tasleem Arif, Sparsh Sharma, Saurabh Singh, Satyabrata Aich , and Hee-Cheol Kim, "Stock Market Prediction Using Machine Learning Techniques: A Decade Survey on Methodologies, Recent Developments, and Future Directions", (2021).
16. Jingyi Shen and M. Omair Shafq, "Short-term stock market price trend prediction using a comprehensive deep learning system", (2020).
17. D. Wei, "Prediction of Stock Price Based on LSTM Neural Network," International Conference on Artificial Intelligence and Advanced Manufacturing (AIAM), (2019).
18. Chen, Sheng & He, Hongxiang. (2018). Stock Prediction Using Convolutional Neural Network. IOP Conference Series: Materials Science and Engineering. 435. 012026. 10.1088/1757-899X/435/1/012026.
19. Wu, J.MT., Li, Z., Herencsar, N. et al. A graph-based CNN-LSTM stock price prediction algorithm with leading indicators. Multimedia Systems (2021). https://doi.org/10.1007/s00530-021-00758-w.
20. Ji, X., Wang, J. and Yan, Z. (2021), "A stock price prediction method based on deep learning technology", International Journal of Crowd Science, Vol. 5 No. 1, pp. 55-72. https://doi.org/10.1108/IJCS-05-2020-0012.
21. Vanukuru, Kranthi. (2018). Stock Market Prediction Using Machine Learning. 10.13140/RG.2.2.12300.7748.
22. A M Pranav, Sujooda S, Jerin Babu, Amal Chandran, Anoop S, 2021, StockClue: Stock Prediction using Machine Learning, International Journal Of Engineering Research & Technology (IJERT) NCREIS – 2021 (Volume 09 – Issue 13).

Hate Classifier for Social Media Platform Using Tree LSTM

Sahil Raut[1], *Nikhil* Mhatre[1], *Sanskar* Jha[1] and *Aditi* Chhabria[1]
[1]Department of Computer Engineering, Ramrao Adik Institute of Technology, DY Patil Deemed to be University, Nerul(E), Navi Mumbai, India

Abstract. Social-media without a doubt is one of the most noteworthy developments ever. From associating with individuals across the globe for sharing of data and information in an infinitesimal of a second, online media stages have enormously altered the method of our lives. This is joined by a steadily expanding utilization of social media, less expensive cell phones, and the simplicity of web access which have additionally prepared for the huge development of social media. To place this into numbers, according to an ongoing report, billions of individuals all over the planet presently utilize web-based media every month, and a normal amount of almost 2 million people new clients are going along with them consistently. While web-based media stages have permitted us to interface with others and fortify connections in manners that were not conceivable previously. Unfortunately, they have additionally turned into the default gatherings for can't stand discourse. Online hate is a wild issue, with the adverse result of disallowing client support in web-based conversations and causing mental mischief to people. Since hate is pervasive across friendly, media stages, our objective was to foster a classifier that is feasible to train classifiers that can identify hateful remarks with strong execution and with the portion of misleading up-sides and negatives staying inside sensible limits.
Keywords. *Hate-speech, Twitter, Feature-extraction, Cyberbullying, Personal attacks.*

1. Introduction

Cyberbullying and forceful language on friendly stages are a portion of the maladies of our cutting-edge time. The right to speak freely of discourse online can without much of a stretch wreck into hostile, outlandish and non-productive analysis towards sexual, political, and strict convictions. AI classifiers and the abundance of information accessible on friendly stages offer a legitimate arrangement to alleviate this issue.[1] Disdain discourse's definition as taken from Cambridge Dictionary: public discourse that communicates disdain or energizes viciousness towards an individual or gathering dependent on something like race, religion, sex, or sexual direction. [2] Notwithstanding this definition, it is likewise generally realized that disdain discourse is more ruinous when it spreads through the media and underlines that disdain discourse is an extreme danger to majority rules system, multiculturalism, and pluralism. The primary objective of this task is to assemble a model equipped for recognizing disdain discourse.[3] In this task, a progression of classifiers, for example, Logistic Regression, SVM, and BERT was prepared on 25000 thousand tweets human-named as hostile or not hostile. The 25000 tweets were gathered by consolidating two unique datasets. A vital test for programmed disdain discourse discovery via online media is the partition of disdain discourse from different occurrences of hostile language.[4] We use public support to present a reference of these tweets into three classes: ones containing disdain discourse, just hostile language, and the other ones with neither or them. Then we build a multiclass classifier in order to understand the various classifications. Close investigation of the expectations and the mistakes depicts when we can isolate disdain discourse from different hostile languages and when this separation is more problematic. Then we track the bigot and homophobic tweets which are bound to be delegated disdain discourse.

2. Literature Survey

Web-based media organizations can straightforwardly report occurrences to the police, yet most badgering is passed on to the casualty to report. Twitter will give a rundown email of connections to messages that can be sent to the police, while Facebook has no such framework set up.[5] Online media stages can be utilized by individuals secretly or with counterfeit profiles, with minimal in method of confirmation. Despite the fact that they regularly give approaches to hailing hostile and scornful substance, as per different studies it is observed that main 17% of all grown-ups have hailed hassling discussion, though just 12% of grown-ups have revealed somebody for such demonstrations.[6] We experience disdain discourse in each part of life, sadly. It is much more testing to manage its damaging impacts in the advanced world. Individuals might act all the more forcefully via web-based media since they can be mysterious, their messages can arrive at a huge openness, and for some different reasons. At the point when we incorporate the messages posted by bots and phony records, disdain discourse turns out to be too normal to ever be distinguished and directed physically.[7] Meanings of online disdain: Instead of one single common meaning, the writing is contained with numerous definitions with particular ways to deal with online disdain.

Definition of Online Hate	Source	Focus
"Language that is used to express hatred towards a targeted group or is intended to be derogatory to humiliate, or to insult the members of the group."	Davidson et al.	Language, target
"Hateful comments towards specific group or target."	Salminen et al.	Target, group
"Hate speech is either 'directed' towards a specific person or entity, or 'generalized' towards a group of people sharing a common protected characteristic."	ElSherief et al.	Target, group
"Comments that are rude, disrespectful or otherwise likely to make someone leave a discussion."	Almerekhi et al.	Individuals, comments, consequences
"An offensive post, motivated in whole or in a part by the writer's being bias against an aspect of a group of people."	Mondal et al.	Language, group, target
"Offensive name calling, purposefully embarrassing others, stalking, harassing sexually, physically threatening, and harassing in a sustained manner."	Wulzyn et al.	Language

Table 1: Definition of Online Hate [8]

The issue of distinguishing disdain discourse has been tended to by different analysts in various ways. As a rule, the issue can be tended to in various ways. One of the potential ways is to foster an unadulterated Natural Language Processing model, which is for the most part a solo model. Thus, the identification turns out to be similarly simpler as there is no requirement for a marked informational index.[9] In this methodology, an NLP model can be planned which orders whether or not a sentence contains disdain discourse. In writing, there are less works that were completed completely dependent on unadulterated NLP-based ideas. One of the likely reasons is the models are relatively slower than the models fabricated utilizing Machine Learning or Deep Learning Models. The AI and profound learning models for the identification of disdain discourse need a named informational collection that is utilized to prepare the model.[10]

A lot of explores have been completed in this space where the analysts made their own dataset. The overall technique is to gather the information from a person-to-person communication site clean the information and afterward get them explained by a group of specialists who physically comment on if a message contains a derisive message or not. Khan et al. led a thorough review of AI models utilized broadly in NLP.[11] Ahmed et al. fostered a dataset that comprises of English and Bengali blended texts and commented on the tweets as disdain discourse or non-disdain discourse. Sahi et al. fostered an administered learning model to distinguish disdain discourse against ladies in the Turkish language. They gathered tweets referencing the apparel selections of ladies and utilized this information to prepare the AI models.[12] Waseem inspected the impact of annotators' information on the order model Waseem et al. given an informational index of 16,000 tweets and they additionally examined which elements give the best presentation with regards to the arrangement of disdain talks. Likewise, there are a lot of works done where scientists take open-source information and attempt to foster models which are utilized to recognize scornful messages on interpersonal interaction locales.[13]

Additionally, the shortfall of comprehensive classifiers suggests that the results across studies and online entertainment stages are not successfully same. Despite the fact that disdain has been seen as an issue in different internet-based web-based entertainment stages, including Reddit, Twitter, YouTube, etc., aside from a couple of exploratory investigations, there is an absence of improvement and testing of models utilizing information from numerous web-based entertainment stages. In aggregate, the fracture of models and component portrayals unnecessarily confounds disdain location across various stages and settings. Additionally, attempting to seem OK of the information with catchphrase-based look doesn't give right outcomes because of the language's design and types of articulation, like incongruity. In a climate where even the greatest news sources on the planet are at times compelled to cripple remarks on delicate recordings they distribute on YouTube, it is beyond difficult to physically battle disdain discourse for organizations and different associations with more restricted assets. Hence, it is unavoidable to depend on strategies that naturally detect disdain discourse.

3. Proposed Methodology

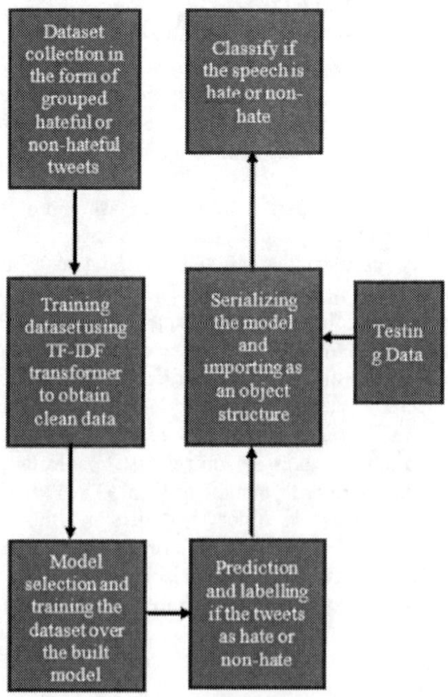

Fig 1: Architecture of Hate Classification Model

The proposed methodology is to examine methods utilized in disdain discourse checking and select the best appropriate procedure for making an altered disdain discourse discovery model to foster an application that will consolidate disdain discourse watchwords for order, allow the client to prepare his decision of dataset on the model and afterward breeze through on the assessment information to actually take a look Sat its level of repulsiveness. To exhibit and test the application while giving investigation on the disdain discourse results from the dataset transferred.

3.1 Word Embeddings

Word embeddings are numerical depictions of words that work with language and understands it by mathematical computations. They rely upon a vector space model that gets the general similarity among person word vectors, in this way giving information on the essential significance of words. Subsequently, word embeddings are by and large used for text portrayal and online hatred acknowledgment. Previous works have shown that unmistakable pre-arranged word embeddings perform well with respect to hate speech. For this undertaking, we picked the Word2Vec model to get the word embeddings.

3.2 Training Models

3.2.1 Logistic Regression (LR)

The choice of logistic regression (LR) is upheld by its straightforwardness and generally expected use for text grouping. Contingent upon the highlights, LR can acquire great outcomes in internet-based disdain identification with low model intricacy. Accordingly, including LR when contrasting various models appears to be sane. Ordinary AI classifiers such as direct relapses models have likewise been utilized to actually identify oppressive web-based language. [14]

3.2.2 Support-Vector Machine (SVM)

Support vector machines (SVM) is one more estimation commonly used in text portrayal. The impulse of SVMs is to view as a hyperplane that helps the insignificant distance between the classes. The cases portraying this plane are known as help vectors. Basically, previous works like Xu et al., Nobata et al., have investigated unique roads in regards to SVM for hatred acknowledgment with great results. [15] The computational multifaceted nature of Support vector machines is lower differentiated with significant learning models, which also gives clearer interpretability.

3.2.3 Bidirectional Encoder Representations from Transformers (BERT) + Convolutional Neural Network (CNN)

Transformers transform one course of action into one more by clearing out any rehash and supplanting it with an extensive part to manage conditions between the data and yield of the structure. With this plan, a model can be arranged all the all the more gainfully in view of the finish of progressive dependence on the past words, extending sufficiency for showing long stretch circumstances. [16] BERT has boundlessly beaten past models, for instance, the GPT and ELMo which stands for Generative Pretrained Transformer and Embeddings from Language Models respectively.

3.2.4 Long Short-Term Memory (LSTM)

These are exceptional kinds of neural organizations which are intended to function admirably when one has arrangement informational index and there exists a drawn-out reliance. These organizations can be helpful when one necessity an organization to recall data for a more extended enough said. This element makes LSTM appropriate for handling printed information. [17] A LSTM is an assortment of comparable cells, though every cell processes the info in a particular methodology. Aside from the contribution from outside sources, every cell likewise gets inputs from its previous cell in the chain.

3.2.5 Tree LSTM

Ordinary type of LSTMs can recollect or allude to the data which it has navigated till now. Be that as it may it doesn't have any proof about the data present after the point crossed till the point. This turns into a significant disadvantage while managing grouping information, particularly text. Tree LSTM is one more form of LSTM which can recall the data from the two bearings. In Tree LSTM we essentially back proliferation in two different ways. Once from the front and once from the back. This cycle makes Tree LSTM a strong apparatus for examining printed information.

As of late, Bisht et al., proposed a solitary LSTM layer as a basic model for distinguishing hostile language and disdain discourse in twitter information. The review utilized pre-prepared word2vec for contribution to one layer LSTM. They observed that word2vec+Tree LSTM performed better contrasted with word2vec+LSTM. It likewise proposed LSTM and Tree LSTM with blend of pre-prepared word vectors as the conveyed word portrayal. In their work, they call attention to that Tree LSTM has a superior F1 score for foreseeing disdain content. [18]

Hence, the justification behind picking Tree LSTM model is that it functions admirably with successive information, where the model requirements to protect the setting of long-grouping. CNN experiences disappearing and detonating slopes issues when the mistake of the angle drop calculation is backpropagated through the organization, which creates CNN cannot recall all input history successfully. Be that as it may, rather Tree LSTM saves long haul conditions in a more powerful manner.

4. Results and Simulation

The overall structure is divided into 6 main parts: Data Cleaning, Training of Models, Displaying results of each model with its accuracy. Testing the built model over input dataset, predicting offensiveness of input statement, classifying statistically all the hate and non-hate tweets present in the dataset.

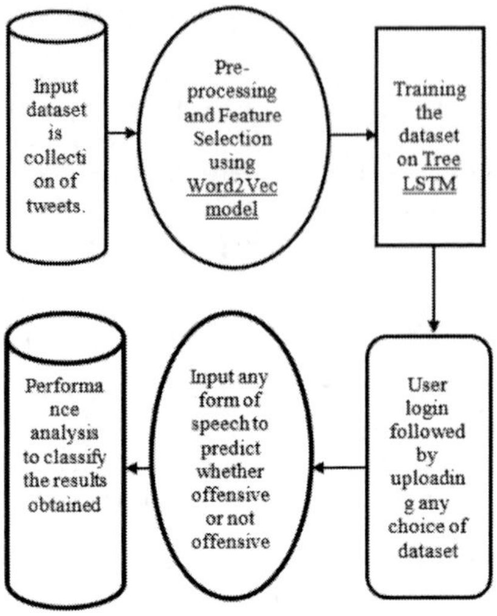

Fig 2: System Design

4.1 Dataset

This dataset is made accessible for naming and gives a definite portrayal of the information assortment standards. The dictionary, gathered from Hate-base, contains words and expressions recognized by web clients as disdain discourse. This dictionary was additionally utilized by the creators as catchphrases to separate the 85.4 M tweets. The assignment was to comment on the tweet with one of three classes: disdain discourse, hostile however not disdain discourse, or neither hostile nor disdain discourse. The tweets with no larger part class were disposed of, making a sum of 24,802 tweets with a doled-out mark of which 5 percent was given the contemptuous name. In this way, we needed to use the Twitter API to remember the dataset. We had the option to get 24,783 tweets (99.9 percent of the first dataset), with 19 tweets either erased or in any case inaccessible.

4.2 Data Cleaning

We as a whole understand that prior to applying any machine learning (ML) model we really want to make our dataset prepared for a possible examination. This progression is especially pertinent when we manage texts. Most words, truth be told, are generally horrible to group forceful sentences.

ID	Count	Hate Speech	Offensive Language	Neither	Tweet
0	3	0	0	3	@Mayasolovely : As a woman ...
1	3	0	3	0	@mleew17: boy dats cold...
2	3	0	3	0	@UrKindOfBrand Dwag: she lo...
3	3	0	2	1	@C_G_Anderson: dat niggug...
...	...	...	...	...	...
...	...	...	...	...	...
25291	3	0	2	1	You's a muthaf***in lie...
25292	3	0	1	2	You've gone and broke the wrong...
25294	3	0	3	0	young buck wanna eat!!...
25295	6	0	6	0	You got wild bitches...
25296	3	0	0	3	Ntac Eileen Dahlia – beautiful col...

Table 3: Dataset Cleaning

4.3 Data Pre-Processing

The following are altogether the preprocessing steps: Lowering of all words in the Tweets, removing of copies, eliminating re-tweets, removing exceptional characters what's more estimating tweets' length, reformatting all spaces and hashtags, removing stop words and additionally words more limited than 3 characters, dropping unnecessary columns and saving last information outline.

4.4 Results

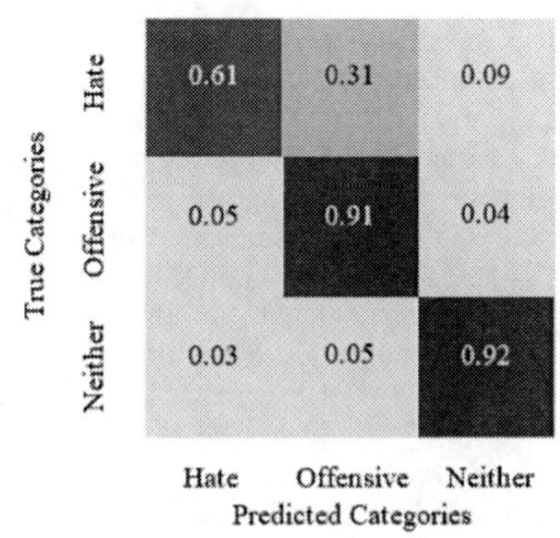

Fig 3: Confusion matrix of Tree LSTM model

As indicated by confusion matrix, we see that practically 40% of disdain discourse is misclassified: the accuracy and recall scores for the disdain class are 0.44 and 0.61 individually. The majority of the misclassification happens in the upper half of this matrix, proposing that the model is one-sided towards ordering tweets as less contemptuous or hostile than the human coders. Far less tweets are named more hostile or contemptuous than their actual classification; roughly 5% of hostile and 2% of harmless tweets have been incorrectly named can't stand discourse.

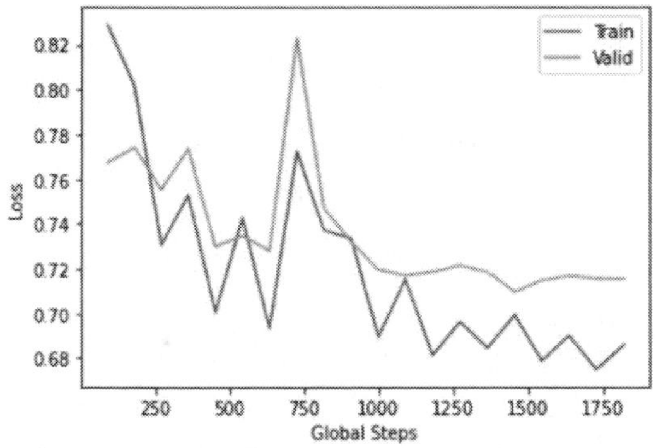

Fig 4: Train and Validation Loss Curve of Tree LSTM model on the dataset

Model	F1 Score	Precision Score	Recall Score	Accuracy
LR Model	0.541	0.66716	0.4564	0.46
SVM Model	0.5545	0.89718	0.7812	0.78
BERT+CNN Model	0.9090	0.93731	0.8850	0.89
LSTM Model	0.8447	0.84479	0.8345	0.83
TREE-LSTM Model	0.9177	0.91893	0.9260	0.92

Table 4: Results of Various Different Models

In all analysis however, the least precision (0.66), recall (0.45), accuracy (45%) and F1 score (0.54) was found in Logistic Regression model utilizing TFIDF highlights portrayal with bigram highlights. Also, the highest recall (0.91), precision (0.92), accuracy (92%) and F1 score (0.91) were gotten by Tree LSTM utilizing TFIDF highlights portrayal with bigram highlights. In feature representation, bigram highlights with TFIDF acquired the best execution when contrasted with Word2vec and Doc2vec.

5. Conclusion

This study utilized computerized text order procedures to distinguish can't stand discourse messages. Additionally, this study thought about two include designing methods and five ML calculations (Logistic Regression, Support Vector Machine, BERT and CNN, LSTM and Tree LSTM) to characterize disdain discourse messages. The exploratory results displayed that the bigram highlights, when addressed through Word2Vec, showed better execution when contrasted with TFIDF highlight designing strategies. In addition, SVM and BERT+CNN calculations showed better outcomes contrasted with LR. The most reduced exhibition was seen in LR. Also, the best execution was found with Tree LSTM as the most effective portrayal of derisive web-based media remarks. The results from this examination concentrate on hold pragmatic significance since this will be utilized as a pattern study to think about impending investigates inside various programmed text characterization techniques for programmed disdain discourse identification. There is still a lot of work to be done in the field of disdain discourse assessment. It is conceivable that a huge improvement in execution would be checked whether word portrayals were utilized rather than character portrayals; a significant part of the jargon of online correspondence and talk includes the utilization of expressions, casual discourse, furthermore allegorical language, which word-based portrayals could maybe better catch. Moreover, twofold grouping in itself can be considered as a restriction. Past examination has shown that disdain has a scope of translations, and understanding the setting of the remarks can be vital for disdain location. Rather than paired order, some specialists have picked distinguishing can't stand targets furthermore more nuanced classes of online disdain. Future improvement endeavors incorporate preparation client explicit information set on the prepared model and afterward permitting the client to input any type of discourse and accumulate its level of obnoxiousness.

6. References

[1] Kovács, G., Alonso, P. & Saini, R. "Challenges of Hate Speech Detection in Social-Media". *SN COMPUT. SCI. 2*, **95** (2021). https://doi.org/10.1007/s42979-021-00457-3

[2] V. Jain, V. Kumar, V. Pal and D. K. Vishwakarma, "Detection of Cyberbullying on Social Media Using Machine learning". (2021). https://ieeexplore.ieee.org/document/9418254

[3] Salminen, J., Hopf, M., Chowdhury, S.A. et al. "Developing an online hate classifier for multiple social media platforms". (2020). https://hcis-journal.springeropen.com/articles/10.1186/s13673-019-0205-6

[4] Venkateshwarlu Konduri, Sarada Padathula, Asish Pamu and Sravani Sigadam, "Hate Speech Classification of social media posts using Text Analysis and Machine Learning". *Paper 5204*, (2020). https://www.sas.com/content/dam/SAS/support/en/sas-global-forum-proceedings/2020/5204-2020.pdf

[5] Velioglu, Riza & Rose, Jewgeni., "Detecting Hate Speech in Memes" *Using Multimodal Deep Learning Approaches: Prize-winning solution to Hateful Memes Challenge.* (2020). https://www.researchgate.net/publication/347880819_Detecting_Hate_Speech_in_Memes_Using_Multimodal_Deep_Learning_Approaches_Prize-winning_solution_to_Hateful_Memes_Challenge

[6] Hou, Y., Xiong, D., Jiang, T., Song, L., & Wang, Q. "Social media addiction: Its impact, mediation, and intervention". *Cyberpsychology: Journal of*

Psychosocial Research on Cyberspace, **13(1),** Article 4. (2019). https://doi.org/10.5817/CP2019-1-4

[7] Kumar, R., Ojha, A.K., Malmasi, S., Zampieri, M, "Benchmarking aggression identification in social media". *In: Proceedings of the First Workshop on Trolling, Aggression and Cyberbullying (TRAC),* (2018) https://aclanthology.org/W18-4401.pdf

[8] Zimmerman, Steven & Fox, Chris & Kruschwitz, Udo. "Improving Hate Speech Detection with Deep Learning Ensembles". (2018). https://www.researchgate.net/publication/326294340_Improving_Hate_Speech_Detection_with_Deep_Learning_Ensembles

[9] Samuel Gibbs. "What can be done about abuse on social media?" *Guardian News & Media Limited*, **12**, (2017) https://www.theguardian.com/media/2017/dec/13/what-can-be-done-about-abuse-on-social-media

[10] Davidson, Thomas & Warmsley, Dana & Macy, Michael & Weber, Ingmar, "Automated Hate Speech Detection and the Problem of Offensive Language". (2017). https://www.researchgate.net/publication/314942659_Automated_Hate_Speech_Detection_and_the_Problem_of_Offensive_Language

[11] Gareth James, Daniela Witten, Trevor Hastie, and Robert Tibshirani. "An Introduction to Statistical Learning". *With Applications in R. Springer Publishing Company, Incorporated, ISBN,* (2014). https://www.ime.unicamp.br/~dias/Intoduction%20to%20Statistical%20Learning.pdf

[12] Bernd Hollerit, Mark Kr¨oll, and Markus Strohmaier. "Towards linking buyers and sellers: Detecting commercial intent on twitter". *Published in WWW '13 Companion 2013 Computer Science*, (2013). http://citeseerx.ist.psu.edu/viewdoc/download?doi=10.1.1.402.3220&rep=rep1&type=pdf

[13] Ostrowski, David, "Semantic Filtering in social media for Trend Modelling". *Proceedings - 2013 IEEE 7th International Conference on Semantic Computing, ICSC. 399-404.10.1109/ICSC.2013*, **78**. (2013) https://ieeexplore.ieee.org/document/6693553

[14] Cheng Xiang Zhai, Charu C. Aggarwal. "Mining Text Data". *Science + Business Media. Springer*, (2012). http://digilib.stmikbanjarbaru.ac.id/data.bc/7.%20Data%20Mining/2012%20Mining%20Text%20Data.pdf

[15] K. Lee, D. Palsetia, R. Narayanan, M.M.A. Patwary, A. Agrawal, and A. Choudhary. "Twitter trending topic classification". *In Data Mining Workshops (ICDMW)*. (2011) http://cucis.ece.northwestern.edu/publications/pdf/LeePal11.pdf

[16] S. Asur and B. A. Huberman, "Predicting the Future with Social Media". *IEEE/WIC/ACM International Conference on Web Intelligence and Intelligent Agent Technology, pp. 492-499, doi: 10.1109/WI-IAT.2010.63*. (2010), https://ieeexplore.ieee.org/document/5616710

[17] SongJie Gong. "A collaborative filtering recommendation system algorithm based on user clustering and item clustering". *Journal of Software*, **vol. 5**, (2010). http://www.jsoftware.us/vol5/jsw0507-9.pdf

[18] Kristin P. Bennett and Erin J. Bredensteiner. "Duality and Geometry in SVM Classifiers". *In In Proc. 17th International Conf. on Machine Learning*, pages 57–64, (2000). http://citeseerx.ist.psu.edu/viewdoc/download?doi=10.1.1.36.9635&rep=rep1&type=pdf

Automated Script Outline Generator using TextRank Algorithm

Urvi Zope [1], Gayatri Thakur [2], Siddhi Karambelkar[3] and Vanita Mane[4]

Department of Computer Engineering, Ramrao Adik Institute of Technology, Dr. D. Y. Patil Deemed to be University, Nerul, Navi Mumbai, Maharashtra, India[1,2,3,4]

Abstract

The process of creating a condensed type of document that retains significant information and thus the general meaning of the source text is known as text summarization. Automatic summarization is a popular method for quickly locating significant information in vast texts with minimal effort. In today's era information is the integral part of knowledge and it is too noisy and not in the precise manner for that both abstractive and extractive summarization techniques are used. The proposed model combines the summarization process for both text and image input type, which makes it easier for user to generate the precise summary regardless of the input type. The proposed model uses Textrank algorithm for the summarization process. Recall oriented understudy of gisting evaluation (ROUGH) score is used by the model to evaluate the accuracy of generated output. The model gives high ROUGH score which assures the accuracy of the generated summary.

1. Introduction

Automatic summarization technology is critical for information retrieval and text classification, and it should give a solution to the problem of knowledge overload [1]. A procedure of lowering the size of a text while keeping its information content intact is known as text summarization.[2] The proposed model uses a sentence clustering-based summarization approach. The model uses textrank algorithm for summarization which works on sentence ranking method. The model compares its results with researches in the field to provide more efficient summary of the data. The proposed model combines the text and image summarization model in a single system which differentiate it from the previous generated models.

1.1 Objectives

This model consists various input type summarization in an integrated form. This model aims at accurate summarization of provided input data. The generated output will be easily understandable and readable. Proposed system will analyze the given input data and remove the irrelevant information and will integrate the central ideas in meaningful way thus the model will focus on the main corpus of information rather than the information which leads to information overload. It will generate a compact yet precise output. Effective text summarization: The model aims to provide the best solution for summarization of the given input data regardless of its type. The proposed system aims to summarize the data in an accurate way regardless of the size of input data.

1.2 Motivation

In this era of information technology information plays a vital role in everyday life of all the individuals.[3] The summarization makes the process of information gathering easy for the users. The proposed model compared its output with the output of previous systems which helped to increase the usability and provide better result to end users of the system.

2. Literature Survey

2.1 Existing Systems

The literature survey for the summarization is briefly explained in the following segment. The main discussion is associated with various algorithms used for summarization and the output generated by all the algorithms respectively.

A. Term Frequency (TF) - Inverse Document Frequency (IDF): TFIDF is the algorithm used in extractive summarization. The terminology Term Frequency (TF) is used to count the occurrence of the words.
The frequency obtained is used to determine the word's importance. The greater the occurrence of a word, the greater its significance in the paper [4]. The simplest way to explain TF is that it counts the number of times a word appears in a document [5]. IDF stands for Inverse Document Frequency, which gives unusual words a greater value and recurrent terms a smaller one. TF sometimes miscalculates the importance of stop words based on their frequency of occurrence. IDF identifies the infrequent occurrence of words in the document to solve TF's problem.

B. Sequence to Sequence Model: The suggested TFRSP approach uses an abstractive summarization algorithm known as a sequence-to-sequence model to create new phrases while keeping the sense of the source content. Google was the first to introduce the Sequence-to-Sequence methodology, which now drives applications such as Google Translate, image captioning, text summarization, online chat bots, and more. It is an encoder-decoder paradigm that maps input and output sequences of various lengths to each other [6]. Long Short-Term Memory (LSTM) is a subcomponent of the encoder-decoder component that is effective for capturing long-term dependencies. There are two steps to the encoder-decoder model: training and inference [7].

C. Text Rank Algorithm: For summarization, the Text Rank algorithm is an extractive and unsupervised learning system. This algorithm is based on the methodology of sentence ranking. As a result, the data in a given input file is first concatenated together in the Text Rank method. After that, stemming is used to separate it into independent sentences [8]. Then, for each sentence, word embeddings are calculated. After that, the algorithm determines sentence-to-sentence similarity and saves the results in a matrix. The similarity matrix is then transformed into a graph, with the sentences serving as vertices and the similarity matrix serving as edges [6]. This graph is then used to rank the sentences, which aids in the creation of the summary. The sentences with the highest scores are given out first.

D. ROUGE score: The ROUGE metric (Recall Oriented Understudy of Gisting Evaluation) is a text summarization metric. It's used to compare the n-gram matches between the system-summary acquired from text summarization and the human-generated reference summary [9]. The ROUGE score is made up of precision, recall, and f-measure values that are added together. The ROUGE-
1 score assesses the overlap of unigrams from the system-generated summary and the reference summary created by humans [10].

2.2 Limitations of Existing System

The existing systems have given good results. But they do not have all the required features. Basically, they are either text input dependent or image input dependent. Following research gaps are identified in the existing techniques:

- Existing model provides a correct summary of only short text. When it came to longer text the summarization performed by these systems was inaccurate and irrelevant.
- The existing systems developed for the summarization were unable to understand the meaning of many vocabulary words. Some of the systems are tested only for specific language datasets. So, the accuracy of generated abstract is not that good when it came to the vocabulary words.
- Many difficulties were faced by the systems while summarizing the long texts.
- As discussed before previously designed systems were unable to generate accurate abstract for the loner texts. It is also observed that when the number of selected sentences exceeds three, the average ROUGE score drops.
- There is no existing model which combined both text and image summarization model into single system.

3. Proposed System

3.1 Problem Statement

"***Automated Script Outline Generator using textrank Algorithm***" is a model that aims to build a system which will be able to perform the summarization of both text and image. Based on the given input the proposed system will extract the important content from that.

3.2 Proposed Methodology

The proposed model uses textrank algorithm for the summarization of given input data. The following diagram shows the detailed working of the proposed system. As shown in the fig. 1 The model first takes the input from the user in either text or image format. If the given input is in image format, then it is first converted into text using OCR. Then Document preprocessing is performed in which stemming and lemmatization is performed on the input data. Then all the data in collected together then textrank algorithm is applied onto that data. The sentences are ranked in a tree manner. User defines the amount of summary required by them. Accordingly, system generates summary with highest ranked sentences. In this way the proposed model works. The ranking algorithms thus provides correctness to the overall output as only the highest ranked sentences are given out as output summary.
The practice of distilling the most significant information from a source into an abridged version for a specific user and task is known as summarizing. A sentence clustering-based summarization strategy is proposed in this model. A system pipeline that utilizes both error detection and correction models.
The summarization process will be performed using textrank algorithm where first the system will take the input data, perform pre-processing on that data and then then model building will be done after which the summarized output data will be generated.

3.2.1 Input:

Input consists of the input data that user is going to provide to the system. For this proposed system BBC news dataset from Kaggle is used. This dataset is made by collecting various news articles gathered from all over the web. Also for the image input the system is provided with an image containing information about android. Thus the system is using the multiple sentence dataset for the evaluation purpose.

3.2.2 Data Pre-processing:

In the preprocessing part of implementation, the system performs stemming and lemmatization on the given input data. In this module first the system will Read the file. After that it will Split the text in the articles into sentences. Then the system will remove punctuations, numbers and special characters. The system will also remove stop words from the sentences. After all this procedure the system will extract word vectors.

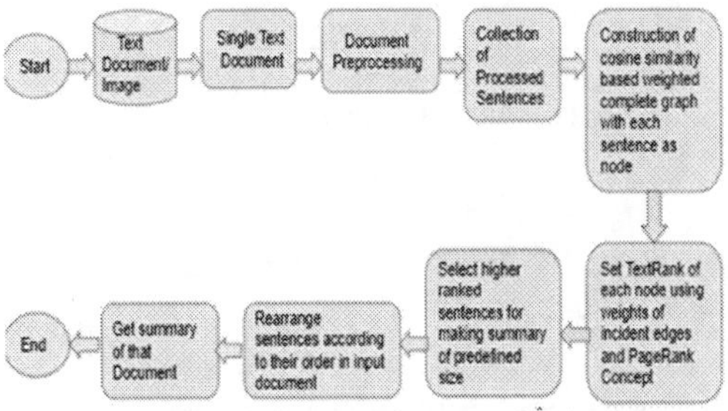

Fig. 1 Block diagram of automated script outline generator

3.3.3. Model Building using TextRank algorithm:

After the preprocessing is performed on the data the next step in the model is model building. In this step basically each sentence is then tokenized into a group of words in the next phase. Then, in order to define sentence similarity, we represent each sentence in the given text document as a word vector. Similarities between sentence vectors are calculated and stored in a matrix using cosine similarity. The similarity matrix is then turned into a graph for sentence rank calculation, with sentences as vertices and similarity scores as edges. Finally, the final summary is made up of a set of top- ranked sentences.
Text Rank algorithm is extractive and unsupervised learning algorithm used for the summarization. This algorithm works on sentence ranking methodology. So, in Text Rank algorithm first of all the data in given input file is concatenated together. Then it is splitted into individual sentences by performing stemming on it. After that word embeddings for each sentence are calculated. Then the algorithm calculates the sentence-to-sentence similarity and stores the calculated similarities in matrix. The similarity matrix is then converted to form a graph where the

sentences act as vertices of graph and the similarity matrix acts as edges of graph. This graph is then used to rank the sentences this ranked sentences helps to generate the summary. The highly ranked sentences are given out as the generated output summary.

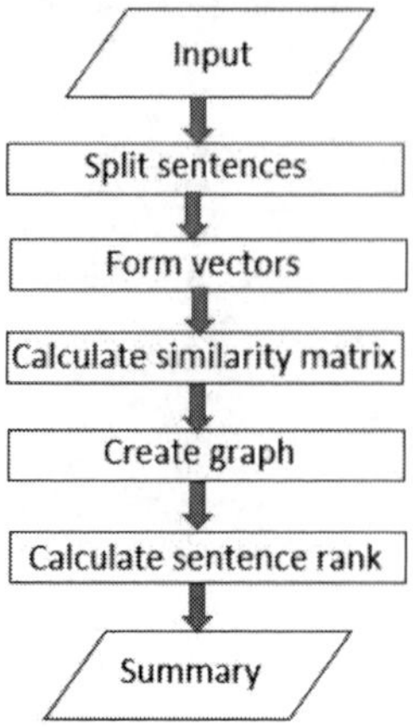

Fig. 2 TextRank algorithm

4. Result and Analysis

4.1 Evaluation parameters

The generated output by the proposed model is compared with other existing summarization models using ROUGH score. Recall-Oriented Understudy for Gisting Evaluation (ROUGE) is an acronym for ROUGE. It consists mostly of a set of measures for assessing automatic text summarization and machine translation. It works by comparing a summary or translation generated automatically with a set of reference summaries. ROUGE score is summary accuracy evaluation metric which compares the human generated summary and machine generated summary. Following are the evaluations done by the proposed system.

Recall (R): Number of overlapping words / total words in reference summary
Precision (P): Number of overlapping words / total words in system summary.
F1_score (F1): (2* Precisions * Recall) / (Precision + Recall)

4.2 Result

The proposed model used BBC news dataset from Kaggle and image input containing the android information as an input. Fig .2 and Fig. 3 show the generated summary for the given input. .Fig.2 shows the output generated by the model when image is given as a input. Fig.3 shows the output generated when text is given as input.

Fig. 2 Output for image input

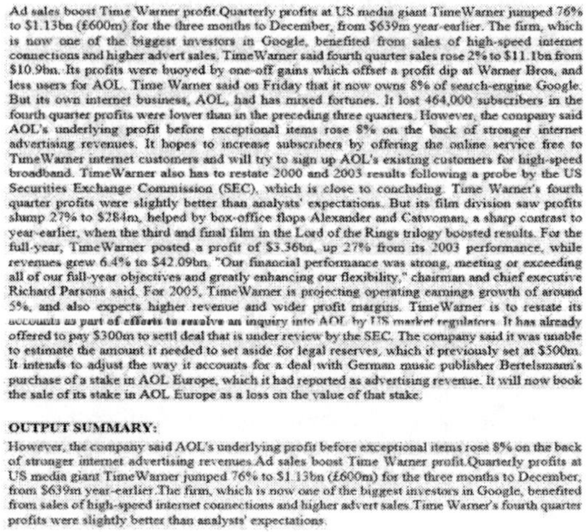
Ad sales boost Time Warner profit.Quarterly profits at US media giant TimeWarner jumped 76% to $1.13bn (£600m) for the three months to December, from $639m year-earlier. The firm, which is now one of the biggest investors in Google, benefited from sales of high-speed internet connections and higher advert sales. TimeWarner said fourth quarter sales rose 2% to $11.1bn from $10.9bn. Its profits were buoyed by one-off gains which offset a profit dip at Warner Bros, and less users for AOL. Time Warner said on Friday that it now owns 8% of search-engine Google. But its own internet business, AOL, had has mixed fortunes. It lost 464,000 subscribers in the fourth quarter profits were lower than in the preceding three quarters. However, the company said AOL's underlying profit before exceptional items rose 8% on the back of stronger internet advertising revenues. It hopes to increase subscribers by offering the online service free to TimeWarner internet customers and will try to sign up AOL's existing customers for high-speed broadband. TimeWarner also has to restate 2000 and 2003 results following a probe by the US Securities Exchange Commission (SEC), which is close to concluding. Time Warner's fourth quarter profits were slightly better than analysts' expectations. But its film division saw profits slump 27% to $284m, helped by box-office flops Alexander and Catwoman, a sharp contrast to year-earlier, when the third and final film in the Lord of the Rings trilogy boosted results. For the full-year, TimeWarner posted a profit of $3.36bn, up 27% from its 2003 performance, while revenues grew 6.4% to $42.09bn. "Our financial performance was strong, meeting or exceeding all of our full-year objectives and greatly enhancing our flexibility," chairman and chief executive Richard Parsons said. For 2005, TimeWarner is projecting operating earnings growth of around 5%, and also expects higher revenue and wider profit margins. TimeWarner is to restate its accounts as part of efforts to resolve an inquiry into AOL by US market regulators. It has already offered to pay $300m to settl deal that is under review by the SEC. The company said it was unable to estimate the amount it needed to set aside for legal reserves, which it previously set at $500m. It intends to adjust the way it accounts for a deal with German music publisher Bertelsmann's purchase of a stake in AOL Europe, which it had reported as advertising revenue. It will now book the sale of its stake in AOL Europe as a loss on the value of that stake.

OUTPUT SUMMARY:

However, the company said AOL's underlying profit before exceptional items rose 8% on the back of stronger internet advertising revenues.Ad sales boost Time Warner profit.Quarterly profits at US media giant TimeWarner jumped 76% to $1.13bn (£600m) for the three months to December, from $639m year-earlier.The firm, which is now one of the biggest investors in Google, benefited from sales of high-speed internet connections and higher advert sales.Time Warner's fourth quarter profits were slightly better than analysts' expectations.

Fig. 3. Output for text input

4.3 Evaluation with ROUGE

For a single document, the proposed method for extractive text summarization is used. For the summarization of input data, the Textrank algorithm is employed. For analyzing generated summaries, ROUGE (Recall Oriented Understudy for Gisting Evaluation) is used. Precision, recall, and the F measure are the performance evaluation metrics used to evaluate ROUGE score.

Table 1 Performance comparison between results obtained by the proposed model and existing model

 https://doi.org/10.1051/itmconf/20224403035

	EVALUATION PARAMETER	Proposed model	Existing model [9]
Total 5 documents tested	P	58.85	55.2
	R	55.05	53.2
	F1	56.87	49.87

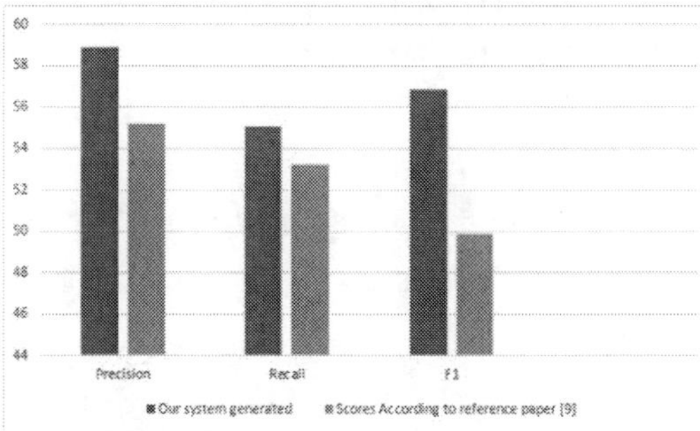

Fig. 5 Graphical comparison between results obtained by proposed model and previous model

Table 1 shows the comparison between the obtained results from proposed model and existing model. Fig.5 graphically shows how the proposed model is working better than the previous model. For total 5 documents the system performance is tested. After comparing the results of proposed model and existing models we came to the conclusion that our model is providing better precision, recall and f1 score than the existing model.

5. Conclusion and Further Work

The textrank algorithm generates a summary for the BBC news dataset combining the text and image summarization model and producing an integrated approach with the increased accuracy of 57.75% when compared with the traditional methods of summarization. The system generating a summary for both text and image input type combined were not existing. Also, the proposed model provides high ROUGH score even after combing the two model together.

Future work:
The proposed model already processes the summarization for text and image input type. The current system can be enhanced by integrating more classification techniques along with adding various functionalities. The proposed model could also be used to test on various different datasets to increase the versatility of the entire model.

References

[1] Li, W., & Zhuge, " Abstractive Multi- Document Summarization based on Semantic Link Network." IEEE Transactions on Knowledge and Data Engineering, (2021)
[2] Pisat, T., Bartakke, M., & Patil, H. "Synonym Suggestion System using Word Embeddings" 4th International Conference on Trends in Electronics and Informatics, (2020)
[3] Singh, P., Chhikara, P., & Singh, J, " An Ensemble Approach for Extractive Text Summarization." International Conference on Emerging Trends in Information Technology and Engineering (2020)
[4] Szucs, G., & Huszti, D. "Seq2seq Deep Learning Method for Summary Generation by LSTM with Two- way Encoder and Beam Search Decoder." IEEE 17th International Symposium on Intelligent Systems and Informatics (2019)
[5] Rahul, Adhikar, S., & Monika. "NLP based Machine Learning Approaches for Text Summarization." Fourth International Conference on Computing Methodologies and Communication (2020)
[6] Aciar, S. V., & Ochs, M. "Classifying User Experience based on the Intention to Communicate." IEEE Latin America. (2020)
[7] Alengadan, B. B., & Khan, S. S. "Modified aspect/feature-based opinion mining for a product ranking system." IEEE IT transactions (2018)
[8] Diaz-Garcia, J. A., Ruiz, M. D., & Martin-Bautista, M. J. "Non- Query-Based Pattern Mining and Sentiment Analysis for Massive Microblogging Online Texts." IEEE Access. (2020)
[9] J.N.Madhuri, Ganesh Kumar.R, "Extractive Text Summarization Using Sentence Ranking" IEEE Access,(2019)
[10] Vrublevskyi, V., & Marchenko, O. "Grammar Error Correcting by the Means of CFG Parser." IEEE International Conference on Advanced Trends in information theory. (2019)

ITM Web of Conferences **44**, 03036 (2022) https://doi.org/10.1051/itmconf/20224403036
ICACC-2022

Predicting students' academic grade using machine learning algorithms with hybrid feature selection approach

Snehal Paddalwar[1,*], *Vanita* Mane[2], and *Leena* Ragha[3]

[1]Department of Computer Engineering, Ramrao Adik Institute of Engineering, Dr. D.Y.Patil Deemed to be University, Nerul, Navi Mumbai, India
[2] Department of Computer Engineering, Ramrao Adik Institute of Engineering, Dr. D.Y.Patil Deemed to be University, Nerul, Navi Mumbai, India
[3] Department of Computer Engineering, Ramrao Adik Institute of Engineering, Dr. D.Y.Patil Deemed to be University, Nerul, Navi Mumbai, India

Abstract. Data plays an important role where any prediction is to be made. Due to the advancement in technology there have been significant ways of collecting students' data. Educators use lots of supportive digital system to teach and collect data, such as Learning Management System, Student Information system, etc. The data collection through these system is huge and nevertheless it can be called as a big data. Lots of research is being done in how to improve the students' performance and overall educational performance and to tap key areas that may help in student progress. Traditional analysis of data includes sampling, whether it be a climate prediction or any performance prediction. In this research, the data collected through classroom observation, academic performance or learning management systems is used to predict the performance of the students. Lots of factors were present in the collected data but the main factors that add relevance for appropriate prediction are only few. This paper proposes how more than one feature selection algorithm results can be combined to improve the predictions with machine learning algorithm.

1 Introduction

The academic or non-academic performance of the students plays a vital role on education system. The performance of students in academics can be affected by various factors. Early predictions of students' performance help in taking prior measures to improve the performance [1] and tap the key areas of students. Data sets related to students can be collected through various means. Some of them are Learning management system, daily observations, quarterly or half yearly assessments results. All these datasets come with a lot of features relating to each student. It is not always all the features that are contributing in prediction accuracy. Some of the features turn out to be irrelevant or might mislead the prediction model. Also, many of the features can act as an outlier that can degrade the accuracy of the model. Selecting the right features for the prediction model is very important to maintain the consistency and accuracy of the results.

Feature selection is selecting related attributes or features to the target results. The features that are selected are from the same dataset. Feature selection plays an important role in data mining and machine learning. On the other hand, feature extraction is different than feature selection. In feature extraction, new variables or attributes are created from the dataset that defines similarity or group of common attributes [3].

The features that are selected from the dataset should improve the accuracy of machine learning models avoiding the overfitting issues. With the help of feature selection, the dataset dimensions can be reduced to a greater extent. Thereby, reducing the time required for computation or training the model. Machine learning is studying the computer algorithms that gets improved with tine and training. Machine learning is classified into supervised, unsupervised, semi supervised and reinforcement learning. This paper focuses impact of feature selection on supervised learning algorithms [4].

2 Background

2.1. Supervised machine learning

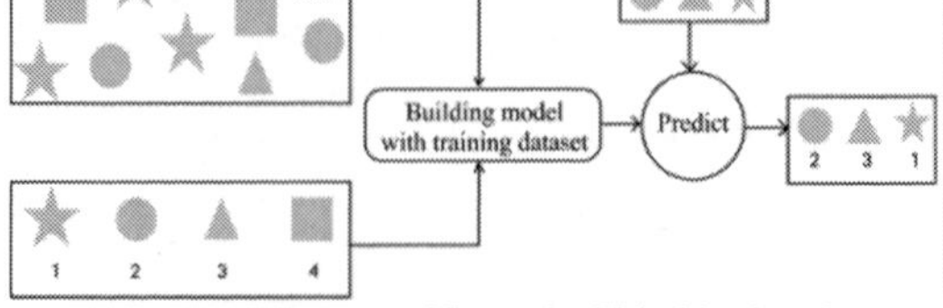

Fig. 1. Work Process of Supervised Machine learning.

* Corresponding author: snehal1335.p@gmail.com

The dataset is labelled in supervised learning. The model that is generated has knowledge of all the types of data in dataset. After the model is trained with the help of some training data, it is then tested using the test data from the same dataset.

2.1.1 Types of supervised machine learning

Supervised learning can be classified into Classification and Regression. Whenever the output of the variable is continuous regression machine learning is used. If the output or the target is discrete then classification machine learning is used [5]. Following are some of the commonly used classification machine learning algorithm in prediction of student academic performance:
- Decision Tree
- Naive Bayes
- K-Nearest Neighbour (KNN)

2.2 Feature selection and its methods

Feature selection is a process in which irrelevant and repetitive features are removed from the dataset. This is done to improve the performance of machine learning algorithms. The factors that defines the performance improvement are accuracy and time required for building the model. Feature selection methods are Filter based, Wrapper based and Embedded based. Out of these methods, filter based methods proved out to perform best when it comes to high dimensional data and time. Filter based methods works independently on each feature and thereby do not look for correlation among the features. In case of Wrapper method, the computational time is little more than filter based method but it looks for correlation between the features while interacting with the classifier [8]. Selection of relevant features is very important when it comes to improving the efficiency and accuracy.

2.2.1. Filter based methods

These methods are based on ranking of features. The ranking technique is easy and provides better results. Filter-based methods used the techniques of Chi Square test, Correlation coefficient, fisher score.

2.2.2. Wrapper based methods

These methods use a greedy search to retrieve the most important and appropriate features that helps in providing the best results. It is basically comprising two types of algorithm sequential and heuristic search.

The sequential search starts by considering an empty set and each feature is then added or remove from the set until the objective function is achieved. Whereas, the heuristic search algorithm basically works on evaluating the dataset to achieve the objective function [10].

3 Literature Survey

In 2019, Kumar, T. R., Vamsidhar, T., Harika, B., Kumar, T. M., & Nissy, R. has used four classification algorithms Naïve Bayes, ID3, C4.5, Random Tree, out of these four Naïve Bayes algorithm performed best in providing good classification of prediction. The main aim of the paper was to compare the classification of the methodology and to predict the student performance [5].

In 2017, Ihsan A. Abu Amra and Ashraf Y. A. Maghari has used two classification algorithm KNN and Naïve Bayes on educational dataset. The agenda was to make early prediction so that it can be used to increase student performance. They have concluded Naïve Bayes performed better than KNN algorithm. As a future work, they have suggested that more classification algorithms can be applied to different education institute datasets. [6].

In 2019, Mohammadi, M., Dawodi, M., Tomohisa, W., & Ahmadi, N. have used KNN, DT and Naïve Bayes classifiers were used on the dataset of 230 students of Kabul University to predict their GPA as high, medium and low. They concluded that KNN showed the highest accuracy rate among others. [7].

In 2016, Danasingh, Asir Antony & Balamurugan, Suganya & Epiphany, Jebamalar Leavline done a literature review on various featured selection based techniques on high dimensional data. They found that filter based methods are more efficient and takes less computational time than wrapper and embedded methods. Also, the filter based method performs much better with classification algorithms [8].

In 2016, Ms.Tismy Devasia1 ,Ms.Vinushree T P2, Mr.Vinayak Hegde3 used Naïve Bayyes algorithm to predict the student performance and proved that it performed better than Decision tree, Regression methods [9].

In 2021, Kaur, A., Guleria, K., & Kumar Trivedi, N. has made a comparative study of different feature selection algorithm and presented advantages and research gaps. They have provided an insight to various feature selection methods that includes filter based, wrapper based and embedded methods. [10] .

4 Proposed Work

The proposed solution aims to overcome the drawbacks of individual feature selection methods. Two feature selection methods can be combine to overcome the limitations of each method. Based on the literature review, we have used filter based and wrapper based feature selection method as a hybrid method to predict the academic performance of students.

 https://doi.org/10.1051/itmconf/20224403036

Algorithm:

Step 1: Select and analyse the dataset.
Step 2: Select features using Filter based feature selection approach.
Step 3: Select features using Wrapper based feature selection approach.
Step 4: Compare the selected features and choose the features that are commonly selected by both the approaches.
Step 5: Select the relevant machine learning algorithm that best fits with the type of dataset.
Step 6: Use the selected features to train the machine learning model and check the accuracy of the model.
Step 7: Compare the results of individual feature selection based approach with hybrid based approach.

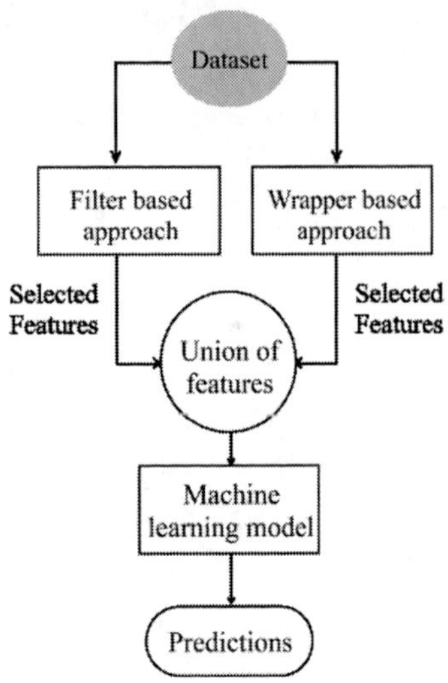

Fig. 2. Generalised Working of Proposed Method

5 Implementations and Results

5.1. Dataset overview and Pre-processing

The dataset is from the UCI Machine Learning repository. The data contains the students' performance in two subjects: Mathematics and Portuguese with additional features that may or may not impact the final grades achieved by the student. Term 1 score and Term 2 score has a strong relation with final grades, but some of the other features shows relevant impact on scoring final grade.

5.2. Feature selection methods used

Two Feature Selection approaches were used. Filter-based and Wrapper based. One of the filter based selection method is Chi-Square test. It is applied to evaluate the correlation in the dataset. In this test, the Chi-square between each feature and the target feature is calculated where the best chi-square feature is selected [10]. Chi-square test is ranking based algorithm and hence works with individual features, without looking out for the dependency on other feature set.

$$X^2 = \frac{(O-E)^2}{E}$$

O - Observed value
E - Expected value

The other approach used was Wrapper-based feature selection. One of wrapper feature selection method RFE (Recursive Feature Elimination) method was used to select the relevant features. RFE deals with the correlation with each feature and the target feature. It is a method that helps in fitting the model by removing the weakest feature until the specific number of features are related to the target.

Initially, the data was processed to categories the final grade into 4 classes. This 'final grade' column was kept as a target feature. The rest 32 features were used for training the model.

Firstly, with 32 features, a model based on decision tree, Naïve Bayes and KNN was build. Corresponding accuracy, precision, recall and f1 scores were also calculated.

Secondly, using Chi-square test and RFE, 10 features were selected by each algorithm. Later, these 10 features of each algorithm was compared and union was formed. It is observed that 15 features were ranked and selected by both the algorithms. Using this 15 features a model based on decision tree, Naïve Bayes and KNN was build. Corresponding accuracy, precision, recall and f1 scores were also calculated.

Later, it was observed that only selection of 15 features by both the algorithm have increased the accuracy, precision, recall and f1 score.

5.3. Results

Accuracy defines how the model have behaved or performed across all classes. It is calculated by the ratio between the correct predictions made out of the total available predictions.

Dataset (No. of Features)	Decision Tree	Naïve Bayes	KNN
Mathematics (32)	80 %	70 %	57 %
Portuguese (32)	79 %	48 %	68 %

Table 1. Prediction accuracy without feature selection.

Dataset (No. of Features)	Decision Tree	Naïve Bayes	KNN
Mathematics (10)	75 %	75 %	50 %
Portuguese (10)	72 %	57 %	59 %

Table 2A. Prediction accuracy with single feature selection method (Chi2)

Dataset (No. of Features)	Decision Tree	Naïve Bayes	KNN
Mathematics (10)	69 %	55 %	53 %
Portuguese (10)	79 %	54 %	47 %

Table 2B. Prediction accuracy with single feature selection method (RFE)

Dataset (No. of Features)	Decision Tree	Naïve Bayes	KNN
Mathematics (15)	87 %	84 %	67 %
Portuguese (15)	83 %	83 %	77 %

Table 3. Prediction accuracy with feature selection (hybrid method).

Dataset: Mathematics, No. Features: 32			
Dataset (No. of Features)	Precision	Recall	F1-Measure
Decision Tree	80	83	81
Naïve Bayes	67	74	68
KNN	69	45	47
Dataset: Portuguese, No. Features: 32			
Dataset (No. of Features)	Precision	Recall	F1-Measure
Decision Tree	68	77	72
Naïve Bayes	45	55	41
KNN	55	41	44

Table 4. Classification report without feature selection.

Dataset: Mathematics No. Features: 10 Feature selected using one method (Chi-2)			
Dataset (No. of Features)	Precision	Recall	F1-Measure
Decision Tree	74	77	75
Naïve Bayes	73	82	75
KNN	73	48	50
Dataset: Portuguese No. Features: 10 Feature selected using one method (Chi-2)			
Dataset (No. of Features)	Precision	Recall	F1-Measure
Decision Tree	69	76	72
Naïve Bayes	67	54	57
KNN	78	55	59

Table 5. Classification report with single feature selection (Chi-2) method.

Dataset: Mathematics No. Features: 10 Feature selected using one method (RFE)			
Dataset (No. of Features)	Precision	Recall	F1-Measure
Decision Tree	68	72	69
Naïve Bayes	65	67	65
KNN	76	50	53
Dataset: Portuguese No. Features: 10 Feature selected using one method (RFE)			
Dataset (No. of Features)	Precision	Recall	F1-Measure
Decision Tree	68	71	79
Naïve Bayes	67	61	54
KNN	49	45	47

Table 6. Classification report with single feature selection (RFE) method.

Dataset: Mathematics No. Features: 15 Features selected by hybrid approach			
Dataset (No. of Features)	Precision	Recall	F1-Measure
Decision Tree	84	87	85
Naïve Bayes	86	84	85
KNN	65	59	59

https://doi.org/10.1051/itmconf/20224403036

Dataset: Portuguese No. Features: 15 Features selected by hybrid approach			
Dataset (No. of Features)	Precision	Recall	F1-Measure
Decision Tree	71	79	74
Naïve Bayes	70	79	74
KNN	82	81	81

Table 7. Classification report with feature selection.

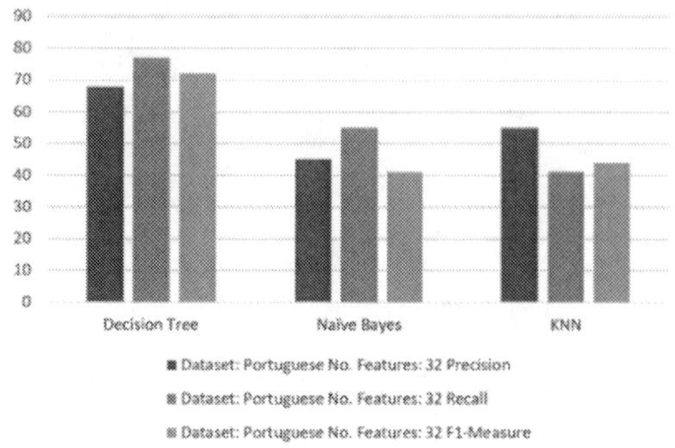

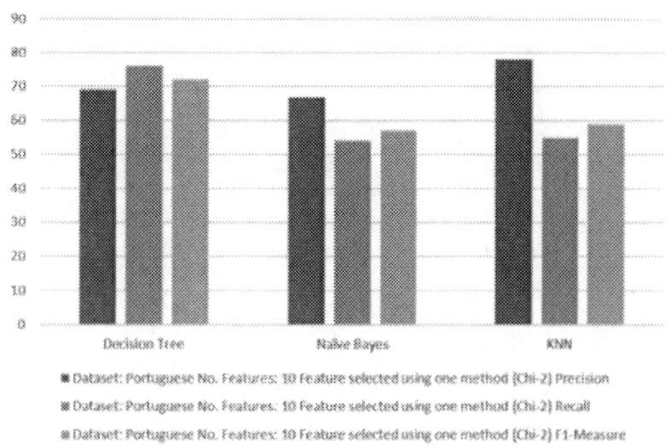

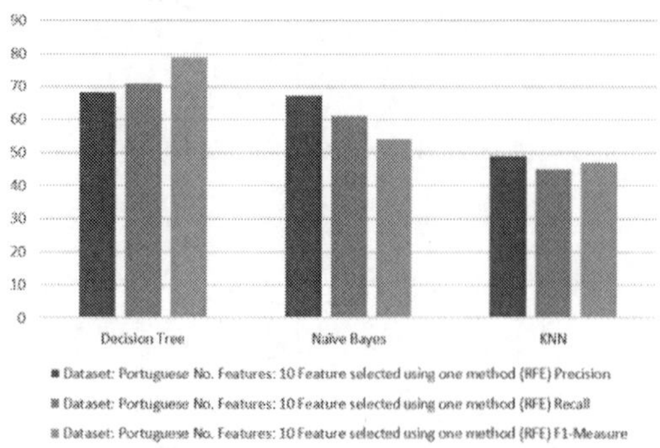

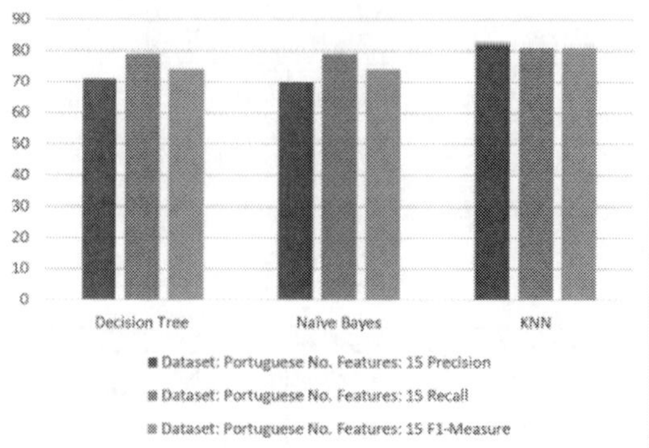

Fig. 3. Classification report comparative graph for Mathematics dataset

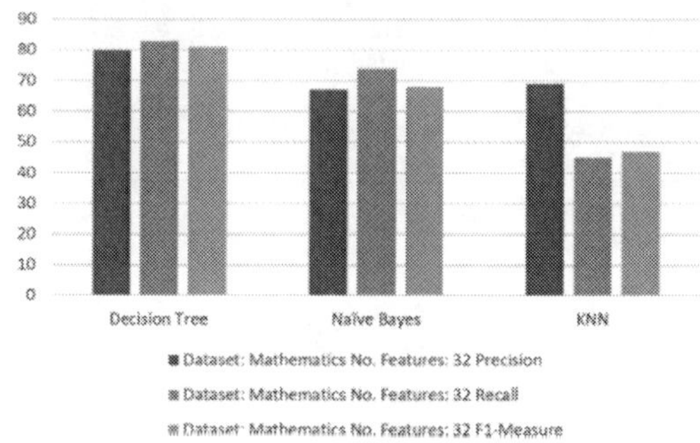

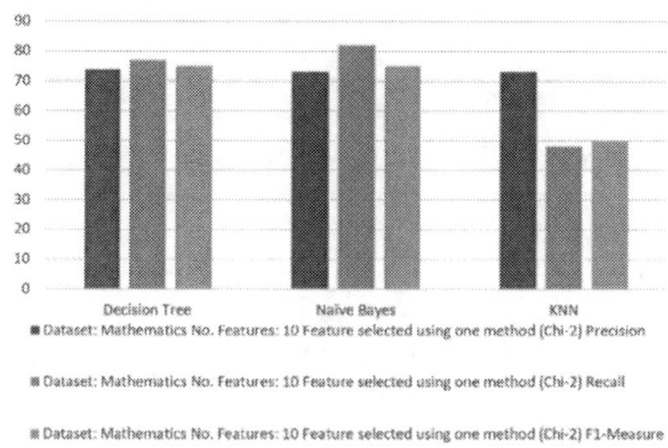

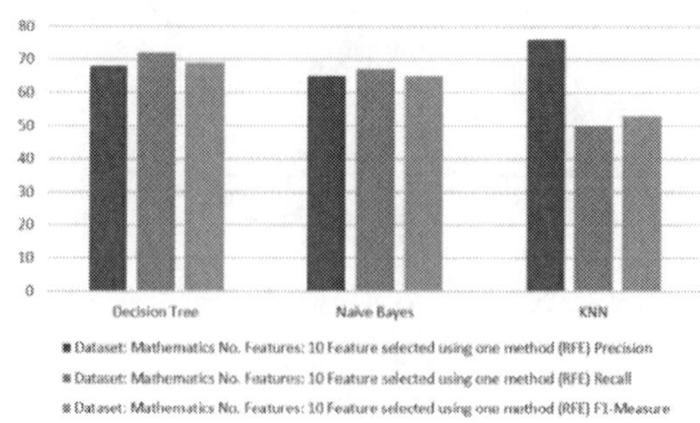

https://doi.org/10.1051/itmconf/20224403036

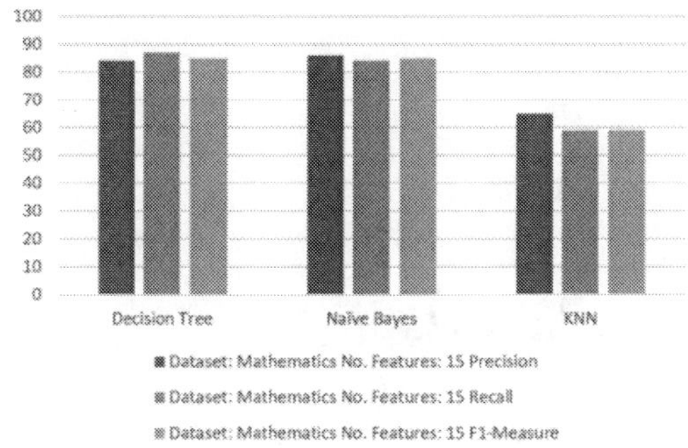

Fig. 4. Classification report comparative graph for Portuguese dataset

Conclusion

Feature selection plays a very important role in machine learning and prediction. With hybrid feature selection the number of features were reduced to more than 50%. Also the accuracy of the different machine learning algorithm prediction has increased to a greater extend. Hence by considering two different feature selection output and providing a union of these output to a machine learning algorithm will increase the prediction accuracy.

References

[1] Gull, H., Saqib, M., Iqbal, S. Z., & Saeed, S. (2020). Improving Learning Experience of Students by Early Prediction of Student Performance using Machine Learning. 2020 IEEE International Conference for Innovation in Technology (INOCON).

[2] Thomas, R. N., & Gupta, R. (2020). Feature Selection Techniques and its Importance in Machine Learning: A Survey. 2020 IEEE International Students' Conference on Electrical, Electronics and Computer Science (SCEECS).

[3] Aparna U.R. and S. Paul, "Feature selection and extraction in data mining," 2016 Online International Conference on Green Engineering and Technologies (IC-GET), 2016, pp. 1-3.

[4] T. R. N and R. Gupta, "A Survey on Machine Learning Approaches and Its Techniques:" 2020 IEEE International Students' Conference on Electrical, Electronics and Computer Science (SCEECS), 2020, pp. 1-6.

[5] Kumar, T. R., Vamsidhar, T., Harika, B., Kumar, T. M., & Nissy, R. (2019). Students Performance Prediction Using Data Mining Techniques. 2019 International Conference on Intelligent Sustainable Systems (ICISS).

[6] Amra, I. A. A., & Maghari, A. Y. A. (2017). Students performance prediction using KNN and Naïve Bayesian. 2017 8th International Conference on Information Technology (ICIT).

[7] Mohammadi, M., Dawodi, M., Tomohisa, W., & Ahmadi, N. (2019). Comparative study of supervised learning algorithms for student performance prediction. 2019 International Conference on Artificial Intelligence in Information and Communication (ICAIIC).

[8] Danasingh, Asir Antony & Balamurugan, Suganya & EPIPHANY, JEBAMALAR LEAVLINE. (2016). Literature Review on Feature Selection Methods for High-Dimensional Data. International Journal of Computer Applications.

[9] Prediction of students' performance using Educational Data Mining. 2016 International Conference on Data Mining and Advanced Computing (SAPIENCE).

[10] Kaur, A., Guleria, K., & Kumar Trivedi, N. (2021). Feature Selection in Machine Learning: Methods and Comparison. 2021 International Conference on Advance Computing and Innovative Technologies in Engineering (ICACITE).

Smart Bus System for Specially Challenged Individuals

Amey Mithbavkar[1,*], *Bhavesh* Salavkar[1,**], and *Anuja* Avhad[1,***] *Supriya.Y.* Bhuran[1,****]

[1]Department of Instrumentation Engineering, Ramrao Adik Institute of Technology, Nerul,Navi Mumbai

Abstract. This paper introduces a system design about bus management system based on Wi-Fi module which implemented the basic functions of the intelligent public transport management system, such as monitoring the time of bus arrival, departing from the bus station and reporting stations name automatically. This system can ensure the punctuality of vehicles to run, improve the automation level of reporting stations, and quality of public transport service. The management system has low cost and thus it is more feasible. This System can be a brilliant helping hand for the individuals who are mentally and physically challenged and help them in reaching their respective destinations on time because, in our respective transportation system, the priority will be to drop the physically and mentally impaired individuals as well as senior citizens safely on time.

Keywords: Wireless Communication, RFID Reader, ESP8266, Cloud Platform.

1 Introduction

The mode of transportation through buses has increased since the pandemic and it is still a safer option than trains. Normally while in public transportation there is a lot of rush which gathers on the Bus Stops or Railway Stations. In these situations, those who are physically or mentally challenged and even senior citizens can't cope with the speed of normal people, to hop in the bus or train and people are not always kind enough to help them. So the creation of this Smart Bus System is basically to help the visually impaired people or the individuals who are physically or mentally challenged.

The main aim of this paper is to promote Wireless Communication. To promote Wireless Communication the best components which the authors could add to their paper are ESP8266 NodeMCU, RFID Tags and Detector and finally output would be displayed on the Cloud Platform to the Operator. The Cloud Platform displays the status of all bus stops and whether or not the bus has arrived at the bus stop. To indicate the arrival of the buses ON the bus stop the "LEDs" are used.

This can be proven to be a really effective system, as the dependency on humans decreases, the authors of this paper have brought up good alternatives to cover up all the man power with effective Wireless Communication. This system can also be in useful to normal passengers but the main priority will always be given to the specially challenged people.

*e-mail: mithbavkaramey44@gmail.com
**e-mail: salavkarb@gmail.com
***e-mail: anujaaavhad2@gmail.com
****e-mail: supriya.bhuran@rait.ac.in

1.1 Literature Survey

The idea of this paper is basically to create a transportation system and keep this transportation through bus prioritized for mentally and physically challenged people, senior citizens, and visually impaired people. D. Adhikari used Cloud Platform to inform the passengers regarding the arrival of the bus [1].

The application and functions of RFID used for reserving seats in transport system and announcements [2]. The NodeMCU is also used for Wireless Communication. The LEDs and buzzer system has been used by P. Herdiansyah for arrival notification of buses. This can act as a huge alerting method to display the impaired or challenged Individuals to understand if their respective bus has arrived or not [3]. The impact of RFID Readers and how accurate they are for detection can be seen in work proposed by M. J. Shah. The Importance of RFID Readers in Vehicles have been discussed. The RFID Reader is used to read the tag of the BUS, then the system will forward this data to the man agement [4].

The amazing system is developed where wireless communication has been used and the ESP8266 is used to convert the signal into data lines and NodeMCU ESP8266 is an integral part of the work developed by G. Suprianto [5]. A major understanding of how buses can be tracked is very well explained by A. Kumar Sharma [6]. The idea to track whether the bus has arrived at the bus stop or not so the implementation of a tracking system has been developed.

The real time data such as location, route, speed, list of passengers can be track and recorded by J. T. Raj [7]. RFID or GPS Technologies connected to a remote server over the Wi-fi Module such as ESP8266. With the help of Wireless Sensors, the alert systems in Smart Bus has been created. In this paper, the authors have mentioned that the Visually Impaired Individuals on the Bus Stop are provided with a mobile application recognized by GPS. Smooth the transportation executed with Wireless Communication by Manali and N.Z. Naqvi. They have executed the tracking system through a GSM module [8] [11]. An android application is used to display all the updates even system is in offline mode [9].The Arduino is used to control the GPS Modules to get the coordinates. This work also displays the major role of Arduino and how with the knowledge of C Language knowledge, any work can be handled efficiently [10] [12].

2 Operation of the Proposed System

The concept behind this work is a basic way to make travel easy for disabled people so the creation of a simple bus stop where the authors have put few buttons which are hosted as an indication to alert the bus driver to halt at a particular stop and then when the driver reaches the exact location an indication bell will indicate the person about arrival of the bus and now the person can board the bus for his destination. The process for this entire work starts with the bus moving from the bus depot to the route which has been assigned to the driver. If the driver gets the notification on its screen that he/she has to halt at some particular stop on their respective route then he/she will halt at that particular stop. Pick-up the passengers and move ahead to their next destination. This is the basic way the system will work.

The communication of the system with other devices is done with the help of a cloud-based service created. When the bus moves towards its assigned route the bus has a system fit inside it with an ESP8266 NodeMcu and an RFID reader which assists the driver at which stop he/she has to halt and will help to understand whether he/she has halted at the right bus stop or not.

This message from the bus depot server will go the driver that the bus has stopped at the right bus stop or not if not the driver gets the notification that, 'Driver has missed the bus stop'. The RFID Card will be installed on each bus stop to know about the particular Bus stop. Here the bus has to be halt when the bus reaches that bus stop. The RFID card is installed at the bus stop and RFID reader in the Bus, will read the RFID Signal and then send it to the ESP8266 Data line. The data line will create a signal and send it to the cloud-based server that the bus has attended the flag host calls of the disabled people from the different bus stops and has a successful journey on defined route. This complete system is visually depicted in Figure.1.

2.1 Block Diagram of the System

Power Supply (for BUS): The power supply block in the diagram is the power source for the system which step downs the voltage from 230V to 5V as essential power required for the system. After the voltage is stepped down to 5V, it is further transferred to the other blocks of the system.

Microcontroller (for BUS): The Microcontroller block consists of the ESP8266 WiFi module chip which

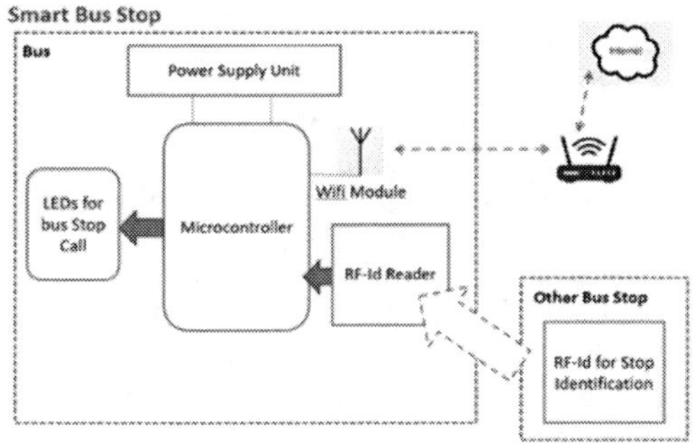

Figure 1. Block diagram of the bus

acts as a communication device between the cloud server and also communicates with the Bus and the Bus Driver with the help of the cloud server data. The Microcontroller controls the functioning of the Bus data and the Bus system.

RFID Reader: The RFID Reader in the system communicates with RFID cards which are placed at the flag host pole on the Bus stops which gives the information that the particular bus has received the particular passenger from the particular bus, and after the entire process it gives the signal to the microcontroller to send it to the cloud server. After the process is done then only the bus moves further to its next destination.

LEDs for Bus Stop call: This LED's give the information to the bus driver to stop at the particular bus stop to pick up the passenger from that stop who is specially challenged passenger who has sent a notification to the particular bus he wants to board it and wants to reach his destination. This way the LED block works in the system.

Figure 2. Block Diagram of System Created for Bus Stop

Power Supply (for BUS Stop): The power supply block in the Figure 2 is the power source for the system which step-downs the voltage from 230V to 5V as essential power required for the system. After the voltage is stepped down to 5v then it is further transferred to the other blocks of the system.

Microcontroller (for BUS Stop): The Microcontroller block consists of the ESP8266 WiFi module chip which acts as a communication device between the cloud server and also communicates with the Bus and the Bus Driver with the help of the cloud server data.

Here they also have an interaction with the pushbuttons which have information of the bus beside them which helps the specially challenged people to board at the right bus which can help them to reach their particular destination. They have also LED's near the push button to know that they have pressed the right button. And the buzzer system is also available which helps them to get notifications about the bus when it reaches the bus stop so that they can board their particular bus to reach their destination.

An RFID card is paced with the push button at the flag host pole near the bus stop with the bus stop identification card which is RFID based.

2.2 Hardware and its functions

In this Smart Bus System the first hardware instrument is RFID Tags. These RFID Smart Cards would be installed on all the respective smart bus stops. The RFID Card Reader will be reading these cards. The role of RFID Card Reader and Detector plays a crucial part in this system.

When the bus arrives at the bus stop, it will detect the RFID card, it delivers the message to the ESP8266 NodeMcu data line, that the bus has arrived at the bus stop. Arriving to the next equipment is ESP8266, The ESP8266 NodeMcu data line will create a signal and send it to the cloud-based server that the bus has attended the flag host calls of the disabled people from the different bus stops and has a successful journey on his defined route. Also another part of a hardware unit is Buzzer.

The buzzer is used in our system to alert disabled individuals about the arrival of the bus. Especially the individuals who are visually impaired will be alerted that their bus is arriving and they can get ready to enter the bus. It can be used as an alarm for visually impaired individuals.

Coming to next part of the Hardware is LED'S. There are LED's attached to the system as well. When the bus arrives at bus stop no 1, LED1 will glow and when the bus will arrive at bus stop no 3 the LED 3 will glow and similarly for other bus stops as well.

2.3 Software used and its functions

Arduino IDE is the software for Arduino. It is a text editor like a notepad with different features. It is used for writing code, compiling the code to check if any errors are there and uploading the code to the Arduino. It supports C/C++ language.

It is open-source software, where the user can use the software as they want it to. They can also make their own modules/functions and add them to the software. Not Exactly a Software but an Cloud Platform is used in this Work to notify the operator whether the bus has arrived at the stop or not.

Cloud Platform is where the final data is received. The Cloud Platform keeps the operator updated on the situation regarding the buses. So the cloud platform is a link between the buses and the operator.

2.4 Connection Diagram

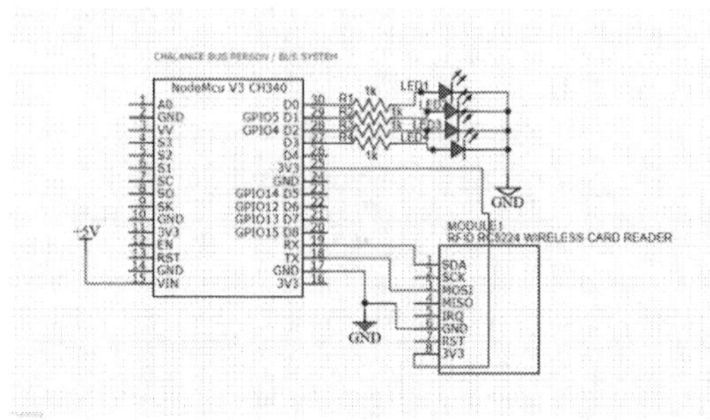

Figure 3. Connection Diagram of Bus

In this circuit the authors have used ESP8266 as the brain of the entire system and RC5224 RFID Card Reader as the heart of the system. Bus Depot with Complete System is clearly shown with Wiring Components in Figure 3 and Figure 4.The RC5224 has a major role in the entire system after the ESP8266 because it helps to collect the data from the bus stops and to send it further to the Bus Depot.

Where RC5224 is placed at the wind screen of the bus and the RFID tags are placed at the bus stop poles where both RFID Card and RFID Card Reader can interact and match their data and verify with the help of cloud server and ESP8266 . The ESP 8266 is also connected with few bus stop where they can get the info of the specially challenged people which gives the bus driver the exact notification of that person and that bus stop which may assets both person on the bus stop and the bus driver to get the exact information of each other . This way the entire Bus System works in the bus with interface of ESP8266 and RC5224 .

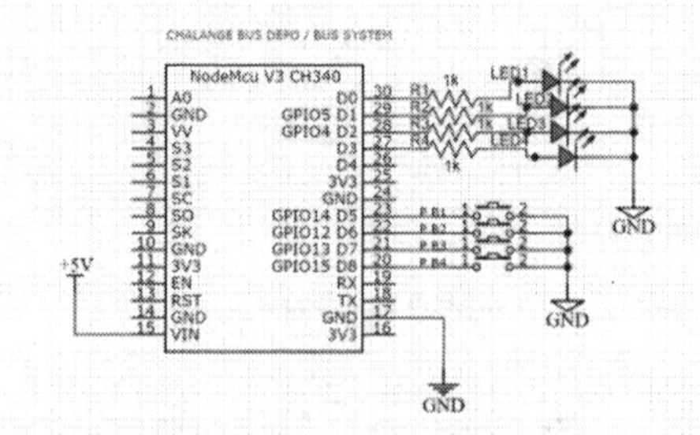

Figure 4. Connection between Bus Depot and Bus System

In this entire circuit an ESP8266 WIFI chip is used which is the heart of the entire system mainly it interacts here with the cloud server and the push buttons. The process starts when the push button is pressed from a certain bus stop location by specially challenged person , it send the signal to the cloud server and after the signal is reaches the cloud server the info of that bus stop is displayed on the cloud server .

This entire cloud server of each bus can be monitored from a single place such as the bus depot. After the interaction of the cloud server is then there will be an interaction of an led blinking at the bus stop and a buzzer notification at the bus stop to let the know that specially challenged person has press the button for the bus driver to pick him from the location and help him to reach at his destination. This way here the circuit at the bus and the bus stop works with the interface of the ESP8266

2.5 Implementation Methodology

The process to implement the entire system is with the help of the ESP8266 , RFID Cards, RFID reader, LED's, etc. These elements can make a management system for the challenged people to make their travel easy and convenient and they can reaches their destination safely and easily. The implementation of this entire system is less complicated but working of system on Cloud platform with Wi-Fi network is challenging task which is the major point and the much essential way of transport necessary of the challenged people.

The system instillation starts with the process of the implementation of the bus stop at the different location for the challenged people. This bus stop contains a **braille script** which informs the blind people with the information of the bus and bus number and a push button which will be indicate in the front of the bus number and the braille instruction sheet where he can push the button and give the info to the specific bus and the bus driver to stop at the particular bus stop.

The bus driver will also get the LED notification of the bus stop where the challenged person is at the particular bus stop. After the driver gets the notification the bus moves towards his pickup and then the driver reaches the bus stop. An RFID reader is placed on the windscreen of the bus and the RFID is paced at the pole of the bus stop as shown in Figure 6 and Figure 7. The RFID reader reads the bus stop data and send a signal to the bus stop buzzer which gives an indication to the special challenged person if he is deaf he gets an blinking LED notification on the bus stop board where the challenged person gets to know that his bus has been arrived and he can board the bus.

The RFID reads the RFID card at a 100 meters of the distance of the bus stop. So the challenged person can get ready to board the bus and start his journey for his destination where he has to reach. After boarding the bus the person will be questioned for his destination and the bus attendant will record to start journey of passenger and when he reaches his destination he get an notification one stop before his destination stop that, 'The destination is going to arrive soon' and the person should be ready to be abort from the bus . There is also an online stop navigation for the bus to know whether the defined person has entered the bus or not. If not he will get notification from the depot and if bus is canceled due to some issues the LED on the bus stop will not glow.

3 Result

Here the creation of a website is done, that displays to the operator whether the bus has arrived at the bus stop or not and otherwise the bus has departed from the respective bus stop. As seen in the image when it's time for the bus to arrive at the bus stop.

A Call signal is sent to the bus to reach the respective bus stop. So accordingly when the bus driver receives a message from the operator the bus driver departs with the bus to reach the bus stop.

On the Server you can be seen the status of the Bus whether it has arrived on the bus stop or not. If the bus is arrived on the bus stop a message will be dropped on the server "Done".

When the bus driver picks up the passenger from the bus stop, the status "Done" is displayed on the server as shown in Figure 5.

4 Conclusion

This paper of Smart Bus Systems will provide a huge benefit to the mentally and physically challenged individuals as well as the senior citizens. They can travel accordingly to their comfort and with the help of RFID Detector and RFID Smart Cards, the passengers waiting at the bus stop can also be alerted of the bus arriving at the bus stop. A Buzzer is used in the system to alert the passengers at the bus stop. In the future, the system can also be updated with an LCD screen to alert the passengers of the buses arriving and departing from the bus stop.

5 Future Scope of the Paper

The safety for physically impaired and mentally impaired people will be high. They will be dropped off to their destination without any hesitation. The whole process of this work can be carried through Wireless Communication. The Demand of Wireless Communication will increase in the future.

The operator and the bus driver of this Smart Bus System will be available for their safe and smooth transportation. This Smart System can be a great alternative for them rather than traveling with the general public in huge numbers. In coming years the injuries and accidents of visually or physically impaired or mentally impaired individuals also would drop down.

References

[1] D. Adhikari and J. H. Park, "Bilateral Bus Information System in Smart City," 2020 IEEE International Conference on Consumer Electronics - Asia (ICCE-Asia), 2020, pp. 1-4, doi: 10.1109/ICCE-Asia49877.2020.9277086.

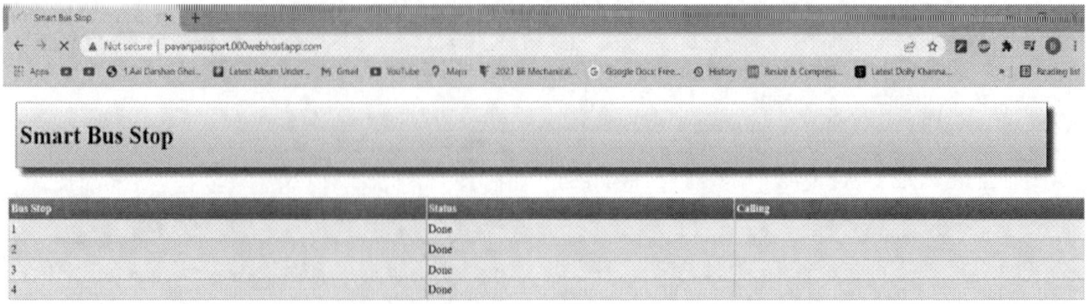

Figure 5. Web Page Showing the Information of the BUS and its BUS Stop

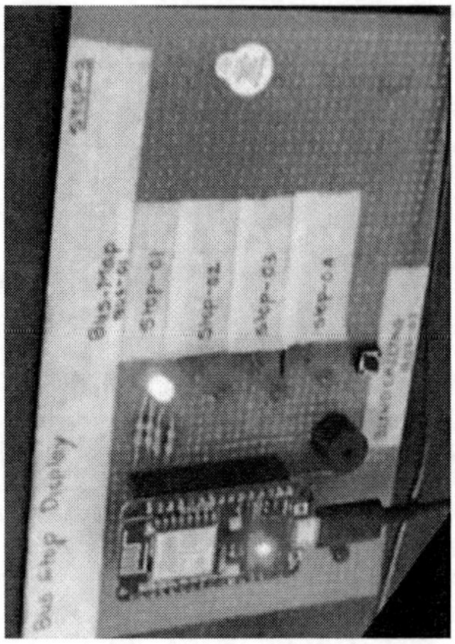

Figure 6. Implementation of the system near the BUS DEPOT

Figure 7. Installation of RFID Reader on each BUS

[2] P. Chowdhury, P. Bala, D. Addy, S. Giri and A. R. Chaudhuri, "RFID and Android based smart ticketing and destination announcement system,".2016 International Conference on Advances in Computing, Communications and Informatics (ICACCI), 2016, pp. 2587-2591, doi: 10.1109/ICACCI.2016.7732447.

[3] P. Herdiansyah, G. I. Hapsari, G. A. Mutiara and L. Meisaroh, "Smart Bus Arrival Notification for Visually Impaired," 2019 International Conference on Electrical Engineering and Informatics (ICEEI), 2019, pp. 614-618, doi: 10.1109/ICEEI47359.2019.8988912.

[4] M. J. Shah, R. P. Prasad and A. S. Singh, "IOT Based Smart Bus System," 2020 3rd International Conference on Communication System, Computing and IT Applications (CSCITA), 2020, pp. 130-134, doi: 10.1109/CSCITA47329.2020.9137816.

[5] G. Suprianto and Wirawan, "Implementation of Distributed Consensus Algorithms for Wireless Sensor Network Using NodeMCU ESP8266,"2018 Electrical Power, Electronics, Communications, Controls and Informatics Seminar (EECCIS), 2018, pp. 192-196, doi: 10.1109/EECCIS.2018.8692952

[6] A. Kumar Sharma, R. Pandey, S. Tarafdar and S. Dubey, "Towards Smart Mobility in Cities - Bus Tracking and Booking System," 2021 9th International Conference on Reliability, Infocom Technologies and Optimization (Trends and Future Directions) (ICRITO), 2021, pp. 1-5, doi: 10.1109/ICRITO51393.2021.9596492.

[7] J. T. Raj and J. Sankar,"IoT based smart school bus monitoring and notification system," 2017 IEEE Region 10 Humanitarian Technology Conference (R10-HTC), 2017, pp. 89-92, doi: 10.1109/R10-HTC.2017.8288913.

[8] Manali and N. Z. Naqvi, "Smart public transport system using mobile phone based sensing," 2015 Annual IEEE India Conference (INDICON), 2015, pp. 1-5, doi: 10.1109/INDICON.2015.7443592.

[9] A. P. Keong, K. G. Smitha and S. Sinha, "Smart Nation: Offline Public Transport Made Easy," 2019 4th International Conference on Intelligent Transportation Engineering (ICITE), 2019, pp. 175-179, doi: 10.1109/ICITE.2019.8880262.

[10] M. T. Kamisan,A. A. Aziz, W. R. W. Ahmad and N. Khairudin, "UiTM campus bus tracking system using Arduino based and smartphone application,"2017 IEEE 15th Student Conference on Research and Development (SCOReD), 2017, pp. 137-141, doi: 10.1109/SCORED.2017.8305406

[11] H.Gull, D. Aljohar, R. Alutaibi, D. Alqahtani, M. Alarfaj and R. Alqahtani, "Smart School Bus

Tracking: Requirements and Design of an IoT based School Bus Tracking System," 2021 5th International Conference on Trends in Electronics and Informatics (ICOEI), 2021, pp. 388-394, doi: 10.1109/ICOEI51242.2021.9452818

[12] R. K. Megalingam, N. Raj, A. L. Soman, L. Prakash, N. Satheesh and D. Vijay, "Smart, public buses information system," 2014 International Conference on Communication and Signal Processing, 2014, pp. 1343-1347, doi: 10.1109/ICCSP.2014.6950068.

https://doi.org/10.1051/itmconf/20224403038

SMAP - A Stock Market Analysis and Prediction Web Application

Ninad Patil[1,*], *Raj* Mutha[2,**], *Veer* Pandey[3,***], *Rohan* Patil[****] , and *Vanita* Mane

[1]Department of Computer Engineering, Ramrao Adik Institute of Engineering, Dr. D. Y. Patil Deemed to be University
[2]Nerul
[3]Navi Mumbai

Abstract. The Stock Market is one of the most rapidly emerging markets in the country as the participation of retail investors has increased significantly in this pandemic.Since there are no specific rules to compute or predict the price of a stock, it becomes very difficult for the first-time investors to invest in the share market. Methodologies such as technical analysis, fundamental analysis, time series analysis, and statistical analysis, etc. are all used to estimate the price of the stocks but since stocks being volatile in nature, none of these methods are proved as a consistent tool to predict the upcoming trends. Machine learning and statistics can be used to predict and reduce the risk factor of loosing money. Apart from predicting future price trends based on the historical data of the company, the web app also show's various decision making factors and fundamental indicators for providing multiple options to users for consideration while making a call. With the help of statistical analysis, the relation between the selected factors and share price is formulated for computing better results. Using facebook Prophet without any hyperparameter's forecast for particular stocks are made. These result's are evaluated by comparing with other models like Last value and prophet with hyperparameter tuning.

1 Introduction

Foreseeing how the financial exchange will perform is quite possibly the most troublesome thing to do. There are such countless elements engaged with the forecast - actual variables versus mental, reasonable and nonsensical conduct etc.Taking factors viable like the most recent declarations about an association, their quarterly income results, and so forth, AI strategies can possibly uncover examples and bits of knowledge we didn't see previously, and these can be utilized to make unerringly precise expectations. In this undertaking, SMAP will perform qualitative analysis on stocks of different companies, their valuation over a period of time along with the sales volume and on that basis, it will predict the risk we put in investing those companies along with the prediction of future stock trends. With the help of a web-based approach, all these information will be available on a website along with other additional information for making better decision while investing in the shares of that particular.

2 Literature Survey

Existing systems were based on particular algorithms such as Artificial Neural Network, Support Vector Machine, Time Series Linear Mode and Hidden Markov Model. It is good for performance, optimization and efficient but failed to provide Real - Time Accuracy[3]. In Hidden Markov Model the advantage of this model is that it uses for optimization purposes. But the model is not able to evaluate, decode and learn from the present data-set[3]. It uses technical indicators as its parameters. Artificial neural network being the main technique in projects. It provided the model with advantages like Better performance compared to regression and lower prediction errors for small noise variation. But as the noise variation increases the prediction got worse. This proves to be the disadvantage of the model[1]. The model where the real information is incorporate to the best straight model. As the customary and occasional patterns came into information, the forecast rate fell down[8]. Model which utilized far reaching profound learning framework it is seen that it didn't lose a lot of exactness when applied to an example from outside the preparation test. Yet, overstate to minor changes in the preparation information which decline the prescient capacity. The parameter used in this model is Consumer investment, net revenue, net income, price per earnings ratio of stock, consumer spending[9]. The ARIMA (Autoregressive Integrated Moving Average) model is one of the most robust and efficient models in our study. But the disadvantage with it is that it is suitable for only short term use[10]. Systems which used supervised machine learning had advantage that previous time points to the input layer contain inputs. Disadvantage is that it is possible to feed those words in through a much smaller set of input nodes[5].

*e-mail: ninadpatil1452@gmail.com
**e-mail: rajmuta1684@gmail.com
***e-mail: pandeyveer45@gmail.com
****e-mail: rohanppp1232@gmail.com

3 Problem Statement

Stock market appears in the news every day. This pandemic has made people realize that just saving money is not enough but investing it can only help in beating inflation. As the number of Demat accounts opened has increased, the rate of investment and business opportunities in the Stock market has increased. If a portal/web application is developed which can predict the short-term price of an individual stock, it can be very helpful for the first-time investors in their investing journey. Existing solutions on the above scenario have either a web interface with all the stock information or there's prediction made on future stock trends on limited companies with no other information about the stock. To eradicate the problems discussed above, a web app is developed with users having the liberty to choose any stock for detailed analysis with stock price trends and other fundamental indicators of that particular stock. The reason behind using Prophet model is that it doesn't require a lot of knowledge or experience forecasting. It performs better compared to others in majority of cases. The forecast is faster as it fits models in stan.

4 Proposed Methodology

With the help of Flask which is a web-framework of python ,which deploy our model for predicting real- time future trends[11]. It provides the web interface which we used to create the web application. The data of the stocks which SMAP uses for the prediction is provided by YAHOO Finance API. Integration of YAHOO Finance API is done for historical data along with other corporate actions and fundamental indicators. The data set is further trained and predicted using a Additive model which is a nonparametric regression method. This methodology is provided by the FB Prophet model[7]. A forecast is generated for the predicted price of the stock which is further plotted using a python library know as Plotly.

As shown in the generalized flow of the system figure 1 here the user selects his/her required stock. The application fetches the data from YAHOO finance and further using the model the predictions are made. The user is later able to see the graph of the future prediction.

In the core of Prophet which is an additive regression which has four main components:

- Logistic or linear piecewise growth curve trend. By selecting change points from the data prophet easily detects changes in the trend.
- A yearly seasonal component modeled using Fourier series.
- A weekly season component using dummy variables.
- A user-provided list of important holidays.

Prophet will provide a components plot which graphically describes the model it has fit as show in Figure 2

- Data is fetched using the Yahoo Finance API. The Adjacent closing values is used by the model for the forecasting purpose.
- FBProphet train's using given data. Further using additive model it fits the non-linear trends and gives the forecast.
- The given forecast and plotting data is given to the Plotly library which is used to plot the graph for visual perspective.
- The non-linear trends which are fit with daily, yearly and monthly effects are shown in Figure 2

5 Implementation and Results

The prophet is integrated to a web application using Streamlit. Streamlit is a app framework which is open source. With the use of the API it automatically updates in the source file[11].
The data is collected for every stock using Yahoo finance API. Every stock is provided with a unique Ticker which is used to refer the company in the database. Using the download method which is provided with fields like the ticker, start and current date, the data is downloaded[7]. A sample input of data is shown in table 1

Table 1. A sample of input dataset

	Date	Open	Close	High	Low	Adj Close
0	Feb 1,2021	133.75	135.38	130.93	134.14	133.93
1	Feb 2,2021	137.25	133.75	137.25	133.74	134.21
2	Feb 3,2021	137.35	137.42	135.86	136.76	136.76
3	Feb 4,2021	137.5	137.42	135.86	136.76	136.76

For the proposed application Facebook prophet is the basic block for prediction. Though there are many methods for predicting the stock price and trend, it was decided to used Facebook prophet model as it is able to generate prediction of a good quality and at scale.
Prophet, is a open-source software which is developed by Meta data Science team. It is based on additive model where non-linear trends are fit with yearly, weekly and daily effects. It performs best when it has a large amount of historical data. It overcomes missing data and shift in trends problems[12].
In this application prophet tool is used for the prediction purpose. The user selects the amount of data to be fetched from the API and feed it into the model. Based on the input data, the model trains the dataset and forecast is done. An example of using Facebook's prophet tool to predicting the price of APPL stock is shown in Figure 3 for next year[1].

To check the effectiveness of the method used, we used the root mean square error (RMSE), mean absolute percentage error (MAPE), and mean absolute error (MAE) metrics. For any metrics, the lower the value the better the prediction.

The MAPE is use to calucalte the overall accuracy of the forecast and it's equation is given as:

$$MAPE = 1/n \sum_{i=1}^{n} \frac{y_{forecast} - y_{true}}{y_{true}}.100\% \quad (1)$$

 https://doi.org/10.1051/itmconf/20224403038

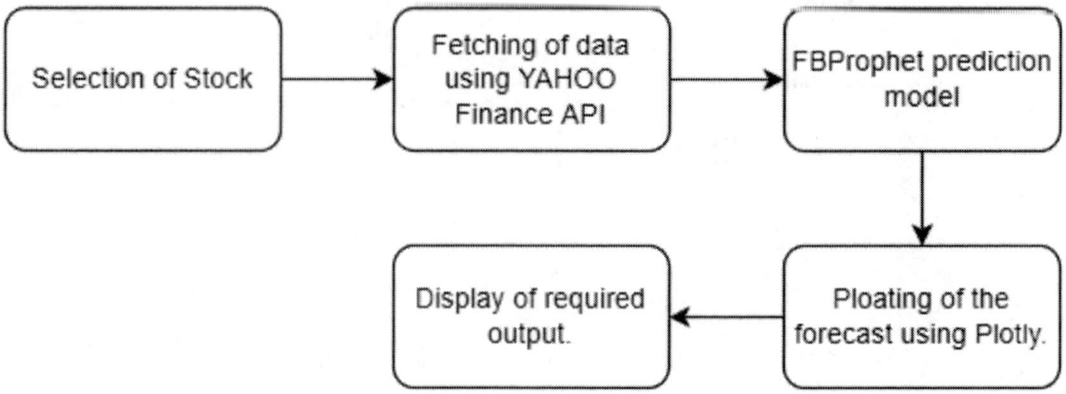

Figure 1. Generalized flow of the SMAP

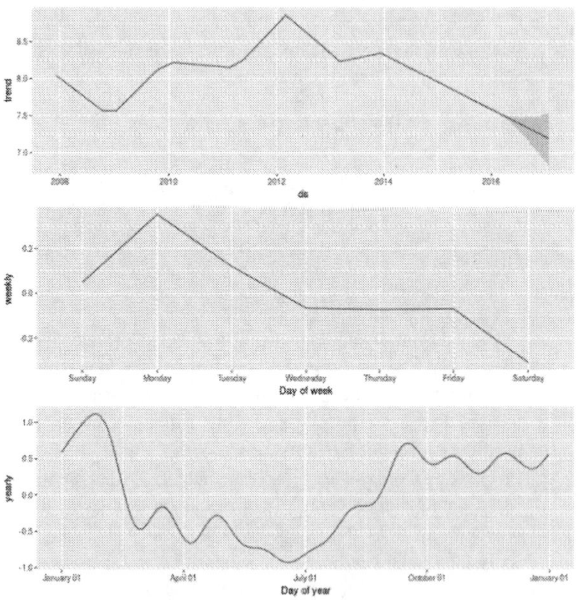

Figure 2. Component plot which describes the model

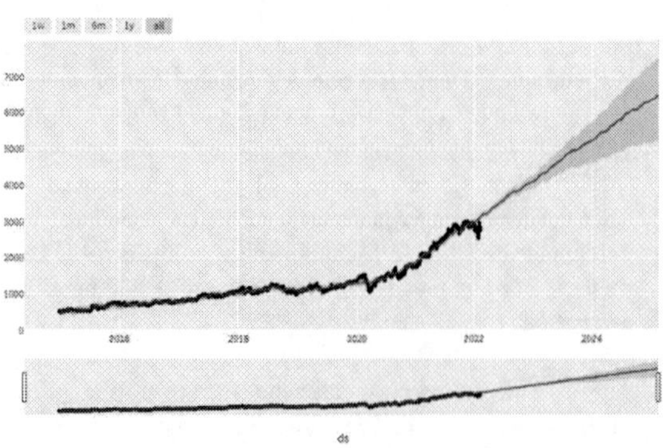

Figure 3. Forecast of the stock using Prophet

Root mean square mistake is perhaps the most usually involved measure for assessing the nature of forecasts. It shows how far expectations tumble from estimated genuine qualities utilizing Euclidean distance.

$$RMSE = \sqrt{\frac{\sum_{i=1}^{n} \|y(i) - y\wedge(i)\|}{y}} \tag{2}$$

The average of all errors is Mean absolute error. It's equation is given as:

$$MAE = (\frac{1}{n}) \sum_{i=1}^{n} |y_i - x_i| \tag{3}$$

The last value method is used as a benchmark in many models. The prediction are set as last observed value. The current adjusted closing price are set as the previous day's adjusted closing price. We would be compared this methods with prophet which we have used in our paper. In SMAP, the data has been taken of 3 years and based on that produced a forecast for 1 year. The forecast gives a RMSE of 1.89, MAPE of 1.59% and MAE of 1.80[7].

Prophet without Hyperparameter tuning gives a forecast with RMSE of 3.4, MAPE of 2.87% and MAE of 3.25. Further we applied hyperpapramets to improve the forecast.

- Prophet with Hyperparameter Tuning of Changepoint. With the used of changepoint_prior_scale the weakness of the default prophet can be overcome. The forecast gives a RMSE of 0.64, MAPE of 0.52% and MAE of 0.58[7].
- Prophet with Huperparameter Tuning of window size. It was observed that by not considering holidays the forecast gave a better result. The forecast gave a RMSE of 3.52, MAPE of 2.98% and MAE of 3.37[7].
- Prophet with Hyperparameter tuining of Fourier order. The number of terms in the partial sum which are used

to estimate seasonality is given by fourier order. The forecast gave a RMSE of 3.4, MAPE of 2.87% and MAE of 3.25[7] .

The results are summerised in Table 2:

Table 2. Performance Evaluation

	Method	RMSE	MAPE(%)	MAE
0	Last Value	1.89	1.56	1.8
1	Prophet with No hyperparameter tuning	3.40	2.87	3.25
2	Prophet tuning changepoint	0.64	0.52	0.58
3	Prophet tuning fourier order	3.52	2.98	3.37
4	Prophet tuning window size	3.4	2.87	3.25

From the summery of Table 2 it is found out that using various hyperparameters the reduction in error is not consistent. Prophet without any hyperparameters shows a stable result but has higher values than last value model.

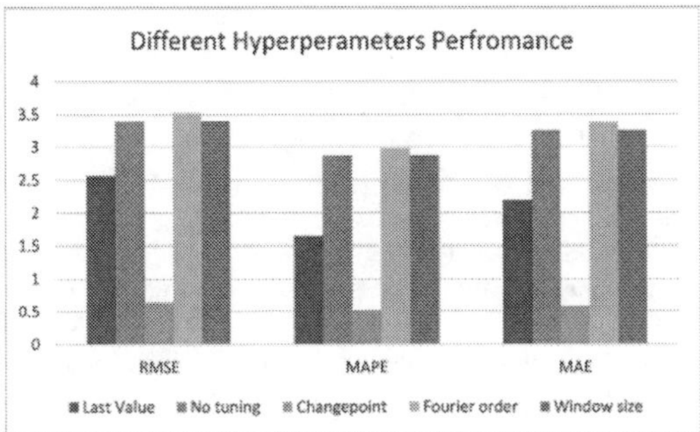

Figure 4. Performance Bar Graph

From the performance Bar Graph in Figure 4 prophet without tuning has close values in all the three parameters with Last value. For window size who's optimum value came as None has a significant high value when compared with Last value. Fourier order is also high when compared to Last value from the Figure 4

6 Conclusion

The application introduced a way for the user to predict the price of a particular stock with ease. The user was able to select the particular stock and generate the required forecast. Thus, Based on real-time analysis, future stock trends will be predicted on the top historical data of that company. To enhance user experience, we have followed a web-based approach, where the user can go through the insights of the stocks along with the future trend prediction and other decisive factors based on real-time analysis. The prophet model was not able to surpass the last value method in terms of performance. When hyperparameters were implemented within the Prophet model, in some instances it gave a better result than the last value but it was inconsistent. Overall prophet without any hyperparameters results are comparable with that of last value.

References

[1] Sharma, Ashutosh, Sanket Modak, and Eashwaran Sridhar. "Data Visualization and Stock Market and Prediction." (2019).

[2] Vijh, Chandola, Tikkiwal, Kumar, (2020). Stock closing price prediction using machine learning techniques Procedia Computer Science.

[3] Nayak, Aparna, MM Manohara Pai, and Radhika M. Pai. "Prediction models for Indian stock market."Procedia Computer Science 2016.

[4] Mruga Gurjar, Parth Naik, Gururaj Mujumdar.(2018) "STOCK MARKET PREDICTION USING ANN", International Research Journal of Engineering and Technology (IRJET).

[5] K. Pahwa and N. Agarwal, "Stock market analysis using supervised machine learning," (2019).

[6] Zunic, Emir, Korjenić, Kemal, Hodžić, Kerim, Donko, Dzenana. (2020). Application of Facebook's Prophet Algorithm for Successful Sales Forecasting Based on Real-world Data. International Journal of Computer Science and Information Technology.

[7] Yibin Ng. Forecasting Stock Prices using Prophet. Towards data science 2019.

[8] Rajat Patil. "Time Series Analysis and Stock Price Forecasting using Machine Learning Techniques" (2021)

[9] Jingyi Shen and M Omair Shafiq. "Short-term stock market price trend prediction using a comprehensive deep learning system" (2020)

[10] Supriti Khanderwal and Debasis Mohanty. "Stock Price Prediction Using ARIMA Model" (2021)

[11] Aslam, Fankar and Mohammed, Hawa and Lokhande, Prashant, "Efficient Way Of Web Development Using Python And Flask"(2015). Volume 6, International Journal of Advanced Research in Computer Science.

[12] Prophet: forecasting at scale. By Ben Letham, Sean J. Taylor

ITM Web of Conferences **44**, 03039 (2022) https://doi.org/10.1051/itmconf/20224403039
ICACC-2022

Optimizing the KYC Process using a Blockchain based approach

Niraj Ratnawat[1*], *Saujanya* Pandey[1], *Rudresh* Paradkar[1] and *Soumi* Banerjee[1]

[1]Ramrao Adik Institute of Technology, Department of Information Technology, Navi Mumbai, India

Abstract—The Know Your Client (KYC) process is an essential part of the financial ecosystem. The KYC process requires banks to validate and verify primary documents. The market these days, though, is flooded with KYC utilities that facilitate this process and share these documents with multiple entities, however they provide very little value addition. Blockchain technology, with its concept of immutable timestamped ledgers and distributed systems, can effectively facilitate banks to improve their KYC methods by allowing near real-time data exchange among various entities for quicker and effective validation ensuring data integrity alongside bringing down the time and costs significantly.

1 Introduction

Know Your Customer (KYC) is a process by which financial institutions validate the identity of customers or businesses with whom they conduct business. The KYC procedure is carried out to prevent institutions from being used/exploited for unethical and illegal activities such as money laundering, whether intentionally or unintentionally. In India, all financial institutions except for regional rural banks are bound to follow the RBI KYC directions, 2016 and update the KYC data once in two years to once in ten years according to the risk associated with the customers. Besides being a legal requirement, KYC is an essential instrument to prevent financial scams and unlawful activities. With the rise of organized criminal activities, terror financing, and money laundering, KYC guidelines are now a crucial tool to tackle illegal transactions in the field of international finance. According to a recent survey by Thomson Reuters[1], major financial institutions spend up to $500 million per year on KYC and customer due diligence. In addition to the processing and underlying expenses, financial institutions around the world must pay substantial fines for non-compliance due to uncertainty, ambiguity, and complexity of the process. This calls for the urgent need to remodel the KYC and due-diligence processes as well as the record-keeping system. It affects the overall transactional-experience for the institutions as well as customers.

Typically KYC and due diligence processes are labor-intensive, recurrent, and sluggish, resulting in higher overhead expenses and irregularities. Below are several points that highlight the issues in the current KYC process.

- KYC verification leads to the expenditure of enormous unnecessary costs, human effort, and time by financial institutions.
- Typically the KYC process is very redundant and repetitive as each financial institution has to perform its own KYC process.
- Clients need to submit the same documents for KYC repeatedly at different financial institutions separately, which leads to an unpleasant experience.
- Sometimes the old systems that store sensitive KYC data fail to provide adequate security assurance.

It is, therefore, a pressing priority to call for a solution that solves the problem of repetition of the process, duplication of records that are stored, and the upsettingly lengthy span of the process. It is crucial to keep into one's view that topics such as this require dealing with sensitive data. The regulations concerning data privacy must not be disregarded.

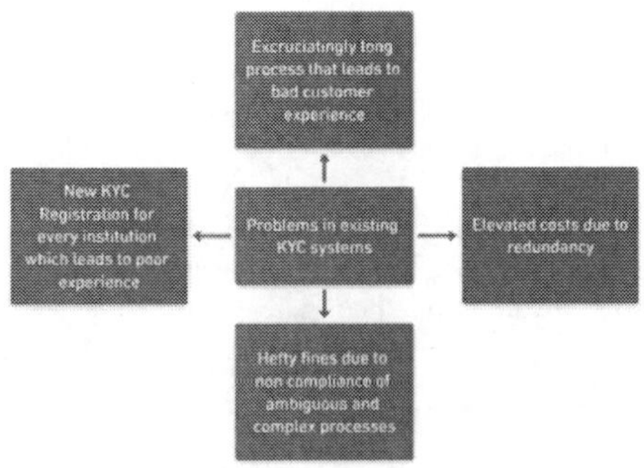

Fig. 1. Problems in current KYC process

2 Literature survey

As a distributed way of storing data, blockchains are currently gaining a lot of consideration. However, because of their irreversibility and openness, blockchains are unlikely to be ideal for personal data. In terms of data privacy, data stored on blockchains can't be altered, which means that personal information in them cannot be deleted, making it crucial to develop blockchain systems with this in mind.

One of the strategies may be to use blockchains solely to provide a timestamp pointing towards data stored elsewhere.

Even if a piece of public information is deleted, it'll still keep a record of the information that once existed. Using blockchains solely as a timestamping system rather than a data store has the added advantage of being more scalable when dealing with massive volumes of data.

"Bitcoin white paper" - Satoshi Nakamoto. Since it was first suggested to account for the Bitcoin cryptocurrency, the fundamental technologies and concepts behind blockchain have extensively developed[9].

"BlockChain Technology (DLT Technique) for KYC in FinTech Domain: A Survey" - R. Kasturi. According to R. Kasturi et al. [4], the data collected from the clients for KYC can be stored on the blockchain and the client can be given an ID to which their data is mapped. The client can then use the ID for future KYC processes. However, this practice is not just expensive, yet often unreliable, since the volume of data that can be processed on a blockchain is either restricted by protocol or requires a large transaction charge. On ethereum, storing a kilobyte of data costs $22.87, which is incredibly costly and unsustainable.

"Double-Blind Consent-Driven Data Sharing on Blockchain" - K. Bhaskaran According to an approach by K. Bhaskaran and others [6], exchanging KYC data via blockchain could help solve duplicacy problems. Hyperledger Fabric is being used in their approach. The customer must provide the personal information required for due diligence to a vendor, which is then sent to the blockchain for validation and storage by the vendor in an encrypted format. Other organizations for which the consumer wishes to conduct KYC may access this information only with the user's permission. The drawback of this approach is that, contrary to the rule, confidential data such as this cannot be kept in an immutable ledger and if the encryption breaks in the future, all of the customers' personal data could be compromised.

"KYC as a Service (KASE)—A Blockchain Approach." - Dhiren Patel Patel et al [8] proposed a method that combines machine learning and blockchain technologies to partially simplify the due diligence phase. According to their method, the customer's data is checked using machine learning techniques, and if approved, the data is encrypted and the transaction is recorded on the blockchain. They plan to use the Ethereum blockchain in conjunction with the proof-of-work consensus method. However, one of the key flaws in this technique is that, because KYC is such a sensitive process, even a small margin of error is costly. And machine learning algorithms are not yet advanced enough to guarantee 100 percent accuracy. As a result, the practical application of this method is still some time away.

3 Proposed system

This section enumerates the steps of the above proposed KYC process utilizing blockchain technology. But before that, a few concepts pertaining to the thorough understanding of the above proposed system are explained below.

3.1 Digital Signature:

A digital signature is a mathematical technique for verifying the authenticity of digital messages or documents. If the prerequisites are satisfied, a valid digital signature gives the recipient a very strong reason to believe that the message was generated by a recognised sender (authenticity) and that it was not altered en route. It employs asymmetric key cryptography as well as a hashing mechanism.

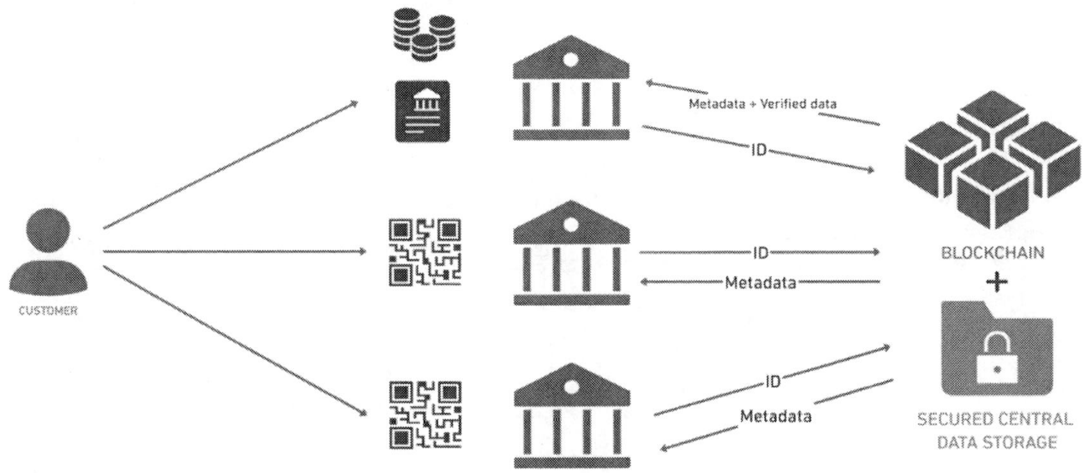

Fig. 2. Proposed KYC System

The sender generates a hash of the document, signs it with their private key, and then sends the signature and hash over. The sender's public key may then be retrieved and compared to validate the signature using the message hash and signature. This method ensures both data integrity and validity.

 https://doi.org/10.1051/itmconf/20224403039

3.2 Smart Contract:

Simply said, a smart contract is a piece of software that runs on Ethereum's blockchain. It's a single address on the Ethereum blockchain that contains a collection of code (its functions) and data (its state). Ethereum accounts come in the form of Smart Contracts. This means they have some balance in their account and can perform transactions through the network. They are not, however, user-controlled. Instead, they are deployed to a network and run according to a set of instructions. User accounts can then engage with a smart contract by sending transactions that cause the smart contract to perform a function. Smart contracts, like conventional contracts, may set rules and have them enforced automatically through programming. Smart contracts can't be erased by default, and their interactions are permanent.

4 Working

The working of the proposed KYC system involves a two-step process - the *addition* and the *verification* process.

4.1 Addition Process (new users):

All the steps listed below for the addition process can be referred to from fig. 3 below.

1. A new client registers on the KYC portal using an existing cryptocurrency wallet such as metamask wallet.

2. The client searches for the desired organization on the portal to perform KYC with, via the search organization functionality.

3. The client visits the organization profile page, selects 'Perform KYC' option and uploads their documents which are then encrypted before being sent to the organization. A new unique request ID is generated during this step.

4. The organization official then downloads the decrypted documents via the portal and performs due diligence to establish the identity of the client.

If the documents are:

a) rejected, the client is notified and asked to initiate a new KYC request with the suggested changes.

b) approved, hash of the documents & hash of the request ID are signed by the organization and sent to the client's account.

The encrypted documents are uploaded to a secure server by the portal once the request is approved.

The client now selects the option to finalize the request, signs the request ID and pays the transaction fees for uploading the digital signatures, public keys and timestamp to the smart contract. The smart contract performs various validations using the digital signatures to check the authenticity of the sender and the data and approves, if all the data is found to be correct.

A new QR code containing the request ID is now generated and the KYC addition is now complete. This QR code can now be used to lookup the details of the KYC request universally.

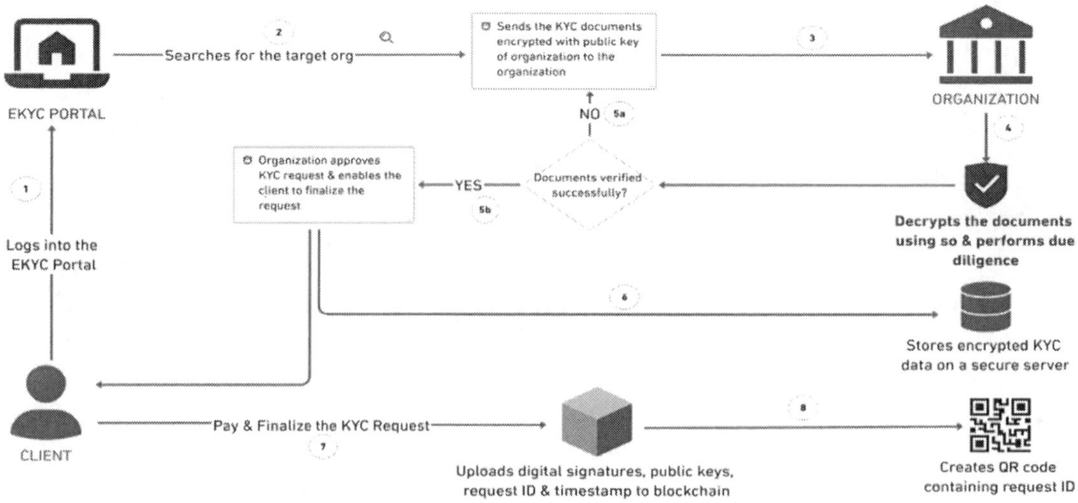

Fig. 3. KYC Addition process for new client

4.2 Verification Process (subsequent KYC verifications):

All the steps listed below for the verification process can be referred to from fig. 4 below.

1. The client who already has a KYC registered with an organization logs into the eKYC portal using their cryptocurrency wallet.

2. The client then searches for the target organization and sends their QR for the registered KYC along with a verification request to the target organization.

3. The organization extracts the *addition* request ID from the QR code, queries blockchain for the details of the registered KYC and validates data to establish the identity of the sender and adding organization.

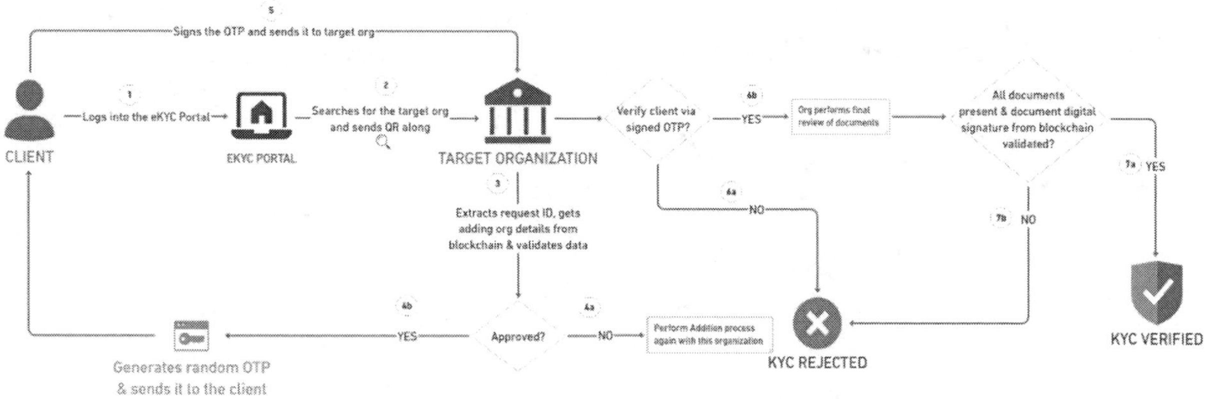

Fig. 4. KYC Verification process for registered client

4. If the target organization:

a) does not approve the previously registered KYC on grounds of not trusting the adding organization or not being able to verify the data, their verification request is rejected with appropriate reasoning.

b) is able to validate the data from blockchain and approves the adding organization, it then sends a randomly generated OTP to the client's registered email-id and phone to re-verify the identity of the client.

The client enters the OTP and signs it using their private key and sends it to the target organization.

If the target organization is:

c) not able to validate the signature, the verification request is rejected.

d) able to verify the signature, the verification request proceeds to the next stage.

The target organization now has to quickly view the documents for their completeness and validate the digital signature of documents from the blockchain.

e) If the documents are complete and the signature is validated, the KYC verification request is successful.

f) If not, the KYC verification request is rejected.

5 Result

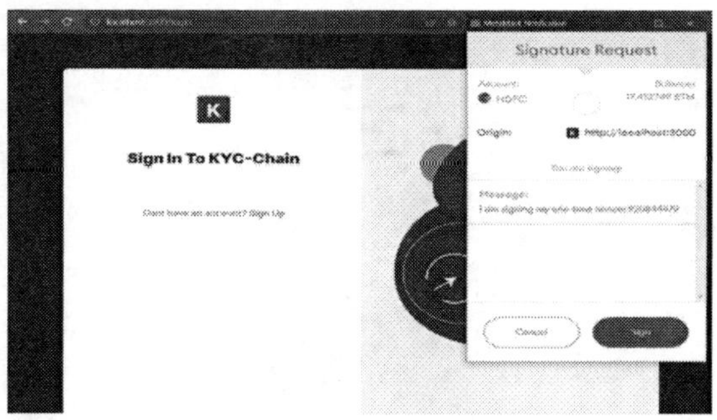

Fig 5. Login screen of the eKYC platform

Fig. 6. Addition request initiated by user

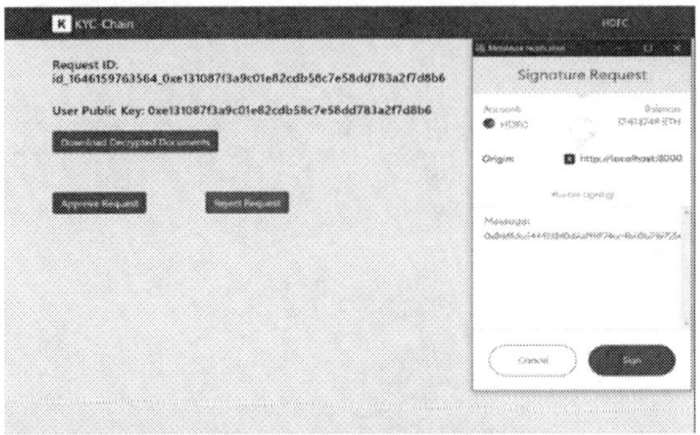

Fig. 7. Organization's view of processing client requests

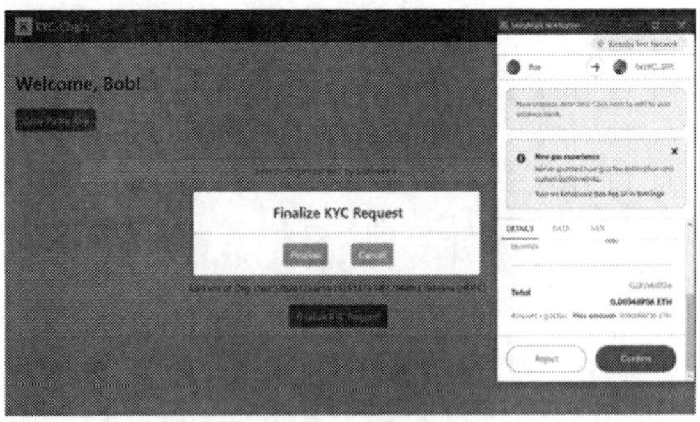

Fig. 8. Organization's view of finalizing client requests

Fig. 9. QR code with relevant information generated after user request approved

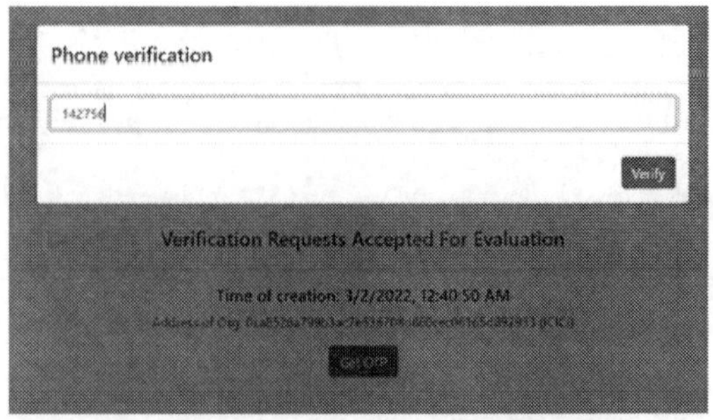

Fig. 10. User identification via OTP verification

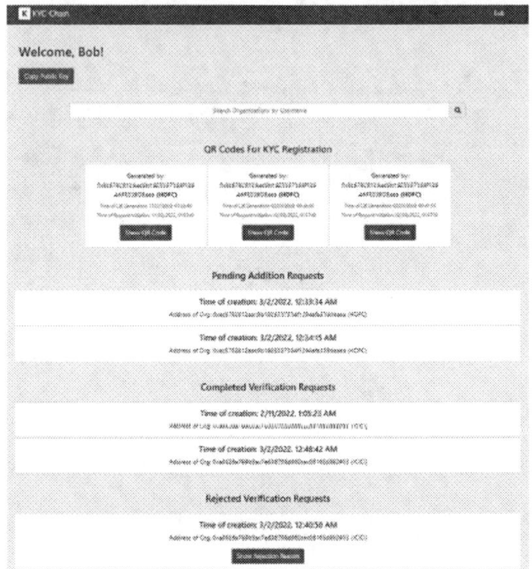

Fig. 11. User's dashboard with all requests data

Fig. 5 shows the login screen of the eKYC portal, which is integrated with Metamask wallet through one-click login functionality, after which the user is redirected to the platform's dashboard.

Fig. 6 shows the interface available to the user after clicking 'Perform KYC' on the organization's profile page.

Users can upload their documents and initiate an addition request.

Fig. 7 shows the interface available to organizations to process the received requests from users.

Once an organization proceeds with processing a user's request, the next step is finalizing the user request as shown in Fig. 8.

A critical step of each user request before finalizing the request is verification of the user's identity to ensure security. This is done by performing an OTP verification from the user's side, as shown in Fig. 10.

A user's KYC request is completed by generating a QR Code for the same, containing all the relevant information, as shown in Fig. 9.

All the ongoing, completed, and rejected requests are available on the user and organization's dashboard as shown in Fig. 11.

6 Comparative Analysis

6.1 BlockChain Technology (DLT Technique) for KYC in FinTech Domain: A Survey [4]

- The proposed system involves leveraging the blockchain for storing users' entire KYC data.
- The KYC data collected from the clients could be stored on the blockchain and referenced with an ID mapped to the data of the respective user.
- In contrast to this solution, our system makes use of a secure centralized server to store sensitive information, which is referenced on the blockchain via unique request IDs.

6.2 Double-Blind Consent-Driven Data Sharing on Blockchain [6]

- This solution presents a system that uses blockchain to encrypt user data and store it in a secure manner.
- The customer must submit personal information to a vendor for due diligence requirements, which is then sent to the blockchain to be validated and stored in an encrypted format by the vendor.
- This approach also involves storing encrypted data on the blockchain as opposed to storing on a centralized secure server in our approach.

6.3 KYC as a Service (KASE)—A Blockchain Approach [8]

- This system proposes storing transaction records after the verification process.
- Machine learning techniques could be combined with blockchain techniques to partially simplify the due diligence process.
- This approach involves performing the due diligence process as compared to manual verification in our approach. Reason being, machine learning algorithms are not yet fully dependable enough to handle such sensitive tasks.

7 Conclusion

The proposed KYC implementation can be used in multiple Banking/Financial services use cases like Lending and Borrowing, Portfolio Management, Stock Exchange, Insurance company, Trade Finance, Mutual Funds and offers a unique yet simple and effective solution to tackle the existing challenges. This approach can help in bringing down the costs, redundancy and duration of the process significantly, thereby, optimizing the current KYC process in terms of non redundant storage and decreased amount of time for overall process. This optimization, however, cannot be measured quantitatively given there is no single standard for KYC processes across institutions. But the overall process remains largely similar, and the proposed system clearly is an optimization over the existing process. However, one of the most worrisome aspects of the presently most popular implementation of blockchain technology is the consensus algorithm it uses, known as Proof of Work, which largely relies on electrical energy usage. This has a harmful influence on the environment and needs to be addressed as soon as possible. Newer consensus algorithms, such as Proof of Stake, are being developed that are more eco sustainable and may be readily adopted by current systems.

8 Future Scope

The current system makes use of a public blockchain. Due to the lack of an authorization system in public distributed ledgers, there's nothing preventing anyone from engaging with and participating in the blockchain's consensus mechanism. As a result, scaling challenges arise, and throughput is relatively lower. To overcome these issues, a consortium blockchain can be utilized, which combines elements from both public and private blockchains. Further, the algorithm used in the smart contract can be optimized to reduce transaction charges.

9 References

[1] https://www.thomsonreuters.com/en/press-releases/2016/may/thomson-reuters-2016-know-your-customer-surveys.html

[2] "'Know Your Customer' (KYC) Guidelines - Anti-Money Laundering Standards". Archived from the original on 2012-08-01.

[3] Master Direction - Know Your Customer (KYC) Direction, 2016 of RBI Circular dated February 25, 2016.

[4] Block Chain Technology (DLT Technique) for KYC in FinTech Domain: A Survey, R. Kasturi and Vimalkumar Pachaiyappan, International Journal of Pure and Applied Mathematics Volume 119 No. 10 2108, 259-265.

[5] Trunomi, E. G. P. P. Gdpr key changes https://www.eugdpr.org/

[6] K. Bhaskaran et al., "Double-Blind Consent-Driven Data Sharing on Blockchain," 2018 IEEE International Conference on Cloud Engineering (IC2E), 2018, pp. 385-391, doi: 10.1109/IC2E.2018.00073.

[7] Buckley, Ross & Arner, Douglas & Barberis, Janos. (2016). The Emergence of Regtech 2.0:From Know Your Customer to Know Your Data. Journal of Financial Transformation. 44. 79-86. DOI=https://dx.doi.org/10.2139/ssrn.3044280.

[8] Patel, Dhiren R. et al. "KYC as a Service (KASE)—A Blockchain Approach." (2020).

[9] Bitcoin white paper. 2008. Satoshi Nakamoto. https://bitcoin.org/bitcoin.pdf

[10] https://www.evry.in/globalassets/india/pdfs---brochures/ekyc-using-blockchain.pdf

[11] https://www.tcs.com/content/dam/tcs/pdf/Industries/Banking%20and%20Financial%20Services/Reimagining%20KYC%20Using%20Blockchain%20Technology.pdf

 https://doi.org/10.1051/itmconf/20224403040

Predicting E-commerce Sales & Inventory Management using Machine Learning

Divyansh Agnani[1*], *Abhishek* Bavkar[1], *Samarth* Salgar[1], and *Sheetal* Ahir[1]

[1] Ramrao Adik Institute of Technology, Department of Computer Science, Mumbai, India

Abstract. Due to prevalent transition from visiting physical stores to online shopping, predicting customer behaviour in the e-commerce market is gaining increasing importance. Over the past few years, e-commerce marketplaces such as Flipkart and Amazon have observed a manifold increase in both sales and market share of retail products sold. During this period, many traditional retailers and wholesalers have set up their e-commerce portfolios to stay relevant in the marketplace. However, due to lack of technology penetration in the Medium, Small, & Micro Enterprises (MSMEs) sector, the technological tools required by them are largely absent. One key issue for small businesses in e-commerce has been to forecast how many products their business will sell, thereby introducing another big issue of predicting how much inventory of a product they need to hold, to match up with demand. This is partly due to the massive scale and opportunities opening for them by participating in an open, large marketplace such as e-commerce. The COVID-19 pandemic has also drastically changed consumer shopping trends and behaviours: consumers have changed their preference to buying online. Thus, Indian small businesses need to have the ability to forecast demand more accurately than ever before.

1 Introduction

Traditional (brick and mortar) small businesses in India have usually differentiated themselves against big retailers by offering products at good value (in terms of excess benefit over cost). This has been possible by a largely two factors: 1) knowing their customers well, helping them price their products competitively, and 2) successfully attaining customer's attention. However, this gap has closed with increasing relevance and participation of customers in the e-commerce marketplace. E-commerce websites such as Amazon and Flipkart have witnessed rapid growth in terms of sales and market share in the past few years. On the other hand, small businesses, termed as Medium, Small, and Micro Enterprises (MSMEs) have been unable to keep up with big retailers partly due to lack of technological tools available to them. Forecasting sales is a challenging, but important task in revitalizing the growth strategy for MSMEs by helping them make informed strategic decisions. It helps the businesses manage inventories, cash flows, and resources by optimizing their supply chain. The benefit of forecasting sales is dependent on how accurately a model predicts sales. If a model is inaccurate and is unable to help a business effectively maintain track of their inventories, it can lead to over-stocking of products, which results in excess overhang, or understocking, which results in potential notional losses or loss of opportunity.

Traditionally, economic forecasts have been used to help businesses predict sales. However, history is a study of surprising events. For years, this reliance on economic forecasts have rendered small businesses inefficient in predicting demand and managing their inventories. This is because, economic forecasts are volatile in nature, changing with ever-coming stream of new information and numbers, giving little time for small businesses to streamline and adapt. Economic forecasts in India are also dated in their approach towards changing consumer trends and growing higher preference for online shopping.

This section of the report is followed by literature review in the subsequent section. Section 3 consists of methodology used to conduct our research, divided into two parts, first containing existing methodology, and second containing proposed changes to existing systems. The methodology also contains an evaluation and comparison of models used in our research. Section 4 involves concluding the paper, with a brief about our findings. The paper ends with references used in the research.

2 Literature Review

E-commerce market in India has dramatically grown in the past few years but is still lower compared to the growth of the market in developed nations, such as the United States (US) and China. Kumar, N. et al. in their research, titled 'Challenges and Opportunities of E-commerce in India' [9] noted that this is due to 1) poor

* Corresponding author: div.agn.rt17@rait.ac.in

telecom infrastructure, 2) poor academic syllabus about ecommerce, and notably for our paper, 3) lack of positive attitude towards technology [9].

The authors recommend policy-driven changes, updated legal frameworks, and infrastructure development for e-commerce. Regarding technological infrastructure needed, Galhotra, B. et al. note in their research to survey and analyse digital platforms preference and changing consumer trends during the COVID-19 pandemic, that machine learning can be applied to predict the changing consumer preferences and trends in the e-commerce space [10].

Traditional sales forecasting methods have used time series analysis techniques, such as autoregressive models, integrated models, and moving average models [1], that forecast sales based on linear functions of historical sales data. Pavlyshenko, B. et al. have compared regression approach against the traditional time series forecasting approach for sales forecasting in their research [4].

Zhao, K. et al. compared autoregressive integrating moving average (ARIMA), a more recent model, with Deep Neural Network (DNN), Transfer Learning (TL), Convolutional Neural Network (CNN) and Weight Decay (WD) models in their research [1]. The author also assesses which model efficiently eliminates the need for manual feature engineering by extracting effective features [1].

Bohanec, M. et al analyse the use of black-box prediction models in a real- world case scenario of B2B sales forecasting. The authors identify the appropriate model based on business understanding by using intelligence and data-driven models [6].

Yu, J. et al. proposes and compares various forecasting approaches namely Winters' Exponential Smoothing, Time Series Decomposition, and ARIMA model [5]. Forecasting sales in e-commerce faces certain challenges regarding non-linear, irregular, and lumpy sales data. Yu, J. et al. in their work on forecasting sales for Amazon also describe necessary data to carry out forecasting as a major obstacle impeding the implementation of the model [5].

Bandara, K. et al. in their research [2] trains a Long Short-Term Memory network (LSTM) to predict sales given non-linear relationships in e-commerce. This inability to record effective data is also reiterated by Holmberg, M. et al. in which the authors recognize restaurants not having required education about forecasting sales and thus not maintaining sales data. The authors of this research have evaluated Extreme Gradient Boosted (EGB) and Long Short-Term Memory Boosted (LSTM-Boosted) models [7].

Another issue that arises from inadequate sales data is explored by Ferreira, KJ. et al. in their research on predicting demand for first exposure items (items that have never been sold before) [8]. The authors have additionally
worked on a price optimization model to help businesses maximize revenue on such products.

E-commerce consumer behaviour studies also showcase a cross-product relationship between different products. Li M. et al. identify different cross-product demand or sales pattern and evaluates their correlation with a pre-processing framework by comparing prediction errors of the NARNNN and ARIMA models [3].

3 Methodology

3.1 Existent Methodology

3.1.1 Data Pre-processing

Data pre-processing is a procedure that describes how the chosen data will be cleaned of any noise and outliers. This entails cleaning out data that has a lot of extra substance but is not relevant to our use case. For example, if the dataset contains missing values for sales and price values, we must manage them appropriately by replacing the missing value with the average value or using mean or median imputation to ensure that the data remains consistent. In this project, the following pre-processing mechanisms have been used.

3.1.1.1 Data Transformation

Data transformation is the process of changing data from one format to another to meet specific requirements. This procedure is also known as the ETL procedure, which stands for Extract, Transform, and Load. As the volume of data has grown exponentially, transformation has become a critical responsibility. As a result, reliable data transformation will allow users to concentrate on data that meets their business requirements. The same will be true for this project; all the data will be robustly transformed, with only the most significant data being integrated into a new data format.

3.1.1.2 Data Imputation

Some of the columns in our dataset have missing values where a default value of 0 should have been placed instead of NaN/undefined/null values. In such cases, we have filled the missing values with 0. In other, more notable cases, columns are missing values which are non-binary in nature and can't be misplaced with arbitrary values. Here, we have taken an approach to fill the missing values with the mode - or the most frequent values present in each column. It is observed that this approach does not consider the correlation between various features and can thus introduce bias in the data by undesirably assigning more values to a specific category. Attention and caution have been undertaken to check for such discrepancies.

3.1.1.3 Feature Extraction

Feature extraction begins with a set of measured data and creates derived values (features) that are aimed to be insightful and non-redundant, facilitating the learning and heuristic steps resulting in better inference and usage by a machine learning model. For example, a feature that contains value count of just one, that is, it contains a unique value for all items in our data frame, it fails to be relevant to the model. Hence, it should be dropped. On the other hand, the unique value feature disposal needs to be dealt with caution. Sometimes, a feature with only two possible value counts, can be flagged as one with a unique value count, if one of the two possible values is filled NaN/undefined/null value. In other cases, features with values that frequently do not deviate from a 'popular' value again deem no relevance. In these cases, such features can be deemed to not be correlated to our target variable. Thus, such columns can be dropped. In some cases, one feature describes the listings in a data frame in a manner that makes other features redundant. This is often in the case of unique identifiers.

3.1.1.4 Feature Selection

Another important and widely used technique in data processing is feature selection, which is used to select appropriate features from noisy data. This method improves the speed of any data processing algorithm, which improves prediction accuracy and reduces the variance in predictions. While seemingly similar, the main distinction between feature selection and extraction is that feature selection retains a subset of the original features while feature extraction generates entirely new ones. Feature selection approaches can be employed if the original features must be preserved, as opposed to feature extraction techniques, which extract useful information from data in order to create a new feature subspace. When model explain ability is a top priority, feature selection strategies are applied.

3.1.1.5 Detecting and removing outliers

Outliers are data points that differ dramatically from the rest of the dataset's observations. It can happen as a result of measurement variability, misinterpretation when filling in data points, measurement errors (fault in measurement instrument), experimental errors, intentional dummy errors to test detection methods, sampling errors, or simply and often visibly, natural outliers. In our project, we have plotted column wise distribution of values in features to detect outliers visually. This plotted distribution can be seen in Figure 3.1 below. Other than that, we have used Grubbs Test function to detect outliers, which is a statistical tool used to detect outliers in a dataset assumed to come from a normally distributed population. This assumes normality. The data value with the highest absolute value is considered first by the algorithm. The considered value is detected as an outlier and excluded from further analysis if the null hypothesis that it is not an outlier is rejected. The value with the second highest absolute value is then considered, and its quality is analysed using the Grubbs test once.

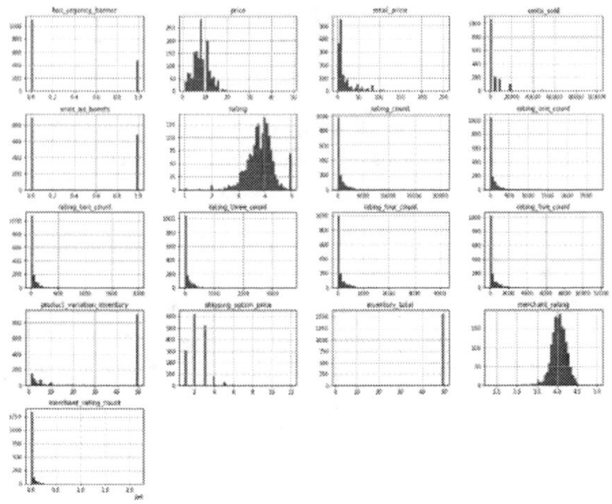

Fig. 3.1 Column wise Distribution of Values

3.1.2 Model Training

Model Training is the process of determining which algorithms will be used to solve a problem. We will be using four distinct regression algorithms for this study, which are Random Forest Regression, Extra Trees algorithm, Gradient Boosting Regression, and AdaBoost Regression.

Regression analysis is a statistical method for modelling the relationship between one or more independent variables and a dependent (target) variable. Regression analysis allows us to see how the value of the dependent variable changes in relation to an independent variable when the other independent variables are held constant.

Random Forest Regression: A random forest is a meta estimator that uses averaging to improve predictive accuracy and control over-fitting by fitting several classifying decision trees on various sub-samples of the dataset. As a result, it can be called a "Forest" of trees, hence the name "Random Forest." The name 'Random' comes from the fact that this algorithm is made up of a forest of 'Randomly constructed Decision Trees.' In other words, the average result of a series of decision trees is called a Random Forest regression. A decision tree is similar to a flow chart in that it asks a series of questions and then makes a prediction based on the answers.

Gradient Boosting: Gradient Boosted Decision Trees constructs an additive model in a forward phased manner, allowing for the optimization of any differentiable loss function. A regression tree is fitted on the negative gradient of the specified loss function at each level. In other words, when a decision tree is the weak learner, the resulting algorithm is called gradient boosted tree. A weak hypothesis or weak learner is defined as one whose performance is at least slightly better than random chance.

Extra Trees Regression: Extra Trees is an ensemble machine learning algorithm that combines the predictions from many decision trees. The Extra Trees approach uses the training dataset to generate a multitude of unpruned decision trees. In the case of regression, predictions are formed by averaging the prediction of the decision trees, whereas in the case of classification, majority voting is used.

Adaptive Boosting: AdaBoost for short, it was the first iteration of boosting algorithms that saw significant success. AdaBoost works by weighing the observations, giving more weight to instances that are difficult to predict and less to those that are already well-predicted. New weak learners are introduced one at a time, with the goal of concentrating their training on the increasingly challenging patterns. This means that difficult-to-predict samples are given increasingly greater weights until the algorithm finds a model that correctly classifies them.

3.1.2.1 Splitting into training and testing subsets

The creation of different samples for training and testing helps us evaluate model performance. Overfitting is a common problem while training a model. When a model performs exceptionally well on the data used to train it, but fails to generalise well to new, previously unseen data points, this phenomenon happens. This can happen for a variety of reasons, including data noise or the model learning to anticipate certain inputs rather than the predictive factors that could help it make more accurate predictions. In general, the more complicated a model is, the more likely it is to be overfitted. Underfitting, on the other hand, occurs when the model performs poorly even with the data that was used to train it. Underfitting happens most of the time when the model isn't appropriate for the problem you're trying to solve. This usually signifies that the model isn't as complicated as it must be in order to learn the parameters that have been shown to be predictive. The most typical strategy for identifying these difficulties is to create separate data samples for training and testing the model. In this way, we can use the training set to train our model and then use the testing set to assess if the model can generalise well to new, unseen data. We have split the data frame in the ratio of 70:30 for training and testing subsets respectively

3.1.2.2 Training or Fitting the Model

A training model is a dataset used to train a machine learning algorithm. It is made up of sample output data as well as the equivalent sets of input data that have an impact on the outcome. The training model is used to process the input data via the algorithm in order to compare the processed output to the sample output. The model is modified based on the results of this association. "Model fitting" is the term for this iterative procedure. The precision of the model is dependent on the accuracy of the training or validation dataset. In machine language, model training is the act of supplying data to an ML algorithm to help it find and learn suitable values for all the variables involved. There are various types of machine learning models, with supervised and unsupervised learning being the most common. When the training data comprises both the input and output values, supervised learning is performed. A supervisory signal is a set of data that contains both the inputs and the intended outcome. When the inputs are fed into the model, the training is based on the deviation of the processed output from the documented result. The process of unsupervised learning entails identifying patterns in data. After then, further data is used to fit patterns or clusters. This is basically an iterative procedure that improves accuracy by comparing the observed patterns or clusters to the expected patterns or clusters.

3.2 Proposed Methodology

We have employed hyperparameter tuning to enhance the performance and computational time of the models used. Parameters are inputs that describe the individual weights between nodes, computed by the algorithm used for training the model. Hyperparameter, on the other hand, are parameters that determine how the model learns, in contrast to the parameters learned by the model.

Hyperparameters are derived by setting different values for parameters, training models using combinations of these values, and evaluating which combination works best. A domain is specified from which these values are taken.

Specifically, we have used Randomised Search hyperparameter tuning for Random Forest and Extra trees models, which randomly transverses parameters from a specified domain, and chooses the optimal combination rand. The benefits of choosing randomised search over other alternatives is the computational times saved, due to its efficiency.

For boosting models, we've employed Grid Search hyperparameter tuning, which creates a grid of possible parameters from a domain and tries every combination to get optimal values. Unlike randomized search hyperparameter tuning, grid search algorithm doesn't transverse through the domain randomly.

3.3 Evaluation

The researchers have conducted an overall evaluation at the conclusion of this study to determine the algorithm which has the highest prediction accuracy, as well as evaluated the effects of hyper-parameter tuning in each of the models used, to arrive at the best algorithm as per our motive. For each algorithm under consideration, the error percentage and prediction accuracy has been assessed. The evaluation metrics used to compare the various algorithms are based on the accuracy of prediction, and error rates. The mean absolute error (MAE), mean squared error (MSE), and root mean squared error (RMSE) of the models used is given in Table 3.1.

Table 3.1 Error rates

Model Used	MAE	MSE	RMSE
Random Forest	0.7434	1.2921	1.1367
Random Forest (Hyper-parameter tuned)	0.7428	1.2554	1.1204
Extra Trees	0.7286	1.2605	1.1227
Extra Trees (Hyper-parameter tuned)	0.7183	1.2275	1.1079
Gradient boosting	0.8146	1.3222	1.1499
Gradient boosting (Hyper-parameter tuned)	0.7730	1.2776	1.1303
Adaptive Boosting	0.9761	1.7546	1.3246
Adaptive Boosting (Hyper-parameter tuned)	0.9627	1.6380	1.2798

The R-squared metric, used to determine the accuracy of the models, are given in Table 3.2.

Table 3.2 Accuracy metrics

Model Used	R-squared
Random Forest	0.9142
Random Forest (Hyper-parameter tuned)	0.9167
Extra Trees	0.9166
Extra Trees (Hyper-parameter tuned)	0.9185
Gradient Boosting	0.9122
Gradient Boosting (Hyper-parameter tuned)	0.9152
Adaptive Boosting	0.8836
Adaptive Boosting (Hyper-parameter tuned)	0.8913

4 Conclusion

We have implemented and evaluated different models to solve one problem of these small enterprises: forecasting sales. This helps them take more strategic decisions when it comes to investments, supply chain management, and inventory management. While working on this project of using machine learning to forecast ecommerce sales, it was observed that while there are many different methods for forecasting ecommerce platform sales, the researcher was able to focus on four regression algorithms that are commonly used when forecasting future sales. All the selected machine learning models were able to be built and tested by the researcher. Furthermore, the researcher were able to hyperparameter tune the models and evaluate the subsequent effects on all four algorithms for comparison. The best algorithm is chosen as the model with the best prediction range, where the predicted value and the actual value are almost identical.

Through surveying existing literature and work, sales predicting is found to be a regression problem, instead of a time-series forecasting problem as it has traditionally been studied. Thus, we have chosen to implement and compare various regression models on algorithms namely Random Forest Regression, Extra Trees algorithm, Gradient Boosting Regression, and AdaBoost Regression. Next, we have tested the mentioned regression models with hyperparameter tuning. Finally, we evaluate and compare the various models and the effects of hyperparameter tuning through R- squared (R2) metric and various error rates.

References

[1] 1. J. Yu, X. Le et al. 'Sales Forecast for Amazon Sales Based on Different Statistics Methodologies'(2016)

[2] 2. K.J. Ferreira, B.H.A. Lee, D. Simchi-Levi et al. 'Analytics for an Online Retailer: Demand Forecasting and Price Optimisization' (2016)

[3] 3. M. Bohanec, M.K. Borstnar, M. Robnik-Sikonja et al. 'Explaining machine learning models in sales prediction' (2017)

[4] 4. Zhao, K. and Wang, C. (2017) 'Sales Forecast in E-commerce using Convolutional Neural Network', (August 2017).

[5] 5. Li, M., Ji, S. and Liu, G. (2018) 'Forecasting of Chinese E-Commerce Sales: An Empirical Comparison of ARIMA, Nonlinear Autoregressive Neural Network, and a Combined ARIMA-NARNN Model'

[6] 6. M. Holmberg, P. Hallden et al. 'Machine Learning for Restaurant Sales Forecast' (2018)

[7] 7. Bandara, K. et al. (2019) 'Sales Demand Forecast in E-commerce using a Long Short- Term Memory Neural Network Methodology'

[8] 8. N. Kumar, A. Jhensanam, Md A. Hossin, B. Chanthamith et al. 'Challenges and Opportunities of E-commerce in India' (2018)

[9] 9. B. Pavlyshenko et al. 'Machine-Learning Models for Sales Time Series Forecasting' (2019)

[10] Bhavna Galhotra and Ayushi Dewan, 'Impact of COVID-19 on digital platforms and change in E-commerce shopping trends' (2021)

A comparative study of fine-tuning deep learning models for MRI Images

Vrushabh Gamare[1,*], *Vaibhav* Kharaje[1,**], *Shreyash* Borole[1,***], and *Saguna* Ingle[2,]

[1]Student, Department of Computer Engineering, Ramrao Adik Institute of Technology, D Y Patil Deemed to be University, Navi Mumbai

[2]Assistant Professor, Department of Computer Engineering, Ramrao Adik Institute of Technology, D Y Patil Deemed to be University, Navi Mumbai

Abstract. A brain tumor is an abnormal development of cells that reproduce uncontrollably and without any external stimulation. If tumors are not found early enough, brain tumors can be fatal to one's health. Specialists and neurosurgeons employ magnetic resonance imaging (MRI) scans to diagnose brain tumors. Several deep learning methods for detecting the existence of brain tumors have been developed to overcome these constraints. The accurate detection of the size and location of a brain tumor is crucial in the diagnosis of the tumor. Medical image processing is a highly complex and tough discipline in which image processing and its methods are an active research topic. There are various technical deep learning and machine learning algorithms which are used to detect brain tumor. We used CNN architecture, ResNet, VGG16 and inception network in this paper and did a comparative study to find out the maximum accuracy for detecting brain tumor. When these algorithms are imposed to MRI images, the prediction of brain tumours is done quickly, and the higher accuracy aids in the treatment of patients. In this paper, after complete procedure and analysis of four different algorithms, we found out that CNN architecture is the most suitable with highest accuracy.

Keyword :

brain tumor, image segmentation, preprocessing, MRI images

1 Introduction

The brain is the most sensitive organ of our body, controlling the core functions and characteristics of the human body. According to the National Brain Tumor Society, around 700,000 persons in the United States have been diagnosed with a brain tumor, with that number predicted to climb to 787,000 by 2020 [1]. The brain is one of the most complex organs in the human body, involving a high number of cells that work continuously. Brain tumors grow when cells divide uncontrollably, resulting in an irregular pattern of cells. The grouping of a cell will have an effect on the normal functionality of brain activity and will harm the healthy cell and cause fatal effects to human life [2] . In MRI image processing, digital image processing is critical. In the restorative profession, X-ray pictures are routinely utilized to evaluate and recognize tumor progression in the body. Brain tumor affect both children and adults. Tumors cause severe brain pressure that spreads throughout the entire brain region. Tumor growth occurs inside the skull, interfering with normal brain function. There could be a tumor that leads to cancer, which is a major cause of death and accounts for approximately 13 percent of all deaths worldwide. The global cancer incidence rate is increasing at an alarming rate.

The key to effective treatment is early diagnosis of tumors. Previously, brain tumors were discovered manually by professional viewing of images, which took time and often resulted in inaccurate conclusions. Detection consists of determining the presence of a brain tumor; segmentation consists determining the size and location of the tumor; and classification consists of determining the stage of brain tumor detection. Many computers now have features that are commonly used in the medical industry. On known MRI images of tumors, image processing processes such as histogram equalisation and sharpening filters are applied. Computer-assisted technology is quite important Role , because the medical sector demands quick and dependable procedures to diagnose life-threatening disorders such as cancer, the world's leading cause of mortality for patients. So, in our study, we used methodologies and a convolutional neural network model to categorize brain tumors as cancerous or non-cancerous using brain MRI scans.

2 Literature Survey

A brief explanation of the literature review done by various scientists and researchers have been discussed below. Researchers has used many algorithms in order to attain the accurate detection of brain tumors. We studied the various methodologies and procedures followed by the experts.

*e-mail: vrushabhcr7@gmail.com
**e-mail: vaibhavkharaje1212@gmail.com
***e-mail: shreyashborole18@gmail.com

Ahuja et al., [3] employed a technique known as transfer learning to detect brain tumors and segment the brain. The VGG transfer learning model was utilized to train the dataset for detection in this model, which came from the BRATS 2019 brain tumor segmentation challenge. Using the Super-Pixel approach, the tumor was separated into LGG and HGG pictures.

Cherguif et al., [4] Medical photos were semantically segmented using U-Net. The U-Net design was used to produce a convoluted 2D segmentation network. The BRATS 2017 dataset was used to test and evaluate the suggested model. With 27 convolutional layers, 4 deconvolutional layers, and a Dice coef of 0.81, the suggested U-Net architecture was created.

Choudhury et al., [5] In this article, deep learning approaches using deep neural networks, as well as merging it with a Convolutional Neural Network model, were employed to produce correct MRI scan findings. The researchers suggested a three-layer CNN design that was subsequently joined to a fully connected neural network by the researchers. The accuracy was 96.05 percent, and the F-score was 97.33.

Habbie et al., [6] Semi-automated segmentation was employed on MRI T1-weighted images to examine the probability of a brain tumor using an active contour model. It looked examined morphological active contours with and without edges and morphological geodesic active contours to see how they performed. MGAC outperformed the other two, according to the data.

In this paper, Neelum et al. [7] A concatenation strategy was used to build the deep learning model, and the likelihood of having a brain tumor was studied. To detect and categorize brain cancers, researchers employed the pre-trained deep learning models Inception - v3 and DenseNet201. The Inception - v3 model was pre-trained to extract features and then concatenate them to classify tumors. After that, a softmax classifier was used to classify the data.

In this paper Nalbalwar [8] An integrated Brain Cancer Detection and Classification System was developed using ANN. Histogram equalisation, picture segmentation, image enhancement, and feature extraction were among the image processing techniques used. When compared to other classifiers, the suggested approach for classifying brain pictures utilizing ANN as a classifier has a high classification efficiency. Additionally, sensitivity, specificity, and accuracy are improved. The proposed method is efficient in terms of computation and yields decent results.

Ahmed [9] presented a research in which a Convolutional Neural Network (CNN) was used to diagnose meningioma, glioma, and pituitary tumor with an overall accuracy of 91.3 percent and recall rates of 88 percent, 81 percent, and 99 percent, respectively. To classify distinct forms of brain tumors from MRI image slices, researchers employed a deep learning architecture based on 2D convolutional neural networks. This work employs techniques such as data collection, data pre-processing, pre-modeling, model optimization, and hyper parameter tweaking. In addition, the model's generalizability was tested using 10-fold cross-validation on the complete dataset.

Fatih [10] The method used in this paper is based on Hough voting, which is a strategy for fully automatic localization and segmentation of anatomical regions of interest. It also used a robust, multi-regional, versatile, and adaptable to diverse modalities segmentation method based on learning techniques. Different amounts of training data and data dimensionality (2D, 2.5D, and 3D) are utilized to predict the final outcomes. Convolutional neural networks, Hough voting with CNN, voxel-wise classification, and Efficient patch-wise evaluation using CNN are used to analyze the image.

Kulkarni [11] Brain tumor detection was demonstrated using thresholding algorithms, and a comparative study on tumor detection was described. The obtained results show that the Sobel edge detection operator is effective for tumor detection and tumor boundary extraction. The size and stage of the tumor are determined. For detecting brain tumors, MRI images are the most effective. In this study, digital imaging techniques are critical for detecting a brain tumor in MR images.

Afshar et al., [12] In this they used capsule networks to classify brain tumors using a Bayesian approach. To improve tumor detection results, a capsule network was used instead of CNN because CNN can lose important spatial information. The Bayes Cap framework was proposed by the team. They used a benchmark brain tumor dataset to test the proposed model.

Arya, et al. [18] presented a review of several image preprocessing and segmentation techniques such as image filtration, denoising, histogram-based segmentation, watershed segmentation, SVM-based segmentation, and MRF-based segmentation that can be used as a module to improve accuracy and reduce error rates.

Baljinder Singh et.al,[19]has previously proposed a pre-processing method for removing noise from pictures using a fuzzy filter and a new mean shift based fuzzy c-means algorithm that consumes less computing time and provides better segmentation output than standard techniques. In the classic fuzzy c-means objective function, the following segmentation approaches have a mean field phrase. Because mean shifts can quickly and readily find cluster centers, all techniques may perform an effective diagnosis of the picture area.

Garima Singh et.al,[20] presents a method for classifying and analyzing image de-noising filters such as the Adaptive filter, Median filter, Un-sharp masking filter, Averaging filter, and Gaussian filter, which are used to remove additive noises such as speckle noise, Gaussian noise, and Salt and pepper noise from MRI images. For comparing the de-noising performance of all the techniques considered, PSNR and MSE are used. A fresh idea is to employ a normalised histogram and segmentation via the K-means clustering method for successful brain tumour identification. For effectively classifying the MRIs, the Nave Bayes Classifier and SVM are used, resulting in precise prediction and classification.

Seetha et al in [21] A CNN system, was presented for the detection and categorization of brain tumours The system uses Fuzzy C-Means (FCM) for brain segmentation, and texture and form features were retrieved from these segmented regions, which were then fed into the SVM and DNN classifiers. The accuracy of the system was determined to be 97.4 percent.

Segmentation of images: The segmentation stage is the most important for properly analyzing images because it affects the accuracy of the subsequent steps. However, effective segmentation is difficult due to the wide range of lesion shapes, sizes, and colours, as well as skin types and textures. Several algorithms have been proposed to address this pervasive issue. [13] Thresholding, edge-based or region-based classification approaches, supervised and unsupervised classification techniques are some of the most common. The following are some examples of common segmentation techniques:

Threshold-based segmentation- It either blackens or whitens the pixels in photographs. The pixel value is compared to a threshold value in this procedure. If the pixel value is less than the threshold value, it is replaced with black; otherwise, it is replaced with white. The threshold value can be adjusted to meet the needs of the situation. It's frequently used to distinguish between foreground and background; however it's always divided into only two classes, which is a drawback .This procedure may be effective if the objects in question have a higher intensity than the background or undesired regions of the image.

Clustering based segmentation- in this method uses a crude initial pixel clustering to produce segmented images. Gradient ascent algorithms are used to refine these clusters, the image until it is segmented .These approaches aim to close the gap between pixels and generated clusters [14, 15]. Techniques such as K-means, SLIC, and watershed clustering are widely used.

Edge-based segmentation- This method identifies image edges and uses them to identify individual items. Sobel and Canny edge algorithms are two popular edge segmentation techniques.

Graph based segmentation- Individual pixels are used as nodes in a graph in network-based segmentation. The degree of similarity between adjacent pixels is determined by the weights of the links connecting these graph nodes. Pixels are grouped into super pixels or various segments using a collection of nodes and edges. Graph cut and Normal cut are two graph-based segmentation techniques that are widely utilized.

3 Proposed Work and Methodology

3.1 Proposed Work

In this research, we applied Image Processing and Data Augmentation techniques on our dataset obtained through kaggle. We evaluated them through a Convolutional Layer CNN model and compared our CNN model architecture accuracy with pre-trained VGG-16, ResNet-50, and Inception networks models. For training, validation, and testing, we separated our dataset into three pieces. The validation data is used to evaluate the model and alter the parameters, while the training data is used to learn the model. Finally, the test data will be used to evaluate our model. Our proposed technique is divided into many steps. Figure 1 depicts a high-level overview of the proposed methodology.

Data loading- We obtained our dataset of MRI images from kaggle and then successfully loaded. The data set is divided into two folders, each labeled Yes or No. Different MRI pictures of the patients can be found in both folders. Patients with brain tumors are in the Yes category, whereas patients without brain tumors are in the No folder.

Image Acquisition- Consider that a patient's MRI scan images are either Gray-scale or intensity images, which are shown with a default size of 220*220 pixels. A gray-scale converted image can be defined if it is a color image, by utilizing a huge matrix whose entries are integer values between 0 and 255, with 0 representing black and 255 representing white.

Image Pre-processing- Before the image is converted, there is a process that takes place. It includes noise filtering as well as a variety of other functions. It could arrive as a result of a thermal effect. To maintain the crucial information in an image, image smoothing involves removing noise and unnecessary elements. Image pre-processing is a step in the converting process that occurs before the image is transformed. Each image was subjected to the following pre-processing steps: Crop the image such that only the brain is visible (the most important part of the image). Because the photos in the collection come in a variety of sizes, resize the image to have a shape of (240, 240, 3). As a result, in order to be provided as input to the neural network, all scans need to have the same structure. To scale pixel values to the 0-1 range, use normalization.

Data Augmentation: Because this is a short dataset, there were insufficient examples to train the neural network in our proposed study. Furthermore, data augmentation proved helpful in resolving the data imbalance problem. The obtained augmentation images are clearly seen in Figure 2.

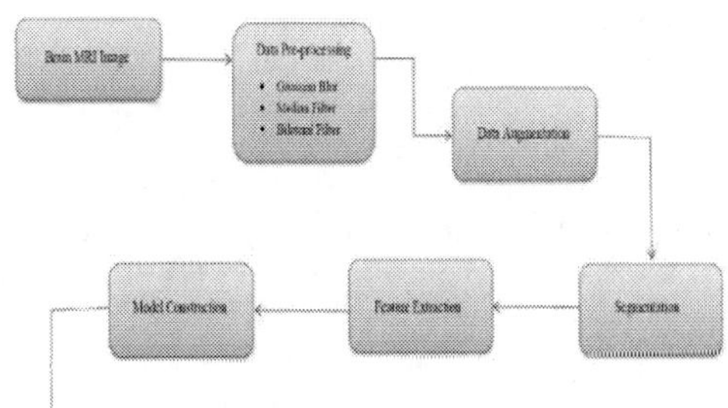

Figure 1. Proposed Methodology

Data Split: The information was divided as follows:70 percentage is used for training,15 percentage for validation of data,15 percentage is used for testing.

 https://doi.org/10.1051/itmconf/20224403041

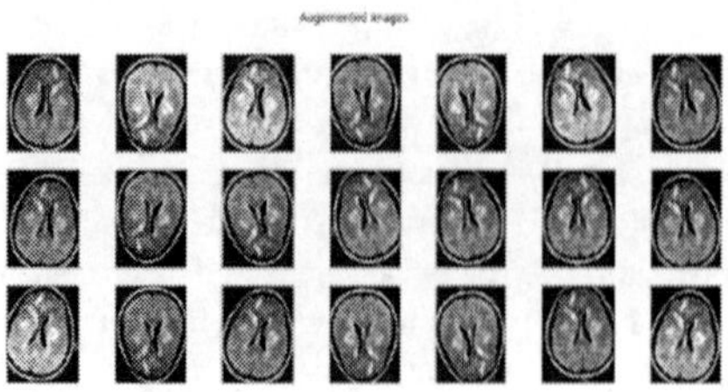

Figure 2. Augmentation Image

3.2 Proposed Methodology

Different classification techniques are employed to classify the brain as normal or consisting of a brain tumor. In Figure 3. We can see the online data-set of images consisting of two classes-yes and no. The images undergo thorough process of four algorithms: CNN, ResNet, VGG16 and Inception v3.

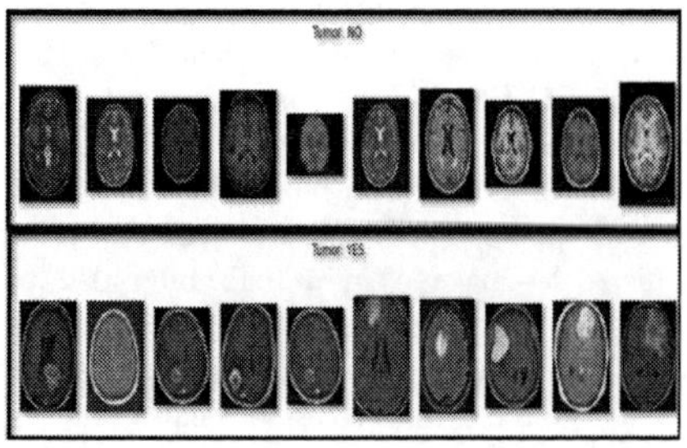

Figure 3. Kaggle Dataset [16]

These classification techniques are described below: - VGG16- a vast visual database project used in the development of visual object recognition software. The VGG16 Architecture was designed and launched by Karen Simonyan and Andrew Zisserman of the University of Oxford in their 2014 paper "Very Deep Convolutional Networks for Large-Scale Image Recognition". The number '16' signifies the number of layers in this design, and the acronym 'VGG' stands for Visual Geometry Group, a group of University of Oxford scholars who devised it. ResNet- An artificial neural network (ResNet) is a residual neural network (ANN). It skips the connections that residual neural networks employ to hop between layers. ResNet models commonly use double- or triple-layer skips with nonlinearities (ReLU) and convolution. An artificial neural network (ResNet) is a residual neural network that may be used to train skip weights using an additional weight matrix (ANN). Skip connections, sometimes known as shortcuts, are used by residual neural networks to go over some layers. ResNet models include batch normalisation and double- or triple-layer skips with nonlinearities (ReLU). Inception Networks- Inception Networks have been shown to be more computationally efficient than VGG, both in terms of the amount of attributes generated and the cost incurred. When modifying an Inception Network, great care must be taken to ensure that computational advantages are not lost. The data was overfit when numerous deep layers of convolutions were utilized in a model. Advantages of Inception- Convolutional neural networks with high performance gain. For the high-performance output of an Inception network, efficient use of computational resources with little increase in computation load is required. The ability to extract features from input data at different scales by using different convolutional filter sizes. Disadvantages- It is very expensive and takes a lot of time to implement.

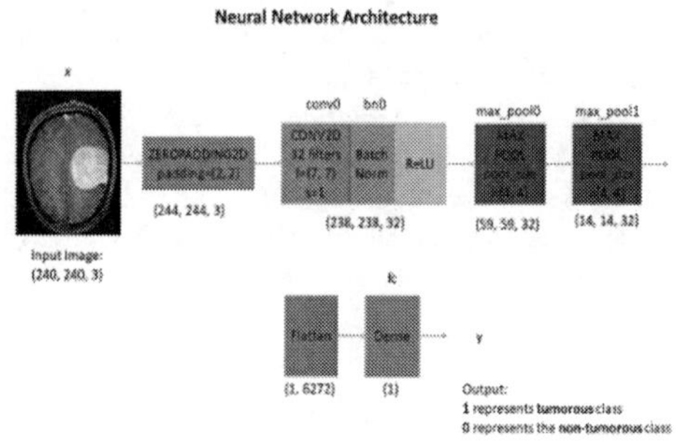

Figure 4. CNN architecture [17]

Convolutional neural network(CNN): - the above Figure 4. depicts the CNN architecture. A deep learning neural network is a sort of machine - learning neural network. Consider CNN to be a machine learning system that can take an input image and apply various filters and parameters in the image, allowing it to distinguish among them. CNN works by recognizing and extracting features from images. A CNN is made up of the following components:

- The grayscale MRI image is used as the initial layer
- Multi-class labels make up the output layer.
- It's vital to realise that ANNs (Artificial Neural Networks) are incapable of grasping.

Advantages of CNN: - They provide translation equivariance, which means that shifting the input data linearly shifts the input in the latent space rather than changing the representation of the input. However, it aids in the development of more robust representations. They make themselves more easily parallelizable, which is why there was a GPU revolution in deep learning research. They provided a significant boost to deep learning research. They can be regarded as backbone function approximates that can be implemented as visual/audio feature extractors for downstream applications that require image/audio data reasoning. They are similar to the mammalian visual cortex, and there are neuroscientific links between biological visual cortex and artificial CNNs. Disadvantages of CNN: - They lack a sense of memory state and are rather limited in sequential modelling (such as language models, speech recognition etc.)

4 Conclusion and Future Work

4.1 Result Analysis

In this study, we used different algorithms for detecting brain tumor after applying pre-processing, augmentation and splitting of images. For each stage of image processing, numerous algorithms have been proposed in the literature. In this paper, we compared four classification algorithms ResNet, VGG16, Inception and CNN which are amongst the most often utilized techniques. From accuracy graphs of respective models in Figure 5. and Figure 6. Comparison analyzing we can conclude that CNN architecture provide best accuracy.

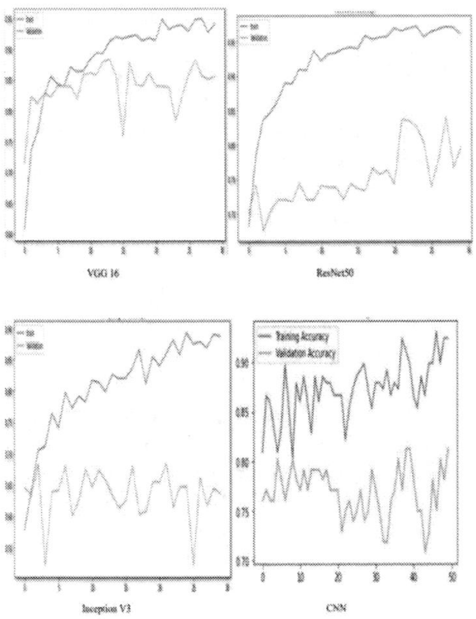

Figure 5. Accuracy graphs of different models

This is the stage where you basically train the algorithms to produce the desired results. The input parameters are available during the learning phase. Basically, you have the model's architecture and the input data. You're actually teaching the algorithm something new. The training parameters are changed by the algorithm during training. It also alters the input data before producing an output. You are assessing once you have received an output. Is that output acceptable, or is it not the expected output? Then you go on to the next step of instruction. When you're pleased with the progress, you'll put the model into execution.

CNN model has the best overall accuracy as well as the best F1 score. CNN is a vital backbone model that is utilized in a wide range of computer vision applications and deep learning models. We compared our CNN model with previous research paper's CNN model and found out that previous model's accuracy was in the range of 72-78 percentage, whereas our accuracy was 82 percentage.To build the model, we divided it into training and test phases, and we trained the model with CNN (Conventional Neural Networks), which increased the accuracy of the model. The model has an 82 percent accuracy, which is better than all other models

Metric	VGG 16	ResNet 50	Inception V3	CNN Model
Train Accuracy	0.940	0.820	0.640	0.80250
Test Accuracy	0.600	0.800	0.500	0.81250
F1 Score	0.714286	0.800000	0.666667	0.81250

Figure 6. Performance of VGG16, ResNet 50, Inception V3 and CNN Model.

4.2 Future Work

Currently, the majority of deep learning algorithms follow the section of tumor area categorization, but the present research is unaware of the tumor region's anatomical location. More research in this area could focus on incorporating this information into the neural network, possibly by feeding the network the entire image. However, due to memory and processing power constraints, training the network on brain tumor images is not possible due to their vast size and high resolution (in the gigapixel range). We can utilise U-Net architecture instead of CNN for more complicated datasets because the max pooling levels are simply replaced with up sampling layers. We eventually want to employ very big and deep convolutional nets on video sequences where the temporal structure provides very useful information that is missing or less visible in static images. In the future, unsupervised transfer learning may gain in popularity. Using VolumeNet in conjunction with the LOPO (Leave-One-Patient-Out) strategy has resulted in high training and validation accuracy. Each iteration of the LOPO test scheme uses one patient for testing and the remaining patients for training the ConvNets; this iterates for each patient. Despite the fact that the LOPO test method is computationally expensive, it allows us to collect more training data, which is necessary for ConvNets training. LOPO testing is reliable and well-suited to our application, which requires us to obtain test results for each individual patient. So, if the classifier incorrectly labels a patient, we can look into it independently.

References

[1] Hassan Ali Khan, Wu Jue, Muhammad Mushtaq and Muhammad Umer Mushtaq "Brain tumor classification in MRI image using convolutional neural network"(2020)

[2] Shraddha S. More; Mansi Ashok Mange; Mitheel Sandip Sankhe; Shwethali Santosh Sahu "Convolutional Neural Network based Brain Tumor Detection" IEEE (2021)

[3] Sakshi Ahuja, B.K Panigrahi, and Tapan Gandhi, "Transfer Learning Based Brain Tumor Detection and Segmentation using Superpixel Technique," in International Conference.

[4] Hajar Cherguif, Jamal Riffi, Mohamed Adnane Mahrez, Ali Yahaouy, and Hamid Tairi, "Brain Tumor Segmentation Based on Deep Learning," in International Conference on Intelligent Systems and Advanced Computing Sciences (ISACS), IEEE, (2019).

[5] Chirodip Lodh Choudhary, Chandrakanta Mahanty, Raghvendra Kumar and Borjo Kishore Mishra, "Brain Tumor Detection and Classification using Convolutional Neural Network and Deep Neural Network," in International Conference on Computer Science, Engineering and Applications (ICCSEA), IEEE, (2020).

[6] Ahmad Habbie Thias, Donny Danudirdjo, Abdullah Faqih Al Mubarok, Tati Erawati Rajab, and Astri Handayani, "Brain Tumor Semi-automatic Segmentation on MRI T1-weighted Images using Active Contour Models," in International Conference on Mechatronics, Robotics and Systems Engineering (MoRSE), IEEEE, (2019).

[7] Neelum Noreen, Sellapan Palaniappam, Abdul Qayyum, Iftikhar Ahmad, Muhammad Imran, "Attention-Guided Version of 2D UNet for Automatic Brain Tumor Segmentation," IEEE Access, Volume: 8.

[8] Rajeshwar Nalbalwar Umakant Majhi Raj Patil Prof.Sudhanshu Gonge 2014 Detection of Brain Tumor by using ANN International Journal of Research in Advent Technology(2014)

[9] Fausto Milletari Seyed-Ahmad Ahmadi Christine Kroll Annika Plate Verena Rozanski Juliana Maiostre Johannes Levin Olaf Dietrich Birgit Ertl-Wagner Kai Bötzel, Nassir Navab Hough-CNN: Deep learning for segmentation of deep brain regions in MRI and ultrasound Elsevier Inc 164 92-102. (2016)

[10] Fatih Özyurt Eser Sert Engin Avci Esin Dogantekin .Brain tumor detection based on Convolutional Neural Network with neutrosophic expert maximum fuzzy sure entropy Elsevier Ltd (2019)

[11] S. S. Hunnur, A. Raut and S. Kulkarni, "Implementation of image processing for detection of brain tumors," International Conference on Computing Methodologies and Communication (ICCMC),Erode,India,ICCMC.2017.8282559.(2017)

[12] Pranian Afshar, Arash Mohammadi, and Konstantinos N.Plataniotis, "BayesCap: A Bayesian Approach to Brain Tumor Classification Using Capsule Networks," IEEE Signal Processing Letters(Volume 27),(2020).

[13] Moumen T.Elmegy, and Khaled M.Abo-El Magad, "A Multiple Classifiers System for AutomaticMultimodal Brain Tumor Segmentation," in IEEE Access, (2019).

[14] Balakumaresan Ragupathy, and Manivannan Karunakaran, "A deep learning model integrating convolution neural network and multiple kernel K means clustering for segmenting brain tumor in magnetic resonance images,"in Int J Imaging Syst Technol. Wiley Periodicals LLC, (2020).

[15] S. Krishnakumar, K. Manivannan, "Efective segmentation and classifcation of brain tumor using rough K means algorithm and multi kernel SVM in MR images", in Journal of Ambient Intelligence and Humanized Computing, Springer (2020).

[16] Navoneel Chakrabarty (2019, November). Brain MRI Images for Brain Tumor Detection. from https://www.kaggle.com/datasets/navoneel/brain-mri-images-for-brain-tumor-detection

[17] NAVONEEL CHAKRABARTY 2017, November). R vs. Python: The Kitchen Gadget Test, Version 1. Retrieved December 20, 2017 from https://www.kaggle.com/datasets/navoneel/brain-mri-images-for-brain-tumor-detection

[18] Arya, Madhvi, and Reecha Sharma. "Brain Tumor Detection through MR Images: A Review of Segmentation Techniques." International Journal of Computer Applications 975 (2016).

[19] Baljinder Singh, Pankaj Aggarwal "Detection of brain tumor using modified mean-shift based fuzzy c-mean segmentation from MRI Images", © IEEE, p.p. 536-545(2017).

[20] Garima Singh, Dr. M.A. Ansari "Efficient Detection of Brain Tumor from MRIs Using K-Means Segmentation and Normalized Histogram", IEEE, (2016).

[21] Seetha, J., and S. S. Raja. Brain Tumor Classification Using Convolutional Neural Networks. Biomedical Pharmacology Journal, 11(7), 1457-1461.(2018)

Implementation of Bird Species Detection Algorithm using Deep Learning

Niyati Jain[1,*] , *Medini* Kamble[1,**], *Amruta* Kanojiya[1,***], and *Chaitanya* Jage[1,****]

[1]Department of Electronics Engineering Ramrao Adik Institute of Technology Nerul, Navi Mumbai 400706

Abstract. Automatically identifying what types of the bird is present in the sound recording using the monitor reading. To distinguishing automatic birds based on their sound patterns.This is useful in the field of ornithology for studying bird species and their behavior based on their sound. Proposed method will be used to distinguish birds automatically using different sound processing methods and mechanical learning methods based on their chirping patterns. We propose a sequential model for audio features within a short interval of time. The model will be used Mel Frequency Cepstral Coefficients to extract features from the audio files and presented it in the model. The proposed work classifies the data set containing three species of bird, and outperform support vector machines.

1 Introduction

An increasing interest in exploring the different impacts of biodiversity is currently taking place around the world. With the rapid decline in wildlife populations around the world due to environmental pollution, there has been an ongoing effort over the years to monitor species that are seen as positive indicators of diversity. Tracking the number of birds in their habitat is one such endeavor, as birds are a good indicator of environmental change. For example, it allows researchers to capture important information such as climate change, migration patterns, pollution and disease outbreaks in the environment. Because birds play such an important role for the environment, a lot of effort has been focused on bird conservation[1]. In recent years, deep-reading strategies have changed the performance of machines in reading, viewing, and processing text. Significant improvements in many classification functions are reported using deep constructs, where deep accumulation neural networks are widely used in computer vision functions.

Human hearing frequency is usually between 20 to 20,000 Hz As we age, we all often lose the ability to hear high waves. Many bird songs have frequencies between 1000 Hz to 8000 Hz, making them readily audible to human hearing. At high altitudes, many warblers, sparrows, warblers, nightingales and many other birds produce sounds up to 8000 Hz or more[2]. When Convolutional Neural Networks(CNN) knows exhausted filters in terms of frequency and time, it faces the limit of deep neurons, lack of time and frequency. The use of deeper and more efficient CNN will also be common and will reflect modern performance in finding objects and image classification challenges. The use of CNN is also popular in voice recognition and voice recognition applications in which audio signals are often converted into spectrum and are taken in the form of inputs on CNN[3].

In audio classification systems, to perform or create models in machine learning, in the first place, we have to extract acoustic features from the audio files. By extracting features, generally the audio files are broken into single frames to make them easy to work. Mel Frequency Cepstral Coefficients(MFCC) is used to extract features from bird audio files. Our main contribution in this paper is create a model that detects a bird on providing an audio file of a bird[4, 5].

We have done an exploratory data analysis (EDA) on the audio files for examining and understanding the data and checking for unlabeled data and errors. To check missing values, we have used Pandas library. There are 2 libraries in Python to work on audio files and they are- Librosa and Scipy. In this paper, we have chosen Librosa because Librosa makes all the files into same sample rate (mono channel) and making the data easy to work on audio processing.

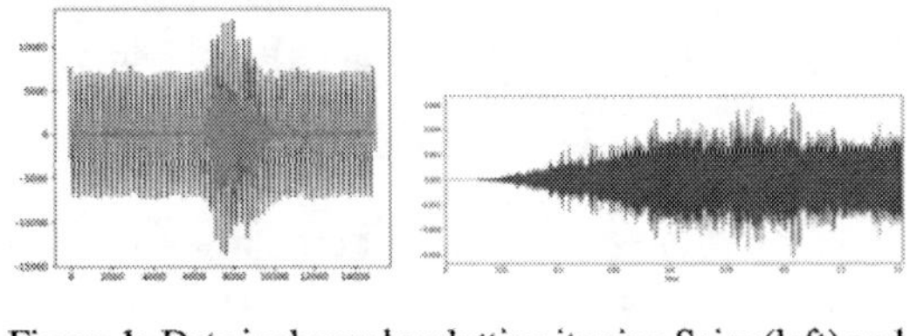

Figure 1: Data is shown by plotting it using Scipy(left) and Librosa(right)

Before creating independent and dependent features, we have to extract features of audio files by using MFCC. By extracting feature, we can uniquely identify the audio signal to predict that in which class it will belong. After features have been extracted, Data frame is created us-

*e-mail: niy.jai.rt19@rait.ac.in
**e-mail: med.kam.rt19@rait.ac.in
***e-mail: amr.kan.rt19@rait.ac.in
****e-mail: chaitanya.jage@rait.ac.in

 https://doi.org/10.1051/itmconf/20224403042

ing those features and the class name. Now that we have both the independent and dependent features, well split the dataset into training and testing dataset for our model.

```
Out[20]: array([[-5.0569110e+02, -4.5887506e+02, -4.3998956e+02, ...,
        -4.1403049e+02, -4.1441818e+02, -4.1519479e+02],
       [ 4.9381628e+00,  2.7463079e+01,  4.1891090e+01, ...,
         2.6102783e+01,  2.4007418e+01,  2.6515617e+01],
       [-2.7153938e+01, -5.2383221e+01, -5.2293556e+01, ...,
        -7.1757324e+01, -7.5127136e+01, -7.2742760e+01],
       ...,
       [-6.6145706e-01,  6.2263441e+00,  5.5203032e+00, ...,
        -3.4690235e+00, -5.3325167e+00, -5.6686220e+00],
       [ 8.7596697e-01,  3.1809101e+00,  2.9348588e-01, ...,
        -5.0395322e+00, -4.3937540e+00, -4.4510584e+00],
       [ 7.5026476e-01, -5.1273794e+00, -1.0170563e+01, ...,
        -1.2300712e+00,  1.1300253e+00,  1.5362844e+00]], dtype=float32)
```

Figure 2: MFCC values in an array

We provide our training x and y dataset to the model and train the model and check the accuracy. Finally, we check our model by providing a testing dataset sample and is being checked if the model is working properly.
This paper is organized as follows: Section 2 comprises of the basic concept of Mel Frequency Cepstral Coefficients has been presented briefly.Section 3 has a detailed explanation of Deep Learning and the model we have used in this work has been written. Section 4 represents the Implementation steps which were performed while designing the algorithm. Finally all the results are given and the conclusions drawn from this work are presented in section 5 and section 6.

2 Mel Frequency Cepstral Coefficients(MFCC)

The first thing for any automated speech recognition system is to extract or remove the features that is identifying the audio signal which are suitable to identify language content and to discard all other data or information which contains elements such as extra background sound.

Therefore the Mel Frequency cepstral coeffiients considers human perception of sensitivity to appropriate frequencies by converting normal frequency into a Mel scale,and thus better suited to speech recognition activities.

MFCCs were introduced by Davis and Mermelstein in 1980 and is the most widely used feature in automated speech recognition system. The feature count is small enough to force us to learn the information of the audio. 12 parameters are related to the amplitude of frequencies. It provides us enough frequency channels to analyze the audio[4, 6–8].

The flow of extracting the MFCC features is given below, As shown in Figure 3:Flow Diagram of MFCC Calculation, MFCC consists of seven computational steps[9?]. Each step has its function and mathematical approaches as discussed briefly in the following:

- Step 1:In this steps it detects signal passing through a filter which will emphasize high waves. This will increase the signal strength at higher frequencies.

$$Y[n] = X[n] - 0.95X[n-1] \quad (1)$$

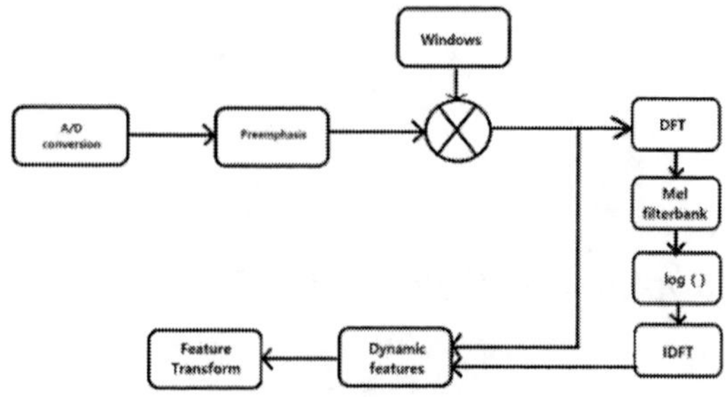

Figure 3: Flow Diagram of MFCC Calculation

Now lets consider a = 0.95 which will make 95% of any one sample is presumed to originate from previous sample.

- Step 2: Framing is the procedure of segmenting the speech samples obtained from analog to digital conversion (ADC) right into a small body with the period in the range of 20 to 40 msec. The voice sign is divided into frames of N samples. adjacent frames are being separated by using M (M < N).

  Normal values used are M=100 and N=256

- Step 3: Hamming window is used as window form through thinking about the subsequent block in function extraction processing chain and integrates all the closest frequency lines. The Hamming window equation is given as: If the window is described as W (n), $0 \leq n \leq N-1$

  Wherein,

  N = numeric values of samples in each frame

  Y[n] = Output signal

  X (n) = input signal

  W (n) = Hamming window,

  then the result of windowing signal is shown below:

$$Y(n) = X(n) * W(n) \quad (2)$$

$$W(n) = 0.54 - 0.46cos[2\pi n/N - 1] \quad (3)$$

$$\text{for } 0 \leq n \leq N - 1$$

- Step 4: Fast Fourier Transform is used to convert N samples from time domain to frequency domain.The Fourier Transform is to convert the convolution of the glottal pulse U[n] and the vocal tract impulse response H[n] in the time domain. This statement supports the equation below:

$$(Y(w) = FFT[h(t) * X(t)] = X(w) * X(w) \quad (4)$$

If X (w), H (w) and Y (w) are the Fourier Transform of X (t), H (t) and Y (t) respectively.

- Step 5: Mel filter bank processing - The frequencies range in FFT spectrum is very wide and voice signal does not follow the linear scale. The bank of filters according to Mel scale as shown in figure 4:Triangular filters of MFCC is then performed.

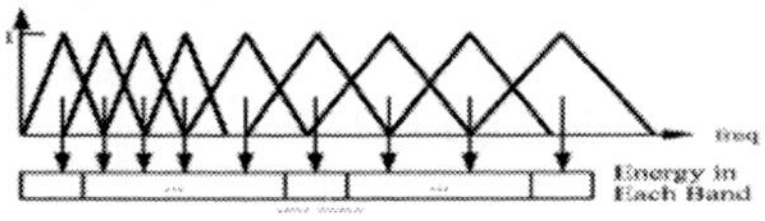

Figure 4: Triangular filters of MFCC

This figure shows a hard and fast of triangular filters that are used to compute a weighted sum of filter spectral components so that the output of system approximates to a Mel scale.Every filters magnitude frequency response is triangular in form and same to unity at the centre frequency and reduce linearly to 0 at centre frequency of two adjoining filters [7, 8]. Then, every filter output is the sum of its filtered spectral components. After that the subsequent equation is used to compute the Mel for given frequency f in HZ:

$$(F (Mel) = [2595 * log[1 + f]700] \quad (5)$$

- Step 6: Discrete Cosine rework this is the process to convert the log Mel spectrum into time domain the usage of Discrete Cosine rework (DCT). The end result of the conversion is known as Mel Frequency Cepstrum Coefficient.
- Step 7: Delta strength and Delta Spectrum The voice signal and the frames changes, along with the slope of a formant at its transitions. Thirteen delta or pace functions (12 cepstral functions plus power), and 39 capabilities a double delta or acceleration function are introduced. The energy in a body for a sign x in a window from time pattern t1 to time pattern t2, is represented at the equation underneath:

$$\text{Energy} = \Sigma X^2[t] \quad (6)$$

Every thirteen delta functions represents the exchange among frames inside the equation 8 corresponding cepstral or strength characteristic, even as every of the 39 double delta capabilities represents the trade among frames inside the corresponding delta capabilities.

$$D(t) = (c(t + 1) - c(t - 1))/2 \quad (7)$$

3 Deep Learning

Deep learning allows computer models composed of multiple processing layers to read data presentations with multiple output levels[10].These methods have greatly improved the state of the art in speech recognition, material recognition, object discovery and many other fields such as drug discovery and genomics. In-depth learning finds complex structures in large data sets using a back-to-back algorithm to show how a machine should modify its internal parameters used to calculate representation in each layer from representation in the previous layer. Deep convolutional networks have brought success to image, video, speech and audio processing, while repetitive networks illuminate successive data such as text and speech[11].

ConvNets or CNNs can process data that comes in many categories, such as speech, text, image, and video. ConvNets is made up of different categories. The first stage is the convolutional layer, where a weight set called the filter bank is combined with the insertion vector. This local measured amount is then transferred to an indirect object such as ReLU called activation function. Two or three stages of conversion, opening operations, and integration layers are packaged, followed by fully integrated layers. Back gradients spread by ConvNet train all the weight of the filter banks. This sequence structure will allow high-level features (details) to be achieved by designing a low level. Convolutional and integration layers in ConvNets are promoted by cells in visual neuroscience

In this work model we use the Keras model. Keras is a neural network application programming interface (API) for python integrated with tensorflow[12].TensorFlow is used to build machine learning models. It gives us a simple and easy-to-use way to define a neural network, which can be built by TensorFlow. Keras facilitates the implementation of complex neural networks with its easy-to-use framework[13, 14].

3.1 Keras Model Overview

Models are a core business that we will be working with when using Keras. Models used to describe TensorFlow neural networks by specifying the features, functions, and layers we want. Keras provides a number of APIs we can use to define your neural network, including:

1. Sequential API, which allows you to create a model layer according to the multi-problem layer. Straightforward (just a simple layer of layers), but limited to single input, single layer output stacks[15].
2. Functional API, which is a full-featured API that supports structural models. It is much more flexible and sophisticated than a sequential API.
3. Model Subclassing, which lets you use everything from scratch. It is suitable for research and very complex application situations, but is rarely used in practice[16].

TensorFlow provides a complete machine learning platform that provides advanced and low-level skills to build and supply machine learning models. The sequential model is suitable for a blank layer stack where each layer has one input tensor and one outgoing tensor. In this model we have used a Dense layer, a function of dropout, a function of Activation and flatten.

1. Dense layer: These layers are used when the connection can be between any element to any other

 https://doi.org/10.1051/itmconf/20224403042

object in the data area. As between two layers of size n1 and n2, there may be a connection of $n1 * n2$ and this is called Dense.

2. Dropout function: Stopping is a way to cut as many connections between the elements by reducing weight (edges) as much as possible. Reducing associations can be used between any layers that stop weight development at the edges. Another important difference here is that it does not have the corresponding weights. There is just throwing things away.

3. Activation function: Opening functions give the model the ability to add compliance to the model. Here a relaxation exercise is performed, which removes a lot of weight.

4. Flatten function: This layer is used when you find a multi-dimensional output and want to edit it to transfer it to a dense layer. Outputs from flat flats are transferred to the phase model or retrieval function you want to achieve.

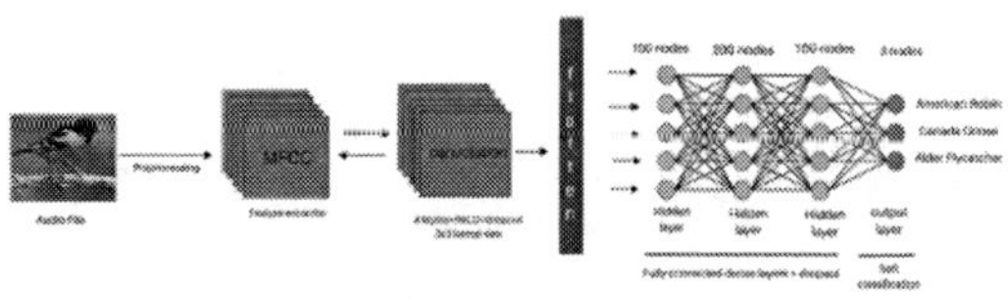

Figure 5: Bird audio detection using Convolutional Neural Network

The first layer by default is the input layer.Afterwards:

Table 1: Layer details of our Model

Layer Index	Layer type	Note
1	Dense	40 Inputs,ReLU activation with 100 dense nodes
2	Dense	ReLU activation with 200 dense nodes
3	Dense	ReLU activation with 100 dense nodes
4	Dense	3 output nodes,Softmax activation

4 Design Algorithm and Result

4.1 Design of Implementation

Some of the functions and libraries used while detecting bird audio files on Jupyter using Python are:

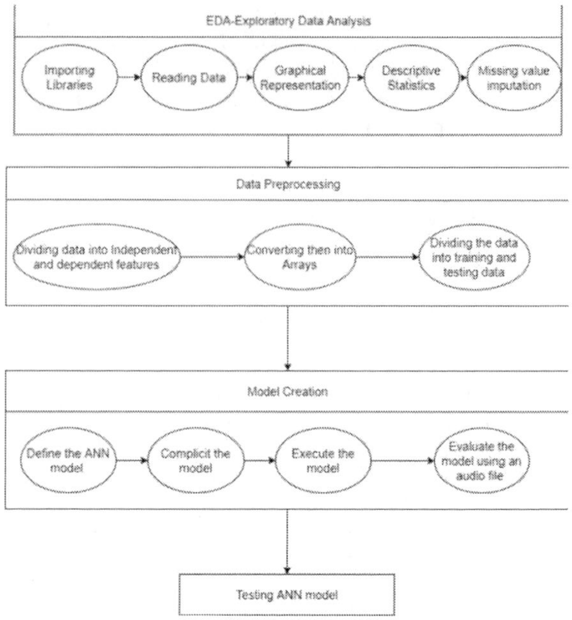

Figure 6: Model workflow on how the project is created

- Librosa: This Library is a python package for music and audio analysis. Libroso is basically used when dealing with audio data such as making music (using LSTM, s), Automatic Speech Recognition. Librosa helps to visualize sound signals and performs the extracts from it using different signal processing techniques [17].
- MFCC: This function comes under Librosa. In short, The mel frequency cepstral coefficients (MFCCs) of the signal are a small set of factors (usually about 10-20) that briefly describe the general shape of the spectral envelope. This feature is one of the most important ways to remove the spectral envelope feature. audio signal and is widely used whenever it works on audio signals[3, 8].
- Tensorflow: This library is uses for data flow graphs to represent calculations, shared status, and functions that change that status. Map graphs of data flow across multiple machines in the collection, and within the machine on all multiple computer devices, including multicore CPUs, standard GPUs, and custom-designed ASICs known as Tensor Processing Units (TPUs). This structure provides flexibility for the application developer[14].
- Pandas: Python library for the rich data structures and working tools and standard data sets mathematics, finance, social science, and many other fields. The library provides integrated, accurate methods for committing common data fraud as well analysis on such data sets. It aims to be the basic layer of the future mathematical computer in Python[12].
- Sklearn: scikit-learn, which is a Python module that integrates a wide range of modern machine learning algorithms for moderate and unregulated intermediate scale

 https://doi.org/10.1051/itmconf/20224403042

problems, is one of the most popular machine learning tools [18].

- Label Encoder: This function is used to make the text of the label simply provide the total value for the total possible variance of the category [19].
- Sequential: It is based on a rotating process between the suggestion of a new hyperparameter configuration for testing and a review of the dynamic model of the relationship between hyperparameter configuration and holdout set performances. So, as the model learns about this relationship, increases its ability to lift better hyperparameter setting and gradually converts to the best solution[15].

Following are the steps to be followed while providing a audio file to the model in Jupyter software:

1. Start with choosing a file from the device. Again, we need to preprocess data where we need to extract the features of that specific data itself with the help of MFCC.
2. Now that we have features of the particular file, we predict the class labels with the help of model that is created. From this, model provides the label.
3. Finally, we have to inverse transform the label to get the class name that will provide us which bird the audio is relating to.

4.2 Result

In this work, we have provided 6342 audio files of 3 birds as the dataset. A random file name is selected from the dataset and then passed to the features-extractor class which returns the class label from the model(numeric value) which is later given out as bird type using inverse transform.

Prediction-feature saves the return extracted features. We have 2 variable x and y , where x represents extracted features and y represents name of the birds. By using the extracted feature, we can predict which type of bird it is by using CNN module. Now we have features by random file module is predicted from this. It is then compared with the given dataset and results the results in array. For a record, if the predicted value is equal to the actual value, it is considered accurate. We then calculate Accuracy by dividing the number of accurately predicted records by the total number of records.Here we get an accuracy of 70 percentage.

5 Conclusion

In this work, We have implemented an algorithm using Sequential model providing the best accuracy at 70 percentage when using the mean values of all features earlier mentioned. By implementing this model, we have classified three birds- Alder flycatcher, Canada Goose and American Robin using only audio data from xeno-canto dataset.This work will really help the researchers to implement the Deep Learning Sequential model to detect bird type using their audio files.

Figure 7: Alder Flycatcher(left) and American Robin(right)

```
test_accuracy=model.evaluate(X_test,y_test,verbose=0)
print(test_accuracy[1])

0.699999988079071
```

Figure 8: Accuracy of 70%

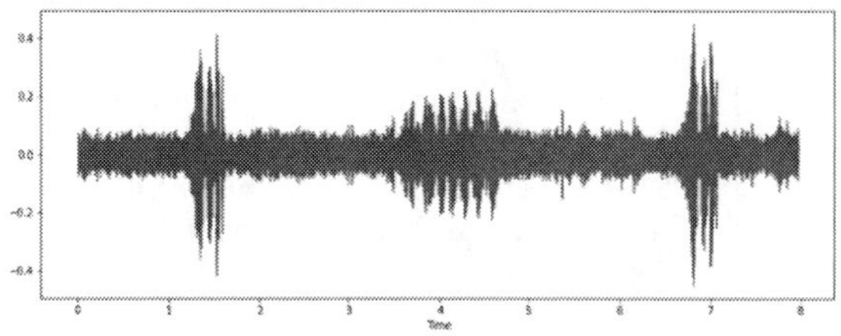

Figure 9: Audio waveform of Alder Flycatcher

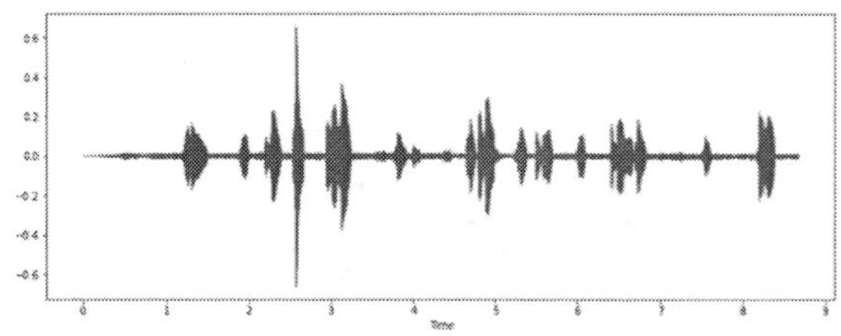

Figure 10: Audio waveform of Canada Goose

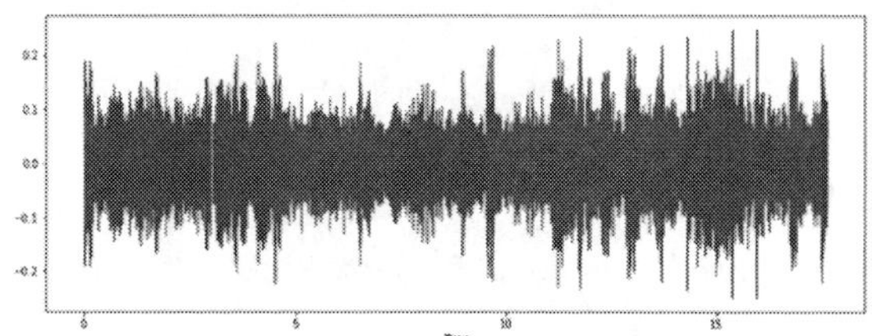

Figure 11: Audio waveform of American Robin

References

[1] J.A. Kogan, D. Margoliash, The Journal of the Acoustical Society of America **103**, 2185 (1998)

[2] D.E. Balmer, S. Gillings, B. Caffrey, R. Swann, I. Downie, R. Fuller, *Bird Atlas 2007-11: the breeding and wintering birds of Britain and Ireland* (BTO Thetford, 2013)

[3] C.H. Lee, Y.K. Lee, R.Z. Huang, Journal of Information Technology and Applications () **1**, 17 (2006)

[4] M. Likitha, S.R.R. Gupta, K. Hasitha, A.U. Raju, *Speech based human emotion recognition using MFCC*, in *2017 international conference on wireless communications, signal processing and networking (WiSPNET)* (IEEE, 2017), pp. 2257–2260

[5] F. Briggs, R. Raich, X.Z. Fern, *Audio classification of bird species: A statistical manifold approach*, in *2009 Ninth IEEE international conference on data mining* (IEEE, 2009), pp. 51–60

[6] A. Coates, A.Y. Ng, in *Neural networks: Tricks of the trade* (Springer, 2012), pp. 561–580

[7] P. Somervuo, A. Härmä, *Analyzing bird song syllables on the self-organizing map*, in *Workshop on Self-Organizing Maps (WSOM03)* (2003)

[8] V. Tiwari, International journal on emerging technologies **1**, 19 (2010)

[9] A. Selin, J. Turunen, J.T. Tanttu, EURASIP Journal on Advances in Signal Processing pp. 1–9 (2006)

[10] Y. LeCun, Y. Bengio, G. Hinton, nature **521**, 436 (2015)

[11] S. Wei, S. Zou, F. Liao et al., *A comparison on data augmentation methods based on deep learning for audio classification*, in *Journal of Physics: Conference Series* (IOP Publishing, 2020), Vol. 1453, p. 012085

[12] W. McKinney et al., Python for high performance and scientific computing **14**, 1 (2011)

[13] L. Breiman, Machine learning **45**, 5 (2001)

[14] M. Abadi, P. Barham, J. Chen, Z. Chen, A. Davis, J. Dean, M. Devin, S. Ghemawat, G. Irving, M. Isard et al., *{TensorFlow}: A System for {Large-Scale} Machine Learning*, in *12th USENIX symposium on operating systems design and implementation (OSDI 16)* (2016), pp. 265–283

[15] Y. Even-Zohar, D. Roth, arXiv preprint cs/0106044 (2001)

[16] C. Kwan, G. Mei, X. Zhao, Z. Ren, R. Xu, V. Stanford, C. Rochet, J. Aube, K. Ho, *Bird classification algorithms: Theory and experimental results*, in *2004 IEEE International Conference on Acoustics, Speech, and Signal Processing* (IEEE, 2004), Vol. 5, pp. V–289

[17] B. McFee, C. Raffel, D. Liang, D.P. Ellis, M. McVicar, E. Battenberg, O. Nieto, *librosa: Audio and music signal analysis in python*, in *Proceedings of the 14th python in science conference* (Citeseer, 2015), Vol. 8, pp. 18–25

[18] F. Yang, X. Wang, H. Ma, J. Li, BMC Medical Informatics and Decision Making **21**, 1 (2021)

[19] J.T. Hancock, T.M. Khoshgoftaar, Journal of big data **7**, 1 (2020)

[20] F. Briggs, B. Lakshminarayanan, L. Neal, X.Z. Fern, R. Raich, S.J. Hadley, A.S. Hadley, M.G. Betts, The Journal of the Acoustical Society of America **131**, 4640 (2012)

[21] P. Somervuo, A. Harma, S. Fagerlund, IEEE Transactions on Audio, Speech, and Language Processing **14**, 2252 (2006)

[22] M.A. Acevedo, C.J. Corrada-Bravo, H. Corrada-Bravo, L.J. Villanueva-Rivera, T.M. Aide, Ecological Informatics **4**, 206 (2009)

[23] J. Pons, X. Serra, *Randomly weighted cnns for (music) audio classification*, in *ICASSP 2019-2019 IEEE international conference on acoustics, speech and signal processing (ICASSP)* (IEEE, 2019), pp. 336–340

[24] J. Devlin, M.W. Chang, K. Lee, K. Toutanova, arXiv preprint arXiv:1810.04805 (2018)

[25] M. Stracy, O. Snitser, I. Yelin, Y. Amer, M. Parizade, R. Katz, G. Rimler, T. Wolf, E. Herzel, G. Koren et al., Science **375**, 889 (2022)

[26] Z. Deng, B. Wang, Y. Xu, T. Xu, C. Liu, Z. Zhu, IEEE Access **7**, 88058 (2019)

[27] M. Aaron, M. Elad, Signal Processing, IEEE Transactions on **54**, 4311 (2006)

[28] T.M. Aide, C. Corrada-Bravo, M. Campos-Cerqueira, C. Milan, G. Vega, R. Alvarez, PeerJ **1**, e103 (2013)

[29] R. Caruana, A. Niculescu-Mizil, *An empirical comparison of supervised learning algorithms*, in *Proceedings of the 23rd international conference on Machine learning* (2006), pp. 161–168

[30] S.S. Stevens, J. Volkmann, E.B. Newman, The journal of the acoustical society of america **8**, 185 (1937)

[31] K. Seyerlehner, G. Widmer, P. Knees, *Frame level audio similarity-a codebook approach*, in *Proc. of the 11th Int. Conf. on Digital Audio Effects (DAFx-08)* (2008), p. 31

[32] P. Minka, Tech. rep., Tech. Rep., Microsoft Research (2003)

[33] R.E. Kass, P.W. Vos, *Geometrical foundations of asymptotic inference* (John Wiley & Sons, 2011)

Body Posture Detection and Motion Tracking using AI for Medical Exercises and Recommendation System

Anuj Patil[1 a)], *Darshan Rao*[1 b)], *Kaustubh Utturwar*[1 c)], *Tejas Shelke*[1d)] *and Ekta Sarda*[1e)]

[1] Department of Computer Engineering, Ramrao Adik Institute of Technology, D Y Patil Deemed to be University, Navi Mumbai

Abstract. Exercises are highly essential in our everyday lives, especially when patients are in the middle of a healing process and need to speed up their body's recuperation. Exercise has become more important in our lives as a result of this. They provide the cornerstone for improving human capacities and extending their lives. Artificial Intelligence and Image Processing can be utilized to improve and supplement the workout process without the need for professional supervision. A software-based motion tracker can keep track of all the exercises you've done and provide you feedback on your posture while you're working out. Through computing data and analysis, the exercise's beneficial efficiency will be increased. The MediaPipe framework could be utilized for this application; in this machine learning model, points are plotted at several joints of the human body posture, and movement is tracked, stored, and analyzed. This detailed analysis of the body tracking could be used in the implementation of an application that could keep a track of the medical exercise of a registered individual. The software could be further improvised in such a manner that the registered user could be mapped to an authentic verified doctor having the access to the diagnosis reports and exercise history of the mapped patient using databases.

Keywords: MediaPipe, BlazePose, BlazeFace, Bicep curls.

1 Introduction

Weighted exercises are one of the most useful and widely used types of physical activity in the world of healthy exercise. Some disorders can be cured by performing workouts that target specific muscle groups. Free weight workouts, on the other hand, can be frightening to beginning users and can have serious bodily consequences if done incorrectly or without professional assistance. Despite these advantages and high-risk activities, little to no effort has been made to use technology to assist new learners in learning new exercises. Our objective is to develop an easy-to-use interface technology that assists users who need to perform workouts to speed up their recovery or healing without the need for active professional supervision [11]. Counting repetitions, tracking your form, offering information on users' sessions, and recommending new and better exercises under passive supervision of health professionals are just a few of the features. When exercising with equipment or weights, this can greatly reduce the risk of injury [8]. Simple biceps curls are an example of this, as your back posture must be perfect to avoid straining your shoulder muscles and spine. [10] For such problems, we intended to create unobtrusive technology that could track users without requiring them to wear anything or place anything on their bodies, and that could subsequently be analyzed by health specialists for future treatment [9].

Push-ups are one of the most frequent bodyweight workouts. This is because it has several health advantages and requires the usage of muscles across the upper body. In terms of how this exercise is carried out, there is a lot of variation.[14] As a result, it was selected as a proof of concept for recording bodyweight exercises and repetitions performed.

The goals of this system are to provide easy access to the software, which can work on any system with x86 or arm-based CPU-s in a for all intents and purposes big way and

[a] Electronic Mail: anujpatil200@gmail.com
[b] Electronic Mail: darshan.rao120501@gmail.com
[c] Electronic Mail: k.utturwar@gmail.com
[d] Electronic Mail: tejasshelke040402@gmail.com
[e] Electronic Mail: ekta.sarda@rait.ac.in

to easily provide a hands-free interface to increase its practicality and usage, which really is quite significant.[13] Preparation of the entire dataset or data hub covering most of the exercises, depending on the diseases or other types of critical health conditions. Apart from this, keeping track of various points on the body to derive an analysis, track and calculate the human body posture percentage could be considered one of the significant objectives.
Lastly, providing tips to improve the body posture for optimal exercise benefits along with recording and preserving the exercise history to provide chronological analytics on your daily/weekly/monthly exercise habits [12].

2 Literature Survey

There is currently no instance of improving health condition-based exercises. Personal trainers and Physiotherapists are the ones that monitor, manually. This has disadvantages for the two types of people. Firstly, the people who live in remote places do not have access to these personal trainers and hospitals. The second type of people is the senior citizens, who need continuous attention for their daily exercise and diet plans.[5] Even if they have their treatment in the hospitals, there is no one to check if their exercise form is correct. [7] There is a need for a person to be allotted for these activities, which increases the additional fees. Sensors such as Xbox Kinect had come up which were able to track body gestures [2]. These sensors were used for games and entertainment and did not enter the field of health. The main reason for this is the expensive build due to which it was not affordable for ordinary citizens. There is a dearth in the exercise to be assigned for the specific types of diseases. Even though some software for detecting gestures, the proper integration with the disease-specific exercise is still in the development stage.[4]

Projects like a real-time on-device hand tracking solution for AR/VR applications that predicts a human hand skeleton from a single RGB camera.[1] These were mainly used for mobile devices and used for hand tracking.[8] Research on metabolic health and nutrition are made which have given a brief idea of biomarkers that can be used the making nutrition plans.[3]

3 Proposed Methodology

The hand tracking method makes use of an ML pipeline comprised of two models that collaborate.:

1. A palm detector that locates palms using an aligned hand bounding box and operates on a whole input picture.
2. A hand landmark model that works with the palm detector's chopped hand bounding box to yield high-fidelity 2.5D landmarks.

Firebase was used as the database framework to store the information about the exercise and easily fetch them as needed. Django framework would handle the front end of the application.

The general architecture of the proposed system is shown in **Fig.1** which shows the generalized flow that includes capturing the live video feed from the patient's camera and displaying it on the dashboard. Further, the exercise tracker tracks the workout performed and pushes the results in the database which are further passed to the doctor's dashboard which is also used to input the instructions for the next exercise set and diet chart and send it back to the patient's dashboard via the database.

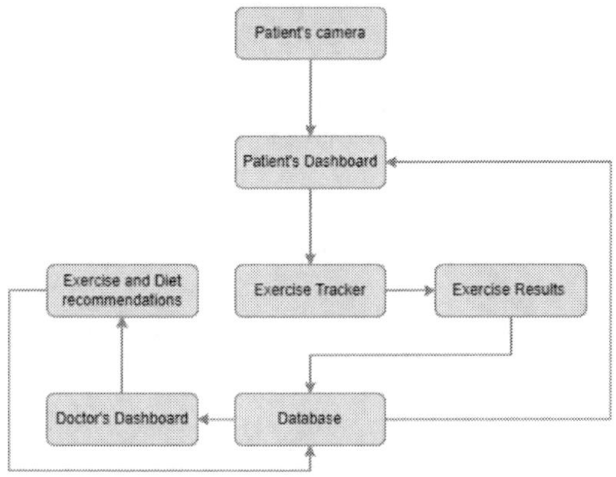

Fig. 1. The generalized flow of the system

There are two kinds of detection models in the MediaPipe framework - BlazePose detector and BlazeFace detector. In BlazePose, there are two machine learning models: a Detector and an Estimator. The Detector removes the human region from the input image, while the Estimator returns key points from a 256×256 resolution image of the detected individual. Similar to BlazeFace, which is also available in MediaPipe, this single shot detector type is suited for mobile real-time applications. Hand gesture detection is a challenging task. The model must detect both occluded and self-occluded hands while functioning with a wide range of hand sizes and a large scale span (20x). The contrast pattern on the face is high around the eyes and mouth, whereas hands do not, making it more difficult to distinguish them just based on their visual characteristics. The system employs three separate validation datasets that span different sectors to compare the quality of the models to other well-performing publicly accessible solutions: Yoga, Dance, and HIIT. There is only one person in each shot, who is 2-4 meters away from the camera. The model simply evaluates 33

https://doi.org/10.1051/itmconf/20224403043

key points from the COCO topology to be compatible with previous solutions.

Fig.2 is an example of how the BlazeFace detector detects the human body posture and can perform segmentation masking over it with very high precision and accuracy.

Fig. 2. Example of MediaPipe Pose real-world segmentation mask. [1]

In **Fig.3**, the algorithmic flow of the MediaPipe framework for hand movement detection is as follows: image snapshots from the given camera are pipelined into the BlazeFace, which then recognizes body points using large datasets. The image is then mapped, cropped, and projected on the screen.

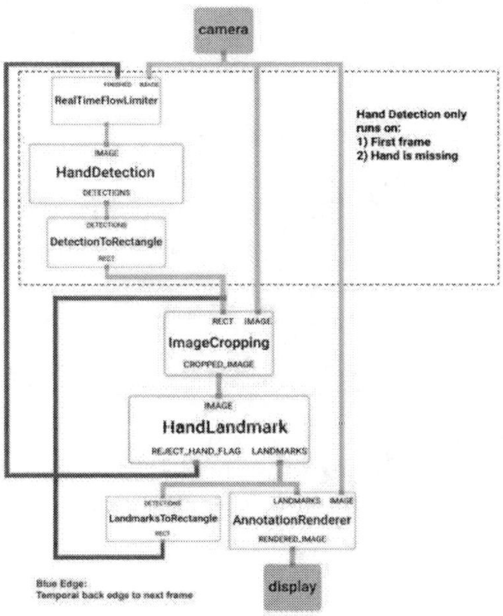

Fig. 3. MediaPipe pipeline flow. [1]

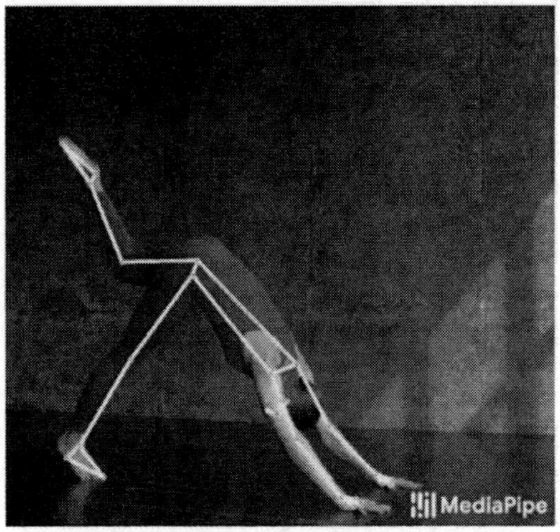

Fig. 4. Example of MediaPipe Poses for pose tracking.[7]

Fig. 5. Hand gesture tracking outputs [7]

Fig.4 and **Fig.5** shows how the MediaPipe model plots the joints and lines on the human body posture and palm for movement detection using the angle between two joints.[6]

4 Implementation and Results.

The Health Tracker is integrated to a web application using Django.

Django is a high-level Python web framework that promotes rapid development and simple, practical design. It's built by professional developers to take care of a lot of the problems of web development, so the user can concentrate on developing your application instead of reinventing the wheel.

The mapping between these two portals would be done using Google's firebase. The patient's portal would consist of a navigation bar along with the profile details. Further, regarding the main components of the portal, the dashboard would consist of a window that is integrated with the system's camera and shows the live feed of the user's environment. Beside the window, allotted daily tasks would be displayed, which would consist of the several exercises allotted by the doctor who is mapped with that particular patient in a list format. The patient needs to follow the daily workout schedule referring to the list and his/her exercise count would be tracked using the machine learning model which is being implemented, and the count would be displayed using a counter container on the corner of the screen for the patient to refer to. Once the allotted instructions are completed by the user, he/she would push the results in the database using the front-end component provided. Apart from the daily exercise recommendations, a section with a detailed description of the daily diet provided by the respective doctor would be also displayed dynamically on the webpage.

The doctor's portal would again consist of the navigation tabs, similar to the patient's portal with the respective profile information which would consist of professional-medical biodata of the doctor. Further, the doctor's dashboard would be mapped with all the patients which are registered to him/her using the database framework, that is, on the webpage, there would be a section where the doctor could view and surf through all the patients mapped with him/her, on the selection of any of the patients, a detailed dashboard consisting of the patient's entire medical journey with that doctor would be displayed in a proper format. The detailed dashboard would consist of several tabular, diagrammatic, and graphical forms of data representations based on the patient's chronological diagnosis to date. On this web page, doctors can dynamically push the daily exercise and diet recommendations in the system's database, which is further pushed on the patient's dashboard.

The Exercise Tracker is the main component of the system. The video captured is further processed in MediaPipe. The MediaPipe uses the BlazePose Detector model for detecting the person. After detecting the person, the landmark model in MediaPipe Pose uses prediction methods to detect the location of 33 pose landmarks as shown in **Fig.6**. The exercise used in this application requires only hand and leg landmarks.

All landmarks are joined, and a function is used to calculate the angle between the joints. The difference in the joint angles caused due to movements is used for calculating the reps of the exercises.

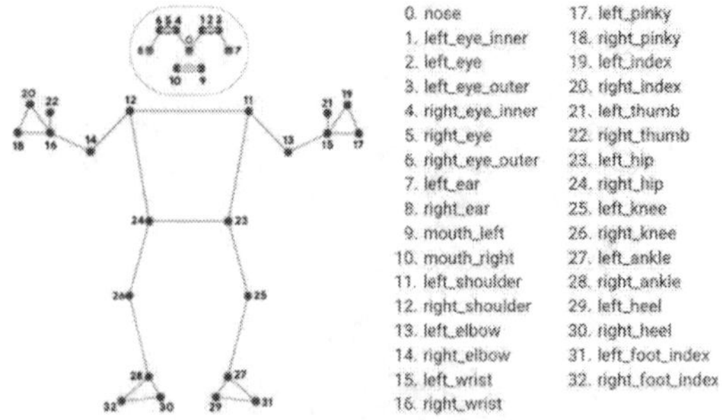

Fig. 6. Landmark identification on the body[7]

For test cases, exercises such as biceps curls and push-ups are used. The biceps curls use angles between the landmarks {16, 14, 12}(right hand) and {11,13,15}(left hand) and angles a difference between timestamps is used for counting reps. Two instances of bicep curls detection and counting are shown in **Fig.7** and **Fig.8**.

Fig. 7. Image showing bicep curls (Instance 1)

Fig. 8. Image showing bicep curls (Instance 2)

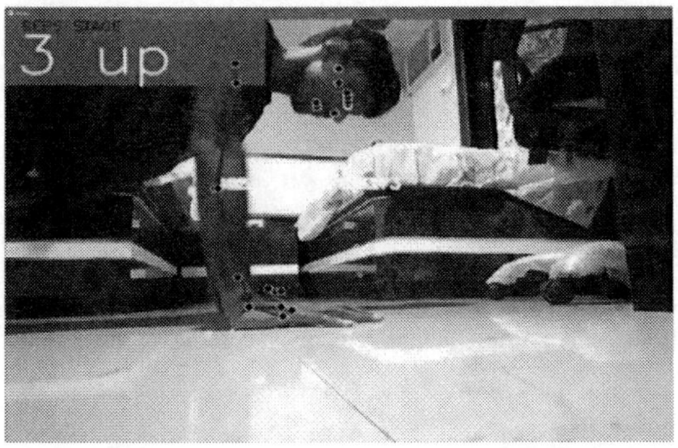

Fig. 9. Prototype User Interface showing repetition and body posture.

Both of the portals, that is the patient's portal and the doctor's portal which are supported by the backend source code needs to be properly mapped together and backed with the collected and stored data in the data warehouse for the application to run and function according to the expected requirements. These requirements could be efficiently and optimally fulfilled by Firebase's Firestore. Google Firebase is an application developed by Google for a development platform that allows creators to create applications for iOS, Android, and the web. Firebase delivers analytics tracking, reporting, and app issue fixes, as well as marketing and product experimentation capabilities.

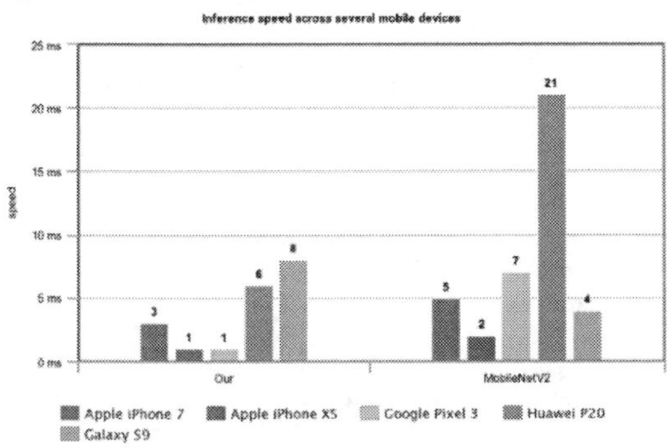

Fig. 10. Graph showing the inference speed across various mobile devices with different operating systems and hardware.

To identify the face or the body, a frontal camera or webcam is used. The model allows the user to keep track of his posture and count the overall number of reps he has completed. Only faces that occupied more than 20% of the image area were considered for the frontal camera model, according to the use case. The 5% criterion was set for the rear-facing camera model. The model was generated after it was trained on a dataset of 66k photos. The developed system has used a 2k image collection from a private, geographically diversified dataset. The mapping of the movements done is of a very high precision because of the usage of the dataset and the robust model, which in comparison to other models on the market is quite accurate **[Fig.9]**. The lighting in the space in which the activities are performed is really important. The tracker was unable to detect the joins while testing in lower illumination settings, resulting in some incorrect findings. The camera's resolution made a difference in the tracking of the workouts; the higher the camera's resolution combined with enough lighting, the closer the precision obtained under such ideal conditions is to perfection.

Model	**Average Precision**	**Inference Time, ms (iPhone XS)**
MobileNetV2-SSD	97.95%	2.1
Ours	**98.61%**	**0.6**

Table 1. Frontal camera face detection performance

The outcomes of certain test cases are shown in **Table.1** above. The tracking which was implemented at first was tested with the existing model MobileNetV2-SSD, but the results were a millimeter off. The model which was developed later on produced better results than the previous model when used. The iPhone XS was the device utilized to test our first case, and the test results were better than the previous ones.

Continuing the testing on the iPhone XS, the testing was conducted on a variety of devices, ranging from phones with lower processing power, such as the iPhone 7, to mobile devices with better computing speed and graphical processing, such as the iPhone XS(iOS) and Pixel 3. (Android).

Accuracy	Exercises
92%	Biceps Curl
84%	Squats
82%	Push-Ups

Table 2. Accuracy of Exercises

The testing of the app for each exercise was done for 50 reps, hence the percentage accuracy of each exercise is the percentage of reps counted per 50 reps. 10 test cases were made for push-up exercise **[Fig.9]** which gave an average accuracy of 92% **[Table.2]**.

5 Conclusion

The system introduced a free weight exercise tracking software that allows users to learn and correct their form in this research. The squat exercise was chosen for this study because it has extremely specific needs that could be tracked and modified easily. Form specifications were used to transmit these requirements, and an indicator of a red X and a green check was used to demonstrate whether they were proper or incorrect. As the user performed the push-up, these parameters, and indicators allowed them to see and adjust their form. In our pilot study, participants were able to enhance their push-ups and correct their form to the point where they felt comfortable performing the exercise.

6 Future Work

The system has a lot of areas for improvement. Further research will be required on the types of exercise used by patients with specific types of diseases. The research requires consulting a physician or Physiotherapists who can verify the exercise is proper or not. Some exercises such as push-ups had lesser accuracy, mainly due to the camera quality used for the application. The less accuracy can be improved by making an android app that can use its higher quality camera for detection. As the system created is only limited to diet plans and health tracking, there is a lot of space for adding more features such as blood tests, vaccine reports. The system can be more versatile by adding vernacular languages, as for countries like India, which has different languages in every state. The application can also be made women-specific that can keep track of women's health, such as checking periods and suggesting women-specific exercise. The UI needs to make it more user-friendly as the targeted patients are the senior citizens and the people living in rural areas who are obtuse to technology.

References

[1] Zhang, F., Bazarevsky, V., Vakunov, A., Tkachenka, A., Sung, G., Chang, C. L., & Grundmann, M. (2020). Mediapipe hands: On-device real-time hand tracking. arXiv preprint arXiv:2006.10214.

[2] Conner, C., & Poor, G. M. (2016, May). Correcting exercise form using body tracking. In *Proceedings of the 2016 CHI Conference Extended Abstracts on Human Factors in Computing Systems* (pp. 3028-3034).

[3] Lee, E. C., Fragala, M. S., Kavouras, S. A., Queen, R. M., Pryor, J. L., & Casa, D. J. (2017). Biomarkers in sports and exercise: tracking health, performance, and recovery in athletes. *Journal of strength and conditioningresearch, 31*(10), 2920.

[4] Van Der Zee, M. D., Van Der Mee, D., Bartels, M., & De Geus, E. J. (2019). Tracking of voluntary exercise behaviour over the lifespan. *International Journal of Behavioral Nutrition and Physical Activity, 16*(1), 1-11.

[5] Shpigelman, L., Lalazar, H., & Vaadia, E. (2008). Kernel-ARMA for hand tracking and brain-machine interfacing during 3D motor control. *Advances in neural information processing systems, 21*, 1489-1496.

[6] Ahmad, A., Migniot, C., & Dipanda, A. (2019). Handpose estimation and tracking in real and virtual interaction: A review. *Image and Vision Computing, 89*, 35-49.

[7] Rahman, S. U., Afroze, Z., & Tareq, M. (2014). Hand Gesture Recognition Techniques For Human Computer Interaction Using OpenCv. *International Journal of Scientific and research publications, 4*(12).

[8] Pedersen, B. K., & Saltin, B. (2015). Exercise as medicine–evidence for prescribing exercise as therapy in 26 different chronic diseases. *Scandinavian journal of medicine & science in sports, 25*, 1-72.

[9] McGloin, R., Embacher, K., & Atkin, D. (2017). Health and exercise-related predictors of distance- tracking app usage. *Health Behavior and Policy Review, 4*(4), 306-317.

[10] Voth, E. C., Oelke, N. D., & Jung, M. E. (2016). A theory-based exercise app to enhance exercise adherence: a pilot study. *JMIR mHealth and uHealth, 4*(2), e4997.

[11] Mohamed Abdur Rahman, Ahmad M. Qamar, Mo- hamed A. Ahmed, M. Ataur Rahman, and Saleh Basalamah. 2013.Multimedia Interactive Therapy Environment for Children Having Physical Disabilities. *In Proceedings of the 3rd ACM Conference on Inter- national Conference on Multimedia Retrieval (ICMR'13). ACM, New York, NY, USA,* 313–314.

[12] Derrick Cheng, Pei-Yu Chi, Taeil Kwak, Björn Hart- mann, and Paul Wright. 2013. *Body-tracking Camera Control for Demonstration Videos. In CHI '13 Extended Abstracts on Human Factors in Computing Systems (CHI EA '13). ACM, New York, NY, USA,* 1185– 1190. *DOI:http://dx.doi.org/*10.1145/2468356.2468568

[13] Han Ding, Longfei Shangguan, Zheng Yang, Jinsong Han, Zimu Zhou, Panlong Yang, Wei Xi, and Jizhong Zhao. 2015. *FEMO: A Platform for Free-weight Ex- ercise Monitoring with RFIDs. In Proceedings of the13th ACM Conference on Embedded Networked Sensor Systems (SenSys '15). ACM, New York, NY, USA,* 141–154.

[14] Kyle Rector, Cynthia L. Bennett, and Julie A. Kientz. 2013. Eyes-free Yoga: An Exergame *Using DepthCameras for Blind & Low Vision Exercise. In Pro- ceedings of the 15th International ACM SIGACCESS Conference on Computers and Accessibility(ASSETS'13). ACM, New York, NY, USA, Article* 12, *8 pages.*

Automatic Number Plate Recognition System for Indian Number Plates using Machine Learning Techniques

Gayatri Hajare[1], *Utkarsh* Kharche[2], *Pritam* Mahajan[3], and *Apurva* Shinde[4]

Department of Computer Engineering, Ramrao Adik Institute of Technology D.Y Patil Deemed to be University, Nerul

[1]gayatriumesh34@gmail.com, [2]utkarshkharche29101998@gmail.com, [3]ironpritam@gmail.com, [4]apurva.karkhanis@rait.ac.in

Abstract. India being a country where the population is above 1.3 billion where each person has at least one car of his/her use. Considering this, the number of cars driven on the roads of India must be greater than the population of the people in the country. India being a diverse country, diversity is not only seen in the language of the number plates but also in size, spacing between the letters on the number plate and the font of the number plate. Diversity differs from state to state. Even though most of the people are using English Number plates, there is no certain law as to how a number plate should be, so some people tend to have number plates according to their preferences. To withstand these problems, we have created a system using You Only Look Once version 5 (YOLOv5) for number plate detection and Google Tesseract for Character Recognition.

Keywords: YOLOv5, Indian Number Plates, Tesseract Engine, Optical Character Recognition.

1 Introduction

Automatic Number Plate Recognition (ANPR) is a tool that can be used not only at toll booths on highways, express ways, etc. to speed up the process of toll collection but also used by car parking managements at malls, movie theatres, etc. This system can also help government authorities to catch and fine individuals who are trying to break traffic rules and regulations as well as help us locate criminals who use vehicles to flee. By detecting vehicle's number plate from CCTV feeds can help to locate stolen vehicles.

India, with a population of 1.3 billion, has the second-largest population [13]and the second largest road network [1] in the world having a road network of 63.72 lakh km has its own needs and requirements for an ANPR system, even though there are certain rules regarding the size of number plate of certain vehicles [2], These rules aren't strictly followed. Different kinds of number plate patterns are found like number plates with different fonts having fancy lettering while number plates are sometimes hand painted on the vehicle especially seen on trucks, etc. The ANPR system along with a combination of sensors is used to track vehicle speed and specifically detect cars that are speeding. Here, a new ticket is generated automatically by estimating the distance between the two cameras. This assists authorities to maintain legislation and rules, which also in return help them to decrease vehicle collisions.

ANPR System provides us an appropriate solution to provide safe parking management in residences. Registered Resident Vehicles are directly allowed to enter and park in their fixed parking spot whereas the non-registered vehicles will be first added into guest list then it will assign them an empty parking spot. In India, around 2 lakh cars are there that are being stolen each year [13]. This number can be decreased if proper precautions are taken. ANPR system can be employed for locating and tracking cars such that if any vehicle is stolen, authorities would be able to detect it from the CCTV feed and locate the path taken by the stolen vehicle and help them to find the find vehicle in less time.

Most ANPR systems are made specific to certain type of Number plates [8,11,12] while some are developed based on certain specific criteria [10,14]. Like [11], Silva and Jung proposed a system specifically for Brazilian number plates by taking a combination of European number plates for training number plate detection model. While in [10], They have given a review of all different ANPR techniques that can be used based on certain criteria.

2 Literature Survey:

Silva, et al. [11] proposed a 4 stage ANPR system in. Unlike other ANPR models where license detection is the first stage, they decided to do vehicle detection first. So, no vehicle with a visible license plate is missed. For vehicle detection, instead of training a model from the start they decided to use an existing model (YOLOv2) based on certain criteria. License detection is done by using a CNN named WPOD-NET which was developed for taking features from YOLO, SSD and STN. For Character Recognition (OCR), They created their own LP characteristics-based YOLO network which can detect LP without any issues.

Shidore and Narote [6] proposed an ANPR Model divided into three sections (1) License Plate Extraction (2) Character Segmentation and (3) Character Recognition. An image is acquired using a good ANPR camera and then various image pre-processing techniques like Gray scaling, Sobel Edge Detection and Thresholding are applied on the image to detect various candidate LPs in the image. Then using Bounding Box Analysis True LP is extracted from the image. For Character Segmentation, Integration of Character Region Enhancement, Connected Component Analysis and Vertical Projection Analysis is used. Character Recognition is done in two parts 1) Character Normalization and Feature Extraction 2) Character Recognition using SVM as classifier. The accuracy level stated in [6] for segmentation is 80% and 79.84% recognition.

Kulkarni, et al. [7] proposed an ANPR system for toll booths for faster toll collection by implementing a 4 stage ANPR model [10] which is a combination of different algorithms specially for Indian LPs. As [10], There is vehicle detection module but by using inductive sensors which then triggers the capture of the rear of the car and then various pre-processing techniques as applied to acquired image and then license localization is done using a 'Feature based number plate localization' which specially developed just for Indian LPs. For Segmentation, they have implemented a method known as Image Scissoring where the LP is scanned and scissored in such a way that there is no white pixel present there and is copied into a matrix. For Character Recognition, statistical feature extraction is implemented.

Kashyup, et al. [4] proposed an ANPR system consisting of 3 modules. Before extracting the number plate, a combination of image processing techniques was applied. Segmentation of Characters is done using Regionprops function. Character recognition is done in two parts, first Feature Extraction and then the Actual recognition is done by Template Matching. Template matching is the process of finding a small image region called a template and comparing it with the same template in the database. The accuracy of the system mentioned here is 82.6%.

Devpriya, et al. [14] proposed a system similar to [4]. The image is pre-processed by converting the RGB image into gray scale image then morphological operation is carried out then the number plate detection is done by applying Canny Edge Detection and Closing and opening morphological analysis. Here, Character Segmentation is done by Connected Component Analysis (CCA) [6]. Then similar to [4] each segmented characters are recognized using Template Matching.

YOLOv3 model which is commonly used for object detection is used for all the stages of ANPR [5]. Tensorflow is used for everything. The input images are converted into grayscale. Here for LP detection annotated dataset is required. For this task, they have used LabelImg for labeling data such that in the input image, a bounded box is around the number plate which helps in detection. The character segmentation and recognition are done using YOLO models. The accuracy of character recognition by the YOLO model is 99% while that of OpenCV is 93%.

Tiwari and Choudhary [12] proposed an ANPR system using Artificial Neural Networks (ANNs) based on Indian Number Plates for the system. As other ANPR systems pre-processing methods like Gaussian Filter, Wavelet Transform and Image Binarization on the acquired image and then the Number Plate is extracted and then Character Segmentation is done and then using Vector generation, Character Recognition is done.

Naren Babu et al. [7] proposed an ANPR system which is divided into two stages License Plate Detection and Character Recognition. They have trained a 37 class CNN single YOLO model for both detecting the license plate as well as recognizing the characters the license plate. They also have achieved the 100% accuracy in detecting the number plate and 91% accuracy in recognizing the characters of the number plate.

3 Proposed System

The proposed system is divided into 3 different stages to understand the working of the system more effectively. A custom YOLOv3 model was created for all the three stages to achieve better accuracy in number plate recognition [7]. In the proposed system, YOLOv5 model used for number plate detection only unlike [7] and further this detected number plate is segmented. We are using YOLOv5 instead of earlier YOLO versions as it can detect smaller objects better while training unlike its predecessor. The system uses Google OCR

Tesseract to recognise and transform the characters on the licence plate into text after segmentation.

3.1 The proposed system is divided into the following steps

1) Acquisition of the image
2) License Plate Detection using the YOLOv5 model.
3) Plate Segmentation
4) Character Recognition of the License plate using Optical Character Recognition (OCR).

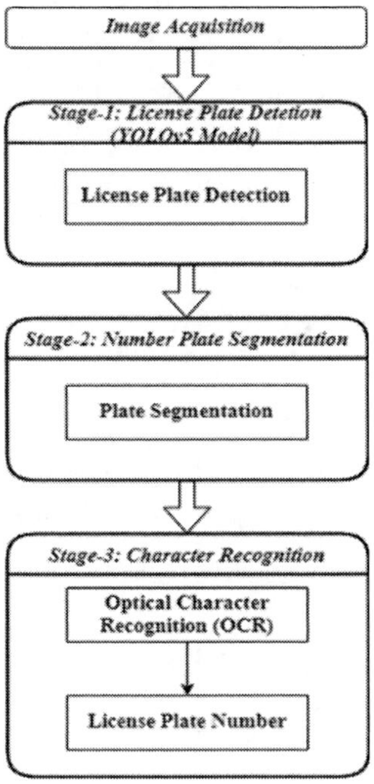

Fig. 1. Architecture of the proposed system

3.2 Stepwise explanation of each module

3.2.1 Image Acquisition

Image acquisition of the vehicles is done using CCTV cameras at the toll booths and parking lots. For our system, we used a dataset consisting of different images of vehicles on the internet as well as we captured images from mobile devices that are shown in fig.1.
Our dataset consists of nearly 500 images of different types of vehicles such as bikes, cars, etc. with both front and rear-view. We also have images of license plates without a vehicle for defining the number plate to our YOLO network for license plate detection. Though our dataset doesn't have any specific dimensions i.e., not all images have the same size, while training the dataset we have converted all images to the same size like 640x640. Also, we annotated the images manually using LabelImg, an open-source API to create a training set in YOLO format.

Fig. 2. Image Acquisition of the vehicle

3.2.1 Stage 1: License Plate Detection

An Automatic Number Plate detection and Recognition system is used to locate the number plate o the vehicle and create the popularity of a plate that will extract text from the image and all others due to calculation modules using local algorithms, plate separation, and character recognition.

We used the YOLOv5 model for license plate detection. It is a Convolutional Neural Network (CNN) used for real-time acquisition where we will use this algorithm to obtain a license plate for a vehicle. CNN can process input images very quickly and detect patterns in them while maintaining its accuracy level.

These types of algorithms are used in AI programs to find specific types of objects based on a selected subject. The YOLOv5 Algorithm is used to locate different types of objects and to distinguish them. YOLOv5 is implemented using the Pytorch library mainly present in Python.

We decided to go with YOLOv5s as its smallest and lightest model with fastest output to cope up with the video frames. Here we can see the difference in different sized models of v5 in accuracy and responsiveness.

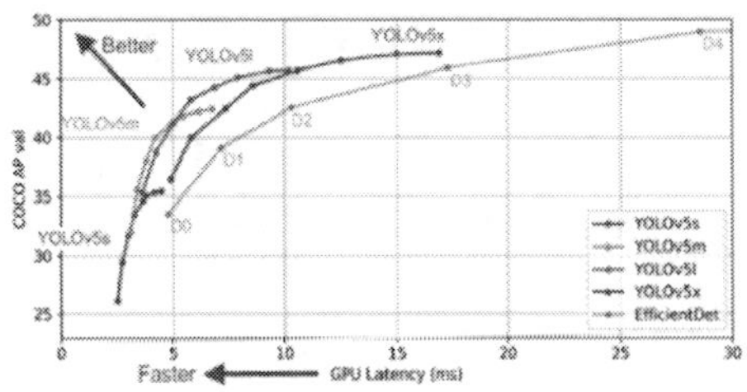

Fig. 3. Accuracy of YOLOv5 compared to other YOLO versions

For the detection of the license plate, The YOLO algorithm uses the following steps:
1)Residual Blocks: In this technique, The whole image is converted in form of square grids and each

grid has an dimension N*N. In this it will highlight the grids which have objects detected in them.

2) Bounding Box Regression: In this technique, A bounding box is formed around the objects found from the whole image. This bounding box has different attributes namely; bounding box width (bb_w), bounding box height (bb_h), Class c (this describes which object is what in. Here License plate) and bounding box centre coordinates (bb_x ,bb_y). Here we want to detect the license plate. In Figure 1, The green outlined bounding box along with its class Plate in green/pink represents the detected license plate from the image using YOLOv5 model.

Fig. 4. Image after the detection of license plate using YOLOv5 Model

Proposed System has achieved the following results after training over 100 epochs.

Precision	0.629
Recall	0.943
map@0.5	0.0958
map@0.5	94% confidence as 0.417

Table 1. Training results of proposed system

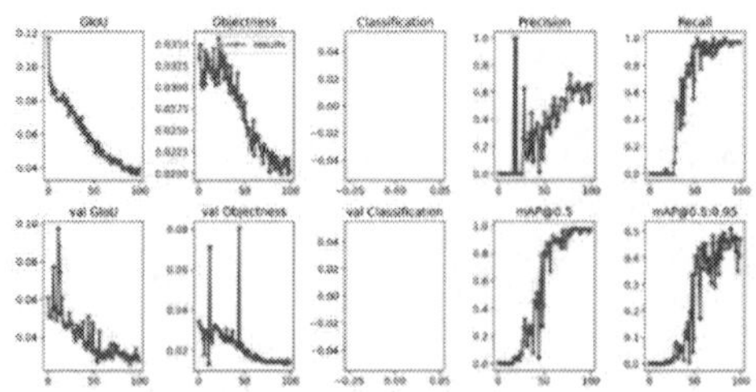

Fig. 5. Accuracy of detection model while training

3.2.2 Stage 2: Plate Segmentation

Plate Segmentation: Segmentation is the next important step in the identification of License Plate that separates the License Plate from the original image and then we apply OCR Tesseract on it for Character Recognition.

Fig. 6. Image of the cropped License Plate

3.2.3 Stage 3: Character Recognition

Character Recognition here is done using OCR Tesseract is the last step of LP recognition that identifies each character separated by a text format [1]. Optical Character Recognition (OCR) is a method that allows a program while otherwise human intervention to identify texts or words written in users' oral communication. Optical Character identification has grown to become the most successful application of the information in the field of pattern acquisition and AI.

As we can see in every stage, we have also tried for one of the regional language number plates in India i.e., for Marathi number plate which should be roughly of following showed format. We used Google Translate API to translate Marathi words to English after passed through the same YOLOv5 model. Though the accuracy we got it is not that great, but it can further improve as more dataset will available in future.

Segmented License plate	Character Recognition of License plate
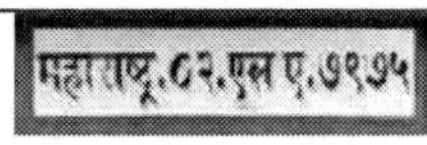	MH02LA7975
	M H 2 0 E E 7 5 9 8

Table 2. Output after performing Optical Character Recognition (OCR)

4 Conclusion

In our proposed program, we have a habit of providing the entire ANPR comprehensive learning program with free results. Our results show that the strategic approach exceeds existing and external strategies in complex data sets, consisting of LPs taken with a strong point where rational retention ends up in many controlled repositories.

Most cars have the same single license plates while bikes, buses, rickshaws, etc., there are usually double license plates. So, this makes LP detection almost impossible. So, to overcome this difficulty, we can create a database with various types of images of all License Plate patterns that can be found in India.

We further wish to work towards creating a dataset containing different Indian number plate types like in different regional languages, different font size and license plate sizes and also there are four types of number plates having different colour indicating the purpose of the vehicle driven e.g., Yellow background number plates are generally used by taxis while White background Number plates indicate that the vehicle is for personal use etc. Hence, we hope to create an ANPR model which can detect all these types of number plates with maximum accuracy and speed.

References

[1] Andrew S. Agbemenu, Jepthah Yankey, Ernest O. Addo, IJCA, *An Automatic Number Plate Recognition System using OpenCV and Tesseract OCR Engine*, **43**, (2018)

[2] Annual Report 21-22, Government of India, Ministry of Road Transport & Highways.

[3] Du S., Ibrahim M., Shehata M., Badawy W., Automatic *License Plate Recognition (ALPR): A State-of-the-Art Review. IEEE Transactions on Circuits and Systems for Video Technology*, **23(2)**, 311–325 (Feb 2013)

[4] Kashyap A., Suresh B., Patil A., Sharma S., & Jaiswal A., ICACCCN, *Automatic Number Plate Recognition,* 6 (2008)

[5] K. Prema, P. Arivubrakan, V. Suganya, TJCME, *Automatic Number Plate Recognition Using Deep Learning,***12**, 3 (2021)

[6] M. M. Shidore, S. P. Narote, IJCSNS, *Number Plate Recognition for Indian Vehicles,* **11**, 4 (2011)

[7] Naren Babu, R., Sowmya, V., & Soman, K. P., ICICICT, *Indian Car Number Plate Recognition using Deep Learning*, (2019)

[8] P. Kulkarni, A. Khatri, P. Banga, K. Shah, ICR, *Automatic Number Plate Recognition (ANPR) System for Indian conditions,* 4 (2009)

[9] Rule 50 & 51 of Central Motor Vehicles Rule 1989

[10] S. Ghadage, S. Khedkar, IJERT, *A Review Paper on Automatic Number Plate Recognition System using Machine Learning Algorithms*, **8**, 4 (2019)

[11] S. Silva, C. Jung, ECCV 2018, *License Plate Detection and Recognition in Unconstrained Scenarios* ,17 (2018)

[12] S. Tiwari, N. Choudhary, IJEAT, *An Efficient Method for Indian Number Plate Recognition*, **7**, 5 (2017)

[13] S.Tripathi, S. Shetty, S. Jain, V.Sharma, IJITEE, *Automatic Number Plate Recognition System (ANPR): The Implementation,*10, 8(2021)

[14]W. Devapriya, C. Nelson Kennedy Babu, T. Srihari, IJCIE, *Indian License Plate Detection and Recognition Using Morphological Operation and Template Matching,* 9, 7 (2015)

Density Based Traffic Control System withConvolutional Neural Network

Mrunali R. Saple, Tejaswini M. Jadhav, Saurabh Y. Kode and Gargi

Phadke Department of Instrumentation Engineering, Ramrao Adik

Institute of Technology, Nerul, Navi Mumbai.

Abstract: - We present a real-time traffic flow controller structure that can keep traffic under control using image processing techniques. In this way, a camera is used in every section of the robot to take pictures of the traffic where traffic jams will appear. The number of vehicles in these images is designed using image processing tools. In the proposed image, green and red signals are represented using LEDs and the diminished green signal supervisor is signified by a specific presentation.

Keywords: - Real time, Traffic Light Control, Vehicle Count

I. INTRODUCTION

Traffic congestion has been a huge problem all around the world. Obstruction senses the delays in worker productivity, trading opportunities, and delays in delivery. Transportation which is current technologies use a hands-on operating system during distribution and require high adjustment during operation. Time ceasing and vehicular traffic increased due to this. The system proposed makes the traffic low and permit the transport built on the density of the lane. This project aims for lowering the traffic flow in the zone wherever the high density of vehicles by using R-pi operating system and Image Processing. Traffic congestion is a very common and increasing problem in upgrading areas, in fact it is constantly evolving gradually environment creates it challenging to catch wherever circulation congestion is in real time, in order to plan restored transportation indication mechanism and efficient traffic flow routes. The source of this could be various conditions such as traffic congestion such as inadequate lane width, road surroundings because of weather, unrestricted claim, and significant red signal delays etc. Although inadequate volume and unrestricted response are related anyplace, light delays have a strong code and are independent on traffic. Indeed, physical control should, therefore, to decrease human potential, the necessity to emulate and advance road traffic control in order to meet the growing demand. Surveillance and safety in image processing is the latest in a series, which is commonly used in vehicle and transportation control with navigation data. The traffic flow congestion resolution is been determined using Image Processing.

II. EXISTING SYSTEM

In existing system, traffic density Monitoring using Raspberry Pi and Open CV. The robots used in India basically have a pre-existing period when the time for each route to have a green signal is adjusted. In four lanes the traffic signal one lane is given a green signal at a time. Thus, the robot allows cars of all directions to pass in a straight line. Therefore, traffic can be either continuous or vertical or rotated by 90 degrees. So even though the traffic congestion on a particular route is very low, it should wait unnecessarily and if it gets an unnecessary green signal, it makes some routes wait even longer with some of the strategies we use. We suggest a method that can be used to control traffic flow using image processing. In line with traffic congestion on all roads, our ideal will intelligently allot green light interval on each road. We selected image processing to calculate traffic congestion as cameras are inexpensive than other procedures like sensors. The planned idea is designed as monitors: We have a Raspberry Pi linked to 4 groups of LEDs representing robots. It is a method of observing traffic congestion on every side and changing the signal rendering to the traffic congestion so it is very beneficial to control traffic.

III. LITERATURE SURVEY

This system proposed is to make some improvement in traffic system that depends on high traffic load. We hear, detecting traffic obstruction using Raspberry-pi, using the python.in libraries. Based on their density the trails priority is maintained. The obstruction calculation algorithm depends on real-time live video frame with citation image and by having cars in the preferred location. In order to control the traffic signal intelligently the traffic congestion can be compared to the other lane. Recently Monitoring and Video surveillance are used for traffic management. To test traffic congestion using image processing a lot of redesigning is done. Various factors like rain; fog and so on is required by these methods for good image with good quality. Various car icons such as radar, ultrasonic, and microwave detector increase the authenticity. Partly, sensors can be expensive, low power, high maintenance, confusing to use and costly to repair. Metal barriers along the road side can affect the radar sensors. During monitoring the field and controlling traffic sensors like Passive acoustic detector array, high temperature, Photosensitive, inductive loop detector, magnetic detector, are used. These sensors have less accuracy. The traffic obstruction that occurs in city areas that cannot be effectively maintained using the endured system of stable signals. When traffic blockage grows beyond the limit on a particular road, it requires the duration of the green signal to lower the traffic. A complication with the lights system is that the timing variable is adjusted in software and as a result valuable time is wasted even if the opposite path is empty.

The goal of resolving signal time control is to determine the sequence of the phase and the length of every part. To explain this problem, formal information labeling the connections of an objective net, traffic information including circulation appeal and rotating vehicle activities, and limitations related to transportation indication constituents are measured. This information is managed according to the ideal design.

IV. PROPOSED SYSTEM

The proposed system main objective is to make the traffic signal more systematic and attainable. So, the traffic is reduced and time gets utilized. The proposed system will be density based so that it will give priority to the lane which has fairly a larger number of vehicles. For evaluating the density Image Processing is used. For Processing Raspberry Pi will be used. Images will be obtained by using the data set. Images obtained will go through steps of image Processing. Then, the differences will be compared and consideration will be assigned.

A. Image Acquisition from Data-Set: An image is taken from the data set of images of road having various amount of vehicle density. Images from the data-set demonstrate the real-time images of traffic.

B. Image Pre-Processing: Pre-processing is basically used for removing unwanted noise and objects from image. The following are the steps that are to be taken in this phase:

Step 1: The first step in which the RGB image is converted into greyscale image for improving performance is Greyscale Conversion. The grey color image consists of pixel intensities between 0-255 where 0 signifies black pixel and 255 signifies white pixel.

Step 2: The greyscale image into the binary image by Threshold. If pixel value is higher than a threshold value it is given one value (white), else it is given the other one (black).

C. Image Processing: This phase aims to serve the processing part on the pre-processed image. Steps taken are as follows: Discontinuities in image are detected by edge detection. Canny Edge Detection algorithm detects all the edges in the image. It provides:

a. Noise Reduction b. Finding Intensity Gradient

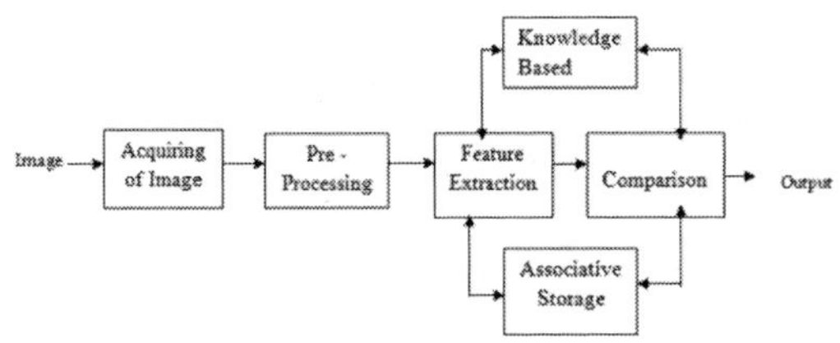

Fig.1 Block Diagram for Image Processing.

 https://doi.org/10.1051/itmconf/20224403045

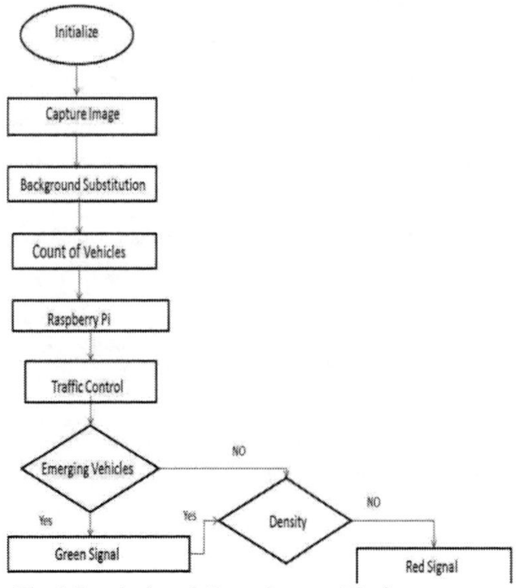

Fig. 2 Density based Control system Block Diagram.

V. IMPLEMENTATION

A. Design Procedure

Raspberry pi, Image Processing using Python and object detection algorithm are used to design this system. To convert raw images to accessible form Image Processing is used. The Raspberry Pi is a main constituent used to switch everything; it acts as a controller. Traffic flow is captured by using the pi camera and this information is received by computer. The computer is connected to the Raspberry pi to perform a hardware launch where it controls the traffic signal using a traffic control system. Crossing the road in the middle of a red light, it will hurt the nails that are attached to the motor. Traffic light-based traffic control system produces and decreases traffic congestion in city areas, time consuming because of traffic congestion.

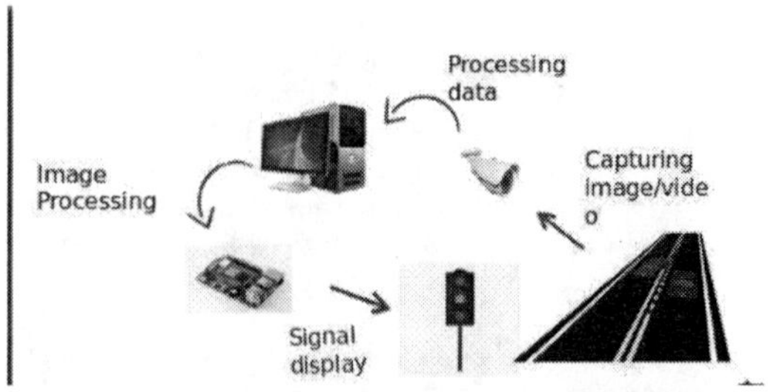

Fig.3 Block Diagram for Density Based Traffic Control Using CNN.

B. Image Segmentation: Image Segmentation helps to gain the region of interest (ROI) from the spitting image. The process of separating an image into different areas is called Image Segmentation. Image Objects are parts in which images are divided. Image segmentation is based on the properties like similarity, discontinuity, etc. Image Segmentation goals to simplify the image for better analysis. It is also the procedure of allocating labels to each pixel in an image. In Machine Learning Image Segmentation is used widely.

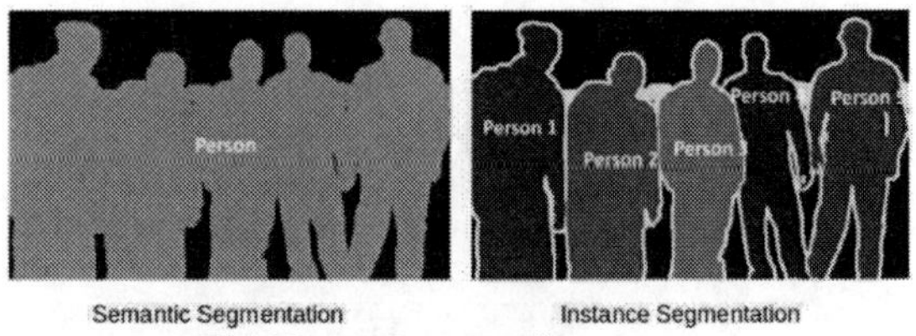

Fig.4 Image Segmentation

C. Image Classification: The process of classifying and classification groups of pixels or vectors inside an image built on certain instructions is image classification. The Separation Act can be designed using one or more spectral elements of text. The two most common methods of classification are 'supervised' and 'unsupervised'.

 https://doi.org/10.1051/itmconf/20224403045

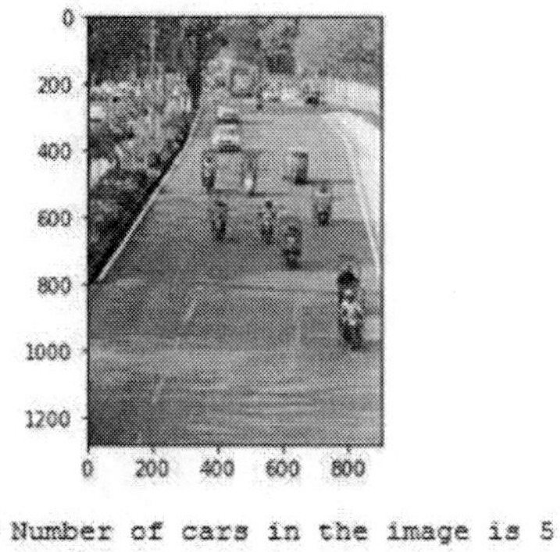

Fig. 5 Image Classification

A. RESULTS AND DISCUSSION

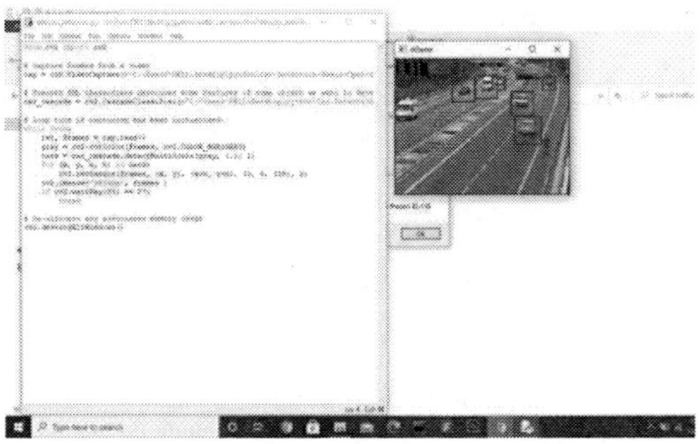

Fig. 6 Output form the image detection

Fig. 6 represent is a new vehicle acquisition and removal system for traffic rentals. Locating moving objects present in complex lanes are involved in this system.

The system implements an advanced background subtraction methodology. The algorithm proposed recreates a background event as needed under any traffic conditions.

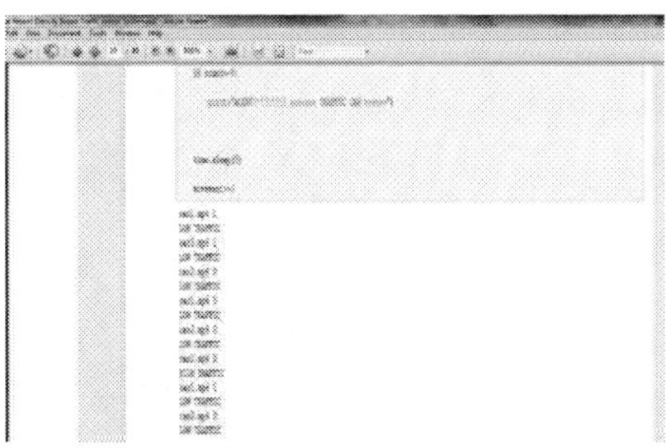

Fig. 7 Video Output

Fig. 7 shows the representation of the output taken from the audiovisual. This indicates the amount of cars existing in the space. It gives the respected signal according to the count of the vehicle.

Output Accuracy:

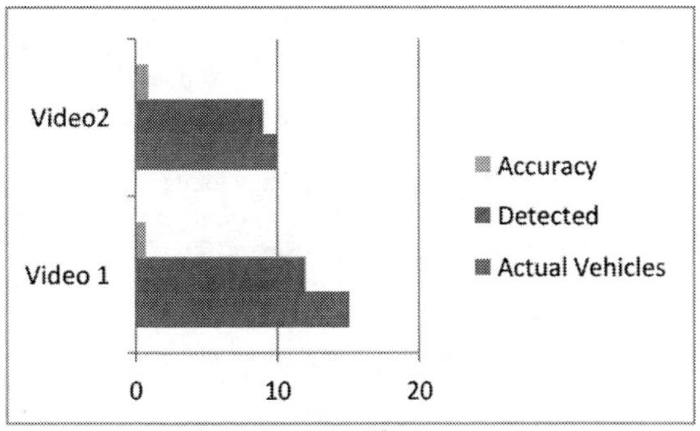

Fig. 8 Accuracy Chart

B. LIMITATIONS

1. Open CV is less flexible compared to MATLAB.
2. Raspberry Pi uses limited amount of memory.
3. In the present system, normal traffic light time is allocated to all the predetermined, defined and modified road routes.

C. FUTURE SCOPE

1. Future work may include a future where we can monitor signals and status can be updated on the server.
2. This will help in the future indicator of producing traffic congestion patterns according specific dates, holidays, time,
Etc.
3. It can also be used prioritize emergency vehicles separated by noise at the top.
4. This may provide a very important feature for ambulances, firefighters, etc.

D. CONCLUSION

We have designed a traffic density measurement control system using Raspberry pi, camera module, and image processing according to traffic congestion in urban areas. The camera module is mounted at the high end of the track. The number of vehicles on the street is given; traffic should be controlled by permitting cars where congestion is higher with the help of that calculation. This system gives good elasticity to keep traffic flow. Indication about the emergency vehicle is also known for transmitting the signal and will be shown a green signal on the route. The traffic light-based traffic control system is highly effective and decreases traffic congestion in urban areas, wasting time because of traffic congestion.

REFERENCES

[1] Somashekhar, G.C, Sarala Shirabadagi, Ravindra S. Hegazi, - High Density Traffic Management using Image background subtraction Algorithm‖, International Journal of Computer Applications (0975 - 8887) Recent Developments in Information Technology, 2014

[2] Angel Serrano, Cristina Conde, Licinio J. Rodr'ıguez-Arago®, Om Computer Vision Application: Real Time Smart Traffic Light -, International Journal of Advances in Engineering and Technology, Nov., 2014

[3] Sachin Grover, Vinay Shankar Saxena, Tarun Vatwani, ‖Developed intelligent traffic control system using image classification‖ International Journal of Engineering and Technology Development, Nov., 2014. cIJAET ISSN: 22311963

[4] Drs. Swapan Kumar Deb, Rajiv Kumar Nathr, ―Carriage Acquisition Based on video traffic monitoring‖, International Journal of Computer Science and Emerging
Technology, IJCSET, ISSN: 2044- 6004

[5] Vidhya, A.Bazila Banu, ―Density BasedTraffic Signal System‖, International
Journal of New Research in Science, Engineering and Technology, pages 2218-2223, 2014 [6] Michael Hobbes, ―Welcome to the Traffic Capital of the World‖ [Online]. Available: https://newrepublic.com/article/118416/what-dhaka-Bangladesh-traffic-capital-world-can teach us. [Accessed Jul. 3, 2014]

[7] D. Li, B. Liang and W. Zhang, Real-time moving vehicle detection, tracking, and counting system implemented with OpenCV, ‖ 2014 4th IEEE International Conference on
Information Science and Technology, Shenzhen, 2014, pp. 631-634.

[8] Bin Tian; Ye Li; Bo Li; Ding Wen, Rear-view vehicle detection and tracking by combining multiple parts for complex urban surveillance, in: IEEE Transactions on Intelligent Transportation Systems, vol.15, no.2, pp. 597–606 (April 2014).

[9] Y. Li, B. Li, B. Tian, F. Zhu, G. Xiong and Kwang, ‖Vehicle detection based on the deformable hybrid image template, ‖ Proceedings of 2013 IEEE International

 https://doi.org/10.1051/itmconf/20224403046

Comparison of Real-Time Face Detection and Recognition Algorithms

*Yasar Borkar**, *Reeve Mascarenhas*, *Shubham Tambadkar*, *and Jayanand P. Gawande*

Department of Instrumentation, Ramrao Adik Institute of Technology, D.Y Patil University, Nerul, Navi Mumbai, Maharashtra, 400706.

Abstract. With the phenomenal growth of video and image databases, there is a tremendous need for intelligent systems to automatically understand and examine information, as doing so manually is becoming increasingly difficult. The face is important in social interactions because it conveys information. Detecting a person's identity and feelings Humans do not have a great deal of ability to identify. Machines have different faces. As a result, an automatic face detection system is essential.in face recognition, facial expression recognition, head-pose estimation, and human–computer interaction, and so on Face detection is a computer technology that determines the location and size of a person's face. It also creates a digital image of a human face. Face detection has been a standout topic in the science field This paper provides an in-depth examination of the various techniques investigated for face detection in digital images. Various face challenges and applications. This paper also discusses detection. Detection features are also provided. In addition, we hold special discussions on the practical aspects of developing a robust face detection system, and finally. This paper concludes with several promising research directions for the future.

1 INTRODUCTION

Face detection and recognition is an integral part of computing systems since its inception in the early 1960s. One of the major areas of application of Face recognition is for security purposes [10]. It may also be used in the medical field to measure biometrics such as heartrate etc. This method is particularly popular for public use since images are the easiest to collect and are the one of the least intrusive biometric methods. The main objective of face detection and recognition is to emulate, and eventually exceed a human being's ability to identify and distinguish a particular face from multiple others. Face detection algorithms usually use universal features like eyes, nose and mouth to distinguish human faces from other objects. Nowadays, with advancement in machine learning and neural networks multiple image processing algorithms are available for face detection and recognition.

1.1 Literature Review

While designing and implementing a face recognition algorithm, many factors, like illumination, ageing and changing physical appearance, unstable placement of the camera, rapid movement of the subject, image resolution, viewing distance, etc. are considered [8]. All these factors are taken into account while choosing a suitable method/algorithm for a specific purpose. Because of the availability of so many algorithms and the many factors that go into selecting a particular method for a specific purpose, it is quite important to analyse these algorithms in reference to each other to efficiently choose a one, or a combination to suit our application.

In this paper, we will be discussing three face detection algorithms. We will conclude on which of the three the most efficient is.

Facial detection and recognition are methods of verifying the identity of an individual through their facial features. It is used for security purposes and falls under the category of biometrics.

The detection part commences with initially collecting either a dataset or a primary reference data through a camera which in turn helps in recognition or identification. Once we have the primary data, an already set program verifies the face through various algorithms and the identity of a person is confirmed.

Face recognition is used in various parts of the industry. The most common example is the smartphone industry. The device initially has to be set up by the owner. Their face is detected and recorded as the primary database. The smartphone unlocks only after recognizing the owner's face. The automation industry has used this technology in an innovative way. Building management systems use facial recognition devices to control access to the restricted areas in a building.

This technology uses various artificial intelligence and machine learning techniques and is rapidly evolving, getting even more accurate by the day. As for security concerns, facial recognition is of great reliability and convenience.

*Corresponding author: reevemascarenhas@gmail.com

1.2 Challenges faced in face detection and recognition

Because it deals with human faces, face recognition has always been one of the most fascinating and intriguing technologies. The Covid-19 outbreak has accelerated the world's transition to touchless facial recognition technology. Because of its contactless biometric features, it is gaining a lot of traction all over the world. Companies are abandoning traditional fingerprint scanners in favour of AI-based facial recognition technology, which is opening up new business opportunities. Security and surveillance, authentication/access control systems, digital healthcare, photo retrieval, and so on are some of the applications where its use has become critical. [6]

As previously stated, opportunities and challenges coexist. Growing commercial interest in face recognition is encouraging, but it also proves to be a difficult endeavour when it comes to problems associated with it, which have continuously hampered its quality of delivery. These difficulties arise when situations are uncooperative, resulting in the various facial appearances/expressions.

The challenges that limit the potential of a Facial Recognition System are listed below.

1.2.1 Illumination

Illumination means light variations. A minor change in lighting conditions poses a significant challenge to automated face recognition and can have a significant impact on its results. If the lighting varies and the same person is captured with the same sensor and almost identical facial expression and pose, the results may appear quite different.

Illumination drastically alters the appearance of the face. It has been discovered that the difference between two identical faces taken under different illuminations is greater than the difference between two different faces taken under the same illumination.

1.2.2 Pose

Facial Recognition Systems are extremely sensitive to variations in pose. When a person's head movement and viewing angle change, the pose of his or her face changes. Head movements or varying camera POVs can invariably cause changes in face appearance and generate intraclass variations, causing automated face recognition rates to plummet dramatically. When the rotation angle increases, identifying the true face becomes more difficult. If the database only has a frontal view of the face, it may result in incorrect or no recognition.

1.2.3 Occlusion

Occlusion is defined as a blockage that occurs when one or more parts of the face are blocked and the entire face is not available as an input image. One of the most difficult challenges in face recognition systems is occlusion.

It occurs as a result of a beard, moustache, or accessories (goggles, cap, mask, etc.), and it is common in real-world scenarios. The presence of such components diversifies the subject, making automated face recognition a difficult nut to crack.

1.2.4 Expressions

Face is one of the most important biometrics because its distinctive features play an important role in determining human identity and emotions. Different situations induce different moods, which result in a variety of emotions and, eventually, a change in facial expressions.

Another important factor to consider is the different manifestations of the same individual. Human expressions, in particular macro-expressions such as happiness, sadness, anger, disgust, fear, and surprise, are examples of human expressions. Micro-expressions are those that exhibit rapid facial patterns and occur involuntarily.

Macro and micro expressions appear on someone's face as a result of changes in one's emotional state, and in the aftermath of such emotions, which are numerous, efficient recognition becomes difficult.

1.2.5 Low Resolution

Any standard image should have a minimum resolution of 16*16 pixels. A picture with a resolution less than 16*16 is referred to as a low-resolution image. These low-resolution images can be obtained from small-scale standalone cameras such as CCTV cameras in streets, ATM cameras, and supermarket security cameras. These cameras can only capture a small portion of the human face area, and because the camera is not very close to the face, they can only capture a face region of less than 16*16 pixels. Because most of the details are lost in such a low-resolution image, it does not provide much information. The process of recognising faces can be difficult.

1.2.6 Ageing

Face appearance and textures changes over time and reflects as ageing, which is another challenge for facial recognition systems. Human face features, shapes, lines, and other aspects change as we get older. It is done after a long period of time for visual observation and image retrieval.

The dataset for a different age group of people over a period of time is calculated for accuracy checking. In this case, the recognition process is based on feature extraction, which includes basic features such as wrinkles, marks, brows, hairstyles, and so on.

The face is the most important part of the human body, and its distinctive features make it even more important in human identification. Various algorithms and technologies are used around the world to improve the accuracy and reliability of face recognition. The applications of this ever-expanding technology are also expanding in healthcare, security, defence, forensics, and transportation, all of which require greater precision. However, some challenges are universal in the development of face recognition technology, such as pose, occlusion, expressions, ageing, and so on, as discussed above in the paper.

1.3 Objectives

One of the main motivations with which we undertook this review is assist to individuals and organizations looking to use face detection algorithms in a specific application to select the method most suitable for their use. This could also be beneficial for researchers intending to design better and more efficient methods than already existing ones. The review and comparison provided in this paper might help researchers combine two or more algorithms mentioned here and develop an algorithm that is more efficient than when those methods are used individually. Another objective we had in mind was to implement a fully functional system to cater a specific application – in this case, a student/employee attendance system. Out of the four methods discussed in the paper we have used two – Haar Cascade (Viola-Jones) and Local Binary Patter histogram (LBPH) in above mentioned implementation.

1.4 Significance of this paper

The information derived from this paper can be used for educational purposes and will help fellow research members. This paper discusses three different algorithms of face detection and recognition. It gives a basic understanding and overview of the algorithms which can be easily understood by not only the industry experts but from a layman point of view as well. Apart from theoretical knowledge, this paper contains an experimental setup in which we have used Haar Casscade and Local Binary Pattern Histogram algorithms to detect and recognize faces. This experiment has helped us with better understanding of the two mentioned algorithms and the provided code can be implemented in the PyCharm IDE to run and check it for yourself to learn more about the selected techniques.

2 METHODOLOGIES

2.1 Haar-Cascade

Haar-cascade algorithm, popularly known as the viola-jones algorithm is used to recognize faces in a real time video [12]. It was initially proposed by Viola and Jones in 2001. The haar cascade algorithm is extensively used in several applications for the same purpose. It uses Haar features to differentiate facial features from the rest of the image to detect a face. Haar features were proposed by Alfred Haar in 1909. They are rectangular black and white regions which help in calculation in a specific rectangular region rather than calculations at pixel level.

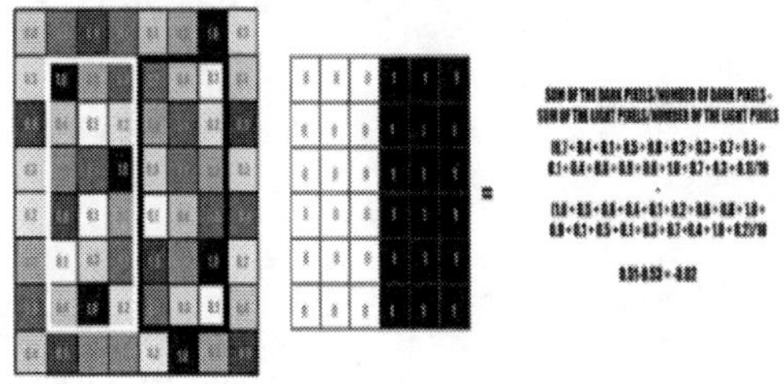

Figure 2

They come in several types, out of which the most common are 2 rectangular features, 3 rectangular features and 4 rectangular features.

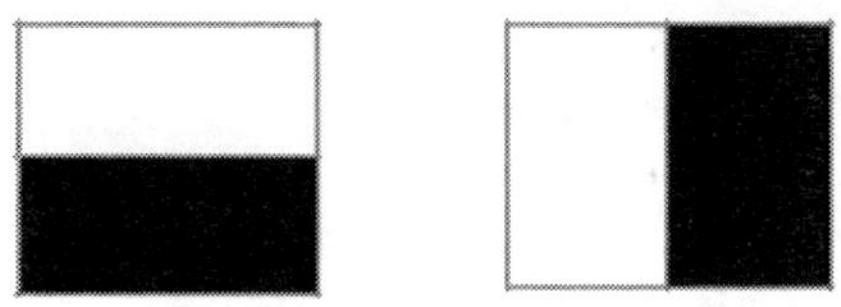

Figure 2: 2 Rectangle Feature

The 2-rectangle feature is used to detect the edge features in a face. The difference between the sum of the pixels is considered to be the value of the 2-rectangular feature.

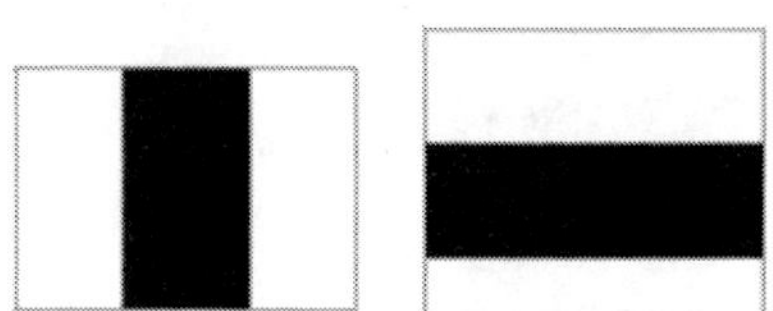

Figure 3: 3 Rectangle Feature

The 3-rectangle feature is used to detect the line features in a face. The sum is computed in the centre region in a 3-rectangle feature.

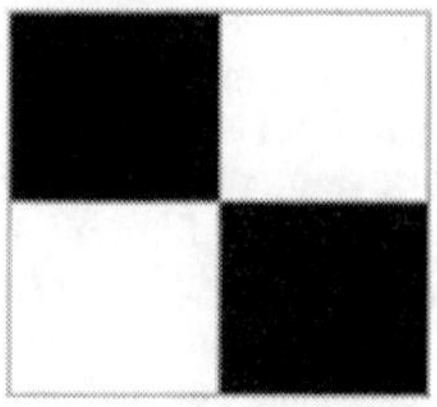

Figure 4: 4 Rectangle Feature

This is used to detect the 4-Rectangle features and computes the difference between diagonal pair of rectangles.

Rectangle features can be computed very rapidly using an intermediate representation for the image which we call the integral image. The integral image at location x,y contains the sum of the pixels above and to the left of x,y, inclusive:

$$ii(x,y) = \sum_{x' \leq x, y' \leq y} i(x',y'),$$

2.2 Fisherface Algorithm

Fisherface is a popular face recognition algorithm that is widely thought to be superior to other techniques such as eigenface due to the effort made during the training process to maximise the separation between classes.

The face, as a part of the human body, is the simplest and most frequently used to determine an individual's identity. Humans can be distinguished and recognised more quickly and easily based on their faces.As a result, the face is used as a means of identifying a person or for face recognition.

In general, image recognition systems are classified into two types: feature-based systems and image-based systems. The first system extracts feature from the eye image components, nose, mouth, and so on, which are then geometrically modelled to determine the relationship between these features. In the second system, image pixels are used to represent them in various methods such as Principal Component Analysis, wavelet transformation, and so on, which are then used for training and image identification classification.

The process of obtaining characteristics that distinguish one face sample from another is known as feature extraction. As the principal component analysis method (PCA) used for face recognition was introduced by Turk and Pentland in 1991, reliable feature extraction techniques are critical in solving pattern recognition problems.

The PCA method seeks to project data in the direction with the greatest variation (indicated by the eigenvector), which corresponds to the covariance matrix's largest eigenvalues. The method's weakness is that it is less effective in class separation.

Cheng et al. introduced the Linear Discriminant Analysis (LDA) method for face recognition in 1991.According to the Fisher Criterion JF, this method attempts to find a linear subspace that maximises the separation of two pattern classes. This is obtained by simultaneously minimising the distance of the within class distribution matrix and maximising the split matrix spacing between the Sb classes, resulting in a maximum Fisher Criterion JF. By maximising the Fisher Criterion JF, the Fisher Linear Discriminant will find subspaces where classes are linearly separated. Sw will be singular if the data dimension is much larger than the number of training samples. This is a shortcoming of the LDA method.

Belheumeur introduced the fisherface method for face recognition in 1997. This method is a hybrid of the PCA and LDA methods. Before performing the LDA process, the PCA method is used to solve singular problems by reducing the dimensions. The disadvantage of this method is that the PCA dimension reduction process results in some loss of discriminant information useful in the LDA process.

The between-class scatter matrix is defined as

$$S_B = \sum_{i=1}^{c} N_i(\mu_i - \mu)(\mu_i - \mu)^T$$

The within-class scatter matrix be defined as

$$S_W = \sum_{i=1}^{c} \sum_{x_k \in X_i} (x_k - \mu_i)(x_k - \mu_i)^T$$

However, in its early stages of development, face recognition with the fisherface method still has some issues, such as computation issues and the condition of the face image into input that will be used for image testing.

The computation problem in face recognition using the fisherface method becomes a problem due to the extremely complicated and complex computation process. While the diversity of the light of the face image, the attributes of the face image, the expression of the face image, and the variation of the position of the image of the face itself are the problems that affect the condition of the face image.

Face recognition systems based on the fisherface method are intended to recognise face images by

 https://doi.org/10.1051/itmconf/20224403046

matching the results of feature extraction. The system is expected to determine whether or not the image to be tested is correctly recognised.

The face image that will be used must first have to go through pre - processing stage. This stage consists of image acquisition as well as RBG image reconfigured into grayscale image. Using a camera, capture images of people's faces. This acquisition's image is a 24-bit RGB JPG file with a resolution of 92 x 112 pixels. Conversion of acquisition face image from RGB to 8-bit grayscale, BMP format, 40 x 40 pixels.

Furthermore, the face dataset must be split up into two (two) parts, with one part of the image serving as the training image (training dataset) and the other serving as the test image (testing dataset).

During this image evaluation phase, the Fisherface technique will be used to create a feature vector of frontal image data that can be used by a system, and then to match the vector of traits of the training sample with the vector of characteristics of the test image through using Euclidean distance formula.

2.3 Local Binary Pattern Histogram (LPBH)

Local Binary Pattern histogram (LBPH) is one of the most common and easily available face recognition algorithms. It was first developed in 1994 and has been since used as an efficient facial recognition algorithm. It uses features extracted from an image and thresholding its neighbouring pixels and getting the result as a binary number. Another advantage about the LBPH algorithm is that it is easily available in the open-source library (OpenCV) and can be used by many programming languages like Python, MATLAB, C++, etc.

The LBPH, like most facial recognition algorithms uses a dataset with the images of the subjects we need to recognize using the algorithm. Usually, the dataset is already pre-processed and only the facial features of the subject are used in the LBPH algorithm. This facial isolation is done using a face detection algorithm – for ex. Haar Cascade.

The LBPH algorithm creates an intermediate image that details and isolates the necessary features for facial recognition. This is done by getting a part of the grayscale image in a matrix format (usually 3x3) where the value of the elements of the matrix are the intensities of each pixel (from 0 to 255). The central pixel is selected and its intensity value is chosen as the threshold value. Now, its 8 neighbouring pixels are compared with the threshold and each are given binary values. The binary values are selected by giving 1 for values greater than or equal to the threshold and 0 for values lower than threshold. Next, the binary values from the matrix are concatenated and a binary number is formed. This binary number is then converted to decimal and set as the central value of the matrix. This procedure is repeated, covering the whole original image and a new LBP result is produced. [13]

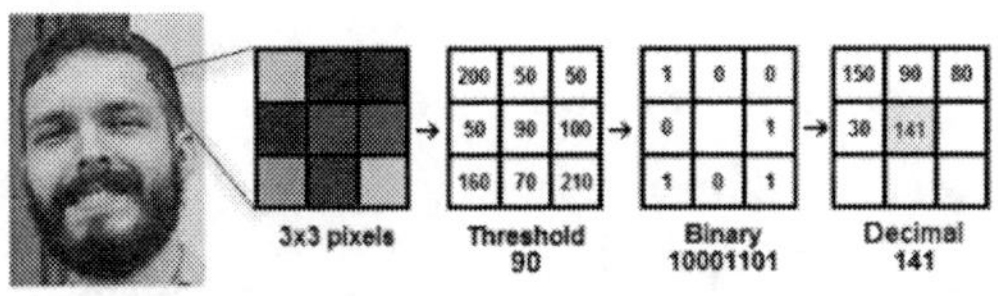

Now, the image generated from LBP procedure is divided into vertical and horizontal grids. The histograms of each grid are created and all the resulting histograms are concatenated to create the final bigger histogram. This final histogram represents the characteristics of the original image.

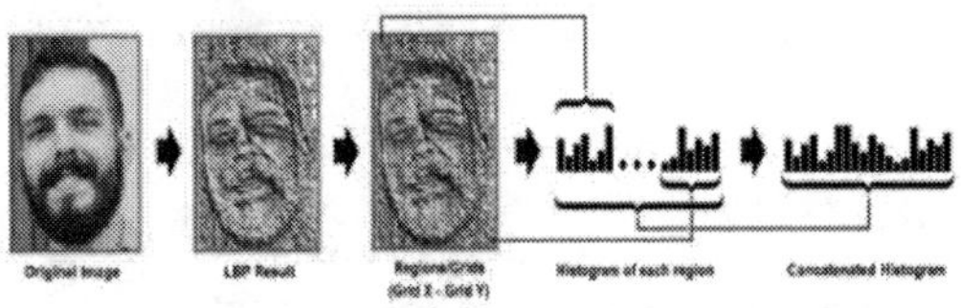

Having obtained the histograms for every image in the training dataset, we repeat the same procedures for the new image obtained from the user input. To find the correct match, simply the two histograms are compared and a threshold value is set passing which the image is concluded to be the same person. One of the most common approaches for comparing the histograms is calculating the Euclidean distance between them. This can be done by the formula:

$$D = \sqrt{\sum_{i=1}^{n}(hist1_i - hist2_i)^2}$$

This distance can also be used as confidence measurement. When a confidence value is lower than the given threshold, it can be assumed that the algorithm has successfully recognized the image.

3 EXPERIMENTAL RESULTS

To conduct the experiment, we used the functions available in the OpenCV library available for Python. We run the program on a laptop computer through PyCharm. The specifications of the computer are:

Company: HP
Windows Edition: Windows 11 Home Single Language
Processor: AMD Ryzen 5 3500U with Radeon Vega Mobile Gfx 2.10 GHz
Installed RAM: 8GB
System Type: 64-bit operating system, x64-based processor

https://doi.org/10.1051/itmconf/20224403046

Four separate programs were written for the four different aspects of the method. One program each was used for:

- Creation of Dataset:

The dataset creation was done using the onboard camera of the computer, running the video input received by the camera through the Haar Cascade algorithm and isolating the grayscaled facial features and storing in the dataset. For every person, an ID is provided so that it can be used in the recognition stage.

- Training of Dataset:

Next, the training is done by collecting the LBP histograms of each image in the dataset. These histograms are stored against their respective IDs so that the recognition code can match the appropriate IDs to its relevant person.

- Input and recognition of image:

In the recognition stage, the camera is once again activated and an input video stream is collected whose LBP histogram is created. This input histogram is compared with all histograms from the dataset. The confidence (Euclidean distance) between the two histograms is calculated. The confidence can be calculated by:

If the confidence is lower than the provided threshold value, that ID is considered to be recognized.

- Writing the results onto a readable file:

Finally, the write to file code is run to save and publish the results onto MS Excel. This result can be monitored by the organization for attendance purposes.

The flow of the experiment is given below for better understanding.

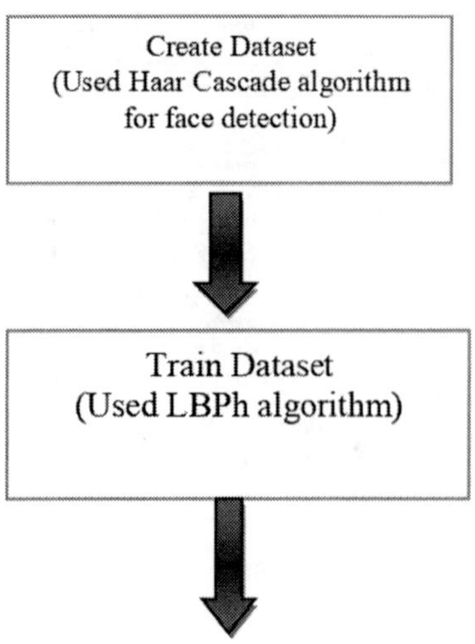

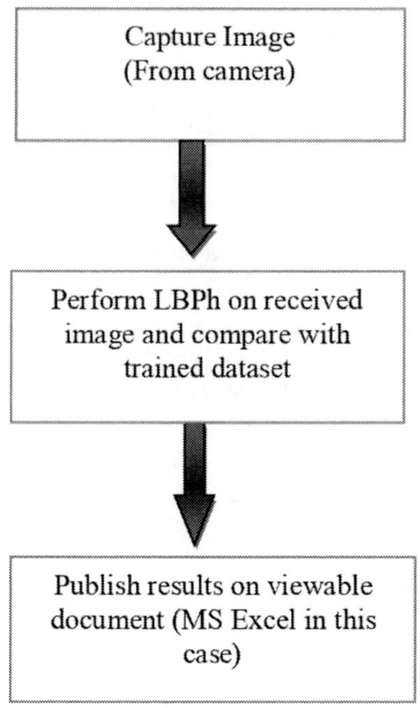

Process Flow Diagram

We tested the Haar Cascade, Local Binary Pattern Histogram and Fisherface algorithms with 3 different subjects for each algorithm. [18]

The experiment was performed three times with varying number of training samples in each case. The first case was conducted with 10 samples, second with 25 and the third with 50 samples. It was observed that the accuracy improves with increasing number of training samples. Accuracy was calculated by comparing the total number of pictures recognized correctly (true positives) with total number of pictures tested. It can be mathematically defined by the following expression:

$$\text{Accuracy (\%)} = \frac{\text{Total no. Of T.P}}{\text{Total no. of T.S}}$$

Where,

T.P stands for True Positive

T.S stands for Training Samples

The results obtained are shown below in tabular form:

	3 Subjects		
	Accuracy in %		
	N = 10	N = 25	N = 50
Haar-Cascade	95	97	99
LBPH	96	98	99
Fisherface	92	95	98

Where N = no. of training samples used.

4 CONCLUSION

This paper gives us a general idea about facial recognition and detection algorithms. It has been written keeping in mind enthusiasts who wish to begin with their research and need a basic understanding of the initially used algorithms for the same. We have discussed some challenges faced in the field of face detection and recognition and how certain algorithms can be used to overcome specific limitations. Three different algorithms have been reviewed in this paper, namely, Haar-Cascade, Local Binary Pattern Histogram (LPBH) and Fisherface. These algorithms have been explained in a simple yet efficient manner so that the reader will understand the concepts with ease.

Apart from the theoretical explanation, we have implemented the above-mentioned algorithms in the PyCharm IDE for an application-based point of view. By real-time implementation, it was observed that out of the three reviewed algorithms, Local Binary Pattern Histogram (LBPH) was the most accurate with an average accuracy of 97.66% followed by Fisherface with an average accuracy of 95.21%. Haar Cascade has proved to be one of the most reliable facial detection algorithms with an accuracy of 97.69%. This program can be implemented by schools, offices or any other organization in the industry that requires a database to keep record of.

Even though LBPH was found to be the most accurate algorithm, various factors are taken in account for various reasons in the face detection and recognition techniques. Every algorithm has its respective benefits and limitations. Hence, it cannot be empirically justified that any one of the existing algorithms is best suited for all kinds of applications. A future avenue for research could be combining these existing algorithms for greater efficiency and minimizing the effects of certain limitations.

5 REFERENCES

[1] NURULHUDA ISMAIL, MAS IDAYU MD. SABRI Department of System and Computer Technology (Multimedia), Faculty of Computer Science and Information Technology, University of Malaya, 50603 LembahPantai, Kuala Lumpur, MALAYSIA, Page 1-11

[2] UMAMAHESWARI, K., SUMATHI, S., SIVANANDAM, S.N., and PONSON, T. (2007). Artificial Neural Network Based Image Data Mining with Hybrid Recognizers. *International Journal of Soft Computing Applications*. Issue 1, Page. 27-49.

[3] PRASEEDA, L. and SASIKUMAR, M. (2008). RBF Based Face Recognition and Expression Analysis. In Proceedings of World Academy of Science, Engineering and Technology. Vol. 35.

[4] MOHAMED, D. (2008). A Simple Face Recognition Algorithm using Eigeneyes and a Class-Dependent PCA Implementation. *International Journal of Soft Computing*. Vol. 3, No. 6, Page. 438-449.

[5] MANJUNATH, B.S. and CHELLAPPA, R. (1993). A Unified Approach to Boundary Perception: Edges, Textures and Illusory Contours. In Proceedings of *IEEE Transactions Neural Networks*. Vol. 4, No.1, Page. 96–107.

[6] Preeti Chatterjee, Subhadeep Jana & Souradeep Ghosh, Global Journal of Computer Science and Technology: G Interdisciplinary
Volume 21 Issue 2 Version 1.0 Year 2021

[7] UMAMAHESWARI, K., SUMATHI, S., SIVANANDAM, S.N., and PONSON, T.(2007). Artificial Neural Network Based Image Data Mining with Hybrid Recognizers. *International Journal of Soft Computing Applications*. Issue 1, Page. 27-49.

[8] K. Delac and M. Grgic, "A survey of biometric recognition methods," in 46th International Symposium Electronics in Marine, vol. 46, pp. 16– 18, 2004.

[9] SINGH,S.K.,CHAUHAN, D.S., VATSA, and SINGH, R. (2003). A Robust Skin Colour Based Face Detection Algorithm. *Tamkang Journal of Science and Engineering*.Vol. 6, No. 4, Page. 227-234.

[10] Sasha Callaway Jeffrey Cheng Alexander Contratti, David Fu Harshika Gelivi Jacob Wachulec, Sonia Purohit, Comparative Analysis of Image Processing Algorithms for Airport Security, New Jersey's Governor's School of Engineering and Technology, July 18, 2020.

[11] TURK, M., and PENTLAND, A. (1991). Face Recognition Using Eigenfaces. In Proceedings of IEEE Conference on *Computer Vision and Pattern Recognition*.Page. 586-591.

[12] Michael Jones, MA 02139 Cambridge, MA 02142, Mitsubishi Electric Research Labs Compaq CRL 201 Broadway, Paul Viola, 8th FL One Cambridge Center Cambridge, Rapid Object Detection using a Boosted Cascade of Simple Features.

[13] Kelvin Salton Do Prado, Facial Recognition: Understanding LBPH Algorithm, November 11, 2017.

[14] J.Brownlee. "How to Control Neural Network Model Capacity with Nodes and Layers." Machine Learning Mastery, 10 Jan. 2020.

[15] ROWLEY, H., BALUJA, S., and KANADE, T. (1998). Neural Network- Based Face Detection. In Proceedings of IEEE Transactions on *Pattern Analysis and Machine Intelligence*. Vol. 1, No. 20, Page.23–28.

[16] W. Koehrsen. "Beyond Accuracy: Precision and Recall." Medium, Towards Data Science, 10 Mar. 2018.

[17] Peter N. Belhumeur, Joao~ P. Hespanha, and David J. Kriegman, Eigenfaces vs. Fisherfaces: Recognition Using Class Specific Linear Projection, IEEE TRANSACTIONS ON PATTERN ANALYSIS AND MACHINE INTELLIGENCE, VOL. 19, NO. 7, JULY 1997.

[18] Amir Dirin, Nicolas Delbiaggio, Janne Kauttonen Haaga-Helia University of Applied Science, Helsinki, Finland, Comparisons of Facial Recognition Algorithms Through a Case Study Application.

Autonomous Hospital Assistant Bot

Pratik Pinto[1*], *Arun Varghese*[1], *Francis Talari*[1], *Shalin Edward*[1], and *Amruta Pabarekar*[1]

[1]Fr. C. Rodrigues Institute of Technology, Vashi, Maharashtra 400703, India

Abstract. In times of outbreaks, pandemics and otherwise there is no system in place to aid the doctors and nurses in their routine work such as monitoring, interacting, and providing necessities to the patient. These tasks tend to overburden the hospital staff and also bring them in proximity to the affected. Thus, The Autonomous Hospital Assistant Bot will act as a middleman through which the hospital staff can remotely monitor, interact and provide necessities like food, water, and medicine to the patient. Thus, reducing burden and proximity risks for the hospital staff. It can also establish communication between the patient and doctor through the bot enabling two-way communication. Bot also has a camera via which doctors can check the patient's health credentials as displayed on the devices connected to them. The addition of the system to the hospital ecosystem will revolutionize the stagnated hospital culture regarding hospital staff and their work & difficulties. The bot can help to cut down these interactions exponentially since it takes care of the three major reasons for the routine check-ups i.e. patient monitoring, interaction, and providing the necessities, thus reducing these physical, in-person interactions down to only the critical patients who need more attention.

1 Introduction

With the exponential rise in the human population, the people at the forefront tending to human health and wellbeing are doctors and nurses. As we know doctors and nurses are the lifelines of any nation. But, a major aspect usually under looked is the effect of overburden and proximity risks implicated in them. The reason for the same is the routine rounds to be made by the staff for patient monitoring and providing the necessary supplements.

To reduce these interactions the Autonomous Hospital Assistant Bot can be used by the hospital staff to remotely check on the patients. This can be done with the help of the website wherein the staff can select which patient they want the bot to go to. The bot will then autonomously reach its specified destination; without knocking over someone as it has an obstacle avoidance system. In this, the staff can also make use of the camera and the two-way communication system to monitor and interact with the patient in the safety and comfort of their remote location. The attached arms allow the staff to supply the patient with required necessities such as food, water, medicine, etc., and when the need arises, they can also measure the temperature of the patient remotely.

2 Literature Survey

A study on the autonomous hospital assistant bot was made. This helped us to learn about how we can build a bot and control the bot. The Literature survey is given in different ways that have been proposed over the years.

2.1 PSLB: Portable Sanitization and Locomotive Bot

In this paper, the portable sanitization bot moves on a track that needs to be placed by the user on the surrounding. On a clean surface like a floor, a black or non-reflective path of considerable width must be placed as shown in Fig. 1. PSLB would follow that path for its motion. The user will control the speed at which the bot will manoeuvre. The infrared sensor comes into play when the user would lay bends in the path, then the sensor would find a non-reflective surface on its path and it correspondingly generates no output signal for which the connected wheel momentarily stops rotating. At higher speed, it would be difficult for the robotic platform to confirm its laid path if the bend angle is considerably high [1].

Fig. 1. Developed PSLB [1]

* Pratik Pinto: pratikpinto2000@gmail.com

https://doi.org/10.1051/itmconf/20224403047

2.2 Line Follower and Obstacle Avoider Robot

It gives the idea that a line follower is an intelligent robot that detects a visual line embedded on the floor and follows it. In this the path is pre-defined. It can be either visible like a black line or a white contrasting color as shown in Fig. 2. IR sensors are used to detect the line that the robot has to follow. The moment of the robot was automatic and could be used for long-distance applications [2].

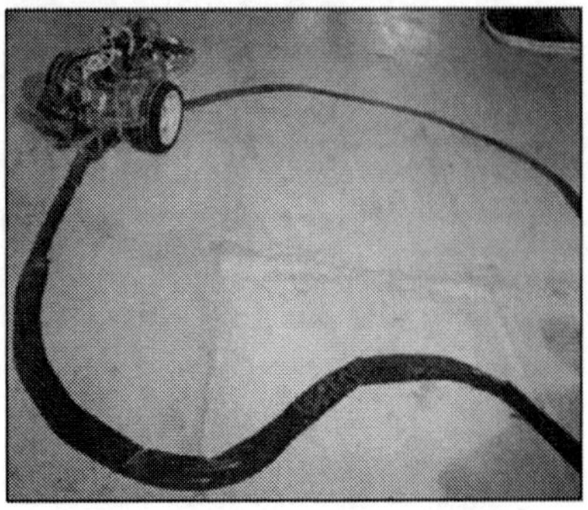

Fig. 2. Line Following Robot Model [2]

2.3 Arduino Controlled War Field Spy Robot using Night Vision Wireless Camera and Android Application

In this paper, a robot is built using a night vision wireless camera run by an android application [3]. A Night vision wireless camera is used which sends videos of the war field to prevent damage to the human beings as shown in Fig. 3.

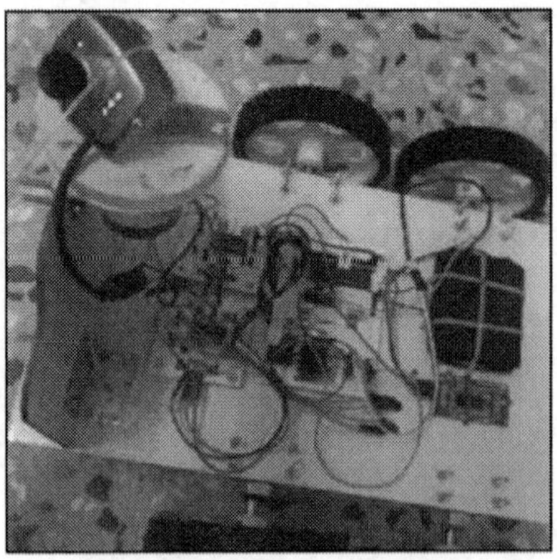

Fig. 3. War Field Robot With Wireless Night Vision Camera [3]

2.4 Smart Surveillance Robot for Real-Time Monitoring and Control System in Environment and Industrial Application

This paper presents the monitoring and controlling of the visuals of a real-time environment. IoT and IoR are the two fields that are growing rapidly and can offer many services for monitoring, manufacturing, security surveillance, etc. in various diverse fields. In this, we saw that the robot was fitted with sensors like DHT11, MQ-6, MQ-3, 3-axis accelerometer, and ESP 8266 module for providing real-time environmental monitoring. Figure 4 shows a smart surveillance bot with real-time monitoring. As shown in Fig. 4[4].

Fig. 4. Smart Surveillance Bot With Real-Time Monitoring [4]

3 System Block Diagram

Fig. 5 represents the block diagram of the Autonomous Hospital Assistant Bot. Herein, Commands can be sent to the RPI via a webpage, and the RPI, in turn, commands the Arduino modules to perform the necessary actions required for autonomous travel. Finally, when the travel is completed, the webpage can be utilized to access the other features of the bot which are aimed at patient-doctor monitoring and interaction.

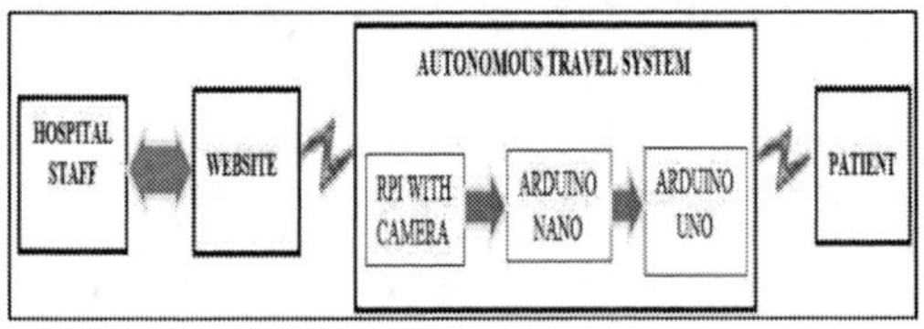

Fig. 5. Block Diagram of the System

3.1 Hospital Staff

The hospital staff are the main focus of this project as the bot aims at reducing their monotonous tasks of checking on the patient, their health credentials, interacting with them, and providing them the necessities may it be eatable, liquid, medicine, etc. The term hospital staff includes nurses, doctors, and any other hospital authority who has the official rights and requirements including but not limited to making rounds, checking on the patients, interacting with them, keeping track of their health progress, their current and past health status as and when the requirement for the same arises.

3.2 Website

The website will be the mode of control given to the hospital staff for enabling them the usage of the bot and its features. The major features of this website will be a camera, audio, and temperature check. For a communication system, a microphone and speaker are available on the bot. With the help of this feature, the hospital staff and the patient will be able to converse with each other. Another major feature of the website is the camera; the hospital staff can select the access to the camera on the bot via the website to see what the camera of the bot is picking up, its mainly in place for viewing the health parameters as displayed on the various and respective devices connected to the patient along with being able to have a general view on the patient and the in-travel areas, if required. Temperature check is also made available on the website.

3.3 RPI with Camera

The RPI is the webpage-bot interface handler as all instructions from the webpage are sent to the RPI and the RPI then segregates these instructions and sends them to the respective devices and modules for their execution. These instructions include Camera access, Communication system access, Temperature Sensing, and Bot's destination selection i.e. the patient bed.

3.4 Arduino Nano

The Arduino Nano handles the Obstacle Detection and Avoidance System of the bot. Herein, if no obstacle is in front of it, it sends a HIGH; and if an obstacle is in front of it will send a LOW. Now these HIGH/LOW signals will be sent to the Arduino UNO and if the signal is HIGH the bot will continue its travel and if LOW it will halt its travel until the signal becomes HIGH again [9].

3.5 Arduino UNO

The Arduino UNO handles the Autonomous Travel of the bot. It achieves this with the help of Rotary Encoder. The rotary encoder is placed in a way that a wheel attached to its dial moves with the bot's wheel. The number of turns made by this dial is sent to the Arduino UNO and for each patient, the point of interest i.e. where it needs to take a turn, is fed to the bot and it moves accordingly when the set rotary encoder value is reached. And thus, it successfully reaches the selected patient [5].

3.6 Patient

The autonomous hospital assistant bot not only aims at assisting the hospital staff in their daily routine responsibility but also provides the patient with a pleasing experience and saves the staff's time thus making it possible for more critical patients to get more attention.

4 CIRCUIT DESCRIPTION AND IMPLEMENTATION

Following are the system that is used for the implementation.

4.1 Obstacle Detection System

The bot to be implementable in a hospital environment must possess some means by which it will avoid collision with people and other objects that will come in the bot's path. The first part of solving this problem is the detection of the obstacle. For this, HCSR04 Ultrasonic Sensor has been utilized to detect obstacle that comes in front of the bot.

To simulate this, the HCSR04 sensor is connected to an Arduino UNO wherein, if the distance of an object in front of the bot is less than 50cm it will glow the LED, and if greater than 50cm the LED remains OFF. As observed in Fig. 6, the LED glows.

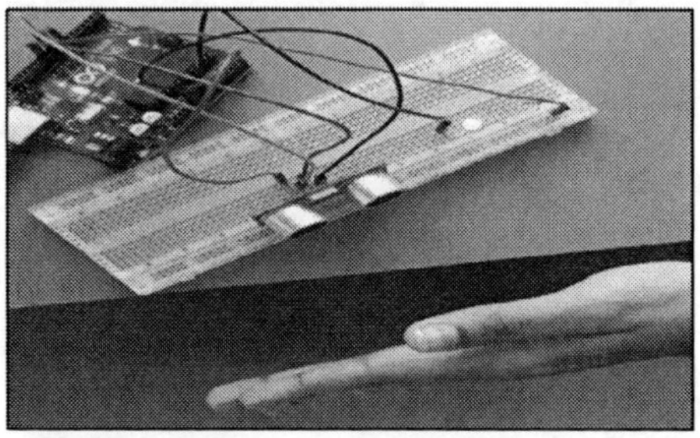

Fig. 6. Ultrasonic Sensor Reaction 1

Thus, with the object being at a distance greater than the set threshold the LED in its corresponding output i.e. Fig. 7 remains OFF[5][6].

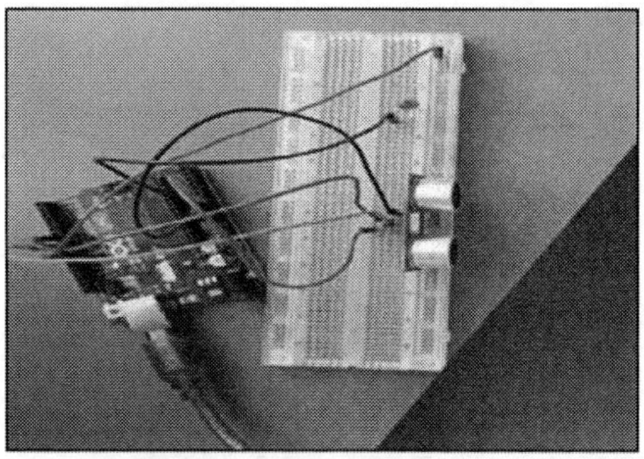

Fig. 7. Ultrasonic Sensor Reaction 2

4.2 Obstacle Avoidance System

In order to implement the Obstacle Avoidance System, simulations were conducted on Tinkercad. The devices and modules used in the implementation were Arduino UNO, IC L293D, Motors, HCSR04 Ultrasonic Sensor, and a 9V power supply.

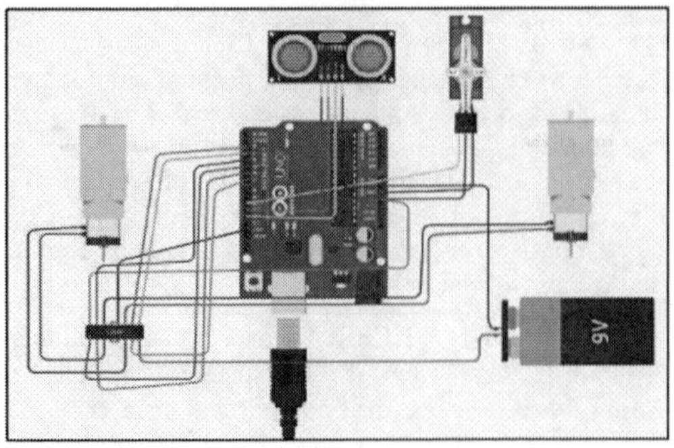

Fig. 8. Obstacle Avoidance System

The layout of the system is described in Fig. 8. Here, the HCSR04 Ultrasonic sensor sends its distance readings to the Arduino UNO. The Arduino UNO then processes the data and checks if the obstacle is at a distance greater than 50cm or not. If this threshold is broken then a signal is sent to the L293D to stop the motors via a LOW to its EN pin. If the obstacle is greater than 50cm then the bot movement is uninterrupted.

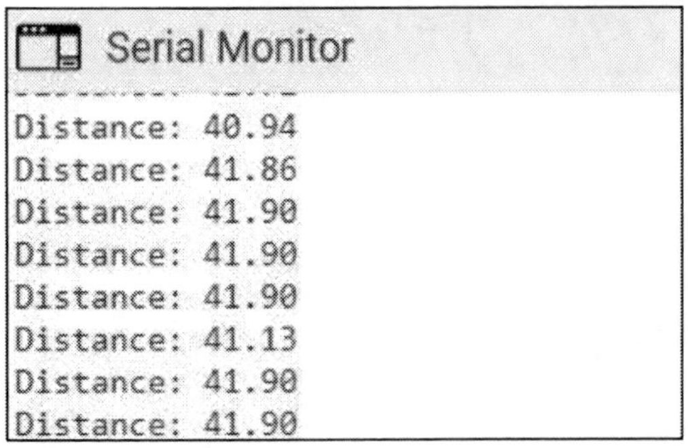

Fig. 9. Obstacle Avoidance System Reading 1

In the 1st test run the obstacle was situated at a distance of less than 50cm, as observed in Fig. 9.
Due to the threshold being broken, the bot movement is halted as observed by the denotation on the motors with 0 RPM.

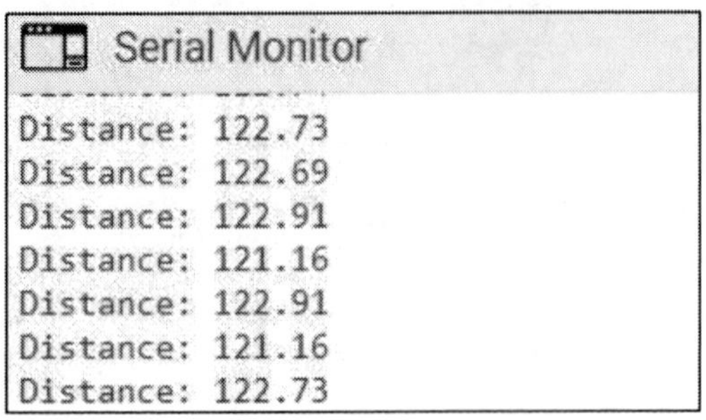

Fig. 10. Obstacle Avoidance System Reading 2

In the 2nd test run the obstacle was situated at a distance greater than 50cm, as observed in Fig. 10.
Due to the threshold not being broken the bot movement is not interrupted as observed by the presence of some RPM values of the motors.

4.3 Temperature Sensing and Display

For the patient's temperature to be checked MLX90614 IR Contactless Thermometer was utilized to measure the temperature without contact. This sensor works on Stefan-Boltzmann Law, according to which the IR energy radiated from a body is directly proportional to its temperature. The temperature thus calculated is displayed on the OLED 128x64 display.
In the test run, there is no body part in front of the sensor and thus the temperature of the room is measured. The room temperature measured was 29°C.

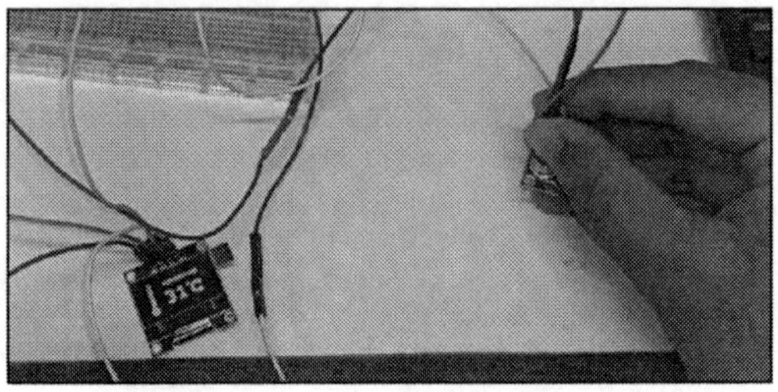

Fig. 11. Temperature Sensor Display

Here, a hand is in front of the sensor and thus the sensor measures the body temperature. As observed in Fig. 11, the reading changes from 29°C to 31°C[5][7].

4.4 Path Processing System

To enable autonomous travel a path is to be fixed. A fixed path is put in place primarily to minimize the obstruction in the hospital corridor by the bot during its movement[8].

Fig. 12. Path Processing System

As observed in Fig. 12, the device made utilizes a Rotary encoder, LCD Display, LED, Wheel, and an Arduino UNO. The system is such, that when it's working the LED glows. From there on the wheel can be traced along the path the bot is planned to be moved. The LCD displays the number of rotations of the Rotary Encoder dial which is noted to plan the path.

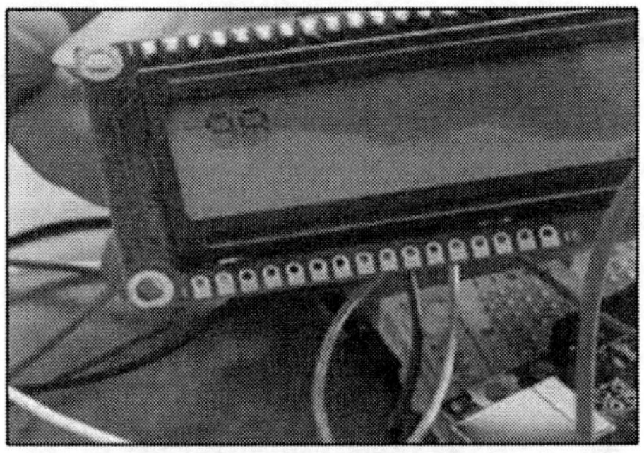

Fig. 13. Path Processing System Reading

As observed in Fig. 13, the LCD is displaying the rotation of the rotary encoder dial, which in this case is 98. These readings are used to put in place points of interest from where changes are required in the bot's path to reach its destination.

4.5 Autonomous Travel Bot

All the systems mentioned above are utilized for enabling the autonomous travel of the bot. The devices and modules used are Arduino UNO, Arduino Nano, Rotary Encoder, HCSR04 Ultrasonic Sensor, and L298N Motor Driver Module.

The Arduino Nano controls the Object Detection System. The Arduino Nano works as a master and sends HIGH or LOW to the Arduino Uno which in turn is controlling the Locomotion system of the bot. If no obstacle is present in the set threshold, then the Arduino Nano sends a HIGH to the Arduino UNO which signals it to resume its travel and if an obstacle arrives it sends a LOW to halt the bot's travel. The Rotary Encoder is attached to a wheel and thus its dial moves along with the bot.

These readings are taken by the Arduino UNO and it compares them to the pre-fed rotary dial turns. When the stipulated Rotary encoder dial turn reading is matched it moves in the required direction. And thus the bot autonomously reaches its destination and stops.

4.6 Battery

The battery that is utilized in this model is a 12V 1.3Ah Lead Acid rechargeable battery. The battery sufficiently powers the bot in its travel and functionality. This battery can be easily recharged via its pins. As the charge in the battery reduces, the voltage output from it reduces too. This functioning of the battery is utilized to act as an indicator for depleted battery charge as before being completely drained out the output voltage will gradually reduce thus resulting in the bot moving slowly as compared to its usual speed. This reduced speed indicates that the charging is depleted and the battery needs to be charged.

4.7 Bot Specifications

The body of the bot was selected to have a wooden framework. The bot body is of the dimensions, 12"x3" of 18mm thickness on the wheel attached portion. 9"x3" of 18mm thickness on its perpendicular end. The pillar are made to be 10"x1.5" of 18mm thickness. The ideal power supply required for this system is 12V. In this, the Arduino UNO, Nano, L298N Motor driver modules are sufficiently powered for their functioning. Upon reduction in the voltage, the system will work at a reduced speed. The battery utilized in this system is a 12V 1.3Ah Lead Acid rechargeable battery. For the Rotary Encoder, 3.3V output of the Arduino UNO is used. And, for powering the HCSR04 Ultrasonic Sensor, 5V output of Arduino Nano is utilized. The L298N motor driver module powers the motors and thus, no external connections are required for the same.

Fig. 14. Bot Body

5 Result

The implementation of the bot initially was conducted with it only possessing the Obstacle Avoidance system along with locomotion capabilities.

The 1st stage of the bot is attached to a stationary power supply which will be replaced by a portable battery. The devices used are Arduino UNO, L298N Motor Driver module, and Ultrasonic Sensor.

Fig. 15. Obstacle Avoidance Bot Test Run 1 with Obstacle (Book)

In the 1st test run, the bot is allowed to move and a book is brought in front of it at random intervals. As observed in Fig. 15, the bot successfully detects the obstacle and halts its movement up until the obstacle is moved out of its path.

 https://doi.org/10.1051/itmconf/20224403047

Fig. 16. Obstacle Avoiding Bot Test Run 2 with Obstacle (Human)

Fig. 16 depicts the 2nd test run wherein a human walks in front of the bot. The bot successfully halts its movement up until the person moves out of its path.

Fig. 17. Autonomous Travel Bot

Fig. 17 depicts the near-completed bot. Here, the bot can autonomously move to the patient as stipulated by the user. The bot performs obstacle detection and avoids the obstacle along with autonomously traveling to its destination.

6 Conclusion

The objective of this project is to make an autonomous movement and provide communication between the hospital staff and patient along with providing the essentials resources to the patient. The bot contains all the gadgets on the top part. These include a Camera, Ultrasonic sensor, temperature sensor, and microphone. The autonomous travel is brought about by the Rotary encoder and Ultrasonic sensor; the interaction is brought about by the camera, microphone, and speaker. The bot will also carry a tray which will be utilized for carrying necessities like food, water, and medicines for the patient. The actual movement of the bot is done with the help of wheels, servo motors, L298N motor driver, and a power supply.

References

1. A. Ray, H. Ray, *PSLB: Portable Sanitization Locomotive Bot*, in proceedings of 4th International Conference on Electronics, Materials Engineering & Nano-Technology, IEMENTech (2020)
2. S.S. Suryawan, S.M. Musamwar, S.R. Kolhe, S.A. Thengane, S.S. Hanumante, Prof. P.H. Sahare, *Line Follower and Obstacle Avoider Robot*, in proceedings of International Research Journal of Engineering and Technology, IRJET (2019)
3. J. Patoliya, H. Mehta, H. Patel, *Arduino Controlled War Field Spy Robot using Night Vision Wireless Camera and Android Application*, in proceedings of 5th Nirma University International Conference on Engineering, NUiCONE (2015)
4. A. Nayar, V. Puri, N.G. Nguyen, D.N. Le, *Smart Surveillance Robot for Real-Time Monitoring and Control System in Environment and Industrial Applications*, in Proceedings of Advances in Intelligent Systems and Computing, AISC (2018)
5. https://www.rs-online.com/designspark/what-is-arduino-uno-a-getting-started-guide. Accessed on: 6th August, 2021
6. https://components101.com/sensors/ultrasonic-sensor-working-pinout-datasheet Accessed on: 6th August, 2021
7. https://robu.in/product/mlx90614-esf-non-contact-human-body-infrared-temperature-measurement-module/efef Accessed on: 8th October, 2021
8. https://howtomechatronics.com/tutorials/arduino/rotary-encoder-works-use-arduino/ Accessed on: 20th October, 2021
9. http://www.rs-online.com/designspark/what-is-arduino-nano-a-getting-started-guide Accessed on: 15th December, 2021
10. https://www.raspberrypi.com/documentation/computers/os.html Accessed on: 27th December, 2021

Logistics & Supply Chain Management Using the Power of Blockchain

Ghrushnesh Rathod
Department of Computer Science
Ramrao Adik Institute of Technology
Mumbai, India
email: ghr.rat.rt18@rait.ac.in

Shubham More
Department of Computer Science
Ramrao Adik Institute of Technology
Mumbai, India
email: shu.mor.rt18@rait.ac.in

Amit Patil
Department of Computer Science
Ramrao Adik Institute of Technology
Mumbai, India
email: ami.pat.rt18@rait.ac.in

Sheetal Ahir
Department of Computer Science
Ramrao Adik Institute of Technology
Mumbai, India
email: sheetal.ahir@rait.ac.in

Abstract: A large portion of supply chain networks have experienced hardships when attempting to incorporate all parts and partners, which make supply chain management network the board framework experiencing an absence of effectiveness and straightforwardness that make a consistent expansion in product misrepresentation and customer's mistake. All data about product creation, stockpiling and transportation should stream obviously during all phases of the production network by following and validating to keep away from item's tampering in the organization. Current software is based on top of brought together engineering and this leaves a hole for altering and bogus data particularly in metropolitan region, yet additionally the current arrangements can't permit the data to be imparted to authorities or buyers to successfully survey and guarantee recognizability and straightforwardness. This paper proposes use of blockchain to store details of product and logistics level to provide assurance to consumer regarding authenticity of product.

Index Terms: Blockchain, Ethereum, Smart Contracts, Decentralized Management

1. Introduction

Throughout the last many years, supply chain and logistics have significantly developed, excused and become a critical variable of seriousness and monetary accomplishment for practically all organizations. Supply chain has not exclusively been put at the focal point of market analysts' consideration, yet organizations spend an ever-increasing number of measures of their turnover on logistics and particularly transportation.

The worldwide exchange and store network, deluged by a developing intricacy of data and actual exchanges, face basic difficulties like consistence, discernibility, administration, higher exchange expenses, and security. In spite of the great money related valuation of the exchanges and the administrative investigation, the center information foundation is at present unfit to stay up with digitalization patterns.

Accomplishing greatness in logistics includes working cooperatively with others to advance the progression of actual products just as the complicated progression of data and monetary exchanges. However, today there is a lot of caught esteem in logistics, generally coming from the divided and serious nature of the logistics business. For instance, in the USA alone, it is assessed that there are more than 500,000 individual shipping organizations. With such countless partners associated with the supply chain, this frequently makes low straightforwardness, unstandardized processes, information storehouses and different degrees of innovation reception.

Many pieces of the logistics esteem tie are likewise bound to manual cycles commanded by administrative specialists. For instance, organizations should frequently depend on manual information section and paper-based documentation to hold fast to customs processes. This makes it hard to follow the provenance of merchandise and the situation with shipments as they move along the supply chain, causing grating in worldwide exchange.

Blockchain might conceivably assist with beating these grindings in logistics and acknowledge considerable increases in logistics process effectiveness. This innovation can likewise empower information straightforwardness and access among applicable supply chain partners, making a solitary wellspring of truth. Also, the trust that is needed between partners to share data is upgraded by the inherent security instruments of blockchain innovation. Besides, blockchain can accomplish cost investment funds by fueling more slender, more computerized, and mistake free cycles. Just as adding perceivability and consistency to logistics tasks, it can speed up the actual progression of products. Provenance following of merchandise can empower capable and manageable supply chains at scale and help to handle item falsifying. Also, blockchain-based arrangements offer potential for new logistics administrations and more creative plans of action. One of the innovations that emerged from this interest was blockchain, a DLT (Decentralized Ledger Technology) that was originally created in 2008 by Satoshi Nakamoto [1], and serves as the basis of Bitcoin and all other cryptocurrencies.

The rest of this paper is coordinated as follows. Section II presents the related work. Section III describes the proposed blockchain Ethereum solution implementation. Section IV gives the testing details and a discussion on the security and testing. Section V finishes up the paper and examines future work.

2. Existing System

Presently, the current frameworks / systems for overseeing records of production network and operations are centralized. A solitary database comprises of the multitude of information and following data of the item. Centralization alludes to an arrangement where whole data and information are put away at the highest point of the organizational structure. There are a couple of drawbacks of the concentrated framework referenced here. You want to enlist an accomplished and technical executive for keeping up with the database administration framework. We really want to purchase costly equipment, programming and generously compensated technical associates. At first, data set size isn't enormous size yet when we store a lot of information, then, at that point, the size of the database increments. Some particularly noteworthy academic studies on system proposals for blockchain and logistics have been done. Hasan and Salah [2] created an Ethereum application for a Proof of Delivery system. A major database can make a ton of issues. For instance, when the data of the database increases too much then the delay time of the result of the database queries can be increased.

Other disadvantages of centralized system in Supply chain use case:

- There are multiple handovers in the current system which leads to error in delivery time estimation.
- There are chances of wrong product delivery due to error in the system.
- A lot of paperwork is involved in the existing system.
- Sometimes, counterfeit products are delivered to the customer.
- Lack of proper authentication across different stages.
- Execution errors—such as mistakes in inventory data, missing shipments, and duplicate payments—are often impossible to detect in real time.

The authors in [3] proposed a simple scheme based on Ethereum blockchain which involves transporting a product between two parties. The scheme depends on a single key that is given to the transporter by the seller [3]. The key is transported along with the item and is handed over to the receiver who is the buyer. The buyer then needs to enter the key for verification. The key hash would already be available in the smart contract which acts as an escrow and holds the buyer's Ether. The Ether would only be placed in the seller's account if the hash of the key entered by the buyer matches the existing hash in the smart contract. A successful verification leads to the transfer of the Ether to the seller [3]. This solution is easy to implement as it is simple and depends on only one key. The method however, depends on trusting the transporter completely that no manipulation of the key would take place before reaching the buyer.

3. Technologies and Methodologies

3.1 Blockchain

The Blockchain is another innovation that tries to make data storage and transmission safer, straightforward, and decentralized. It's an appropriated information base, or public/private shared record of every advanced event, that has been run and shared by the blockchain's representatives. As far as innovation, Blockchain is a circulated framework that stores all exchanges as squares from its initiation. These are comprised of information that clients send in. Cryptographic calculations protect these information moves.

Blockchains can be summed up and used to build a bunch of settled upon decides that nobody can break, including clients and framework administrators. For multi-part applications requiring minimal common trust, they depend on a solitary framework compositional stage. The decentralization of information is a critical component of blockchain innovation, which likewise gives a more significant level of information assurance.

While the blockchain isn't the main type of appropriated record innovation, it is the most normal structure among digital currencies from which the modern applications talked about in this article are drawn. The conveyed design of the blockchain gives extra security looked at conventional "client server" models.

Blockchains can possibly assemble another age of transactional applications that set up trust, responsibility, and straightforwardness at their center while smoothing out business processes and legitimate requirements [4][5].

3.2 Application of Blockchain:

The mechanical insurgency, just as the expansion of information assortment parts, has raised worries about information security and correspondence dependability. The capacity, unwavering quality, recognizability, and realness of data gave by blockchain, just as insightful authoritative arrangements for a trust less climate, predict a key change in production network the executives. Moreover, pursuing the direction toward savvy coordinated operations and the utilization of correspondence and data innovation at all levels of the worth chain, the Blockchain offers critical potential for upgrading the strategies cycle. Quite possibly the most urgent part of a blockchain application is its connection with this present reality, which requires the utilization of pertinent apparatuses and advances.

Blockchain uses decentralized, conveyed record to work with the development of products, data and funds. It incorporates utilization of smart contracts to automate the supply chain capacities.

Blockchain Technology may not replace legacy systems or old applications soon. However, Blockchain can certainly be a complementary application to legacy systems and may even lead to the development of new systems in the near future [6].

3.3 Ethereum

The Ethereum Network gives an adaptable improvement environment. It is a P2P network that can interaction any sort of smart contract, which can be handily made with a couple of lines of code, and without the need of making your own unique reason blockchain foundation. Instead of Bitcoin and other single-reason blockchains, Ethereum decoupled this savvy contract layer, which presently runs on top of the fundamental Ethereum blockchain, making it simple to make smart contracts with only a couple of lines of code.
As indicated by Dr. Gavin Wood, prime supporter of Ethereum association, Ethereum, taken overall, can be considered a transaction-based state machine: we start with a beginning state and steadily execute exchanges to transform it into some last state. It is this last state which we acknowledge as the canonical "version" of the world of Ethereum [7]. It permits everybody to set their own guidelines in exchanges and state progress capacities. This is finished by including shrewd agreements, a bunch of cryptographic guidelines that are executed under the conditions that the maker of the agreement has set up.

3.4 Smart Contracts

The idea of "Smart Contracts" was presented by N. Szabo in 1994 in an unpublished composition and afterward officially in 1997 [8]. Smart Contracts are basically blockchain-based applications, briefly, self-executing programs which contain the terms and arrangements between the parties involved. A smart contract is addressed by its 20-byte address, like an EOA address, and it tends to be created in high level languages, like Solidity. At the point when the contract is conveyed on the organization, the code is assembled as bytecode. The component that separates Smart Contracts is that they consequently check whether or not the particulars of the understanding have been satisfied. Furthermore, to guarantee unwavering quality, adaptation to internal failure, and straightforwardness, the codes (Smart Contracts) are recreated on numerous hubs in the blockchain. There is an endless measure of contracts that can be made in the organization. Nonetheless, there are some "layouts" that are ordinarily utilized, as they give a few primary functionalities to gets that need to accomplish comparative practices. These are token norms that characterize a typical rundown of rules and are called ERC (Ethereum Request for Comments). Ethereum smart contracts empowered blockchain by allowing the execution of code [9].

4. Proposed blockchain solution

In this section we describe our Ethereum blockchain solution that makes use of security provided by this technology to create a solution between different levels of supply chain and logistics. Our solution can handle multiple entities at the same time. Our solution can be extended to multiple products and multiple levels of supply chain. Product details are also included in this solution which can be verified before the transaction.

4.1 System Overview

The proposed blockchain solution prioritizes proof-of-delivery of physical commodities being transferred across different entities in supply chain. Figure 1 represents multiple entities interacting with smart contract.

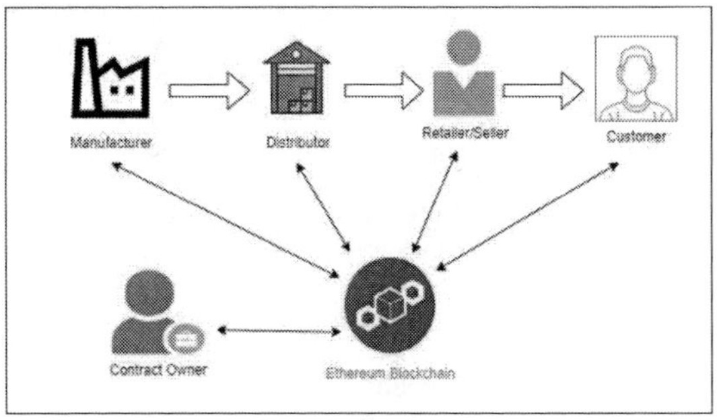

Fig. 1. Different Entities and their interaction with Smart Contracts stored in Blockchain

Each entity can interact with smart contracts stored in blockchain if they have required permissions. All the data of product is stored in smart contracts.
The roles of the participants can be summarized as follows:

- **Owner:** Owner is the one who deploys the contract to the blockchain network.
- **Manufacturer:** The manufacturer manufactures the product and add the details of product to the contract in the chain. He can put the product on sale.
- **Distributor:** The distributor buys the product from Manufacturer and gains ownership of the product to sale it to next level of people on supply chain.
- **Retailer/Seller:** Seller buys the product from distributor and then passes it to customer finally to complete the transaction process.
- **Customer:** Customer gains the ownership of the product and is free to pass on the product through resale as well.

Blockchain provides transparency and the ability to log the events on public ledger. Ethereum provides ability to store and fetch events which helps in tracking the product across the supply chain and helps to know the current owner of product easily. To provide verification of product across the supply chain product details are hashed and stored along with it. This helps in transfer of product across the chain easily and validation is provided. Blockchain is used to store details of products and transaction details between any two entities.
In order to fulfill these requirements of transparency and traceability of the product, smart contracts consist the following [10]:

- **Modifiers:** Modifiers are utilized in the smart contracts to make a prerequisite before the execution of a function. Modifiers were likewise

https://doi.org/10.1051/itmconf/20224403048

used to confine the execution of a function in view of the Ethereum address of the function caller.

- **Methods:** Methods are used in Smart contracts to create function calls. Functions are responsible to perform specific set of actions. In this project methods are used to implement various functions to avoid repetition of code.
- **Events:** Events go about as notifications and are utilized as logs which can help in tracing back a product if there should be an occurrence of issues. Hence, any work that is executed makes an event that refreshes all entities about the status of product.
- **Variables:** Variables are used to store data that can be updated as they move across the chain. In this project variables are used to store product details and hash of the product.

4.2 System Design

The process of supply product from manufacturer and consumer consists of various entities. This contract is designed to store data of product across the entities and provide details of current and previous owners. To avoid the handling multiple contracts across different entities single contract was designed to store the entire information of supply chain. In multiple contracts are used to store data which makes the ecosystem confined to a particular product. Contracts are designed in such a way that all the data is stored in single contract and only the owner of product can transfer the product to next entity.

Once the manufacturer adds the details of product in contract the data is encrypted to create a hash will be used later during transfer of ownership between any two entities. During the process the buyer will have to verify the product details and hash of details to get the complete ownership. In this the system relies on buyer to make sure he verifies the details properly before making payments through ether. The general view and code structure of the smart contract was developed using Truffle [11], a testing tool for smart contracts, and Ganache-cli [11], an Ethereum blockchain simulator. Fig 2 gives clarity about the use of blockchain to store information of transaction and details of products in smart contracts.

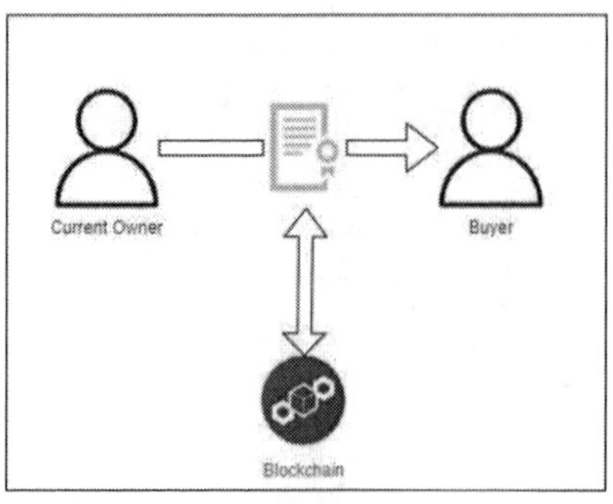

Fig 2: Transfer of ownership and transaction details are stored in blockchain

5. Implementation Details

In a Software, where blockchain is one of the parts, they may have to utilize the information put away on blockchain and the shrewd agreements running on blockchain to really take a look at specific conditions. ID of the exchanges or blocks on blockchain is a piece of information that can be effortlessly incorporated into the current information. Approval of the information can be executed by smart contracts running on blockchain.

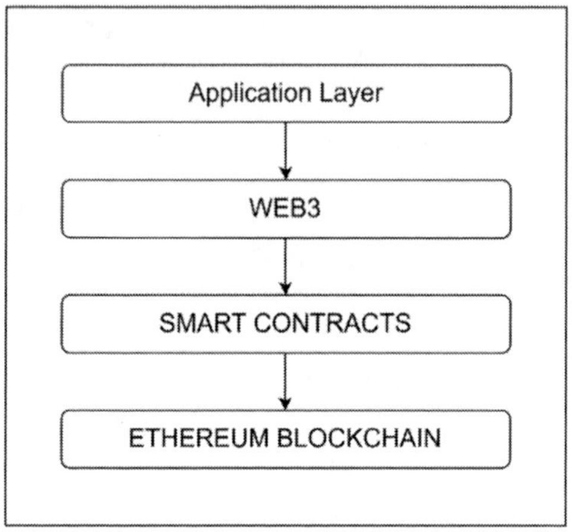

Fig. 3. General Implementation of System

Fig 3 gives information about overall architecture of the application. The system is divided into four layers: application layer, service layer, contract layer and data layer.

The application layer provides the user interface, and the service layer provides Node.js Web services needed by the system. The contract layer completes business logic control and access control of blockchain data and database data by smart contracts. The data layer is responsible for storage of blockchain data and database data. The invocational connection between service layer and contract layer is realized through RPC interface exposed by Ethereum.

Fig 4 gives information about the working and coordination between smart contracts across levels to generate a data layer for information to be stored in blockchain.

The main working routine of the distributed network is as follows:

The node for every substance produces new exchange information, and engenders the information through the shared organization.

Every one of the hubs create the new squares as per the Ethereum agreement instrument.

Every one of the hubs compute the mark and hash, and the principal that completes the estimation and broadcasts the outcome can get the option to record this exchange.

 https://doi.org/10.1051/itmconf/20224403048

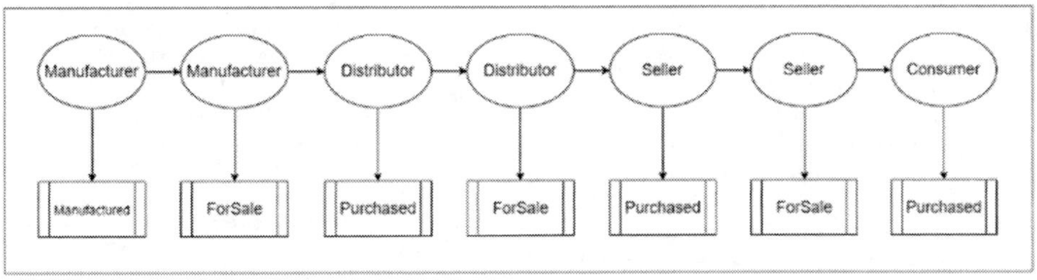

Fig. 4. Overall flow of a project

The exchange data, past hash, timestamp, difficulty and different fields would be stored with the new block;
The hub that gets the option to record the exchange communicates the new square to the whole organization.
The wide range of various hubs that get the new block will do the data check and put the new block to the tail of the blockchain.
The contracts are made and tried utilizing the Remix IDE which gives the fundamental devices to testing and investigating. Thus, making it simpler to change the code on a case-by-case basis while programming. This segment centers around the execution calculations utilized in the savvy contracts. The language utilized for composing the savvy contracts is Solidity. Three sorts of agreements are made as referenced in the Design segment which are the fundamental contract, CS gets that rely upon the quantity of carriers and the BT end of chain contract. One of the fundamentals viewpoints in the execution is that each agreement across the chain has the agreement locations of both the parent and the youngster contract.
Figure 5 gives information about the details to be stored regarding the product in the blockchain. Some set of details will be used for creating of product hash which will be constant throughout the supply chain. Refer figure 6 for flow of contract between 2 entities.

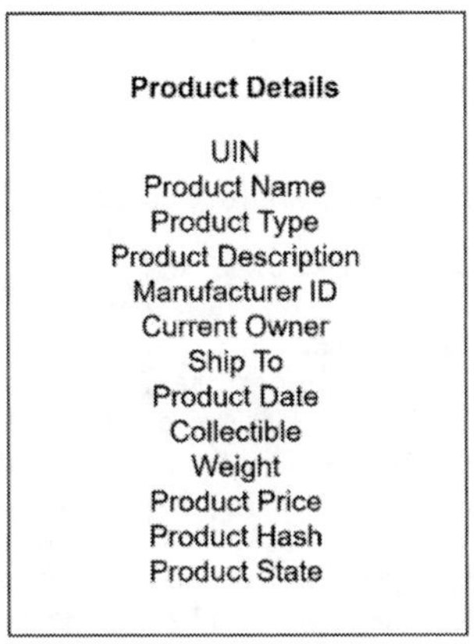

Fig. 5. Item Details stored in Blockchain

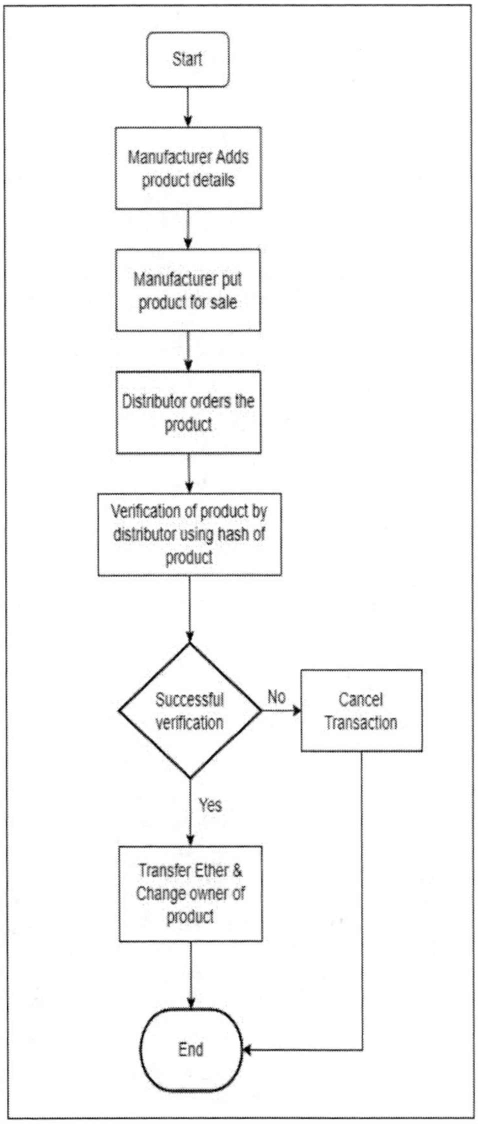

Fig. 6. Shows transaction and transfer of ownership between 2 entities.

Algorithm 1: Create product by Manufacturer

Input: Product name, Description, Type, collectible, weight, manufacturer address

1. If(address belongs to manufacturer)
2. uin = randMod() // to generate UIN from algorithm 2
3. item = new Item(Product name, Description, Type, collectible, weight, manufacturer address) ;
4. update product hash from algorithm mentioned in algorithm 3.
5. Make manufacturer owner of product
6. Add item to manufacturer products list
7. Emit event indicating product creation in blockchain
8. Else
9. Show error

5.1. Creating product by Manufacturer

Algorithm 1 shows how a product can be created by Manufacturer. For manufacturer to be able to add a new product in supply chain, it is mandatory he needs to have the required rights. Address of manufacturer can be added as an entity in Supply chain by the owner of main supply chain contract. Manufacturer needs to enter the product details to be able to create the product. Once the product is created, manufacturer is the owner of product by default. Events are used to indicate creation of product in blockchain.

Algorithm 2: Generate UIN of product

Input: address of sender, currentTimeStamp

1. randNonce = randNonce + 1
2. pack address of sender, currentTimeStamp and randNonce
3. Use keccak256() to convert into 256-bit hash.
4. Convert hash to uint
5. return remainder of uint by % 82589933

5.2. Keccak

Keccak is a family of sponge function instances, encompassing capacity values ranging from 0 to 1599 bits. All these instances are well-defined and so are their security claim. Keccak is the result of using the sponge construction on top of the Keccak-f permutations and applying the multi-rate padding to the input. Using multi-rate padding causes each member of the Keccak family (and in particular for each value of the capacity) to act as an independent function. As a native feature, Keccak provides variable output length, that is, the user can dynamically ask for as many output bits as desired [12]. Keccak is used for hashing purpose during verification in Algorithm 2, 5 and 6 mentioned here.

Algorithm 3: Sale by Manufacturer

Input: address, uin (product id), address of Receiver, price

1. if address is of manufacturer
2. if product is owned by address
3. update product visibility to true
4. update product state to for Sale by Manufacturer
5. update product price to price
6. emit event on blockchain network
7. else
8. fail transaction
9. else
10. fail transaction

5.3. Sale by Manufacturer

Once the product is manufactured by manufacturer it is up to him to decide the price and add the product for sale. The contracts first verify the address mentioned is of the manufacturer and is mapped with the current owner of the product. In case any of these conditions fail the transaction is cancelled. Once the conditions is satisfied the product visibility is changed to true and state is updated to "For Sale By Manufacturer". Default product price is updated to price decided by the manufacturer. Event is emitted regarding the product in blockchain.

Algorithm 4: Shipped by Manufacturer

Input: manufacturer address, distributor address, uin (product id)

1. if transaction address is manufacturer
2. if receivers address is distributor
3. if product state is sale by manufacturer
4. change product state to "shipped by Manufacturer".
5. change "Ship to Address" to distributors address
6. update product hash using algorithm 5
7. emit event of product to "Shipped by Manufacturer".
8. Else
9. error
10. Else
11. error
12. Else
13. Error

5.4. Shipped by Manufacturer

Off the chain discussion regarding sale will take place between manufacturer and distributor which will be executed on blockchain. Address of entity who called the function will be verified if it belongs to manufacturer. The manufacturer then adds the address of the user he wants to ship the product to. The address will be verified if it belongs to distributor and will be processed for further steps. The product state is updated to "Shipped by Manufacturer". Using the steps mentioned in Algorithm 5 you can update the product hash which will be used for verification during algorithm 6. The event will be emitted to blockchain network.

Algorithm 5: Generate Product Hash using Keccak 256

Input: address1 (owner), address2 (receiver), uin (product id), productName

 https://doi.org/10.1051/itmconf/20224403048

1. "pack" uin, address1, address2, product name using abi.encodePacked.
2. convert this packed data to hash using keccak 256
3. convert the obtained hash to integer.

5.5. Received Product by Distributor

In this part the distributor i.e., the receiver will be confirming the product transfer and will transfer the amount in the form of ether to previous owner of the product. Algorithm 6 carries out the execution of this part and transfers ownership of the product to distributor.

Algorithm 6: Received Product by Distributor

Input: uin, address1, address2, amount
1. if productPrice <= amount
2. if product state is shipped by manufacturer
3. if ("product Hash" is equal to hash generated by algorithm 5 from "uin", "address1", "address2", "product Name")
4. update product state as "Received by Distributor"
5. update visibility of product to false
6. transfer price of product to previous owner
7. update current owner to distributor address
8. add product to distributors product list
9. emit an event "Received by distributor" to network
10. else
11. cancel transaction
12. else
13. error
14. else
15. error

5.6. Hash Verification:

Hash verification occurs every time the product is transferred to next entity. Receiving entity verifies the hash value of the product details with the hash value stored in the contract to complete the transaction. Once the verification is completed ownership of product is transferred to receiver's address. Refer figure 7 for verification process.

By changing the entities and roles mentioned in the algorithm, the same algorithm is used by distributor to sell product to retailer and retailer to customer. Depending upon the type of commodity customer can also sell this product to another consumer with ease.

6. Testing and Validations

This section describes the testing and validation of smart contracts. Manufacturing of product by manufacturer and sale of the product to distributor will be tested in this section. Transaction will take place between manufacturer and distributor in this section. Multiple entities across the chain will be go through the process of transfer of product before reaching the customer.

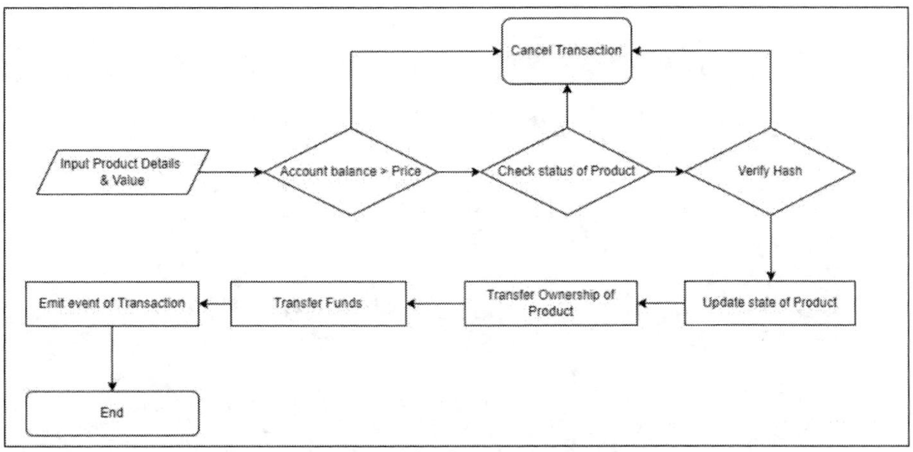

Fig. 7. Verification of Product and Receiver

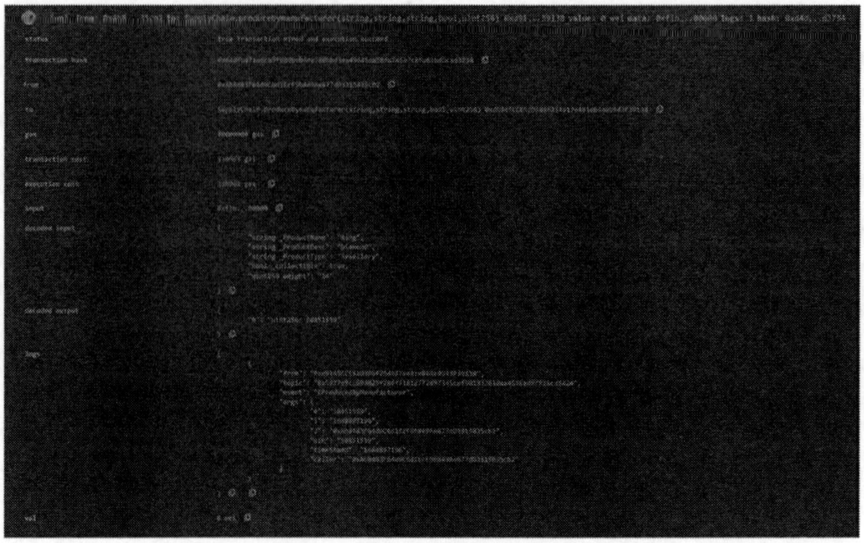

Fig. 8. Product is successfully added to the blockchain.

6.1. Add Product to Supply chain

Only the manufacturer can add the product to supply chain. Figure 8 shows that the product is successfully added to chain.
Once the product is successfully added to the chain the contract returns a unique identification number for the product which remains constant throughout the chain. Logs are generated which gives information about the transaction. Event is emitted in blockchain.

6.2. Sale product by Manufacturer:

Once the state of product is updated to "For sale by manufacturer" event in emitted in blockchain.
Manufacturer depending upon his convenience he can add the product to sell at the specific price. Figure 9 shows how the product can be updated for sale purpose.

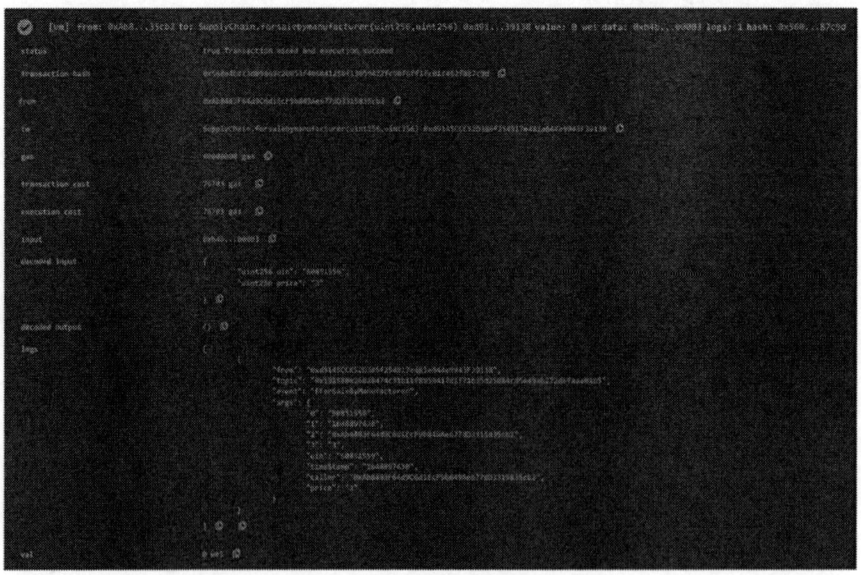

Fig. 9. Product on sale by manufacturer

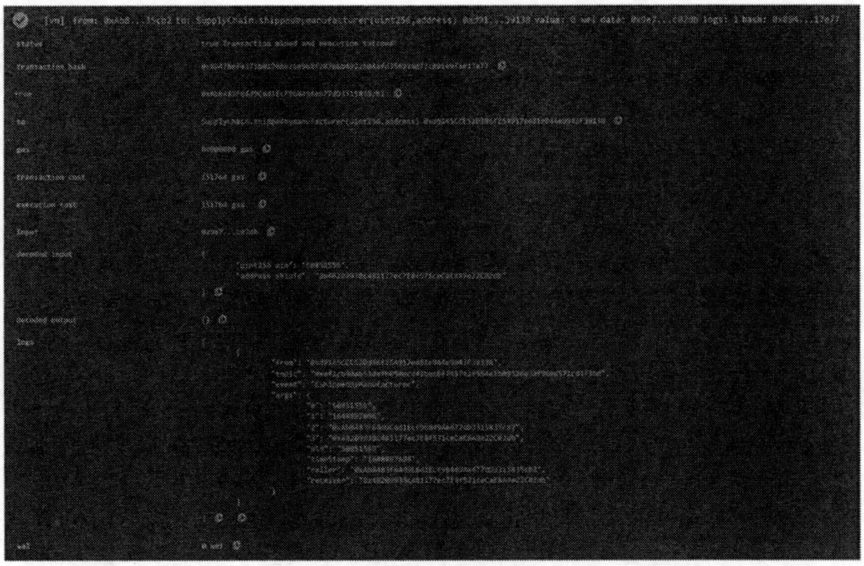

Fig. 10. Product is shipped to receiver's address

6.3. Ship Product to Next Entity (Distributor)

Off chain communication takes place between two parties for sale of product. Once the product is sold and shipped the event is emitted in blockchain. Sender will add the address of the receiver which will be encrypted in product hash. Figure 10 shows that the product state is updated and is currently in shipping state to receiver.

6.4. Delivery of Product to Wrong Address

During the process of process of shipping the product can get delivered to wrong address. In such scenario during verification if would be found and transaction won't be completed. Since the new hash generated during receiving purpose, will be compared to product hash added during shipping.

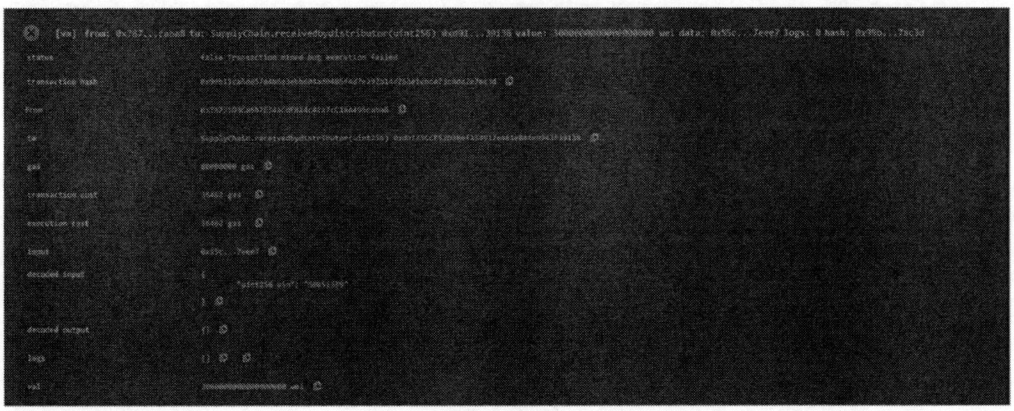

Fig. 11. Shows the failed transaction of product and ownership is retained by previous owner.

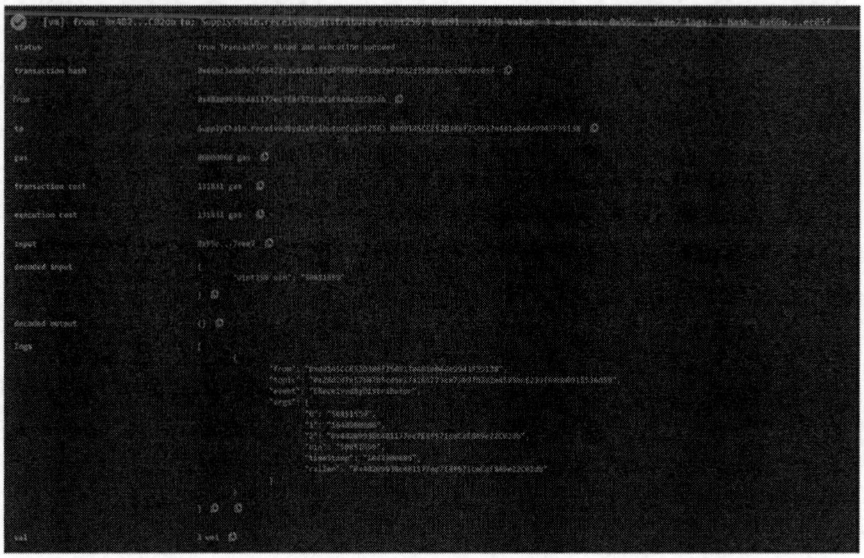

Fig. 12. Successful verification of Hash

6.5. Delivery of Product to Correct Address

Once the product is received by correct recipient of product the transfer of funds from receivers account are automatically transferred to the previous owner resulting in transfer of funds and ownership of product refer Figure 12.

7 Conclusion

The developed method works for both multiple and single transporters, and includes a punishment and incentive mechanism to encourage all players to act honestly. The system also does away with the requirement for a trusted third party, instead using smart contracts as an escrow to automatically settle payments and distribute each participant's agreed-upon share once the item is delivered to the buyer. To meet the demand of delivering across multiple transporters, our decentralized solution uses a chain of contracts with no cyclic dependencies. For future work, depending on the product of application changes can make it compatible for multiple products. E.g., Vaccine, Agriculture products, Medicines

8 References

[1] S. Nakamoto, "Bitcoin: A peer-to-peer electronic cash system," 2008.
[2] H. R. Hasan and K. Salah, "Blockchain-based solution for proof of delivery of physical assets," in International Conference on Blockchain. Springer, 2018, pp. 139–152.
[3] Two party contracts. [Online]. Available: https://dappsforbeginners. wordpress.com/tutorials/two-party-contracts/
[4] K. Toyoda, P. T. Mathiopoulos, I. Sasase, and T. Ohtsuki, "A novel blockchain-based product ownership management system (poms) for anti-counterfeits in the post supply chain," IEEE Access, vol. 5, pp. 17 465–17 477, 2017.
[5] K. Biswas and V. Muthukkumarasamy, "Securing smart cities using blockchain technology," in 2016 IEEE 18th International Conference on High Performance Computing and Communications.
[6] A Review of Blockchain Technology and Its Applications in the Business Environment by Thomas Kitsantas Athanasios Vazakidis Evangelos Chytis at International Conference on Enterprise, Systems, Accounting, Logistics & Management July 2019
[7] https://ethereum.github.io/yellowpaper/paper.pdf
[8] http://myinstantid.com/szabo.pdf.
[9] K. Christidis and M. Devetsikiotis, "Blockchains and smart contracts for the internet of things," IEEE Access, vol. 4, pp. 2292–2303, 2016.
[10] https://ethereum.org/en/smart-contracts/
[11] A. M. Antonopoulos and G. Wood, Mastering ethereum: building smart contracts and dapps. O'Reilly Media, 2018.
[12] The Keccak reference by Guido Bertoni, Joan Daemen, Michaël Peeters Gilles Van Assche, https://keccak.team/index.html.

Plant Disease Detection Using Cnn

Nishant Shelar[1], *Suraj* Shinde[2] , *Shubham* Sawant[3] , *Shreyash* Dhumal[4] , and *Kausar* Fakir[5]

[1,2,3,4,5]Department of Electronics and Telecommunication, Ramrao Adik Institute of Technology, Navi Mumbai, India

Abstract. Plants and crops that are infected by pests have an impact on the country's agricultural production. Usually, farmers or professionals keep a close eye on the plants in order to discover and identify diseases. However, this procedure is frequently time-consuming, costly, and imprecise. Plant disease detection can be done by looking for a spot on the diseased plant's leaves. The goal of this paper is to create a Disease Recognition Model that is supported by leaf image classification. To detect plant diseases, we are utilizing image processing with a Convolution neural network (CNN). A convolutional neural network (CNN) is a form of artificial neural network that is specifically intended to process pixel input and is used in image recognition.

1. INTRODUCTION

Agricultural production is a very old means of obtaining food. It is a vital source of income for people all around the world. No one can exist in our world without food. Plants are crucial not only for humans, but also for animals who rely on them for food, oxygen, and other necessities. The government and experts are taking significant initiatives to enhance food production, and they are working successfully in the real world. When a plant becomes afflicted with a disease, all living organisms in the environment are affected in some way. This plant disease can affect anywhere on the plant, including the stem, leaf, and branch. Even the types of illnesses that impact plants, such as bacterial and fungal diseases .etc can differ. The illness that impacts the crops will be determined by factors such as climate. There are a large number of people that are food insecure. This occurs as a result of insufficient food crop output. Even significant climate changes will have an impact on plant development. This type of natural tragedy is unavoidable. Early detection of plant disease aids in the prevention of large-scale crop losses. Farmers must apply the appropriate insecticides for their crops. Too many pesticides are harmful to crops and farmland. Getting expert advice will help you avoid misusing chemicals on plants. Plants have been the focus of many researchers to aid farmers and others involved in agriculture. When a disease is visible to the naked eye, it is straightforward to detect. The illness may be discovered and treated early if the farmer has sufficient information and monitors the crops on a regular basis. However, this phase only exists when the disease is extreme or crop output is low. Then there are the different innovations. Farmers will benefit from the introduction of automated disease detection tools. This approach yields outcomes that are suitable for both little and large-scale agricultural cultivation. Importantly, the results are precise, and the disorders are detected in a very short amount of time. These technologies rely heavily on deep learning and neural networks to function. Deep Convolutional Neural Network is utilized in this study to identify infected and healthy leaves, as well as to detect illness in afflicted plants. The CNN model is designed to suit both healthy and sick leaves; photos are used to train the model, and the output is determined by the input leaf.

2. LITERATURE REVIEW

K.Muthukannan and colleagues discovered spot infections in leaves and categorized them according to the diseased leaf categories using various machine learning algorithms. LVQ - Learning Vector Quantization, FFNN - Feed Forward Neural Network, and RBFN - Radial Basis Function Networks were utilized to diagnose diseased plant leaves by analyzing the collection of form and texture data from the afflicted leaf picture. The simulation showed that the proposed system is effective. With the support of this work, a machine learning-based system for improving crop quality in the Indian economy can be developed. [1]

The study of plant leaf disease detection by Malvika Ranjan and colleagues starts with image capturing. Color data, such as HSV features, are retrieved from the segmentation results, and an artificial neural network (ANN) is then trained by selecting feature values that can effectively discriminate between healthy and sick samples. Using a combination of image data processing methods and ann, the current study suggests a method for identifying cotton leaf illnesses early and reliably.[2]

The goal of Syafiqah Ishakais and colleague's study of Leaf Disease Classification using Artificial Neural

Network is to acquire and analyze data from leaf photos in order to determine healthy or diseased leaves of medical plants using image processing methods. To extract pictures and get data, an algorithm of adjusted contrast, segmentation, and features extraction is employed from the image processing approach. The Artificial Neural Network was used to analyze the findings of the experiment. The architecture of the network used to classify healthy or unhealthy leaves is multilayer feed-forward Neural Networks, which are multilayer perceptron and radial basis function RBF. The end outcome of the experiment demonstrates that the RBF network outperforms the MLP network. [3]

Srdjan Sladojevic and colleagues present Deep Convolutional Neural network Supported Identification of Crop Diseases by Plant Image Classification, a new method for the construction of a crop diseases recognition model based on plant image classification and deep convolutional networks. The methodology employed and the novel technique of training allow for a quick and painless system set up in practice. With the ability to identify crops from their surroundings, the built model can recognize thirteen types of plant illnesses from healthy leaves. All of the necessary processes for applying this diseases recognition model are detailed throughout the study, beginning with the collection of photographs in order to establish a database that is evaluated by agricultural experts. Caffe, a deep learning framework developed by Berkley Vision and Learning Centre, was used to perform the deep CNN training. The experimental results on the developed model achieved precision between 91% and 98%, for separate class tests, on average 96.3%.[4]

CNN and Modeling Adversarial Networks were used to classify plant diseases. Others, like Emanuel Cortes A deep neural network and semi-supervised algorithms were trained to distinguish crop species and disease status of 57 different classes using a publicly available dataset of 86,147 photos of ill and healthy plants. rs-net was the unlabeled data experiment that functioned successfully. With a detection rate of 1e-5, it was able to score more than 80% in the training phase in less than 5 epochs.[5]

Plant disease identification and treatment using neural network models, Konstantinos P. Ferentinos and colleagues built CNN models to conduct crop disease identification and diagnosis using basic leaf pictures of healthy and sick plants. The models were trained using an open collection of 87,848 photos, which included 25 kinds of plants in 58 various classes of [plant, illness] pairs, including non-affected plants. Multiple model architectures were developed, with the top-performing one achieving a success rate of 99.53 percent. The model's high success rate makes it a valuable or early detection tool.[6]

In the study Soybeans, Crop Disease Detection Using Cnns, Serawork Wallelign, and the others The viability of CNN for crop diseases identification in leaves pictures captured in the natural surroundings is presented in this study. To accomplish the soybeans plant disease classification, the model is built using the LeNet architecture. The PlantVillage collection yielded 12,673 samples tested green photos from four types, including healthy leaf images. The photos were taken in an unstructured setting. The built model obtains an accuracy of classification of 99.32 percent, demonstrating it a Convolutional neural network effectively extract significant features and diagnose plant diseases from photos captured in the wild. [7].

A Deep-Learning-Based Detection for Real-Time Recognition of Tomato Plant Pest and Diseases Alvaro Fuentes and colleagues look at three types of detectors: the Faster Region-based Cnns (Faster R-CNN), the Area Convolutional Neural Network (R-FCN), and the Single Action Multibox Detector (SSD), all of which are referred to as "deep learning meta-architectures" in this paper. We use "deep feature extractors" like VGG net and Residual Network to merge every one of these meta-architectures (ResNet). We show how deep morpho and feature extractors perform, and we also suggest a way for locally and globally category labeling and feature extraction to improve accuracy and reduce false positives throughout training. We train and test our systems end-to-end on our large Tomato Diseases and Pests Dataset, which contains challenging images of diseases and pests, including several inter-and extra-class variations, such as infection status and location in the plant.[8]

This paper outlines a method for accurately identifying apple leaf diseases. Building enough unhealthy photos and unique architecture of a deep CNN based on AlexNet are required to identify apple leaf infections. Using a database of 13,689 pictures of sick apple leaves, the suggested deep CNN model is meant to detect four common apple leaf disorders. The total accuracy of the suggested illness detection model is 97.62 percent. When compared to the AlexNet model, the parameters of the suggested model were reduced by 51,206,928 and the model's accuracy was enhanced by 10.83 percent with produced pathological pictures. According to this research, the deep learning model for disease management may be more accurate and have a faster convergence rate, therefore enhancing disease control.[9]

Prasanna Mohanty and colleagues developed a deep convolutional neural network using deep learning to detect 14 different crops and 26 illnesses. On a held-out test set, the training set model obtained an accuracy of 99.35 percent, illustrating the practicality of this strategy. The model still obtains a 31.4 percent accuracy when tested on a collection of photographs acquired from reputable web sources - i.e. images shot under settings distinct from those used for training. While this accuracy is substantially greater than the one based on random selection 2.6%, a larger collection of training data is required to increase overall accuracy.[10].

To diagnose plant leaf illnesses, Ashwin Dhakal and colleagues created a model that includes feature extraction, segmentation, and classification of collected leaf patterns. Yellow Leaf Curl Virus, Bacterial Spot, Late Blight, and Healthy Leaf are the four classifier labels employed. With 20 epochs, the retrieved characteristics are fitted into the neural network. Various neural network-based topologies are used, with the greatest accuracy of 98.59 percent in predicting plant disease. [11]

S. Khirade and colleagues used digital image processing algorithms and BPNN - backpropagation neural networks to solve the problems of detection of plant diseases in 2015. Different techniques for identifying plant disease using photographs of leaves have been developed by the authors. To segment the contaminated section of the leaf, they used Otsu's thresholding, followed by border detection and spot detection algorithms. They then extracted properties such as colour, texture, morphology, edges, and so on in order to classify plant diseases. The BPNN algorithm is used to classify or identify plant diseases. [12]

In 2017, Peyman Moghadam and colleagues proved the use of hyperspectral imaging in the diagnosis of plant diseases. In this research, the VNIR - visible and near-infrared and SWIR - short-wave infrared spectrums were employed. For leaf segmentation, the authors employed the k-means clustering approach in the spectral domain. To remove the grid from hyperspectral pictures, they suggested a unique grid removal technique. The accuracy of vegetation indices in the VNIR spectral range was 83 percent, and full-spectrum accuracy was 93 percent. Despite the fact that the suggested technique achieved improved accuracy, it necessitates the use of a hyperspectral camera with 324 spectral bands, making the solution prohibitively expensive. [13]

Sharath D. M. and colleagues created a Bacterial Blight detection method for Pomegranate plants in 2019 utilizing variables including colour, mean, homogeneity, SD, variance, correlation, entropy, and edges. Grab cut segmentation was used by the authors to segment the image's region of interest. The edges of the photos were extracted using the Canny edge detector. The authors have succeeded in developing a system that can forecast the degree of infection in the fruit.[14].

The convolutional neural network was used by Garima Shrestha and colleagues to identify plant disease in 2020. With an accuracy of 88.80 percent, the authors were able to classify 12 plant diseases. Experimentation was carried out by using a collection of 3000 high-resolution RGB photographs. The convolutional layer and pooling layer have 3 blocks in this network. Eventually, the network becomes very expensive as a result of this. Additionally, the model's F1 score is 0.12, which is extremely poor due to the significant amount of erroneous negative predictions.[15]

3. PROPOSED SYSTEM

We are building a neural network model for image classification. this model will be deployed on the android application for live detection of plant leaf disease through an android phone's camera. The recognition and classification procedures are depicted in Fig. 1

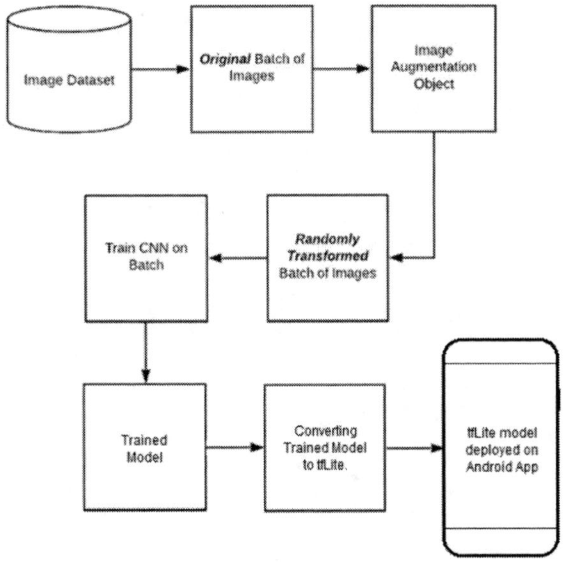

Fig. 1. Block Diagram Of Proposed System

(1) The first step is to collect data. We are using the PlantVillage Dataset, which is widely available. This dataset was released by crowdAI.

(2) Pre-processing and Augmentation of the collected dataset is done using pre-processing and Image-data generator API by Keras.

(3) Building CNN(Convolutional Neural Network) Model (Vgg-19 architecture) for classification of various plant diseases.

(4) Developed model will be deployed on the Android Application with help of TensorFlow lite.

4. CONVOLUTIONAL NEURAL NETWORK ARCHITECTURE (VGG-19)

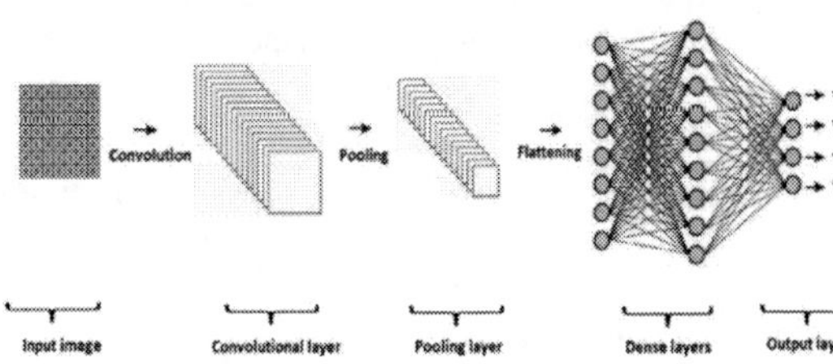

Fig. 2. CNN Architecture

A Convolutional Neural Network has three layers: a convolutional layer, a pooling layer, and a fully connected layer. Fig 2 shows all layers together.

4.1 Convolution Layer

Convolutional layer: produces an activation map by scanning the pictures several pixels at a time using a filter. Fig 3 shows the internal working of the convolution layer.

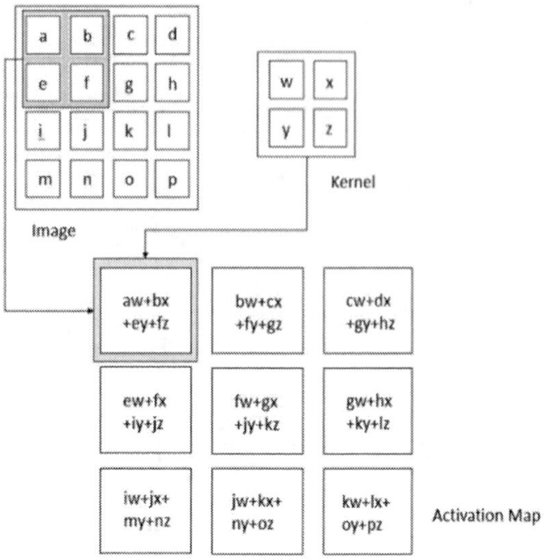

Fig. 3. Convolution Layer

4.2 Pooling Layer

Pooling layer: reduces the amount of data created by the convolutional layer so that it is stored more efficiently. Fig 4 shows the internal working of the pooling layer

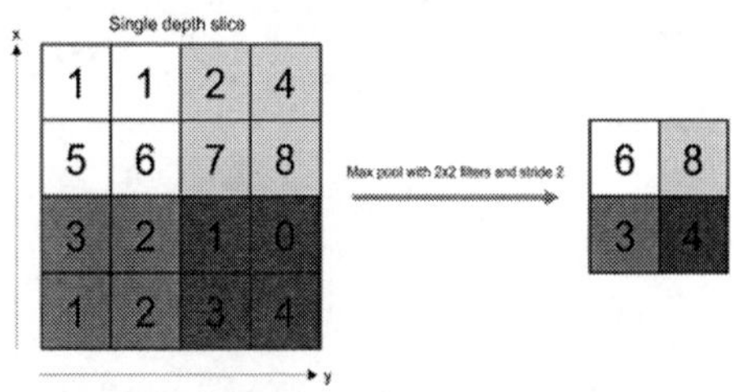

Fig. 4. Pooling Layer

4.3 Fully Connected Layer

Fully connected input layer – The preceding layers' output is "flattened" and turned into a single vector which is used as an input for the next stage.

The first fully connected layer – adds weights to the inputs from the feature analysis to anticipate the proper label.

Fully connected output layer – offers the probability for each label in the end.

Fig 5 shows the internal working of fully connected layer

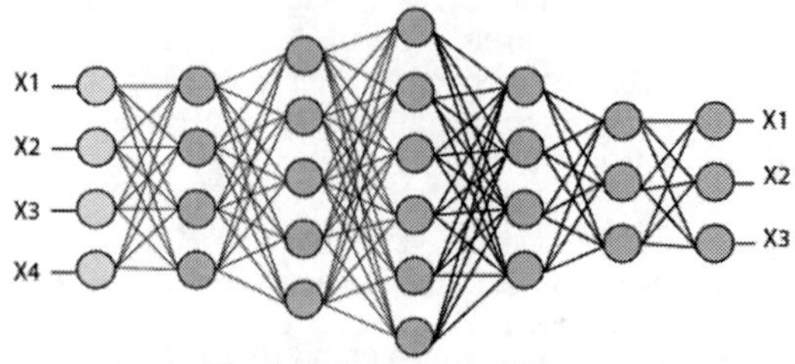

Fig. 5. Fully Connected Layer

VGG19 is a sophisticated CNN with pre-trained layers and a thorough grasp of how an image is defined in terms of form, color, and structure. VGG19 is a deep neural network that has been trained on millions of photos with challenging classification problems.

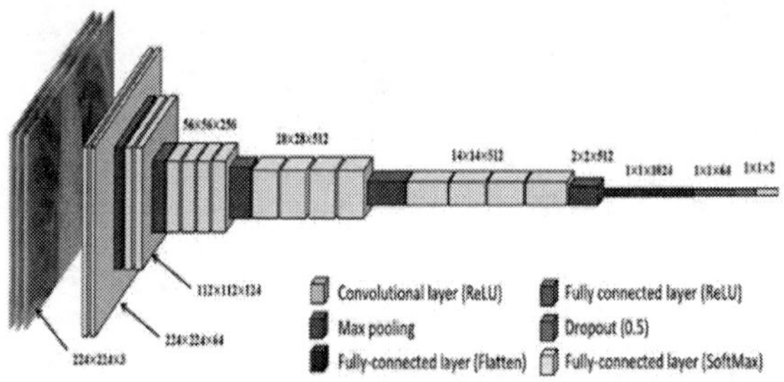

Fig. 6. vgg-19

5. RESULT

A 95.6% accuracy rate was achieved using early stopping while Training the model on 50 epochs. Figure 7 depicts the visualization of training and validation accuracy. The result of detecting and recognizing a strawberry plant is shown in Figure 8. On the left, a healthy plant leaf, and on the right, a sick infected plant. The result of detecting and recognizing a potato plant is shown in Figure 9. On the left, a healthy plant leaf, and on the right, a sick infected plant.

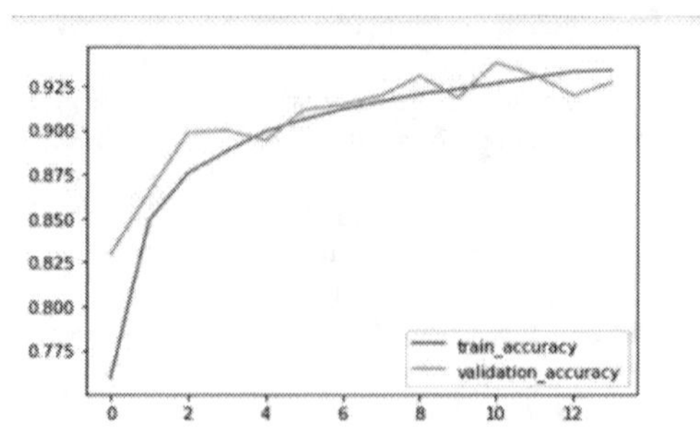

Fig. 7. Training vs Validation

Fig. 8. Result of detection and recognition of a strawberry plant.

Fig. 9. Result of detection and recognition of a Potato plant.

6. CONCLUSION & FUTURE WORK

We are successful in creating disease classification techniques used for plant leaf disease detection. A deep learning model that can be used for automatic detection and classification of plant leaf diseases is created. Tomato, strawberry, soybean, raspberry, potato, corn, Pepper bell, peach, orange, grape, cherry, blueberry, apple are 13 species on which the proposed model is tested. 38 classes of plants were taken for identification through this work. We were successfully able to work with the image data generator API by Keras. Through this, we were able to do image-processing tasks. We were also able to create the vgg-19 model which is an advanced convolution model and train the model with the data for prediction. The prediction is done by our model is almost correct. We have successfully deployed these model on the android app and are trying to increase the accuracy of the android app as well as the model.

7. REFERENCES

[1] Diseased plant leaves using Neural Network Algorithms K. Muthukannan1, P. Latha2, R. Pon Selvi1 and P. Nisha1 1Department of ECE, Einstein College of Engineering, Anna University, Tirunelveli, India 2Department of CSE, Government College of Engineering, Anna University, Tirunelveli, India ARPN Journal of Engineering and Applied Sciences, VOL. 10, NO. 4, MARCH 2015, ISSN 1819-6608

[2] Ranjan, Malvika, and others "Detection and classification of leaf disease using artificial neural network." International Journal of TechnicalResearch and Applications 3.3 (2015): 331-333.

[3] Ishak, Syafiqah,and others. "Leaf disease classification using artificial neural network." Jurnal Teknologi 77.17 (2015).

[4] Sladojevic, Srdjan,and others. "Deep neural networks based recognition of plant diseases by leaf image classification." Computational intelligence and neuroscience 2016 (2016).

[5] Cortes, Emanuel. "Plant disease classification using convolutional networks and generative adversarial networks." (2017).

[6] Wallelign, Serawork, Mihai Polceanu, and Cedric Buche. "Soybean plant disease identification using convolutional neural network." The thirty-first international flairs conference. 2018.

[7] Fuentes, Alvaro,and others. "A robust deep-learning-based detector for real-time tomato plant diseases and pests recognition." Sensors 17.9(2017): 2022.

[8] Liu, Peide, Tahir Mahmood, and Qaisar Khan. "Multi-attribute decision making based on prioritized aggregation operator under hesitant intuitionistic fuzzy linguistic environment." Symmetry 9.11 (2017): 270.

[9] Mohanty, Sharada P., David P. Hughes, and Marcel Salath´e. "Using deep learning for image-based plant disease detection." Frontiers in plantscience 7 (2016): 1419.

[10] Dhakal, Ashwin, and Subarna Shakya. "Image-based plant disease detection with deep learning." International Journal of Computer Trends and Technology 61.1 (2018): 26-29.

[11] Ramcharan, Amanda, et al. "Deep learning for image-based cassava disease detection." Frontiers in plant science 8 (2017): 1852.

[12] S. D. Khirade and A. B. Patil, "Plant Disease Detection Using ImageProcessing," 2015 International Conference on Computing CommunicationControl and Automation, 2015, pp. 768-771, DOI: 10.1109/ICCUBEA.2015.153

[13] P. Moghadam, D. Ward, E. Goan, S. Jayawardena, P. Sikka andE. Hernandez, "Plant Disease Detection Using Hyperspectral Imaging,"2017 International Conference on Digital Image Computing: Techniques and Applications (DICTA), 2017, pp. 1-8, doi:10.1109/DICTA.2017.8227476

[14] Sharath, D. M., et al. "Image-based plant disease detection in pomegranate plant for bacterial blight." 2019 international conference on communication and signal processing (ICCSP). IEEE, 2019.

[15] Shrestha, G., and M. Deepsikha. "Das, and N. Dey,"." Plant Disease DetectionUsing CNN," 2020 IEEE Applied Signal Processing Conference(ASPCON). 2020.

Disease Detection Using Artificial Intelligence

Sumeet Shah[1], *Akshata Tattu*[2], *Ashna Sheregar*[3], *Sneha Mane*[4]

[1]Electronic and Telecommunication Engg,Ramrao Adik Institute of Technology,Navi Mumbai,India

[2]Electronic and Telecommunication Engg,Ramrao Adik Institute of Technology,Navi Mumbai,India

[3]Electronic and Telecommunication Engg,Ramrao Adik Institute of Technology,Navi Mumbai,India

[4]Electronic and Telecommunication Engg,Ramrao Adik Institute of Technology,Navi Mumbai,India

Abstract

Currently, artificial intelligence is widely used to aid humans in a variety of ways. One area where artificial intelligence is particularly beneficial is medical image detection where diagnostic procedures require the collection and processing of large amounts of data for particular diseases. The topics covered in this paper include Pneumonia, Lung Cancer, and Brain Tumors. Early detection and treatment are crucial when treating these types of diseases. The paper describes the use of a convolutional neural network algorithm in order to process medical images so that it can aid in decision making process and help save time. Machine learning is useful in increasing consistency and accuracy in detection. The highest accuracy that the model has achieve 94.297%.

Keywords

pneumonia detection; lung cancer; brain tumor; machine learning; image processing; artificial intelligence; convolutional neural network.

1.INTRODUCTION

Pneumonia is a disease that generally affects the human lungs. This disease can cause illness in all age groups. This disease can be caused by contagion, bacteria as well as fungi. In today's world with new advancements in medical fields we have been successful in reducing the number of deaths caused by pneumonia. Every year 1 million people have been reported to suffer from this disease by the hospital and 50,000 people die due to this disease alone. Children who are under the age group of 5 and older adults who are over the age of 65 are generally more prone to this disease. It is one of the largest causes of infectious death in children under the age of 5. Pneumonia related complications resulted in 1 million children dying in 2015. The main way to detect pneumonia is by using chest x-rays which can be a grueling task. In general, the x-rays depict increase in viscosity of the infected area which is a mixture of connection and atelectasis. Air bronchogram can be seen due to the presence of air column girdled by connection and atelectasis when the disease becomes severe. This area appears as patchy and invariant throughout the area. With significant advancements in computer vision and deep learning, effective ways have been introduced in image bracket, recognition, and segmentation.

Lung cancer, additionally regarded as lung carcinoma, have contributed about 98–99% of all lung cancers, is a malignant lung tumor characterized through uncontrolled growth in lung tissue. Lung cancers is generally the result of long-term tobacco smoking. About 10–15% of cases manifest in humans who haven't smoked. These cases are frequently caused by an aggregate of hereditary factors and vulnerability towards gases such as radon, asbestos, passive smoke and different forms of air pollution. Lung cancer is generally detected by computed tomography (CT) scans. The analysis is proven via biopsy, which is commonly carried out by using bronchoscopy or CT-guidance. 2.2 million people had reported the presence of lung cancer and 1.8 million people lost their lives due to it. This disease causes death irrespective of gender. This disease is generally prognosed at the age of 70 years. The mortality rate can be increased by early detection and treatment of the disease. The most popular and effective imaging technique (CT scanning) is used in the diagnosis as it can discover every suspected and unsuspected lung cancer nodule.

According to WHO brain tumor is the 10th leading cause of death. The survival rates in brain tumor cases vary depending upon several factors like the type of spinal cord or brain tumor. The survival rate in women is 36z% where as in men it is 34%.308,102 people had been diagnosed with primary brain tumor. 251,329 people lost their lives due to primary cancerous brain tumors. Brain tumor is caused by the growth of abnormal cells in the brain. There are several types of brain tumor. Some are benign which means that they are noncancerous, and some are malignant that is they are cancerous. Brain tumors can originate within the brain or it can be present in other parts of the body and then spread to the brain as a secondary brain tumor. The speed at which a brain tumor grows can vary and its severity depends upon its location, size, etc.

Deep Learning methods can be deployed in medical sector for analysis and detection purpose from biomedical images. A lot of chest x-rays, MRIs and CT scans are taken daily in hospitals across the nation which in turn creates a large unexplored image dataset. Hence, we have proposed a convolutional neural network algorithm which is capable of classifying the above-mentioned diseases with the help of CT scans, MRI and x - rays.

2. Related works: -

In this paper [1], machine learning algorithm is used to support the decision-making process so that it determines the correct diagnosis with help of chest x-rays. They have concentrated on only one disease that is pneumonia and have used a CNN algorithm to classify the chest X-ray images. Pneumonia is a disease which affects the lower part of respiratory tract. The main libraries that were used in this paper are: Keras, Matplotlib, Seaborn and NumPy. Their model had an accuracy of 88.90%.

In this paper [2], multiple variants of convolutional neural networks namely, InceptionResNetV2, DenseNet201 and VGG19 are trained using different methods of feature extraction and fine-tuning. Due to the low amount of data, a higher detection accuracy is attained on the chest X-rays pneumonia detection task. The highest accuracy of their model can reach up to 94.20%.

In this paper [3], they have created two different models which includes detection and segmentation of Brain tumor. The first model was used to segment the tumor with the help of FCM and then classify the images with the help of ML algorithms. CNN is widely used for medical image processing. They came up with an exemplary that correctly classifies tumor from the MRI images. Their paper had some drawbacks such as sparsity of connection and parameter sharing.

In this paper [4], they have used CT scan images as their input. Their methodology has various steps that is pre-processing where the noise and other disturbances present in the image are removed and the next step is image segmentation. The important during the diagnosis of this disease. They selected a Fully Convolutional Network in order to gain more accuracy. They used FCN to segment the lung nodules. Their method was able to provide 100% accuracy.

In this paper [5] they have included a CNN-based brain tumor diagnosis system which helps in detecting brain tumor. This helps in rapid treatment planning, to carry out pre-processing on images, the process of feature extraction, reduce the feature space, and lastly classify the images for diagnosis of brain tumor. They improved the performance of their model with the help of geometrical and statistic data augmentation on the images that they had acquired. Their model achieved an accuracy of 100%.

In this paper [6] consists of a computerized method for identifying the tumor region in the brain with the help of the MRI images. Firstly, they have classified the brain into healthy brain or a brain that has tumor and after that they have further classified them into benign or malignant tumor. The algorithm that they have used consists of pre-processing, feature extraction, image segmentation and classification of images. They have created a user-friendly MATLAB GUI program.

3. Methodology

2.1. Input Biomedical images: - We take CT images for lungs, Xray for pneumonia and MRI images for brain tumor detection. The main advantage of biomedical images is that they have better clarity, less noise and distortion. The mean and variance and the calculated value is very close to the original value.

2.2. Pre-processing: - Pre-processing is the first process in disease detection. It consists of steps that format the images before they can be used for the model training. It converts image data into a form that will allow machine learning algorithm to solve it. This technique improves the detection accuracy and also eliminates the background noise or disturbance. There are several techniques which are used to pre-process an image like image resizing, converting the images to grayscale images, noise reduction and image augmentation. For the dataset that we have acquired we have only resized the images.

2.3. Feature extraction: - Feature extraction is a decisive step in image processing and machine learning that uses various algorithms and techniques for detecting and isolating the required features of an image. It is a section of the dimensionality reduction process, in which, a preliminary set of the raw information is divided and reduced to extra manageable groups. So, when we choose to process it will be easier. The most important attribute of these massive datasets is that they have a vast range of variables. These variables require a lot of computing assets to process. So, feature extraction helps to get the satisfactory characteristic from those large datasets by choosing and combining variables into features, thus, efficiently lowering the amount of data. These facets are effortless to process, however nonetheless capable

 https://doi.org/10.1051/itmconf/20224403050

to describe the real data set with accuracy and originality. Hence, we can conclude that, feature extraction in image processing and machine learning helps in increasing the accuracy and reducing the computational time which is required for the learning algorithm.

2.4. Train ML model: -Training a model is the most vital step in machine learning process. It is a process that allows a machine learning algorithm to automatically learn various patterns based on the given data. These patterns are statistically learned using either supervised learning or unsupervised learning. We first divide our dataset into two parts as training and testing dataset. While training the model, we use the training dataset which helps in the learning process of the model. Once the model is trained, we then test our model using the testing dataset to check whether the predictions made by our model are accurate. Once the model is trained, we can use save this model and use the model for future predictions.

2.5. Detection of Disease: - Once the machine learning model is trained and saved. The user can input the medical image which will be properly resized. After this the inputted image will be classified into a particular disease by the machine learning model.

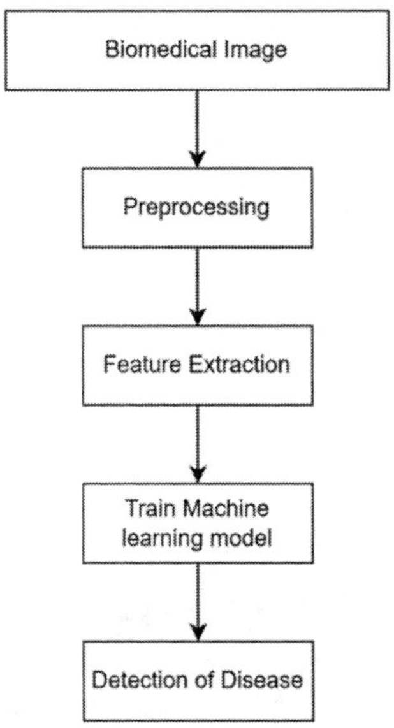

Fig.1. Steps involved in Detecting a disease

4. RESULTS AND DISCUSSION

Phase 1: - Here figure (2) represents some of the biomedical images for the particular diseases from the dataset. The images taken are a combination of Chest X-rays, MRI and CT scans. The data set contains 7182 images of lung cancer, brain tumor and pneumonia.

Phase 2: - The size of all the images is resized to a standard (100x100) dimension size with the help of resize function as all the images had different sizes.

Phase 3: - All the images were converted into an array with the help of np.array function so that the features can be extracted properly before passing on the data for training purpose.

Phase 4: - Here in table (1) represents all the parameters that we have added in order to train the machine learning model. Once all the parameters and filters have been applied the machine learning model is trained.

Phase 5: - Here in figure (4) represents all the predicted results along with the actual results.

Phase 6:- Here in table (2) shows the accuracy rate for the model that has been trained.

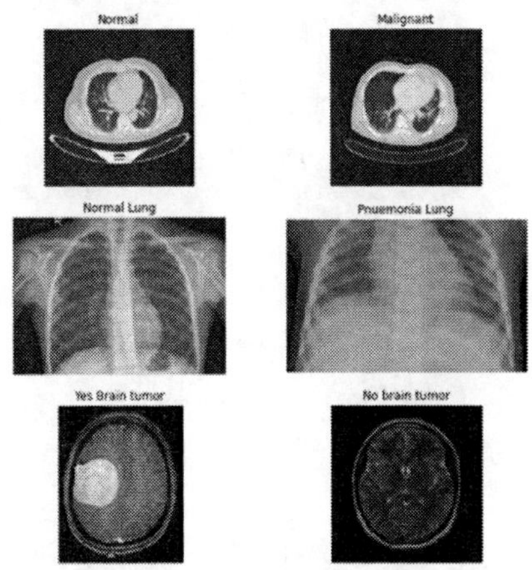

Fig.2. Images for combined Dataset (Lung Cancer, Pneumonia and Brain Tumor)

Layer(type)	Output Shape	Parameter
conv2d(Conv2D)	(None, 99, 99, 32)	416
max_pooling2d (MaxPooling2D)	(None, 49, 49, 32)	0
dropout (Dropout)	(None, 49, 49, 32)	0
conv2d_1 (Conv2D)	(None, 48, 48, 16)	2064
max_pooling2d_1 (MaxPooling2)	(None, 24, 24, 16)	0
dropout_1 (Dropout)	(None, 24, 24, 16)	0
conv2d_2 (Conv2D)	(None, 23, 23, 8)	520
max_pooling2d_2 (MaxPooling2)	(None, 11, 11, 8)	0
flatten (Flatten)	(None, 968)	0
dense (Dense)	(None, 128)	124032
dense_1 (Dense)	(None, 6)	774
Total Parameters	127806	
Trainable Parameters	127806	
Non-trainable Parameters	0	

Table 1. Table represents the total number of parameters and types of filters applied

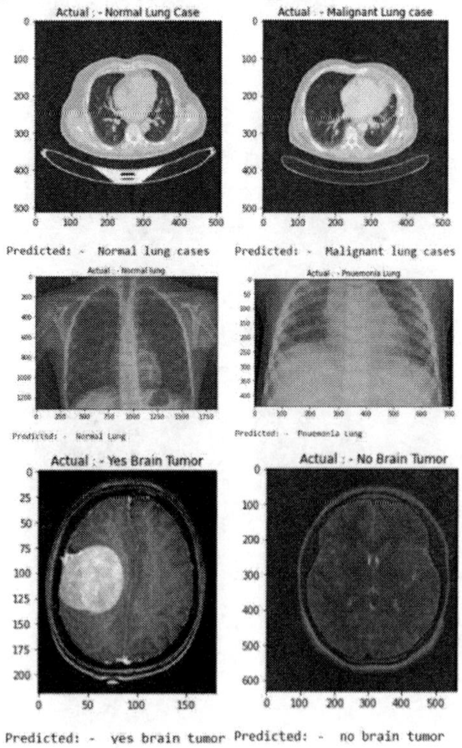

Fig.4. Predicted Disease Results

```
Confusion Matrix
[[ 53   0   0   1   0   0]
 [  0  10   0   0   0   1]
 [  0   0 152   0  10   0]
 [  1   0   0  46   0   0]
 [  0   0  22   0 407   0]
 [  0   5   0   0   1  10]]
```

Fig.5. Confusion Matrix for trained model

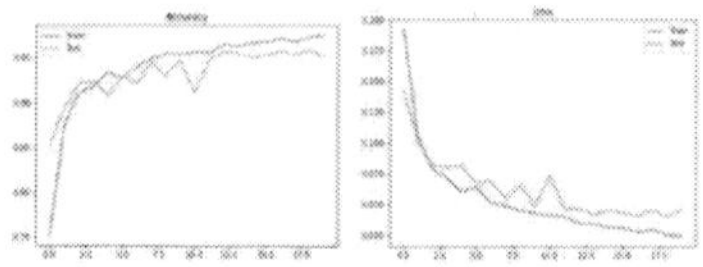

Fig.6. Graphs depicting the Accuracy and Loss for the trained model

Model	Accuracy
CNN + Sequential	94.297%

Table 2. Accuracy of trained model

5.CONCLUSION AND FUTURE WORK

We have implemented a machine learning model using convolution neural network to distinguish the presence of the diseases. Images with variable size have been used to train the machine learning model. Our model is able to detect whether the person is suffering from the above-mentioned diseases with an accuracy of 94.297%.

In the future, we will try to expand this application for other diseases which involve the use of image classification. We will also try out various other machine learning models to try and improve our model's accuracy.

ACKNOWLEDGMENT

We acknowledge the Principal and Head of Department of Electronics and Telecommunication Engineering, Ramrao Adik Institute of Technology for their motivation and support. We would also like to thank our guide for guiding us along the way.

REFERENCES

[1] Račić, L., Popović, T. and Šandi, S., 2021, February. Pneumonia Detection Using Deep Learning Based on Convolutional Neural Network. In 2021 25th International Conference on Information Technology (IT) (pp.1-4). IEEE.

[2] Jiang, Z., 2020, June. Chest x-ray pneumonia detection based on convolutional neural networks. In 2020 International Conference on Big Data, Artificial Intelligence

and Internet of Things Engineering (ICBAIE) (pp. 341-344). IEEE.

[3] Rajendra Tharu1 Ranjan KC2, Md. Irfan musalman3, Shibam Mallick4, Ganesha M5, 2020, July, Brain Tumor Detection and Classification Using Machine Learning In the 2020 International Journal of Innovative Research in Science, Engineering and Technology (IJIRSET) Volume 9, Issue7

[4] Suresh Babu, Jennifer.J, Hesima.C, Preethi.S 2020 March Lung Cancer Detection using GLCM and

Convolutional Neural Network In the 2020 International Journal of Innovative Research in Science, Engineering and Technology (IJIRSET) Volume **7**, Issue 3

[5] Hossain, T., Shishir, F.S., Ashraf, M., Al Nasim, M.A. and Shah, F.M., 2019, May. Brain tumor detection using convolutional neural network. In 2019 1st international conference on advances in science, engineering and robotics technology (ICASERT) (pp. 1-6). IEEE.

[6] Badran, E.F., Mahmoud, E.G. and Hamdy, N., 2010, November. An algorithm for detecting brain tumors in MRI images. In The 2010 International Conference on Computer Engineering & Systems (pp. 368-373). IEEE.

[7] Rajpurkar, P., Irvin, J., Zhu, K., Yang, B., Mehta, H., Duan, T., Ding, D., Bagul, A., Langlotz, C., Shpanskaya, K. and Lungren, M.P., 2017. Chexnet: Radiologist-level pneumonia detection on chest x-rays with deep learning. arXiv preprint arXiv:1711.05225.

[8] Sharma, D. and Jindal, G., 2011. Identifying lung cancer using image processing techniques. In International Conference on Computational Techniques and Artificial Intelligence (ICCTAI) (Vol. **17**, pp. 872-880).

[8] Mukherjee, S. and Bohra, S.U., 2020, December. Lung Cancer Disease Diagnosis Using Machine Learning Approach. In 2020 3rd International Conference on Intelligent Sustainable Systems (ICISS) (pp. 207-211). IEEE.

[9] Rahane, W., Dalvi, H., Magar, Y., Kalane, A. and Jondhale, S., 2018, March. Lung cancer detection using image processing and machine learning healthcare. In 2018 International Conference on Current Trends towards Converging Technologies (ICCTCT) (pp. 1-5). IEEE.

[10] Radhika, P.R., Nair, R.A. and Veena, G., 2019, February. A comparative study of lung cancer detection using machine learning algorithms. In 2019 IEEE International Conference on Electrical, Computer and Communication Technologies (ICECCT) (pp. 1-4). IEEE.

[11] Paliwal, G. and Kurmi, U., 2021, December. A Comprehensive Analysis of Identifying Lung Cancer via Different Machine Learning Approach. In 2021 10th International Conference on System Modeling & Advancement in Research Trends (SMART) (pp. 691-696). IEEE.

[12] Sravya, V. and Malathi, S., 2021, January. Survey on Brain Tumor Detection using Machine Learning and Deep Learning. In 2021 International Conference on Computer Communication and Informatics (ICCCI) (pp. 1-3). IEEE.

[13] Jahan, R. and Tripathi, M.M., 2021, June. Brain Tumor Detection Using Machine Learning in MR Images. In 2021 10th IEEE International Conference on Communication Systems and Network Technologies (CSNT) (pp. 664-668). IEEE.

[14] Someswararao, C., Shankar, R.S., Appaji, S.V. and Gupta, V., 2020, July. Brain tumor detection model from MR images using convolutional neural network. In 2020 International Conference on System, Computation, Automation and Networking (ICSCAN) (pp. 1-4). IEEE.

[15] Gaikwad, S., Patel, S. and Shetty, A., 2021, May. Brain Tumor Detection: An Application Based on Machine Learning. In 2021 2nd International Conference for Emerging Technology (INCET) (pp. 1-4). IEEE.

[16] El-Feshawy, S.A., Saad, W., Shokair, M. and Dessouky, M., 2021, July. Brain Tumour Classification Based on Deep Convolutional Neural Networks. In 2021 International Conference on Electronic Engineering (ICEEM) (pp. 1-5). IEEE.

[17] Siar, M. and Teshnehlab, M., 2019, October. Brain tumor detection using deep neural network and machine learning algorithm. In 2019 9th International Conference on Computer and Knowledge Engineering (ICCKE) (pp. 363-368). IEEE.

[18] Labhane, G., Pansare, R., Maheshwari, S., Tiwari, R. and Shukla, A., 2020, February. Detection of pediatric pneumonia from chest X-ray images using CNN and transfer learning. In 2020 3rd International Conference on Emerging Technologies in Computer Engineering: Machine Learning and Internet of Things (ICETCE) (pp. 85-92). IEEE.

[19] Panwar, A., Yadav, R., Mishra, K. and Gupta, S., 2021, July. Deep learning techniques for the real time detection of Covid19 and pneumonia using chest radiographs. In IEEE EUROCON 2021-19th International Conference on Smart Technologies (pp. 250-253). IEEE.

[20] Nag, T., Rajawat, S.S. and Rana, A., 2021, September. Detection of Pneumonia using Chest X-Ray Images and Convolutional Neural Network. In 2021 9th International Conference on Reliability, Infocom Technologies and Optimization (Trends and Future Directions)(ICRITO) (pp. 1-4). IEEE.

Convolutional Neural Network Based Traffic Sign Recognition System for Simultaneous Classification of Static and Dynamic Images

Mrunal Padalkar[1, a], *Tejal* Shirsat[1, b], *Snehal* Vekhande[1, c], and *Ekta* Sarda[2, d]

[1]Department of Computer Engineering, Ramrao Adik Institute of Technology, D Y Patil Deemed to be University, Navi Mumbai, India
[2]Asst Prof, Department of Computer Engineering, Ramrao Adik Institute of Technology, D Y Patil Deemed to be University, Navi Mumbai, India

Abstract: Traffic symbols are crucial part of the road infrastructure which are erected at the side of the roads that communicates basic instructions of the road with the help of simple visual graphics which can be understood in no time. As compared to previous decades, traffic congestion is a major issue faced in densely populated cities. Unfortunately, drivers may not notice these traffic signs due to adverse traffic conditions or ignorance which may cause accidents. Therefore, building an intelligent traffic sign recognition model is the need of the time. Besides contributing to the safety and comfort of drivers, traffic symbols recognition has important benefits for autonomous vehicles. In this paper, we have used simple CNN technique to recognize static as well as dynamic traffic symbols on the German Traffic Symbol Recognition Benchmark (GTSRB) dataset which has more than 40 classes of traffic symbols in different orientation and lightning condition. Further, a comparison of performance of CNN on static and dynamic input was done and the efficiency was compared. The experimental results show that the detection rate of the CNN model on static images is 96.15% which significantly higher than that of on dynamic images, which resulted in 95% accurate.

Keywords— CNN, GTSRB, detection, classification.

[a] Corresponding author: mrunalvilas@gmail.com
[b] Corresponding author: 10tejalshirsat@gmail.com
[c] Corresponding author : Snehalvekhande10@gmail.com
[d] Corresponding author : Ekta.sarda@rait.ac.in

1 INTRODUCTION

Road signs or Traffic symbols play extremely pivotal role in facilitating road traffic and increasing road safety through uniform adoption of traffic rules. They also provide special road conditions like curvature, no passing of heavy vehicles, speed limits etc. Unfortunately, drivers may not notice these signs due adverse traffic conditions or ignorance which may cause accidents. Therefore, building an intelligent traffic symbols recognition model is the absolute necessity. The main intention of these models is to provide crucial information for the drivers, in order to minimize their effort and hence result in safe driving. For this, various advance technologies like ML and Artificial Neural Networks are used. Moreover, with advancements and research in the field of Artificial Intelligence various giant companies are developing self-driving cars. For achieving full accuracy and success in this the car needs to be able to interpret the traffic symbols and make proper decisions. Speed Limit, Traffic Signals, No Entry, Direction Indication are some of the many traffic signals we come across usually. Apart from this Traffic sign Recognition has many other real time applications like Self driving cars, Traffic surveillance, Road Network Sustenance, etc. so to build a model that will be able to identify these signs is the motivation of this paper. In this Paper, we have built a Deep Neural Network model that classifies given static and dynamic traffic symbols into different classes. In addition, comparison is made based on efficiency of model on two different types of inputs i.e., Static Images and Dynamic Real-Time Images.

2 LITERATURE SURVEY

Traffic symbol detection and recognition is a system that is created to reduce accidents [3]. The system uses camera to capture the image, which is then processed by an image processor, classifier, and detector before being converted to audio. They used German datasets to classify them using faster R-CNN algorithm. [1] This paper has presented an experimental comparison analysis of eight deep neural network-based traffic sign models. They looked at the most important features of specific detectors, like detection time, precision, accuracy, the amount of floating-point operations in CNN and SPP, as well as the area of the workspace. They developed a real-time system for traffic road sign recognition using SVM algorithm for autonomous vehicles in their work "India Traffic Road Sign Recognition for Intelligent Driver Assistance System using SVM" [4]. In daylight settings, the SBIV function of their system performed effectively in identifying traffic road signs. In the published work [6], the authors Amara Dinesh Kumar R. Karthika Latha Parameswaran has created a model that uses a capsule network to detect pose and spatial variances more effectively than CNN. Even on blurred, rotated, and distorted images the detection performed better. The model presented in the paper [5] enables real-time, high-resolution video processing. This model was developed on the Nvidia K1 CPU, which had CUDA for performance acceleration. A parallel window searching based on GPGPU, which is a real time traffic symbol recognition system for input images with high resolution, was presented in the paper [2]. For stable recognition irrespective of illumination variation, this model includes Byte-MCT. SVM and CNN classifiers were used. In the paper "Traffic Sign Recognition with Convolution Neural Network Based on Max Pooling Positions" [7], a CNN model was created to develop a compact but containing discriminative feature characterization. They have also developed a recognition approach based on MPPs that enhances classification performance and speed. Using MPP's, a unique strategy for improving classification performance and speed has been developed.

3 PROPOSED METHODOLOGY

This paper has proposed a traffic symbol recognition system that will classify both static and Real-Time images using a Deep convolution neural network. An efficient Traffic symbol interpretation system based on CNN is proposed. First, the CNNs a GTSRB ("German Traffic Sign Recognition Benchmark") dataset, which had images that are labelled, rotated and in different lighting conditions. Later, our method performs classification of traffic symbols using CNN; finally (to make it work more efficiently) this model is additionally enhanced to recognize and classify Dynamic Real-Time Images. In comparison to these two kinds of inputs, Static Images when tested by the model provide more efficient outcomes.

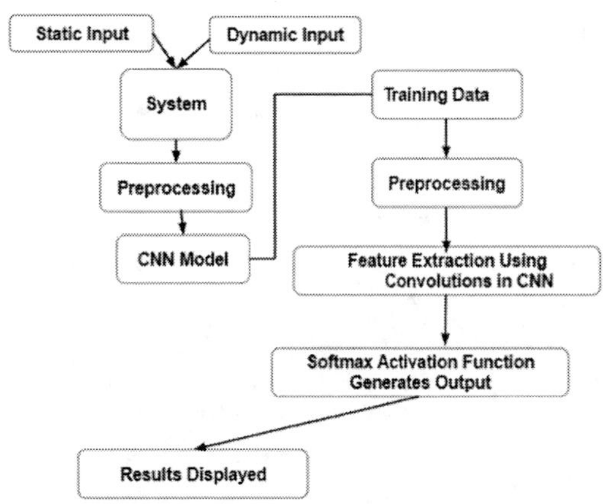

Fig. 1. Proposed Methodology

Fig. 1 Represents Proposed Methodology diagram. In this proposed system, Data Pre-processing is performed for making input images suitable for analysis and operation. Data Pre-processing is different for static and dynamic input. CNN is used for feature extraction, SoftMax layer of CNN is used for classification results.

Paper began by gathering accurate and appropriate datasets for this model. The GTSRB dataset, which included approximately 50,000 images and 43 classes, was chosen. Various data exploration techniques were then applied to this data, which was then visualized to gain visual insights. In addition, we used CNN to create a classification model. The model is then validated and trained, and a test is run on the developed model. CNN is one of the main technologies used in Neural Networks to perform image identification and image segregation. The classification of images in CNN follows a simple logic: Create hierarchical representations of the data, finding features from a number of images provided and then identifying these features to classify unseen images in specific classes. In the case of Deep Learning CNN models for testing and training of input images, each image is processed through a series of convolutional layers with Kernels, fully connected layers (FC), Pooling, and SoftMax function is applied to these images to classify an object. Deep learning CNN models use probabilistic values ranging from 0 to 1 to express the likelihood of an object pertaining to a particular class. The evaluation of the convolutional layer is expressed in the following equation:

$$a^k = f\left(\sum_{i.j=1}^{3} W^k{ij} \times X_{ij} + b^k \right) \quad (1)$$

Where,

W_{ij} ^{k} is the weight value filter of convolution,

X_{ij} is the pixel value convolution filter used previously,

b ^{k} is the bias

and the activation function is f(x)

Each step of Proposed methodology is discussed below in brief detail.

3.1. Data Pre-processing

Data Pre-processing is a procedure that is performed on data to format it in the best possible format for subsequent processing. Preprocessing ensures that data from resources is in a common and appropriate format. Because our input data is of two different kinds i.e., Static and Dynamic, it is essential for us to ensure that each image in the dataset has a common dimension; dealing with images of varying sizes is not a feasible choice. As a result, we resized the entire input to a fixed size of 30 x 30. Furthermore, we appended the appropriate labels to the image and converted the image content into an array for submitting to the model. Hence, the data dimension is (39208,30,30,3), and there are 39208 images with dimensions of 30 x 30 and they are colored with RGB (3) channels. To gain a better understanding of the dataset, we used exploratory data analysis techniques to visually summarize it. Because the dataset contains many attributes, the set of correlation values between pairs of its attributes is represented in the correlation matrix, and a heat map of this correlation matrix was plotted later to visually understand the attribute linearity. Data preprocessing simplified subsequent processing and gave us an understanding that the data was enough for applying any deep learning algorithm.

3.1.1. The human brain does not interpret an image by pixel, but rather processes visual signals through a structure of multiple layers. As neural networks represent these structures well, we opted for neural networks in this model. An architecture with insufficient depth may necessitate the addition of many more computational elements. Therefore, CNN algorithm with increased depth and layer(pooling) for transition of invariant features is built. Because the fully connected layers include the majority of the network's parameters, they are frequently prone to overfitting. To prevent this dropout layer is added. Finally, as there are 43 different classes to categorize, the model is built using cross entropy measures.

3.1.2. After successfully building a hierarchical CNN model, our model was trained on training set, validation set, multiple epoch values, batches sizes, and 2 activation functions for the classifier. In case of Static images, the batch size of 32 outperformed other batch sizes with 10 epochs and on the other hand batch size of 50 with 10 epochs worked best for Real-Time images; the accuracy was stable. Training set gave an accuracy of 96.15 percent on Static input and 95% on Dynamic input. An analysis of the accuracy and the loss was conducted.

3.1.3. Finally, now that the model has been trained. A GUI model using Tkinter was created, which provides effective interface to classify static and Real-Time Traffic sign images. It takes as input static images and classifies the traffic sign present on them and also allows user to start Camera equipment to capture Real-Time scene and recognize Traffic Symbol from it.

3.1.4. In case of Dynamic Images, The live stream Detects the Traffic Symbol and Displays the Meaning and probability of Classification of that Traffic Symbol on the screen itself.

3.2 System Diagram for CNN Model

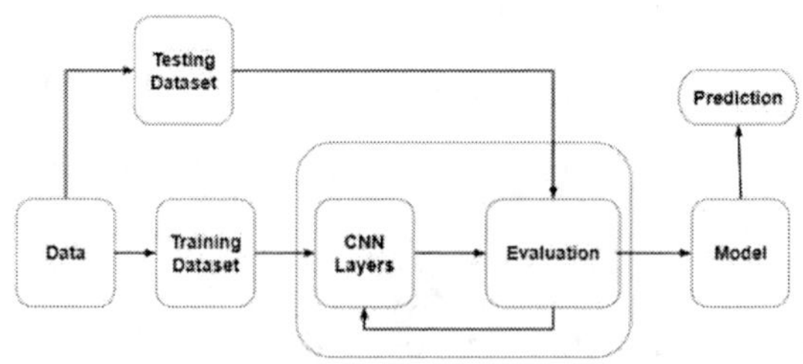

Fig. 2. System Diagram of CNN Model

Fig. 2 depicts an overview of the simple CNN model that was used. The initial dataset is divided into train, test and validate folders, and then images in the train folder are each passed through the CNN model layers, and the model with the help of convolution identifies the features and develops a feature map. It extracts these features, and finds interrelationships between them, before passing this feature extracted and reduced dimensional data through the activation function for final decision making. Having many convolution layers in the network allows the network to Develop representations of objects in the images from simple features to more complicated features and up to sensitivity to distinct categories of objects. When the data from the test folder is tested against the trained model, the activation function evaluates the probability of each test image belonging to one of the many categories of traffic symbols, and the one with the highest probability is displayed in the frontend screen and a prediction is made.

3.3 System Diagram for Classification of Static Images

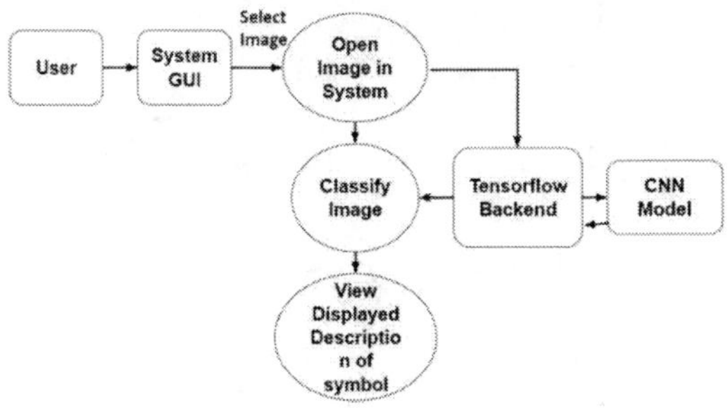

Fig. 3. Proposed System Diagram For Classification of Static Images

The Fig. 3 depicts our system's diagram for classification of static images. To use the system, the user must first start up the TensorFlow backend. Once the backend is activated from the command line, the GUI loads and the user can see an upload button and a button for Real-Time classification, upon pressing upload image button the user can input image through file system in his/her system and can upload any image in .png or .jpg format upon loading the input image in the system classify button will appear upon pressing which the user activates the CNN Model through TensorFlow backend and the image is resized to 30 x 30 image and then fed to the model. After processing the image, the model generates a list of decimal values indicating the probability that the image belongs to one of the 40+ classes. The SoftMax function then chooses the class with the highest probability, and thus a static traffic symbol image is successfully interpreted by our simple CNN Based system in no time.

3.4 System Diagram for Classification of Dynamic Images

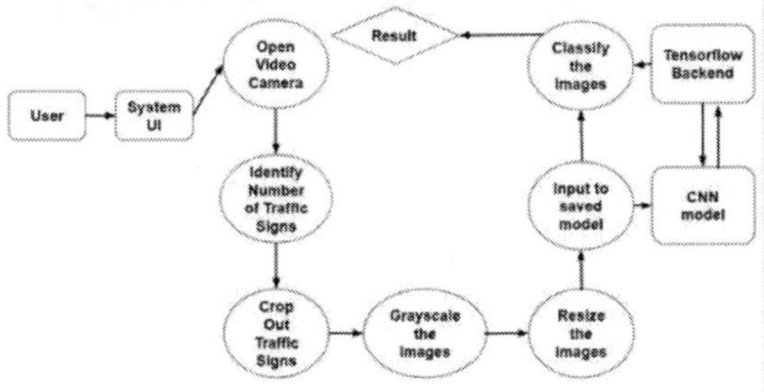

Fig. 4. Proposed System Diagram for Classification of Dynamic Images

Fig. 4 represents details For Dynamic or Real-Time Recognition, Classify Video button provided on System User Interface must be hit for dynamic or real-time recognition, which Turns the camera Equipment of the System the user is using, allowing a live scene to be taken. Red Rectangles indicate which region is identified as traffic sign. The class the Probability value of that particular image is displayed on the screen. The camera will continue to collect the live footage until the user tells it to stop. What happens in the background for real-time traffic sign classification is, when user starts the live image capturing, First Identification of Number of Traffic Signs present on the screen is performed, which is followed by cropping out the required region containing Traffic sign is done, which is then followed by preprocessing of that required region and feeding that image frame to the model is performed, this entire process results in detection of the traffic symbol from live stream and displaying the meaning and probability of classification of that traffic symbol on the screen.

4 RESULTS AND SIMULATIONS

4.1 Stages of Implementation

Stage 1:

Gathering the Input data to work on is done in this stage 1.

Stage 2:

In stage 2, for Static Input we use PIL library to open image content into array. For Dynamic Input OpenCV library is used. Now that we have images in the form of a list i.e., data and labels, we convert it to array for further processing. The system then summarizes and analyze the data using Data Analysis techniques. The images are in colored form. The variable explorer shows that we have 39208 (pictures) with a height of 30 px, a width of 30 px, and in RGB color format.

	Width	Height	Roi.X1	Roi.Y1	Roi.X2	Roi.Y2	ClassId	Path
0	27	26	5	5	22	20	20	Train/20/00020_00000_00000.png
1	28	27	5	6	23	22	20	Train/20/00020_00000_00001.png
2	29	26	6	5	24	21	20	Train/20/00020_00000_00002.png
3	28	27	5	6	23	22	20	Train/20/00020_00000_00003.png
4	28	26	5	5	23	21	20	Train/20/00020_00000_00004.png

Fig. 5. Head Function on GTSRB Dataset

	Width	Height	Roi.X1	Roi.Y1	Roi.X2	Roi.Y2	ClassId	Path
39204	52	56	5	6	47	51	42	Train/42/00042_00007_00025.png
39205	56	58	5	5	51	53	42	Train/42/00042_00007_00026.png
39206	58	62	5	6	53	57	42	Train/42/00042_00007_00027.png
39207	63	69	5	7	58	63	42	Train/42/00042_00007_00028.png
39208	68	69	7	6	62	63	42	Train/42/00042_00007_00029.png

Fig. 6. Tail Function on GTSRB Dataset

Fig. 5 and 6 show the details of the dataset by revealing the starting and ending 5 rows, respectively, as well as its characteristics. The dataset has 39208 rows and 8 columns.

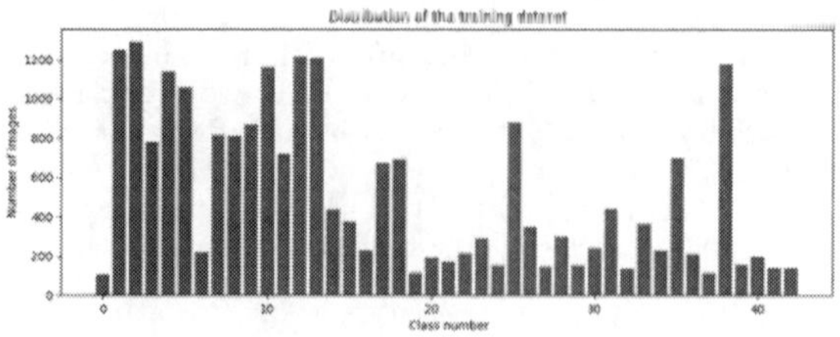

Fig. 7. Bar graph representing number of images per class

Fig.7 Represent the Distribution of Images among 43 classes. It can be seen that the distribution is Un-even and hence we might get good classification for one class and bad classification of other. It is Observed that the class with minimum 500 images had had great classification accuracy. Classes with less than 500 images receive poor classification accuracy.

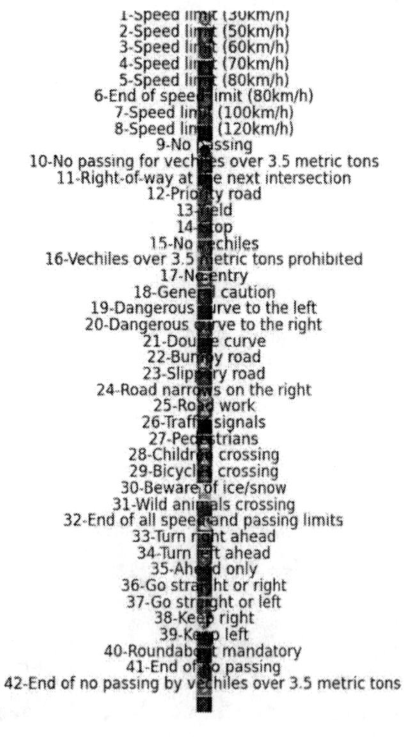

Fig. 8. Classes with their corresponding images

Fig. 8, gives a brief Idea about all the Images Present in the Dataset with Their corresponding classes marked by Class-ID. Division of data into train, test and validate set is performed. Neural networks expect the labels of classes in a dataset to be organized in a one-hot encoded manner. Conversion of labels into one hot encoding is also executed.

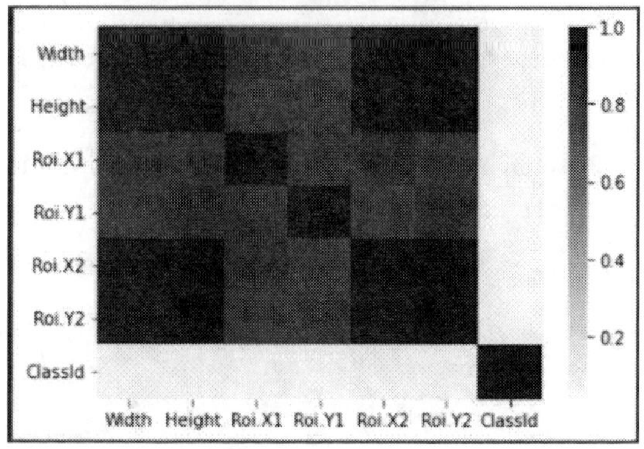

Fig. 9. Relationship Analysis using Heatmap

Summarization of the data for more advanced analysis is provided by correlation matrix. Fig. 9 depicts a heatmap-based visual representation of the correlation matrix. The data in a Two-Dimensional form is represented using heatmap. Heatmap is serves as a way of representing the data values in colored graph.

Stage 3:

Data Preprocessing is performed at stage 3. First Images are converted to Grayscale to reduce the dimensionality. As we all know grayscale images have only one channel this helps the model to focus on important parameters. Gray scaling is followed by Equalization phase. Equalizing is important to ensure that all images have standard Lighting. Normalization is last step in Preprocessing, here Values are normalized by dividing each image by 255, so resulting values are between 0 and 1 instead of 0 to 255.

Fig. 10. Augmented Images

Stage 4:

Augmentation of Image is performed in this stage 4, to ensure that the images are generic and not plain. Images are Shifted left and right, zoomed in and rotated. Fig.10 Represents Augmented Traffic Sign Images.

Stage 5:

In Stage 5, CNN Model is Built from Various Layers. Keras was used to create and train the neural networks to classify the images. This model will be of Sequential Type, meaning output of one layer will be fed as input to next layer. Convolutional networks for classification are constructed for two purposes: for image processing and for readout from a sequence of convolutional layers and fully connected layers. Few Convolution layers, few pooling layers and dropout layers and stacked and at the end a Dense layer is used as an Output layer with 43 possible outputs.

Model: "sequential"

Layer (type)	Output Shape	Param #
conv2d (Conv2D)	(None, 28, 28, 60)	1560
conv2d_1 (Conv2D)	(None, 24, 24, 60)	90060
max_pooling2d (MaxPooling2D)	(None, 12, 12, 60)	0
conv2d_2 (Conv2D)	(None, 10, 10, 30)	16230
conv2d_3 (Conv2D)	(None, 8, 8, 30)	8130
max_pooling2d_1 (MaxPooling2	(None, 4, 4, 30)	0
dropout (Dropout)	(None, 4, 4, 30)	0
flatten (Flatten)	(None, 480)	0
dense (Dense)	(None, 500)	240500
dropout_1 (Dropout)	(None, 500)	0
dense_1 (Dense)	(None, 43)	21543

Total params: 378,023
Trainable params: 378,023
Non-trainable params: 0

Fig. 11. Summary of CNN Model

Fig. 11 represents the summary of the CNN model that was created. A deep convolutional neural network is a network that has multiple layers. Each layer in a deep network receives its input from the layer before it, with the very first layer receiving its input from the images used as training or test data. We used a stack of 11 neural networks for extracting features and for making model learn their inter relationships.

Stage 6:

The model is trained and validated with multiple batch sizes in stage 6, training set and validation set. The model performs better with 32 batch size for static images and 50 for dynamic images than with other batch sizes, and the accuracy is stable when the epoch is set to 10.

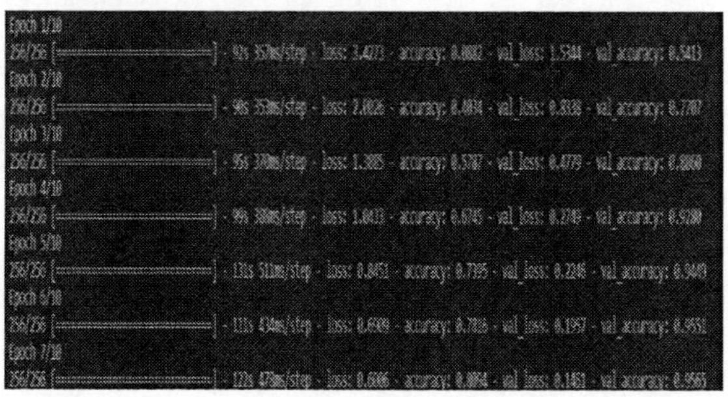

Fig. 12. Each Iteration Details

Fig.12 Represents the Accuracy and loss of the model at each Iteration. It can be seen that accuracy is increasing with each epoch.

Stage 7:

Now that the data is trained, In stage 7 we tested our model. We begin by resizing the height and width to 30x30 and creating a NumPy array of the test images. The accuracy score is imported. Lastly, we save the model. Data visualization of accuracy and loss is done. During learning process, the network observes change in weights, and as weights change, the network becomes more adjustable to the features of the images that differentiate between classes. This means that the used loss function becomes smaller and smaller. Observing the change in the loss with learning can be useful to see whether learning is progressing as expected. For progressive learning loss function must go down.

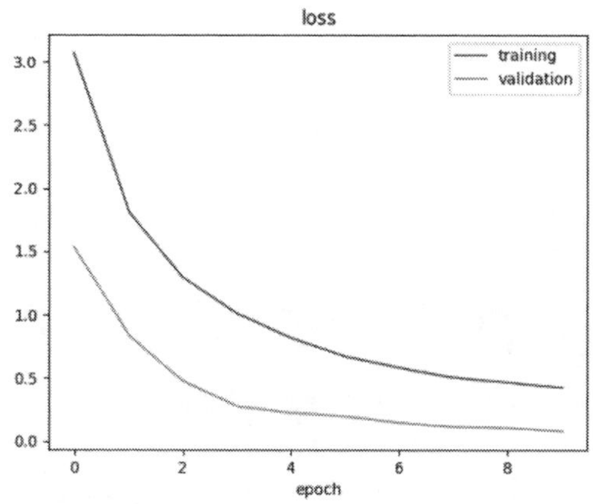

Fig. 13. Plot of training set loss and validation set loss

Fig. 13 is a curve showing the categorical cross-entropy loss in a network that is learning to classify different types of traffic signs, measured on the training set and validation set. You can see that there are rapid decreases in loss in the first few epochs of training, after which learning slows down. This is a sign that learning is going rather well. For validation loss learning is progressing in similar fashion and it never flattens out, this means when validated against a separate set of data, loss becomes better and smaller.

https://doi.org/10.1051/itmconf/20224403051

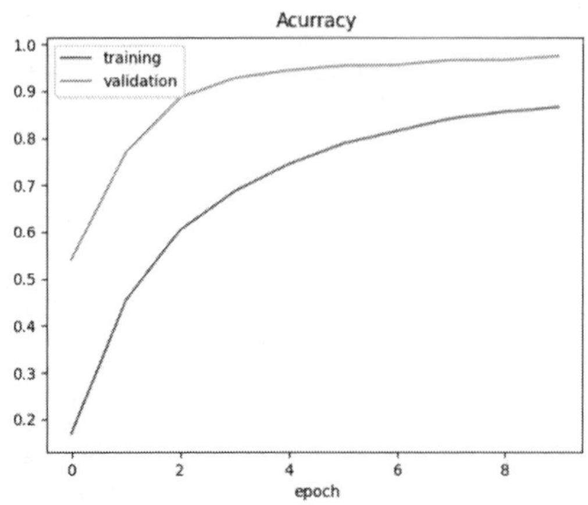

Fig. 14. Plot of training set accuracy and validation set accuracy

Given Figure 14 is a curve showing the accuracy in a network that is learning to classify different types of traffic signs, measured on the training set and validation set.

Stage 8:

Stage 8 represents the GUI application for the model. This model is built using Tkinter module.

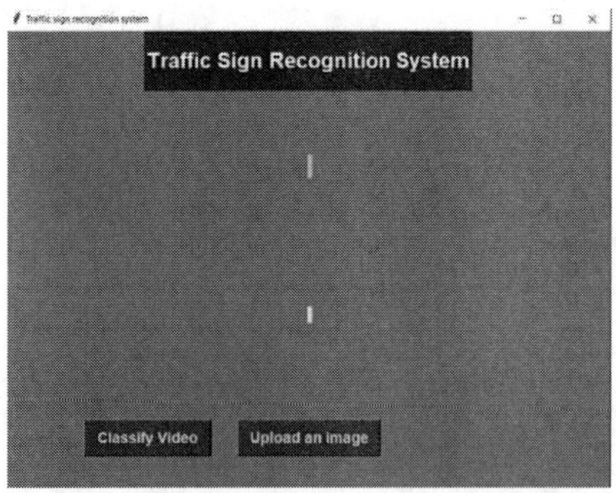

Fig. 15.1 GUI Of proposed System

The GUI of the deep learning model is shown in figure 15 and its sub-figures. As shown in Fig. 15.1, The system has two buttons: upload an image and classify video. The Upload an Image button will take us to the test folder, where we can select any random image from a pool of over 40 classes. After uploading the image, when you click the classify image button, the image is correctly classified. For example, if we have uploaded a general caution road sign in the GUI the system will correctly predict its label. If user opted for Classification of Video option, Camera Equipment starts to capture the live scene and Traffic Symbol is detected and classified in Real-Time Environment. OpenCV Library is Used to set the Camera Parameters for inputting live feed. The accuracy fluctuates in the training set, but it is stable in the validation set.

Fig. 15.2 Classification Result of Dynamic Image

Fig 15.2 Displays the results of Dynamic Image Classification, the traffic sign is captured live from camera feed and the class of that particular traffic sign is shown with its probability. Hence, we can conclude that proposed system is able to detect traffic sign in a live video and the traffic sign is highlighted by red rectangle. The class of that image is displayed on top left corner along with its probability percentage are displayed.

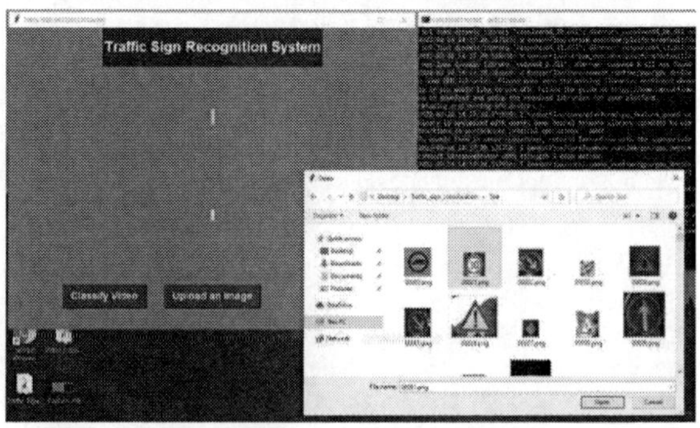

Fig. 15.3 Input static image from proposed system

Fig 15.3 Displays how system allows for selection of static images from system. User can upload one image at a time. The image formats supported by system are .png and .jpg

 https://doi.org/10.1051/itmconf/20224403051

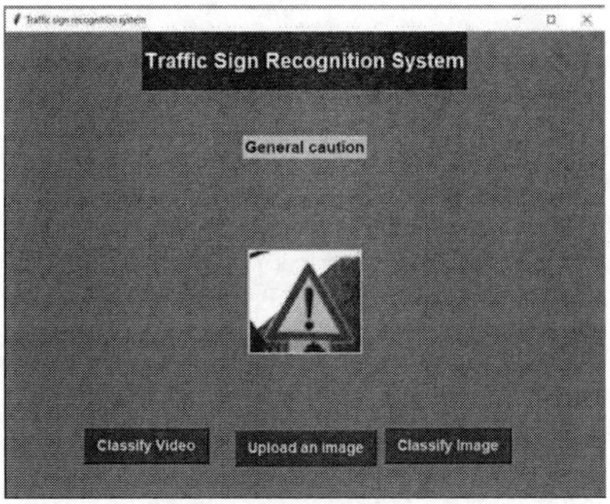

Fig. 15.4 Classification Result of Static Image

Fig 15.4 shows result of static image recognition. The chosen traffic sign is displayed on the screen along with the class it belongs to.

4.2 Results:

Table 1. Classification Accuracy of CNN Model on static and Dynamic input

Input Type	Accuracy
Static Image	96.15%
Real-Time Dynamic Images	95%

As shown in Table 1, Classification accuracy for static input is 96.15% and, Classification accuracy for Dynamic input is 95%. Accuracy for static input is more than that of dynamic image input.

5 CONCLUSION

Thus, This Paper has been successful in Development of an efficient Deep CNN Based traffic sign recognition system for classification of both Static and Dynamic Input. The color information present in the image, as well as the dimensional property of the road symbols, are used to classify the identified traffic symbols. According to the implementation, the system can grab a high recognition rate of 96.15 percent for static images and 95% for Real-Time Images. The system delivers accurate results in different weather, lighting, and daylight conditions This System is relevant for static as well as dynamic images, but accuracy of dynamic image classification can be improved, which becomes the Future scope for this paper.

6 REFERENCES

1. Dewi, C., Chen, R. C., & Tai, S. K. (2020). Evaluation of robust spatial pyramid pooling based

on convolutional neural network for traffic sign recognition system. Electronics, 9(6), 889.

2. Lim, K., Hong, Y., Choi, Y., & Byun, H. (2017). Real-time traffic sign recognition based on a

general purpose GPU and deep-learning. PLoS one, 12(3), e0173317.

3. Anju, C. P. Traffic Sign Detection and Recognition.

4. Sathya, R., Thiruvenkatasuresh, M. P., Arieth, R. M., Revathy, G., & Babu, M. D. V. (2020).

INDIAN TRAFFIC ROAD SIGN RECOGNITION FOR INTELLIGENT DRIVER ASSISTANCE SYSTEM USING SVM. Journal of Critical Reviews, 7(9), 3177-3185.

5. Shustanov, A., & Yakimov, P. (2017). CNN design for real-time traffic sign recognition. Procedia engineering, 201, 718-725.

6. Amara Dinesh Kumar R.Karthika Latha Parameswaran (2018) .Novel Deep Learning Model for Traffic Sign Detection Using Capsule Networks.

7. Qian, R., Yue, Y., Coenen, F., & Zhang, B. (2016, August). Traffic sign recognition with convolutional neural network based on max pooling positions. In 2016 12th International conference on natural computation, fuzzy systems and knowledge discovery (ICNC-FSKD) (pp.578-582). IEEE.

8. Sheikh, M. A. A., Kole, A., & Maity, T. (2016, October). Traffic sign detection and classification using colour feature and neural network. In 2016 International Conference on Intelligent Control Power and Instrumentation (ICICPI) (pp. 307-311). IEEE.

9. Kale, Amol Jayant, and R.C. Mahajan. "A road sign detection and the recognition for Driver Assistance Systems", 2015 International Conference on Energy Systems and Applications, 2015.

10. Fang, C.Y.. "An automatic road sign recognition system based on a computational model of human recognition processing", Computer Vision and Image Understanding, 200411.

11. Abderrahmane Adoui El Ouadrhiri, Jaroslav Burian, Said Jai Andaloussi, Rachida El Morabet, Ouail Ouchetto, Abderrahim Sekkaki. "Chapter 34 Fast-Tracking Application for Traffic Signs Recognition", Springer Science and Business Media LLC, 2018.

12. Jharna Majumdar, N R Giridhar, P E Gagan. "Autonomous Mobile Robot Navigation on Identifying Road Signs using ANN", 2019 10th International Conference on Computing, Communication and Networking Technologies (ICCCNT)

Copy Move and Splicing Image Forgery Detection using CNN

Devjani Mallick[1,*], *Mantasha* Shaikh[2], *Anuja* Gulhane[3] and *Tabassum* Maktum[4]
[1,2,3,4]Ramrao Adik Institute of Technology, D. Y. Patil deemed to be University, Navi Mumbai, Maharashtra, India.

Abstract. The boom of digital images coupled with the development of approachable image manipulation software has made image tampering easier than ever. As a result, there is massive increase in number of forged or falsified images that represent incorrect or false information. Hence, the issue of image forgery has become a major concern and it must be addressed with appropriate solution. Throughout the years, various computer vision and deep learning solutions have emerged with a purpose to detect forgery in case of digital images. This paper presents a novel approach to detect copy move and splicing image forgery using a Convolutional Neural Network (CNN) with three different models i.e. ELA (Error Level Analysis), VGG16 and VGG19. The proposed method applies the pre-processing technique to obtain the images at a particular compression rate. These images are then utilized to train the model and further the images are classified as authentic or forged. The paper also presents the experimental results of the proposed method and performance evaluation in terms of accuracy.

Keywords- *Image, forgery, CNN, ELA, VGG16, VGG19, CASIA v2.O, NC2016*

1 Introduction

Image forensics is field in technology which aims to categorize authentic and forged images using various techniques like CNN etc., which is otherwise impossible by naked human eye to make out the difference. The recent growth of research in this field is mainly due to number of free image manipulation software available to common man, which can lead to various crimes, riots, misunderstandings, etc. that might create turbulence in the smooth conduct of public order and safety of citizens. We have witnessed an unprecedented growth of digital images, a trend which has been made possible by the numerous devices around, such as smart phones and tablets along with cheap and fast internet. Moreover, the development of user-friendly image manipulation software that is available at reasonable prices, has made the manipulation of such content easier than ever. Three of the most common manipulations in literature are: (1) Copy-move forgery, in which a specific region from the image is copy pasted within the same image. (2) Removal, in which an image region is removed and the removed part is then in-painted. (3) Splicing, in which a region from an authentic image is copied into a different image.

There have already been numerous approaches in the literature that address the task of image forgery detection. The traditional computer vision approaches have focused mainly on developing algorithms that tackle some of the aforementioned sub-tasks in isolation. The more recent research has focused on applying deep learning techniques for detecting image forgeries. Furthermore, the Convolutional Neural Network can learn how to extract the necessary features irrespective of the manipulation method. However, while some of these methods seem to solve the image forgery detection problem, the accuracy achieved raises suspicions about their robustness. Studies have shown that CNN network architecture achieves an impressive accuracy of approximately 70% on the CASIA v2.0 dataset for detection of copy move forgery. Following that, the same network architecture is trained and tested on the Media Forensics Challenge 2016 dataset whose tampered images are significantly more difficult to recognize. Finally, the effect of hyper parameter tuning and data augmentation on the network performance is analyzed in the paper.

The main goal of this study is to evaluate how the performance of an image forgery detection using CNN varies based on the sample difficulty. In order to address this issue the classification pipeline is developed in the proposed method. The existing approaches achieve high accuracy as the CAISA datasets used to test their network are manipulated in a way that is easy to recognize by humans. Therefore, the behavior of such a CNN on a more challenging dataset is analyzed in this paper. The performance of the CNN model may degrade significantly and to validate this intuition, two datasets are selected which are used to train the CNN. The CASIA v2.0 dataset is chosen over the v1.0 as it is considered more challenging and also offers nearly seven times more samples. This is particularly important given that previous studies have shown that a high number of training samples is a requirement for a reasonably well-trained CNN.

The rest of the paper is organized as follows:
The section 2 gives the survey of existing systems for detection of image forgeries. The section 3 demonstrates the proposed system to detect copy move and splicing image forgery presented. The results of the proposed system are presented in section 4. Finally, the conclusion and future work is expounded in Section 5.

* Corresponding author: anuja.gulhane@gmail.com

2 Related Work

In the paper [1], a reliable deep learning-based approach for detecting image forgeries in the context of double image compression is presented. To train the model, the difference between the original and recompressed versions of an image is used. The forged image is recompressed, and the difference between it and the original image is determined. The forged part is now highlighted due to the difference in the source of the forged part and the original part of the image. The proposed approach is compact, and its performance reveals that it outperforms current techniques. The experiment's results are promising, with a validation accuracy of 92.23 percent overall.

The paper [2] proposes a system architecture based on ResNet50v2 as basic convolutional model with five stages initialised with YOLO CNN weights for the purpose of object detection and transfer learning. In this paper, it first performs batch normalisation and then it applies activation function to update and optimise the weights. This paper addresses the degradation problem by means of deep learning framework.

The paper [3] proposes a blind image splicing detection system that uses a backbone of deep convolutional residual networks, followed by a fully connected classifier network that distinguishes between genuine and altered images. The paper [4] presents "a robust copy move forgery classification using end to end convolution neural network", which provides a deep neural network-based technique with good outputs for categorizing photos based on whether they include any copy move forgeries. The following work seeks to classify all photos that have copy move forgeries and are scaled, rotated, or compressed at different levels. They proposed a novel CNN model that achieved a 93-95 percent accuracy with datasets of the same type or hybrid datasets.

A novel image splicing detection and localization approach based on deep convolutional neural networks is presented in [5]. The first layer of the CNN model is initialized using an optimal combination of the 30 fundamental high-pass filters used in the spatial rich model (SRM) for image steganalysis to suppress the effects of picture contents and extract more diversified and expressive residual features for RGB color images. The paper [6] proposes using a convolutional neural network to extract information from the spliced image automatically. After that, the discrete wavelet transform (dwt) is applied. Later, for classification, a support vector machine is used. Additional tests are carried out, in which the discrete cosine transform is used instead of dwt, and principal component analysis is used.

To deal with tampering and segmentation of visual data, paper [7] suggests a CNN model. To identify tampering, the model is designed to preserve local and global properties of data. Their approach is an inception-based module CNN network that uses mask r CNN to detect tampering. We demonstrated that our proposed technique can find out how to identify various photo tampering methods without relying on pre-selected highlights or any pre-preparing and segment out modified region through a series of tests. CAISA v1.0 and v2.0 datasets yielded 98.76 percent and 97.92 percent accuracy, respectively.

The goal of the paper [8] is to provide an efficient splicing detection and CMFD pipeline architecture whose major goal is to detect the traces left by various splicing and copy-move forgery post-processing operations, such as jpeg compression, noise addition, blurring, contrast adjustment, and so on. The accuracy of the COMOFOD and BOSSBASE datasets is 95.97% and 94.26 percent, respectively.

The paper [9] provides a unique strategy based on neural networks and deep learning to improve copy-move forgery detection, concentrating on the convolutional neural network architectural approach. To achieve satisfactory results, the suggested method uses a CNN architecture with pre-processing stages. The paper [10] uses a fusion processing technique that combines a deep convolutional model and an adversarial model to detect copy-move forgeries. A total of 4 datasets are used. The results show that the deep learning CNN and discriminator forgery detectors have a considerably high detection rate (95 percent). A two-branch design and a fusion module are used to build the network. Through CNN and GAN, the two branches are used to locate and identify copy-move forgery regions.

The paper [11] provides a new deep learning-based image fraud detection system for automatically learning hierarchical representations from input RGB color photographs. Image splicing and copy-move forgeries can be detected using the suggested CNN. The fundamental high-pass filter set employed in the spatial rich model (SRM) is utilized to establish the weights at the first layer of our network, which serves as a regularizer to efficiently suppress the effect of picture contents and capture the subtle artefacts created by the tampering operations. The pre-trained CNN is used as a patch descriptor to retrieve dense features from the test images, and the final discriminative features for SVM classification are obtained via a feature fusion technique. The suggested CNN-based model outperforms various state-of-the-art approaches, according to experimental results on multiple public datasets.

The authors in [12] offer a shallow convolutional neural network (SCNN) that can identify manufactured region boundaries from original edges in low-resolution images. SCNN was created to make use of chroma and saturation data. Two techniques based on SCNN, termed sliding windows detection (SWD) and fast SCNN, have been developed to detect and identify image forgery regions.

3 Proposed Work

The overall system architecture is shown in Figure 1. The major components of the proposed methodology are, pre-processing, error level analysis and CNN. These modules are elaborated in detail next in this section.

(a) Data pre-processing: pre-processing involves normalization of images. the purpose of normalization is to make sure that all the images have similar data distribution. For normalization, the whole dataset is resized into 128*128 pixels.

(b) Error level analysis: To convert an image to ELA, the pre-processed photos must first be resaved at a certain quality level. The image is whitened or brightened as a result of this technique. In order to resave the images, consider both genuine and falsified images that have been pre-processed and then resave photos at a specific compression level. Finally, Images that have been pre-processed and images that have been re-saved are compared to see how much of a difference there is. The modified parts of the image in the forged ELA-generated image are brighter than the corresponding original components. Having ELA in the preprocessing tends to have a huge advantage in further processing of the neural network as the ELA converted image contains only non-redundant information and has similar intensity contrast as compared to the nearby pixels. Due to this reason, our neural network training optimizes in only 8-9 epochs with a learning rate of 0.0001.

The next step involves changing the image size. We want each RGB value to lie between 0 and 1 so we divide each cell value by 255.0 to normalize the training of the neural network so that it converges faster. After that each image is categorically labelled where 0 stands for authentic image and 1 stands for forged image. Subsequently, the images are divided into two sets – 80% of the images are taken for the training set and the remaining 20%for the validation set for the working of the convolutional neural network.

(c) Convolutional neural network: Convolutional layer is used as feature extractor that learns feature representation. This is from the image that is input to CNN. Meanwhile, the pooling layer shrinks the convolution layer's output map and prevents overfitting. In general, there are stacks of numerous convolutional and pooling layers before the fully connected layer that serve to extract a more abstract feature representation. Pooling layer usually decodes the image (such as 2x2) in the aggregation into a single unit. The first layer of the CNN is a convolutional layer with a kernel size of 5x5 and a total of 32 filters. The second layer CNN consists of a convolutional layer with a kernel size of 5x5 and 32 filters, as well as a max pooling layer with a size of 2x2. After that, the max pooling layer adds dropouts for 0:25 to avoid overfitting. The flatten layer changes the feature map by flattening it into the feature vector and passing it to a fully connected layer after images are trained from a finetuned pre-processed model. In the last dense layer of models, the fully connected layer is employed for pattern recognition, and the SoftMax activation function is used to convert the feature vector in a probabilistic manner. The training set is compared to the test set using SoftMax activation, and a probability distribution on actual and forged images is returned.

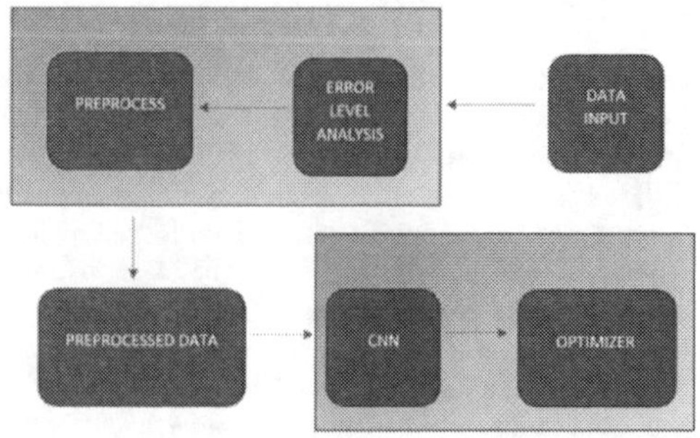

Figure 1. System Architecture

Optimization applied during the training is RMS Prop Optimizer in which the learning rate for each parameter is automatically adjusted without our intervention which helps in optimizing several parameters like number of features, number of training samples, target MSE, number of hidden layers, etc.

For the training of the neural network, two popular neural architectures have been utilized mainly VGG16 and VGG 19.

A. VGG16

VGG 16 is a CNN consisting of 22 neural layers out of which the end one is a SoftMax output classifier (4 dense layers at the end). The ELA converted training set images are then passed to finetuned VGG-16 model consisting of 16 layers to generate the model which is trained on a training set of 13,000 images out of which 6,500 are authentic and the rest are forged images.

VGG16 architecture consists of two convolutional layers which consists of mainly 64 channels of size 3 by 3 kernel, one maxpool layer of size and stride 2 by 2, two convolutional layers of 128 channels, one maxpool layer of size and stride 2 by 2, three convolutional layers of 256 channels, one maxpool layers of 2 by 2 pool, three convolutional layers of 512 channels and finally maxpool layers. RELU layers are added at each step so as to filter all the negative values from previous layers, which is finally passed on to dense layer to flatten the output and then add dense layers of 4096 units and dense softmax layers of 2 units. The SoftMax layer will show values between 0 and 1 based on the model that which category the input belongs to. So, with learning rate as 0.0001 and 12 epochs, from the confusion matrix we get the training accuracy for VGG16 as 94.4%.

B. VGG-19

VGG 19 is neural network having 24 layers with similar chronology distribution as compared to VGG 16. Pre-processed output after ELA is then passed on to VGG19 model which has total of 19 layers. Afterwards the images are classified based on SoftMax activation where training set id utilized to make comparison with the test set and return the probability distribution of different classed. The VGG19 architecture consists of two convolutional layers, one pooling layer, two more convolutional layers, again pooling layer, two more convolutional layers, and two more stacks of three convolutional layers separated by maxpool layer, followed by SoftMax layer at output. So, with learning rate as 0.0001 and 12 epochs, from the confusion matrix we get the training accuracy for VGG19 as 95%.

4 Results & Discussion

For training images, for every model, a mix of CASIA-2.0 and NC2016 dataset was considered. The dataset consists of RGB images from two these two datasets, which are split into training and testing datasets, and then passes on to our CNN models to classify them into two classes i.e. authentic and forged. First step is to divide the dataset into 2 categories: original and fake images. Therefore a total of 13000 images were trained using ELA, VGG16 and VGG19 model with ratio of authentic -tampered images being 50-50.And similarly for testing, a total of 1000 images with 500 authentic and 500 tampered were considered for each model. VGG16 Training result is shown in Figure 2.

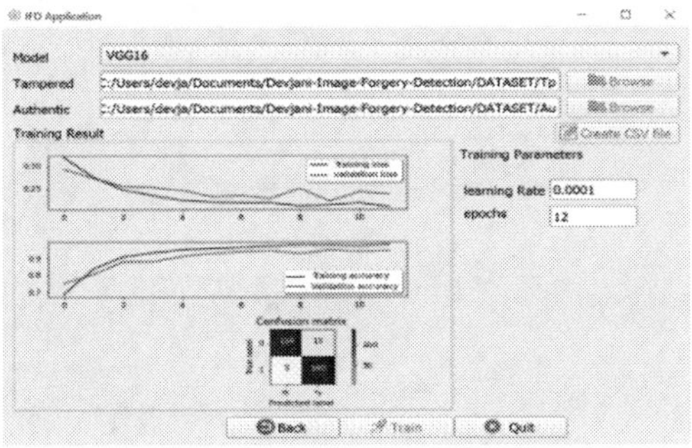

Figure 2. VGG16 Training Curve

So, with learning rate as 0.0001 and 12 epochs, from the confusion matrix we get the training accuracy for VGG16 as 94.4%. VGG19. The training result is shown in Figure 3.

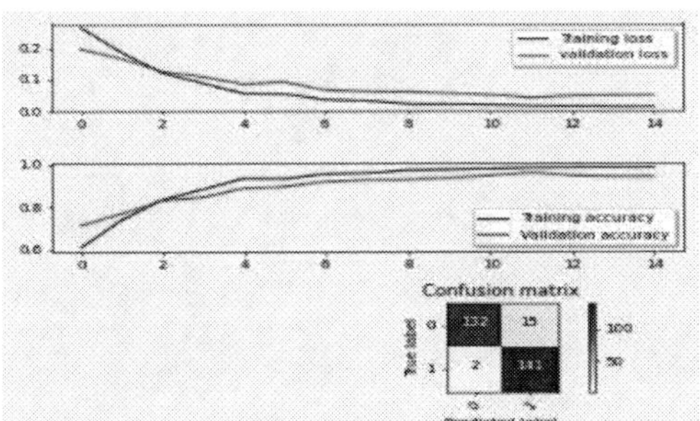

Figure 3. VGG19 Training Curve

So, with learning rate as 0.0001 and 12 epochs, from the confusion matrix we get the training accuracy for VGG19 as 95%. After training is completed, testing part begins on all three models i.e. ELA, VGG16 and VGG19. Also the threshold for all three models was kept as 50% , i.e. if CNN model predicts accuracy of authenticity as greater than 50% , then it will be labelled as authentic else forged. The example of one such test image on VGG16 model has been shown in Figure 4.

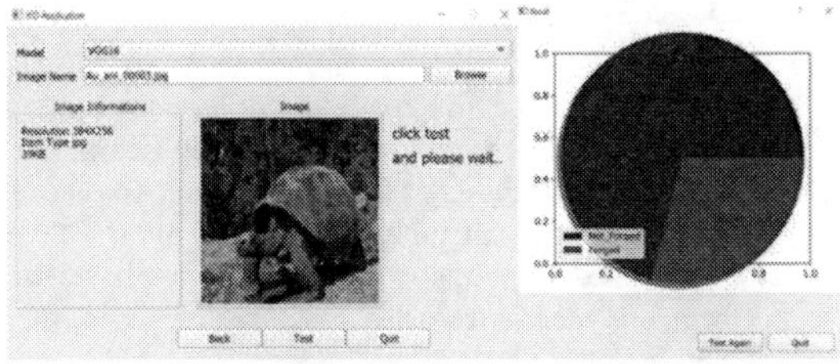

Figure 4. VGG16 Test Result

The overall comparison of different models along with their accuracy has been summarized in the Table 1. The accuracy of different models with respect to separate authentic and tampered images and hence overall accuracy have been presented with the help of bar chart in Figure 5.

Table 1:Comparison of Different Models

model	ELA	VGG 16	VGG 19
true positive	315	347	329
true negative	185	153	171
false positive	391	369	400
false negative	109	131	100
accuracy	70.6%	71.6%	72.9%

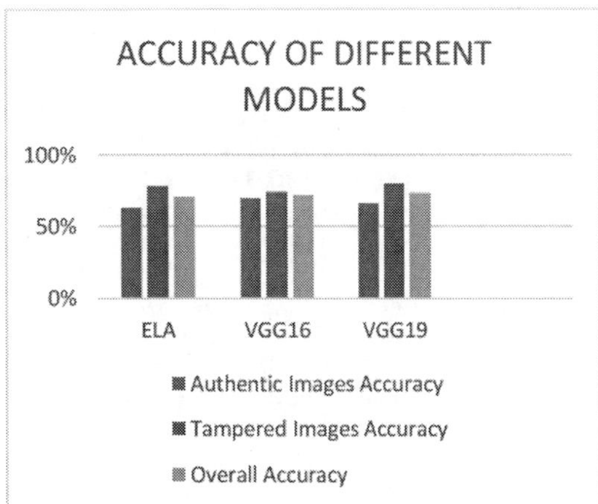

Figure 5. Accuracy comparison of different models

As can be inferred from the above bar-chart, VGG19 is the best model when it comes to accuracy of predicting forged images as forged whereas the same is true for

VGG16 in case of authentic images. The overall accuracy VGG19 comes out to be slightly better than VGG16 if not similar.

5 Conclusion

In this paper a method to detect copy move image forgery using CNN is presented. The proposed method uses CNN network to extract features from two datasets of varying difficulty, namely CASIA v2.0 and NC2016. The experimental results validate that the classification performance decreases when the samples are more challenging. However, the implemented architecture does not easily generalize to datasets with different underlying distributions. The proposed model is based on three different models i.e. ELA, VGG16, and VGG19 is able to achieve good results, with their respective accuracies as 70.6%, 71.6% and 72.9% , on a set of images from CASIA2.0 and NC2016 datasets. Nevertheless, there is surely a lot of work still to be done in the image forgery detection domain and neural networks will be able to detect tampered images regardless of their difficulty. In future there is scope to improvise the VGG19 training model by increasing the dataset and using high computational power systems.

References

[1] S.S. Ali, I.I. Ganapathi, N.S. Vu, S.D. Ali, N. Saxena, N. Werghi, *Image Forgery Detection Using Deep Learning by Recompressing Images*, Electronics 11(3), 403 (2022).

[2] E.U.H. Qazi, T. Zia, A. Almorjan, *Deep Learning-Based Digital Image Forgery Detection System*, Appl. Sci., 12, 2851 (2022).

[3] S. Nath, R. Naskar, *Automated image splicing detection using deep CNN learned features and ANN-based classifier*, Springer-Verlag London ltd., part of springer nature (2021).

[4] S. Kumar, S.K. Gupta, *A robust copy-move forgery classification using end to end convolution neural network*, 8th international conference on reliability, Infocom Technologies and optimization (Trends and Future Directions) (ICRITO) (June 2020).

[5] Y. Rao, J. Ni, *Deep learning local descriptor for image splicing detection and localization*, IEEE access, v**ol. 8** (2020).

[6] Eman I. Abd El-Latif, A. Taha, Hala H. Zayed, *A passive approach for detecting image splicing using Deep Learning and Haar Wavelet Transform*, Arabian journal for science and engineering, **vol. 6** (2020).

[7] S. Saleem, A. Dilawariand U.G. khan, *Multimedia Forensic: An approach for splicing detection based on deep visual features*, 2019 international conference on robotics and automation in industry (ICRAI) (2019).

[8] R. Thakur, R. Rohilla, *Copy-Move forgery detection using residuals and convolutional neural network framework: a novel approach*, 2019 international conference on robotics and automation in industry (ICRAI) (2019)

[9] Y. Abdalla 1, M.T. Iqbal, M. Shehata, *Convolutional Neural Network for copy-move forgery detection*, conference: computer vision and pattern recognition Doi: 10.11.19/ICPR (Nov 2019).

[10] Y. Abdalla 1, M.T. Iqbal, M. Shehata, *Copy-Move forgery detection and localization using a generative adversarial network and convolutional neural network*, article - department of computer science, math, physics, and statistics, university of British Columbia, Kelowna, BC v6t 1z4, Canada (2019).

[11] Y. Rao, J. Ni, *A deep learning approach to detection of splicing and copy-move forgeries in images*, 2016 IEEE international workshop on information forensics and security (WIFS) (2016).

[12] Zhang, Zhongping & Zhang, Yixuan & Zhou, Zheng & Luo, *Boundary-based image forgery detection by fast shallow CNN*, conference: computer vision and pattern recognition Doi: 10.11.09/ICPR (2009).

Smart Navigation for Visually Impaired people using Artificial Intelligence

Rajvardhan Shendge[1*], *Aditya* Patil[2], and *Siddhi* Kadu[3]
[1]Department of Computer Engineering, Ramrao Adik Institute of Technology, D Y Patil Deemed to be University, Navi Mumbai
[2]Department of Computer Engineering, Ramrao Adik Institute of Technology, D Y Patil Deemed to be University, Navi Mumbai
[3]Department of Computer Engineering, Ramrao Adik Institute of Technology, D Y Patil Deemed to be University, Navi Mumbai

Abstract. This research introduces a blind aid that uses a live object recognition system. People who are blind or partially sighted rely significantly on their other senses, such as touch and auditory cues, to comprehend their surroundings. There is a need to deploy a technology that assists visually impaired persons in their daily routines, as there is now very little aid. Existing solutions, such as Screen Reading software and Braille devices, assist visually impaired individuals in reading and gaining access to numerous gadgets. However, these technologies are rendered worthless When the blind need to perform basic activities like recognizing the situation before them, such as recognizing people or objects, technologies become ineffective. This method will benefit blind or visually challenged all across the world. The goal is to help a person with total or partial blindness obtain a second set of eyesight without the assistance of a guardian, allowing them to live a better and more independent life. This project outlines working to create a more welcoming and inclusive environment, focusing on assistive technology that provides services, resources, and information to those with visual disabilities.

1 Introduction

One of the essential human senses is eyesight, and it plays a massive role in human life and helps in the perception of the surrounding environment. India is believed to include around 60 percent of the world's visually handicapped population. As our society progresses technologically, many new tools for people with disabilities have developed. The assurance of mobility for visually impaired persons is one of the many urgent supports. This innovative program will aid blind and visually impaired people all around the globe. The main goal is to let someone who is partially blind gain a second pair of vision without the assistance of a guardian, allowing them to live a worry-free life. This project expands on how it will aid in the creation of a more welcoming and inclusive environment, focusing on assistive technology that delivers services, resources, and data to those with visual difficulties. Independent living is a need in today's environment, yet visually impaired people confront significant limitations. The visually impaired are at a disadvantage since they need physical assistance to get information about their surroundings. As a result, there is a need to deploy technology that aids visually impaired people in their daily routines, as there is now very little help accessible. Screen Reading software and Braille devices are examples of existing solutions with different scopes. These technologies, on the other hand, are made useless. When visually impaired persons need to do essential fundamental functions like identifying the scene before recognizing people or things, such technologies are rendered ineffective. As a result, current technologies are more or less inaccurate, and if they fail, they may injure the user.Consequently, the user would be guided by this sophisticated program without the need for a personal human guide. This application would be self-contained, autonomous, and stand-alone since employing a fulltime guide may be costly and jeopardize one's privacy. It is straightforward for a vision-impaired person to carry out fundamental tasks since they can see items in their surroundings and can readily overcome any obstacles in their path. To avoid a mistake, visually impaired persons must be aware of their environment and their interactions with items. Finally, as an output of the object recognised within the frame with the highest confidence score among all other things present, the system will emit audio. The frame is chosen at a specific time interval to ensure that the audio output is not hampered.

In this paper, an ensemble approach is proposed to detect an object detected in a live camera frame and the audio output of the detected object's name and its actual position.The major contribution of the paper are providing smart android app for blind people which includes object detection with audio as output of

* Corresponding author: author@email.org

detected object in live camera frame as well as position of detected object.

The remaining of the paper is organized as follows: Section 2 is the survey of existing systems for detecting objects for a blind person—section 3 includes the analysis gap, followed by section 4 of system design. Section 5 demonstrates the proposed system along with the ensemble approach. The implementation details in 6 and the proposed system results are presented in section 7. Finally, the conclusion and future work is expounded in Section 8.

2 Literature Survey

There are various scholarly publications such as journal articles, research work etc. that are done in the field related to mobile navigating app for blind people. Many such articles and research reports were referred before making the project report. The reports/articles that have influenced the project, the most are :

WafaM.Elmannai, et al. [1] proposes a method intended to assist the visually impaired. The system combines sensor-based techniques with computer vision concepts to achieve an economically viable solution. Based on fuzzy logic and deep image information, the system uses an algorithm called New Obstacle. Using these techniques, visually impaired people can avoid obstacles by detecting objects in front of them. This system helps to support six blind people indoors and outdoors. System hardware requirements include a camera, GPS, Wi-Fi, microphone, compass, gyroscope, and microcontroller. This requires multisensory data and computer vision approaches at the software level. 96% detection accuracy and 100% the system offers obstacle avoidance. It helps to pass safely and gives high performance. The system is considered more dependable, straightforward, transportable, inexpensive, and accessible. This system also has some limitations when it comes to working. The system is compatible while detecting large objects because their size plays an essential role as they may not be detected in the frame. So it can be not easy to detect by finding the difference between background and background.

Qi-Chao Mao, et al. [2] proposes a method focuses on the autonomous mobility of the visually impaired. This is done by designing a wearable assistive device limited to blind people detecting traffic lights. The system is designed based on the AdaBoost algorithm. This approach is faster and more powerful in detecting objects. This system is enhanced with a flexible parallel architecture on the FPGA (Field Programmable Gate Array) platform. The main image and the confidence of the weak classifier are calculated in parallel. The parameters of the weak classifier are trained by the AdaBoost algorithm with MATLAB software and then configured on the FGPA platform. FPGA is said to be more flexible and consume less power. Testing shows that the system will detect traffic lights in videos at a frame rate of around 30 fps.

ZhenchaoOuyang, et al.[3] proposes a method for providing orientation and navigation capabilities, this paper proposes an electronic device called NavCane. This device helps visually impaired people move indoors and outdoors in a barrierfree journey. This device sends advance information about obstacles to its user without information overload, and the information is sent to that person through operational and auditory methods. It has many components such as an ultrasonic sensor and wet floor, GSM and GPS module, gyroscope, radiofrequency recognition vibrator motor, global module positioning system, and battery. The system has been evaluated by 80 visually impaired people and successfully in various situations. This NavCane device detects obstacles in known indoor environments, unlike other electronics. It is considered a low power consumption device in the vehicle and a low power consumption system. Analysis indicates that this NavCane improves barrier-free performance much more than white cane.

Wei Fang, et al.[4] This research focuses on knowledge-based data training and projected regions of interest to present a technique for developing a high-speed deep neural network for real-time video object recognition. The method creates a framework for training datasets across deep neural networks using limited sampling and cross-network knowledge prediction, which increases performance while reducing processing complexity. The training process is regulated by learn projection matrices by projecting knowledge and images representation of the teacher-level network from its middle layer to the middle layer sub-network layer. Experiments are being carried out to show that this system decreases the network's computational complexity by a factor of 16 and increases network performance significantly.

VidulaV.Meshram, et al.[5] proposes a method for embedded devices which have limited memory and processing power, making real-time object detection extremely difficult. Therefore, this paper proposes a design in which detection accuracy is not reduced. This task uses a lightweight object detection method called MiniYOLO v3. For darknet-based backbones, there are depthseparable folds and point-by-point group folds to reduce the size of parameters in the network. The process reducing the dimensions and then increasing the dimensions is adopted with the structure. The boundary effect of pointwise group convolution is suppressed by channel shuffle.

WenmingCao, et al.[6] proposes a method You Only Look Once, i.e., YOLO is one of the most well-known and widely used high-speed real-time object detection methods. To get better object detection results in a particular environment, this document proposes using the Tinier YOLO, which focuses improving better detection of results, performance and accuracy by reducing the size of the system . Focuses on network performance, detection speed, and accuracy to overcome all TinyYOLO issues. One of the biggest challenges in deploying fire modules in TinyYOLO V3 is to see the total number of these modules and their location in the system. Second, there is a problem with the style of connectivity between the modules to achieve better results in detection, accuracy, and performance. The system employs tight connections between launch modules to enhance mobility and maximize the flow of information between networks. However, beyond that, the model size cannot be reduced, as recognition accuracy can be significantly reduced. Therefore, this system uses a continuity component to address the issue of reducing model size. These path layers help you combine features extracted from the front layer to get the right features. Finally, the system suggests removing batch normalization from the module and overall performance to reduce computational cost factors.

3 Gap Analysis

Above mentioned papers propose a method of detecting the object in a live frame or detecting an object on the captured image by providing an audio output of the detected object in the captured image. At the same time, our system proposes a system in which an object is detected in the live frame, and the audio output of that detected object in the live frame and the position of the detected object in provided as output to the user.

4 System Design

In figure 1 the overall design is described as follow which is an MVC design Pattern and some people also termed it as architecture as well.

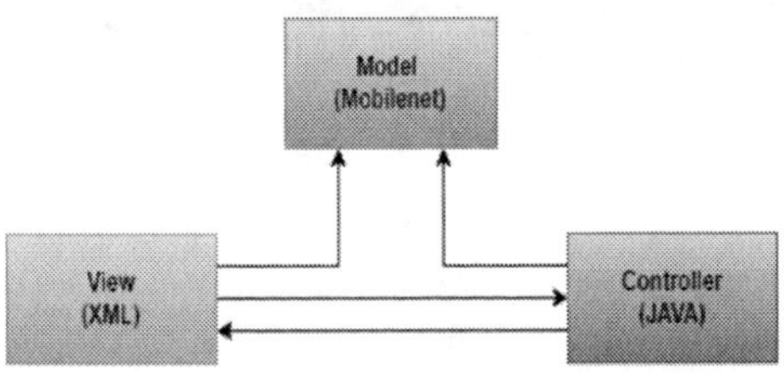

Fig. 1. System Design

4.1 Architecture Diagram

In figure 2 the system detects various objects in realtime producing real-time voice output. This system is divided into two parts: the user and the android application layers. Android Application receives input from the camera as input then the captured image is stored in the android application as a captured image then the pre-processing of captured image takes place , this process is called as pre-processing of the image the algorithm which identify the object from the image from the captured image, after detection of an object from the captured image it labels the object according to maximum accuracy and the application generates an audio signal for the identified object these audio signals are transferred to user layer as an audio output.

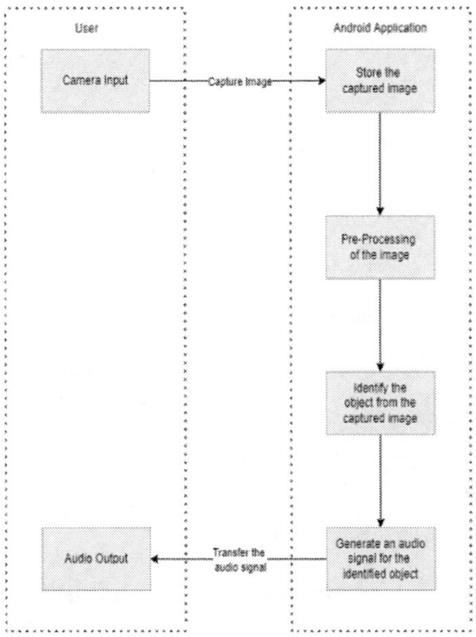

Fig. 2. Architecture Diagram

4.2 Object Detection

The particular thing detection plus recognition have in order to be accurate, because well as the particular detecting speed through the object ought to be high to ensure that the navigation regarding blind folks might be more simple. A fast R-CNN algorithm has already been implemented[10]. The particular system does not work on several regions; the R-CNN algorithm processes the particular image into the bundle of containers and checks when some of these types of boxes contain any object. The quick R-CNN algorithm offers to use the picky search methodology in order to extract these containers from a picture. The varying weighing scales, colors, textures, and enclosure would become the four areas that form a suitable object. The picky search identifies the particular patterns within the particular image, and based on those designs, they split the particular image into numerous regions. Object recognition is a software program related to pc vision and picture processing that detects the particular existence of items in digital pictures and videos along with a limitation package and types or even classes of items [2].

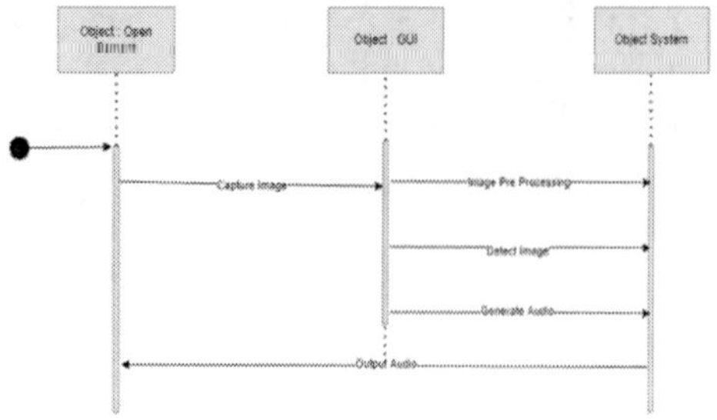

Fig. 3. Sequence Diagram for Object Detection

figure 3 shows the obejct is captured live through mobile camera and this captured image gets passed through object gui where image preprocessing process takes place, where object is detected and audio is generated for that particular detected object along with posiiton of that detected object in object system and this generated audio is given as ouput to user.

Aesthetically impaired users may understand their environment without difficulty plus remain independent simply by utilizing object recognition. Input: a picture or an image with one or even more objects. A single or more restricting boxes (e., Gary the gadget guy, defined by the dot, width, plus height) and a course label for every restricting box are created as output. Bounding boxes are produced, which predict the confidence score's particular certainty. This particular score lets us know that the bounding package will contain some items. For every bounding box, the cellular predicts a course of that item which provides the probability distribution amongst all the accessible classes in a given model. The particular confidence score plus the probability calculated gives all of us the final rating, which lets the particular consumer know exactly how likely it is that the particular bounding box consists of some specific item.

4.3 Tensorflow

In Figure 4, the TensorFlow Lite design is represented in a specific efficient transportable format known as FlatBuffers (identified by the. flite document extension), which offers several advantages over TensorFlow's protocol buffer design format, including reduced size (small programme code footprint) and faster inference (data is accessed directly without an additional parsing/unpacking step), allowing TensorFlow Lite to execute efficiently. For an automated era of pre- and postprocessing pipelines during on-device inference, a TensorFlow Lite model may optionally incorporate metadata with the human-readable model description and machine-readable information. This item detection approach makes use of the COCO mobile net SOLID STATE DRIVE v1 model, which has datasets for 80 different object categories that are regularly encountered in our world.

Fig. 4. Tensorflow Architecture

4.4 Image Pre-Processing

The first goal of Image and Video pre-processing is to drop frames from the video stream. A contemporary camera can record videos that contain at least 20 - 30 frames per second. Since the individual is not moving quickly, we may drop frames from the video to simplify the processing. We will transform the images/frames into a grayscale format to further reduce redundant processing. A picture comprises pixels, and every pixel stores its RGB color values. Reducing the number of values held by our pixels may minimize our processing by a substantial margin. YOLO (You Only Look Once) realtime object detection algorithm is one of the most successful object detection algorithms that comprises many of numerous concepts coming out of the computer vision research field. Object detection is a fundamental component of autonomous vehicle technology. It is an area of computer vision that's incredibly popular and functioning so much better than only a few years ago [8].

4.5 Android Studio

The particular Android SDK is used to create the application , in which aesthetically impaired users certainly discover objects and understand their environment. This platform is used for the front-side conclusion and backend of the program form. This system includes all the libraries and deals required to put this system together.

4.6 Dataset

The SSD MobileNet model was trained using the Common Object in Context (COCO) dataset, which recognises 80 different categories in this study.

5 Proposed System

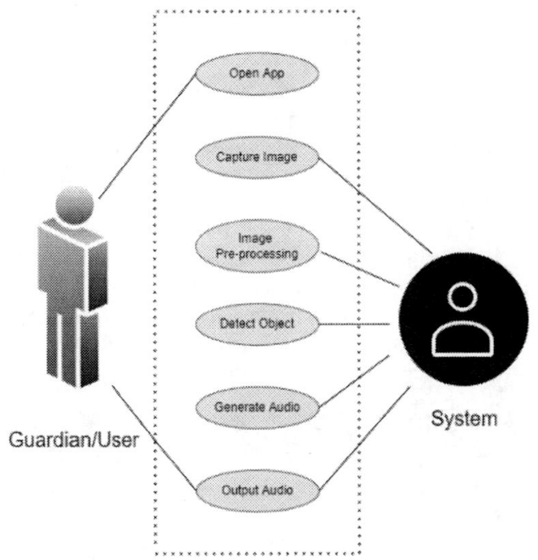

Fig. 5. Proposed System

In figure 5 Many difficulties are faced by the partially visually impaired person while performing basic tasks and most commonly during self-navigation in an unknown environment. They cannot detect the object around them, and in most cases, they cannot identify their objects accurately. So self-navigation difficulties can be overcome by this application, which will help them live everyday life like an average person. In an ordinary case, object detection cannot achieve high accuracy as other applications only detect objects in the captured image rather than object detection in live streaming video. So, the proposed system detects objects continuously in the live stream and not by taking photos every time. For detecting and recognizing objects in the live stream, the processing speed should be very high to detect all objects within the frame, but object detection and recognition are more accessible in the case of captured images. In this system, the android smartphone captures an image from the camera, and it is processed through an algorithm, and the object detection and voice output module produces the required output for the system. In object detection, the detected or recognized objects are first segmented, and where multiple segmented regions are formed, then the image is classified and identified using a fast RCNN algorithm. The voice module converts the detected object constraint area into text and produces output using speech processing. For conversion of text to speech, the system uses Google API. system proposes a system in which an object is detected in the live frame, and the audio output of that detected object in the live frame and the position of the detected object in provided as output to the user.

6 Implementation Details

In this system, the TensorFlow model is used, which in the backend uses the SSD algorithm and can work by speed, balance, and accuracy. This system accurately detects 81 objects approximately. The Mean Average Precision of this system is 74.3 mAP and is the highest among real-time processing models. This system produces speech feedback for the object detected in the frame. The same thing, on the other hand, may be called out multiple times when it is discovered. Even if the detection result is the same, repeating the same object's name is inconvenient. It's also improper if the names of two items are said at the same time, overlapping, or the user is unable to distinguish between them. To overcome this problem, even if one object is recognised in the first frame and is talking, the software will not mention that class for the next two seconds. At the same time, the problem of multiple detection of the same object has been fixed. In this way you get an idea of the exact performance of the model when the objects are correctly detected. Viable objects can be detected immediately, but only objects with an accuracy value more significant than the fixed threshold value will be reported to visually impaired users using output feedback by equals output. Voice. Multiple objects can even be detected precisely at the same time. The system is built by combining some of the technologies described below. Since Android Studio is a full-fledged integrated development environment (IDE) specially designed for Android applications, it is used to create apps. The Android framework allows taking photos and videos with the phone. Camera intent or Hardware.camera2 API. It's a video recording application that recognises objects and interprets text in real time. To create object identification patterns, the Tensorflow library is employed, which allows for rapid numerical computations. It offers a general-purpose design that allows calculations to be used across several platforms with ease. The SSDMobileNetCOCO paradigm allows for real-time video processing. The SSD is made up of a single composite network that learns to predict bounding box positions before the object that was recognised as bounding boxes [7].

The application detects objects using the MobileNetCOCO SSD model. The entire input image uses only one neural network. The network then divides the input image into multiple regions and predicts the boundary regions within the squares based on their probability scores [7]. This object parses images and extracts text from them. It can detect text in all types of images after it has been initialized. Google TexttoSpeech is used to implement the TTS feature, which talks items that have been discovered and identified, as well as providing guidance for those objects.

6.1 Object Detection Pipeline

In Figure 6 the camera will operate in a loop and take an image at a rate of about 2Hz. The picture will run through face detection, and any face boundbe drawn on the screen as a teaser. The system then selects if it merely detects the largest face in the frame, and the complete image will be given to CNN. If the user opts to detect the largest face, then the head and shoulder region of the biggest face bound will be cropped out, and the new photo will be submitted to CNN for classification. CCN will produce an N dimension vector of confidences of detected features. Feature vector will be input into the logistic regression prediction function, then deliver probability/score for each of the classes.

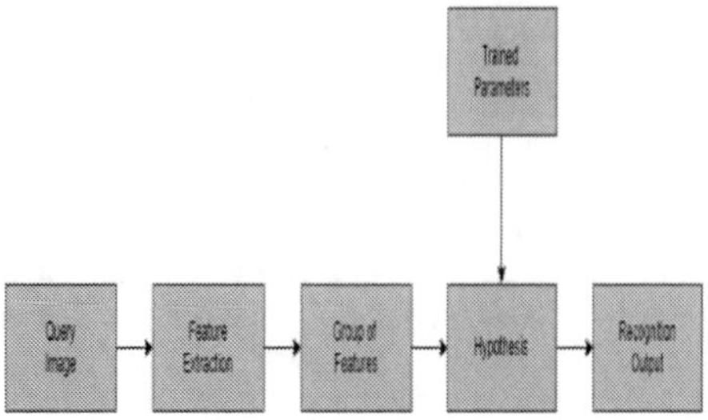

Fig. 6. Object Detection Pipeline

6.2 Convolutional Neural Network

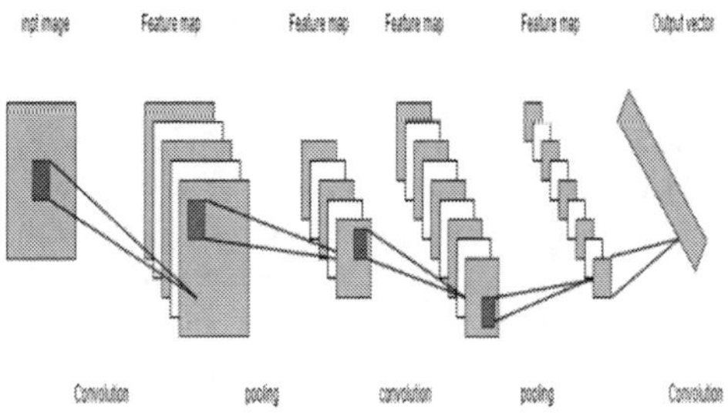

Fig. 7. Convolutional Neural Network [13]

Figure 7 displays a CNN that can recognise 2D shapes regardless of object rotation, scale, or other characteristics. The basic form of convolutional neural network, according to LeCun, is neurobiologically driven and comprises of three key activities:

- Extracting Feature.
- Mapping Feature.
- Subsampling.

7 Results and Analysis

In figure 8 and figure 9 the Tensorflow's object detection model was utilized in this application with the SSD method in the backend, and it works by balancing accuracy and speed. This sample successfully identified about 81 items. This model has the highest mAP value of 74.3 (mean accuracy) among models designed for real-time processing. After completing this endeavor, he was meant to respond to the identified item with a voice. However, the same item is recognized several times throughout time. Even though the detection findings are the same, it is not preferable to mention the same object name. Furthermore, it was unfavorable if two-item names were reported as identical or so near that the user could not tell the difference. If an item is spotted in the first picture and spoken about, this difficulty may be solved. The application will not speak about its class for the following five seconds, even if discovered. As a result, the difficulty of detecting an item many times has been addressed. Here are some examples of forecasts about an object's precise observed value. Please give us an idea of the model's accurate performance when detecting things. Multiple items can be recognised at once, but only those with a precision value better than a threshold value will be speech output feedback to visually impaired users. It is even possible to accurately identify many things at the same time.

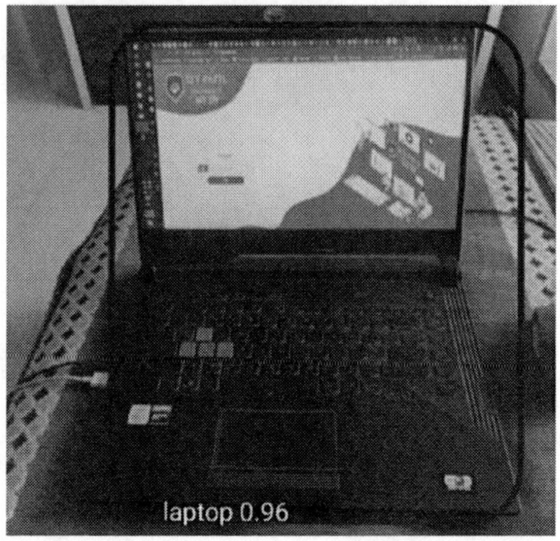

Fig. 8. Detection Of Objects

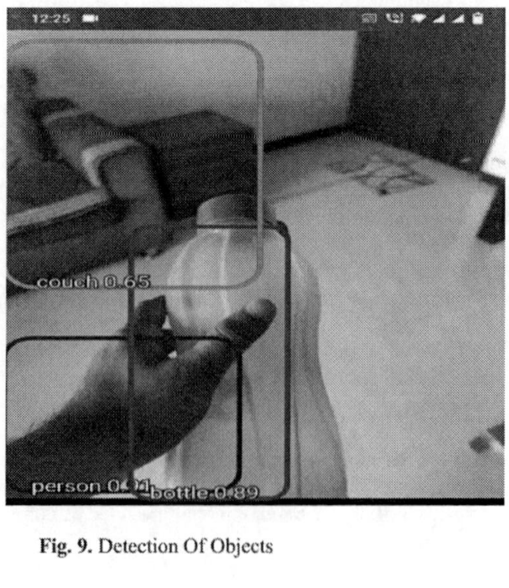

Fig. 9. Detection Of Objects

8 Conclusion

This study presents a computer vision-based navigation system for blind people and attempts to solve the problems presented in the introduction section. The suggested method is put to the test in various scenarios and against comparable apps on the App Store. The findings show that the application runs well under various conditions and is quicker and more efficient than its competitors. The program is also somewhat accessible to blind users since it enables them to launch it simply by putting an earphone jack into their mobile phones, which is not available in many other apps. This program can be used for navigation in its present version, but it has a few restrictions. This paper proposes a system that assists partially blind persons in navigation and detects the object surrounding of user. The proposed system is tested with many conditions and against applications of similar interest. It is noticeable from the observations that the applications perform pretty under the circumstances and is efficient, balanced, and fast. The application is easy to access for a partially blind person, and as the application provides output in the form of audio, it makes it more efficient for a partially blind person who is missing in other applications. In the current state of the application, this application can be used for navigation but has a few limitations. Being optimistic that it would cause people to be sympathetic with visually impaired people and they understand that even they should be able to live life like a normal person.

In the future, GPS can be added so that the device can guide the user where they want to travel and the device would direct them there. Use of the proximity sensor that will inform the visually impaired person of the object's or person's distance. An integrated reading mechanism, in which we can upload a book to the gadget and have it read by the device. Aside from that, device's battery consumption optimization can be imporved.

References

1. WafaM.Elmannai,KhaledM.Elleithy. "A Highly Accurate and Reliable Data Fusion Framework for Guiding the Visually Impaired". IEEE Access 6 (2018) :33029-33054.
2. Qi-Chao Mao,Hong-Mei Sun,Yan-Bo Liu,RuiSheng Jia. "MiniYOLOv3:Real-Time Object Detector for Embedded Applications".IEEE Access 7 (2019) :133529-133538.
3. ZhenchaoOuyang,JianweiNiu,YuLiu,MohsenGuiz an i. "Deep CNN-Based Real-Time Traffic Light Detector for Self-Driving Vehicles".IEEE Access 19 (2019):300-313.
4. Wei Fang,LinWang,Peiming Ren. "Tinier-YOLO:A Real-Time Object Detection Method for Constrained Environments".IEEE Access 8 (2019) :1935-1944.
5. VidulaV.Meshram,KailasPatil,VishalA.Meshram, Fe lix Che Shu. "An Astute Assisstive Device for Mobility and object Recognition for Visually Impaired People".IEEE Access 49 (2019) :449-460.
6. WenmingCao,JianheYuan,ZhihaiHe,ZhiZhang,Zhi q uan He. "Fast Deep Neural Networks With Knowledge Guided Training and Predicted Regions of Interests for Real-Time Video Object Detection".IEEE Access 6 (2018): 8990-8999.
7. QiankunLiu,BinLiu,YueWu,WeihaiLi,Nenghai Yu. "Real-Time Online Multi-Object Tracking in Compressed Domain".IEEE Access 7 (2019): 76489-76499.
8. Meimei Gong,Yiming Shu. "Real-Time Detection and Motion Recognition of Human MovingObjects Based on Deep Learning and Multi-Scale Feature Fusion in Video".IEEE Access 8 (2020):25811- 25822.
9. Kiruthika, U., Somasundaram, T. S. (2018). Efficient agent-based negotiation by predicting opponent preferences using AHP. Journal of applied research and technology, 16(1), 22-34.
10. N. Senthil kumar, A. Abinaya, E. Arthi, M. Atchaya, M. Elakkiya, "SMART EYE FOR VISUALLY IMPAIRED PEOPLE", International Research Journal of Engineering and Technology, Volume: 07 Issue: 06, June 2020.
11. Liang – Bi Chen, Ming-Che Chen, "An implementation of an intelligent assistace system for visually impaired/blind people, "IEEE, 2018.
12. Shahed Anzarus Sabab, Md. Hamjajul Ashmafee, "Blind Reader: An IntelligentAssistant for Blind", 19th International Conference on Computer and Information Technology, December 18-20, 2016, North South University, Dhaka, Bangladesh.
13. M. Z. Alom, T. M. Taha, C. Yakopcic, S. Westberg, P. Sidike, M. S. Nasrin, M. Hasan, B. C. Van Essen, A. A. Awwal, and V. K. Asari, "A state-of-the-art survey on deep learning theory and architectures," Electronics, vol. 8, no. 3, p. 292, 2019.
14. Shreyash Patil, Oshin Gawande, Shivam Kumar, Pradip Shewale,"Assistant Systems for the Visually Impaired", International Research Journal of Engineering and Technology (IRJET), Volume: 07 Issue: 01 — Jan 2020.
15. Gagandeep Singh, Omkar Kandale, Kevin Takhtani, Nandini Dadhwal, "A Smart Personal AI Assistant for Visually Impaired People", International Research Journal of Engineering and Technology (IRJET), Volume: 07 Issue: 06 — June 2020.
16. Mingmin Zhao, FadelAdib, Dina Katabi Emotion Recognition using wireless signals.

ITM Web of Conferences **44**, 03054 (2022) https://doi.org/10.1051/itmconf/20224403054
ICACC-2022

Image Super Resolution using Enhanced Super Resolution Generative Adversarial Network

Raj Sarode[1], Samiksha Varpe[2], Omkar Kolte[3], Leena Ragha[4]

[1,2,3,4]Ramrao Adik Institute of Technology, D Y Patil Deemed to be University, Navi Mumbai, India

Abstract: Aside from enhancing the accuracy and speed of single picture modification utilizing fast and in-depth convolutional emotional networks, one significant challenge remains mostly commonly unaddressed, namely how do we recover soft texture details when we concentrate too much on exceptional improvement features? The resultant evaluations offer greater transmission ratings, but the high frequency data is non-existent and unsatisfactory mostly in sense that now it fails to meet the consistency anticipated in high resolution. The resulting ratings have higher signal-to-audio ratings, but the high frequency data is non-existent and unsatisfactory in the sense that it fails to match the consistency expected in high resolution. Introducing ESRGAN, an Advanced Optical Genetically Modified (GAN) network of high-resolution image (SR). To our knowledge, it is aframework capable of identifying immature real-world images up to 4x points rising. To achieve this, we propose a function of loss of vision that combines the loss of content with the loss of content (Mean Squared Error Loss). Controversial Loss has our solution for many uncooked pictures utilizing a discriminatory network which is taught to distinguish between high resolution images and realistic images. We have built a structure that contains several RRDB blocks (Residual in Residual Dense Block) outside the Batch Normalization layers. Our deep residual network can find realistic image texture in very low sample images. Additionally, we used techniques that included residual measurement and a small implementation to train a deeper model. We also introduced the relativistic GAN as a racist, who learns to judge whether one image is more realistic than another, directing the generator to return a detailed texture. In addition, we have improved vision loss by using features before activation, which provides greater security and thus restoring more precise light and texture.

Keywords: Machine Learning, classification, feature selection, prediction, heart disease, diabetes

1. Introduction

For generating realistic textures during single picture super- resolution, super-Resolution Generative Adversarial Network (SRGAN) is the best pioneering study. Nevertheless, hallucinogenic features are usually paired by unsettling arte-facts. Comprehensively analyze the three essential aspects of SRGAN – network design, adversarial loss and perceptual loss, and improve each of them to achieve an enhanced SRGAN(ESRGAN) which helps to enrich the visual quality. Specifically, as the core network construction unit, we offer residual-in-residual dense blocks (RRDBs) without batch normalization. Furthermore, we adopt the notion of relativistic GANs to enable the discriminator to predict relative reality rather than absolute value. Features before activation improves the perceptual loss and it gives more adequate supervision for brightness stability and texture recovery. Consistently superior visual quality with more realistic and natural textures is delivered ESRGAN comparatively better than SRGAN. In machine learning, image super resolution refers to produce an upscale, clear, high resolution as the output when low resolution image is given as input. In more technical terms, when we apply a degradation function to a high-resolution (HR) image, we get a low-resolution (LR) image - i.e. LR = degradation (HR). We propose a deep learning method for single image super-resolution (SR). With HR image as our target and LR image as our input we train our convolutional neural network. • We will primarily focus on NOISE, RGB and other quality factors. Moreover, for showing better overall reconstruction quality, we extend our network to cope with three color channels simultaneously. For retreiving the texture details in the reconstructed SR images our primary target is to reconstructhigh-resolution images by up-scaling low-resolution images. Satellite and aerial image analysis, medical image processing, compressed image/video enhancement are some of the numerous applications. The scope of this is

Surveillance: Security cameras can be used for detecting, identifying, and performing facial recognition on low- resolution images.

Medical: In medical, obtaining high-resolution MRI pictures may be challenging when scanning, spatial coverage, and signal-to-noise ratio (SNR) (SNR) are also some applications. Super-resolution helps to overcome issue by creating high- resolution MRI from otherwise low-resolution MRI pictures.

Media: Super-resolution may be used to minimize server costs, since media can be supplied at a lower resolution and be up-scaled on the fly.

Objectives :

Combination of low resolution (noisy) sequences of

images of a scene can be used to generate a high-resolution image or image sequence which gives an outcome in the form of Super-resolution. Reconstruction of the original scene image with the high resolution given a set of observed images at lower resolution is attempted by the model.

- The general approach considers the low-resolution imagesas resulting from resampling of a high-resolution image.
- The goal is then to recover the high-resolution image, which, when resampled based on the input images and the imaging model, will produce the low resolution observed images.
- Thus, the accuracy of the imaging model is vital for super-resolution, and incorrect modeling, say of motion, can actually degrade the image further.

2. Literature Survey:

Survey of Existing System:x

Up sampling Methods: Up sampling (Various methods include raising the spectral resolution of photographs or even just increase the quantity of pixel rows/columns or both in the image).

1. Interpolation-based methods – Resizing of digital pictures is image interpolation which is commonly utilized by image-based systems. Nearest-neighbor interpolation, linear, bilinear, bicubic-interpolation areconventional approaches.

   a. Nearest-neighbor Interpolation – The nearest-neighbor interpolation is a simple and obvious approach. It determines the closest pixel value for each region to be interpolated irrespective of any other pixels.

   b. Bilinear Interpolation – Bilinear interpolation (BLI) first conducts line translation on one image axis and then another on another axis. As a result of quadratic translation with a two X 2 reception field size, it demonstrates substantially larger performance than adjacent neighbor rendering while retaining reasonably consistent speed.

   c. Bicubic Interpolation: (BCI) executes interpolation on each of two axes. Unlike BLI, the BCI nto in cts ed.

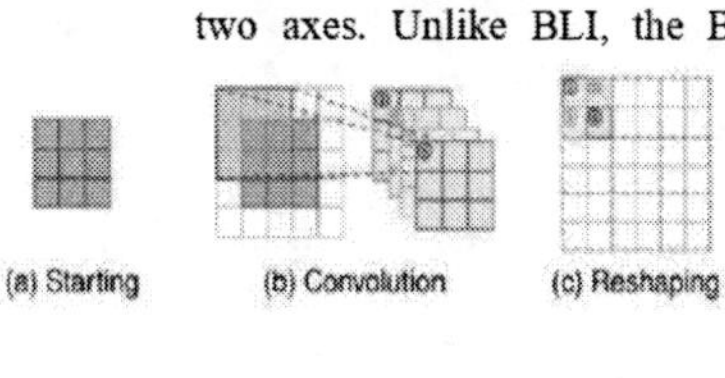

Shortcomings Interpolation-based approaches frequently cause side effects such as computing intricacy, noise escalation, blurring outcomes

2. Learning-based up sampling: A modified-convolution layer and a tiny pixel layer are inserted inthe SR field, to alleviate the shortcomings of methods based on interpolation and sampling learning from the end method.

   a. **Transposed convolution:** Layer,
   i.e. deconvolution layer, aims to transform a standard convolution, i.e., the probable input is forecasted based on the feature maps scaled like convolution output. By introducing zeros and conducting convolution, the imaged is extended to enhance picture resolution.
   Transposed convolution layer: The blue boxes signify the input, and the green boxes indicate the kernel and the convolution output.

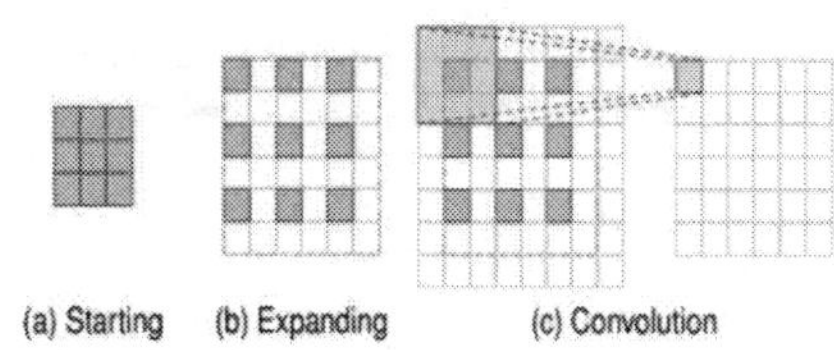

Figure 1

   b. **Sub-pixel Layer:** The sub pixel layer, one other readable layer from end to end, makes sampling by producing multiple channels with convolution and bending the shows. Within this layer, conversions are performed primarily to create results with s2 channels, where s is a measurement factor. Assuming the input size is h × w × c, the optimal value will be h × w × s2c. Subsequently, a rearrangement procedure was undertaken to yeild results of size SH × SW × C.

3. Pre-Up sampling Super - We do not undertake direct mapping of LR images on HR images as it is viewed a complex process. Using old sampling techniques to obtain high quality images, then enriching ourselves using deep neural networks is a basic option. For example, LR photos are taken from top to transform HR images to the desired size using bicubic rendering. Then deep CNNs are delivered to these photos to merge high quality images.

4. Post-Up sampling Super - In order to enhance mathematical performance by fully utilizing in-depth learning technology to automate the solution dynamically, researchers recommended to perform multiple calculations in a low-resolution area by substituting predictable predictions with end-to-end readable structures installed at the end of the models. In the pioneer research of this approach, namely post- upsampling, without any further modification SR, LR input files are uploaded to deep CNN, and at the end of the network easy- to-read layers are injected.

2. Problem Statement:

Our purpose is to develop a convolutional neural network that would intake low-resolution photographs, and we can train it to output Higher Resolution images that resemble the actual ones the finest. Here we are going to employ the Generative Adversarial Networks (GANs) technique. Particularly ESRGAN (Enhanced Super-Resolution Generative Adversarial Networks). HR (high resolution images) is analyzed to obtain down sampled LR images. Here, we have LR and HR images for training samples. LR images are given as input to generator that upsamplesand provides SR images. Utilization of discriminator is done to seperate the HR image and back-propogate GAN laws for training the discriminator and generator. Here we are utilizing the super-resolution dataset DIV2k dataset.

3. Proposed Methodology:

In this study, we propose Advanced Networks for Advanced Maintenance Networks. It basically uses the GAN principle for large-scale operation, an audio image generator will try to create and its testing will be done by discriminator. Generator may o/p images that can surpass accurate training data, for this both will preserve training.

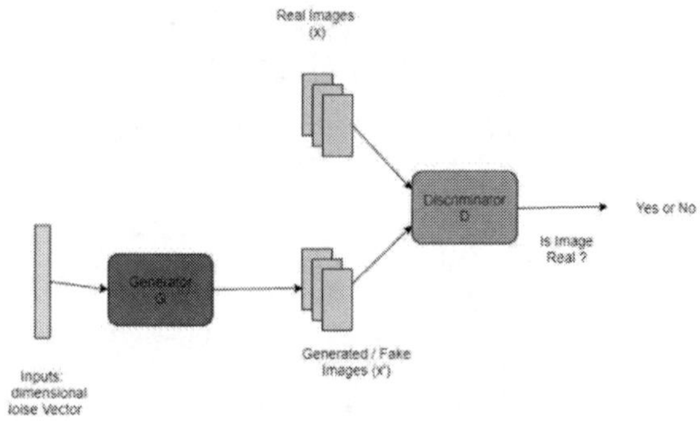

Figure 3: Architecture of Generative Adversarial Network

Key aspects of the approaches –

- Post up sampling kind of framework
- Sub-pixel layer for up sampling
- Includes residual blocks
- Utilizes Perceptual loss

Sometimes we need models that can detect what is in the input and generate new samples; such models are labelled generative models. For example, if we could have a discriminative model that discriminates between an image of a dog and a cat, we would not be capable of creating the cat picture, although it might have a sense of what a cat looks like. In generative adversarial networks, we may create a model that can produce samples comparable to the ones in the supplied dataset. Generative Adversarial Networks, or GANs for short, comprises of two components that compete against each other during training. One network aims to produce realistic samples that have never been seen before. The other one attempts to determine if its inputs are legitimate or fraudulent. The first network is designated the generator, while the second is called the discriminator. The generator's purpose is to trick the discriminator, while the goal of the discriminator is not to be misled. The discriminator receives either a randomly selected picture from a dataset or a synthetic sample made by the generator. Given this input, it produces the probability that the input is a true training example rather than a false sample created by the model. After training both the networks in this situation, both the generator and discriminator grow better towards attaining their antagonistic objectives. In the beginning, both the generator and discriminator perform badly. The generator may create terrible samples, however the discriminator still does not detect it. The generator and discriminator both become better and better as they continuously learn to defeat a superior opponent. Ideally, they would get to an equilibrium where the generator provides flawlessly realistic samples and the discriminator always outputs probability. Once the training is complete, you may toss away the discriminator and utilize the generator to produce fresh samples. In adversarial training, the model is driven to work harderwhen it is failing. The

 https://doi.org/10.1051/itmconf/20224403054

discriminator consistently examines for the inadequacies of the generator to identify whether it is a phonysample, and the generator is driven to correct such deficiencies. Once those vulnerabilities are rectified, the discriminator attempts to discover the next problems and so on. In generative adversarial networks, the generator and discriminator functions minimize their loss functions. What ESRGAN (Enhanced Super-Resolution Generative adversarial networks) achieves is that they calculate the weighted average of two models, one trained using Mean Squared Error and the other fine-tuned using adversarial training. Blending the parameters this manner enables finding the correct balance between the two models without retraining them. The network structure of the generator is enhanced by adding the Residual-in-Residual Dense Block (RRDB), that enhances the network capacity as well as makes the training more accessible. To enhance the quality of generated images SRGAN, mainly two modifications are made in network architectureRemoval of all Batch Normalization (BN) layers Replacing the original basic block with the RRDB.

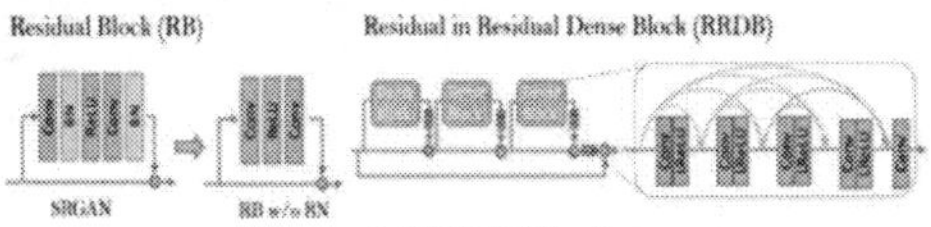

Figure 4: RRDB Block

In the above image, the BN layer is removed, and in this right image, the RRDB block is employed in a deeper model where β is the residual sizing factor. It has been observed that the removal of BN layers leads to an increase in performance and reduction of computation complexity and memory usage in many network architectures. Comparatively the fundamental residual block in SRGAN has less deeper and sophisticated structure for the Generator Network than RRDB which ultimately boosted the performance of the network. The residual scaling parameter is kept constant between 0 and 1 to prevent the instability of the network. By utilizing the idea of Relativistic average GAN (RaGAN), the improvement of the discriminator is witnessed in second upgrade, which helps the discriminator assess "whether an image is more realistic than the first one" rather than "whether an image is genuine or fake". Above is the distinction seen between the regular discriminator and relativistic discriminator. Instead of the usual discriminator, which offers the probability that a picture is genuine or false, a relativistic discriminator seeks to foresee the possibility that an actual image is relatively more realistic than a false one.

3.1. System Architecture:

Firstly, we are using the DIV2k dataset which contains images. We are passing these Images to Generator which consists of Transposed CNN and Sub-pixel CNN and then we are getting super-resolved images. After getting those SR images, we are passing them to discriminator CNN and here the discriminator will discriminate whether it is a real or less real image and depending on that it will calculate total loss.

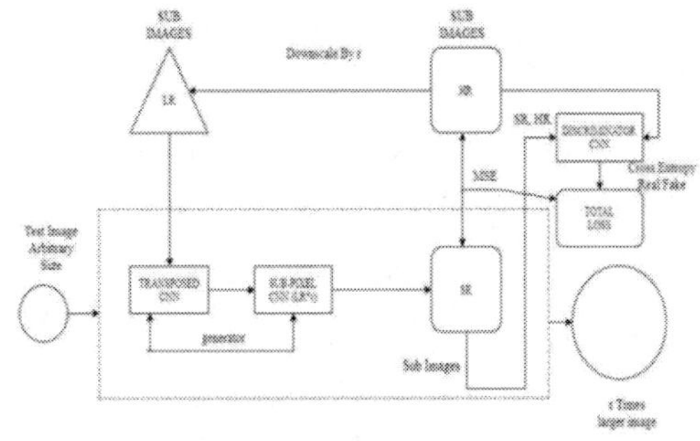

Figure 5: System Architecture

This loss now will be passed on to generator and generator will address those losses and will try to minimize the loss and will again generate image and will pass on to discriminator and this process will go on till generator will generate perfectly realistic image and discriminator will detect with 0.5 probability.

4. Implementation Details:

The proportion between the greatest potential power of an image to the strength of corrupting noise that influences the fidelity of its representation is Peak signal-to-noise ratio (PSNR). Equating the picture to an ideal clean image with the highest potential power is necessary to evaluate the PSNR of a picture.

$$\text{PSNR} = log_{10} 10 \frac{(L-1)}{MSE} = log_{10} 20 \frac{(L-1)}{RMSE}$$

$$\text{MSE} = \frac{1}{mn} \sum_{i=0}^{m-1} \sum_{j=0}^{n-1} [O(i,j) - D(i,j)]^2$$

Where,

O - matrices values of original image.

D - matrices values of degraded image.

m - number of rows of pixels.

i - index of that row of the image.

n - number of columns of pixels.

j - index of that column of the image.

RMSE – root mean squared error.

4.1. Model Analysis:

- DIV2K
- Set5
- Set14
- BSD100

 https://doi.org/10.1051/itmconf/20224403054

- Manga109
- Decision Tree
- Urban 100

5. Implementation Details & Results:

We can improve the traditional SRGAN method from three aspects.

1. A deeper model is trained in which Residual in Residual Dense Block (RRDB) is used without batch normalization.
2. Relativistic average GAN is employed instead of the normal GAN.

In contrast to SRGAN deeper ESRGAN model exhibits its higher performance with simple training. Balancing the visual quality and PSNR can be done by network interpolation strategy. Smooth animation with the interpolation parameters was demonstrated which has changed from 0 to 1. In this smooth control of the RRDB_PSNR model and fine-tuned ESRGAN model is given by network interpolation strategy.

PSNR and the perceptual index calculated by various methods are shown below:

i) Here we are using Set14 dataset for testing and the PSNR ratio calculated for different method is shown in below image. PSNR ratio for one image is shown below:

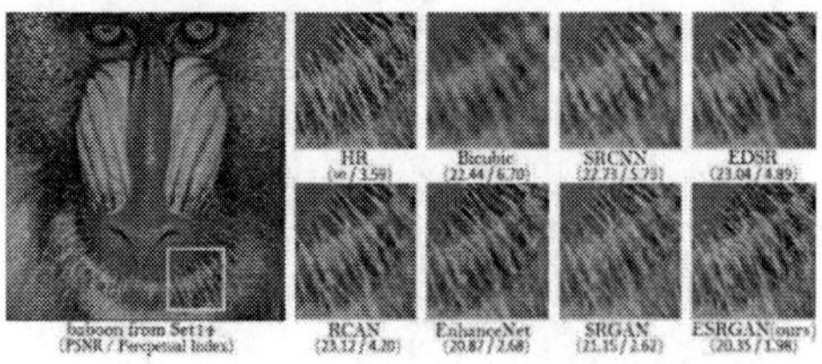

Figure 6

ii) Here we are using BSD 100 dataset for testing and the PSNR ratio calculated for different methods is shown in below image. PSNR ratio for one image is shown below:

Figure 7

The influence of training patch size (Dataset):

It is observed that training a deeper network benefit from the greater dataset size. Furthermore, it is noted that the greater improvement can be detected in the schematic one rather than deeper model. Benefit of larger training dataset size can be takenby bigger model as it is more efficient. This is assessed on set5 dataset with RGB channels as shown in the graph below.

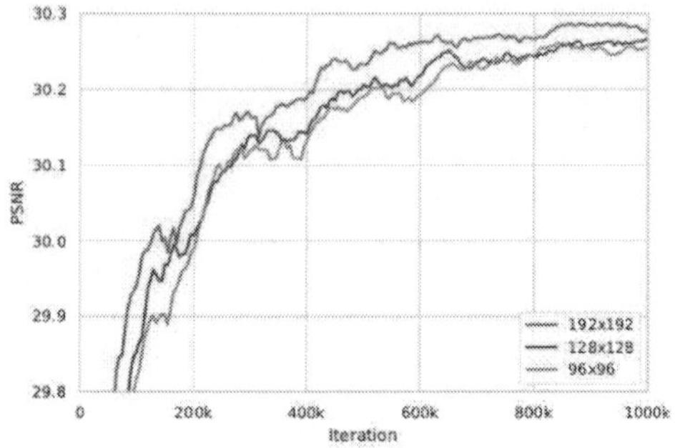

Figure 8: (a) 16 Residual Blocks

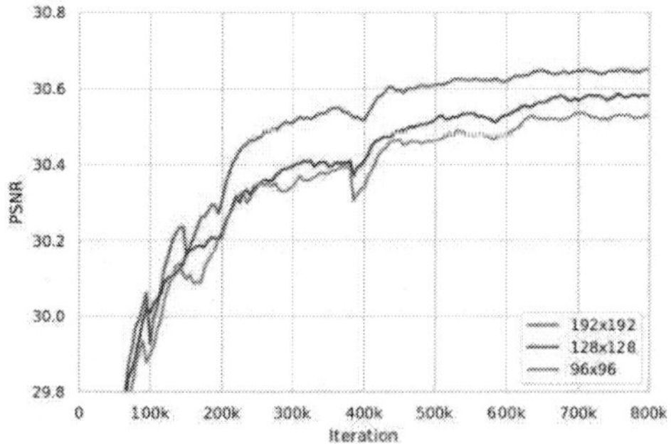

Figure 9: (b) 23 RRDBS

The below table summarizes the PSNR ratios of different methods keeping training dataset constant against varying testing datasets:

Method	Training dataset	Set5	Set14	BSD100	Urban100	Manga109
SRCNN	DIV2K	30.48/0.8628	27.50/0.7513	26.90/0.7101	24.52/0.7221	27.58/0.8555
EDSR	DIV2K	32.46/0.8968	28.80/0.7876	27.71/0.7420	26.64/0.8033	31.02/0.9148
RCAN	DIV2K	32.63/0.9002	28.87/0.7889	27.77/0.7436	26.82/ 0.8087	31.22/ 0.9173
ESRCNN	DIV2K	32.73/0.9011	28.99/0.7917	27.85/0.7455	27.03/0.8153	31.66/0.9196

Project Outcomes:

Following is the GUI interface of our application. Hereuser will input an image and the output will be high resolved image.

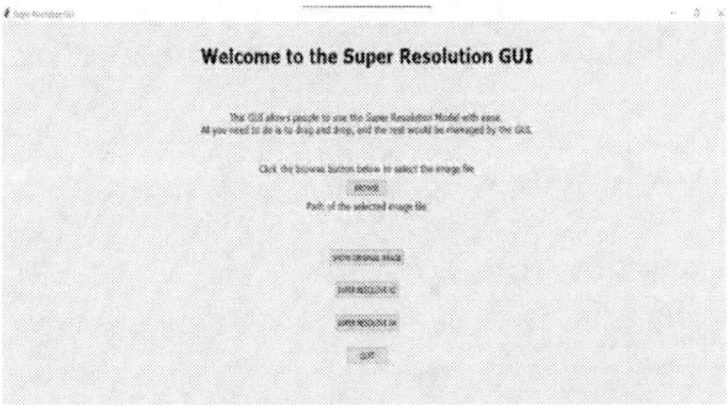

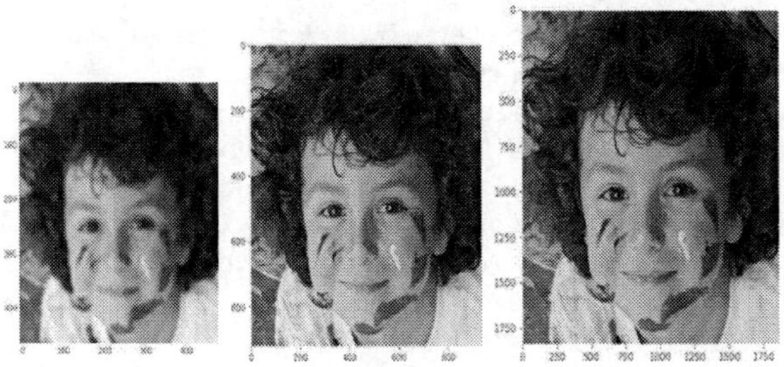

6. Conclusion & Future Work:

This study represents a image enhancing technique for enhancing the lower pixel images and attempts to solve the problems presented in the introduction section. This suggested method is put to test in various scenarios which includes getting images from CCTV surveillance, satellite, distant cameras and also from the live videos as well as recorded videos. The finding shows that the application is running well under static images collected from dataset. Till now we are able to achieve to get higher resolved images from the larger and variety of datasets. This paper proposes a system that will take highly resolved images from the dataset and will enhanced it's pixels using ESRGAN method and the output image will be compared with the original one and the mean square error will be calculated and based on MSE the PSNR ration will be calculated and more the value of PSNR the more efficient will be that method. Formulation of the architecture contains several RRDB blocks without BN layers. In addition, the proposed deep model is trained using residual scaling and smaller initialization. For learning how to determine whether an image is more realistic than other which will help the generator to recover more detailed textures, introduction to use of relativistic GAN as the discriminator is also shown. The ESRGAN, a PSNR oriented model trained with DIV2K dataset can achieve high PSNR performance as compared to SRGAN, EDSR, RCAN trained model. The application has user friendly GUI interface which allows user to take an image of a lower resolution and it outputs a highly resolved image. But, this application has its own limitations such as we cannot capture image from a live streaming video or a surveillance camera but this can be the future scope of this project. Currently in our application, we are providing an option to increase the resolution of an image by the scaling factor of 2 and 4, what we can do further is that we can take users input and then our model will fulfill the required resolution need of the user.

References:

[1] Jianchao Yang, Wright, J., Huang, T. S., & Yi Ma. (2010). Image Super-Resolution Via Sparse Representation. IEEE Transactions on Image Processing, 19(11), 2861–2873.

[2] Capel, D., & Zisserman, A. (2003). Computer vision applied to super resolution. IEEE Signal Processing Magazine, 20(3), 75–86.

[3] Sung Cheol Park, Min Kyu Park, & Moon Gi Kang. (2003). Super-resolution image reconstruction: a technical overview. IEEE Signal Processing Magazine, 20(3), 21–36.

[4] Chao Dong, Chen Change Loy, Kaiming He, and Xiaoou Tang. Image super-resolution using deep convolutional networks. IEEE TPAMI, 38

[5] Manuel Fritsche, Shuhang Gu, and Radu Timofte. Frequency separation for real-world super-resolution. In ICCVW, 2019.

[6] Yue, Linwei & Shen, Huanfeng & Li, Jie & Yuan, Qiangqiang & Zhang, Hongyan & Zhang, Liangpei. (2016). Image super-resolution: The techniques, applications, and future. Signal Processing. 128.10.1016/j.sigpro.2016.05.002.

[7] Christo Ananth, G.Gayathri, I.Uma Sankari, A.Vidhya, P.Karthiga, "Automatic ImageSegmentation method based on ALG", International Journal of Innovative Research in Computerand Communication Engineering (IJIRCCE), Vol. 2, Issue 4, April 2014,pp- 3716-3721

ITM Web of Conferences **44**, 03055 (2022)
ICACC-2022
https://doi.org/10.1051/itmconf/20224403055

Facial Emotion Classifier using Convolutional Neural Networks for Reaction Review

Makarand Madhavi[1,*], *Isha* Gujar[2], *Viraj* Jadhao[3], and *Reshma* Gulwani[4]

[1]Department of Information Technology, Ramrao Adik Institute of Technology
[2]Department of Information Technology, Ramrao Adik Institute of Technology
[3]Department of Information Technology, Ramrao Adik Institute of Technology
[4]Department of Information Technology, Ramrao Adik Institute of Technology

Abstract. Applications of facial emotion classification is gaining popularity in the world. There are many ways to train a model to classify human facial expressions by use of existing technologies. The strategy to order and recognize feelings of an individual conveyed by his facial expression is done by contrasting it to a gathered set of labelled experiences of feelings. In this paper, we propose the making of an intelligent system that will recognize and classify facial emotions. A multi-layer Convolutional Neural Network model is proposed. Another method of training using pretrained ResNet50 Model is explored. A basic live video streaming application is developed to showcase the use case of our model which will be capable of monitoring and recording facial emotions in real time from a live video stream and subsequently summarize the overall reactions at the end of the stream.

1 Introduction

By definition, A facial expression is one or more developments or positions of the muscles under the surface of the face. Facial expressions are form of nonverbal communication.[1] These muscles create a specific visual structure on the face which relates to the specific emotion. Humans recognize an emotion by identifying the structure of the facial expression and relating to an emotion by past experiences. The way humans recognize emotions comes very naturally by the way their brains are coded. In order for machines to be able to recognize and classify emotions it needs to replicate neuron structures of the brain in a similar fashion. Convolutions Neural Networks is a technology that is widely used in artificial intelligence in image classification. It can be used to train a model to classify human emotions based on labelled set of data of images of facial expressions.

Such a system that can identify and classify a facial emotion has a wide range of applications. It can be used is psychological studies and surveys with huge sample size. An automatic emotion recognition system will efficiently be able to summarize survey data with very less time requirement. It can be used in patient monitoring systems. An intelligent emotion recognition system has a high usage in advertising and customer feedback. In movie review analysis consisting of a large number of reviewers and film duration in hours, an emotion classification system will be able to summarize audience reaction in a matter of minutes after a reviewer finish watching. A live webcam image can be obtained to quantify real time emotions which additionally have various applications.

2 Background

The most common applications of artificial intelligent systems are mimicking human behaviours in one form or another.[2] To be able to mimic such behaviours a computer needs to understand the behaviours themselves, how they look, how they act, how they feel. The very soul of why humans act the way they do is driven more by emotions than by logic. Decades of artificial intelligence research fuelled by the development and availability of high amounts of compute power enabled by GPUs have made it possible to make artificially intelligent systems capable of classifying images to classes. Such systems can be used to recognize emotions from facial images. If computers can keep track of psychological state of a user they can better serve and act accordingly.[3] Emotion recognition is therefore vital to human computer interaction.

3 Literature Review

3.1. Use of Neural Network technologies

* Corresponding author: makarandmadhavi99@gmail.com

Neural Networks can be used to classify facial expressions like happy, sad, angry, surprise etc. using MATLAB, and Neural Network Toolbox technologies as per research conducted by Dilbag Singh. They are however difficult to implement with complex high-density code and sub optimal processing.[4] Research done by Nithya Roopa.S1 elaborates the use of open-source technologies such as Inception Net, Tensor Flow, Kaggle with the help of Karolinska directed emotional faces (KDEF) Datasets and transfer learning algorithms, however they only achieve an accuracy of 35%.

3.2 Movie Review Analysis: Emotion Analysis of IMDb Movie Reviews

Movie reviews can be also analysed from IMDb reviews however this research argues that the reviews are highly user taste specific containing user biases with halo and horn effects.[5] A more direct reaction capturing system will be more accurate to get un-biased and organic reviews.

3.3 Reaction Review

Movie test screening involves three phases:

1. In Before Phase: The audience is informed about the movies and is advised about a few things.
2. In Watching Phase: The facial emotion recognition system views and records the facial emotional reactions of the audience. The audience watches the movie and their emotional reactions are recorded and saved. The system takes reviews from the audience or provides feedback. The audience is given a questionnaire to review the film which gives clarity of the plot and other general information.
3. In After Phase: After collecting feedback, the recorded emotional reactions and reviews from the audience are analysed to prepare a report.

3.4 Supporting Technologies

A web-based user interface system will be most suitable for a facial emotion recognition system as it can be further scaled up for specific applications as per the need.[6] Python programming language is widely used in data science and artificial intelligence; we will be using python library Flask supported by other open-source libraries to enable us to train and use a facial emotion recognition model. Having a central backend for the highly computational emotion recognition module will be efficient in order to scale up in future scope.

4 Methodology

First, we are preparing dataset for training and testing models. We will prepare two models and pick the best among the two. The first is the CNN model and the following is transfer learning model with a pre-prepared Resnet50 model.[7] Then we implement a module for prediction emotions from images utilizing trained models which are used for recognizing emotions from frames of video and then visualizing predicted emotional data.

4.1 Datasets

Neural networks require large amounts of data for training and validation. The selection of high labelled data is directly responsible for the performance of the model. Thus, we require high quality and quantity of data for training and validation. There are quite a few datasets available for research like the KDEF dataset and FERC-2013 dataset. The Facial Expression Recognition Challenge (FERC-2013) dataset is more suitable for this task. This dataset contains cleanly labelled 32000 in number 48x48 resolution images. [8,9,10] The FERC-2013 set shows emotions from all genders, races and age groups. This dataset might be difficult to interpret however the diversity of the data will be beneficial for our model. We thus hope to achieve good results from this dataset. FERC-2013 dataset is employed for training the model, the dataset contains 48x48 grayscale images categorized into these emotions – happy, sad, angry, neutral, fearful, surprised and disgust. We will be training a neural network model using deep learning with the assistance of the TensorFlow library. Using Keras module that uses deep learning to train a neural network model.

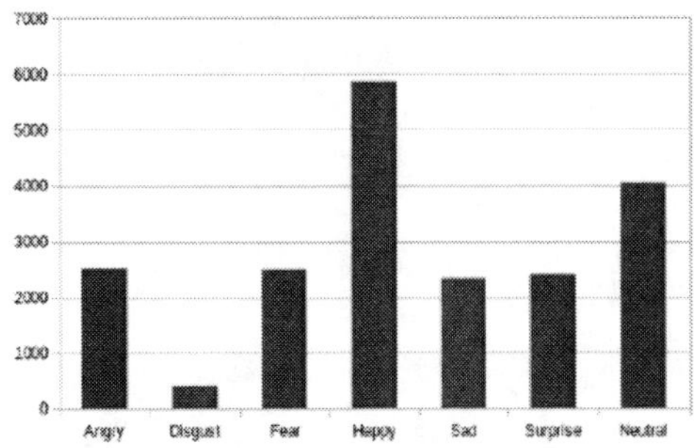

Fig. 1. FERC2013 dataset summary.

4.2 Convolutional Neural Network Model

We are going to construct a CNN model with the following architecture. Having four convoluted layers, a dense layer and a final seven output SoftMax layer denoting the expressions - angry, happy, sad, disgust, fear, surprise and neutral. [11,12] The filters range from 2x2 to 3x3 filters with 64-128 filters on each convolutional layer. We also add dropout and max-pooling to improve training efficiency and avoid overfitting of the model. The images before being fed to the model are pre-processed by rescaling them to a factor of 1.0/255 for each pixel value so that they are in 0-1 range. This is done since we are training the model from scratch and it is generally good for computation.[13] This also eliminates bias for darker and lighter images such that now every image had equal impact on the model. If images are not rescaled, the brighter images have more impact on the model than lighter images. This also eliminates racial biases in training. So, a pale complexion of skin colour is given the same weightage as a darker complexion of skin colour. We also implement some image augmentation techniques to prevent overfitting the model and ensure features learned are not highly positional but relative in nature. Image augmentation also virtually increases the number of images in the dataset as each augmented image is treated as a different image. We randomize width shifting, height shifting and zooming by a factor of 10%. We also randomly flip images horizontally. The loss function used is categorical cross entropy and optimizer for training used was Adam optimizer with a relatively slow learning rate. [14,15] The model was tried to train with Stochastic gradient descent algorithm but the training turned out to be too slow with a lower learning rate, and unstable with a higher learning rate. The training and validation loss and accuracy curves followed each other for a few epochs after which the validation loss was seen to be incrementally increasing which denoted that the model is overfitting beyond this point and couldn't get any better. We are achieving 66.23% accuracy with this model.

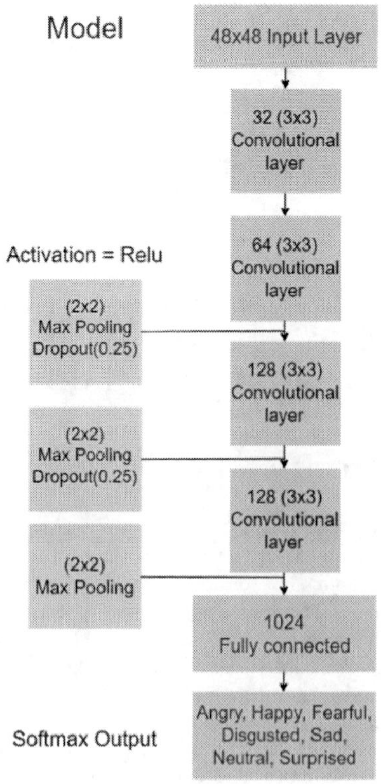

Fig. 2. Structure of CNN model used for training.

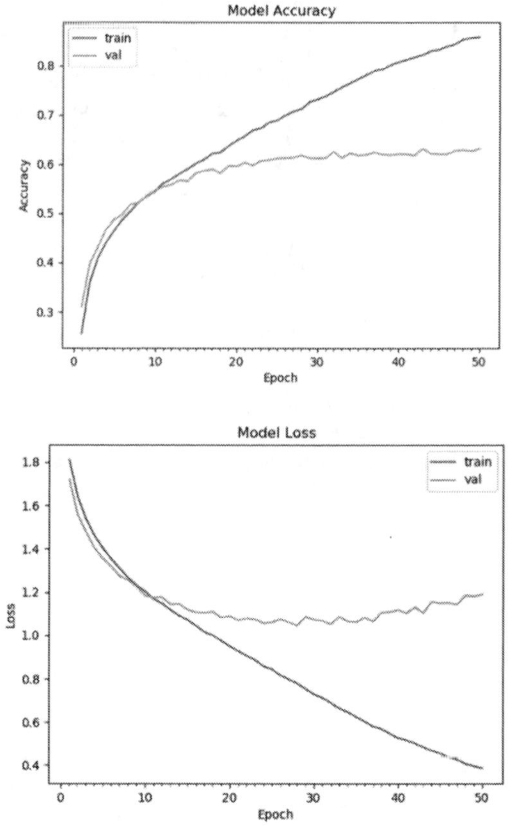

Fig. 3. Accuracy and loss curve after training.

4.3 ResNet-50 Model

ResNet, which stands for residual networks, is widely used CNN model for computer vision tasks. This model was awarded the ImageNet Challenge in 2015.[16] This model has 50 convolutional layers. ResNet solved the problem of vanishing gradients which resulted for easier training. We are using the ResNet50 variation of this model for our application. We use pretrained "ImageNet" weights loaded into the model before our training and fine tuning. We add 2 more dense layers on top of this model. The first dense layer is equipped with a L2 class layer weight regularizer. This is added to apply penalties and regularize layer output. Lastly, we add a 7 neuron SoftMax activation output layer representing our 7 emotion classes. Dropout layers with a factor of 0.2 are added in-between the dense layers as well to prevent overfitting. All the layers in this model are set to non-trainable except for last four layers for initial training. [17,18] Same image augmentation techniques are used on the images that was discussed in the previous model consisting of rescaling, width shifting, height shifting, zooming and horizontal flip. The ResNet50 model is designed to take RGB images as input. It takes 3 layers of RGB as input but our dataset images are grayscale having only one grayscale layer.[19] To overcome this issue, we send the same

 https://doi.org/10.1051/itmconf/20224403055

grayscale layer to all 3 RGB input layers. Once again, the Adam optimizer seems to be working the best here. The loss function is categorical cross entropy. We train for 50 epochs. The accuracy gradually increases and loss gradually decreases for a few epochs after which we don't see any improvement in training.

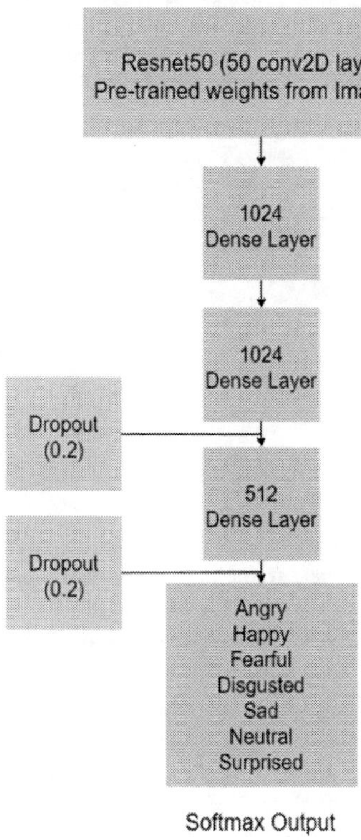

Fig. 4. Resnet50 model by Transfer Learning.

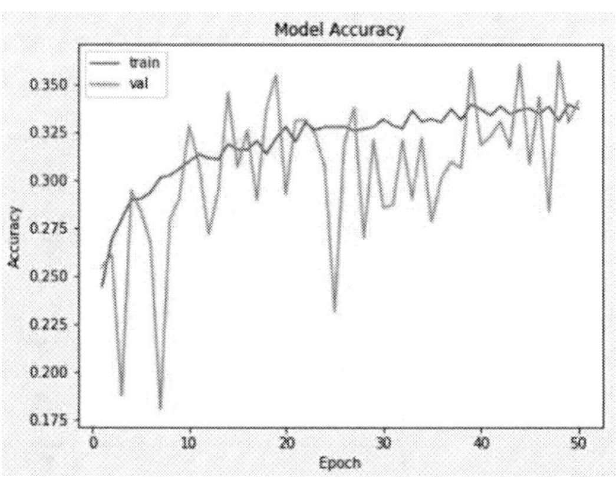

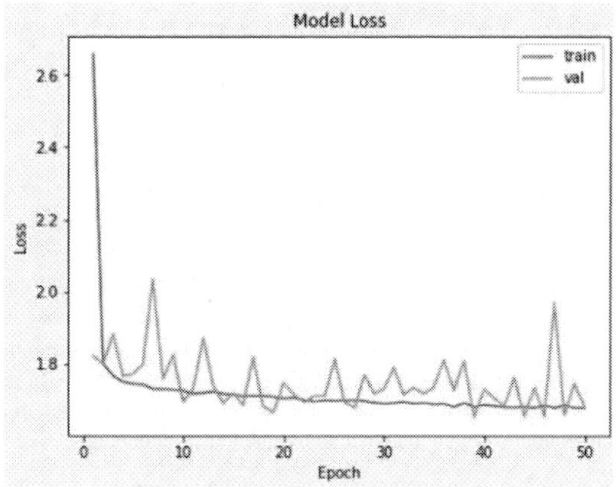

Fig. 5. Accuracy and loss curve after first training ResNet.

We get 33.25% accuracy from this training. We then try to fine tune this model by setting all layers of the model to trainable. Using Adam optimizer and a very low learning rate we train for another 50 epochs.

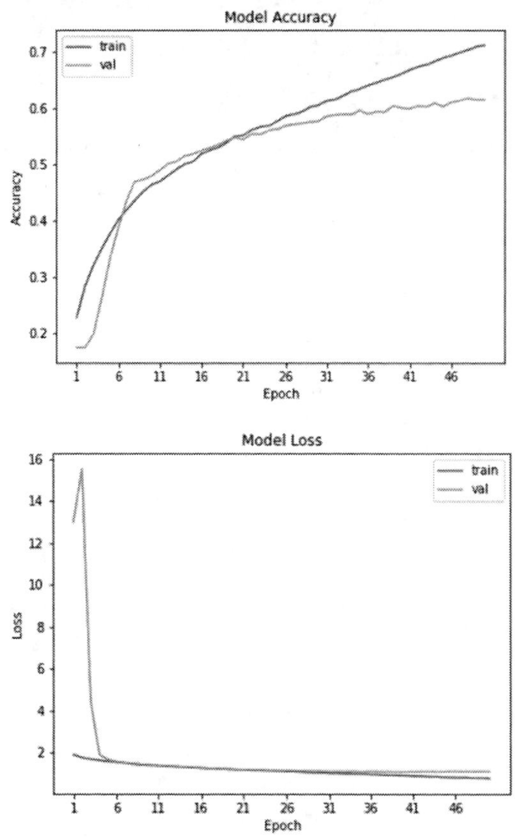

Fig. 6. Accuracy and loss curve after fine tuning ResNet50 model.

The accuracy gradually increases and loss gradually decreases for both training and validation. The training and validation curves follow each other for a few epochs after which training curves continue to steadily improve with no improvement in validation curves which denotes the model is overfitting beyond this point.[20] We achieve an optimum validation accuracy of 61.47% after fine tuning.

4.4 Predicting Emotions

For prediction, input images are converted to grayscale since our model is trained on grayscale images. All the faces inside an image frame are detected using the OpenCV Haar cascade classifier. Detected faces are scaled to 48x48 resolution images which is the same resolution of images our model was trained on. Scaled images are run through the model to classify emotions.

Fig. 7. Faces and emotions classified by CNN model.

5 Application

For detecting faces and classifying emotions from a video stream we sample the video into image frames at regular intervals. These sampled images are run through a processing pipeline that classifies emotions as described above. Sampling every frame of the video stream and processing it to detect faces and classify emotions requires high memory and compute power. Given that in video consecutive frames are very similar in nature [21], processing a similar frame which has a high probability of giving similar results in use cases where precision isn't important is a waste of resources. Thus, we can sample image frames at an interval of 0.5 seconds or 1 second depending on the use case. For e.g., movies that have a runtime of over 120 minutes can choose to sample audience reactions at every 2 or 5 seconds.

The web-based GUI is created using flask at the backend for handling requests and connecting with our modules for the prediction of emotions from input images, videos and livestreams. [22] Using OpenCV library we were able to take input videos and images stored in any format. Web based GUI's however can only display images of PNG, jpeg or SVG format and videos with X264, 3GP or MPEG4 format with MP4, WebM, or Ogg containers. We therefore create predicted and processed video in mp4/H.264 format using opensource openh264 from windows utility; predicted images in PNG format.

We are also able to predict emotions live on the go and record them from user's video capture device or screen share device. This is made possible by capturing images at regular intervals from the capture device. The user is given the option to set the interval at which frames are sampled. Sampled images are sent to the prediction server via HTTP post requests. [23,24] The images are processed by the emotion recognition system in a similar fashion as stated in prediction. The images are stacked in a mp4/H.26 container to generate the output.

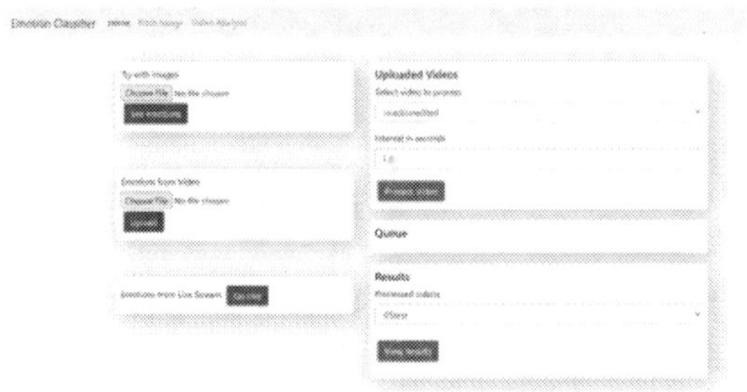

Fig. 8. Main Page.

5.1 Application Results

The videos are passed frame by frame into the model to predict emotions. The output format we get from video processing module is as follows: -

Table 1. Output from video processing module.

Timestamp	0.5	1
Happy	1	0
Disgusted	0	0
Fearful	0	0
Neutral	0	0
Sad	0	0
Surprised	0	1
Angry	0	0

Emotion probability is plotted against time for each emotion for analysis. Prediction data is aggregated in a csv file for recording and further analysis as required.

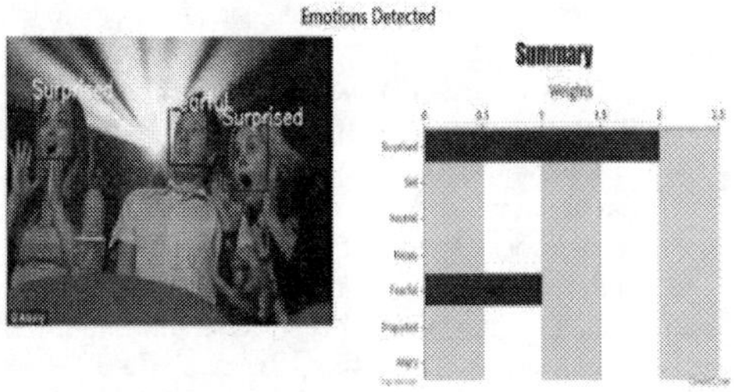

Fig. 9. Emotions detected in image.

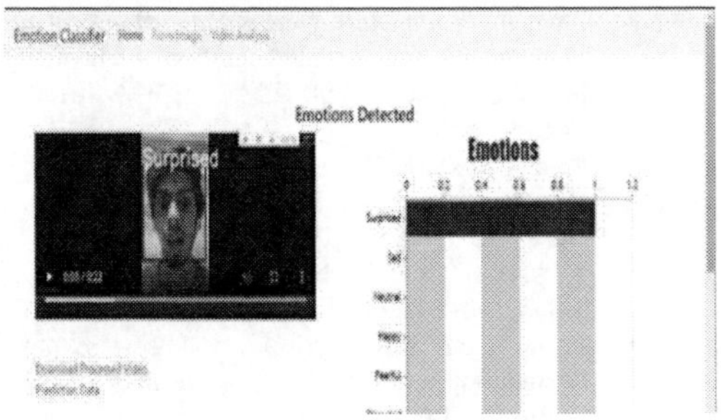

Fig. 10. Summarising reactions from live stream.

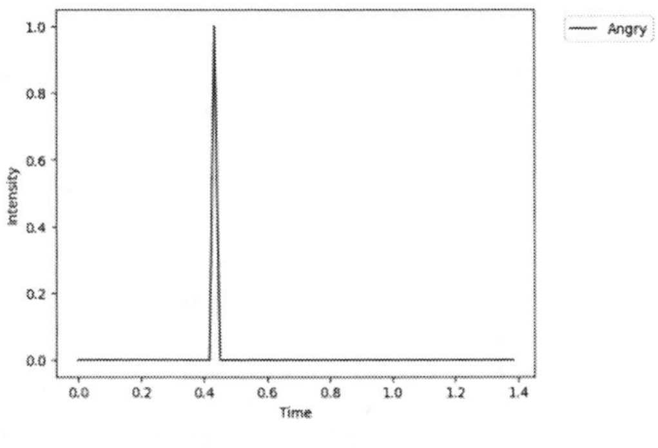

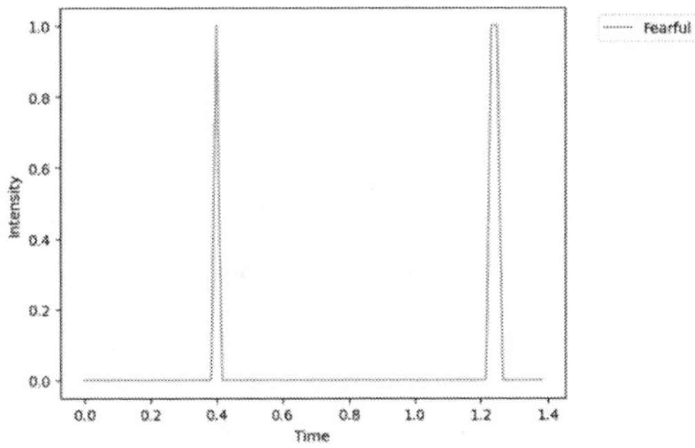

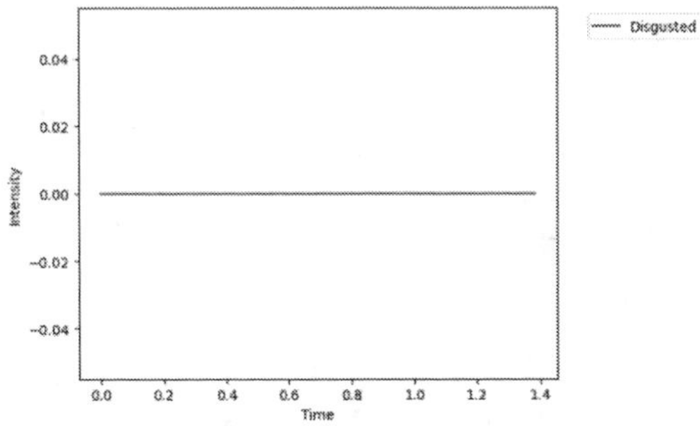

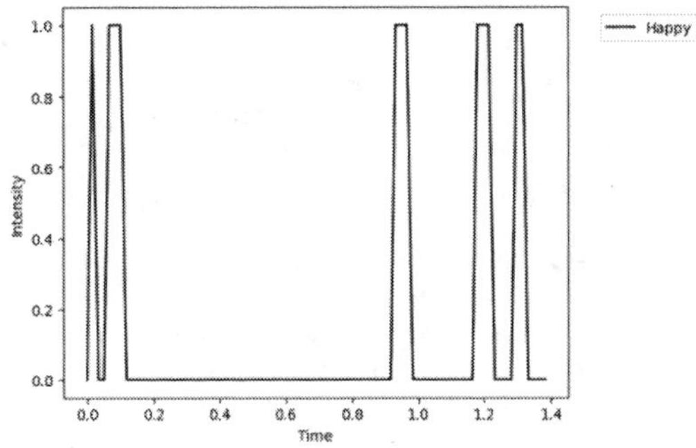

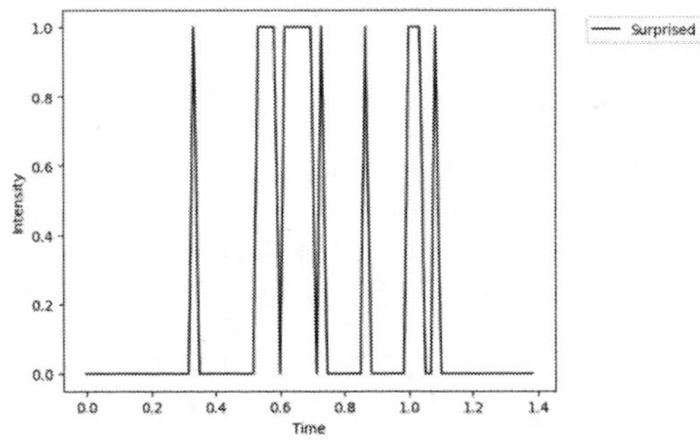

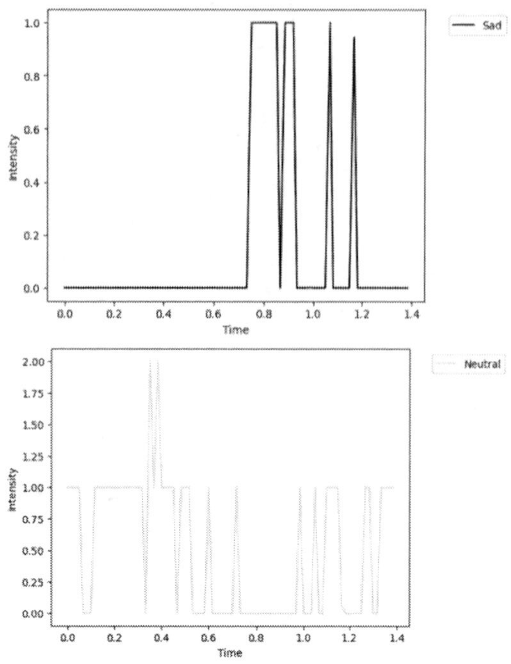

Fig. 11. Emotions Detected vs Time.

6 Conclusion

Emotion Recognition by facial expressions is one of the most challenging human-computer interaction tasks. There has been extensive research on in this topic in the past decade owing to its possibilities of numerous applications. In this project, we have trained and compared a traditional convolutional neural network model to ResNet50 transfer learning model. The results from both the models were similar denoting we achieved optimum accuracy for the dataset we used. Further research can be done to create more accurate and reliable models from a much larger dataset and compute power. We have created a system that will automatically detect the face data of individuals watching a movie or any other type of content and capture various emotions of those individuals.[25] This system will help take honest movie reviews with a good accuracy percentage. This project highlighted how current technologies and tools can used to build a primitive emotion recognition system. The system we build is capable of quantifying human facial emotions in real time and record the same. This system can be used as a tool for numerous real-world applications in various fields like entertainment, advertising, market research, psychology etc.

References

[1] Nithya Roopa.S ,"Emotion Recognition from Facial Expression using Deep Learning," in August 2019 International Journal of Engineering and Advanced Technology,Volume-8,Issue-6S.

[2] Dilbag Singh "Human Emotion Recognition System," in August 2012 MECS(http://www.mecs-press.org/) DOI10.5815/ijigsp.2012.08.07).

[3] Zhiwei Deng, Rajitha Navarathna, Peter Carr, Stephan Mandt, Yisong Yue, Iain Matthews, "Factorized Variational Auto encoders for Modelling Audience Reactions to Movies", Greg Mori Simon Fraser University, Disney Research, Caltech.

[4] S. P Khandait, Dr.R.C. Thool & P.D. Khandait, "Automatic Facial Feature Extraction and Expression Recognition based on Neural Network", (IJACSA) International Journal of Advanced Computer Science and Applications. 2, No.1, January 2011.

[5] Octavio Arriaga, Paul G. Ploger, Matias Valdenegro, "Real-time Convolutional Neural Networks for Emotion and Gender Classification".

[6] John Gideon,Soheil Khorram,Zakaria Aldeneh,Dimitrios Dimitriadis2,Emily Mower Provost, "Progressive Neural Networks for Transfer Learning in Emotion Recognition", University of Michigan at Ann Arbor, IBM T. J. Watson Research Centre.

[7] Prathap Nair, Andrea Cavallaro, "3-D Face Detection, Landmark Localization, and Registration Using a Point Distribution Model", IEEE TRANSACTIONS ON MULTIMEDIA, VOL. 11, NO. 4, JUNE 2009.

[8] Jayalekshmi J, Tessy Mathew, "Facial Expression Recognition and Emotion Classification System for Sentiment Analysis", 2017 International Conference on Networks & Advances in Computational Technologies (2017).

[9] Kamil Topal, Gultekin Ozsoyoglu, "Movie Review Analysis:Emotion Analysis of IMDb Movie Reviews", 2016 IEEE/ACM International Conference on Advances in Social Networks Analysis and Mining (ASONAM).

[10] Anurag De , Ashim Saha, "A Comparative Study on different approaches of Real Time Human Emotion Recognition based on Facial Expression Detection", 2015 International Conference on Advances in Computer Engineering and Applications (ICACEA),IMS Engineering College, Ghaziabad, India.

[11] Andrew Ryan, Jeffery F. Cohn, Simon Lucey, Jason Saragih, Patrick Lucey, Fernando De la Torre, Adam Rossi, "Automated Facial Expression Recognition System", 43rd Annual 2009 International Carnahan Conference on Security Technology,5-8 Oct. 2009.

[12] N Mehendale, "Facial Emotion Recognition using convolutional neural networks (FERC)", SN Appl. Sci,446 (2020).

[13] C. Shetty, A. Khan, T. Singh and K. Kharatmol, "Movie Review Prediction System by Real Time Analysis of Facial Expression," 2021 6th International Conference on Communication and Electronics Systems (ICCES), (2021).

[14] Almeida, João, Luís Vilaça, Inês N. Teixeira, and Paula Viana. "Emotion Identification in Movies through Facial Expression Recognition" Applied Sciences 11, no. 15: 6827, (2021).

[15] Ijaz Ul Haq ,1 Amin Ullah ,1 Khan Muhammad ,2 Mi Young Lee ,1 and Sung Wook Baik, "Personalized Movie Summarization Using Deep CNN-Assisted Facial Expression Recognition",

Complexity, vol. 2019, Article ID 3581419, 10 pages, (2019).

[16] Mariya A. Ali1, Dr. Sonali B. Kulkarni2, "Emotion Detection and Sentiment Analysis for Hindi Movie Reviews", International Journal of Emerging Trends & Technology in Computer Science, Volume 10, Issue 1, (2021).

[17] Priyanka Nathawat, Dr. Vivek Chaplot, "A Review of Facial Expression Recognition", European Journal of Molecular & Clinical Medicine, Volume 7, Issue 4, (2020).

[18] S Minaee,A Abdolrashidi, "Deep-Emotion: Facial Expression Recognition Using Attentional Convolutional Network", Sensors (SENSORS-BASEL),(2019).

[19] Singh, Chandra Bhushan and Sarkar, Babu and Yadav, Pushpendra, "Facial Expression Recognition",SSRN, (May 25, 2021).

[20] K. Lekdioui, Y. Ruichek, R. Messoussi, Y. Chaabi and R. Touahni, "Facial expression recognition using face-regions," 2017 International Conference on Advanced Technologies for Signal and Image Processing (ATSIP), (2017)

[21] H Nguyen, S Yeom, G Lee, H Yang, I Na and S Kim, "Facial Emotion Recognition Using an Ensemble of Multi-Level Convolutional Neural Networks", International Journal of Pattern Recognition and Artificial Intelligence Vol. 33, No. 11, (2019).

[22] In-kyu Choi, Ha-eun Ahn and Jisang Yoo, "Facial Expression Classification Using Deep Convolutional Neural Network", J Electr Eng Technol, (2018).

[23] Damir Filko, Prof. Goran Martinović, "Emotion Recognition System by a Neural Network Based Facial Expression Analysis", Automatika, (2013).

[24] Milan Tripathi, "FACIAL EMOTION RECOGNITION USING CONVOLUTIONAL NEURAL NETWORK", Ictact Journal on Image and Video Processing, Volume: 12, Issue: 01, (2021).

[25] Wafa Mellouk, Wahida Handouzi, "Facial Emotion Recognition using Deep Learning: Review and Insights", Procedia Computer Science, Volume 175, (2020).

https://doi.org/10.1051/itmconf/20224403056

Crop State Analysis Using Image Processing

Pratik Khandagale[1,*], *Dharmaraj* Eranjikal[1,**], *Siddhi* Parab[1,***], and *Bhawana* Garg[1,****]

[1]Ramrao Adik Institute Of Technology, Nerul, Navi Mumbai

Abstract. Agriculture is the primary source of livelihood for the majority of India's population. In the traditional agriculture system, farmers have to manually check their crops in every section of the farm. This may at times leave some sections unchecked. Due to this, crop health may deteriorate considerably. If may not be treated in time, the farmer will face a huge loss and the crop yield will also reduce. To overcome this problem, a system can be employed to monitor the crops and out the sections where crop condition is worsening. For this, an image processing algorithm can be implemented. Through this system, the received data from the algorithm will be manipulated to enhance the crop's health.

1 Introduction

For around 60 percent of India's total population, agriculture is their primary source of income. Agriculture and forestry had a Gross Value Added of Rs. 17 lakh crore (US dollars 225 billion) in Financial Year 20. In Financial Year 20, agricultural and allied sectors accounted for 17.8 percent of India's gross value added (GVA) at current prices. Consumer spending in India would grow by as much as 6.8 percent in 2021, following a pandemic-driven drop.

Due to its enormous potential for value addition, notably in the food processing industry, the Indian food industry is poised for massive expansion, increasing its contribution to world food commerce every year. The Indian food and grocery market is the sixth largest in the world, with retail accounting for 70 percent of total sales.

The Indian food processing sector, which accounts for 32 percent of the country's overall food market and is rated fifth in terms of production, consumption, export, and predicted growth, is one of the country's largest industries.

In Financial Year 21, total agriculture and allied product exports totalled 41.25 billion dollars.

In the case of plants, illness is defined as any disturbance of a plant's normal physiological function that results in recognisable symptoms. A symptom is a phenomena that occurs in the presence of something and is used to prove its existence. A pathogen, or any agent that causes disease, causes disease. The majority of the time, pests or illnesses are visible on the plant's leaves or stems [1].

As a result, identifying plant leaves, as well as determining diseases, percentages of disease occurrence, and symptoms of disease assault, is critical to successful crop cultivation.

*e-mail: pkhandagale02@gmail.com
**e-mail: mandareranjikal@gmail.com
***e-mail: siddhiparabs00@gmail.com
****e-mail: garg.bhawana@rait.ac.in

Traditional farming is difficult because extreme weather conditions or insects can harm crops, reducing overall crop development. As a result, a drop in crop yields might result in a significant loss of population. Image processing algorithms can be used to solve this challenge. The received data from the algorithm will be altered using this system to improve the crop's health.

1. Determines a crop's condition in early stage.
2. Generates data about farm's every area.
3. It predicts approximate yield of crops.
4. Helps in saving farmer's time.

The next section covers the block diagram of the rover. The third section gives technical insights into all the hardware and software used for the rover. In section 4, the complete design and the components implementation are reviewed. The fifth section will consist of the snaps taken of the constructed rover. The final section consists of the concluding remarks. The references are also mentioned at the end.

2 Proposed Work

The system consists of Motors controlled by a Motor Driver, ESP-8266 micro-controller, ESP-32 Camera installed on Rover Chassis. The rover connects with the Computer via an Internet of Things application. The ESP-8266 micro-controller connects to the internet by connecting to a Jio-Fi portable hotspot which creates a portable WiFi range. The Computer system runs an application which is responsible for controlling the rover. The application is also responsible for processing the images of the farm, handling the database and representing data in understandable format. When the system is turned on, the rover and the application waits for each of them to connect to the IoT platform. Once they both are connected, the application ask the user to feed the dimensions of the area of farm to be scanned. On this, the application push

https://doi.org/10.1051/itmconf/20224403056

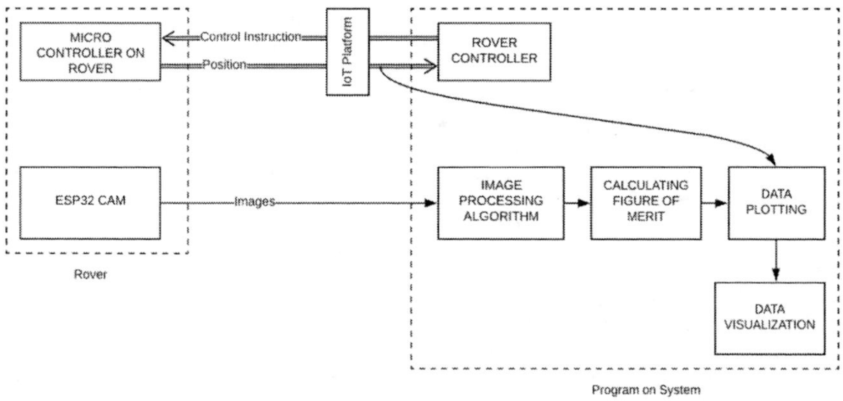

Figure 1. Block Diagram of System

the user-fed data to rover in-order to set-up the rover. The application waits for set-up process to be completed [2] [3] [5]. On completion of the set-up process, the application send the control signal to the Rover. The control signals are produced respective to the arrow keys pressed by the operator. The rover reads the control signals transmitted by the host computer. The micro-controller on the rover, drives the motors to move the rover in accordance to the keys pressed by the operator. Simultaneously, the rover calculates the time elapsed and finds the distance travelled by the rover. This value of distance travelled is sent back to the host computer. The application on the computer, estimates the location of the rover in the field. After travelling for a particular distance, the application commands the rove to halt and capture image of the farm. The image of the farm is sent to the application where, it calculates the intensities of the Red, Green and Blue light reflected by the leaves of the crops. Depending upon the intensities of colour bands, the application predicts if the crop is health or is going down in terms of health. The application generates a figure of merit for the health condition of the crop. It then enlists this value along with it's respective co-ordinates in the farm in a database [4]. After completion of scanning the said area of the farm. The application generates an image which highlights the area where crop health might be deteriorating. This report can help the farmer to pay attention to these areas and protect the crops from worsening.

In this paper, a system is proposed which can identify the areas where crop health may be affected. Thus, farmers can produce more yield.

3 Technical Insights

3.1 Hardware Implemented

Frame: The frame of a rover is the main structure, or the skeleton upon which the rest of components will be attached.

ESP8266: It is a microcontroller which helps us to connect with the wifi, or send or receive data. With the respective of the project, the ESP is employed as a receiver. It produces PWM signals similar to the signals produced by the receiver output for corresponding values sent by the transmitter.
Motors: The motors are the main drain of battery power on your drone.
Motor Drivers: It acts as an interface between motors and the control circuits. It takes low current signal and converts it into higher current signal that can drive a motor.
Battery: LiPo batteries are the power sources of the rover. LiPo is used because of the high energy density and high discharge rate.
JioFi Portable Hotspot: It is a device which creats the wifi signal. It is used to connect our project to the internet and the cloud-based server.
ESP32 Cam: It will be used to capture images for processing.

3.2 Software Implemented

PyCharm: IDE for developing Python program which controls the motion of the rover and also performs image processing and data visualization.
pythonanywhere.com: It is used for creating an IoT platform by using python which helps the rover to connect with the computer.

4 Design and Implementation Details

4.1 Design Methodology

The overall system can be classified into two main sections: (a)Rover and (b)Application.
The rover has a 2200mAh Li-Po battery which powers the entire system. The rover has an ESP8266 micro-controller as the main controlling element which is responsible for all

 https://doi.org/10.1051/itmconf/20224403056

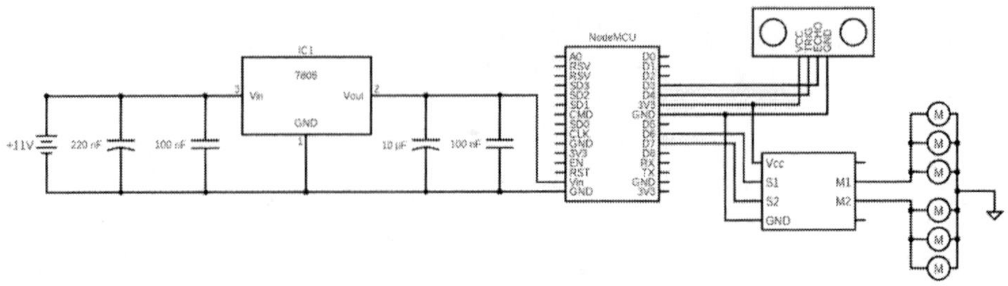

Figure 2. Circuit Diagram for Rover

the movement of the rover. The ESP8266 receives 5V regulated power supply from the 5V pin of the motor driver. 5V from the motor driver circuit is connected to the Vin pin of the micro-controller. The D1 and D2 pins of the micro-controller are assigned to generate signals for controlling the set of motor on the left of the rover. Whereas, pins D12 and D13 are assigned for producing controlling signals for motors which drive the wheels on the right side of the rover. The movement of the rover is controlled by driving the set of wheels in respective pattern to produce left turns, right turns, forward and reverse drive. The D1, D2, D12 and D13 pins are connected to respective pins on the motor driver. The motors are then connected to the respective output pins of the motor driving circuit.
Another important component installed of the rover is an ESP-32 Camera. The ESP-32 camera helps the application running on the host computer to capture images of the farm. The ESP-32 camera and the ESP8266 micro-controller connect to an Internet-of-Things(IoT) platform via a Portable hotspot employed on the rover. The Portable hotspot provides internet connectivity to the rover.
The process flow of the application can be explained with help of following Flow Diagram:

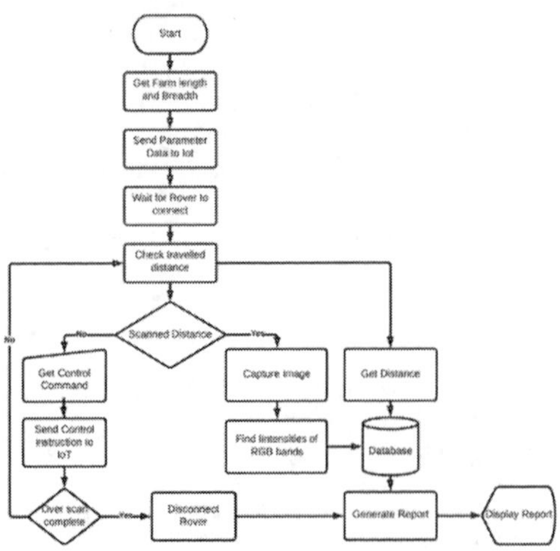

Figure 3. Flowchart

On start, the application asks the user to enter a reference name for the session. On next page, the application requires two mandatory fields which represent the length and breadth of the area on the farm to be scanned. On entering the required data the application sends the necessary data to the rover via IoT platform for completing the set-up process of the rover.
After successful set-up of the rover, the application checks for the successful communication between the rover and the application [6].
The application send control commands to the rover in-accordance to the arrow keys pressed by the operator. The micro-controller on the rover controls the movement of the rover according to the control commands received by the same. Simultaneously, rover transmits the distance travelled by it. Depending on these distance values, the application predicts the co-ordinates of the rover on the farm. On travelling as predefined amount of distance, the application commands the rover to halt and captures the image for processing it [7].
The captures image is processed to find the intensities of red, blue and green light. The application, generates a figure if merit for health condition of the crop in the area scanned. It then stores the co-ordinates and figure of merit in a database [8]. On scanning the said area, the application processes the data in database to generate a report which highlights area with their respective health condition. By reading the generated report, one can predict the areas where crop health might be going down.

4.2 Selection Criteria for Components:

To cut down the cost, an 11.1V 2200mAh Li-po Battery was employed in the system. ESP8266 has an on-board WiFi Module. Thus, it proves to be efficient and economical for building a system requiring to connect with the internet. ESP32 CAM is an economical, low power consuming camera which can be easily interfaced with our desired circuitry. Jio-Fi Portable Hotspot was implemented in the system to provide strong internet connectivity while in motion as it was compact and portable.

Figure 4. The Rover with ESP32 CAM implemented

5 Results and Discussion

As aimed, the rover scans the defined area of the farm, and calculates the figure of merit which represents qualitative status of health condition of the crops. The report generated by the system highlights the areas on the farm where crop health might be going down.

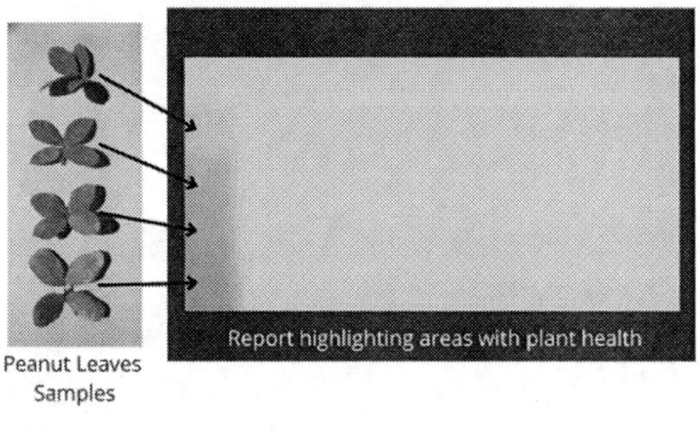

Figure 5. The Report generated by the system

Figure 5 demonstrates working of the developed system. The report shows colour shades which represents the health condition of the crops in particular area. The red colour represents crops with least qualitative value of crop health, whereas, green colour represents highest qualitative value of crop health.
On comparison with sample leaves which represent the crops, a conclusion can be drawn that the generated report image has highlighted the area with wilting leaves as red. On other hand, the area where crops are in good condition is marked with green shade. Thus, the designed system can predict the area where crop health is going down.
This report can be used by the farmer to pay more attention to the crops in highlighted areas where crop health is affected. This will in-turn, assist the farmer to save the crops from being damaged before it is late. By saving the crops from being spoilt, farmers can avoid any plant disease from spreading to other area and can increase their crop yield.

6 Acknowledgements

With great pleasure, we avail this opportunity to express our profound gratitude and deep regards to our Head Of Department of Instrumentation Engineering, RAIT, Dr. Sharad P. Jadhav, for his generous support and constant encouragement. It helped us to work in the right direction to attain the desired objective.

7 Conclusion

Image Processing can contribute to the foregoing numerous aids a farmer uses to tend his farm. A slight change in a leaf's colour, i.e. a small infection to a crop which could go undetected by naked eye will get detected easily. The use of a rover in the farm will consequently become more present in the near future when more of this systems are bought into action. A farm's crops are the most important aspect of a farmer's living, which will be saved by the implemented machine. The farm's dimensions when entered will get the system connected and the whole farm area will get examined. The farmer once alerted, can keep an eye over the area where the damage is detected. Hence, it will result in crops getting saved for the sake of farmer's convenience. Moreover, the annual yield produced by the farmer will increase. The model could be used by the government that helps farmers with their benefits. In some cases, rover companies can sell the model directly to the farmer. Image Processing is a evolving field with many opportunities for affecting future research.
Considering the rover's future scope, a GPS will be added to trak the location of the rover and to recover the rover from a lost location in case of any system failure. As this system works over the surface, measurement of soil moisture level could be done.

References

[1] Jaware, T. H., Badgujar, R. D., Patil, P. G., Crop disease detection using image segmentation. World Journal of Science and Technology **2**, 190-194 (2012)
[2] M. Jhuria, A. Kumar and R. Borse, Image processing for smart farming: Detection of disease and fruit grading **71**, 521-526(2013)
[3] J. R. Fiona and J. Anitha, Automated Detection of Plant diseases and Crop Analysis in Agriculture using Image Processing Techniques: A Survey **71**, pp. 1-5(2019)
[4] Vibhute, Anup S.K.Bodhe, Application of Image Processing in Agriculture: A Survey. International Journal of Computer Applications **1**, 34-40(2020)

[5] Shibghatallah,Muhammad Abdul Hakim and Khotimah,Siti Nurul and Suhandono,Sony and Viridi,Sparisoma and Kesuma,Teja, Measuring leaf chlorophyll concentration from its color: A way in monitoring environment change to plantations **1**, 210-213 (2013)

[6] H. A. M. Tran, H. Q. T. Ngo, T. P. Nguyen and H. Nguyen, Design of Green Agriculture System Using Internet of Things and Image Processing Techniques **37**, 28-32 (2018)

[7] T. McDonald, Y. R. Chen, Transactions of the ASAE **4**, 784–791 (1990)

[8] Vijai Singh, A.K. Misra, Detection of plant leaf diseases using image segmentation and soft computing techniques, Information Processing in Agriculture, **4**, 41-49(2017)

Diabetes & Heart Disease Prediction Using Machine Learning

BhaveshDhande[1], Kartik Bamble[2], Sahil Chavan[3], Tabassum Maktum[4]

[1,2,3,4]Ramrao Adik Institute of Technology, D Y Patil Deemed to be University, Navi Mumbai, India

Abstract One of the root causes of mortality in today's world is the culmination of several heart disease and diabetes illnesses. In clinical data analysis, predicting multiple diseases is a significant challenge. The machine learning approach has proved to be functional in assisting in the decision-making and governing of large amounts of data generated by the healthcare field. The various experiments scratch the surface of machine learning to predict different diseases. The papers present a novel method for identifying significant features using machine learning techniques, which improves the diagnosis of multi-purpose disease prediction. The different features and many well-known classification methods are used to implement the prediction model to predict the heart disease and diabetes. The proposed method utilizes ensemble approach for achieving a higher degree of accuracy rates for by using classification algorithms and feature selection methods. The proposed method implements voting classifier that has sigmoid SVC, AdaBoost, and Decision tree algorithms. The paper also implements the traditional classifiers and presents the comparison of different models in terms of accuracy. The web application is also developed for users to avail its services very easily and make it convenient for their use, particularly in the prediction of heart and diabetes collectively.

Keywords: Machine Learning, classification, feature selection, prediction, heart disease, diabetes

1. Introduction

The answer to calculate risk through diseases via model-based prediction is very difficult. Due to this, the examination of several medical datasets and their forecasting using soft computing is very handy and a cheaper way for professionals in the healthcare industry. These techniques include exploratory analysis and constructive models that support the professional in statistical decision-making, which is a massive requirement of the medical industry. The high blood pressure, high cholesterol, diabetes, peculiar pulse rate, and various other risk factors make it harder to detect illnesses [1]. The severity of heart disease in humans was determined using various data mining and neural network methodologies. The decision trees, genetic algorithm, Naive Bayes, and K-nearest neighbor algorithm are all used to classify the severity of the disease. Because the features of their problems are complex, the diseases must be treated with caution. Failure to do so can reduce the effectiveness of organs or lead to early death. The various metabolic diseases are identified using a medical science approach and raw data mining. The data mining techniques with classification plays a vital role in predicting heart disease and data research [2]. The random forest is an ensemble machine learning algorithm. It is perhaps the most popular and widely used machine learning algorithm with excellent performance across a wide range of classification and regression predictive modeling problems. Also, random forest approach has a brute way of tuning parameters that helps in easier feature selection [3]. The process of aggregating the votes for class labels from individual models and predicting the class with the most votes is called a hard voting ensemble. Whereas, in a soft voting ensemble, the predicted probabilities for class labels are added up, and the class label with the highest sum probability is predicted. The voting ensemble method assumes that all the models have equal contributions in predictions, which is a drawback because some models give better results in some scenarios and poor in others. The Adaptive Boosting (AdaBoost) algorithm is a boosting technique in machine learning that is employed as an ensemble method. The weights are re-allocated to each instance, with higher weights applied to improperly identified instances. This is termed as adaptive boosting. In supervised learning, boosting is used to reduce bias and variation. It is based on the concept of sequential learning [4]. The linear SVM algorithm is used for linearly separable data, which implies that if a dataset can be categorized into two classes using only a single straight line, it is called linearly separable data, and the classifier employed is called linear SVM. Under the supervised learning approach, one of the most prominent machine learning algorithms is logistic regression. It is a method for predicting a categorically dependent variable from a set of independent factors and to describe data and explain the connection between one dependent binary variable and one or more independent variables. As a result, the result must be a discrete or categorical value such as Yes or No, 0 or 1, true or false, and so on. But, instead of giving exact values like 0 and 1, it delivers probabilistic values that are somewhere between 0 and 1. In order to predict and diagnose the recurrence of cardiovascular disease, ensemble learning incorporates the model approaches of five classifiers, including support vector machine, artificial neural network, Nave Bayesian, regression analysis, and random forest [7]. Dataset Cleveland's cardiovascular data records were retrieved from the UCI repository, and for Diabetes, the PIMA dataset has been used. The findings of the experiments showed that an ensemble model is a superior strategy in terms of diagnostic performance predictability and accuracy.

In this paper an ensemble approach is proposed to detect the heart disease and diabetes. The different algorithms combined include ADA-boost, decision tree and sigmoid SVC. The classifiers are combined by varying their weights. The major contribution of the paper is as follows:

- Provide a new approach to concealed patterns in the medical data.
- To predict the chance of heart disease with the highest accuracy of prediction.
- To predict the chance of diabetics with the highest accuracy of prediction.
- Error rate compression for the results found to make it relatively exact in accuracy.

The rest of the paper is organized as follows:
The section 2 gives the survey of existing systems for predicting heart diseases and diabetes. The section 3 demonstrates the proposed system along with the ensemble approach. The results of the proposed system are presented in section 4. Finally, the conclusion and future work is expounded in Section 5.

2. Literature Survey

The most general causes of mortality on the planet have been heart and diabetic problems. Furthermore, today, the prediction of the same or even hinting at a minute probability of it is a problem that needs a solution. In the medical field, machine learning has paved its purpose by helping make choices and predict by training over large amounts of data existing in the form of datasets.

The study in [1] represent that diabetes mellitus and hypertension were moderately associated while cardiovascular diseases are strongly associated with severity and mortality for COVID-19. The paper helps to gain relation between diabetics and heart diseases and create a link or gain experience to handle data for both diseases at the same time as it gives an idea of immunity prediction of COVID through the data of diabetics and heart. The quantitative estimate of severity outcomes and or deaths in COVID-19 patients was performed with Comprehensive Meta-Analysis Software (CMA) version 3.0. In paper [2] the K-means clustering algorithm is used for predicting heart diseases and analysis is carried out using visualization tool tableau. The Cleveland heart disease raw dataset with 76 features of 303 patients was pre-processed with exploratory data analysis which narrowed down the dataset to 209 records and 7 important features. The study includes 4 types of chest pain with age, maximum heart rate, and chest pain type which are considered as vital features in prediction. In paper [3] HRFLM method is proposed that stands for union of Linear Method (LM) and Random Forest (RF), which boosts efficiency by improving selection. The study involves pre-processing of Cleveland UCI repository with use of R rattle GUI (Feature Selection and Classification modelling) which provides an easy-to-use visual graphics, working environment for the user of the dataset, and building the predictive analytics. The several approaches are presented in [4] for predicting heart diabetes. The methodology with logistic regression provides 96% precision. This was the first paper that observed the study of more than one dataset and competes between algorithms, with pipeline affected to 98.8% fidelity using Adaptive Boost classifier. In paper [5] the difficulties in the diabetic analysis were convened in relation to the COVID-19 rate. The conclusion derived from the study is that different categories of diabetes have a unique effect on the percentage of mortality rate. The paper [6] Bhavesh Dhandeproposes an approach to predict diabetes mellitus by applying machine learning techniques. The paper concludes that minimum redundancy maximum relevant approach is better than principal component analysis. It cements random forests as a better algorithm than others. The two datasets, Luzhou and Pima were utilized, with 80.84% and 77.21% accuracies fetched respectively. The ensemble approach with various classification algorithms such as KNN, Adaptive boost (AB) Gradient boost (XB), decision tree and random forest is proposed in [7]. Based on the analysis of different algorithms, it can be concluded that the proposed system on this research edged are under cover (AUC) promisingly. The perfect couple for prediction turned out to be an ensemble of (AB+XB) classifiers. In paper [8], for predictive data mining for medical diagnosis various techniques such as KNN, Neural Networks, Bayesian classification, Classification based on clustering, Decision Tree, etc. are used. An overview of all prediction models is studied in this research paper. According to a performance study of data mining algorithms Naïve Bayes, Decision Tree, KNN provide the highest accuracy rate. The Weka 3.6.0 tool was used for conducting the research. In paper [9], the methodology is for finding out the best algorithm to extract best features from the medical dataset. Whenever, data collection is followed by data pre-processing, data mining, and pattern evaluation, the suitable and highest accuracy is achieved. The data extraction was performed with the WEKA software tool then compared using predictive accuracy, ROC curve, ROC value. The approach for prediction of heart disease by applying various algorithms like ANN, random forest, SVM is presented in [10]. By using 3-fold cross-validation along with SVM algorithm maximum accuracy achieved was 83.17%. The application of decision tree algorithm with 37 splits and 6 leaf nodes led to an accuracy of 79.12%, and when used with 5-fold cross-validation technique accuracy achieved was 79.54%. By using random forest algorithm, the accuracy achieved was 85.81%, which is maximum as compared to all other algorithms. The method presented in [11] utilizes XGBoost, AdaBoost, gradient boosting, extra trees, light gradient boosting Lightgbm, SGDC, Nu SVM algorithms for prediction of cardiovascular effects. The data pre-processing was performed on UCI repository and Framingham dataset. The data pre-processing using the Multiple Imputation Chain Equation model for filling inexistent values proved to be efficient way of data pre-processing and the accuracy of 95.83% was obtained by using the stacking algorithm. In [12] research was carried out using three methods KNN, Neural Networks and SVM on real dataset of Algerian people. Neural Network algorithm gave highest accuracy of 93%. [13] paper predicted diabetes based on different human body attributes. Study showed that Body Mass Index (BMI) and growing age are major factors in the development of risk for diabetes.

The limitations of existing system are as follows: The prediction of possibilities is not accurate for disease aggregated inputs and hence thereby cannot handle enormous datasets for patient records effectively. As the machine learning approach is based on predicting outcomes using existing data, we cannot be sure of whether it will apply to the current specimen on which it would be experimented, because evolution is a constant. To add up, poor results on very small datasets causes frequent overfitting occurrences. To conclude, these previous studies focus on the particular impacts of specific machine learning techniques and not on the optimization of these techniques using optimized methods.

3. Proposed Methodology

The paper presents a methodology to predict heart diseases and diabetes by applying machine learning techniques and perform classification by using ensemble classifiers over the datasets involved. This will provide an easy glance at the data involved to build analysis. Thus, the proposed methodology will then materialize into a model evaluation phase based on performance, only after bits like feature selection and data pre-processing. By considering the features extracted, the proposed method will frame an idea of the presence of either disease. Furthermore, all the basic models will be analyzed by ensemble techniques for the prediction. Innovative models like the Random Forest and Adaptive Booster will be added, which will further be evaluated using voting classifiers with weights and without weights.

3.1. System Architecture

The process of making new observations or categorizations from the given available data with the help of supervised learning approach is known as classification. The training dataset is hatched within several classifiers and passed through them.
These 1-n classifications are then used through a combined powerful algorithm. This algorithm then further analyzes the data by considering the ensemble considered model. This ensemble model has a build of various boosters/classifiers/outliers to improve results. Then, after building a model with the available test dataset, the prediction of any random data set can be performed. The system architecture of the proposed method is shown in Figure 1. The working of the proposed method is as follows:

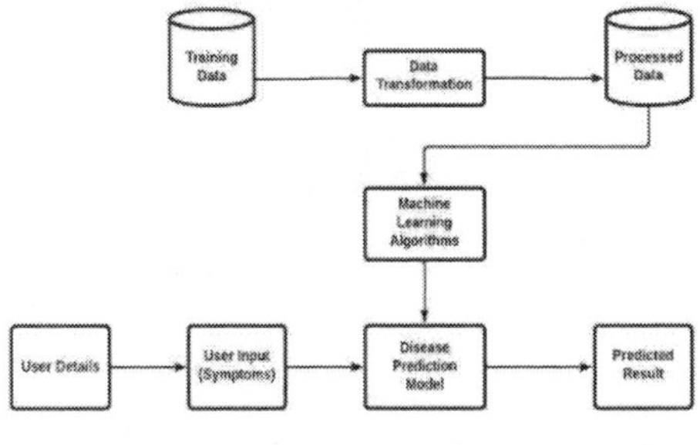

Figure 1: System Architecture

Training Data is undergoing the transformation to generate useful information i.e., processed data by various data analysis and preprocessing using various basic methods. After obtaining processed data, the algorithms can be implemented on the data involved. User then interacts by giving input which is symptoms to the application as a feed to predict results. Then, the algorithms work upon comparisons through various models in the disease prediction model to fetch the required Predicted Result.

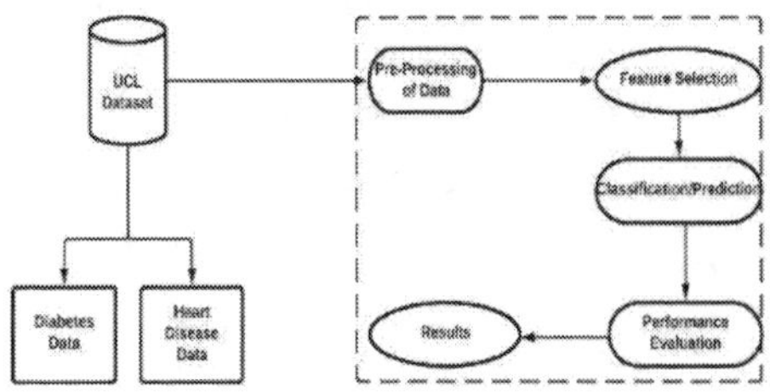

Figure 2: Dataset Handling

The paper considers the two datasets i.e., Heart and Diabetics Datasets. The advantage here is that since both have been taken from the UCL repository and belong to the same result, we can club their results.

The data from these datasets is preprocessed as per the need of exploratory data analysis and the right information is extracted. The process of feature selection is then applied, where the most important features and the impactful ones are thoroughly selected by using required classifiers and outliers. Further, the different classification models are built using random forest, Ada boost, KNN, logistic regression and decision tree approach to predict the disease. Further ahead, these results are then performance evaluated to check their accuracies and precision, to obtain the final declarative results with their legitimacy decided. After calculating the accuracy of all the above-mentioned models, we selected the top-performing models which are sigmoid SVC, AdaBoost, and Decision tree, and combined them into a single voting classifier.For diabetes prediction after calculating all model accuracies we created a voting classifier with the following weights follows KNN - 2, Decision Tree – 1, Logistic Regression – 2, and Random Forest – 2.

4. Implementation Details& Results

The proposed method is implemented by the use of the environment provided by Google Colab and key Python libraries. The various ways in which data has been analyzed are listed as follows: Plots of Attributes for Detailed Analysis, Voting Classifier: Pair, Scatter Plots, KDE/Target Visualization Plot, Correlation Matrix, Box/Pair Plotting, Plotting Comparison of Models Before & After Standardization. The various models in which data has been analyzed are listed as follows: Ensemble Classifiers, XGBoost, Decision Tree, Logistic Regression, KNN, Random Forest, AdaBoost, Sigmoid SVC, Polynomial SVC, RBF SVC & Linear SVC.

Before beginning EDA, we double-checked the dataset's dimension and variable data types, as well as looked for null values. Diabetes affects about a third of the persons in the dataset, and this division of the 'Outcome' will help our algorithms forecast more accurately for both classes (1 and 0). Co-relational Matrix is plotted to compute relation between the features used for prediction. Co-relation between the features is directly proportional to the accuracy of prediction. The co-relation between the features is improved by scaling all the feature in a specific range of values.

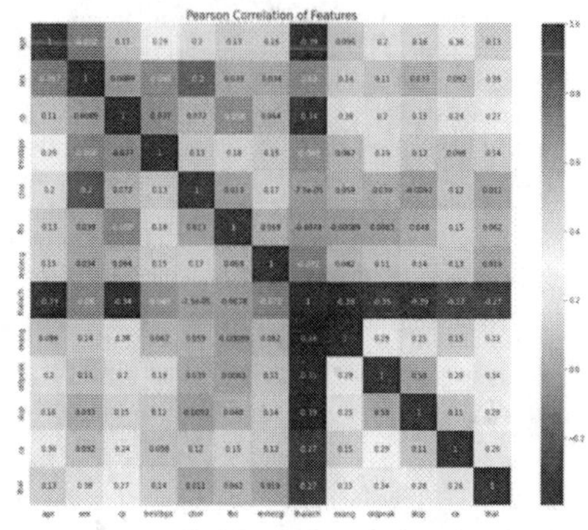

Figure 3: Heart Correlation Matrix

Out of the raw data of 303 patients, 6 had null values in either ca or thal; for this being so few instances of missing data, those data points were just dropped, making a total of 297 data points. The original labels, ranging from 0, no heart disease, to 4, the most advanced stage of heart disease, was redesigned to range from 0, no heart disease, and the original values of 1, 2, 3, and 4 were squished into a single category, 1, "presence of heart disease." In these figures, we are extracting data from the dataset using dat.info so that the dataset can be studied in an analytical manner.

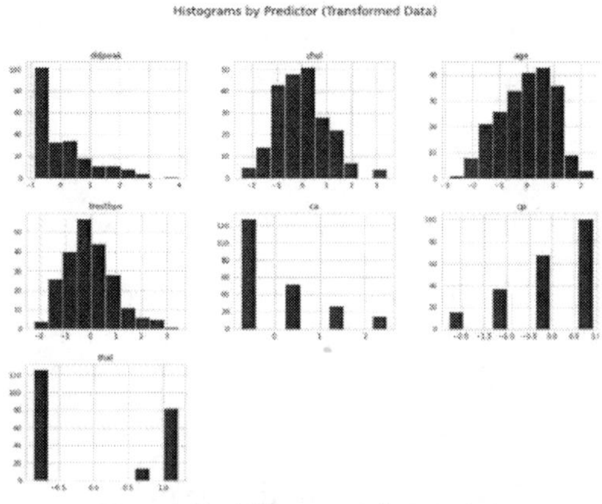

Figure 4: Heart Histogram Attribute Analysis

After selecting the top 7 features for prediction using the random forest classifier. From correlation matrix (Figure 3) and the histogram (Figure 4), we can observe that the features are not correlated to each other. Hence, we scaled all the values between -2 to 2. After data preprocessing the accuracies and recall of most of the algorithm has increased significantly.

Table 1: Model Accuracies for Heart (Before Parameter Tuning)

Algorithm	Accuracy	Recall
Random Forest	84.00%	0.8571
Logistic Regression	85.33%	0.8857
XGBoost	80.00%	0.8000
Decision Tree	74.67%	0.7714
AdaBoost	76.00%	0.8285
KNN	84.00%	0.8285
SVC	81.33%	0.8571

Table 2: Model Accuracies for Heart (After Parameter Tuning)

Algorithm	Accuracy	Recall
Random Forest	84.00%	0.8571
Logistic Regression	85.33%	0.8857
Decision Tree	85.33%	0.8571
AdaBoost	84.00%	0.8285
KNN	84.00%	0.8285
Linear SVC	85.33%	0.8571
RBF SVC	82.67%	0.8857
Sigmoid SVC	88.00%	0.8571
Polynomial SVC	81.33%	0.8571

Comparing Table 1 and Table 2, most models are seeing a lot of improvement all models are up on recall score; the KNN model's accuracy increased from 59% to 84%; the SVC's accuracy increased from 55% to 81% with the exception of XGBoost, which saw a 4% decrease.We have selected the three top-performing models: the sigmoid kernel from the SVCs, AdaBoost from the boosters, and the decision tree over the random forest, for a total of 3 individual models. We combined them into a single voting classifier below. The accuracy which we achieved was 88.57 %. This ensemble method does as well as two of the individual models, the decision tree and AdaBoost, but not as well the sigmoid SVC

To demonstrate the link between different variables, we produced a correlation matrix shown as Figure 5. This graph enables us to see which characteristics have a high correlation with others and hence eliminate them from the model; however, none of the variables had a high correlation with any of the others.The dataset has 8 features which are Pregnancies, Glucose, blood pressure, skin thickness, Insulin, BMI, Diabetes Pedigree Function, Age.

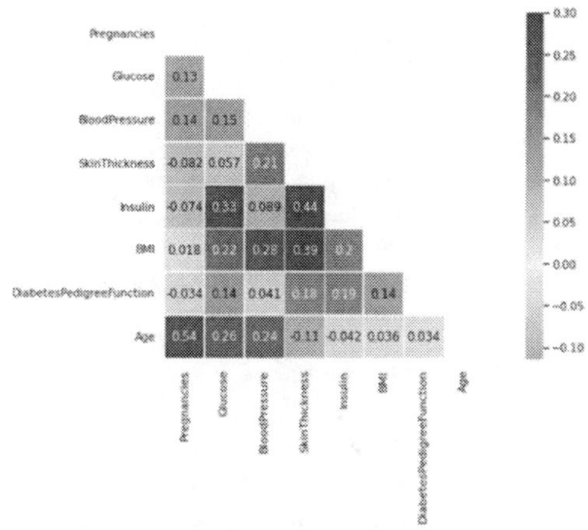

Figure 5: Diabetes Co-relational Matrix

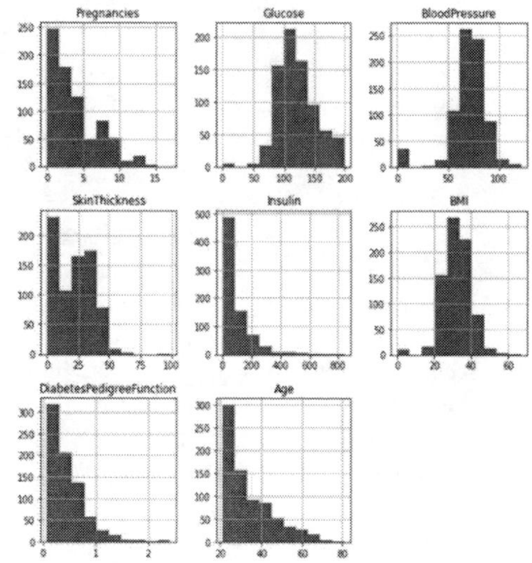

Figure 6: Diabetes Histogram Attribute Analysis

We have developed prediction models using the following classifiers Random Forest, Adaptive Boosting, KNN, Logistic Regression and Decision Tree.

Table 3: Model Accuracies for Diabetic Prediction

Model	Accuracy
Random Forest	72.73%
KNN	77.93%
Logistic Regression	77.92%
Decision Tree	74.46%
AdaBoost	72.73%
VC (without weights)	80.52%
VC (with weights)	80.95%

In Table 3, the two ensemble models used are Adaptive Booster (Boost) & Random Forest (Mean)

as the foundation. By the combination of different classifiers using Voting Classifier with and without weights, precision can be enhanced. The precision boosts to 80.95% and 80.52% for voting classifiers with or without weights respectively. The same weights were applied on other foundations with the hefty weights delegated for the effective models. The weights used in voting classifier are as follows KNN - 2, Decision Tree – 1, Logistic Regression – 2 and Random Forest – 2.

5. Conclusion

The heart disease and diabetes can be synonymously aggregated to drive a patient's conclusions. In this paper sufficient exploratory analysis and pre-analysis of normalized models has been carried out to understand the need of these predictions using ensemble technique. The system promises to handle and correlate both events of heart and diabetes to drive to quicker prediction using machine learning concepts .For heart disease, it can be concluded that the Voting Classifier of Decision Tree, Sigmoid SVC, and Adaboost has the highest accuracy of 88.57 % and for diabetes, the voting classifier has an accuracy of 80.95 %. The proposed methodology can be extended since it has a scope to conclude immunity of a patient from COVID through the study conducted. The development of a robust model with the help of automated feature selection to work on possibility of COVID through the analysis of both diseases can be carried out in future.

References

[1]Bianca de Almeida-Pititto, Patrícia M. Dualib, Lenita Zajdenverg, JoanaRodrigues Dantas,Filipe Dias de Souza , Melanie Rodacki and Marcello Casaccia Bertoluci on behalf of Brazilian Diabetes Society Study Group (SBD), "Severity And Mortality Of COVID 19", In Patients With Diabetes, Hypertension And CardiovascularDisease: A Meta-analysis, Diabetology & Metabolic Syndrome Research, (2020)

[2] R. Indrakumari, T. Poongodi, Soumya Ranjan Jena, "Heart Disease Prediction using Exploratory Data Analysis, International Conference" on Smart Sustainable Intelligent Computingand Applications under (ICITETM 2020)

[3] Senthilkumar Mohan 1, Chandrasegar Thirumalai1, And Gautam Srivastava 2,3, (Member,IEEE), "Effective Heart Disease Prediction Using Hybrid Machine Learning Techniques",(IEEE ACCESS)

[4] Aishwarya Majumdar, Dr. Vaidehi , "Diabetes Prediction using Machine Learning Algorithms, International Conference" On Recent Trends In Advanced Computing (2019, ICRTAC2019)

[5] Emma Barron, Chirag Bakhai, Partha Kar, Andy Weaver, Dominique Bradley, Hassan Ismail, Peter Knighton, Naomi Holman, Kamlesh Khunti, Naveed Sattar, Nicholas J Wareham,Bob Young, Jonathan Valabhji, Associations of type 1 and type 2 diabetes with COVID-19 related mortality in England: a whole-population study, (Lancet Diabetes Endocrinol 2020)

[6] Quan Zou, Kaiyang Qu , Yamei Luo, Dehui Yin, Ying Ju,Hua Tang, "Predicting Diabetes Mellitus With Machine Learning Techniques," Bioinformatics and Computational Biology, (Frontier Genetics Journal, 2018)

[7] Md. Kamrul Hasan , Md. Ashraful Alam , Dola Das, Eklas Hossain, (Senior Member, IEEE), And Mahmudul Hasan, "Diabetes Prediction Using Ensembling of Different MachineLearning Classifiers", (IEEE ACCESS 2020)

[8] Jyoti Soni, Ujma Ansari, Dipesh Sharma, Sunita Soni, Predictive Daa Mining for Medical Diagnosis", International Journal of Computer Applications, (Volume 17, #8,2011)

[9] Himanshu Sharma, MA Rizvi, "Prediction of Heart Disease using Machine Learning Algorithms: A Survey" ,International Journal on Recent and Innovation Trends in Computing and Communication (2016)

[10] Baban U. Rindhe, Nikita Ahire, Rupali Patil, Shweta Gagare, Manisha Darade, "Heart Disease Prediction Using Machine Learning", IJARSCT, (2021)

[11] Nabaouia Louridi, Samira Douzi, Bouabid El Ouahidi, "Machine Learning-Based Identifcation Of Patients With A Cardiovascular Defect", Journal of Big Data (2021)

[12] Dhai Eddine Salhi, Abdelkamel Tari, M-Tahar Kechadi, "Using Machine learning for heart disease prediction", researchgate.net.

[13] Minakshi R. Rajput ,Sushant S. Khedgikar, "Diabetes prediction and analysis using medical attributes: A machine learning approach", Journal of Xian University of Architecture and Technology.

Deep Learning based Facial Emotion Recognition

Swayam Badhe[a], *Sangita Chaudhari*
Department of Computer Engineering, Ramrao Adik Institute of Technology,
D. Y. Patil Deemed to be University, Nerul, Navi Mumbai India

Abstract. Human Facial expression is the mirror of human emotions playing very significant role in nonverbal communication. In many applications like human machine interface, medical diagnosis, AI based games, market research etc. needs facial expression recognition. Although it is very easy task for humans, its bit challenging for machines to detect and recognize correct emotions from series of human facial expressions. Since decade, many researchers have tried different image processing, machine learning and deep learning-based approaches to correctly identify human emotions. Some of them could able to identify emotions but with more complexity. WE have proposed CNN model with 4 convolution layer and 2 FC layers which is giving good accuracy over the existing models with less complexity. It is performing better for all classes except fear and disgust.

Keywords: Emotion Recognition, Facial Expressions, Emotion Classification, CNN

1. Introduction

In recent years, there is huge outbreak in Image processing and pattern recognition domain with the introduction of machine learning and deep learning approaches. It gives opportunities to researchers to avail automated feature extraction instead of hand-crafted feature extraction with better accuracy. Human emotion detection is always an important issue to be solved in many human computers interface-based application. Human visual system was so developed that it could easily detect and identify various emotions from facial expressions of human. However, to do so with machine is very challenging task and it needs perfect learning for machines to correctly classify the human emotions from facial expressions. It involves separation of various facial states from static human face images or from dynamic video sequence to determine the emotions of the object (face).

In today's era, there is huge demand of automated human emotion recognition in various applications like identification and access control, human machine interface, automated surveillance, medical diagnosis, AI based games etc. There are 7 classes of human emotions which can be identified and used in various applications to make out decisions. There emotions classes are Neutral, Angry, Disgust, fear, happy, sadness and surprise (figure 1). Each of these emotions results in changes in human facial components like lips, eyes, chin, cheeks, forehead, eyebrows. There orientation, colour, size, shapes may get changed. In automated human emotion recognition system, most of the approaches tries to extract features based on this and classify them in correct emotion. The objective of this work is to develop Automatic Facial Expression Recognition System which considers human face images/video and recognize and classify the emotions on human face into seven different expression classes as shown in figure 1.

1. Happy
2. Surprise
3. Neutral
4. Sad
5. Anger
6. Disgust
7. Fear

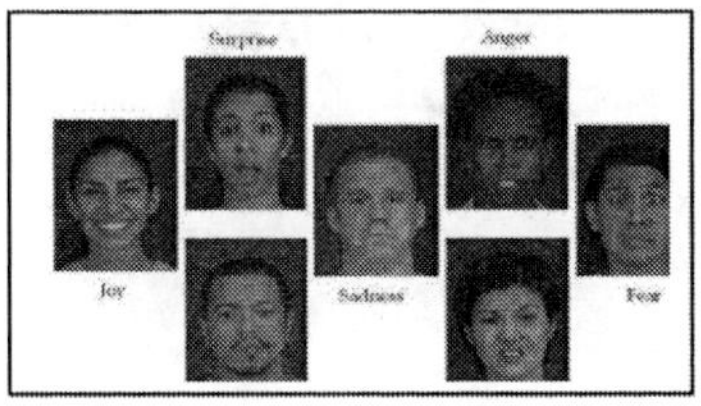

Fig 1. Basic Facial Expressions

Our goal will not only be to develop anAutomatic Facial Expression Recognition System but also improving the accuracy of this system compared to the other available systems for various emotion classes.

2. Literature Survey

Over the years, many attempts have been made to detect and recognize emotions from facial expression as it has its own importance in human computer interaction used in variety of applications. Various Techniques like PCA, LDA, Curvelets, Gabor filters, SVM, AdaBoost, Local binary patterns, Histogram based descriptors, Neural network and its variants have been adopted in automated human emotion detection and recognition. Table II shows summary of comparison of few of the existing techniques based on PCA, SVM, AdaBoost, Local binary patterns and CNN [1-6].

[a] Electronic Mail: badheswayam@gmail.com

Table I. Comparison of Existing Systems

Paper	Technique Used	Accuracy (%)	Emotions considered	Limitations
Mutsugu et al. [1]	CNN Curvelet face +LDA Curvelet face+PCA	97.6 83.5 83.9	Happy, Neutral	Individuality of facial expression not detected
Zhang et al. [2]	Patch based 3D Gabor filters + SVM and AdaBoost	92.93	Happy, Neutral, Sadness,Surprise	JAFFE DB requires larger sizes of patches
Hayat [3]	SVM+ clustering	94.34	Anger, Disgust, Happiness, Fear, Sadness, Surprise	Less accuracy
Hablani [4]	Local binary patterns	94.44	Happy, Neutral, Sadness, Surprise, Anger, Fear, Disgust	Manual feature extraction, does not perform well for temporal behavior of facial expressions
Zisheng [5]	Pyramid HOG	96.33	Happy, Neutral, Sadness, Surprise and Anger, Fear and Disgust	less accuracy
Yu [6]	Deep CNN	61.2	Happiness, Disgust, Fear, Angry, Surprise and Sadness	Less accuracy, complex network

Mutsugu et al. used CNN with curvelet, PCA and LDA for emotion detection, but could recognize only Happy, Neural and lip movement capturing. The system proposed by them is insensitive to individuality of facial expressions [1]. In SVM based approaches, the classes considered are 7 but these approaches could not give good accuracy and it require larger database with larger patch size [2, 3]. In most of the approaches, emotions can be easily recognized from static facial images, but results out in poor accuracy in case of temporal changes in the facial expressions [4, 5]. In deep learning-based approach proposed by Yu, 7-layer CNN is used. Although majority of the emotion classes are considered, it does not give good accuracy at the cost of complexity in the used model. Similar approaches based on VGG, AlexNet and Resnet are also studied and it was found that although they tend to give moderate accuracy for temporal changes in human emotions, their complexity is very high [7-9].

3. Proposed Methodology

Human being is a gifted species with many emotions to have effective communications with humans and other species. Most common emotions emanated through human facial expressions are neutral, happy, sad, surprise, fear, anger and disgust. All these emotions are well expressed by contraction and expansion of various facial muscles. They are reflectance of human mind and changing dynamically with variety of thoughts in mind representing very complex signal with ample amount of information about our state of mind. One of the challenging tasks in artificial intelligence is recognize these emotions in various applications to make it more user friendly and to study the impact of contents and services on user. For example, customer's interest can be very well measured using facial recognition. Patients' facial expression can be monitored in response to the treatment and accordingly change in line of treatment can be followed in healthcare system. Human visual system is so developed that such expressions can easily identified and used in our day-to-day activities. However, machine is not yet so intelligent to identify/recognize emotions and use it in decision making. There are many attempts made by researchers to devise solutions to recognize human emotions. We are proposing simple deep learning neural network model to make machine to recognize emotions and make inference out of it. Figure 2 shows general structure of facial recognition system.

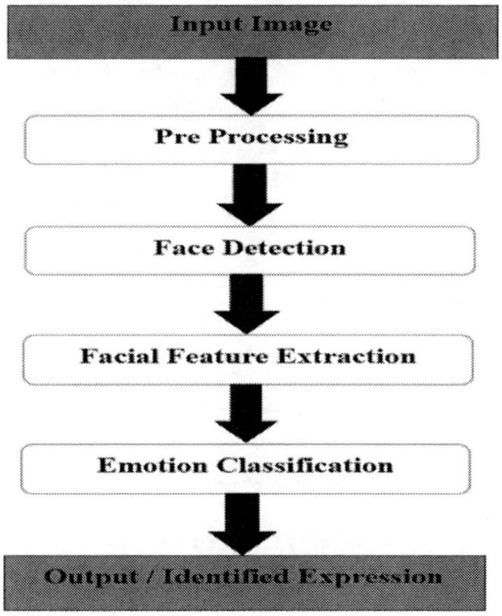

Fig. 2. Generalized Face Recognition System

In most of the systems, preprocessing, face detection, facial feature extractions and classification are common steps. Locating the face in the given scene is one of the crucial tasks in the crowded face scene. Once, face is detected, hand crafted facial features like shape, skin texture, face parts orientation, colour etc. are extracted and finally classification step will help to categories facial expression in one of the categories based on features. To improve the accuracy, deep convolutional neural network model is used in our proposed facial emotion recognition system. Figure 3 shows simple layer structure of convolution neural network.

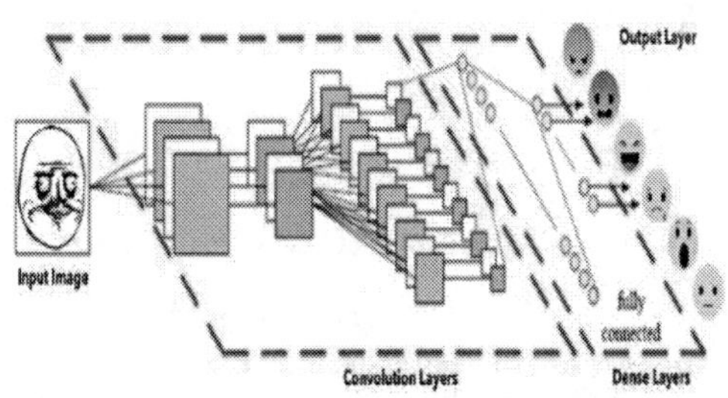

Fig. 3. Convolution Neural Network Layers

It is feedforward network with series of convolutional and pooling layers. Convolutional layer is performing the task of feature representation of the input image. Pooling layer tries to reduce spatial resolution of feature maps with maximum value pooling from input feature map. Fully connected layer followed by these layers interprets these resultant feature maps to perform high level reasoning with SoftMax operator on top of it to achieve classification. Algorithm 1 shows the series of steps performed to recognize emotions from facial expressions.

Algorithm1: Facial Recognition

1. Resize input images Image to 48x48 pixels with labeling in to 7 classes (anger, disgust, fear, happiness, sadness, surprise, and neutral)
2. Apply preprocessing to remove noise and enhance images using median filtering and contrast enhancement.
3. Detect the face from each image
4. Detected face conversion to gray scale image
5. Pass the input image [1x48x48] to convolution2D layer.
6. Perform convolution to generate feature maps.
7. Apply Maxpooling2D using (2, 2) window across the calculated feature map by keeping maximum pixel value only.
8. Training with fully connected neural network with forward/backward propagation on the pixel values.
9. Use SoftMax function to represent probability of emotion class.

4. Results and Discussion

The proposed model is trained using 35887 images from FER2013 dataset. It consists of 48x48 grayscale face images. All images are pre-cropped with same region occupancy in image. The dataset consists of 8989 images for Happy, 6077 images for Sad, 6198 images for *Neutral,* 4002 images for *Surprise,* 5121 images for *Fear,* 4593 images for *Angry and* 547 images for *Disgust.* The facial images are to be classified in one of these seven classes: Happy, Sad, Neutral, Surprise, Fear, Angry and Disgust. Figure 4 shows sample images from FER2013 dataset.

Fig 4. FER2013 Emotion Data [10]

The model is performing really well for Happy, Sad, surprise, angry and neutral and able to predict the emotions of real time dynamic expression variation in face. Figure 5 shows visual results of Happy expression recognition with the prediction array. Similarly figure 6-9 shows visual results with the prediction array for Sad,

 https://doi.org/10.1051/itmconf/20224403058

surprise, angry and neutral respectively. The proposed models' results are compared with the existing Model used in many other face recognition systems (Table II). It is observed that proposed model is giving good accuracy compared to other models for 5 classes and results out in less complexity as well. It could classify all emotions correctly except fear and disgust misclassified occasionally.

Fig 5. Prediction of Happy Emotion

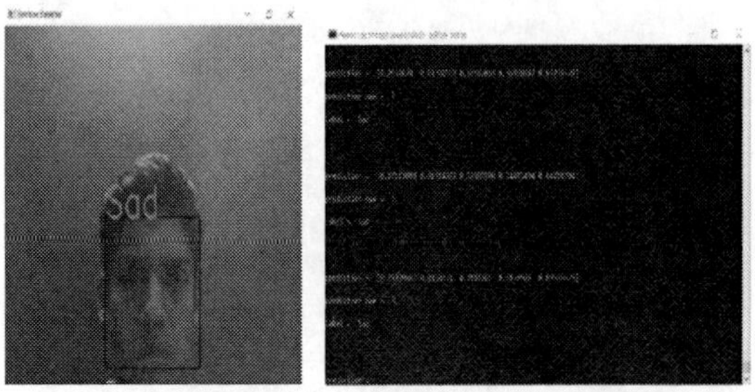

Fig 6. Prediction of Sad Emotion

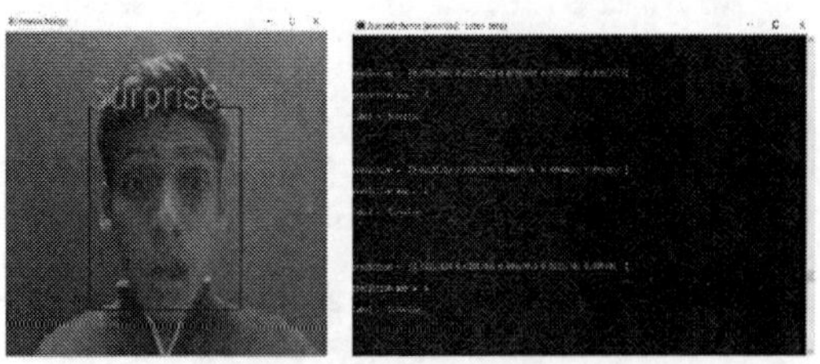

Fig 7. Prediction of Surprise Emotion

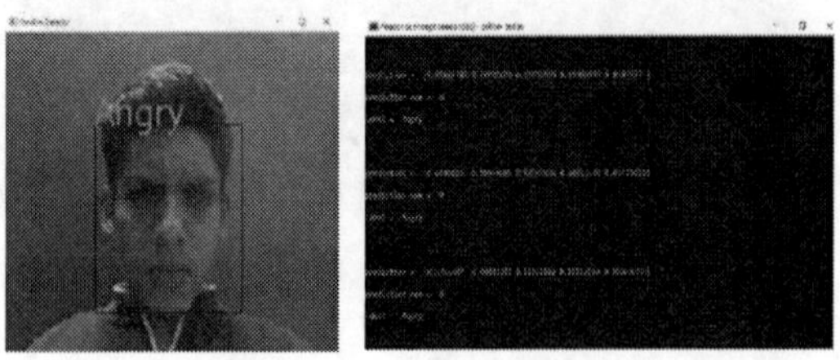

Fig 8. Prediction of Angry Emotion

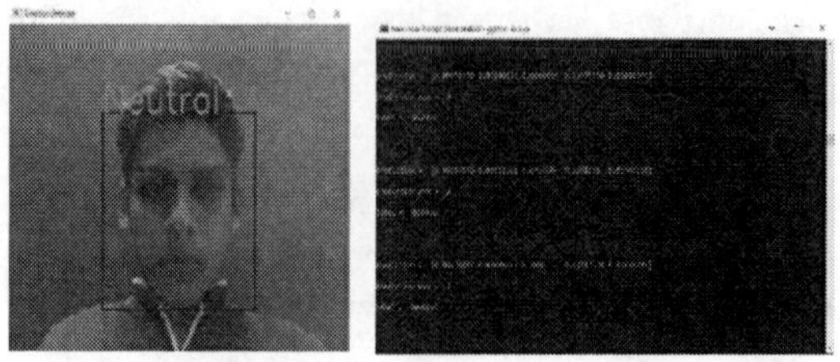

Fig 9. Prediction of Neutral Emotion

Table II Comparison with other DL Models

Algorithm	**Accuracy (%)**	**Computational Complexity**
Alexnet	84	O(4)
VGG	81	O(9)
Resnet	88	0(16)
Proposed CNN	94.27	O(4)

5. Conclusion

Human facial expressions are playing important role in effective communication as well as to understand state of the mind. In recent years, many intelligent human computer interaction systems are trying to recognize human emotions through facial expressions and make effective use of it in various applications. The proposed convolutional neural network model is giving good results by recognizing emotions from real time video of human with dynamically changing emotions. Currently, the system is able to give correct predictions for all classes except disgust emotion. We are trying to add more images for all emotions in the existing pool of images to get correct recognition of all emotions from static as well as dynamic facial expressions.

References

[1] Matsugu, Masakazu, et al. "Subject independent facial expression recognition with robust face detection using a convolutional neural network." Neural Networks 16.5-6 (2003): 555-559.

[2] Zhang, Ligang, and Dian Tjondronegoro. "Facial expression recognition using facial movement features." IEEE transactions on affective computing 2.4 (2011): 219-229.

[3] Hayat, Munawar, and Mohammed Bennamoun. "An automatic framework for textured 3D video-based facial expression recognition." IEEE

Transactions on Affective Computing 5.3 (2014): 301-313.

[4] Hablani, Ramchand, Narendra Chaudhari, and Sanjay Tanwani. "Recognition of facial expressions using local binary patterns of important facial parts." International Journal of Image Processing (IJIP) 7.2 (2013): 163-170.

[5] Li, Zisheng, Jun-ichi Imai, and Masahide Kaneko. "Facial-component-based bag of words and phog descriptor for facial expression recognition." 2009 IEEE International Conference on Systems, Man and Cybernetics. IEEE, 2009.

[6] Yu, Zhiding, and Cha Zhang. "Image based static facial expression recognition with multiple deep network learning." Proceedings of the 2015 ACM on international conference on multimodal interaction. 2015.

[7] Sekaran, Sarmela AP Raja, Chin Poo Lee, and Kian Ming Lim. "Facial emotion recognition using transfer learning of AlexNet." 2021 9th International Conference on Information and Communication Technology (ICoICT). IEEE, 2021.

[8] Hussain, Shaik Asif, and Ahlam Salim Abdallah Al Balushi. "A real time face emotion classification and recognition using deep learning model." Journal of Physics: Conference Series. Vol. 1432. No. 1. IOP Publishing, 2020.

[9] Verma, Monu, Santosh Kumar Vipparthi, Girdhari Singh, and Subrahmanyam Murala. "LEARNet: Dynamic imaging network for micro expression recognition." IEEE Transactions on Image Processing 29 (2019): 1618-1627.

[10] https://www.kaggle.com/msambare/fer2013

Analysis of Multiple Factors Influencing COVID-19 Induced Mental Health

Neha Chauhan[1,*], *Niranjan Dubey*[1], *Dr. R. Kamatchi Iyer*[1] *and Nilima Zade*[2]

[1]School of Management and Entrepreneurship, Atlas SkillTech University, Mumbai, Maharashtra, India
[2]School of Digital Technology, Atlas SkillTech University, Mumbai, Maharashtra, India

Abstract. The pandemic has adversely affected human lives. People have lost their loved ones, their jobs and even their homes in some cases. People have suffered from anxiety, frequent frustration, distress and loneliness due to the pandemic. With many of us facing challenges and battles in our lives, it is natural to feel frustrated and distressed. The spread of rumours and misinformation has led to anger, disorder and stress among people. In a country like India, where the population has suffered immensely due to the lack of healthcare infrastructure, Mental Health conditions are also deteriorated considerably due to different factors, we can confirm this by looking at the rise in suicide cases not only in India but also all around the world. We surveyed a few people across India including the working class and students from various age groups as the Covid-19 pandemic situation has brought everyone into renewed focus. The result indicates the level of suffering and the factors that have influenced their recovery.

1 Introduction

Mental health has been a growing concern for a while now, the importance of mental health has increased in recent times especially due to Covid-19 where people went through frustrations, isolation, distress and a lot more. Even though there are multiple factors influencing mental health, the following factors are considered to have a major impact during the pandemic scenario where people are isolated from each other, lost their loved ones, and deteriorated in financial condition; cases of the same have boosted [1-3]. The objective of the study was to assess the reasons for an increase in mental illness due to the covid-19 outbreak in India and to investigate people's attitude towards the pandemic, how people have been affected mentally, what activities are they doing to feel relaxed as well as to come up with a recommendation that is effective for people to stay stress-free [4-5].

1.1 Background

The alarming rate at which the covid-19 cases have increased in India has led to an increase in mental illness among the people. According to the Indian Psychiatry Society, as shown in Fig.1, there has been an increase in mental disease cases, with at least one out of every five Indians affected. There has been a big rise in suicidal cases, with many of these being unreported [6]. India reported 381 suicides daily for the year 2019, a 3.4 percent increase in deaths due to suicide as compared to 2018. A study by the International Journal of Mental Health Systems found a 67.7 percent increase in the online news report about Suicidal behaviours [7]. The purpose of the study was to understand various factors leading to diminishing mental health among people due to covid-19 and how they can cope with the same [8].

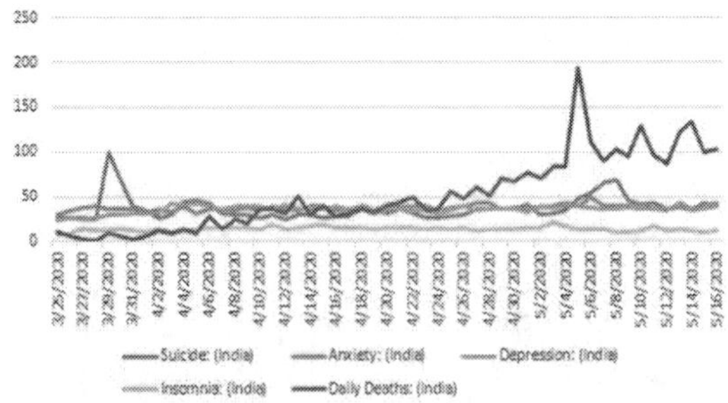

Fig. 1. Int Psychogeriatr. 2020 Jun 3: 1–2. Published online 2020 Jun 3. doi: 10.1017/S1041610220001052 (nih.gov). Source: PubMed Central

2 Methodology

A self-administered web-based questionnaire was created using Google Forms and circulated via social media among the people of Maharashtra, India. Participation in the survey was voluntary and participants were not offered any incentives, the participants were a self-selected sample. Our research was mainly divided into two parts, the first part was mainly designed to understand the demographics of our respondents followed by a study designed to understand various factors leading to diminishing mental health among people due to covid-19 and how they can cope

* Corresponding author: nehachauhan3129@gmail.com

with the same. This research has been performed using a qualitative method wherein the target audience are students and working professionals from all age groups wherein major responses are from the age group of 18 to 45. The tools used to carry out the analysis were Microsoft Excel and Power BI. To know from which demographics our respondents belong to we asked them about their Demographic data like age, gender, city, state, profession and the sectors they work in, which helped us understand whose mental health is affected by the pandemic the most. The research was further narrowed down to the mental illness of the respondents it was necessary to know the root cause behind the same, for which the questions were asked whether or not the family members/loved ones were affected by the virus? There were also certain questions related to assessing their mental health. As the research is based on how the Covid-19 virus affected people's mental health, narrowing down to know the root cause of it, it is necessary to know if they or any of the direct family members have been affected by the Covid-19 virus and if they were affected by the virus did, they notice any change in their mental behaviour. Noticed any change in their mental behaviour. There are many other changes which Covid-19 pandemic brought in people's lives, which also introduced to new normal like working or studying from home but as everything comes with a drawback, working/studying from home comes with additional performance pressure. Due to the deterioration in the economy, the people also faced job loss, salary cuts, and it even led to the shutdown of many businesses. These are the main parameters for this research. To understand how people cope with the new normal the research aimed to understand various activities they practice to relax. Stress being a normal part of life can sometimes be difficult to manage on our own for which the research also tried to find out whether or not people consulted someone to battle their stress.

2.1. Discussion points and results

The following observations are based on the survey:

Finding 1: To find which gender experienced an increase in performance pressure during the pandemic.
Outcome: Of all the respondents, 60percent of women and 40percent of men as shown in Fig.2, felt an increase in performance pressure due to working from home as compared to working from the office / visiting college campuses before the pandemic.
This has been due to the Women's facing difficulties in managing work-life balance while working from home. Along with their work, they also had to meet and manage the demands of their family who were at home due to the pandemic which led to overburdening them with work.
Finding 2: To find which gender experienced mental issues while they were affected by covid19 during the pandemic.
Outcome: Through our analysis, we observed that of all the respondents who were affected by the covid19 virus, 53.1 percent of women claimed to have suffered from mental health issues whereas only 18.4percent of men were affected as shown in Fig.3.

Gender ▲	No	Yes
Female	44.00%	60.00%
Male	56.00%	40.00%
Total	**100.00%**	**100.00%**

Fig. 2. Increase in Performance Pressure

Gender	No	Yes	Total
Female	46.88%	53.13%	**100.00%**
Male	81.58%	18.42%	**100.00%**

Fig. 3. Mental Health

Finding 3: Reason for being frustrated too often.
Outcome: 30.2percent of respondents claimed to be frustrated too often mainly due to Loss of freedom/Boredom and 37.2percent of people were frustrated because of performance pressure as shown in Fig.4.
Humans are social animals and since most people are not used to being in isolation, they are facing frustration. Also, while working/studying from home they are expected to deliver more as compared to working from the office/studying on campus.

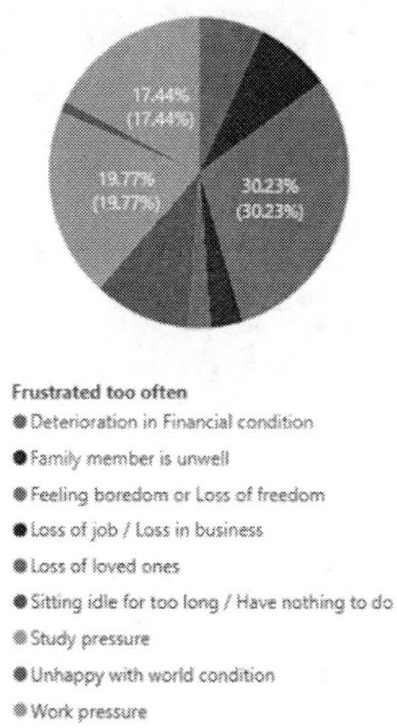

Fig. 4. Reasons for being frustrated

Finding 4: Professionals from which sector faced financial deterioration.
Outcome: The survey results also highlighted that the Education sector faced a major salary cut or loss in income i.e., almost 50percent of professionals working in that sector. Professionals working in Digital

Marketing, and IT had no salary cut as these industries were at a booming stage. Also, in the BFSI industry, only 6.2percent of people faced this issue as shown in Fig.5.

It was interesting imperative to discover that the healthcare sector which is one of the most demanded sectors whose employees worked almost 24*7 at risk of getting infected by covid-19 had to face a salary cut.

Sector	No	Yes	Total
BFSI (Banking, Financial Service and Insurance)	93.75%	6.25%	100.00%
Digital Marketing	100.00%		100.00%
Education	50.00%	50.00%	100.00%
Healthcare	66.67%	33.33%	100.00%
IT (Information Technology)	100.00%		100.00%
Not Applicable (I'm a Student)	82.35%	17.65%	100.00%

Fig. 5. Professionals from different sectors facing the financial crunch

Finding 5: Change in eating and sleeping pattern during the pandemic

Outcome: Change in eating and sleeping patterns can occur due to various reasons and it has been an area of concern for people suffering from mental illness. From the responses, 34percent of people claimed that they didn't notice any change in their eating or sleeping patterns whereas 27.4percent and 17.9percent as shown in Fig.6, claimed that there was a change in their eating and sleeping patterns during the pandemic since they failed to make a schedule of their day and due to the anxiety/stress respectively.

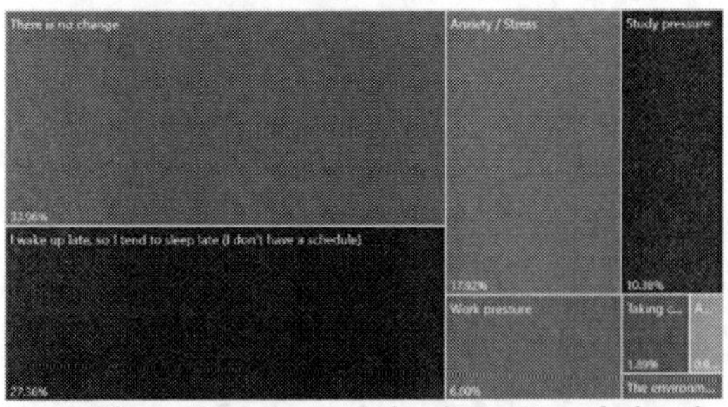

Fig. 6. Change in eating and sleeping pattern during the pandemic

Finding 6: What kind of relaxation practices do people follow?

Outcome: Relaxation plays a crucial role in handling mental health especially when it comes to adapting to a new normal. It aids decision making, improves focus and overall mental stability. While adapting to the new normal, 22.9percent of the respondents watched movies/series, 18.1percent of people spent time with their family/loved ones and 33.3percent i.e.,1/3rd of people preferred doing physical activities which were essential during covid-19 to boost immunity as shown in Fig.7. Also, sleeping has been observed as a most helpful relaxation activity.

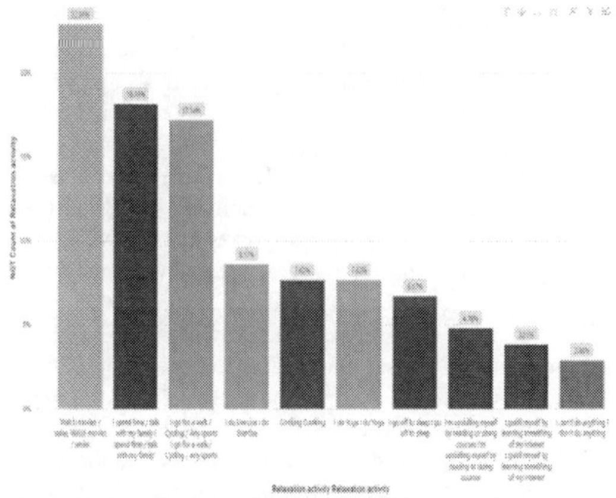

Fig. 7. Relaxation practices followed by people

Finding 8: Do people talk to or consult someone when they feel mentally unstable?

Outcome: Consultation/talking to someone has been known to increase mental stability and battle against stress, anxiety and other mental health conditions. From our analysis, we found that 37.7percent of people do not consult/talk to someone while facing mental health issues of which 0.9percent of them don't have anyone to talk with. The data also showed that of all those who preferred to talk to someone when they suffered from mental health issues, 20.75percent of people are between the age bracket of 19-25, followed by the 26-45 age bracket which is 17.92percent respondents as shown in Fig.8.

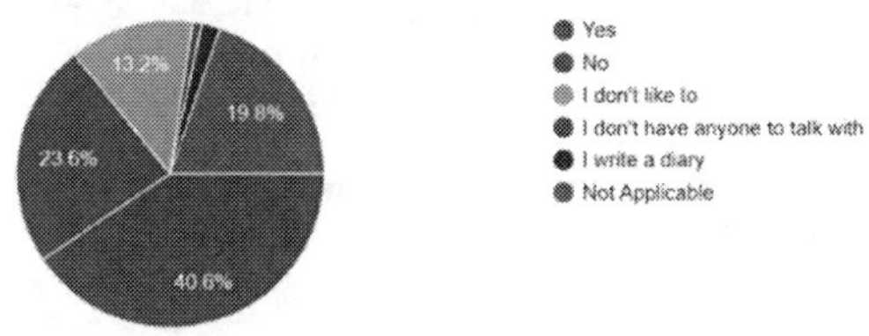

Fig. 8. Percentage of people who talk or consult someone when they feel mentally unstable

Finding 9: How would people rate their mental health on a scale of 0-5?

Outcome: While rating themselves on a scale of 5 on how mentally healthy they feel, 43.4percent of respondents rated themselves at 3 or below as shown in Fig.9. From this, it can be inferred that the majority of people feel mentally healthy.

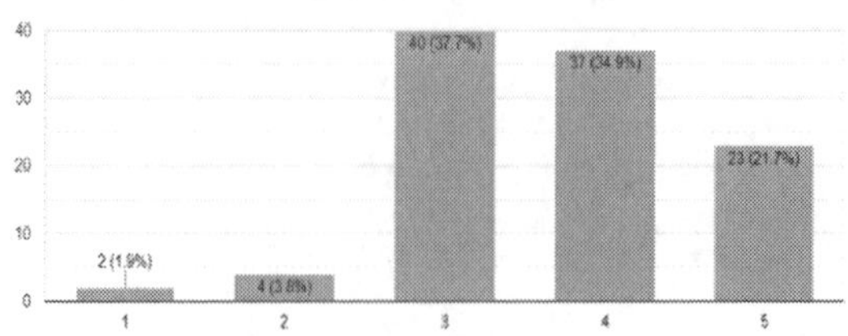

Fig. 9. People rating their mental health

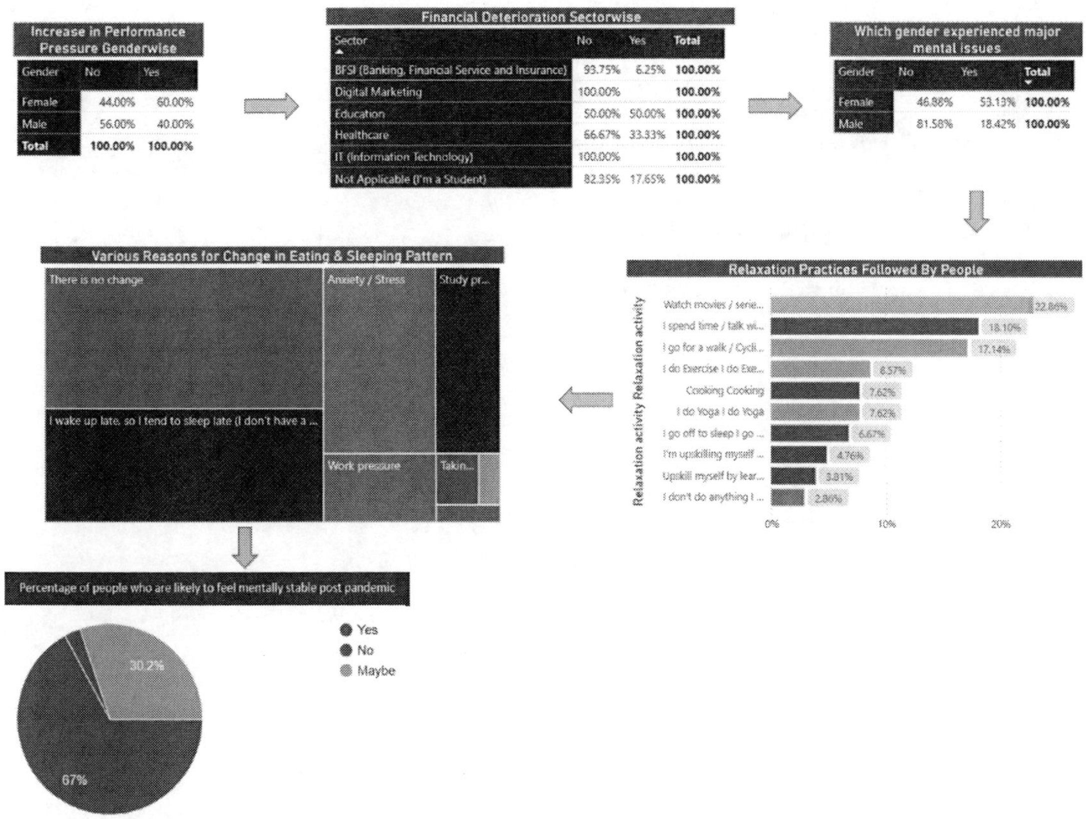

Fig. 10. Summarization of whole analysis

2.2 Recommendations

As shown in Fig.10, to cope with mental health issues, it is recommended to do some relaxation activities to not fall victim to mental fatigue. The majority of people during the pandemic liked watching movies/series, spending time with family/loved ones and going for a walk/any other sport to overcome their stress. Consulting someone while suffering from a mental health issue can help to upscale one's morale and psychological condition. Maintaining a proper routine during lockdown/pandemics has been difficult for people, but those who have been able to follow a routine have experienced mental stability.

3 Conclusions

The purpose of this research was to understand various factors affecting mental health issues during Covid19 and how people can cope with the same. We observed that the major factors affecting mental health during the pandemic were Performance pressure, loss of freedom/boredom, failing to consult someone, not doing any relaxation activity and following Covid-19 news too often. Based on our analysis we can conclude that the majority of people i.e., 67 percent of people feel that they are likely to feel healthier once the pandemic is over whereas 30.2percent of people are still unsure and the remaining 2.8percent are deeply affected by mental illness and need a proper consultation from experts.

Keywords

Mental health, Stress, Covid-19, Pandemic, Anxiety

References

1. Gupta S., "How social distancing may affect mental health," ScienceNews, 2020. [Online]. Available: https://www.sciencenews.org/article/coronavirus-covid-19-socialdistancing-psychological-fallout. [Accessed: 06-June-2021].
2. "Anxiety and physical illness," Harvard Health, 2018. [Online]. Available: https://www.health.harvard.edu/staying-healthy/anxiety_and_physical_illness. [Accessed: 07-June-2021]
3. Rubin G. J. and Wessely S., BMJ, **368**, no. January, pp. 1–2, (2020).
4. Brooks S. K. et al., Lancet, **395**, no. 10227, pp. 912–920, (2020).

5. BBC, "Coronavirus : India defiant as millions struggle under lockdown." [Online]. Avalable : https://www.bbc.com/news/world-asia-india-52077395. [Accessed: 30- Mar-2020].
6. Schmall E. and Saaliq S., The Washington Post, (May2020).[Online].Avalable: https://www.medrxiv.org/content/10.1101/2020.05.03.20081380v1.full.pdf
7. Lancet Psychiatry, **vol. 7**, pp. 148–161, (2020). [Online].Avalable:https://pubmed.ncbi.nlm.nih.gov/31879245/
8. Mina S., Jabeen M., Singh S., and Verma R., **60**, no. 2, p. 211, (Mar. 2015).

Stock Price Prediction using Facebook Prophet

Sumedh Kaninde[*], *Manish Mahajan*[*], *Aditya Janghale*[*] *and Bharti Joshi*[*]

Abstract. Estimating stock prices has always been a challenging task for researchers in the financial sector. Although the Efficient Market Hypothesis states that it is impossible to accurately predict stock prices, there is work in the literature that has shown that stock price movements can be predicted with the right level of accuracy, if the right variables are selected and appropriate predictor models are developed. those that are flexible. The Stock Market is volatile in nature and the prediction of the same is a cumbersome task. Stock prices depend upon not only economic factors, but they relate to various physical, psychological, rational and other important parameters. In this research work, the stock prices are predicted using Facebook Prophet. Stock price predictive models have been developed and run-on published stock data acquired from Yahoo Finance. Prophet is capable of generating daily, weekly and yearly seasonality along with holiday effects, by implementing regression models. The experimental results lead to the conclusion that Facebook Prophet can be used to predict stock prices for a long period of time with reasonable accuracy.

1. INTRODUCTION

Proper stock forecasting can lead to great profits for both the investor & the trader. Often, it proves that forecasting is more complex than random, which means that it can be predicted by cautiously analysing the history of the relevant stock market. ML is an effective way to represent such an operation. It predicts market value close to carrying value, thereby raising accuracy. Introduction of ML in the stock market has demanded a lot of research since the efficient and precise measurements. An important part of ML is the data taken. Databases should be as tangible as possible because small changes in data can further drastic changes in results. In this project, supervised ML is based on a database gained from Yahoo Finance. This database contains the following 5 variants: open, closed, low, high and volume. Open, closed, low & high are different stock bid prices at different times with almost specific names. Volume is the no. of shares passed from 1 owner to different over a period of time. The model was then tried on sample data.

Stock market is specified as non-linear, fluctuating and unpredictable in nature. Predicting stock prices is a demanding task as it depends on different factors including but not limited to political atmosphere, world economy, company's financial conditions and performance etc. Hence to maximise the profit and minimise the loss, technique to predict the amount of the stock in advance by analysing the trends over the last some years, could prove to be extremely useful for stock market prediction.

The most important motivation for attempting to predict the stock market prices is financial earning. The capability to uncover a mathematical model that can predict the path of the future stock prices will make the owner of the model very rich. Thus, researchers, investors and brokers are always trying to find stock market models that would give them better returns than their counterparts.

2. Related Work

There are multiple systems existing to provide a solution to the above problem. The existing systems indeed help us to come to a conclusion, but that is not the final outcome in the current situation. The conclusion from the above problem should be accurate enough to predict the decision. The judgement is very sensitive and hence it needs to be very precise. Below is the list of the existing systems that have been researched, the methodologies and the parameters they had taken into consideration as in table 1.

Table 1 Various Techniques

Techniques	Advantages	Disadvantages	Parameter used
ANN	Better efficiency compared to regression. Lower prediction mistakes	Increase in noise variation makes prediction worse	Stock closing-price

[*] Corresponding author: sumedhkn10@gmail.com

Techniques	Advantages	Disadvantages	Parameter used
SVM	Does not lose accuracy when applied to specimen from the training specimen	Amplified to the small flexibility of training data that lowers the capacity to predict	Consumer funding, nett revenue, nett income, PE ratio, consumer-spending
Hidden Markov model (HMM)	Utilised for optimization purpose	Assessment, decoding & studying	Technical indicator
ARIMA	strong & efficient	capable of short term predictions only	Open, high., low, close prices and moving average price
Time series linear model	Integrates the real data to the best possible linear model	conventional and the seasonal-trends included in the data	Data & no. of month

2.2 Limitations of Existing System

The existing systems like lstm, arima doesn't work well on robust data or inconsistent data so predictions are hard .They are suitable for short term only • Increase in noise variation makes prediction worse .Existing systems doesn't work well with traditional and seasonal trend present in data

3. Proposed Methodology

To have the completion of the proposed work, a proper methodology is supposed to be followed. Firstly, the dataset needs to be fetched and cleaned which can then be used for training purposes.

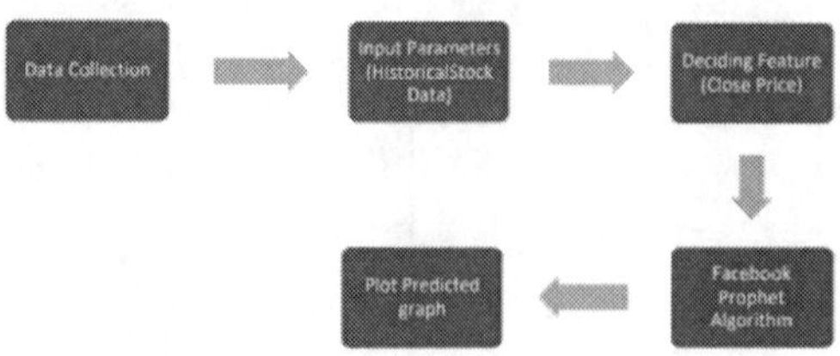

Fig. 1 Proposed System

The first step is to collect data from Yahoo finance. Then taking appropriate input parameters in consideration. After this, deciding the features for predictions is done. Finally using facebook prophet the predictions are made and plotted on the graph

3.1 Dataset

Downloading a set of data from Yahoo finance.Yahoo! Finance is the media asset which is a member of Yahoo! network. Provides financial information, data and comments including stock quote, media release, financial details, and actual material.

3.2 Parameter

The stock closing price has been chosen as our parameter. Whatever move the stock makes throughout the day due to volatility in the end the closing price is something of utmost importance. The closing stock price is remarkable for many reasons. Investors, traders, financial institutions, regulators and other stakeholders use it as a reference place for calculating performance over a particular time such as a year, a week and over a smaller time frame such as 1 minute or less. In fact, investors and stakeholders make their settlement on closing stock prices. Institutional investors check stock's closing price to make decisions in their investment assets. Hence the close price is our target attribute used for predictions.

3.3 Facebook Prophet

Prophet is a process of predicting time series data derived from add-on models where nonlinear trends correspond to seasonal, weekly, and daily periods, as well as holiday outcomes. It works best with a series of time periods with strong seasonal results and a few seasons of historical data. The prophet is strong on lost data & shifts in trend, and often treats outsiders. The prophet makes it clearer to make a more accurate forecast that is very fast compared to other time series forecast strategies. This model requires very little calculation time compared to other models. The prophet is equal to the models in Stan to get predictions in a few seconds. It allows us to get accurate weather forecasts with dirty data without any manual effort. The Prophet has numerous "human" seasons of the week and time of year.

4. IMPLEMENTATION AND RESULTS

This section contains implementation and results of all the features used in the proposed system. Yahoo finance has been used to collect data for the last 20 years i.e. from 1 jan 2000 to current date to forecast the predicted price for the next five years using facebook prophet. Streamlit framework has been used for this work as it is very easy to use and has several features in it and helps to make an interactive application. The period for prediction is taken as 5 years. The dataset is trained using the prophet fit function. In prophet it is required to rename the major columns to ds and y. The Date column is taken as 'ds' and the 'Close' column as y and hence predicted the future prices and plotted them on the graph.

Fig. 1 Proposed System

The first step is to collect data from Yahoo finance. Then taking appropriate input parameters in consideration. After this, deciding the features for predictions is done. Finally using facebook prophet the predictions are made and plotted on the graph

4.2 Results

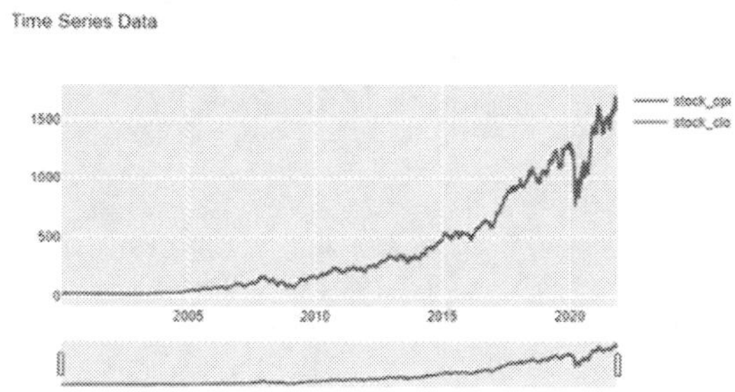

Fig 2. Graph before predictions

In Fig 2 On X-axis the closing price of stock is taken and On Y-axis Time (years) is taken. Graph is plotted for the closing price of HDFC Bank from 1 Jan,2000 to the current date.

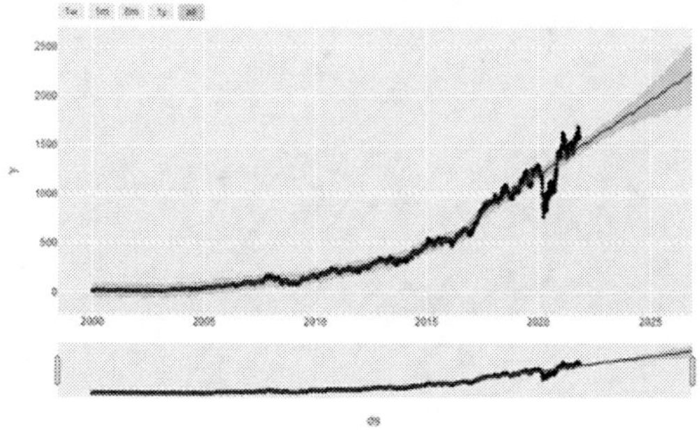

Fig 3. Graph after Predictions

In fig 3 On X-axis the closing price of stock is taken and On Y-axis Time (years) is taken. Graph is plotted for the closing price of HDFC Bank from the 1 Jan,2000 and also the predictions for the next five years represented by the blue line.

Table 2. RMSE and Calculation time

Model Name	RMSE	Time (s)
ARIMA	0.796109	1.63
LSTM	0.228731	13.28157353
FB PROPHET	**0.935556**	**0.659962893**
FAST RNN	0.202456	3.337492943

In table 2 RMSE and calculation time studied for the state of the art and the put forward models for Apple Inc. stock price. [1]

5.Conclusion

System is designed for predictions of the future prices of stocks for next 5 years using Facebook Prophet that can be used for better investments. This makes it easy to determine which stock to choose for investment based on the predictions giving the highest percentage of returns in a given period of time.The prediction accuracy can be increased by using several other features of Facebook Prophet and also make the application interactive and easy to use. In future, the stock market prediction system can be further increased by utilising a larger dataset than one being used presently. This will help to rise the accuracy of prediction models

References

[1] Mehar Vijh,Deeksha Chandola,Vinay Anand Tikkiwal,Arun Kumar."Stock Closing Price Prediction using Machine Learning Techniques" in International Conference on Computational Intelligence and Data Science (ICCIDS 2019),167(4):599-606,January 2020,DOI:10.1016/j.procs.2020.03.326

[2] Hu Z, Zhao Y, Khushi M. "A Survey of Forex and Stock Price Prediction Using Deep Learning" in Applied System Innovation.4(1):9,February 2021,DOI: https://doi.org/10.3390/asi4010009

[3] Jaydip Sen,Sidra Mehtab."Stock Price Prediction Using Convolutional Neural Networks on a Multivariate TimeSeries".February2020. https://arxiv.org/abs/2001.09769

[4] Subba Rao Polamuri,Kudipudi Srinivas,A. Krishna Mohan."A Survey on Stock Market Prediction Using Machine Learning Techniques" in International Conference on Data Science, Machine Learning and Applications, Springer Link. pp.923-931,May 2020,DOI:10.1007/978-981-15-1420-3_101

[5] ARAVIND GANESAN,ADARSH KANNAN."Stock Price Prediction using ARIMA Model" in International Research Journal of Engineering and Technology (IRJET).Volume: 08, August 2021

[6] Wu, Jimmy Ming-Tai & Li, Zhongcui & Herencsar, Norbert & Vo, Bay & Lin, Chun-Wei."A graph-based CNN-LSTM stock price prediction algorithm with leading indicators" in Multimedia Systems.February 2021,DOI:10.1007/s00530-021-00758-w.

[7] Jaydip Sen,Sidra Mehtab,Abhishek Dutta."Stock Price Prediction Using Machine Learning and LSTM-Based Deep Learning Models" in Second Symposium on Machine Learning and Metaheuristics Algorithms.October 2020,DOI:10.13140/RG.2.2.23846.34880

[8] Jiake Li."Research on Market Stock Index Prediction Based on Network Security and Deep Learning" in Security and Communication Networks.vol:2021,April 2021,DOI: https://doi.org/10.1155/2021/5522375

[9] Jaideep Singh,Matloob Khushi."Feature Learning for Stock Price Prediction Shows a Significant Role of Analyst Rating",March 2021. Appl. Syst. Innov. 2021, 4(1), 17; https://doi.org/10.3390/asi4010017 - 01 Mar 2021

[10] Karim Elbahloul."Stock Market Predictions Using Various Statistical Methods Volume II", August 2019,DOI:10.13140/RG.2.2.19926.40005

[11] Yadav, Konark & Yadav, Milind & Saini, Sandeep."Stock values predictions using deep learning based hybrid models" in CAAI Transactions on Intelligence Technology.Volume:7,June 2021,DOI:10.1049/cit2.12052

[12] Liheng Zhang,Charu Aggarwal,Guo-Jun Qi."Stock Price Prediction via Discovering Multi-Frequency Trading Patterns" in KDD '17, Halifax, NS, Canada. Pages 2141–2149,August 2017,DOI:https://doi.org/10.1145/3097983.3098117

[13] Zunic, Emir & Korjenić, Kemal & Hodžić, Kerim & Donko, Dzenana."Application of Facebook's Prophet Algorithm for Successful Sales Forecasting Based on Real-world Data" in Info Studio d.o.o. Sarajevo, Bosnia and Herzegovina.Vol 12,May 2020.

Product Review Based on Facial Expression Detection

Chirag Bera
Department of Electronics Engineering,
Ramrao Adik Institute of Technology,
Navi Mumbai, India.
berachirag123@gmail.com

Prathmesh Adhav
Department of Electronics Engineering,
Ramrao Adik Institute of Technology,
Navi Mumbai, India.
prathmesh.adhav22112000@gmail.com

Shridhar Amati
Department of Electronics Engineering,
Ramrao Adik Institute of Technology,
Navi Mumbai, India.
amatishridhar@gmail.com

Navin Singhaniya
Department of Electronics Engineering,
Ramrao Adik Institute of Technology,
Navi Mumbai, India.
navin.singhaniya@rait.ac.in

***Abstract*—The research presents a method for assessing public acceptability of items based on their brand by analysing the facial expression of a consumer who intends to purchase the product at a supermarket. In such circumstances, face expression detection is crucial in product evaluation. Emotions are conveyed through facial expressions. Sentimental analysis is a type of natural language processing that may be used for a variety of purposes. As a result, several techniques to classifying human emotional states have been proposed. The extraction of feature points via a cascade classifier is used to identify facial expressions, which minimizes the time complexity. The owner can view the feedback of the the reviewed product. This product ranking will assist the business owner in increasing product sales while also ensuring that the top products are available for the customers.**

I. INTRODUCTION

Human facial expressions have the power to communicate sentiments and emotions, which in turn determine how people react. Facial expression is an adaptive reaction that reveals a person's mental state while doing a task (e.g. expression of a consumer while purchasing a product). Machines that analyze human facial expressions have been a hot topic of study. Variations in expressions, the necessity for quick processing, and product-specific applications are the key issues in facial expression detection. Some products require more precise and quicker algorithms, and balancing these two characteristics is a key challenge [3].

In practice, analyzing a customer's facial expression and voice at that time may be quite useful. For instance, statistics on consumer curiosity might provide insight into a consumer's level of interest in a certain product [2].

Facial expression recognition is one of the most significant applications of image processing. The looks on our faces disclose our emotions. In interpersonal communication, facial expressions are crucial.

It is a non-verbal scientific gesture in which our emotions are reflected on our faces. Face recognition plays a major part in artificial intelligence and robotics, and is hence a generational requirement. The goal is to create an Automatic Facial Emotion Recognition System that can detect and categorize human facial photos comprising various expressions into seven different expression classes, including Neutral, Angry, Disgust, Fear, Happy, Sadness, and Surprise.

Identify applicable funding agency here. If none, delete this.

II. LITERATURE SURVEY

In this paper, a method is suggested for detecting facial emotion of a customer for a given prodect. The facial expression is categorised as angry, disgust, fear, happy, sad, surprise and neutral.

Vikrant Chaugule, Abhishek D, Aadheeshwar Vijayakumar, Pravin Ramteke, and Shashidhar Koolagudi: They suggest a technique to evaluate public acceptability of items based on their brand by studying the facial expression of a client planning to purchase the product from a supermarket or hypermarket in this article. The extraction of feature points is used to detect facial expressions using a modified Harris algorithm [1]. The suggested approach is compared to current techniques in terms of time complexity. To minimize computational complexity, the Harris method for feature point extraction has been updated.

Preeti Thakre and Pankaj Agarkar: In this paper, the impact of semantic categorization on NLP tasks is investigated in this research. They look at why emotion and sentiment analysis might help models be more accurate [2]. In word vector space, emotional words have good emotional semantic discrimination, whereas feelings have greater discrimination in emotional semantic space.

They present a Python-based novel Emotion-Semantic Enhanced Convolutional Neural Network Model that generates the emotional space using vectors corresponding to sentiments for Emotion Recognition on Amazon Product Reviews. In comparison to other models, the ECNN model is better at capturing emotional semantics. The system's future capabilities include recognizing emotion from user-uploaded material.

III. Methodology

The proposed approach takes a real time video or recorded video as input and extract frames (images) at regular intervals. Once the face has been retrieved, the bottom one-third of it is utilised to identify the mouth and lips, while the top one-third is used to detect the eyes. Using the suggested technique, feature points are recovered from both the lips and the eyes. The curious ratio, which is determined from the feature points of the eyes and mouth, is used to identify facial emotions. Curiosity, indifference, contentment, and a person's degree of enthusiasm may all be determined from facial faces.

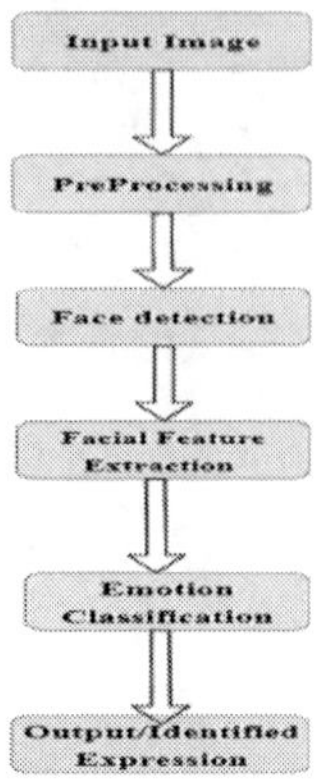

Fig. 1. Problem formulation

1) **Pre-processing:**
   Pre-processing is a common name for operations with images at the lowest level of abstraction both input and output are intensity images. Most pre-processing steps that are implemented are –
   - Reducing the noise.
   - Converting the image To binary/grayscale.
   - Pixel brightness transformation
   - Geometric transformation
2) **Face Registration**
   Face Registration is a computer technique that recognises human faces in digital photographs and is used in a range of applications. Faces are initially found in the picture using a series of landmark points known as "face localization" or "facial detection" in this face registration process. In a procedure known as "face registration," these discovered faces are subsequently geometrically normalised to match a template image.
3) **Facial Feature Extraction:**
   The technique of finding certain areas, points, landmarks, or curves/contours in a given 2-D picture or 3D range image is known as facial features extraction. The resultant registered picture is used to produce a numerical feature vector in this feature extraction stage. Common features that can be extracted are
   - Lips
   - Eyes
   - Eyebrows
   - Nose tip
4) **Emotion Classification:** The activation of certain groups of unique muscles on the human face is used to portray facial emotions. The signals in our expressions, which can be subtle yet complicated, frequently carry a wealth of information about our mental state. Humans are well-trained in reading other people's emotions; in fact, kids as young as 14 months old can determine the difference between joyful and sad. Can machines, on the other hand, access emotional states more effectively than humans? In order to address the issue, we created a deep learning neural network that allows robots to infer our emotional states. To put it another way, we give them eyes to view what we do. Facial expression recognition is a human or computer-based procedure that includes the following steps:
   a) Facial detection in a scene (e.g. in image; this step is also referred to as face detection)
   b) Facial characteristics are extracted from the identified face region (e.g. face feature extraction is the process of recognizing the form of facial components or characterizing the texture of the skin in a facial region.)
   c) Analyzing the motion of facial features and changes in their appearance, and categorizing this information into facial-expression interpretative categories such as facial muscle activation (smile or frown), emotion (affect) categories like happiness or anger, attitude categories like (dis)liking or ambivalence, and so on.

   The algorithm attempts to categorise the provided faces depicting one of the seven fundamental emotions in the third phase of categorization.

Fig. 2. Classification of facial expressions

According to numerous research studies, there are four essential procedures that must be completed in order to undertake this work.

A. DeepFace

In this paper, we are going to detect the facial expression of an already existing image using OpenCV, Deepface, and matplotlib modules in python. Deepface is a Python face recognition and facial attribute analysis module that is extremely lightweight. The open-source DeepFace library comprises all cutting-edge AI face recognition models and conducts all facial recognition operations in the background. When you use DeepFace for facial recognition, you receive access to the following features:

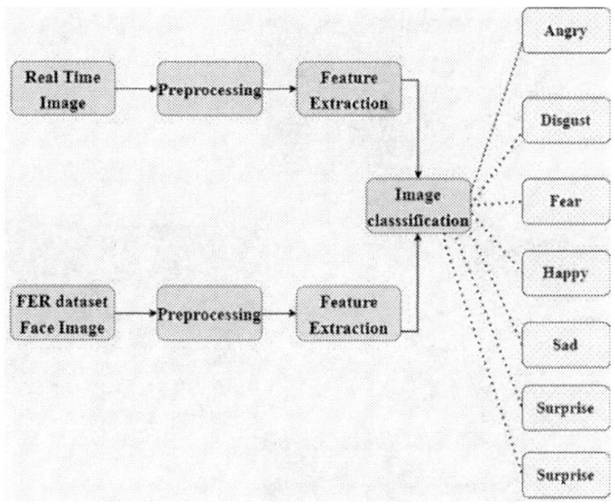

Fig. 3. Image Processing System Architecture

- **Emotion labels in the dataset:**
  1) **Face Verification:**
  Face verification is the process of comparing one face to another to see if they are same. As a result, face verification is frequently used to match a candidate's face to that of another contender. This may be used to verify if a person's physical appearance matches that of an ID document.
  2) **Face Recognition:**
  The process includes searching a picture database for a face. Face recognition necessitates a large number of runs of face verification.
  3) **Facial Attribute Analysis:**
  Facial attribute analysis is a task that involves evaluating the visual features of face photographs. Face characteristics analysis is therefore used to extract features including age, gender categorization, mood analysis, and race/ethnicity prediction.
  4) **Real-Time Face Analysis:**
  With the real-time video stream from your camera, you may test face recognition and facial attribute analysis.

B. System Implementation

Steps:

1) Collection of real time data (48-by-48-pixel grayscale images of faces each labeled with one of the 7 emotion classes: anger, disgust, fear, happiness, sadness, surprise, and neutral).
2) Images Pre-processing.
3) Detection of a face from each image.
4) Grayscale photographs of the cropped face are produced.
5) The pipeline ensures every image can be fed into the input layer as a (1, 48, 48) numpy array.
6) The numpy array gets passed into the Convolution2D layer.
7) Feature maps are created through convolution.
8) The feature map only keeps the maximum pixel value.
9) The pixel values were subjected to forward and backward propagation.
10) The model can predict which emotion will be the dominating emotion in a given situation.

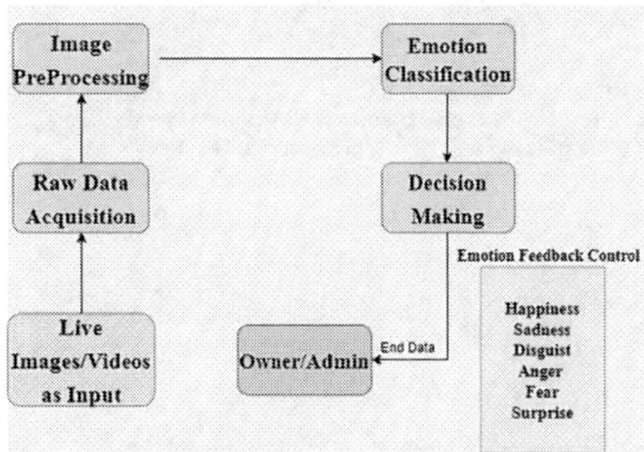

Fig. 4. Proposed System Architecture

IV. Result and discussion

The dominating emotion is predicted by reshaping an image obtained from a live video source. After 10 seconds, a timer array keeps track of the feelings and returns the most shown/captured emotion.

Step 1: The commodities for sale are added to the shop by the owner (admin) for reviewing purpose.

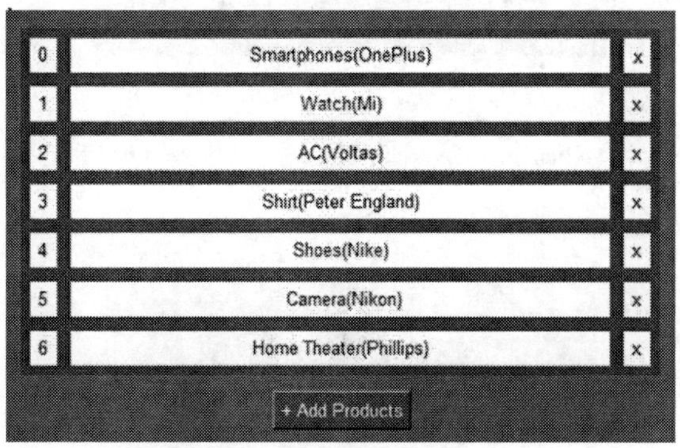

Fig. 5. Products Added by Admin

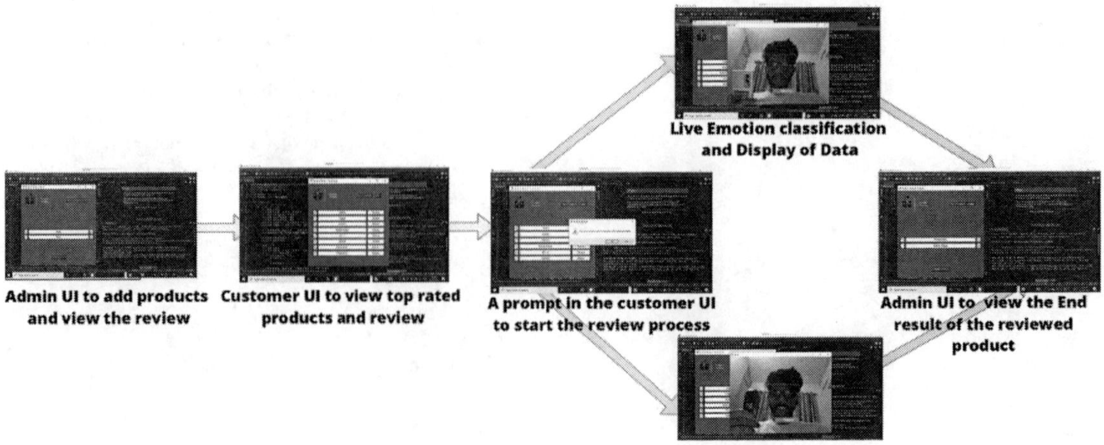

Fig. 6. Work flow

Step 2: The owner's (Admin) products are now visible to the customer (User) with the option of reviewing.
Step 3: Customers can now leave product reviews by using the CLICK ME button.
.

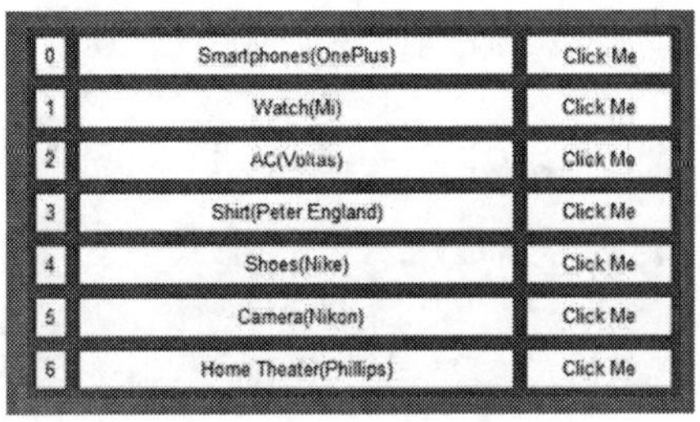

Fig. 7. List of product available

Step 4: Customer feedback "happy" will be recorded in the system.
.

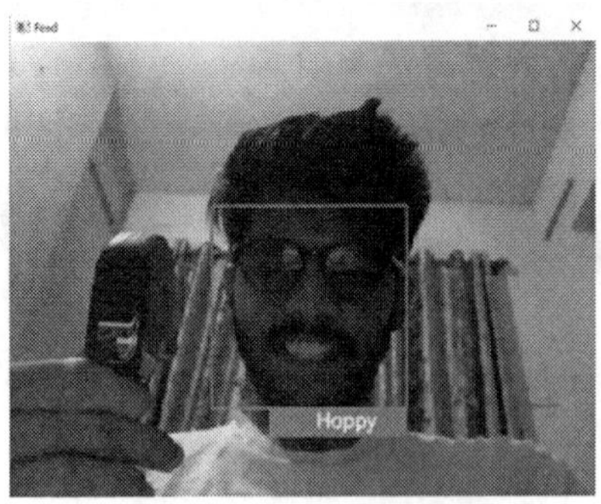

Fig. 8. Happy feedback by customer

Step 5: Customer feedback "Neutral" will be recorded in the system.
.

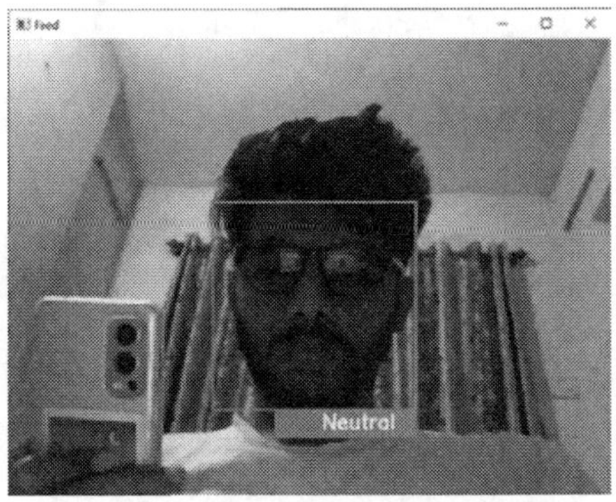

Fig. 9. Neutral feedback by customer

Step 6: The owner can now see the reviewed product. This product ranking will assist the business owner in increasing product sales while also ensuring that the top products are available for his customers.
.

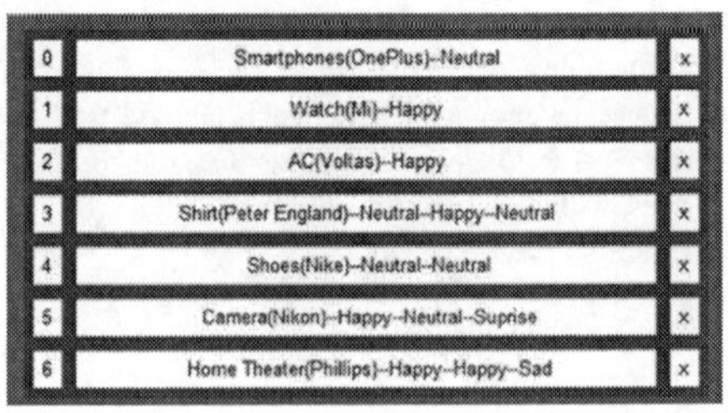

Fig. 10. Product feedback visible to owner

Owner of several shops with a range of items may gain customer feedback by collecting real-time emotion analysis of the customer at the same time, and the gathered data can be evaluated by the owner for better decision-making.

V. Conclusion and Future Work Plan

The facial expression recognition system described in this paper presents a robust face recognition model based on the mapping of behavioural and physiological biometric variables. The physiological properties of the human face that are relevant to various expressions such as pleasure, sorrow, fear, anger, surprise, and disgust are linked to geometrical structures that are reconstituted as the recognition system's basis matching template. Our work focuses on analyzing live facial expressions of consumers who are viewing a certain product, allowing us to conduct a real-time assessment of that product and score it based on the customer's facial expression analysis results.This product rating will assist the business owner in increasing product sales while also ensuring that the top items are available for his clients. This feature is significantly more accurate and quicker than previous techniques, which had a greater margin of error.

Further,this work might be expanded to look at other elements for product evaluation outside emotions, such as the amount of time a consumer spends looking at a product (which would necessitate face recognition in supermarkets) and product review in online purchasing. Because everyone nowadays is lured to internet buying, such reviews may aid in the early detection of product failure, minimizing future production losses.

References

[1] Vikrant Chaugule, Abhishek D, Aadheeshwar Vijayakumar, Pravin Ramteke, and Shashidhar Koolagudi, Optimized Facial Expression Detection ,(Aug 13, 2016)

[2] Preeti Thakre and Pankaj Agarkar,Customer Emotions Recognition Using Facial and Textual Review (Feb 14, 2020)

[3] P. Viola and M. J. Jones, "Robust real-time face detection," International journal of computer vision, vol. 57, no. 2, pp. 137–154, 2004.

[4] Zhaolong Li, "Analyzing Emotion on Amazon Product Review for User Modeling", Master's Thesis, October 2013.

[5] Y. Zhang and B. C. Wallace, "A Sensitivity Analysis of (and Practitioners' Guide to) Convolutional Neural Networks for Sentence Classification", ArXiv preprint arXiv:1510.03820v4, 2016.

[6] L. C. De Silva, T. Miyasato, and R. Nakatsu, "Facial emotion recognition using multi-modal information," in Information, Communications and Signal Processing, 1997. ICICS., Proceedings of 1997 International Conference on, vol. 1. IEEE, 1997, pp. 397–401.

[7] K. Durand, M. Gallay, A. Seigneuric, F. Robichon, and J.-Y. Baudouin, "The development of facial emotion recognition: The role of configural information," Journal of experimental child psychology, vol. 97, no. 1, pp. 14–27, 2007.

[8] D. Bamman and N. A. Smith, "Contextualized Sarcasm Detection on Amazon Product Review", in Proc. of the 9th Int. AAAI Conf. on Web and Social Media. Citeseer, 2015

[9] Niko Colneric and Janez Demsar, "Emotion Recognition on Amazon Product Review: Comparative Study and Training a Unison Model", IEEE TRANSACTIONS ON AFFECTIVE COMPUTING, FEBRUARY 2018.

[10] A. Radford, R. Jozefowicz, and I. Sutskever, "Learning to Generate Reviews and Discovering Sentiment", 2017.

[11] B. Nejat, G. Carenini, and R. Ng, "Exploring Joint Neural Model for Sentence Level Discourse Parsing and Sentiment Analysis",Proc .of the SIGDIAL2017.

[12] N. Nodarakis, S. Sioutas, A. Tsakalidis, and G. Tzimas, "Using Hadoop for Large Scale Analysis on Amazon Product Review: A Technical Report", arXiv preprint arXiv:1602.01248, 2016.

[13] X. Liu, J. Gao, X. He, L. Deng, K. Duh, and Y. Y. Wang, "Representation Learning Using Multi-Task Deep Neural Networks for Semantic Classification and Information Retrieval", Proc.of the Conf. of the North American Chapter of the Association for Computational Linguistics: Human Language Technologies, 921, 2015.

[14] S. M. Mohammad and S. Kiritchenko, "Using Hashtags to Capture Fine Emotion Categories from Tweets", Computational Intelligence, vol. 31, no. 2, pp. 301326, 2015.

[15] M. Dixit, P. Rai, and S. Silakari, "A review on smiley face recognition," in Communication Systems and Network Technologies (CSNT), 2013 International Conference on. IEEE, 2013, pp. 137–139.

N-gram models for Text Generation in Hindi Language

Tejashree Ghude, Roshni Chauhan, Krushna Dahake, Atharv Bhosale and Tushar Ghorpade*

Department of Computer Engineering
Ramrao Adik Institute of Technology
D Y Patil Deemed to be University

Abstract - Native language plays a vital role for communication. Hindi is preferred by most Indians and it is the fifth most spoken language in the world. Hence, to make User Experience more effective while interacting with Software Applications , we aim to build a Model using Natural Language Processing which takes a specific word as an input and predicts the subsequent words for completing the sentence.It will act as a tab- complete function in Hindi language. This will also pave a way as a use case for building chatbots, personal emails, applications which are content based like cooking recipes, blogs, etc. in Hindi Language.

Keywords :Natural Language Processing, N-Gram, Language Model, Text-Generation.

1 . Introduction

To overcome the language barrier in software applications, a NLP based model will play an effective role to enhance the user experience while interacting with Software Applications. Natural language processing (NLP) is a branch of artificial intelligence that helps computers understand, interpret and manipulate human language [1]. NLP draws from many disciplines, including computer science and computational linguistics, in its pursuit to fill the gap between human communication and computer understanding. NLP combines computational linguistics rule-based modeling of human language with statistical, machine learning, and deep learning models. Together, these technologies enable computers to process human language in the form of text or voice data and to interpret its meaning. This system addresses the problem of predicting the subsequent words, given an initial fragment of text. This approach of autocompleting will prove to be effective for error detection in Hindi language. Instance-based learning and N-gram models can be used to conjecture completions of sentences.

Further the paper is divided as follows: - in the section 2, Literature Survey is discussed, section 3 includes the problem definition of the system, section 4 contains Proposed System, section 5 and section 6 shows the Results and Analysis respectively for the system and conclusion along with future at end of the paper.

2 . Literature Survey

Literature survey provides an idea of the current condition and background information of the topic. In these papers various existing technologies and algorithms are mentioned. Following papers are one of the important papers.

In a research paper by Sharmila Mani et. al.[2], Optimized N-gram is proposed for faster Word Completion and Next Word Prediction (NWP) for mobile phones in real time. N-gram model is trained in various languages like English, Marathi, Hindi, Kannada, Tamil etc and optimised using Stupid Backoff approach.

Mrinalini K et. al. [3] explained a S2ST system for travel expressions which can perform translation between English and Hindi. This is based on the statistical machine translation approach. Bilingual Corpus is used as a training dataset.

Nikita P. Desai et. al.[4] has done a taxonomic survey of resources for Hindi Language NLP systems on all the available research and work done till 2021. Various available NLP resources for Hindi Language like corpora and lexical resources, language scripts, preprocessing such as tokenizer, parser, etc and about other toolkits like inltk, indic nlp, StanfordNLP etc are listed. It provided brief information about research done in Hindi NLP systems and their approaches and resources.

M Mittal et. al.[5], has worked on correction of grammatical errors using a hybrid approach in hindi language. Both statistical and rule based approaches are used to detect grammatical errors in Hindi sentences. Indian language Corpora Initiative (ILCI) corpora is used to detect part of speech using Hidden Markov Model Based technique. Hand written Hindi grammar rules are used to detect the errors and then suggestions are provided for correcting it.

From a brief literature survey, it is discovered that a hybrid approach of text generation using multiple predicting algorithms of text generation using multiple predicting algorithms is the best way to make NLP systems. Preparing, processing and cleaning Hindi datasets and methodologies related to it are discovered.

3 . Problem Statement

Native language plays a vital role for communication. The tab-completion feature for Hindi Language is not effectively developed in existing software applications. Natural Language Processing will prove to be an effective

*Corresponding author: tejashreeghude@gmail.com

 https://doi.org/10.1051/itmconf/20224403062

approach for predicting the subsequent words which can be used in various applications.

4 . Proposed Methodology

When it comes to text generation, following linguistic grammar is necessary. Every spoken language has its own grammar rules. These grammar rules can be applied in text generation using context free grammar. For English there already exists technologies which can provide these grammatical rules at one place. While for Hindi there aren't many resources which can help us in defining and implementing these grammar rules.

Although defining grammar rules and generating sentences using them is an ideal way, it does not give any assurance that the generated sentence will be meaningful. Here instead of using a rule based approach, a probability based approach was followed for text generation. As one cannot possibly cover all grammar rules for a particular language, a probabilistic approach could be more reliable for text generation. So for a probabilistic approach data was collected, processed and sorted based on the probability of their occurrence before using it for prediction. In this way the accuracy of phrase generation of the text generation model is ensured.

4.1 Architecture/ Framework

The aim is to implement an N-gram model for prediction of text generation . An N-gram model is a type of probabilistic language model for predicting the next item following a similar approach as $(n-1)$ order Markov model. The Markov model predicts that the state of an entity at a particular position in a sequence depends on the state of one entity at the preceding position. Here instead of depending on the probability of occurrence of the whole sentence this system chooses to consider the last few words. By considering the probability of the whole sentence will restrict the possible patterns for text generation. Considering the probability of word using Markov assumption would reduce the complexity and give the text prediction process more flexibility. It doesn't only make it flexible but also makes it more efficient . The Data Cleansing process is first carried out on the data before passing it data to the model. Data cleaning is rather a very crucial process for this model. Without the cleaning process, the dataset is just a bunch of words that the computer doesn't understand.

4.2 Processing and Prediction :

The initial phase of processing involved data cleaning for the model. As mentioned before, the processing of all the text data was carried out in this phase. First break the paragraph data into list sentences. As the text generation is carried out in Hindi language which is in Devnagri script , all the English latin characters were removed from the dataset during the data cleaning process. Along with these characters there are often some numeric characters which are of no use to us. These types of characters were removed along with all the punctuation signs from each sentence. After data cleaning ,the received data is in the form of a list of sentences . After tokenizing each sentence of the previous list is converted into a list of words which is passed to our model. The model counted the occurrences of each n-gram (i.e. a sequence of n words) in the whole data set. After this process, the probability for each combination is calculated. For calculating the probability , the occurrence count of the whole n-gram taken was divided with the occurrence count of the first (n-1) word in the dataset which gave its value between 0 and 1 . This probability is then used for prediction .This n-gram model checks and adds the next word to the sentences from data based on its probability of occurring with last (n-1) words . For example if input is given as " हम ये " these two words i.e. (n-1) words to the trigram (n-gram) model will predict the third word "नहीं" accordingly which is described in figure no. 4.1.

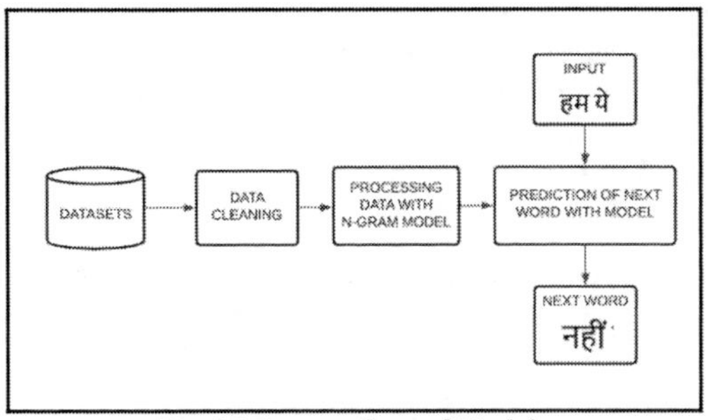

Fig 4.1 Process flow of Model

Text generation could be given as a task of generating text with the goal of appearing indistinguishable to human-written text. This task is also known as "natural language generation". Text generation or natural language generation produce meaningful phrases and sentences in the form of natural language. In essence, it automatically generates text based data in a human-like manner at high speed. Text generation can be addressed with Markov processes or with deep generative models. Here the implemented model is based on the N-gram algorithm. Here different Hindi datasets were used. The data from these dataset is first processed before passing to the model. This model could suggest the next word from the previous word based on the probability of their occurrence.

4.4 Data Preprocessing

The two dataset used here are Hindi Wikipedia Articles - 172k [12], HindiEnglish Corpora (HindEnCorp 0.5 by Charles University) [13]. Hindi Wikipedia Articles - 172k is actually a collection of text files of hindi Wikipedia articles . At first , each article was extracted from the text file and converted to a list of articles ,there were about 1,72,279 articles which were then further broken into a list of sentences to get 26,61,107 sentences. HindiEnglish Corpora (HindEnCorp 0.5 by Charles University) was in csv file form with Hindi and English sentences, Hindi sentences were extracted from this dataset

https://doi.org/10.1051/itmconf/20224403062

which were about 97,662 sentences. In total there were **27,88,714** sentences for training our model.

Along with this a small Hindi dataset was created using essays and stories in Hindi to show which was used in the initial stage of model building . This Hindi dataset which was initially in the form of a text file was also cleaned and processed before passing to the model. As mentioned before, the data cleaning process involved removing numbers, special characters and latin characters and at the end a collection of sentences(list of sentences) was generated. These sentences were first tokenized before feeding them to our model .

4.5 N-gram model

An N-gram language model is a technique of counting sequences of characters or words that allows us to support rich pattern discovery in text. For comparison of the accuracy two n-gram models : trigram and bigram were implemented. Bigrams are a set of two consecutive words from the sentences whereas trigram is three consecutive words in a sentence. So first all the possible trigram (i.e. consecutive three word combinations in the corresponding sentence)were extracted.

In the Trigram model the occurrence of each trigram was also counted during extraction. In the model the association of these words is stored based on the occurrence of third with respect to the first word. After getting the count of each pair i.e group of first two words and third word, its count is divided with the count occurrence of first two words. This gives us the probability which was used for prediction. Following figure 4.2 shows how sentences are divided and then the relation in which they are stored .

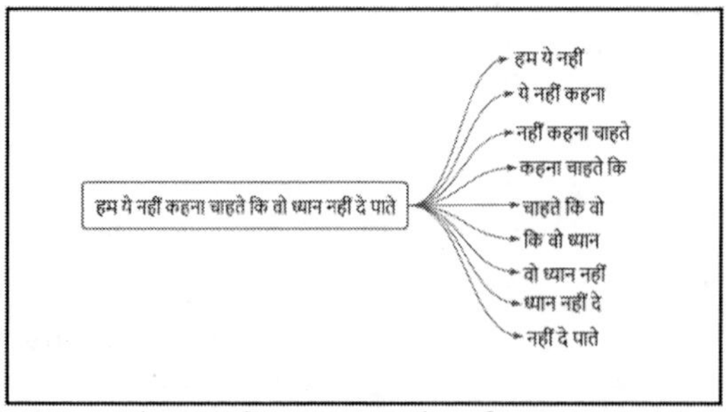

Fig 4.2 Trigram Generation of sentences

After getting the probabilities for each combination (n-gram) in the dataset, text generation was carried out shown in figure 4.3. For text generation ,input of the first two words to the model was given .The model checks all the possible third words and then appends the third word to the sentence if its probability of occurrence is more than or equal to threshold probability i.e 0.1 .The last step is repeated until it reach to condition were the last two word don't have further third word combination, at this condition te process terminates.

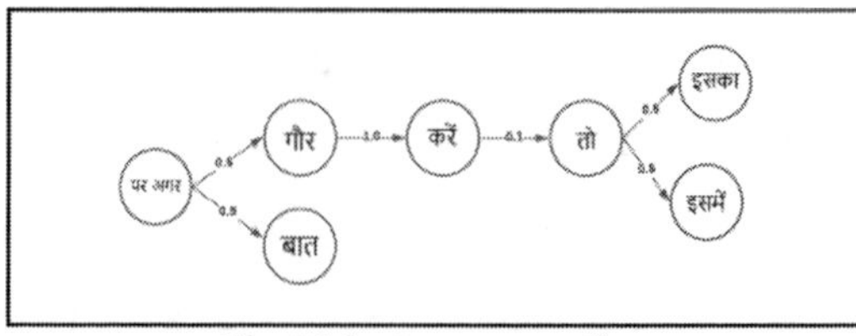

Fig 4.3 Sentence generation

Using these N-grams and the probabilities of the occurrences of certain words in certain sequences could improve the predictions of auto completion systems.

5 . Results

Different inputs were given to the models to analyze their outputs. The trigram takes input of a group of two words while bigram takes only the last word as input. Because of this reason we got different outputs for trigram and bigram .

The input " पर अगर " was given to the Trigram model and it predicted the sentence shown in the next fig 5.1 . The Trigram model stopped after words " पाते हैं " as there would have been no combination with " पाते हैं " as input variable in the model .Whereas when the input " पर " given to the Bigram model it predicted the sentence shown in the fig 5.2 . Similar to the Trigram model, the Bigram model terminates the sentence at " हैं ".

Fig 5.1 Output of Trigram model

Fig 5.2 Output of Bigram model

To show working of the model a small dataset was used to show how the model takes the decision on the basis of probability. After adding each word , the model repeats the last step and checks the possible third word and their probability. The following figure 5.3 is the output of that experiment, where the probability and the third word option model at each step was also printed.

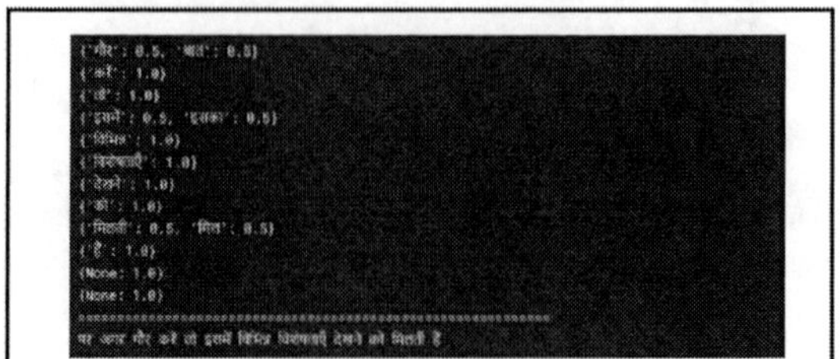

Fig 5.3 Output of with probabilities

The accuracy for this Trigram model was calculated by considering true positive (grammatically correct and present in the dataset) and true negative (grammatically correct and not present in the data set as our data set . Here TP = True Positives, TN = True Negatives, FP = False Positives, and FN = False Negatives. Following is the output we got for the accuracy of the Trigram model which is shown in Figure 5.4.

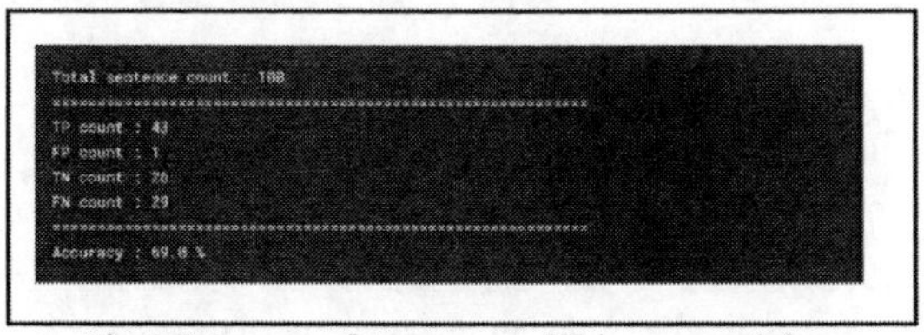

Fig 5.4 Output of Accuracy of Trigram model

6 . Analysis

The N-gram method gives the assurance of accurate phrase generation. To analyze the model and to find the efficiency of the model, accuracy of the model was checked. The accuracy of our model depends on whether the model is generating grammatically correct sentences or not. The model was generating sentences by predicting the next word for a group of words based on probability of their occurrence in the dataset. There were chances of the model generating a grammatically wrong sentence as only upto Trigram model implemented but nevertheless the accuracy was enough for smooth functioning of text generation model.

To find the accuracy of the model True vs. False and Positive vs. Negative method was used. Here the True vs False condition was set for grammatical correctness of the model. Sentence was assigned True if it is meaningful and grammatically correct, while it was assigned False if it is not meaningful or if it is not grammatically correct. And Positive vs. Negative condition was applied on the basis of whether the sentence is present in our data set or not.

A true positive is an outcome where the model correctly predicts the positive class. Similarly, a true negative is an outcome where the model correctly predicts the negative class. A false positive is an outcome where the model incorrectly predicted the positive class. And a false negative was an outcome where the model incorrectly predicted the negative class.

Accuracy is the fraction of predictions the model got right. Formally, accuracy has the following definition in (1) :

$$Accuracy = \frac{Number\ of\ correct\ predictions}{Total\ number\ of\ predictions} \quad (1)$$

For binary classification, accuracy can also be calculated in terms of positives and negatives as follows in (2) :

$$Accuracy = \frac{TN + TP}{TP + TN + FP + FN} \quad (2)$$

Where TP = True Positives, TN = True Negatives, FP = False Positives, and FN = False Negatives. In this way we have evaluated the accuracy of our model . By calculating the accuracy of sentence generation we have estimated the performance of the model.

In n-gram two models were built. The first is the Bigram model and the second is the Trigram model. The sentence generation in both models is quite accurate but it's more accurate in trigram as the trigram considers the last two words while bigram only considers last words . Usually, n-grams more than 1 is better as it carries more information about the context in general. However the result might differ for different datasets. In bigram one past word is considered and in trigram two past words are considered. It can happen that past two words itself occurred less time and when it happens it contains all those probable words in the same, more or less frequency.This can highly affect the accuracy of the model . So we can say that the accuracy will depend on the training and testing datasets. A testing dataset is created for the training. The testing dataset consisted of pairs of words which we fed to the Bigram and Trigram model to generate sentences. After generating sentences and classifying the result with True vs. False and Positive vs. Negative method as mentioned before the final accuracy we got was 69% for the same Trigram model and 61% of Bigram model.The confusion matrix of the result is given in fig 6.1 .This accuracy test was done on the small dataset. So by increasing the size of the dataset we can surely increase the range of accuracy of our model.

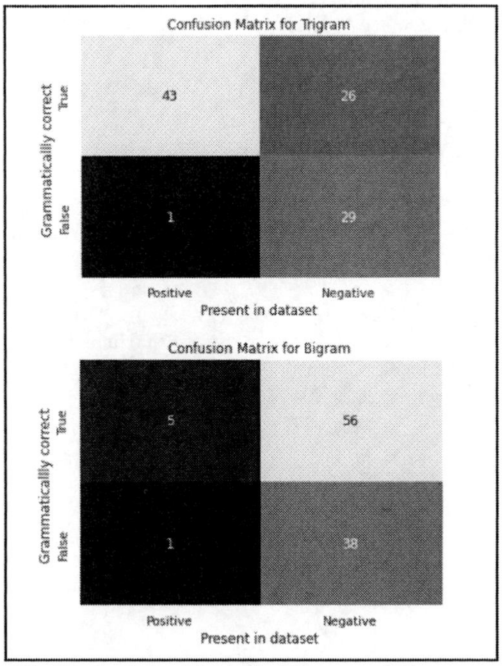

Fig 6.1 Confusion Matrix for Trigram and Bigram

There are many other algorithms which can be used as

support for this model to improve its efficiency and accuracy. Here no other technology was used for grammar structuring for text generation. In spite of using a basic probabilistic approach for the text generation, 69% and 61% accuracy was achieved by the Trigram and Bigram model respectively.

7 . Conclusion

The proposed N-gram approach in this paper ensured the accuracy of the language model. As language is creative and new sentences are generated every minute, It is very difficult to count all possible sentences. But other ways by which this task can be carried out are not as efficient and accurate as probability based prediction. So it is observed that the accuracy of these existing models is not adequate to our current accuracy needs. The N-gram language model could be relatively more accurate .With support of other technologies we can improve its performance in future. Hence, this model can be used in various software applications for tab completions, emails, chatbots, etc.

References

[1] Diksha Khurana, Aditya Koli, Kiran Khatter, Sukhdev Singh, "Natural Language Processing: State of The Art, Current Trends and Challenges", Researchgate, 2017.

[2] Sharmila Mani, Sourabh Vasant Gothe, Sourav Ghosh, Ajay Kumar Mishra, Prakhar Kulshreshtha, Bhargavi M and Muthu Kumaran, "Real-Time Optimized N-gram For Mobile Devices", IEEE 13th International Conference on Semantic Computing (ICSC) 2019.

[3] Mrinalini K,Vijayalakshmi P , "Hindi-English Speech-to-Speech Translation System/or Travel Expressions", ICCPEIC, India 2015.

[4] Nikita P. Desai, Mr.(Dr.) Vipul K. Dabhi, "Taxonomic survey of resources for Hindi Language NLP systems", Arxiv, 2021.

[5] M Mittal, SK Sharma, A Sethi, "Detection and Correction of Grammatical Errors in Hindi Language", International Journal of Computer Sciences and Engineering, 2020, 2347-2693.

[6] Ankur Sonawane, Sujeet Kumar Vishwakarma, Bhavana Srivastava, Anil Kumar Singh, "Generating Inflectional Errors for Grammatical Error Correction in Hindi", Proceedings of the 1st Conference of the Asia-Pacific Chapter of the Association for Computational Linguistics and the 10th International Joint Conference on Natural Language Processing, 2019.

[7] Tobias Scheffer, Peter Haider, Steffen Bickel, "Using Hybrid Approach Predicting Sentences using N-Gram Language Models", International Conference on Advanced Information and Communication Technology, Dhaka, Bangladesh, 2020.

[8] Jayashree Nair, K Amrutha Krishnan, R Deetha, "An Efficient English to Hindi Machine Translation System Using Hybrid Mechanism", ICACCI, Sept. Jaipur, India 2016.

[9] Tom Young, Devamanyu Hazarika, "Recent Trends in Deep Learning Based Natural Language Processing",Arxiv, 2018.

[10] B. S. Harish , R. Kasturi Rangan, "A Comprehensive Survey on Indian Regional Language Processing", Springer, 12 June 2020.

[11] T. Brants, A. C. Popat, P. Xu, F. J. Och, and J. Dean, "Large language models in machine translation," in Proceedings of the 2007 Joint Conference on Empirical Methods in Natural Language Processing and Computational Natural Language Learning (EMNLP-CoNLL), 2007.

[12] https://www.kaggle.com/aiswaryaramachandran/hindienglish-corpora

[13] https://www.kaggle.com/disisbig/hindi-wikipedia-articles-172k

[14] https://www.clarin.eu/resource-families/parallel-corpora

https://doi.org/10.1051/itmconf/20224403063

Business Meeting Summary Generation Using NLP

Department of Electronics, Ramrao Adik Institute of Technology, Nerul

Aryan Jha[*], *Sameer* Temkar[**], and *Preetam* Hegde[***] *Navin* Singhaniya[****]

Abstract. The effort of providing a brief and fluent summary of a business meeting while preserving vital information content and overall meaning is known as business meeting summarizing.Initially, we convert recorded meetings, interviews, speeches, lectures, and other audio streams into text documents in the proposed work. After that, speaker text is divided into segments. Minutes of the Meeting are generated automatically using the proposed work.We also focused on using the Rev-AI Speech-to-Text API to better the voice to text conversion of a specific recorded audio file while making sure that the summarised text that has been converted from speech to text provides a specific and accurate meaning that is not only understandable to everyone but also covers all of the recorded file's important points.

1 Introduction

Nowadays, most data distributed is through Internet due to which summarizing the distributed data becomes really essential. text synopsis turns out to be effectively significant[1]. The earliest numerical data points in this topic date back to 1958. Analysts recommended that the recurrence of words be utilized as a numerical measure in this interaction, which is still valid for certain strategies [1].To save time during these lengthy Business Meetings, it is necessary to summarize the meetings. Text Summarization can be used to perform this summarization of Long Business Meetings. Business Meeting Summary Generation Using NLP can be used to listen to the doubts of the users and give answers to them in form of results which of late is something that Internet service providers are doing from past few years. For business summarizing, there are essentially two methods:

- Abstractive Summarization :The Abstractive approach attempts to give an outline by interpreting the text utilizing progressed regular language calculations to build a new, more limited message that fetch the most data - portions of which may not present in the first record.This is in contrast to the extractive method, which uses just the sentences found in the original text
- Extractive Summarization :The Extractive summarization approach will be used in our research .It uses just the sentences found in the original text.

1.1 Motivation

Because of the growing availability of documents, extensive study in the field of text summarization is required. A summary is a text that is created from one or more texts, contains crucial information from the original text, and is no more than half the length of the original text, and frequently much less. The principle of Text Summarization is followed in "Business Meeting Summarization Using NLP."

1.2 Objective

The job of developing a brief and fluent summary while keeping the important data content and overall meaning is known as business meeting summarizing. The goal is to employ Business Summarization in Business Meetings to assist us summarise a recorded meeting while maintaining critical information and ensuring that the summarized meeting has the right context and meaning. Investigate various Business summarizing strategies. Compare the summaries that were generated.

Determine the best summary parameters (for example, k in extractive business meeting summarization). To scale an algorithm to different languages, identify or make adjustments (if possible). Also, distinguish between the various applications of business meeting summarization.

2 Organization Of Report

The report is organized as follows:

- The introduction is given in Section-1. It describes the fundamental terms used in this project. It motivates to study and understand the different techniques used in this work. This Section also presents the outline of the objective as well as defines the problem of the report.Motivation,Objective and Novelty of the project is also mentioned in this section.
- Section 3-presents the System Implementation in which algorithms and techniques are explained .

*e-mail: ary.jha.rt18@rait.ac.in
**e-mail: sam.tem.rt18@rait.ac.in
***e-mail: pre.heg.rt18@rait.ac.in
****e-mail: navin.singhaniya@rait.ac.in

https://doi.org/10.1051/itmconf/20224403063

The,architecture workflow of the system implemented in the research paper is discussed in detail.Flowchart of the algorithm that is used in the project is also mentioned in this section.

- Section 4-presents the proposed method for summarization. .The flowchart,workflow of Extractive summariation method that we have used in this project is also explained in this section.
- Section 5-Applications of the summary generator that we have build are mentioned in this section.
- Section 6-All the Experimental results of the summary generator that we have build are included in this section.
- Section 7-Output screenshots are included in this section.
- Section 8-The conclusion and the result are presented in this section.

3 System Implementation

3.1 Techniques and Algorithms

3.1.1 TextRank Algorithm

The TextRank approach is based on PageRank, a popular algorithm for ranking web pages in search results. It makes a m x m lattice, with every cell having the probability that the client will visit that site, which is 1/(number of special connections in website page wi). The qualities are then intelligently refreshed and helps in briefing of text.

TextRank performs in the same way. It creates a m x m adjacent matrix, where m is the number of sentences in the text. It is compared to mj (where i!= j) for each sentence mi (where i is a record). This comparison is made using cosine similarity or another approach for comparing two sentences. This is how the entire matrix is filled. The full row is added to decide the score for each sentence mi (where i = 1, 2, 3, and so on). A greedy search is used to select the top k sentences based on this score. These are the needed summary sentences. While processing the graph, the adjacency matrix increases the time complexity to O(m power 2), but the adjacency list reduces the difficulty to O(v + e)[1]. TextRank is more efficient at identifying the links between sentences. It is simple to apply cosine similarity when vectors are used.A chart connect between the two sentences shows that both are fundamental for an essential condition. As a result, TextRank performs brilliantly .Steps involved are :

- ConvertToVector():Convert every one of the sentences to vector representation..
- SentencesSimilarities():Construct the graph with sentences as hubs(nodes) and register the hub edges (sentence likenesses)
- kTopRanked(): Return K top words for generation of summary.

	1	2	3	4	5	6	7	8
1		3	0	0	0	0	0	0
2	3		12	2.7	5	0	0	0
3	0	12		0	0	0	0	0
4	0	2.7	0		3.5	3	0	0
5	0	5	0	3.5		0	0	0
6	0	0	0	3	0		3	8
7	0	0	0	0	0	3		0
8	0	0	0	0	0	8	0	

Figure 1. Adjacency Matrix

3.1.2 PageRank Algorithm

PageRank is a Google Search algorithm that ranks webpages in search engine results. Larry Page, one of Google's founders, was the inspiration for PageRank. PageRank is a metric for determining how important a website's pages are. The PageRank algorithm produces a probability distribution that is used to indicate the possibility of a random user clicking on links ending up on a specific page. PageRank can be estimated for any large set of documents. Several research publications assume that at the start of the computational process, the distribution is evenly distributed among all documents in the collection. The PageRank computations require several passes through the collection, referred to as "iterations," in order to alter approximation PageRank values to more nearly represent the theoretical actual value.

3.1.3 Cosine Distance – Text Similarity Metric

The degree to which two documents are related in terms of context or meaning is measured by text similarity. The Cosine distance-text similarity metric measures the cosine of the point between two m-layered vectors projected in a multi-faceted space. The cosine similarity between two papers can range from 0 to 1.[3] When the Cosine similarity score reaches 1, two vectors are aligned in the same orientation. The lower the value, the less comparable the two papers appear to be. The mathematical equation of Cosine similarity between two non-zero vectors is:

$$\cos x = \frac{A \cdot B}{|A||B|} \tag{1}$$

4 Proposed method for Summarization

4.1 Extractive Summarization

An extractive summarization method involves extracting key sentences or paragraphs from the original text and compressing them into a shorter text. As the name suggests, extractive summarization focuses on calculating

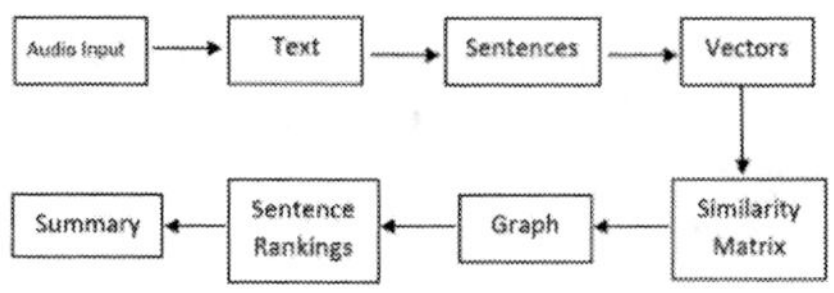

Figure 2. Flowchart

sentence weights and selecting (top k sentences) them for the summary. Extractive summarization techniques generate summaries by selecting a subset of the original text's sentences. To understand how summarization system actually works, we'll go over three fairly independent tasks that all summarizers must finish.

- Generate an intermediate representation of the input text which expresses the text's key points.
- The sentences are selected depending on their representation.
- Choose a summary composed of a few sentences.

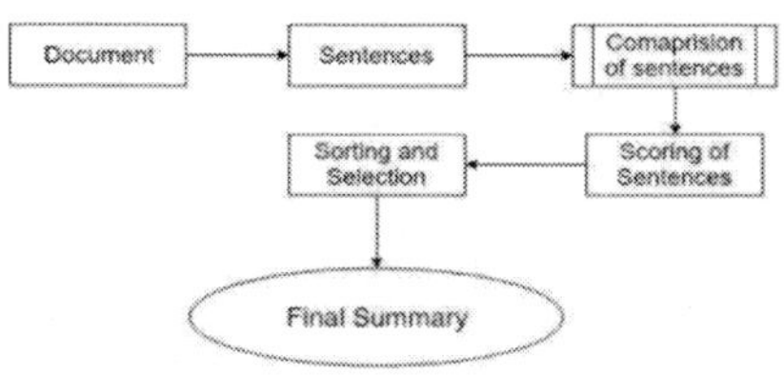

Figure 3. Extractive Summarization Flowchart

4.1.1 Intermediate Representation

Every summarizing system creates an alternative representation of the text it aims to summarize and uses it to identify essential content. There are two techniques to representation based on the representation: subject representation and indicator representation. Topic representation techniques reads the text's themes by converting it into an intermediate form. Each sentence is described as a list of significant attributes such as sentence length, position in the document, the existence of certain phrases, and so on in indicator representation approaches.

4.1.2 Sentence Score

After generating the intermediate representation, we assign an importance score to each sentence. The score of a sentence in topic representation approaches demonstrates how well the sentence explains some of the most important topics in the text. The score is computed in the most of of indicator representation methods by aggregating evidence from multiple indicators.

4.1.3 Summary Sentences Selection

To produce a summary, the summarizer algorithm selects the top k most important sentences. Some approaches use greedy algorithms to select important sentences, while others convert sentence selection into an optimization problem where a set of sentences is chosen with the constraint of maximising overall relevance and coherency while minimising redundancy. The meaning in which the summary is generated, for example, could be useful in determining its importance. Another factor which may impact sentence selection is the type of document (newspaper article, email, scientific paper, etc.).

5 Applications

This method is used by well-known social media platforms to generate summaries for postings that are categorized into subjects. These subjects are utilised to keep consumers interested on the internet. Instead of only offering links, today's search engines provide immediate answers to the user's inquiry.Technologies for voice-based associates that responds to the users inquiries utilizes a similar methodology.Many weekly emails start with a brief introduction before presenting a well curated selection of relevant articles.

- One of the most significant marketing channels is video. People are now uploading videos on professional networks like LinkedIn, in addition to video-focused venues like YouTube or Vimeo. Scripting may or may not be required, depending on the type of video. When it comes to writing a script that integrates research from a variety of sources, summarization can be a valuable ally.
- For example,[1] Google's home feed makes synopses on basis of the users likes and dislikes . Instead of only offering links, today's search engines provide immediate answers to the user's inquiry.
- Many weekly emails begin with an introduction and then feature a carefully curated selection of relevant content. Summarization would allow companies to supplement newsletters with a stream of summaries (rather than a list of links), which is a more mobile-friendly format.
- One of the most significant marketing channels is video. People are now uploading videos on professional networks like LinkedIn, in addition to video-focused venues like YouTube or Vimeo. Scripting may or may not be required, depending on the type of video. When it comes to writing a script that integrates research from a variety of sources, summarization can be a valuable ally.

6 Experimental Result

6.1 ROUGE(Recall-Oriented Understudy for Gisting Evaluation)

Our summary generator correlates highly with human judgment and has high recall and precision significance test with manual evaluation results.We choose ROUGE(Recall-Oriented Understudy for Gisting Evaluation)precision, recall as the measurement of our experiment results.It is a set of metrics for calculating automatic Business Meeting summarization as well as machine translation.It compares an automatically generated summary or translation to a set of reference summaries (typically human-produced).To get a good quantitative value, we can actually compute the precision and recall using the overlap.

6.1.1 Recall

Recall (in the context of ROUGE) refers to how much of the reference summary the system summary is recovering or capturing .Recall can be measured as:

$$\frac{number\_of\_overlapping\_words}{total\_words\_in\_reference\_summary}$$

6.1.2 Precision

In precision, you're essentially determining how much of the system summary was actually relevant or actually needed.When trying to generate brief summaries, precision becomes extremely important .Precision can be meaured as :

$$\frac{number\_of\_overlapping\_words}{total\_words\_in\_system\_summary}$$

6.1.3 F-Measure

It is test accuracy metric. It contains the test's precision and recall, where the precision is the number of true positive test result divided by the overall number of positive test result, including those that were, by mistake grouped, and the recall is the number of true positive test result divided by the overall number of measurements that should have been identified as positive[5].The formula for f-measure is given by:

$$2 \times \frac{precision \times recall}{precision+recall}$$

Table. 1 ROUGE-1 Precision, Recall, f-measure values of samples

Samples	Precision	Recall	f-measure
Sample(1)	70.09	68.18	69.12
Sample(2)	65.48	67.27	66.36
Sample(3)	87.12	100	93.12
Sample(4)	83.72	80.00	81.80
Sample(5)	62.50	86.20	72.46
Sample(6)	58.06	100	73.46
Sample(7)	72.51	100	84.07
Sample(8)	77.98	73.91	75.89
Sample(9)	68.31	100	81.17
Sample(10)	49.56	82.60	61.95
Average	70.00	85.81	75.94

Here we are comparing the model-generated summary of our input text with the reference summary.The reference is a human-made summary of our input text.Here we used ROUGE-1 values because it helps to measure the fluency of the summaries much better than ROUGE-2 and ROUGE-l. We are getting accuracy of 70 percent ,recall of 85.81 percent and f-measure of 75.94 percent. The data for this experiment comes from the kaggle bbc news summary dataset.

7 Output

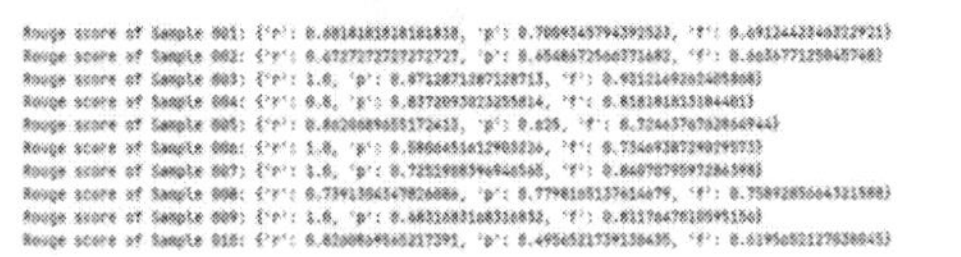

Figure 4. ROUGE-1 scores of the sample dataset

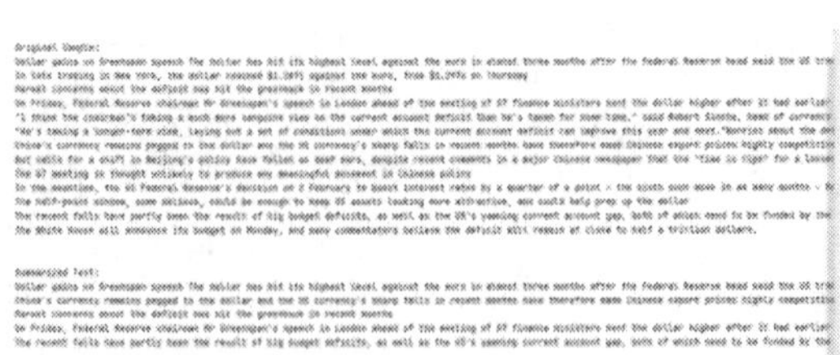

Figure 5. Sample input and its summarized output

8 Conclusion

We have successfully converted 30 minutes of meeting into text using Rev-Ai API and then successfully summarized the text using our model.For the evaluation of our results we have used ROUGE-1 score values(Precision,recall and f-measure) because it helps to measure the fluency of the summaries much better than ROUGE-2 and ROUGE-L.In our experimental results we have used bbc news summary kaggle dataset and we are getting the accuracy of 70% ,recall of 85.81% and f-measure of 75.94%. Our summary generator correlates highly with human judgment and has high recall and precision significance test with manual evaluation results. We also focused on improving the voice to text conversion of a specific recorded audio file using the Rev-AI Speech-to-Text API while ensuring that the summarised text that has been converted from speech to text offers a precise and concise meaning that is not only understandable to everyone but also covers all of the recorded file's important points.

References

[1] Virender Dehru, Pradeep Kumar Tiwari, Gaurav Aggarwal, Bhavya Joshi and Pawan Kartik "Text Summarization Techniques and Applications", IOP Conference Series: Materials Science and Engineering, 2021

[2] Ivona Milanova, Ksenija Sarvanoska, Viktor Srbinoski and Hristijan Gjoreski "Chapter 1 Automatic Text Generation in Macedonian Using Recurrent Neural Networks", Springer Science and Business Media LLC, 2019"

[3] Verma, Jaiprakash Vinodkumar and Tanwar, Sudeep and Garg, Sanjay and Rathod and Abhay Dinesh "Crowdsourced Social Media Reaction Analysis for Recommendation"

[4] V Gupta and G.S Lehal, "A survey of text summarization extractive techniques", Journal of Emerging Technologies in Web Intelligence, vol. 2, no. 3, pp. 258-268, 2010

[5] Schwenke, Carsten "Diagnostic Studies"

[6] J Goldstein, J. Carbonell and M Kantrowitzt, Multiple-document summarization by sentence Extraction, pp. 40-48, 1998

[7] F. Liu and Y. Liu, "Towards abstractive speech summarization: Exploring unsupervised and supervised approaches for spoken utterance compression", IEEE Transactions on Audio Speech and Language Processing, vol. 21, no. 7, pp. 1469-1480, 2013.

[8] E. Collins, I. Augenstein and S. Riedel, A supervised approach to extractive summarisation of scientific papers, 2017.

A Novel Method of Deepfake Detection

Rutuja Hande[1,*], *Sneha* Goon[1,**], *Aaditi* Gondhali[1,***], and *Navin* Singhaniya[1,****]

[1]Department of Electronic Engineering, Ramrao Adik Institute of Technology, Nerul, Navi Mumbai

Abstract. Deep-Fake is a novel artificial media technology that uses the likeness of someone else to replace people in existing photographs and films. Deep Learning, as the name implies, is a type of Artificial Intelligence that is used to create it. It is critical to develop counter attacking approaches for detecting fraudulent data. This research examines the Deep-Fake technology in depth. The Deep-Fake Detection discussed here is based on current datasets, such as the Deep-Fake Detection Challenge (DFDC) and Google's Deep-Fake Detection dataset (DFD). The creation of a bespoke dataset from high-quality Deep-Fakes was utilised to test models. The results of both with and without Transfer Learning were analysed. Finally, the trained models were used to spot well-known deep-fakes of former US President Barack Obama and well-known actor Tom Cruise. A comparison study was performed on all three models. The findings show that the detection are generally domain-specific tasks, however that using Transfer Learning considerably improves the model performance parameters, whereas convolutional RNN gives sequence detection advantage.

1 Introduction

The internet and social media platforms have brought people from all corners of the globe closer together. In a world where recording and sharing are an inevitable part of everyday life, it's critical to be aware of the far-reaching consequences that Deep-Fakes can have [1]. However, This is no longer the case, thanks to the appearance of Deep Fake Video. Technology has the ability to convince others that something is real when it isn't." Deep-Fakes is a combination of the words "deep" and "fake." "Deep learning" and "fake" are two words that come to mind.

In this technology, autoencoders and GANs are utilised to create visual and auditory data. The resulting information is rife with deception [2]. Autoencoders are neural networks that retrieve and change the original image dimension by extracting crucial facial traits.Tradionally in the latent space The latent space is used to depict data analysis in a more realistic way. In a transparent way The de-noising process will overlay the original image's features [3]. This image was coded by a coder who had been specifically trained for it. Incorporation of a Generative The addition of an adversarial network to the decoder side helped to increase the technique's generation.

If unmanaged, this cutting-edge technology can be disastrous, as the era of believing what one sees is over. However, malevolent Deep-Fakes dominate the productive applications of Deep-Fakes, such as film dubbing, special effects, and instructional objectives [5]. Information

Figure 1. Frames from Deep-Fake video of former US President Donald Trump

Warfare, Celebrity Defamation, Political Propaganda, Scams and Fraud, Harassment, and Blackmailing are all examples of how this weapon of mass misinformation can be utilised. With the enhanced quality of Deep-Fakes created with deep learning algorithms like GANs and Autoencoders, it's getting increasingly difficult to tell them apart with human eyes.The project's vision is to develop Deep-Fake video detection frameworks that are trained on the most recent video datasets and then assess the results. Convolution Neural Network architecture was used to detect deep fakes [9]. A hybrid approach of Convolutional Recurrent Neural Network is used to diverse video datasets, taking into account the added temporal nature of video. LSTM or GRU can be used as the RNN's later extension network. Comparative analysis of the results is performed on some of the most recent DeepFake datasets, providing insight into the types of data

*e-mail: rutujahande2000@gmail.com
**e-mail: snehagoon2000@gmail.com
***e-mail: aaditi.gondhali28@gmail.com
****e-mail: navin.singhaniya@rait.ac.in

dealt with by each of them on all three model architectures as discussed, and is also represented using tabular and graphical representation for better visual understanding [13].

2 IMPLEMENTATION

APPROACHES BASED ON CNN+RNN NETWORKS

The extraction of features from video frames is done using a Convolutional Neural Network. RNN is trained on these features in order to classify videos into one of two categories: fake or authentic. Another model based on the ConvLSTM hybrid architecture [2] examines the minute visual markers on the faces using CNN and then evaluates the visual data using an RNN network based on the features. The widespread availability of resources such as GPUs has resulted in the widespread distribution of these videos. RNN is well-known for its ability to analyse sequential data. We already know that the temporal information extracted from video frames is sequential. As a result, RNN can be used to process it. The work presented by Sabir et al. [3] used the best model for recognising facial alterations, which uses the RNN network. The FaceForensics++ dataset was used for testing, and efficiency was enhanced by 4.55 percent above the prior state-of-the-art. Some approaches focus on the faces in the films to reliably detect modifications, extract- ing visual and temporal information from the faces. These methods employ a recurrent neural network-based architecture to investigate temporal trends. D. M. Montserrat et al. [4] explored one such strategy. They compared the results with previous methodologies using the DFDC dataset. The proposed method produces more accurate findings than existing methods. MTCNN is the CNN model employed, and the model is used for temporal feature extraction. The proposed model has a validation accuracy of 92.61 percent and a test accuracy of 91.88 percent.

APPROACHES BASED ONLY CNN NETWORKS

Face warping is used in Deep-Fake generation, as are numerous morphing techniques like as scaling, translation, and affine deformations. Deep-Fake generation's warping method causes resolution discrepancy [7] between the warped face area and the surrounding back- ground area. Visual artefacts [8] are created in the video frames as a result of this. This type of technique takes advantage of the errors that occur throughout the generation process. Convolutional Neural Network models are used in these methods to detect the artefacts. Convolutional Neural Networks could be quite effective for computer-assisted detection difficulties. H.-C. Shin et al. present a paper [9] that uses deep CNN factors to solve Computer-Aided Detection (CADe) problems. Another CNN-based effort on image modification detection [10] use a high pass filter in conjunction with a CNN network to detect the image's hidden features.

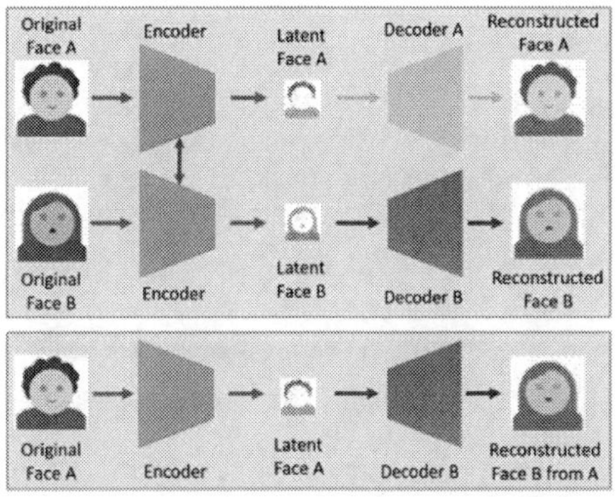

Figure 2. deepfaken creation model using two encoder-decoder pair [3]

2.1 ALGORITHM

For the detection of Deep-Fake films, a variety of existing models have been applied. A comparison of CNN-based video deep fake detection with CNN followed by RNN-based Video DeepFake Detection is discussed in this paper. The performance of two RNN variants, LSTM and GRU, on temporal feature exploitation is compared. The algorithm of the method presented in this paper is discussed in this part. The major goal of the system under discussion is to determine whether the video input is authentic or Deep-Fake. This goal can be accomplished in three different ways. Three different models are compared based on their performance for Video Deep-Fake Detection. The following is the algorithm for these processes:

STEP 1: Frames from the input video are extracted. The input video is preprocessed and turned into frames in this step. The CNN architecture is fed a predetermined amount of frames as an input.

STEP 2: Each frame's features are extracted. For an unseen test sequence input, CNN extracts a collection of facial traits for each frame. The feature set is concatenated when this operation is performed on a preset number of frames at the same time.

STEP 3: Analyze Temporal Sequences[1] This stage is different in each of the three models covered in the paper. To acquire a single probability value for the CNN model, the concatenated features are input to dense, followed by a global average pooling layer. Concatenated features are given to LSTM and GRU layers for temporal

Dataset	**realvideos**	**fakevideos**	**totalvideos**	**rightcleared**	**agreeingsubject**	**total subjects**	**synthesismethods**
UADFV	49	49	98	NO	0	49	1
DeepFakeTIMIT	640	320	960	NO	0	32	2
FF++	1000	4000	5000	NO	0	N/A	4
CelebDF	590	5639	6229	NO	0	59	1
GoogleDFD	363	3000	3363	YES	28	28	5
DeeperForensics|&50000	10000	60000	YES	100	100	1	
DFDC	23654	104500	128154	YES	960	960	8
KoDF	62166	175776	237942	YES	403	403	6
FkeAVCeleb	500	19500	20000	NO	0	500	4

Table 1. Quantitative comparison of fakeAVCeleb to existing publically available deepfake dataset

feature analysis in CNN followed by LSTM and GRU models, and a single probability value is obtained.

STEP 4: Classification of video as real video or Deep-Fake video From the probability values obtained in step 3, the unseen test input video is classified as either manipulated i.e. DeepFake video, or non-manipulated video i.e real video on the basis of decided threshold

2.2 Mathematical Modelling

Each layer of CNN can be as follows:

- Convolutional Layer
- Pooling Layer
- Fully connected layer
- Convolutional Layer

Convolutional product is applied to this layer using many filters followed by an activation function .

- Pooling Layer The pooling layer's goal is to down sample the input's features without reducing the.the following notations are taken into account:
- Fully Connected Layer It is a finite number of neurons. It takes input as a vector as well as returns a vector.

Maths Behind LSTM-

$$Ct = Ct + (ItC't)$$
Calculating output
$$Ht = tanh(Ct)$$

Maths Behind GRU-

Consider time step t and input is a minibatch $XtRnxd$, where n is number of examples and d is number of inputs. Hidden state of earlier time step is $Ht1Rnxh$. From this we can calculate reset gate $RtRnxd$ and update gate $ZtRnxd$ as follows:

$$Rt = (XtWxr + Ht1Whr + br)$$
$$Zt = (XtWxz + Ht1Whz + bz)$$

Here W are the weights and b are the biases. Candidate Hidden State $H\tilde{}tRnxd$ state is given as: $H\tilde{}t = tanh(XtWxh + (RtJHt1)Whh + bh)J$ is the Hadamard Product Operator.

Proposed Model Architecture

More preciously, at the **1st layer**, we denote:

- INPUT: $a^{[l-1]} with size \left(n_H^{[l-1]}, n_W^{[l-1]}, n_C^{[l-1]}\right), a^{[o]}$ being the image in the input
- PADDING: $p^{[l]}, Stride : s^{[l]}$
- NUMBER OF FILTER: $n_C^{[l]}$ where each $K^{(n)}$ has the dimension $\left(f^{[l]}, f^{[l]}, n_C^{[l-1]}\right)$
- BIAS OF nth CONVOLUTION:$b_n^{[l]}$
- ACTIVATION FUNCTION: $\psi^{[l]}$
- OUTPUT:$a^{[l]}$ with size $\left(n_H^{[l]}, n_W^{[l]}, n_C^{[l]}\right)$

And we have

$$\forall n \in [1, 2, ...n_C^{[l]}]$$

$$\text{conv}(\ a^{[l-1]}, K^{(n)})_{x,y} =$$

$$\psi^l(\sum_{i=1}^{n^{l-1}} \sum_{j=1}^{n^{l-1}} \sum_{k=1}^{n^{l-1}} K_{i,j,k}^n a_{x+i,y+j-1,k}^{[l-1]} + b_n^l)$$

$$\dim(\text{conv}(a^{[l-1]}, K^{(n)})) = (n_H^l, n_W^l)$$

Thus,

$$a^{[l]} = [\psi^{[l]}(conv(a^{[l-1]}, K^{[n]})),$$
$$\psi^{[l]}(conv(a^{[l-1]}, K^{(2)})), ..., \psi^{[l]}(conv(a^{[l-1]}, Kn_c^l))]$$
$$dim(a^{[l]}) = (n^{[l]}H, n^{[l]}W, n_C^{[l]})$$

with,

$$n_{H/W}^{[l]} = \left[\frac{n_{H/W}^{l-1}}{s^{[l]}} + 1\right]; s > 0$$

$$= n_{H/W}^{l-1} + 2p^l - f^l; s = 0$$
$$n_C^l = \text{number of filters}$$

The learned Parameters at the lth layer are:

 https://doi.org/10.1051/itmconf/20224403064

- FILTERS with $\left(f^{[l]} \times f^{[l]} \times n_C^{[l-1]}\right) \times n_C^{[l]}$ Parameters
- BIAS with $(1 \times 1 \times 1) \times n_c^l$ Parameters

Pooling Layer Parameters

$$a^l_{x,y,z} = pool(a^{l-1})x, y, z = \phi^l(a^{l-1}_{x+i-1,j-1,z})i, j\varepsilon\lfloor 1, 2, 3 ... f^l \rfloor$$

$$dim(a^l) = n^l H, n^l W, n^l_C$$

$$\mathrm{n}^{[l]}_{H/W} = \left[\frac{n^{l-1}_{H/W}}{s^{[l]}} + 1\right]; s > 0$$

$$= \mathrm{n}^{[l-1]}_{H/W} + 2p^{[1]} - f^{[l]}; s = 0$$

$$\mathrm{n}^{[l]}_C = n^{[l-1]}_C$$

Fully Connected Layer Parameters

- INPUT: $a^{[l-1]}$ with size $\left(n_H^{[l-1]}, n_W^{[l-1]}, n_C^{[l-1]}\right)$,$a^{[0]}$ being the image in the input
- PADDING: p^l rarely used, (stride): $s^{[l]}$
- SIZE OF THE POOLING FILTER: $f^{[l]}$
- POOLING FUNCTION: $phi^{[l]}$
- OUTPUT: $a^{[l]}$ with size $(n_H^{[l-1]}, n_W^{[l-1]}, n_C^{[l-1]})$

2.3 DATA SET

Various already existing as well as unique datasets were used to train and test the models. In this section, we'll go through a quick overview of each dataset we'll be using. The datasets utilised for Video Deep-Fake Detection are as follows:

Figure 3. Deep-Fake Detection Challange Dataset [4]

1. Deep-Fake Detection Dataset (DFD): This is the Google/Jigsaw Deep-Fake Detection Dataset. It has 3068 Deep-Fake videos, which were created from 363 original videos of 28 persons of various genders, ages, and cultural backgrounds.

2. Deep-Fake Detection Challenge Dataset (DFDC): This is the Deep-Fake Detection Challenge Dataset from Facebook. There are two versionsO f it. 5K films made by two facial modification algorithms are included in the preview edition. There are 124k videos made by eight facial modification algorithms in the entire dataset version.

3. Custom Dataset: A custom dataset is generated by mixing 936 from existing Video Deep-Fake Detection Datasets, such as DFDC and DFD datasets, to test the generalizability of an algorithm.

3 PERFORMANCE PARAMETER

Performance Parameter To evaluate the outputs of three deep neural networks: CNN, CNN-LSTM, and CNN-GRU, the following metric parameters are used to compare their performance:

• Accuracy:- The most common comparative metric is accuracy. The ratio of the total number of correctly classified examples to the total number of examples classified is known as the correct classification rate. It is the most commonly used metric for comparing models.

• Precision:- The ratio of successfully categorised positive examples divided by the total number of anticipated positive examples is the precision value. A high precision number indicates that an example classified as positive is, in fact, positive.

• recall:- The ratio of the total number of correctly categorised positive examples di- vided by the total number of positive examples is known as recall. A high recall rate shows that the class has been appropriately identified.

• F1 score:- The combined effect of precision and recall value is represented by the F1 score. It's the sum of recall and precision in a harmonic form.summarise the results of the CNN model, CNN-LSTM model, and CNN-GRU model trained and tested on the DFD dataset, DFDC dataset, and Custom dataset, respec- tively. The variation of the metric parameters described above is obtained over a number of epochs. It's also worth noting that the outcomes of the models with and without finetuning are nearly identical.

4 OUTPUTS AND RESULT ANALYSIS

In this sections results of each model based on their performance on various datasets, various testing parameters, and training parameters is discussed by graphical means. This section also includes some popular video Deep-Fake detection tested using the models described in the report and

https://doi.org/10.1051/itmconf/20224403064

comparative analysis between three models: CNN, CNN-LSTM, CNN-GRU. The results of CNN model, CNN-LSTM model and CNN-GRU model trained and tested on DFD dataset , DFDC dataset and Custom dataset. The variation of above mentioned metric parameters are obtained over number of epochs. And it is observed that with and without finetuning the models results are similar.

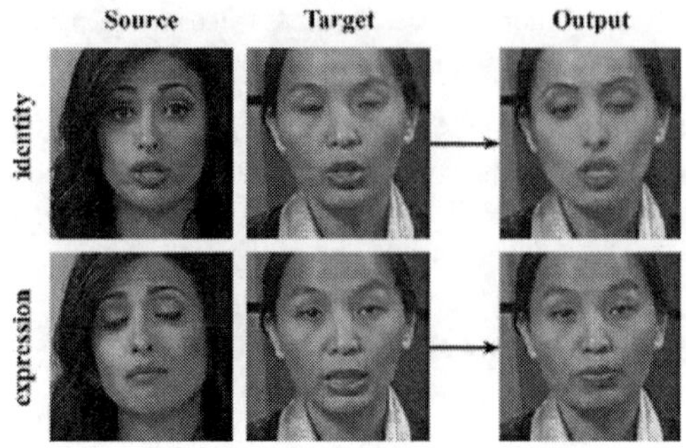

Figure 4. Output

• Deepfake Video ft. Former President Barak Obama

Figure 5. Output

5 CONCLUSION

Seeing is Believing' was formerly thought to be true. It's no longer true, given the emergence of Deep-Fake content. Deep-Fake has begun to erode people's faith in media information. Deep-fake films have the potential to wreak havoc on politics, slander individuals, disseminate hate speech, and the list goes on. As a result of the spread of this, detecting it effectively becomes critical.

This paper describes a method for detecting Deep-Fake videos that employs Transfer Learning. By leveraging pre-trained netwoks, the work provides valuable insights into the detection (with comparative aspect) of AI-generated bogus videos.

Frame Extraction and detection are used to successfully achieve Deep-Fake detection of videos in this work. Frame detection involves obtaining a sequence of frames, which are then analysed for temporal feature changes using CNN, LSTM, and GRU. Using the CNN model, we retrieved face features from numerous consecutive frames and passed them to the network. The hybrid framework of

• Deepfake Video ft. Tom Cruise

Figure 6. Output

CNN layer followed by LSTM proves to be relevant for the task of Video Deep-Fake Detection, according to the findings obtained by performing tests on different datasets. In a Deep-Fake Video, the CNN-LSTM pipeline works in the domain of temporal discontinuities in successive frames. The CNN model is followed by the CNN-LSTM model, which also does well in detecting Fake frames since the detection pipeline collects frames from videos and feeds them into the model.

6 FUTURE WORK

According to the literature review, Deep-Fake detection is one of the most popular research topics in the AI sector. To detect alterations in the Deep-Fake media, a variety of methods based on neural networks and biological signals have been used.

So far, the research has concentrated on the visual and audio components separately. So, in order to do the Deep-Fake Video as well as Audio Detection at the same time on the same video, future work will entail implementing Deep Temporal Convolutional Model (TCN) for deeper analysis and testing. Also, rather than using interpretability methods to those models as the methodology utilised, work on developing models that operate as a generalising architectural foundation for the majority of sorts of Deep-Fake techniques.

Although much research has been done on the development and detection of Deep-Fake, not enough has been done on the reversal of those manipulations.
As a result, a network may be created that can undo the Deep-Fake video or image, allowing us to reconstruct the original video or image from the Deep-Fake one.Seeing the consequences that Deep-Fakes have on human life it is a good lesson to double check the content before sharing it on any internet platforms

References

[1] Mitra, Alakananda, Saraju P. Mohanty, Peter Corcoran, and Elias Kougianos. "A novel machine learning based method for deepfake video detection in social media." In 2020 IEEE International Symposium on Smart Electronic Systems (iSES)(Formerly iNiS), pp. 91-96. IEEE, 2020.

[2] Mitra, Alakananda, Saraju P. Mohanty, Peter Corcoran, and Elias Kougianos. "A machine learning based approach for deepfake detection in social media through key video frame extraction." SN Computer Science 2, no. 2 (2021): 1-18.

[3] Güera, David, and Edward J. Delp. "Deepfake video detection using recurrent neural networks." In 2018 15th IEEE international conference on advanced video and signal based surveillance (AVSS), pp. 1-6. IEEE, 2018.

[4] Zhao, Tianchen, Xiang Xu, Mingze Xu, Hui Ding, Yuanjun Xiong, and Wei Xia. "Learning self-consistency for deepfake detection." In Proceedings of the IEEE/CVF International Conference on Computer Vision, pp. 15023-15033. 2021.

[5] Zhao, Hanqing, Wenbo Zhou, Dongdong Chen, Tianyi Wei, Weiming Zhang, and Nenghai Yu. "Multi-attentional deepfake detection." In Proceedings of the IEEE/CVF Conference on Computer Vision and Pattern Recognition, pp. 2185-2194. 2021.

[6] Khalid, Hasam, Shahroz Tariq, Minha Kim, and Simon S. Woo. "FakeAVCeleb: a novel audio-video multimodal deepfake dataset." arXiv preprint arXiv:2108.05080 (2021).

[7] Jafar, Mousa Tayseer, Mohammad Ababneh, Mohammad Al-Zoube, and Ammar Elhassan. "Forensics and analysis of deepfake videos." In 2020 11th international conference on information and communication systems (ICICS), pp. 053-058. IEEE, 2020.

[8] Rana, Md Shohel, and Andrew H. Sung. "Deepfakestack: A deep ensemble-based learning technique for deepfake detection." In 2020 7th IEEE international conference on cyber security and cloud computing (CSCloud)/2020 6th IEEE international conference on edge computing and scalable cloud (EdgeCom), pp. 70-75. IEEE, 2020.

[9] Deshmukh, Anushree, and Sunil Wankhade. "Deepfake detection by exposing ai-generated fake face video." In Proceedings of Integrated Intelligence Enable Networks and Computing, pp. 673-679. Springer, Singapore, 2021.

[10] Tolosana, Ruben, Ruben Vera-Rodriguez, Julian Fierrez, Aythami Morales, and Javier Ortega-Garcia. "Deepfakes and beyond: A survey of face manipulation and fake detection." Information Fusion 64 (2020): 131-148.

[11] Cozzolino, Davide, Andreas Rössler, Justus Thies, Matthias Nießner, and Luisa Verdoliva. "Id-reveal: Identity-aware deepfake video detection." In Proceedings of the IEEE/CVF International Conference on Computer Vision, pp. 15108-15117. 2021.

[12] Pu, Jiameng, Neal Mangaokar, Lauren Kelly, Parantapa Bhattacharya, Kavya Sundaram, Mobin Javed, Bolun Wang, and Bimal Viswanath. "Deepfake videos in the wild: Analysis and detection." In Proceedings of the Web Conference 2021, pp. 981-992. 2021.

[13] Wodajo, Deressa, and Solomon Atnafu. "Deepfake video detection using convolutional vision transformer." arXiv preprint arXiv:2102.11126 (2021).

[14] Lu, Yuhang, and Touradj Ebrahimi. "A New Approach to Improve Learning-based Deepfake Detection in Realistic Conditions." arXiv preprint arXiv:2203.11807 (2022).

Performance of fault classification on Photovoltaic modules using Thermographic images.

Ashwini Raorane[1], Dhiraj Magare[2], Yogita Mistry[3]

[1]Department of Electronics Engineering, Ramrao Adik Institute of Technology,Sector 7, Phase I, D. Y. Patil Deemed to be University,
Nerul, Navi Mumbai, Maharashtra, India, E-mail: ashwini.raorane@rait.ac.in
[2]Department of Electronics Engineering, Ramrao Adik Institute of Technology,Sector 7, Phase I, D. Y. Patil Deemed to be University,
Nerul, Navi Mumbai, Maharashtra, India, E-mail: dhiraj.magare@rait.ac.in
[3]Department of Electronics Engineering, Ramrao Adik Institute of Technology,Sector 7, Phase I, D. Y. Patil Deemed to be University,
Nerul, Navi Mumbai, Maharashtra, India, E-mail: yogita.mistry@rait.ac.in

Abstract. In this paper, presents thermal image analysis on Fault Classification (FDC) of Photovoltaic (PV) Module. The traditional manual approach of PV inspection is generally more time-consuming, more dangerous, and less accurate than the modern approach of PV inspection using Thermography Images (TI). The benefit in using (TI) images is that it can be used to quickly establish problematic areas in PV Module and provide various measurement details. Thermal image analysis conducted in this research will contribute to inspect PV module by providing a more accurate and cost-efficient diagnosis of PV faults. To maintain the long-term reliability of solar modules and maximize the power output, faults in modules need to be diagnosed at an early stage. In this research, thermographic images were used to detect faults in PV Module using traditional methods and Deep learning methods are mainly used to identify and classify the type of faults that can happen in PV Module. This method will present and discuss on the fault classification and its performance parameters. The fault detection stage determined whether the PV module has an abnormal condition. In this research, performance metrics of fault classification using Deep Neural Networks (DNNs) models is analyzed, which offers high accuracy for detecting abnormalities in image classification tasks.

Keywords--- Thermal Imaging, fault detection, classification, deep learning, pretrained models

1 Introduction

Photovoltaic (PV) systems, also known as solar panels, are gaining more attention nowadays than ever. The global PV market has grown exponentially from 1992 to 2019, and the market is expected to grow faster in the coming years. According to a market report, Renewables 2019 by IEA (n.d.), the maximum amount of energy produced by renewable energy is estimated to increase by 50% between 2019 and 2024. The IEA addresses that Solar PV accounts for about 60% of the increase.

Utilizing Thermography Images has become an emerging solution for a PV system inspection. (TI) inspect more faster than traditional handheld methods by lying over PV modules and also guaranteeing safety by avoiding dangerous working conditions. Additionally, PV inspection using thermographic images promotes accuracy in detection and classification of anomaly [1]. TI for PV inspection allows us to "assess performance of photovoltaic modules, superseding time-consuming traditional manual methods. This chapter introduces an overall introduction to the research subject. It includes the problem, significance, the purpose, research questions, assumptions, delimitations, and limitations.

Deep learning algorithms have recently been utilised in fault diagnostics of photovoltaic (PV) systems to identify and classify the types of problems that can be detected. As public awareness of climate change grows, the globe is turning its attention to renewable energy sources. Solar energy has managed to come out as a major non-polluting and never ending source of energy in this scenario. Solar energy installations have grown at an exponential rate during the last 10 years [2]. As the world's reliance on solar energy is increasing, researchers and engineers are naturally concerned about providing reliable and long-lasting solar energy production systems. Because solar modules have a long life span (25-30 years), it is vital to maintain solar energy production systems on a regular basis in order to get the most out of them. Furthermore, fault diagnosis is required in order to maintain an uninterruptible power supply and minimise solar power plant shutdowns.

Traditional methods of PV module's fault detection with the help of electrical performance measurement are well known. However, these methods have significant drawbacks. Despite an irregular IV characteristic curve of PV panel is a clear sign of problem, its difficult to locate the source or actual location of the fault (Particularly when number of modules installed are large). Therefore, another procedure is required to determine specific location and the nature of issue. The use of IR imaging for PV plant has a number of benefits. A thermal image can clearly show anomalies. Due to the fact that IR imaging is a non-invasive technology, PV panels can be evaluated without disrupting routine operations. Thermal imaging allows you to scan a large number of panels in a short amount of time. In addition to that, IR imaging allows for early fault detection before the system fails completely. As a result, IR imaging provides an easy, quick and dependable means of scanning flaws in solar panels to ensure fault free operation. In this research an automatic detection of anomaly using thermographic images is analysed. In this study a detail analysis of fault detection and classification is studied methodology using deep learning techniques. In this Four different pre-trained of deep Convolutional Neural Network (CNN): Resnet 101, ResNet50, Alexnet and VGG16 were used for transfer learning. In this work, a novel PV thermal image dataset of photovoltaic-thermal-images-dataset based on PV system has carried out for classification of fault [20]. During this project, we have reported three schemes of classifications: Single anomaly, multiple anomaly and contagious string detection is done. The validation accuracy with the help of pre trained model is studies initially. This is often a moderately high accuracy rather than using traditional techniques of the accuracies reported within the literature. Therefore, this research is going to be useful in addition to quickly diagnosing fault using thermographic images and might be helpful to diagnose faults in PV modules. To ensure the reliability and accuracy with high computational speed for long term operation of a PV panel, even minor faults or hotspot formation on a PV panel need to be detected. Out of many kinds of PV module faults, the study focuses on the environmental phenomenon. The performance of PV modules was easily affected as the weather changes and the amount of sunlight was not constant.

The traditional PV system inspection is inefficient, inaccurate, and dangerous. Therefore, the new approach of PV system inspection using thermography images has become a solution in the industry. The thermal image analysis method introduced in this work will help PVS inspection by finding and analyzing the thermal appearances of PV faults and eventually provide meaningful factors for PV inspection using thermography images [10]. The key strengths of these research methods are reliability and accuracy in hot spot detection. This research takes three main steps to conduct a PV module inspection: (i) Thermal data collection for PV module, (ii) Thermal image preprocessing, and (iii) Thermal image data analysis for PV fault detection and classification (FDC). In this research, classification of fault based on deep learning method and traditional machine learning methods will be analysed using imaginary datasets. An intelligent fault detection using thermal image dataset with experimental faults detection and classification with machine learning concept using CNN is studied.

1.2. Organization of paper

We use the CNN architecture for defect detection in this work. In this research, the work is organized as follows. We firstly provide some theoretical background information about the CNN network needed for the Image Analysis in Section II. Literature survey is discussed in section III Then, we briefly describe the thermal images used in our work in Section III. Further on vast literature a solution has come up with proposed methodology in section IV. A deep CNN framework of pretrained model is studied in detail in section V. Based on label dataset a network has analysed using pretrained model in section VI and followed by results and conclusion.

2 Literature Survey

Object-based computer vision (CV) has become a significant technological challenge. There are a few frameworks in this industry, and many developers are looking for a system that can recognize objects quickly and accurately. It is vital to highlight that computer vision is quite beneficial in a variety of fields. Key concepts of the literature study involve PV module inspection, FLIR thermography, fault detection, and classification. The key concepts are present in the concept map in Figure 2. Traditional defect detection approaches need precise operating conditions, which take time, effort, and money. In today's fast-paced environment, innovative tactics and technology breakthroughs are expected to provide rapid results. Advanced and automated fault diagnosis is a procedure that provides immediate results and ensures a longer lifespan for a variety of important PVM components. Hypothesis: This research uses convolutional neural networks (CNN) to conduct fault identification in PVM, successfully classifying different errors based on photos taken by FLIR camera. Pierdicca et al. [20] stated that the utilization of RES, whose application is rising in necessity because of environmental and climatic concerns, as well as the global economy's recovery, will be a major challenge. Renewable energy sources will be the sole option for reducing fossil fuel use and pollution. The amount of allocated PV plants that generate energy has expanded dramatically, and the problem of checking and providing a PV plant has become critical, posing several issues such as efficiency, dependability, safety, and stability. This research describes a unique method for estimating PV cell degradations using DCNNs. Amaral et al. [14] stated that PV power plants currently play a significant part in the production of electricity from renewable sources. The PV modules of these power plants are mounted in trackers to achieve optimal efficiency. However, the trackers' movable construction is prone to failure, putting the required perpendicular position among the PV modules and the smartest moment in the sky in jeopardy. As a result, diagnosing a tracker failure is critical to ensuring optimal energy output. Sensor-based and statistical-based approaches have been investigated; however, they are costly and inefficient. To deal with these issues, a unique technique based on machine learning is presented for defect diagnostics in PV trackers.

3 Design Methodology

Positioning, module identification, defect detection, and defect categorization are the primary hurdles of these tec hniques. The problem of data interpretation still needs to be tackled with these approaches. The high amount of images to be examined is one of the most significant challenges. Mechanized data management intelligent system, using modern signal processing algorithms for

computer vision and capable of identifying single panel frames could be the winning answers for making monitoring more economical and dependable. Deep Convolutional Neural Networks (DCNNs) have been shown to outperform state-of-the-art methods in image classification when compared to widely used machine learning approaches (e.g. Support Vector Machine (SVM), Neural Networks (NN), k Nearest Neighbor (KNN), Decision Tree (DT), Random Forest (RF), Nave Bayes) when dealing with large image collections analysis [26].
They pave the way for total picture interpretation, outper forming existing approaches in terms of accuracy and, in some cases, efficiency, by uncovering many levels of re presentation that can capture the data's more abstract sem antics. One important feature of deep learning's success in image categorization is the usage of convolutional networks. The convolutional layers are the core of DCNN since they learn the feature representations of their input images. By integrating low/mid/high-level features and classifiers in an end-to-end multi-layer, DCNNs have led to a series of advancements for image categorization. DCNNs can automatically extract visual features and categorize them from a vast amount of picture data.

This work describes a novel approach to estimating PV cell degradations using DCNNs, according to research trends. While many scholars have accomplished image classification in the literature, the thermal image dataset is used to illustrate the research work using thermal infrared sensor based on a dataset survey. The tests on the "Photovoltaic Images Dataset" are presented to demonstrate the degradation problem and to thoroughly evaluate the classification approach presented in this study. The DCNN that we choose is based on the VGG19 network architecture, and a typical machine learning technique based on SVM is evaluated and compared for better parameter assessment.

3.1. Infrared Thermograph

The low resolution of IR cameras limits its ability to identify microscopic faults such as microcracks, soiling effect, and hotspots. It has not had an impact on a solar module's photoelectric conversion efficiency. These methods are unable to capture the deep interactions between several parameters, such as particle size, thickness, and coverage, as well as environmental factors (such as sun irradiation and humidity), that are necessary for studying the impact on power loss [2].

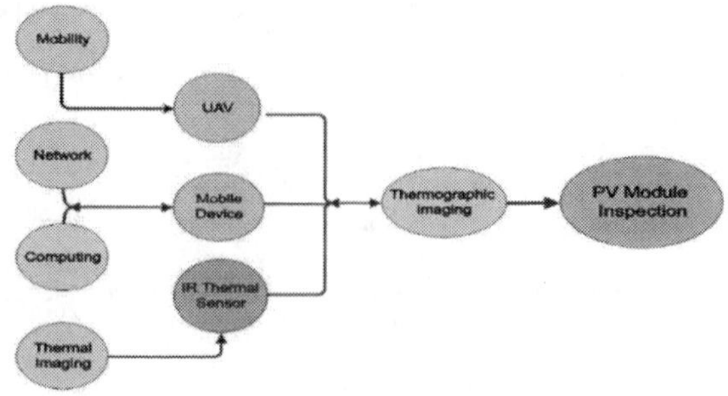

Figure 1. Source of Thermal images for PV Inspection Module

3.2. PV module defects

Identifying defects in PV modules is crucial because these faults can prompt severe power losses and degenerate performance. A single fault in a cell has the potential to spread to other modules near it and result in a complete failure in its functioning. In this work module faults and string and system faults are discussed. For the module faults, there are "hot spots on the cells and for the string and systems faults, "Wiring troubles such as diode defect, frayed wires, charge controller issues, inverter and fuse failures" are among the problems. This research focuses on Hot Spot (HS) phenomenon, which is the primary defect of the PV modules. This phenomenon is a sign of energy loss in PV modules and it is considered as one of the critical faults in PV modules. The leading causes of the hot spots are PV cell failures, partial shadowing, PV cell mismatch, or connection failures in cell links.

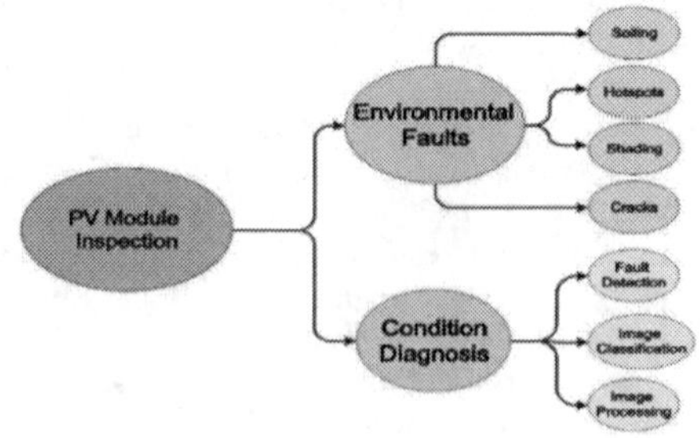

Figure 2. Key concepts of PV Module diagnosis.

3.3. Fault Detection and Classification

This research used a visual method of Fault Detection and classification (FDC) to inspect PV modules. FDC was realized by using Matlab 2021 using deep learning tool box. In this fault detection and classification of thermal images is studied using traditional machine learning techniques and deep learning model. Further comparative analysis is carried out to study deep learning model in future.

Throughout the FDC procedures, the hot spot phenomenon was identified by using image segmentation carrying out binary mask images. With the help of binary mask using image analysis, hotspot are identified and classified in three classes as single hotspot, multiple hotspot and string. This fault are Classifier used in reseach method are support vector machine (SVM) and Pretrained model VGG19. Using this traditional learning and deep learning performance evaluation is carried out. The performance metric like accuracy, precision, recall and F1 score parameters are studied.

3.4. Deep learning

A convolutional neural network (CNN/ConvNet) is a deep neural network used to interpret visual imagery. In neural networks, multiplication is the norm, but in ConvNet, it is the exception. Using a special technique called Convolution, a third function can be generated that describes how one function is modified by another. In mathematics, convolution is the process of combining two functions. One of the most important aspects of deep learning's effectiveness in image categorization is the usage of convolutional structures. The convolutional layers are the core of DCNN since they learn the feature representations of their input images. The DCNN is trained using image labels showing the anomaly in the images in order to provide information regarding cell degradation.. The training is performed with a VGG19 network.

3.5. Data Annotation

A thermographic inspection of a ground-based PV system was conducted on a PV plant with a capacity of around 66 MW in Tombourke, South Africa, for its collection. The "Pv thermal images Dataset" is composed of 3336 images collected as follows: 1007 images with damaged PV cells. This work mainly focuses on damage cell and classification of faults based on three labelled classes is used.

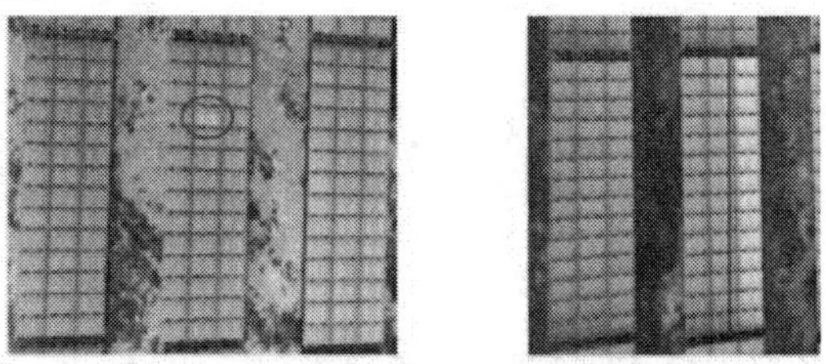

Figure 3. Deep CNN image classification Pipeline

In this Figure no. 3, the thermographic images are studied to identify the hotspot captures on pv array. Based on the mask region using segmentation approach following faults are labeled. Here in this work Three fault are considered to classify the actual predicted class and comment on its model performance.

IV. System Implementation

1. Thermal image input is taken from the FLIR camera from existing data set.
2. Then the Model detects the spot in binary mask using Image segmentation, Identify the Faulty class.
3. After the detection of thermal images, using the Convolution Neural Network (CNN) classifier, the system detects whether single hotspot, multiple hotspot or string features are extracted.
4. Using SVM classifier fault classification is analyzed.
5. Using pretrained model fault classification is analyzed

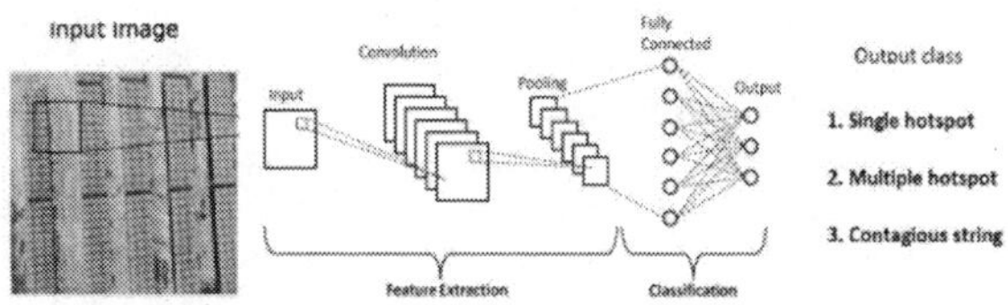

Figure 4. Deep CNN classification Pipeline

Pre-processing:

1. The dataset is divided into two batches: training and validation.
2. Color preprocessing is done with batch normalization.
3. Machine learning and deep learning model has been set up for image classification purposes.
4. A 3x3 convolution filter is used to classify images.
5. Fully connected with softmax layer faults are classified with set of 1000 images

4.1. Performance Evaluation Metrics

The following measures are used to assess the algorithms' performance:

Accuracy: shows the probability of the true predicted value of the class label, which approximates the algorithm's effectiveness.

$$\text{Accuracy}= \frac{Tp+Fn}{Tp+Tn+Fp+Fn} \quad \text{..................} \quad (1)$$

where tp is the number of true positives and fn the number of false negatives.

Recall is a function of the number of cases that were successfully classified (true positives) and the number of examples that were mistakenly classified (false negatives).

$$\text{Recall}= \frac{Tp}{Tp+Fn} \quad \text{........................} \quad (2)$$

Precision is a function of true positives and examples categorized as positives when they aren't (false positives).

$$\text{Precision}= \frac{Tp}{TP+Fp} \quad \text{.........................} (3)$$

F1-score: is a measure of a test's accuracy.

$$\text{F1-score}=\frac{(\beta^2+1)*precision*recall}{\beta^2*precision+recall} \quad .... (4)$$

The F1-score is a weighted average of the precision and recall scores. The evaluation of performance parameter will help in greater extend to analyses the model used in classification task.

V. Result and Discussion

In this paper various literature work based on various fault classification methods are discussed and more research based on DNNs have continued to advance, with numerous breakthroughs and advances, and they now offer significant advantages in picture categorization. Computational hardware and capacity have also improved significantly, supporting DNN requirements have been met, and their application has been expanded, thanks to major improvements in computational hardware and capacity. Only the training set will be used to train each classifier. Similarly, the test set is established at the start and is utilized for all tests. The dataset is split into 70 percent training and 30 percent test images. The performance of the traditional learning ad deep learning is studied for the faulty dataset is reported in Table 1. In this comparative analysis of performance parameter on the model has studied.

Learning Model	Accuracy (%)	Precision	Recall	F1-score	Run Time (s)
Machine learning (SVM)	72.54	0.6931	0.562	0.6816	4253
Deep learning (VGG19)	92.0068	0.9694	0.5256	0.6534	3625

Table 1. Evaluation parameter of PV Diagnosis

Table 1 compares the evaluation outcomes for each model, using Equations (1) – (3) to calculate accuracy, precision, recall, and F-measure. A good prediction model requires not only high accuracy but also generalizability. From the network analysis, it is observed that deep learning gives the better performance compared to machine algorithm. Following observations have done as shown below.

1. This has also proved thar training time required to execute process takes longer time compare to machine learning algorithm.
2. Evaluation metrics are usually derived from the confusion matrix) to evaluate classification results, where true positive (TP) means both actual and predicted results are anomaly; true negative (TN) means both actual and predicted results are normal; false positive (FP) means actual results are normal but predicted to be anomaly; and false negative (FN) means actual results are anomaly but predicted to be normal.

6 Conclusion

This research work is studied in detail with experimental dataset to analyse faulty thermal images. The models created for defect detection and classification using Machine learning and Deep learning techniques. The performance of the pretrained models is assessed against the labelled dataset, which holds a set of 1007 images with 3 class labels. The experimental results of the deep learning model are examined in terms of four performance measures such as precision, recall, accuracy, and F1-Score. In the first objective, a comparative study of machine learning and deep learning models for fault classification has been studied and presented.

7 References

[1] HIGUCHI and Tadatoshi BABASAKI, Failure detection of solar panels using thermographic images captured by drone, 7th International Conference on Renewable energy Research and application.

[2] A. Triki-Lahiani, A. Bennani-Ben Abdelghani, and I. Slama-Belkhodja, "Fault detection and monitoring systems for photovoltaic installations: a review," *Renewable and Sustainable Energy Reviews*, vol. 82, pp. 2680–2692, 2018.

[3]An On-Board Remaining Useful Life Estimation Algorithm for Lithium-Ion Batteries of Electric Vehicles, Faculty of Transportation Engineering, Kunming University of Science and Technology, Kunming 650500, China.

[4] Aicha Djalab1, Mohamed Mounir Rezaoui1, Lakhdar Mazouz1, Ali Teta1 , Nassim Sabri2Robust Method for Diagnosis and Detection of Faults in Photovoltaic Systems Using Artificial Neural Networks, Periodica Polytechnica Electrical Engineering and Computer Science, 64(3), pp. 291–302, 2020

[5] M. Sabbaghpur Arani and M. A. Hejazi, "The comprehensive study of electrical faults in PV arrays," *Journal of Electrical and Computer Engineering*, vol. 2016, Article ID 8712960, 10 pages, 2016.

[6]Köntges, M., Kurtz, S., Jahn, U., Berger, K., Kato, K., Friesen, T., et al.: Review of failures of photovoltaic modules. In: IEA PVPS Task, p. 13 (2014).

[7] Haque, A., Bharath, K.V.S., Khan, M.A., Khan, I., Jaffery, Z.A.: Fault diagnosis of photovoltaic modules. Energy Sci Eng 7(3), 622–644 (2019).

[8] hao, Q., Shao, S., Lu, L., Liu, X., Zhu, H.: A new PV array fault diagnosis method using fuzzy C-mean clustering and fuzzy membership algorithm. Energies 11(1), 238 (2018).

[9] Dhimish, M., Badran, G.: Photovoltaic hot-spots fault detection algorithm using fuzzy systems. IEEE Trans. Device Mater. Reliab. (2019).
[10] DeepSolarEye: Power Loss Prediction and Weakly Supervised Soiling Localization via Fully Convolutional Networks for Solar Panels, Sachin Mehtal, Amar P. Azad2, Saneem A. Chemmengath2, Vikas Raykar2, and Shivkumar Kalyanaraman, University of Washington, Seattle, WA, USA 2 IBM Research Lab, India.

[11] Power loss due to soiling on solar panel: A review Mohammad Reza Maghami a,b,n , Hashim Hizam a,b , Chandima Gomes a , Mohd Amran Radzi a , Mohammad Ismael Rezadad c , Shahrooz Hajighorbani.

[12] Masoud Alajmi, Sultan Aljahdali, Sultan Alsaheel, Mohammed Fattah, and Mohammed Alshehri Machine Learning as an Efficient Diagnostic Tool for Fault Detection and Localization in Solar Photovoltaic Arrays, College of Computers and Information Technology, Taif University Al-Hawiya, Taif 21974, Saudi Arabia.

[13] J. A. Tsanakas, D. Chrysostomou, P. N. Botsaris, A. Gasteratos, "Fault diagnosis of photovoltaic modules through image processing and Canny edge detection on field thermographic measurements," International Journal of Sustainable Energy (T&F), vol.34, no. 6, 2015.

[14] Amaral, Tito G., Vitor Fernão Pires, and Armando J. Pires. "Fault Detection in PV Tracking Systems Using an Image Processing Algorithm Based on PCA." Energies 14, no. 21 (2021): 7278.

[15] A. Chouder, S. Silvestre, "Automatic supervision and fault detection of PV systems based on power losses analysis," Energy Conversion and Management, vol. 51, pp. 1929-1937, 2010.

[16]Jaffery, Z. A., Dubey, A. K., Irshad, & Haque, A. (2017). Scheme for predictive fault diagnosis in photo-voltaic modules using thermal imaging. Infrared Physics & Technology, 83, 182–187. doi:10.1016/j.infrared.2017.04.015.

[16] T. Takashima, J. Yamaguchi, K. Otani, T. Oozeki, K. Kato, M. Ishida, "Experimental studies of fault location in PV module strings," Solar Energy Materials and Solar Cells, 93 (2009) 1079-1082 [17] D. Riley, J. Johnson, "Photovoltaic prognostics and heat management using learning algorithms," IEEE 38th Conference (PVSC), 2012, pp. 001535- 001539.

[18] P. Ducange, M. Fazzolari, B. Lazzerini, F. Marcelloni, "An intelligent system for detecting faults in photovoltaic fields," 11th International Conference on Intelligent Systems Design and Applications (ISDA), 2011, pp. 1341-1346.

[19] A. Coleman, J. Zalewski, "Intelligent fault detection and diagnostics in solar plants," IEEE 6th International Conference on Intelligent Data Acquisition and Advanced Computing Systems (IDAACS) 2011, pp. 948-953.

[20] Pierdicca, R.; Malinverni, E.; Piccinini, F.; Paolanti, M.; Felicetti, A.; Zingaretti, P. Deep Convolutional neural network for automatic detection of damaged photovoltaic cells. In Proceedings of the International Archives of the Photogrammetry, Remote Sensing and Spatial Information Sciences, 2018 ISPRS TC II Mid-term Symposium "Towards Photogrammetry 2020", Riva del Garda, Italy, 4–7 June 2018; Volume 42.

[21] Christopher Dunderdale1 ;Warren Brettenny1 ; Chantelle Clohessy1 E. Ernest van Dyk2, "Photovoltaic defect classification through thermal infrared imaging using a machine learning approach" Nelson Mandela University; South Africa Statistical Association, DOI: 10.1002/pip.3191, Accepted: 20 August 2019.

[22] Baba M., Shimakage T. and Takeuchi N., "Examination of fault detection technique in PV systems," 35th International Telecommunications Energy Conference, 2013.

[23] Chen Z, Wu L, Cheng S, Lin P, Wu T, Lin W. Intelligent fault diagnosis of photovoltaic arrays based on optimized kernel extreme learning machine and I-V characteristics. Appl Energy. 2017;204:912-931.

[24] Deitsch S, Christlein V, Berger S, et al. Automatic classification of defective photovoltaic module cells in electroluminescence images. Solar Energy. 2019;185:455-468.

[25]. Jaffery ZA, Dubey AK, Haque IA. Scheme for predictive fault diagnosis in photovoltaic modules using thermal imaging. Infrared Phys Technol. 2017;83:182-187.

[26] Pierdicca R, Malinverni ES, Piccinini F, Paolanti M, Felicetti A, Zingaretti P. Deep convolutional neural network for automatic detection of damaged photovoltaic cells. Int Arch Photogramm Remote Sens Spat Inf Sci. 2018;42:893-900.

ITM Web of Conferences **44**, 03066 (2022) https://doi.org/10.1051/itmconf/20224403066
ICACC-2022

A novel feature extraction and mapping using convolutional autoencoder for enhancement of Underwater image/video

Jitendra Sonawane[1,*], *Mukesh Patil*[1], and *Gajanan Birajdar*[2]

[1]Ramrao Adik Institute of Technology, Department of Electronics and Telecommunication Engineering, Nerul, Navi Mumbai, Maharashtra, India - 400706
[2]Ramrao Adik Institute of Technology, Department of Electronics Engineering, Nerul, Navi Mumbai, Maharashtra, India - 400706

Abstract. Marine resources known to human are very limited and as 71% world is surrounded by ocean, we are yet to discover the many of the species and the enriched resources. Often the Underwater scenery collected are poorly illuminated, degraded, and distorted due to light propagation model underwater, water molecules and impurities as well. Counting on to these factors images/videos collected in underwater environment are in need of enhancement. We propose a method of utilizing convolution autoencoder, which can be able to collect the features of underwater images and enhanced image and then the feature mapping of this can be used in testing of the other underwater images/videos. The method utilizes the technique, which combines benefits of unsupervised convolution autoencoder to extract non-trivial features and utilized them for the enhancement of the underwater images. In order to evaluate the performance, we have used both subjective as well as objective evaluation method. Evaluation parameters used represent the results of the proposed method are significant for enhancement of underwater imagery. With the proposed network, we expect to advance underwater image enhancement research and its applications in many areas like in study of marine organism, their behaviour according to the environment, ocean exploration and Autonomous underwater vehicle.

1 Introduction

Underwater imagery collected using various approaches like underwater photography or with the help of autonomous underwater vehicle often used for collecting data, but they face various issues as, underwater images are collected at larger depth in ocean are often impacted with blue wavelength, they have inappropriate illumination, particle in water often corrupt the images captured. As the underwater environment is a challenging environment, so we need to mitigate some of the problems like blurring of the images, contrast adjustment of the images, fogging etc. [1,2]

These factors can impact on the performance of underwater image enhancement applications which can significantly influence on various applications such as marine exploration, aquatic robotics and underwater surveillance. As we require the images with high resolution and lowly impacted by all the interfering parameter that are generally faced while capturing underwater images or videos, and obviously it can never be fully eliminated so the necessity of the algorithms, which can then be able to recover the important data from the footage of these underwater scenery. The different methods employed for the image enhancement should be faster, optimized and adaptive.

Underwater image enhancement process is often divided into two approaches; one is Underwater Image Enhancement and other is Underwater Image Restoration. Out of which the Underwater Image Enhancement technique does not follow the principal of image formation model. Here in this paper we too have considered the approach of underwater image enhancement. In underwater image/video enhancement generally we consider the approaches like contrast enhancement, colour correction and even there are few fusion techniques are developed by the few researchers.

Various researcher is recent few years used variety of the underwater image enhancement techniques such convolutional neural network(CNN), autoencoder, deep learning etc. In 2017, Sun et al. [3] suggested the use of deep CNN architecture for detecting underwater objects using data augmentation. Even if the data imbalance is there with dataset this approach leads to the appropriate results. Also created a weighted probabilities decision mechanism improve objects from frames in underwater video. Honnutagi [4], in later year introduced Fusion-Based Underwater Image Enhancement by Weight Map Techniques, in which it was used to resolve low contrast problem usually encounters in underwater images is resolved also the MSE, PSNR, entropy were improved by

* Corresponding author: jitendra.sonawane@rait.ac.in

the proposed system. In 2019, liu et al. [5] created an underwater imaging system and proposed the RUIE benchmark. for visibility degradation, color cast, and higher-level detection/classification. Liu also benchmarked some of the important challenges of the Underwater enhancement. Jamadandi et al.[6] in 2019, implemented deep learning framework to enhance underwater images by augmenting network with wavelet corrected transformations, which resulted in helps in recovering highly degraded images and helps achieve low noise and overall better global contrast while retaining the sharp features which are obfuscated by the backscattering of light underwater and results were characterized by the PSNR and SSIM values. In the same year Tang et al. [7] put forward method to enhance the image and video based on Retinex. The method pre corrects the colour and reduces the dominant colour. Also, the improved multiscale Retinex is performed with the intensity channel in order to calculate the reflectance and the illumination component. The image is restored from logarithmic domain, and the dynamics is compensated at the same time. at the end he was able to preserve the colour by this way.

Hashisho et. al [8] proposed a network that estimates the accuracy and the computation cost for ensuring real-time deployment on underwater visual tasks using autoencoder network. Irfan et. al [9] used classification convolution autoencoder which is designed as a hybrid network. classify large size underwater images with higher degree of accuracy and classification. Yadav et. al [10] proposed a technique using the principle of histogram equalization for enhancement of underwater images and the method utilized for the same was able to retain the colours using convolution neural network model. Wang et. al [11] presented stacked convolutional sparse denoising autoencoder model which utilized parse denoising autoencoders and convolution neural network for denoising the underwater heterogeneous information data. The algorithm proposed by the Mello et. al [12] feeds the output image to a degradation block as per image formation model that reinforces its degradation. The degraded and input images are then matched using a loss function. Algorithm can restore image from the decoder after the training process.

Autoencoders is the one of the important machine learning algorithm which can regenerate the images on just by training itself and if the image is mapped properly, it would be difficult to tell visually to observe the difference between original image and the regenerated one. In this paper we propose an algorithm designed with the convolutional autoencoder architecture for feature extraction and mapping. Which later used in enhancement of the Underwater image/video. Experiments are performed on the EUVP Dataset [13] contains separate sets of paired and unpaired image samples of poor and good perceptual quality to facilitate supervised training of underwater image enhancement models.

The rest of this paper is organized as follows. In Section 2, we review Convolutional autoencoder, in which we have presented detailed description of encoder and decoder model for Underwater and enhanced image a. In Section 3, we present details of the proposed technique. Experimental results are discussed in Section 4. Experimental results are presented in subjective as well as in objective manner and discussed. In section 5, Conclusion is drawn.

2 Convolutional autoencoder

Autoencoder is comprised of encoder part and decoder part [12,13]. Neural network of the type autoencoders are used for dimensionality reduction [13], also capable of finding a structure within the image or data in order to develop a compressed representation of the input.
The encoder acquires a knowledge to interpret the input and compresses it to an internal representation which is defined by the bottleneck layer [14].

Output of the encoder is provided to the decoder which then tries to regenerate the input.

For a input image X the encoder function is defined as :

$$Y_i = \text{encoder}(X) = \sigma(X * W^i + b) \tag{1}$$

Where, b is encoder bias, Y_i is encoding of the input X in a low dimensional space, W^i is 2-D convolutional filter, $*$ is 2-D convolution and σ denotes activation function such as ReLU [15,16].

In decoding phase, Y_i is the input of the decoding function, which can be defined as

$$G = \text{decoder}(Y_i) = \sigma\left(Y_i * \tilde{W}^i + \tilde{b}\right) \tag{2}$$

Where $\tilde{W}^i$ is 2-D convolutional filter in decoder and $\tilde{b}$ is bias of decoder [17].

3 Proposed Method

The proposed technique utilizes the benefit of the convolutional autoencoder which adds up to two parts encoder and decoder combined with the mapping network. In the proposed method we first train the underwater image for feature extraction as shown in Figure 1. Similar approach is followed for the Enhanced image feature extraction through the convolution autoencoder as shown in Figure 2.

https://doi.org/10.1051/itmconf/20224403066

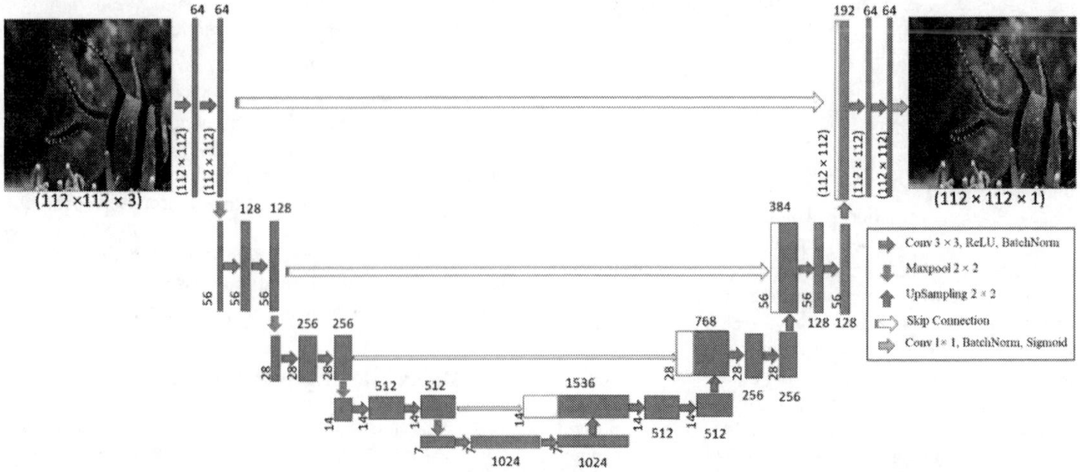

Fig. 1. Autoencoder for Underwater image feature extraction.

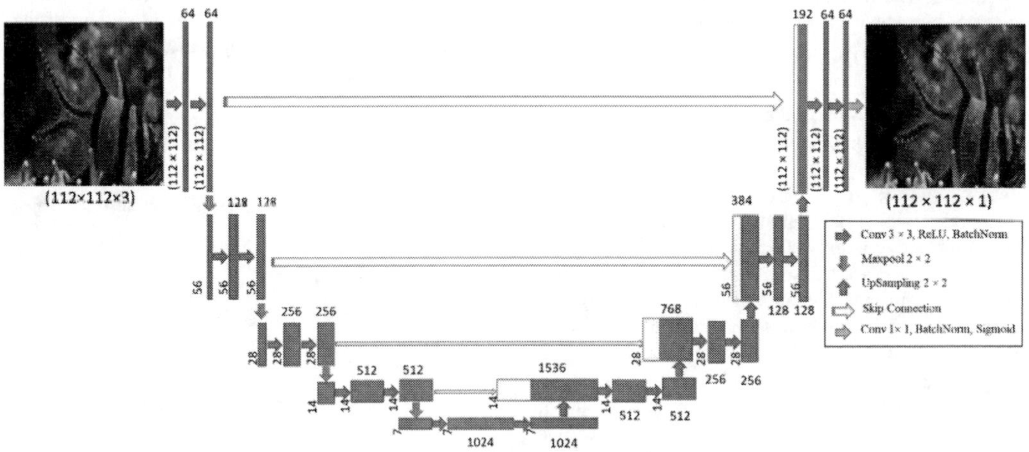

Fig. 2. Autoencoder for Enhanced image feature extraction.

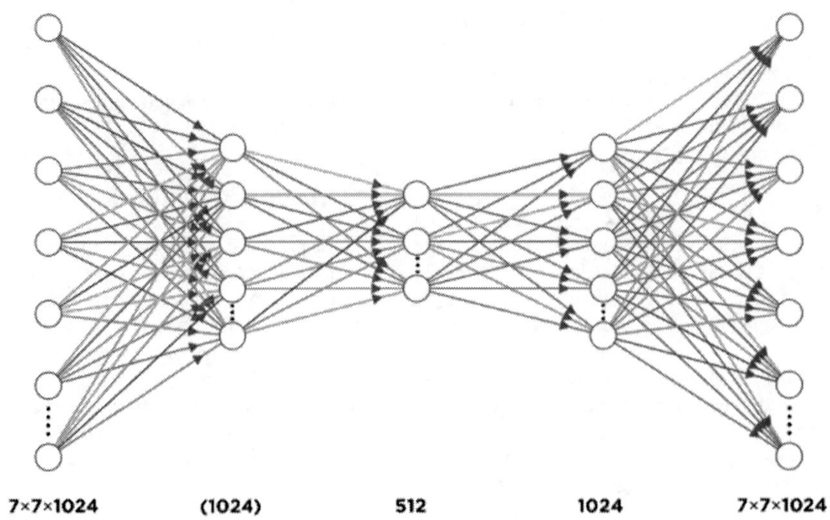

Fig. 3. Mapping Network.

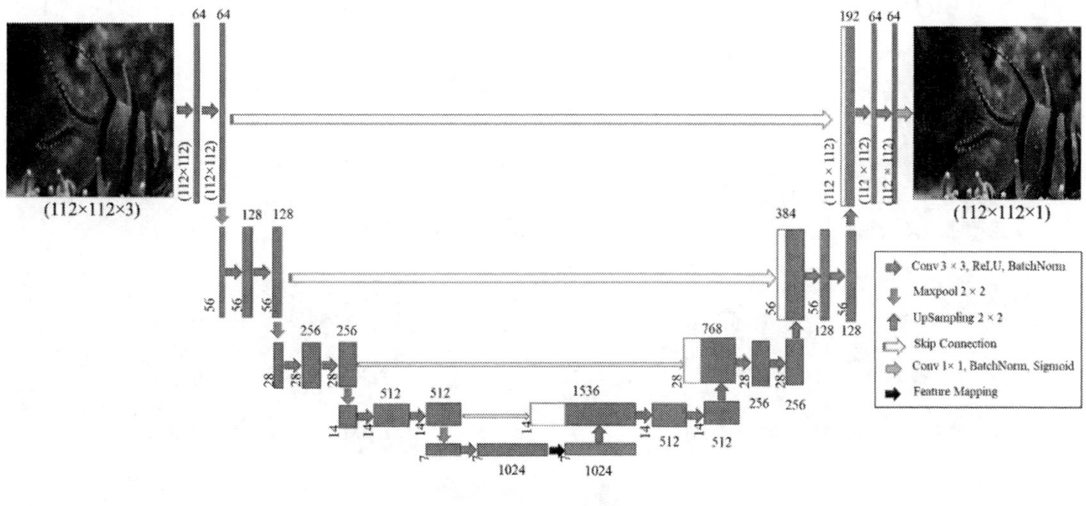

Fig. 4. Modified architecture of the autoencoder for testing the images.

All the features are then mapped using the mapping network shown in Figure 3. In lateral phase of experimentation while testing we had modified the architecture as shown in Figure 4, wherein the architecture presented is only used for the testing purpose.

The autoencoders generally designed in such a way that architecture follows the bottleneck at the centre of the model, which can be used for the reduction of the dimensionality of input for the representation, later that the reconstruction is followed.

There are various types of autoencoders, and their use varies depending upon application, but more common use is as a learned or automatic feature extraction model. Following Table 1. shows the exact configuration followed for the proposed method.

Table 1. Autoencoder layer configuration used in proposed method.

Encoder			
Layer name	**Filter Size (Filter)**	**Output size**	**Connected to**
input	-	112×112×3	-
conv (2)*1	3×3 (64)	112×112×64	input
maxpool	2×2	56×56×64	conv
conv (3)*2	3×3 (128)	56×56×128	maxpool
maxpool	2×2	28×28×128	conv
conv (3)*3	3×3 (256)	28×28×256	maxpool
maxpool	2×2	14×14×256	conv
conv (3)*4	3×3 (512)	14×14×512	maxpool
maxpool	2×2	07×07×512	conv
conv (3)*5	3×3 (1204)	07×07×1024	maxpool
mapping network			
Decoder			
upsampling	2×2	14×14×1024	mapping network
skip connection, concat	-	14×14×1536	upsampling [output, conv*4]
conv (3)	3×3 (512)	14×14×512	skip connection, concat
upsampling	2×2	28×28×512	conv
skip connection, concat	-	28×28×768	upsampling [output, conv*3]
conv (3)	3×3 (256)	28×28×256	skip connection, concat
upsampling	2×2	56×56×256	conv
skip connection, concat	-	56×56×384	upsampling [output, conv*3]
conv (3)	3×3 (128)	56×56×256	skip connection, concat
upsampling	2×2	112×112×256	conv
skip connection, concat	-	112×112×192	upsampling [output, conv*2]
conv (3)	3×3 (64)	112×112×64	skip connection, concat
conv	1×1 (1)	112×112×1	conv
conv (3)	3×3 (64)	112×112×64	skip connection, concat
conv	1×1 (1)	112×112×1	conv

4 Experimental Results and discussion

In this section we have discussed the results in subjective as well as in objective manner to present the results of the proposed technique.

4.1 Subjective Evaluation

As shown in Figure 5. represents the results obtained on different algorithms and the proposed method. Subjective evaluation is the process where assessment of the quality

 https://doi.org/10.1051/itmconf/20224403066

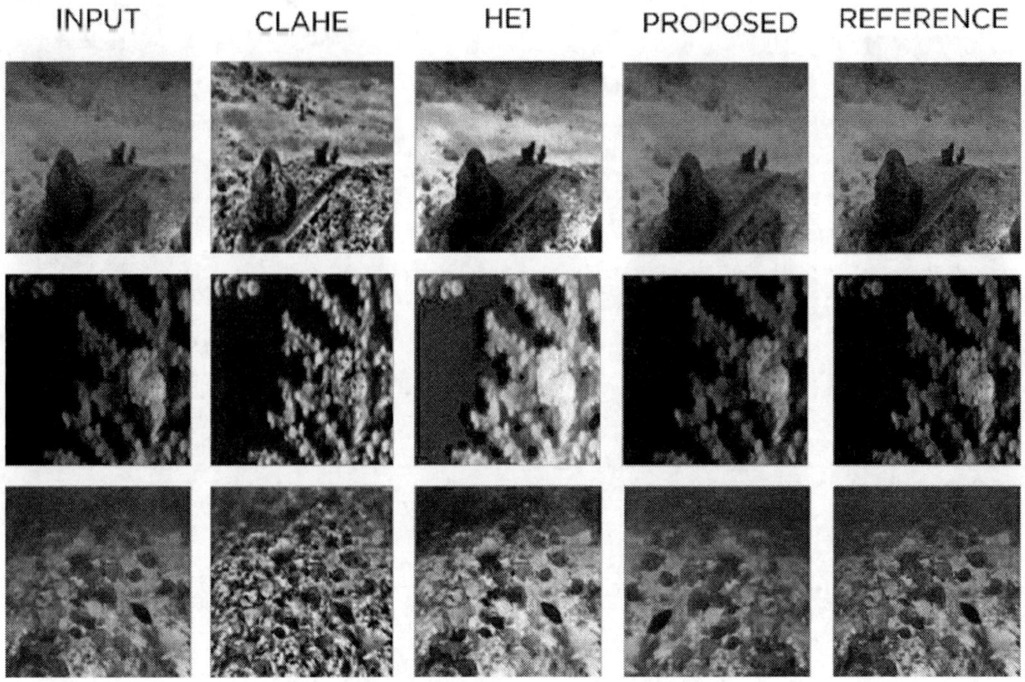

Fig. 5. Subjective Evaluation for the proposed method.

of the results can be experienced visually, results represented in this paper can help us analyse the outcome.

4.2 Objective Evaluation

Objective evaluation can be done based on the statistics of the results, and for that we have used performance evaluation parameter [18] such as peak signal-to-noise ratio (PSNR) [19,22], structural similarity index measure (SSIM) [20,22], underwater image quality measure (UIQM) [20,21,22].

Table 2. Performance evaluation parameter on input, output and

U-UIQM	:	2.34 ± 0.43
E-UIQM	:	2.44 ± 0.50
U-E-PSNR	:	17.52 ± 1.96
U-E-SSIM	:	0.72 ± 0.066

U-UIQM: Underwater (input image) UIQM
E-UIQM: Enhanced (output image) UIQM
U-E-PSNR: PSNR between Underwater & Enhanced image
U-E-SSIM: SSIM between Underwater & Enhanced image

Table 3. Performance evaluation parameter on input, output and

	HE	**CLAHE**	**Proposed**
P-UIQM	2.51±0.341	2.921±0.34	3.412±0.18
U-P-PSNR	11.82±2.40	12.45±1.12	18.77±2.84
E-P-PSNR	12.6 ±2.57	12.81±1.41	18.65±2.35
U-P-SSIM	0.567±0.21	0.48±0.03	0.633±0.07
E-P-SSIM	0.55±0.112	0.47±0.056	0.71±0.064

P-UIQM: Proposed algorithm UIQM
U-P-PSNR: PSNR between Underwater & Proposed (resulted) image
E-P-PSNR: PSNR between Enhanced & Proposed (resulted) image
U-P-SSIM: SSIM between Underwater & Proposed (resulted) image
E-P-SSIM: SSIM between Enhanced & Proposed (resulted) image

It is evident from the results that the proposed technique outperforms in both the evaluation mechanism, subjective and objective.

5 Conclusion

In this paper, convolutional autoencoder based approach proposed to improve the subjective as well as objective quality of the underwater images, same can be utilized on the underwater videos for enhancement. Proposed model uses the power of unsupervised deep convolutional autoencoder extracts the useful features from underwater images and also from enhanced underwater images and based on the features extracted from both the phases of proposed algorithm, later mapping network will map the features which then shall be able to regenerate the underwater image. From experiment performed on underwater image data sets demonstrate that proposed model has significant outcome.

References

1. Hashisho, Yousif, Mohamad Albadawi, Tom Krause, and Uwe Freiherr von Lukas. "Underwater color

restoration using u-net denoising autoencoder." In 2019 11th International Symposium on Image and Signal Processing and Analysis (ISPA), pp. 117-122. IEEE, 2019.

2. Irfan, Muhammad, Jiangbin Zheng, Muhammad Iqbal, and Muhammad Hassan Arif. "A novel feature extraction model to enhance underwater image classification." In International Symposium on Intelligent Computing Systems, pp. 78-91. Springer, Cham, 2020.
3. Sun, Xin, Junyu Shi, Lipeng Liu, Junyu Dong, Claudia Plant, Xinhua Wang, and Huiyu Zhou. "Transferring deep knowledge for object recognition in low-quality underwater videos." Neurocomputing 275 (2018): 897-908.
4. Honnutagi, Pooja, V. D. Mytri, and Y. S. Lalitha. "Fusion-based underwater image enhancement by weight map techniques." In Recent Developments in Machine Learning and Data Analytics, pp. 327-339. Springer, Singapore, 2019.
5. Liu, Risheng, Xin Fan, Ming Zhu, Minjun Hou, and Zhongxuan Luo. "Real-world underwater enhancement: challenges, benchmarks, and solutions." arXiv preprint arXiv:1901.05320 (2019).
6. Liu, Risheng, Xin Fan, Ming Zhu, Minjun Hou, and Zhongxuan Luo. "Real-world underwater enhancement: challenges, benchmarks, and solutions." arXiv preprint arXiv:1901.05320 (2019).
7. Tang, Chong, Uwe Freiherr von Lukas, Matthias Vahl, Shuo Wang, Yu Wang, and Min Tan. "Efficient underwater image and video enhancement based on Retinex." Signal, Image and Video Processing 13, no. 5 (2019): 1011-1018.
8. Yadav, Anushka, Mayank Upadhyay, and Ghanapriya Singh. "Underwater Image Enhancement Using Convolutional Neural Network." arXiv preprint arXiv:2109.08916 (2021).
9. Wang, Xingmei, Yixu Zhao, Xuyang Teng, and Weiqi Sun. "A stacked convolutional sparse denoising autoencoder model for underwater heterogeneous information data." Applied Acoustics 167 (2020): 107391.
10. Mello, Claudio D., Paulo L. Drews, and Silvia C. Botelho. "Degradation-Driven Underwater Image Enhancement." In 2021 Latin American Robotics Symposium (LARS), 2021 Brazilian Symposium on Robotics (SBR), and 2021 Workshop on Robotics in Education (WRE), pp. 186-191. IEEE, 2021.
11. M. J. Islam, Y. Xia and J. Sattar, "Fast Underwater Image Enhancement for Improved Visual Perception," in IEEE Robotics and Automation Letters, vol. 5, no. 2, pp. 3227-3234, April 2020, doi: 10.1109/LRA.2020.2974710.
12. Liou, Cheng-Yuan, Wei-Chen Cheng, Jiun-Wei Liou, and Daw-Ran Liou. "Autoencoder for words." Neurocomputing 139 (2014): 84-96.
13. Tschannen, Michael, Olivier Bachem, and Mario Lucic. "Recent advances in autoencoder-based representation learning." arXiv preprint arXiv:1812.05069 (2018).
14. Wang, Yasi, Hongxun Yao, and Sicheng Zhao. "Auto-encoder based dimensionality reduction." Neurocomputing 184 (2016): 232-242.
15. Badrinarayanan, Vijay, Alex Kendall, and Roberto Cipolla. "Segnet: A deep convolutional encoder-decoder architecture for image segmentation." IEEE transactions on pattern analysis and machine intelligence 39, no. 12 (2017): 2481-2495.
16. Agarap, Abien Fred. "Deep learning using rectified linear units (relu)." arXiv preprint arXiv:1803.08375 (2018).
17. Qin, Hongwei, Xiu Li, Zhixiong Yang, and Min Shang. "When underwater imagery analysis meets deep learning: A solution at the age of big visual data." In OCEANS 2015-MTS/IEEE Washington, pp. 1-5. IEEE, 2015.
18. Irsoy, Ozan, and Ethem Alpaydin. "Autoencoder trees." In Asian conference on machine learning, pp. 378-390. PMLR, 2016.
19. Yang, Miao, and Arcot Sowmya. "An underwater color image quality evaluation metric." IEEE Transactions on Image Processing 24, no. 12 (2015): 6062-6071.
20. C. Li et al., "An Underwater Image Enhancement Benchmark Dataset and Beyond," in IEEE Transactions on Image Processing, vol. 29, pp. 4376-4389, 2020, doi: 10.1109/TIP.2019.2955241.
21. Panetta, Karen, Chen Gao, and Sos Agaian. "Human-visual-system-inspired underwater image quality measures." IEEE Journal of Oceanic Engineering 41, no. 3 (2015): 541-551.
22. Raveendran, Smitha, Mukesh D. Patil, and Gajanan K. Birajdar. "Underwater image enhancement: a comprehensive review, recent trends, challenges and applications." Artificial Intelligence Review 54, no. 7 (2021): 5413-5467.

Plant Disease Identification – A portable mobile application system

Diksha Nigam[1,*], *Anurag* Patil[1], *Pratik* Murari[1] and *Puja Padiya*[1]

[1]Ramrao Adik Institute of Technology, Computer Engineering Department

Abstract. Plant Disease Classification System is software that takes the image of the plant leaf and identifies the possible plant disease at an early stage. In plants, disease indications usually occur on leaves, fruit, buds, and young branches. This situation causes fruit to be wasted (to drop) or be damaged. In addition, these diseases lead to the formation of new infections and the spread of the disease for reasons such as seasonal conditions. For this reason, it is very important to determine the disease in advance and to take the necessary precautions before it spreads to other trees. This project proposes a user-friendly, portable, scalable, and accurate way for identifying, and treating plant diseases at an early stage using Convolution Neural Networks. This will help in the timely and accurate diagnosis of plant diseases, which plays an important role in preventing the loss of productivity and quantity of agricultural products.

1 Introduction

Food safety and plant health are closely related. Pests and diseases threaten food security by causing the loss of 20–40% of global food production. Plants can be protected from these infestations by using pesticides. However, the use of such substances is environmentally harmful. In plants, the indication of disease is typically found on leaves, buds, fruits, and young branches. As a result, fruit is wasted or damaged when it falls from the plant. Plant diseases are typically detected manually by botanists or agricultural engineers, first by visual inspection and then by testing in a laboratory. It is often time-consuming and complicated to use these traditional methods. This is why machine learning and image processing have become important for automatically identifying diseases. Using a visual inspection to diagnose plant diseases can be of great benefit to users who have little to no experience in farming.

The main objectives of the proposed system are early and easily accessible detection of plant disease and suggestion of precautions to avoid the loss of agricultural goods.

1.1 Related Work

1.1.1 Survey of existing systems

In Zhang, K.; Wu, Q.; Liu, A.; Meng, X. [1], experimental result shows that multilayer convolutional neural networks (MCNN) are effective in identifying tomato leaf disease, and they could be used to identify other plant diseases.

In M, TURKO ¨ GLU.; D, Hanbay. [2], the results ˘ show that deep feature extraction and SVM/ELM classification produced better results than transfer learning but both of them had complex computations not suitable for the type of system we are aiming for.

Singh, U.P.; Chouhan, S.S.; Jain, S.; Jain, S.[3] validated on a real time dataset of mango leaves using Multilayer Convolutional Networks(MCNN).

J, Chen; Q, Liu.; L, Gao.[4] used accuracy and mean accuracy indices to evaluate the algorithm. In order to extract the disease-related characteristics of tea plants automatically from images, a CNN model named LeafNet was developed using different sized feature extractors.

Krizhevsky, Alex; Sutskever, Ilya; Hinton, Geoffrey E.[5] introduced AlexNet which achieved a winning top-5 test error rate of 15.3% compared to 26.2% achieved by the second-best entry in the ImageNet LSVRC-2010 contest. A large dataset of 1.2 million high-resolution images was used for training.
The rest of the papers referred are the review papers which helped as gain an insight into various possible algorithms and their uses.

1.1.2 Limitations of existing systems

Study of existing systems have shown that Existing software is complicated and does not provide an easy, intuitive experience. It is not reliable because it is prone to human error. Accurate ML implementations require high processing power, they are not scalable hence not commercial. The disadvantage of current systems is that they require a lot of resources and do not give fast performance. Another method is with the help of experts

* Corresponding author: nigamdiksha02@gmail.com

but not everyone has access to resident experts in this field.

1.2 Problem Statement

To develop a user-friendly and portable system that uses image analysis to identify the plant disease using AlexNet (CNN) and suggest the precautions that need to be taken in order to save the plant at an early stage. Proposed method uses one of the Convolution Neural Network's variation, AlexNet to analyze and visualize imagery of 38 different plant diseases.

1.2 Scope

The system for Plant Disease Detection involves training the AlexNet model on a dataset of 300,000 plant leaf images consisting of 26 diseases and 11 healthy samples. To make the system a user-friendly, cross-platform application made using Flutter to run on Windows, Linux, MacOS and Web. Build a Simple UI to upload the image of the plant leaf and get the results which include the identified disease, probability of the identified disease compared to other diseases, essential resources, etc. Make the application multilingual to make it accessible across the globe. Provide an in-app chat option for users to make networks with other users and plantsman(https://en.wikipedia.org/wiki/Plantsman).

2 Proposed Work

Proposed method uses one of the Convolution Neural Network's variation, AlexNet to analyze and visualize imagery of 38 different plant diseases. An artificial neural network, such as a convolutional neural network, is used primarily for applications involving image recognition. Several layers called perceptron's are used to learn the features present in images with great detail. The term "convolution" refers to the process of understanding an image's features. Convolution is required to extract these features. Filters are used to extract these features, we specify how many filters to use (Kernels). Alexnet contains a total of eight layers with weights; where five of them are convolutional and three are maxpooling layers. These are followed by three layers that are fully connected. A 38-way softmax layer produces distribution over 38 class labels using the output of the last fully-connected layer.

2.1. Proposed system

The system proposes to develop a user-friendly and portable system that uses image analysis to recognize the plant disease using AlexNet (CNN) model and suggest the precautions that need to be taken in order to save the plant at an early stage. Proposed method uses one of the Convolution Neural Network's variation, AlexNet to analyze and visualize imagery of 38 different plant diseases

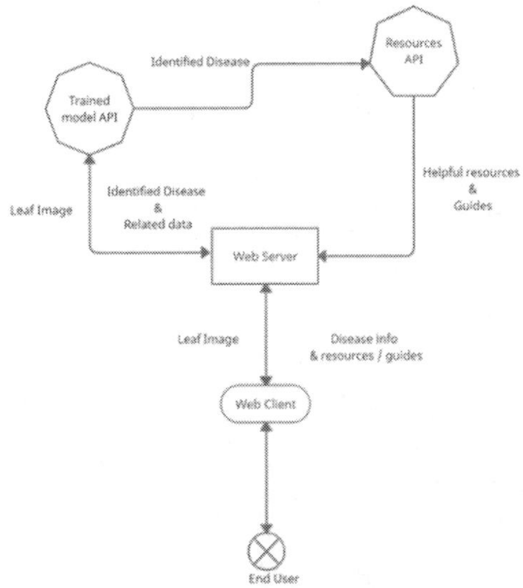

Fig. 1. The system architecture is depicted in the figure. The end user inputs a leaf image through the user interface. This input is processed and sent to trained model for prediction. The model predicts the disease and sends it to resources api which in turn fetches helpful resources and guides.

2.2 Details of Hardware/Software Requirement

The surveyed papers show that most of the good performing techniques are either computationally expensive to implement or have a huge error margin. Thus we wanted to implement such a technique which can be scaled and provides better results while requiring less complex computations. We used SQLAlchemy as the database. To make the application scalable we wrapped the backend written in python with starlette api.

Table 1. Hardware Specifications

Processor	Intel Core i5
RAM	4GB
Operating System, Server	Windows 10,Linux,Unix

Table 1. Software Specifications

Database	SQLAlchemy
Client	Google Chrome, Microsoft Edge or any other web browser
Development Tools	Microsoft Visual Studio
Programming Language	Python, HTML, CSS, Java, Dart, Kotlin

2.3 Implementation Details

In the back-end, Python API written in Starlette is used to serve the CNN Model. Client facing API to be written in Java (Kotlin) using SpringBoot. PostgresSQL will be used to store the backend side data of users. Redis is used for cache. Nginx server is used for load-balancing. We built a user-friendly cross-platform application with Flutter to work with Windows, Linux, MacOS and Web. SharedPrefrences is used as clientside database. Dart is used as the language because of its null-safety, robust and strict nature. We implemented AlexNet CNN model for the disease prediction system. We compared the performance of AlexNet, GoogleNet and VGG models as these are most widely used high performing models in competitions.

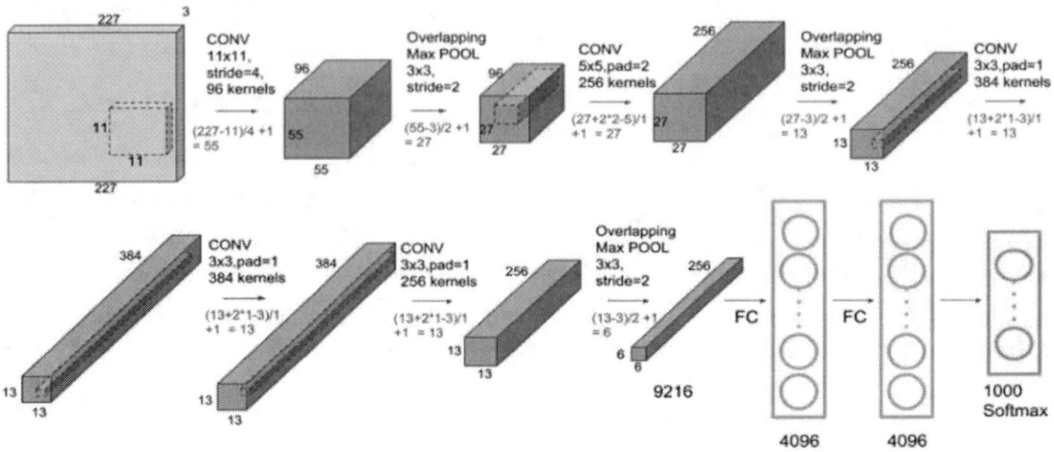

Fig. 2. AlexNet Architecture[10]. The figure depicts the architecture of AlexNet with it's 5 convolution layers, 3 max-pooling layers and 3 fully connected layers.

As depicted in Fig. 2, there are 11 layers in AlexNet architecture as:
Layer C1: Convolution Layer (96, 11*11)
Layer S2: Max Pooling Layer (3*3)
Layer C3: Convolution Layer (256, 5*5)
Layer S4: Max Pooling Layer (3*3)
Layer C5: Convolution Layer (384, 3*3)
Layer C6: Convolution Layer (384, 3*3)
Layer C7: Convolution Layer (256, 3*3)
Layer S8: Max Pooling Layer (3*3)
Layer F9: Fully-Connected Layer (4096)
Layer F10: Fully-Connected Layer (4096)
Layer F11: Fully-Connected Layer (36)

In the first convolutional layer, 96 kernels of 11*11*3 sizes with a stride of 4 pixels filter the 224*224*3 input image (this represents the distance between the receptive field centers of neighboring neurons). 256 kernels of size 5*5*48 are used to filter the output of the second convolutional layer. It takes the normalized and pooled output of the first convolutional layer as input. There are no intervening layers between the third, fourth, and fifth convolutional layers. There are 384 kernels in the third convolutional layer of 3 * 3 * 256 with the outputs of layer two connected (normalized, pooled). There are 384 kernels in the fourth convolutional layer, and 256 kernels in the fifth convolutional layer. Each of the 5 fully connected layer has 4096 neurons. Alexnet has around 60 million parameters.
Therefor the overall system consists of a user interface, a machine learning model and a resource API. All these components are programmed in Python language. The system is also integrated with an in-app chat system using web sockets which allows user to interact with fellow users in the system. We built a scalable backend, which can even be hosted on low end machines as prediction is very fast and memory efficient.

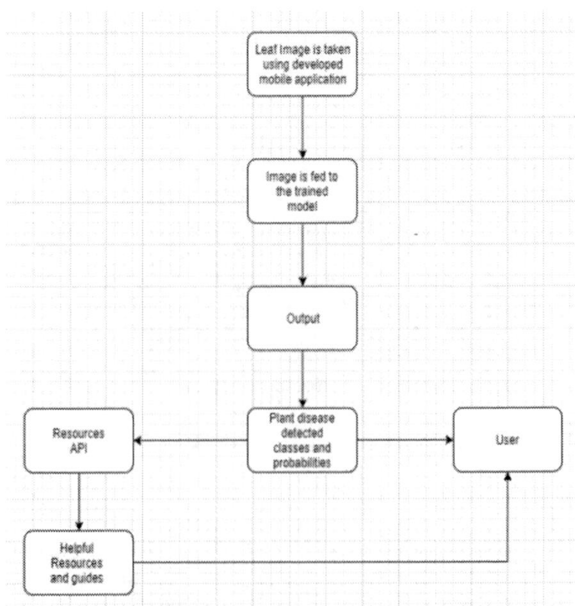

Fig. 3. Flow Chart for the system

Figure 2 shows a flowchart for the overall disease identification system implemented. A leaf image is

https://doi.org/10.1051/itmconf/20224403067

provided as an input through the mobile application. Image is fed to the trained model. Output of the system is in the form of array of key-value pairs of plant disease detected classes mapped to their probabilities. This information is displayed to the user in the form of bar graph with predicted classes probablitites. These probabilities is again fed to the resource api to derive helpful resources and guides for the highest probability predicted disease.

The dataset used is Plant Village Dataset [9]. It consists of images of leaves of diseased plant with their corresponding labels. It has 300,000 plant leaf images consisting of samples of 26 plant diseases and 11 healthy samples. Plants included in the dataset are commonly found all over the world, hence the application can be used globally.

3 Result Analysis

We Operated with 3-channel images that were (224*224*3) in size. We used max pooling and ReLU activations when subsampling. We used ReLu activations for convolutions. We used (3*3) kernels for max pooling. We used either (11*11), (5*5), or (3*3) kernels for convolutions. All this was done to classify images into one of 38 classes. We built a user friendly mobile application, in which we added support for multiple languages in the application. It also includes a realtime chat feature with other users. We added better and improved resources when the user is shown predicted disease and other results. It also shows news / posts for gardners from internet.

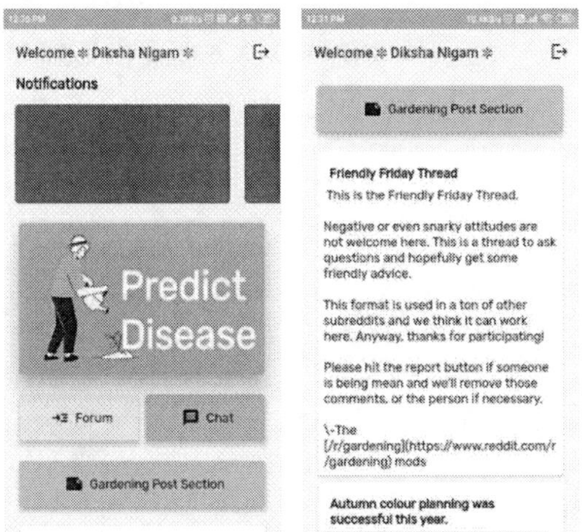

Fig. 4. The first screen shows the start of main page which contains notifications feature, disease prediction option and chat and forum options. The second screen shows the below section of main page containing the daily newsletters and posts.

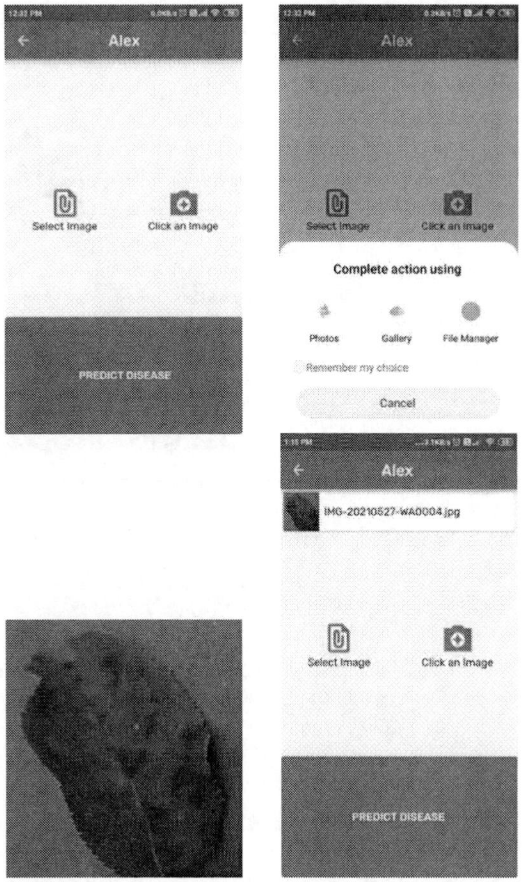

Fig. 5. The process of disease prediction is depicted in the picture. The image can be taken through a camera or from the image gallery. The prediction shows the top 5 probabilities of diseases.

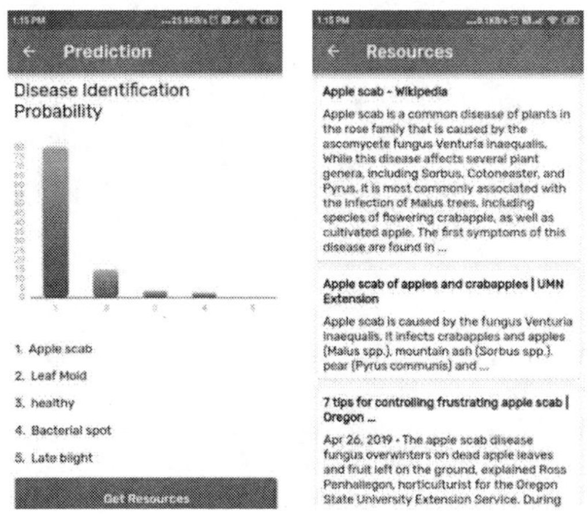

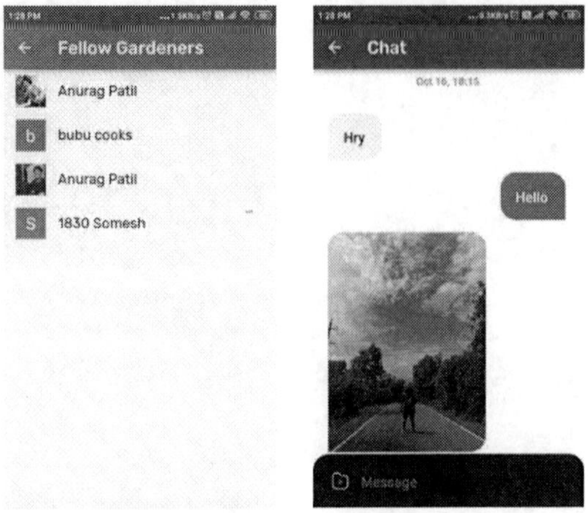

Fig. 6. Resources related to the predicted disease is made available to the user. The user can also interact with other users through personal chats and chat forums.

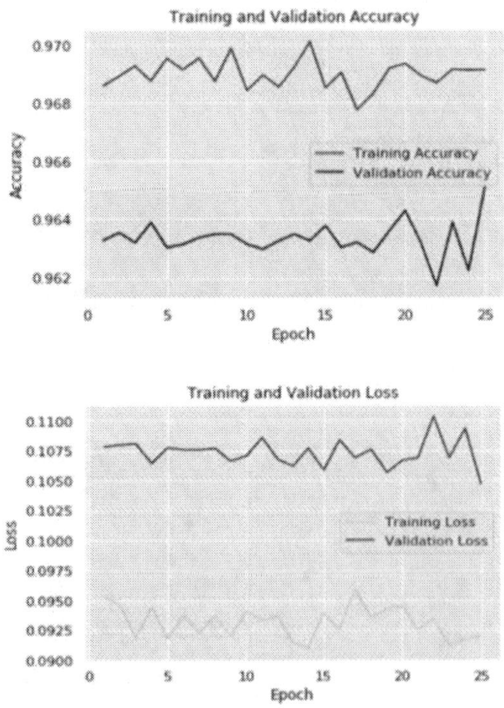

Fig. 7. Graphs showing the trend of accuracy over epoch and validation loss vs epoch.

In order to avoid overfitting, we used a technique called cross-validation to test whether the model is over-fitting. The data is divided into two parts - a training and a validation set. A model is trained on a training set, and then it is evaluated on a validation set. With this in mind, loss and accuracy are calculated on the training set and validation set. As seen from figure 5, the training accuracy and validation accuracy achieved by the model are 97% and 96.6% respectively.

References

1. K. Zhang, Q. Wu, A. Liu, and X. Meng, "Can deep learning identify tomato leaf disease?", Advances in Multimedia, vol. **2018** (2018)
2. M. Turko¨ glu and D. Hanbay, "Plant disease and pest detection using deep learning-based ˇ features," Turkish Journal of Electrical Engineering & Computer Sciences, vol. **27**, no. 3, pp. 1636–1651 (2019)
3. U. P. Singh, S. S. Chouhan, S. Jain, and S. Jain, "Multilayer convolution neural network for the classification of mango leaves infected by anthracnose disease," IEEE Access, vol. 7, pp. 43721–43729 (2019)
4. J. Chen, Q. Liu, and L. Gao, "Visual tea leaf disease recognition using a convolutional neural network model," Symmetry, vol. 11, no. 3, p. 343 (2019)
5. A. Krizhevsky, I. Sutskever, and G. E. Hinton, "Imagenet classification with deep convolutional neural networks," Communications of the ACM, vol. 60, no. 6, pp. 84–90 (2017)
6. K. Simonyan and A. Zisserman, "Very deep convolutional networks for large-scale image recognition," arXiv preprint arXiv:1409.1556 (2014)
7. J. Kaur, R. Chadha, S. Thakur, and E. R. Kaur, "International journal of engineering sciences & research technology a review paper on plant disease detection using image processing and neural network approach" (2016)
8. S. Kumar and R. Kaur, "Plant disease detection using image processing-a review," International Journal of Computer Applications, vol. 124, no. 16, (2015)
9. Plant Village Dataset- Dataset of diseased plant leaf images and corresponding labels https://github.com/spMohanty/PlantVillage-Dataset
10. https://neurohive.io/wp-content/uploads/2018/10/AlexNet-1.png

Sports Injury Prediction System using Random Forest Classifier

Akshay Shringarpure, *Ronak* Shetty, *Ajinkya* Surve, and *Amarsinh* Vidhate[1]
[1]Dept. of Computer Engineering, Ramrao Adik Institute of Technology, Navi Mumbai Maharashtra, India

Abstract. One of the largest growing industries in the modern-day world is the sporting industry. Currently valued at around 500 billion USD, with a growth scope of exponential potential, its ability to attract investors is incredible. And just like any other investment. It is part andparcel of the investor's fiscal responsibility to take good care of their assets. The biggest assets in the sporting industry are of course the players, and the greatest threat to said assets is injuries. We take into consideration said factors and deem it important to solve said issues, and understanding the money involved, the industry sides with us too. We seek to solve the said problemby taking into account all previous injury records and datasets of various players and predicting the kind, number, and severity of the injuries in the future. We seek to create a methodology for such prediction, which applies to all and any sports, being one of the only such models of its kind.

1. Introduction

Sports Injuries are common in competitive sports as players always have to push themselves to peak conditions and often end up causing harm to themselves or opponent players. Today, a player earns money for every minute spent on the field and a retiring injury can neither be afforded by the players nor the team management. According to a census, about 11 million injuries are sustained each year, most cases being damage to limbs and head injuries. Head injuries are the major cause of death in competitive sports. Accordingto a survey of major sports leagues in the U.S., the financial loss due to sports injuries was around $29 M in the year 2017, which includes loss of sponsorships due to failed performanceof teams and medical expenses. Our project isof great benefit for asset protection and management. Simply put, our project can save millions of dollars.

1.1. Objective

Our project can help sports organizations in evaluating players based on their history of injuries. The various clubs and team management can decide whether to sign a player or not based on the predicted data. Another medical aspect of this software is warning the players about their predicted injuriesso that they can be cautious during the game and protect themselves from serious harm or permanent damage. The team players can also strategize their game plans based on these predictions such that every player in the game can function at their highest form.

1.2. Motivation

All members of our team are big fans of sporting events of all kinds. One of the worst pains that you can feel as a fan, is when the club you support buys a player who develops an injury record and never flourishes into the greatness he or she was pegged for. Besides the monetaryloss of the club, the players themselves have to suffer terrible mental and physical torment. We seek to eliminate such trauma, making the sport even more exciting.

1.3. Organization of Paper

The remaining part of the report is organized in the following manner. The prerequisite for the study is covered in Chapter 2. It contains thesurvey of existing systems, limitations of existingsystems which constitutes the literature survey, problem statement, and scope. Chapter 3contains the project proposal which includes proposed work, methodology, and hardware andsoftware requirements.

The planning and formulation part of the project is covered in chapter 4. Chapter 5 contains the Design of the System. Chapter 6 includes the necessary implementation details, project outcomes, and results. Finally, Chapter 7 contains the Conclusion and Future Work.

2. Literature Survey

It has been observed that sports science and its techniques are benefitting sports persons a lot. A literature survey is given hereunder.

2.1. Survey of Existing System

- Liu, Guangying at. El [1] has proposed a learning-based model to predict sports injuries based on the available data from various sources. The authors have reduced attributes that have a significant impact on injury risk and have provided an algorithm based on the Random Forest method to test with real-world data. The better achievement is low error rates.

- Chen Huang, Lei Jiang [2] .Data monitoring and sports injury prediction model based on embedded system and machine learning algorithm[6]. The authors here propose

https://doi.org/10.1051/itmconf/20224403068

the use of an Artificial Neural Network (ANN), whose purpose is to develop and use early-doing ability and exercise load data to validate a hierarchical machine learning prediction system with accurate detection of player injuries.

- Bittencourt NFN, et.al [3]. Complex systems approach for sports injuries: moving from risk factor identification to injury pattern recognition—narrative review and new concept 2016. This paper is based on the demonstration of the implementation of a relatively complex system. This model seeks to identify the regularities and patterns of relationships amongst determinants and characterizes the formation of a pattern that arises from a complex web of afore-mentioned determinants.

- Alejandro López-Valenciano,(2018) et.al has proposed a preventive Model for Muscle Injuries. It's a novel approach based on Learning Algorithms [4]. Here, their motive was to analyze and juxtapose the functioning of multiple MachineLearning (ML) methods to pick the best performing injury risk factor model. The model focused on lower extremity muscle injuries of male football and handball athletes.

- David L. Carey, et. al (2017). had mentioned predictive modeling of training loads and injury in Australian football[5]. The data that the authors worked on, was collected from an elite Australian football club, over 3 seasons. The data of the first two seasons were used to build their injury prediction model, and the predictions were generated for the third season.

These are the most notable works in this venture that comprises, and sum up our understanding so far. Our collective research has led us to a basic, and in turn, an in-depth understanding of the problem statement, further accentuating our comprehension of the need, and lack of a solution to the problem [6,7,8].

Multiple models exist, respectively efficient in their sports, but the efficiency of these models could still be further improved, and the need for a more comprehensive model still exists, keeping in focus the algorithm used and its efficiency.

2.2. Limitations of Existing System

Existing systems are currently minimal in the application, and even when in use, have major issues. Some are very complex to implement for wide usage whereas some have less accuracy, or tend to be time-consuming. To overcome these limitations, first, we had to understand these issues and their significance. We have compared all the different methodologies implemented by fellow researchers based on their limitations as presented below.

2.3. Problem Statement

Competitive sports are a multibillion-dollar industry, and the most prized assets of said industry are the players that participate. Sportspersons undergo a lot of training and physical exertion, which causes extensive wear and tear to the body, and in accordance,

Author	Method Used	Limitations
Bittencourt NFN, Meeuwisse WH, Mendonça LD, Nettel-Aguirre A, Ocarino JM, Fonseca ST (2016) Complex systems approach for sports injuries: moving from risk factor identification to injury pattern recognition-narrative review and new concept.	Complex Web of Determinants	Complex model causes increased implementation cost. Model based on too many uncertain factors
David L. Carey, Kok-Leong Ong, Rod Whiteley, Kay M. Crossley, Justin Crow, Meg E. Morris (2017) Predictive modelling of training loads and injury in Australian football	Multiple classification algorithms	SVM and logistic regression giving low accuracy for certain datasets
Alejandro López-Valenciano, Francisco Ayala, José Miguel Puerta, Mark De Ste Croix, Francisco Vera-García, Sergio Hernández-Sánchez, Iñaki Ruiz-Pérez, Gregory Myer (2018) A Preventive Model for Muscle Injuries: A Novel Approach based on Learning Algorithms	Alternating Decision Tree	Multiple ADTrees required to achieve good accuracy and avoid underfitting.
Chen Huang, Lei Jiang (2020). Data monitoring and sports injury prediction model based on embedded system and machine learning algorithm	Artificial Neural Network (ANN)	Neural networks depend a lot on training data. This leads to the problem of over-fitting and generalization.

Fig. 1. Methodology Comparison

of injuries. Suppose, a million-dollar player gets injured during a game and cannot continue, It can costa fortune to his/her team owner, as well as affectsaid player's career drastically in case of prolonged injury. We seek to predict the injury patterns of various players after taking into account past data, and hence, predicting theseverity and duration of the injuries that a player might incur in the future.

2.4. Scope

We believe that our project is of great significance to a multi-billion-dollar industry. We primarily focus on major asset care, and as a result, our work is indispensable. What is also incredible is that our project works for multiple sports, making its appeal more universal.

3. Proposed Solution

3.1. Proposed Work

To develop a system that accurately predicts player injuries with the help of previous data and specialized algorithms, so that players and clubs can take necessary precautions.

3.2. Proposed Methodology

Injury prediction, in its nature, is a pattern of the non-linear format. What we understand from current cumulative research, is that injury classification is to partition various patterns into groups and that, there are (or were) many different models which can be categorized into traditional time series methods and Intelligent methods and semi-recently, artificial neural network models. The drawback of such traditional systems (like linear regression, time series model), is the very first fact that we learn i.e.; Injury prediction, in its nature, is a pattern of the non-linear format. Traditional systems only consider time sequence features, and it is difficult for them to deal with the nonlinear nature of patterns. ANN, on the other hand, gives better performance with nonlinearity issues, but in most (if not all) of the NN models, the error on the training data set is good, butthey perform badly when out-of-sample data is presented to the network. As a result, after a lot of research, we have concluded that the random forest algorithm is the best as, even in comparison to Support Vector Machine (SVM)and decision trees, it gives us the most accurateresults, with comparative time delay.

4. Planning and Formulation

4.1. Schedule and flow:

The idea was conceived and worked on by the start of June-July 2021. First ensured the commencement of the research phase, which lasted for approximately a month, for the duration of which, the problem statement and current solutions were explored in explicit detail. Then began our work on the formulation and implementation of our ideology and solution. And in accordance, we could reach a model of our solution, which works as expected.

4.2. Detail Plan of Execution

The first phase of the proposal spanned 4 months. To effectively design and develop a cost-effective model, the Waterfall model was practiced.

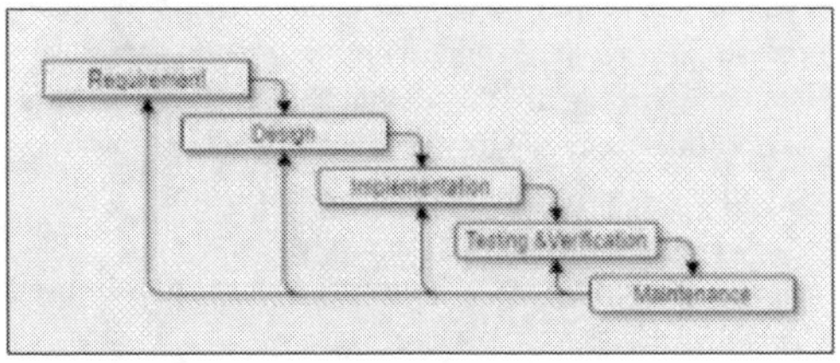

Fig. 2. Workflow diagram

5. Design of System

5.1. Flowchart:

The diagram that we have used to better our understanding and ease our ideological implementation is a flowchart. We have included the standard procedure of employing a standard machine learning algorithm which involves data preprocessing, data splitting, training and ultimately testing the model with the testing data. It summarizes the design and working of our system eloquently, as is evident by the diagram below.

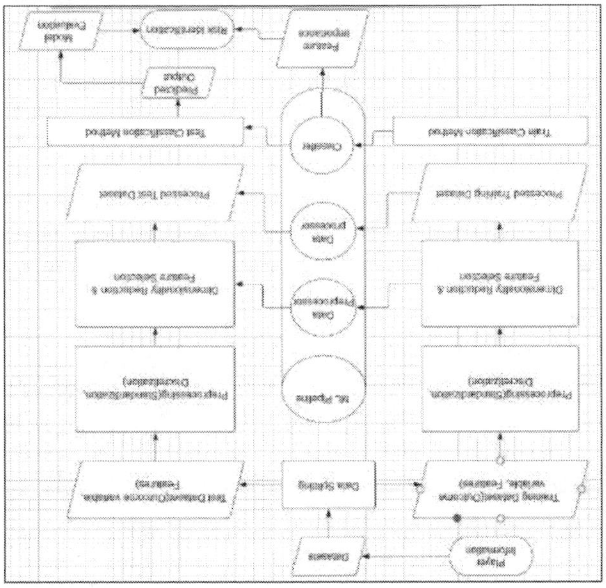

Fig.3. Flowchart of the proposed system

6. Expected Results

6.1. Implementation Details

For implementation, we have used a Jupiter notebook. In accordance, with datamanipulation and visualization, we have usedvarious python libraries like pandas, matplotlib, and seaborn. For input, we use CSV files, namely - NHL_Injury_Database.csv, NHL_Bio.csv, and NHL_Time_on_ice.csv. We then perform data cleaning and manipulation on these CSV files, combining all the data into one file, named-'NHL_alldata.csv'. With the help of python libraries, we can then perform data visualization, like:

- Checking % of injured and non-injured players
- Checking the injury type of the players
- Checking injury count of different types
- of injury
- Checking the injury of the player
- Checking injury chances along with age

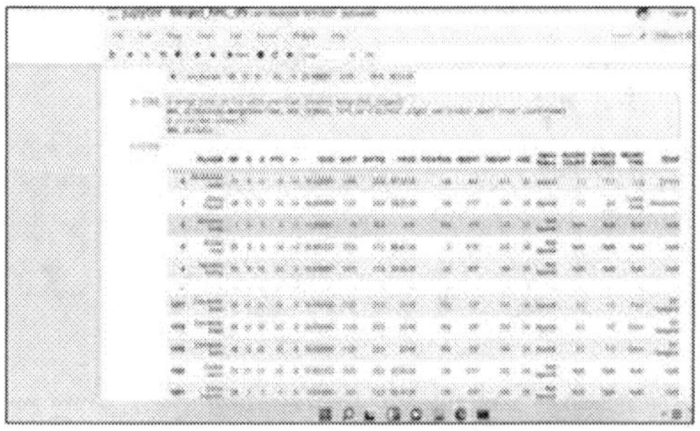

Fig 3. Loading the data from NHL_Injury_Database.csv, NHL_Bio.csv, NHL_Time_on_ice.csv

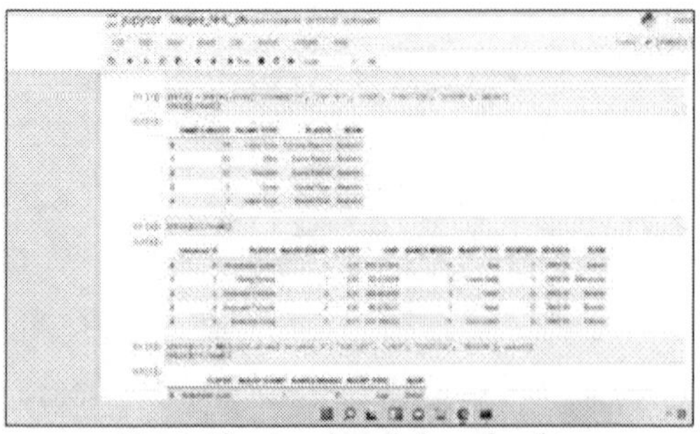

Fig. 4. Cleaning player information data

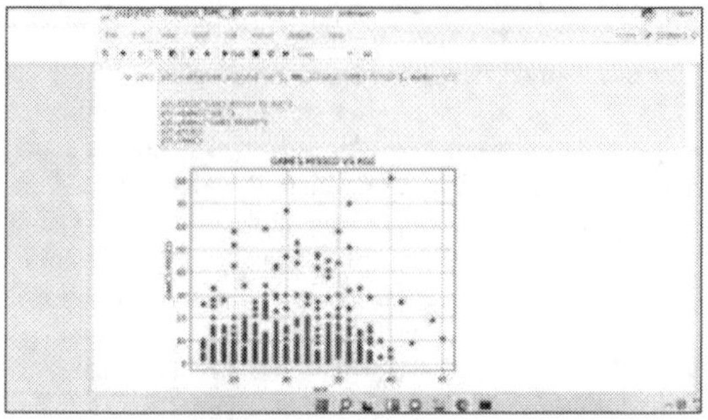

Fig. 5. Scatterplot of Games missed by age

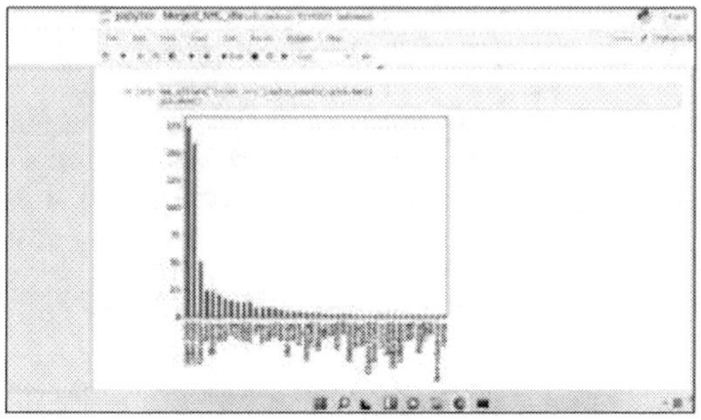

Fig. 6. No of Injuries Vs body parts

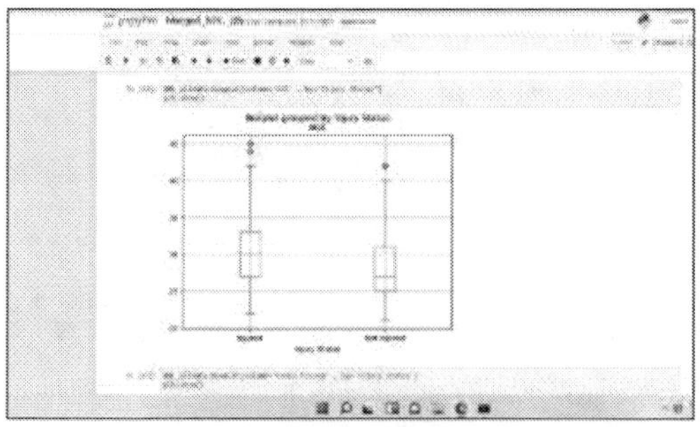

Fig. 7. No of the times players are injured (1, 2, 3)

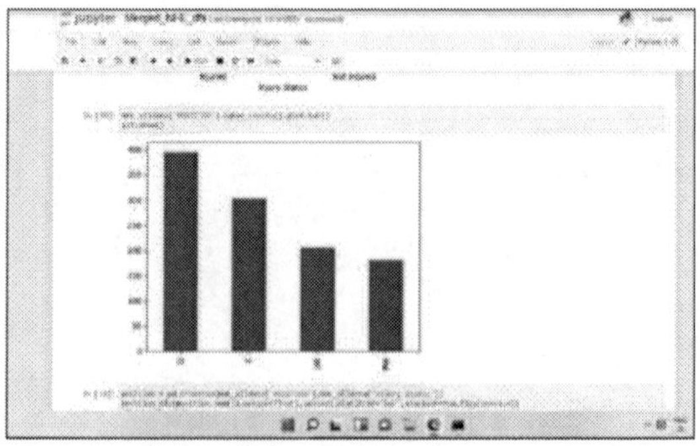

Fig. 8. Boxplot of Age Vs Games missed

https://doi.org/10.1051/itmconf/20224403068

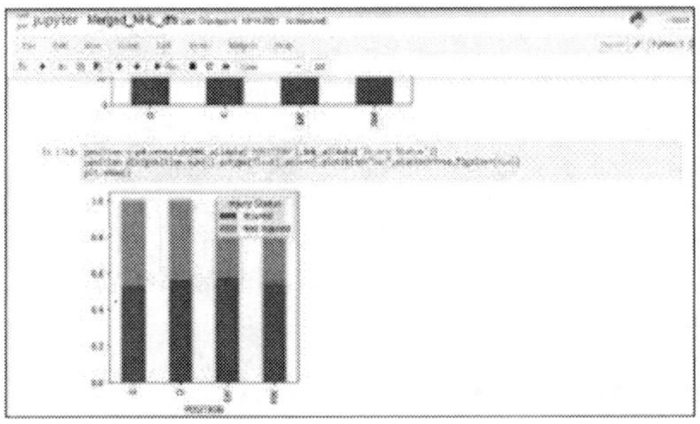

Fig. 9. Injury Status by Position of play

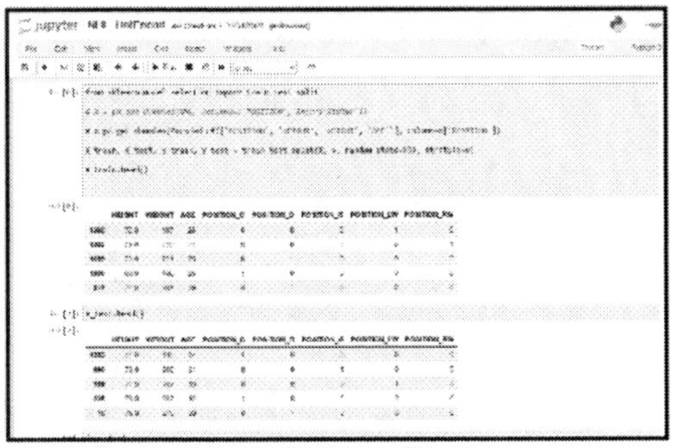

Fig. 10. Splitting of Data

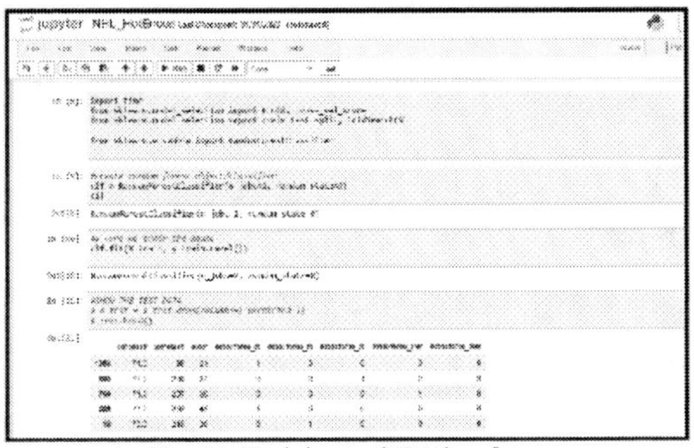

Fig. 11. Training and Testing data

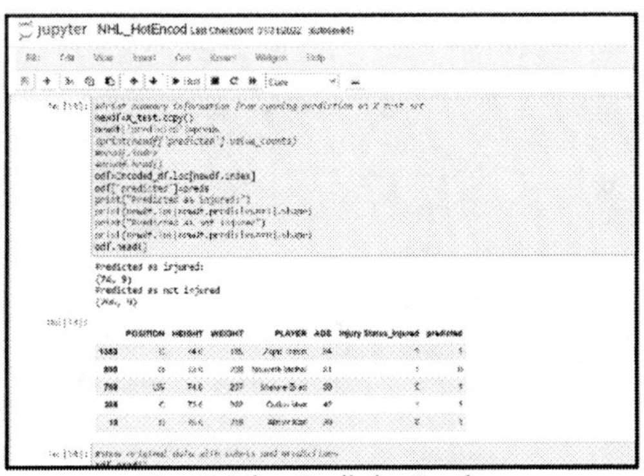

Fig. 12. Injury Prediction Results

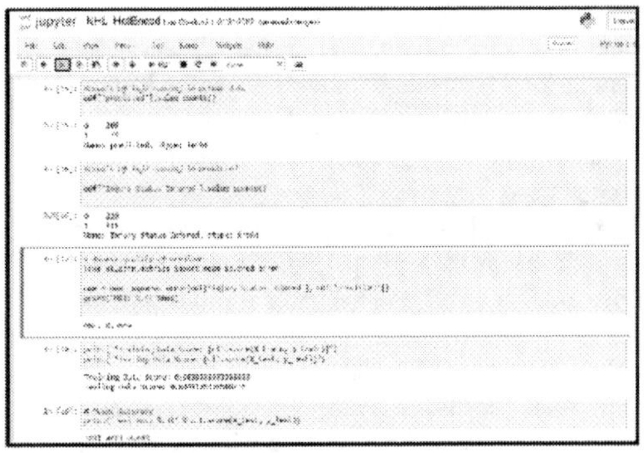

Fig. 13. Accuracy Score

6.2. Project Outcomes

As can be inferred from the data and graphs, the correlation between age and injuries is evidentialin the sense that, as age increases, so does the frequency of injuries. Along with this, it can also be inferred that players who play position d,get injured very often, in comparison to other positions. What is also very noticeable, is the correlation between weight and injuries, further adding another aspect to the prevention of injuries.

The amalgamation of various such considered criteria could potentially affect how a club looks after its players and could also determine optimum utilization of their players. As the player ages, or gains more weight, the club could plan to play him in another position, which may better suit him, and help the club get the most out of him through consistent use.

Our model is designed to potentially shape the transfer policies of clubs, in the sensethat clubs can determine what players are worth the money, and could perform consistently in the long-term and whether they should bet on older talents or not.

Our model could also help players elongate their careers, as they too will grasp a better understanding of their ideal positions, weight, and overall understanding of their injuries. They could plan their careers by our predictions, maximizing their efficiency and prowess in their sport. This model provides insightful readings regarding a player's fitness, bankability, and vulnerabilities.

7. Conclusion and Future Work

An ideology was put in motion with the inauguralwork of our project. The ability to predict injuries,and assess and determine patterns in said prediction is a novel concept.

In addition, our approach is further newer, and hence has a lot of potential for further research. Our current approach gives the most accurate results available, and can truly change the modern sports-entertainment industry.

Our project could change the multi-million-dollar sporting industry as we know it. But we are very well aware that our idea and its execution are in a state of infancy, and a lot ofwork has to be done for the full potential of the idea to be realized. A very raw concept of it has already been displayed to you last semester. Weseek to further strengthen our idea with the help of algorithmic hybridization. Our base working model uses the Random-Forest algorithm, which wedeemed to best fit us as it provided the best accuracy and was comparatively quick. But given the severity of our project, we decided to work on the suggestions put forth by our mentors, and enact hybridization. This will severely improve the accuracy and speed of our application, and in turn, improve efficiency.

Throughout our research, which is still ongoing, we have come across multiple algorithms, out of which we have shortlisted some of them such as linear regression, logistic regression, naive Bayes, and linear discriminant analysis. We might also add a more user-friendlyGUI to increase the reach and ease the use of our project.

References

1. Wang Jian Bai, Hwa Sun, Guangying Liu, Hongmei Li "A learning-based system for predicting sports injuries," MATEC Web of Conferences **189**, 10008 (2018)

2. Huang, Chen and Jiang, Lei, "Data monitoring and sports injuryprediction model based on embedded system and machine learning algorithm, " MAM,**81**, 103654 (2020)

3. Bittencourt, N F N and Meeuwisse, W H and Mendon, "Complex systems approach for sports injuries: moving from risk factor identification to injury pattern recognition," British Journal of Sports Medicine, **50**, 21, 1309-1314, (2016).

4. López-Valenciano A, Ayala F at. el , "A Preventive Model for Muscle Injuries: A Novel Approach based on Learning Algorithms," Med Sci Sports Exerc., **50**(5):915-927, (2018)

5. David L. Carey, Kok-Leong Ong, Rod Whiteley, Kay M. Crossley, Justin Crow, Meg E. Morris "Predictive modeling of training loads and injury in Australian football", IJCSS,**17**, 49-66,(2018).

6. Song, Hesheng and Han, Xiu-Ying and Montenegro-Marin, Carlos and Krishnamoorthy, Sujatha, "Secure prediction and assessment of sports injuries using deep learning-based convolutional neural network," JAIHC,**12**,1-12,(2021)

7. Brooks JH, Fuller CW, Kemp SP, Reddin DB. Incidence, risk, and prevention of hamstring muscle injuries in professional rugby union. Am J Sports Med.,**34(8)**:1297-306(2006)

8. Gregory Ornon, Jean-Luc Ziltener, Daniel Fritschy, Jacques Menetrey, "Epidemiology of injuries in professional ice hockey: a prospective study over seven years" *J EXP ORTOP* **7**, 87 (2020)

https://doi.org/10.1051/itmconf/20224403069

Computer Control Using Vision-Based Hand Motion Recognition System

Anshal Varma[1], *Sanyukta Pawaskar*[2], *Sumedh More*[3], *and Ashwini Raorane*[4]

[1]Department of Electronics Engineering, Ramrao Adik Institute of Technology, D. Y. Patil Deemed to be University, Navi Mumbai, ans.var.rt18@rait.ac.in

[2]Department of Electronics Engineering, Ramrao Adik Institute of Technology, D. Y. Patil Deemed to be University, Navi Mumbai,san.paw.rt18@rait.ac.in

[3]Department of Electronics Engineering, Ramrao Adik Institute of Technology, D. Y. Patil Deemed to be University, Navi Mumbai, sum.mor.rt18@rait.ac.in

[4]Department of Electronics Engineering, Ramrao Adik Institute of Technology, D. Y. Patil Deemed to be University, Navi Mumbai, ashwini.raorane@rait.ac.in

Abstract. In our day-to-day communication and expression, gestures play a crucial role. As a result, using them to interact with technical equipment requires small cognitive data processing on our part. Because it creates a large barrier between the user and the machine, using a physical device for human-computer interaction, such as a mouse or keyboard, obstructs the natural interface. In this study, we created a sophisticated marker-free hand gesture detection structure that can monitor both dynamic and static hand gestures. Our system turns motion detection into actions such as opening web pages and launching programs. This system will bring a revolution in various industries, which has the potential to replace traditional devices and time-consuming computer handling methods.

1 Introduction

Today's world demands Human Control Interaction. The easiest and most natural way of communication is Hand gestures. Gesture recognition is a fast and accurate means of interacting with the computer. It is used in human-computer interaction for text input. The user's hand gestures are captured and decoded to determine the desired outcome coded with that gesture. Researchers in the IT industry have been looking for many intuitive ways to interact with computers and other devices. They are looking for an easy way to interact with these devices through measly communications between the user and the device. Besides the conventional way (i.e., keyboard and mouse), nowadays there are other means such as voice commands (Alexa, Google Assistant), and gestures (virtual reality) that are getting popular in particular sectors, such as remotely controlling equipment at dangerous places.

Gesture recognition is a subset of pattern recognition technology that identifies specific human gestures via sensors in our computer devices. The idea is to provide a more natural and intuitive way for people to interact with their devices. No mouse, no trackpad, no keyboard. Modern gesture recognition technologies are made up of two main types. A sensor-based (or haptic-based) and camera-based (or image-based) system. Wearable equipment, positioned on the arms or hands or both, involves techniques involving the use of additional devices, which should be worn by the user. Through devices capable of estimating position and movement, such as accelerometers and gyroscopes, these devices can map the positioning of the user's members in a way to extract information about their movement.

On the other hand, vision data-based groups employ cameras to capture the user's movement without the need for extra equipment. The data captured by the camera may include depth pictures, infrared signals, and other means if different types of cameras are used in addition to the more standard RGB frame capture. The data captured should be by the means of a camera and not with any sensor or device attached to the user. Vision-based approaches are deemed less invasive to the user, allowing for broader use.

Machine learning approaches gained prominence in academia and business as a result of advancements in the computer sector to handle natural language processing issues, e.g., voice recognition and vision computational challenges. The complexity of computing is one of the major obstacles to employing these video classification approaches to accomplish the issue of gesture recognition. This is related not just to the volume of data contained in a video, but also to the complicated techniques employed in data processing employing neural networks, such as 3D convolution networks (3DCNN), and LSTM (Long-Short Term Memory) networks, Optical Flow processing, and so on.

This article proposes a convolutional neural network model gesture recognition tool designed for use in systems shipped. More specifically, gesture recognition, that is, making predictions for each new frame of image obtained using only video capture in RGB in order to not require a special camera for use. This technique was developed based on high-accuracy models aimed at generic applications, noting which structures would be feasible to implement as their own versions for systems embedded, with a view to minimizing the complexity of computation and keeping accuracy compatible with these techniques.

2 Literature Survey

Research in the fields of human-computer interaction (HCI) and its use in virtual environments are done a lot lately. Researchers have attempted to use video devices for HCI to identify virtual objects in order to control the system environment. Various natural motions can be identified, tracked, and analyzed by utilizing web cameras as the input device. Sarita Gavale et al. [7] worked on a system that uses the ultrasonic sensor HCSR04 to detect the motion of the hand with certainty. This sensor's range is around 1 to 13 feet. A hand gesture controller, also known as human activity gesture, detects hand movement. During the sensor testing for this project, the Sensor Kit UNO Arduino kit was used, which is required for processing raw data with microcontrollers.

[2] Aashni Haria et al. built a robust and engaging gesture-based system. These motions are then utilized to handle and control various desktop elements. To detect fingertip, this system use the Graham scan technique and the contour detection algorithm. Chua, S.N.D., et. al. [4] For hand gesture identification and localization, their project uses a single RGB camera. Due to the lack of standard gestures intended for human–computer interaction, the authors created their own dataset comprising all gestures used for training and their accompanying commands. Bedregal, Benjamin, et al. [13] introduced a fuzzy rule-based method for hand gesture identification. The method is extremely reliant on the preceding extensive examination of the properties of the gestures to be detected, as well as the manual transfer of those results to the recognition system. Experiments with users were conducted by Rios-Soria et al. [9] to assess the suggested gesture-detection algorithm's functionality. We asked participants to make a set of motions and actions in front of the camera and then measured whether or not the movements were detected. An observer kept track of whether the algorithm's output matched the actual gesture being performed.

Hojoon Park [10] used the index finger to move the cursor and the angle between the index and thumb to click events. Chu-Feng Lien [11] controlled the mouse cursor with only his fingertips, and his clicking approach was based on image density, requiring the user to hold the mouse cursor at the target place for a brief time. Fingertip tracking was used by A.Erdem et al. [12] to control mouse motion. A mouse button click was implemented by designing a screen such that when a user's hand passed over the region, a click occurred. Another way of clicking was utilized by Robertson et al. [13]. To mark a clicking event, they employed the thumb's motion (from a "thumbs-up" posture to a fist). The mouse cursor was moved by moving the hand while making a certain hand sign. Without the use of special color objects or gloves, Shahzad Malik devised a real-time system that can track the 3D position and 2D orientation of each hand's Index finger and thumb. Nasser H. Dardas et al. created a 3D game control system based on finger gesture detection.

3 Problem Formulation

Gesture-based engagement is becoming more common in our daily lives, yet we are still not taking advantage of its full potential. Gestures provide the user with a new form of interaction that closely resembles their participation in the real world. This seems natural and does not necessitate the use of an additional device or interruption. Furthermore, instead of a single input, they provide the user with multiple options. This project helps to overcome the disadvantages of standard input techniques by allowing the user to intuitively give commands. Devices have become an increasingly crucial element of our daily lives as the information society develops. In recent decades, the most common form of human-computer interaction (HCI) has been based on basic input devices such as the keyboard and mouse. This is effective, but the technique's underlying flaw is that there is an inherent lack of naturalness and physical connection with the computer.

With the introduction of new display technology such as virtual reality, this constraint becomes even more critical. As a result, some novel solutions for overcoming the HCI bottleneck have recently been devised. This situation has been thoroughly investigated, and numerous viable options have been identified. People frequently rely on their hand movements to engage with the environment, hence gesture interaction is believed to be superior to other ways. As a result, it gives users a natural and three-dimensional experience. As a result, the system that must be designed must meet the following requirements. i.e.

- The system will look for human behavior

https://doi.org/10.1051/itmconf/20224403069

- Process the action and convert it to input
- Lastly, perform a user-defined action

4 Design Methodology

4.1 3D CNN

This review article takes a look at the fundamentals of CNN and how it may be used to a variety of radiological jobs, as well as its challenges and future directions in the field of radiology..This paper also covers two difficult conditions in using CNN for radiological tasks: short datasets and overfitting, as well as techniques to avoid them. Knowing the standards and benefits of CNN, as well as its limits, is essential for maximising its potential in diagnostic radiology, with the objective of improving radiologists' overall performance and patient care.

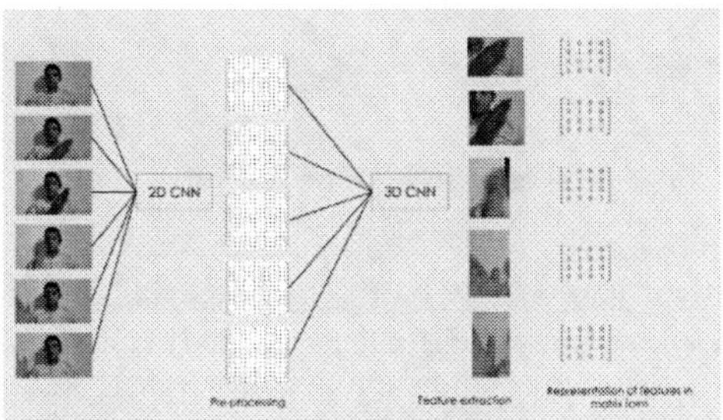

Fig 1. Work flow in CNN

In 2D CNNs, convolutions are implemented at the 2D characteristic maps to compute functions from the simplest spatial dimensions. When implemented to video evaluation problems, it's suitable to seize the movement statistics encoded in a couple of contiguous frames. To this end, we advise to carry out 3-d convolutions withinside the convolution levels of CNNs to compute functions from each spatial and temporal dimensions. The 3-d convolution is executed through convolving a 3-d kernel to the dice shaped through stacking a couple of contiguous frames together. By this construction, the characteristic maps withinside the convolution layer is hooked up to a couple of contiguous frames withinside the preceding layer, thereby shooting movement statistics a 3-d convolutional kernel can simplest extract one form of functions from the body dice, for the reason that kernel weights are replicated throughout the complete dice. A standard layout precept of CNNs is that the quantity of characteristic maps need to be elevated in past due layers through producing a couple of varieties of functions from the identical set of lower-degree characteristic maps.

4.2 RT3D 16F

For the development of the proposed network, the structures of the network that compose the state of the art were analyzed. Most of these networks do not have the purpose of executing in embedded systems, and certain structures that make up these networks end up not being good candidates for this purpose. One example of this is the processing of flow frames using algorithms of Optical Flow, which is widely present.

In preliminary tests, the use of this type of processing has proven computationally expensive (high computation time), even in simplified versions of the algorithm. The 3D convolution structure is present in most state-of-the-art gesture recognition techniques. It is on this that the models proposed in this article are based. The level of computational complexity of a layer of a 3DCNN network depends on the dimensionality of the input tensor. A tensor, in this context, serves as a generalisation of dimensional numerical representations such as vectors (1D) and matrices (2D), however, being able to represent any representation containing N dimensions. In order to reduce the complexity of video frames directly on a 3DCNN network, the model recommends having a 2DCNN network that receives the frames. preprocesses these, thus reducing the dimensionality of the data, in addition to aiding in the extraction of spatial features (being able to identify elements present in the frames, such as hands, arms, etc.)

Layer	Input Channels	Output Channels	Filter	Pooling
2D Conv	3	16	[3,3]	No
2D Conv	16	32	[3,3]	No
2D Conv	32	64	[3,3]	[2,2]
2D Conv	64	128	[3,3]	[2,2]
2D Conv	128	256	[3,3]	[2,2]
3D Conv	256	256	[3,3,3]	No
3D Conv	256	256	[3,3,3]	[2,2,2]
3D Conv	256	256	[3,3,3]	[2,1,1]
3D Conv	256	256	[3,3,3]	[2,2,2]

Table 1. 2D and 3D Convolution Layers for the proposed model

The layers that make up the blocks of 2DCNN and 3DCNN are listed in Table I. These are the main network parameters. In addition, there are activation layers using ReLU [14] (an activation function that maintains the signal identical for positive values and zeroes for negative values) and batch normalization layers, which were omitted for simplicity. The linear block of the network consists of 3 layers. This block has the function of reducing the output tensor of the 3DCNN network to progressively transform the 6144 nodes into 3072 nodes, then 1024 nodes, and finally into the number of classes existing in the model, depending on the data set.

5 System Implementation

https://doi.org/10.1051/itmconf/20224403069

This project proposes a convolutional neural network model gesture recognition tool designed for use in modern systems. The adaptation of 2D convolution networks to operate in one more dimensionality is the 3D convolutional network (3DCNN). Video frames are images with two-dimensional representations. When several frames of video are combined, the information is now represented by three dimensions: the height of the frames, the width of the frames, and the different frames themselves.

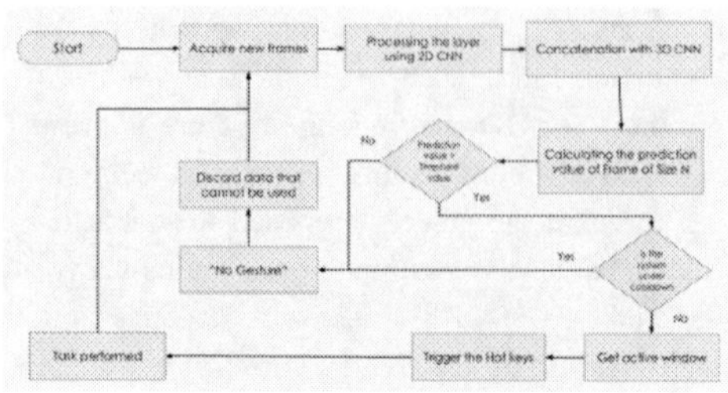

Fig 4. Architecture Diagram

The model will be trained on a dataset consisting of 148,092 sample images containing gestures divided into 26 classes. Here, the input stream of webcam video is set at 12 to 16 frames per second, as the greater the number of frames per second, the greater the amount of movement information contained in the images and, therefore, the greater the expectation of the highest accuracy for the technique. The size of frame will have a height of 100 px and a width of 200px for processing the image.

Webcam-based analysis is based on the user's distinguish data about their surroundings. The complete gesture interactive mechanisms that act as a building block for vision-based hand gesture recognition systems are composed of three fundamental phases. Hand detection and segmentation of related picture areas are the first steps in hand gesture recognition systems. A vast number of strategies have been presented that makes use of a variety of visual cues, as well as their combination in many circumstances. It is a quick approach to detect and operate at an image acquisition frame rate, but it may also be used for tracking if the detection method is fast enough to work at an image acquisition frame rate. Tracking hands, on the other hand, is notoriously difficult since they move quickly and their look might vary dramatically within a few frames.

The frame-to-frame connection of the segmented hand regions or characteristics towards interpreting the observed hand movements is known as tracking. Gesture recognition is a technique that allows users to conduct essential orders with their hands and other actions by providing devices with real-time data via motion sensors. As the major source of data input, the device's motion sensors recognise and analyse human motions.

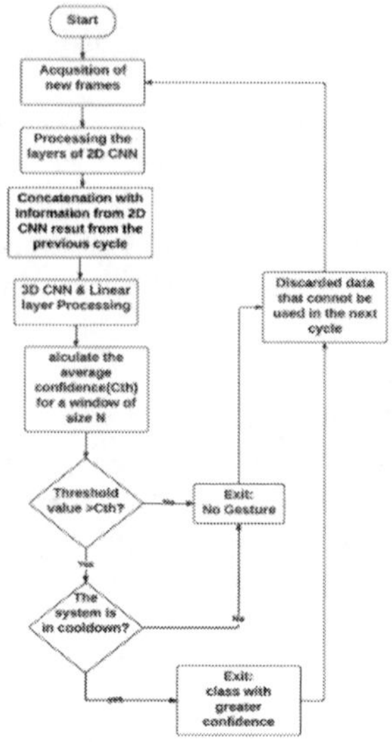

Fig 5. Flow of the Project

As a result, the identification of human hand movements by artificial intelligence (AI) systems has been a significant advancement in the previous decade, with applications in high-precision surgical robots, health monitoring devices, game systems, and other applications.Despite the fact that gesture recognition is still in its infancy, the creative market environment that has seen Virtual Reality become a global phenomenon forces us to recognise that its application domains may continue to expand. Finally, the entire undertaking proved to be a beneficial learning experience.

6 Result and Discussion

The developed project is a gesture detection based computer operating system which is expected to take gesture as the data from the user and take action given to that gesture. The image is trained using RT3D algorithm and a ckp file is created out of these trained images. The trained data is compared with the tracked images and the algorithm recognises the gesture and executes the action.

6.1 Initialization

The main file and the learned model are retained for initialization. We type "conda activate dante" to activate the conda environment for the machine, then "python main.py" to access the main file where the application is written, after launching cmd.

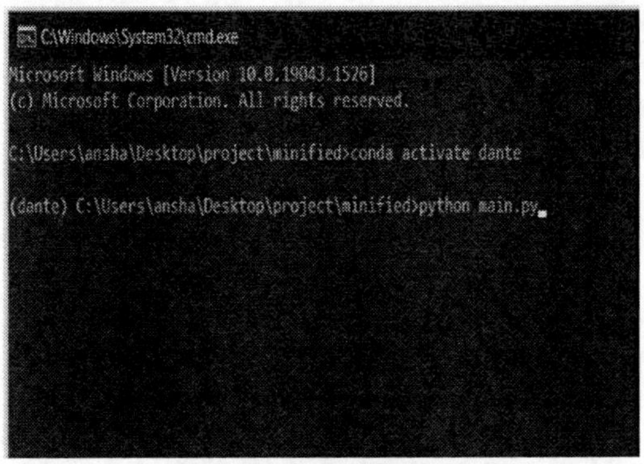

Fig. 6(a). Initialization

6.2 Gesture Detection

Fig. 6(b). Gesture gets Recognised

Following the program's initialization, all of the models are loaded, and the system begins predicting motions using a webcam. As soon as the user makes the motion, the trained model begins to forecast and displays the top three predictions.

6.3 Gesture Application

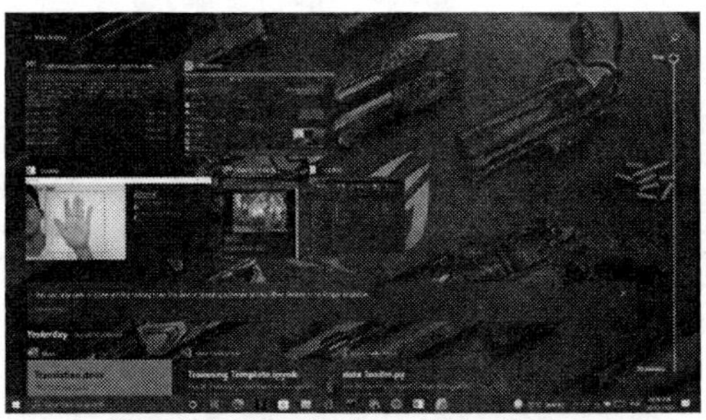

Fig. 6(c). Application of the Gesture

The model reacts to the gesture when it has been identified by the model. Depending on the active window in which the gesture is done, each gesture has a distinct action. For example, in the photo, the gesture was a stop sign, and the action key attached to it was win + tab, which displayed all open tabs.

6.4 Threshold Value

The value of the gesture made by the user is displayed in the cmd. The value for the threshold is 0.45. If the gesture value is greater than 0.45, the action associated with the gesture for the given active window will be performed. If the value falls below the threshold, it will pause for 0.1 seconds before predicting again.

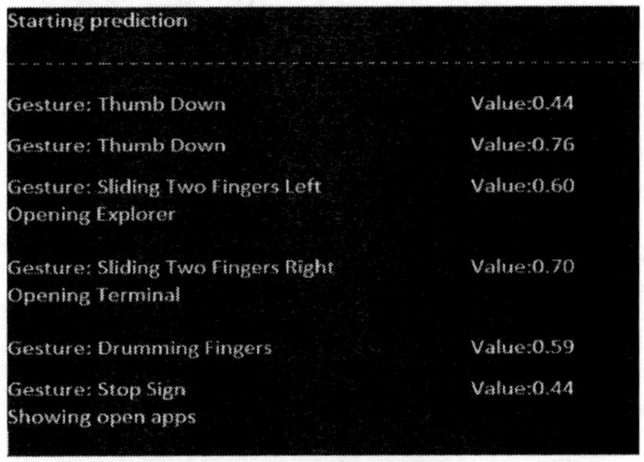

Fig. 6(d). Threshold Value

Given below (fig) is a list of the 20 actions performed by the system. You can find all the gestures and its action assigned to the system in the figure. In a similar way, various gestures can be added for performing different actions.

Implemented Gestures

Gesture	Desktop	Chrome	File Explorer	VLC
Stop Sign	Recent Tabs	Switch to another tab	Opening Chrome	Play/pause
Sliding two fingers Right	Opening Terminal	Open new tab	Taking Screenshot	Forward 10 seconds
Sliding two Fingers Left	Opening Explorer	Closing Tab	Closing Explorer	Rewinding 10 seconds
Sliding two fingers up	Virtual Keyboard	Maximize Window	Navigate to previous folder	Increase volume
Sliding two fingers down	Open YouTube	Minimize Window	Opens selected folder	Decrease Volume
Swiping up	-	Open Chrome	-	-
Swiping down	-	Minimize all	-	-
Zooming in	-	Zoom In	-	-
Zooming out	-	Zoom out	-	-

Fig 7. Gestures and their respective action

Accuracy & Analysis

To analyze the accuracy of the gesture, we tried to evaluate the prototype 20 users. Each of them performed six gestures with right and left hand, resulting in 240 observations. We calculated an average of observations for each gesture. Given below is the study.

Hand Used	Gesture Performed					
	Swiping Left	Swiping Right	Two Fingers To The Right	Two Fingers To The Left	Drumming Fingers	Shaking Hands
Right	98%	97%	95%	99%	100%	91%
Left	95%	99%	91%	95%	96%	92%
Total	96.5%	98%	93%	97%	96%	91.5%

Table 2. Analysis of the gesture accuracy

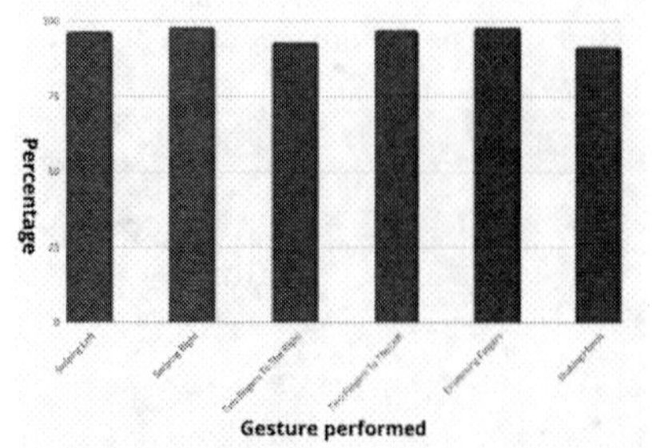

Fig 8. Analysis of the gesture accuracy

7 Conclusion

Thus, the project aims to provide control over the computer without using traditional interfaces like mouse, and keyboard which somewhere becomes a bottleneck that depends on heavy machine user interaction. from reading lots of related articles we have learned that we can easily reduce this bottleneck with computer vision and make it scalable and be used at a large scale in various fields. By using a 2DCNN network to preprocess the frames, this study builds a hand gesture detection system for engaging with various applications such as picture browsers, games, and so on and gives a successful solution towards a user-friendly interface between human and machine. Using a low frame rate minimizes the quantity of data to be processed and allows us to execute our model with more time between frames. Finally, the flickering of the gesture-controlled mouse is decreased using the smoothening approach. Moreover, the proposed system gives more performance and uses a system of cooldown to avoid multiple computations of the same gesture.

8 Future Scope

Gesture recognition makes it easy to use a variety of devices, including personal navigation devices, computers, laptops, and mobile phones.The growth trend of touchless gesture recognition in niche applications, including games, could boost the growth of the touchless sensor market during the forecast period.Medical applications have additional gesture recognition capabilities that nurses and doctors may not be able to touch the display or trackpad for health and safety reasons, but still need to control the system.Proper gestures, such as swiping by hand or using your finger as a virtual mouse, are a safer and faster way to control your device.

References

1. S. Albawi, T. A. Mohammed and S. Al-Zawi, "Understanding of a convolutional neural network," *2017 International Conference on Engineering and Technology (ICET)*, 2017, pp. 1-6, doi: 10.1109/ICEngTechnol.2017.8308186.
2. Aashni Haria, Archanasri Subramanian, Nivedhitha Asokkumar, Shristi Poddar, Jyothi S Nayak, Hand Gesture Recognition for Human Computer Interaction, Procedia Computer Science, Volume 115, 2017, Pages 367-374, ISSN 1877-0509
3. S. M. A. Hoque, M. S. Haq and M. Hasanuzzaman, "Computer Vision Based Gesture Recognition for Desktop Object Manipulation," 2018 International Conference on Innovation in Engineering and Technology (ICIET), 2018, pp. 1-6, doi: 10.1109/CIET.2018.8660916.
4. Chua, S.N.D., Chin, K.Y.R., Lim, S.F. et al. Hand Gesture Control for Human–Computer Interaction with Deep Learning. J. Electr. Eng. Technol. (2022).
5. S. Song, D. Yan and Y. Xie, "Design of control system based on hand gesture recognition," 2018 IEEE 15th International Conference on Networking, Sensing and Control (ICNSC), 2018, pp. 1-4, doi: 10.1109/ICNSC.2018.8361351.
6. F. Brandolt Baldissera and F. L. Vargas, "A Light Implementation of a 3D Convolutional Network for Online Gesture Recognition," in IEEE Latin America Transactions, vol. 18, no. 02, pp. 319-326, February 2020, doi: 10.1109/TLA.2020.9085286.
7. D. Xu, Y. Chen, C. Lin, X. Kong and X. Wu, "Real-time dynamic gesture recognition system based on depth perception for robot navigation," 2012 IEEE International Conference on Robotics and Biomimetics (ROBIO), 2012, pp. 689-694, doi: 10.1109/ROBIO.2012.6491047.
8. Gavale, Sarita & Jadhav, Yogesh. (2020). HAND GESTURE DETECTION USING ARDUINO AND PYTHON FOR SCREEN CONTROL. International Journal of Engineering Applied Sciences and Technology. 5.271-276. 10.33564/IJEAST.2020.v05i03.043.
9. Chaudhary, Ankit & Raheja, Jagdish & Das, Karen & Sonia, Raheja. (2011). Intelligent Approaches to interact with Machines using Hand Gesture Recognition in a Natural way: A Survey. International Journal of Computer Science and Engineering Survey. 2. 10.5121/ijcses.2011.2109.
10. Rios-Soria, D.J. & Schaeffer, S.E. & Garza-Villarreal, S.E.. (2013). Hand-gesture recognition using computer-vision techniques. 1-8.
11. H. A. Jalab and H. K. Omer, "Human computer interface using hand gesture recognition based on neural network," 2015 5th National Symposium on Information Technology: Towards New Smart World (NSITNSW), 2015, pp. 1-6, doi: 10.1109/NSITNSW.2015.7176391.
12. Gupta, Aviral & Sharma, Neeta & Scholar, M. (2020). A REAL TIME CONTROLLING COMPUTER THROUGH COLOR VISION BASED TOUCHLESS MOUSE. 9. 5077.
13. Lenman, Sören & Bretzner, Lars & Thuresson, Björn. (2012). Computer Vision Based Hand Gesture Interfaces for Human-Computer Interaction.

ITM Web of Conferences **44**, 03070 (2022) https://doi.org/10.1051/itmconf/20224403070
ICACC-2022

A Robust Home Alone Security System Using PIR Sensor and Face Recognition

Harsha Saxena[1,*], and *Leena Ragha*[2,**]

[1]Ramrao Adik Institute of Technology, Nerul, Navi Mumbai
[2]Ramrao Adik Institute of Technology, Nerul, Navi Mumbai

Abstract. CCTV-based video monitoring technology is one of the fastest growing security technologies markets. The existing video monitoring systems are, however, still not in a position to be used to prevent crime. For public safety purposes, large networks of cameras are increasingly deployed in public places like Residential Buildings, College Campus, offices, airports, railway stations, and shopping malls. Such systems are primarily dependent on human observers and are therefore limited over long periods by factors such as exhaustion and monitoring. In order to overcome this constraint, "intelligent" systems are required, which can highlight the critical data and remove normal conditions that are not a safety hazard. We propose a model utilizing machine learning techniques in order to build these smart systems. This research aims to create an application in real time, which is necessary for labs, places of work or homes where human detection and Recognition will be done for human safety

1 Introduction

Globalization and societal liberalization and search of new opportunities makes a trend among young people to move to urban areas is increasing A 2015 research article published in Demographic Research, 'One-person households in India' by Premchand Dommaraju[1] places the increase in India at 9.04 million one-person households in 2011 from 6.8 million according to the 2001 census. This reflects a 0.1% increase in such households along with an overall propensity towards the nuclear family setup.

On the other hand, the crime incidents targeting home alone senior citizens or kids with the intention to cause harm to the people staying is also increasing alarmingly. Because of many reshaping living arrangements, in majority cases, small kids, single girls, and elderly people stay alone at home and are vulnerable to this kind of attack. Most of these cases could be prevented from becoming a tragedy if timely help is provided. In this project, we propose to address this problem faced by all those who care for the loved ones staying alone in distant places. The remote connection of the people not only keeps them closure, but it also connects them emotionally. One can keep a watch on the dear ones as they get alert when any suspicious event occurs. A smart monitoring system as shown in Figure 1 can help in an investigation in case of problems.

The system intends to overcome the drawbacks of the past surveillance systems and to enhance security, adaptability, and efficiency. The main aim of this research is to automatically detect the humans using the PIR sensor mounted on the front door and activate the CCTV which should be active as long as the visitors are in the house. The system identifies and recognizes the visitors as family, friends, relatives, service provider and Unknown and continuously tracks the movement of the people to recognize the abnormality.

The system is developed using PIR sensors to detect the presence of humans, Haar cascade for face detection and Local Binary Pattern Histograms (LBPH) for Face Recognition. The Setup of CCTV to capture the human movement at a door using the PIR Sensor. The system recognizes the person at the door and categorizes the people as a family, relative, friend and an unknown person. As soon as the system recognizes the unknown person and sends an alarm and picture to the registered email to prevent the tragedy from happening.

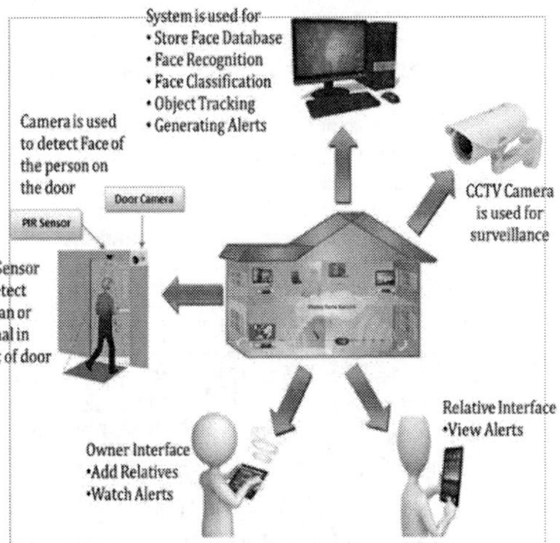

Figure 1: Home safety measures using face recognition and PIR.

There are few problems faced by the face recognition process are change in illumination, low-resolution camera, occlusion due to objects like eye-glasses, hair styling, and makeup is solved in the paper.

Using this kind of solution, the monitoring is discrete based on the accuracy of the system. The monitoring person need not be observing the camera footage continuously. The alert message will trigger his attention to observe the live CCTV footage and take necessary

* email: harsha.saxena@rait.ac.in
**email: hodce@rait.ac.in

action quickly like calling neighbors, police, etc. If missed, then the messages will give the time details and searching for abnormal events in the stored CCTV footage will be facilitated.
The further organization of the paper is listed in this section as follows: The literature survey and the related work done before-in this field of research for our study-is covered in Section-2. Section-3 contains the analysis of performances of the various algorithm used in Face recognition. Further, Section-4 and Section-5 contain our proposed methodology to solve the problem and the expected results. Finally, we conclude the paper by stating the conclusion of the entire paper in Section-6. The references of papers we used for working have been annotated in Section-7.

2 Literature Survey

Security is one of the most challenging house requirements. The goal of this study is to identify a person using face recognition and alert them of safety in danger [2]. One of the applications for image processing is facial recognition. The image processing method involves transforming an image into a digital form and performing some image processing operations, producing an improved image, or collecting some useful data. In our CCTV cameras, we will benefit from image processing and face recognition. Video recording and review of the images collected is a process that requires a significant amount of memory. CCTV video tracking is used in every part of the world today [3]. However, there is no effective video surveillance implemented yet. Video surveillance is usually used to install a camera and review the captured video. However, we can do something better at the same cost [4]. Our device is fitted with a camera in the safe room. A PIR sensor is used along with the camera so that the camera is not turned on every time. If the sensor detects the human presence, the camera activates and begins to record the footage. Human facial characteristics are observed from the frames extracted from the captured video. The image is compared to the picture stored in the data set [2]. When the face is recognized [12] it will open the door for family members and relatives. When the face is not recognized as an alarm/doorbell. It will also track the person inside the house and give an alert message if any suspicious activity happens.

There are many Door lock security systems [13][14] are classified based on technology used GSM based[6][7], smart card-based, Password-based, Biometric based[8][9], Social networking sites based, RFID based, Bluetooth based, Door phone-based, OTP based[10], Motion detector based using PIR[11], VB based, Combined system.

3 Analysis of Literature Survey

A literature survey shows that a lot of work is still going on face recognition in videos to improve its accuracy. Table 1 shows the surveys.

Table 1: Literature Survey Analysis

Title	Advantages	Limitations
Human Detection Using HOG [17].	low complexity	High dimensional feature Vector
HOG feature human detection System [15]	The multi-layer classifier produced better results than individual	Did not overcome SVM classifier
pedestrian detection using variant scale blocks-based HOG features [16]	1.The hybrid method produced a better result than traditional SVM and Adaboost. 2. The block size is not limited to a wide range of functionalities that allow highly classified features to be extracted in the interests of accuracy and time savings for detection systems.	There have been some wrong results that can happen when a pedestrian is shown in crowded scenes
Robust Real-Time Face Detection [18][19][20]	minimizes computation time while achieving high detection accuracy	1.A Face detector fails on significantly occluded faces. 2. harsh Backlighting may cause problems sometimes. 3. The detector becomes unreliable with more rotation than given
Open CV Face detection and recognition [21]	1.LBPH analyses feature of training set individually. 2. changes in luminosity of the image don't affect results	Not invariant to rotations
Face Recognition using Discriminative Feature Learning [22]	Do not depend on the label prediction	Need very high computation cost
Class Specific Linear Projection used in Eigen vs. Fisher faces Recognition [5]	1.Eigenface and Fisher's face are easy to implement. 2. Fisher face method has error rates that are lower then those of the Eigenface technique	Not invariant to illumination

 https://doi.org/10.1051/itmconf/20224403070

Analyzing the literature, we came to know some limitations and scope to work on the face detection problem as follows: Even though still image face recognition results are excellent, face recognition in video frames is still an unsolved problem as the frames are influenced by real-time effects like the motion, lighting, no of people, and occlusion [2]. The conventional face recognition such as Eigen and Fisher facing problems in the detection of side views. Besides, face recognition on Fisherfaces is not ideal for the shift in lighting in different real conditions [5].

Facial recognition based upon LBPH, consisting of an array of histograms and blocks measuring distance. The scheme detects the right faces, even in slanting pictures, by showing a rectangular region on the face [5]. With the use of a locally normalized histogram with a gradient orientation similar to SIFT descriptors [17], it produces very good results in the detection of persons, which reduces false-positive rates.

The Authors in the paper [23] proposed the method in which the aim was to detect and recognize the face in Real-time. They achieved an overall system accuracy of 96.8% by using MTCNN for face detection and Inception-Res Net network for Face Recognition. The author created the database of 100 faces, if the face matched with the face in the database, then it is considered as a known person else it will be recognized as unknown. They detected face under various circumstances and achieved average confidence value under illumination 93%, Head Pose 94.15%, Occlusion by hand 89%, Face Expression 91% and Makeup 95%. But the technology used was time-consuming, it takes more time for recognition when multiple faces occur in the environment. Few faces were not tracked when video moves faster. In this paper face recognition is done by using CNN, the accuracy obtained is very accurate but it needs very high computation cost as it passes through many hidden layers. There is a need to look into the simplest and fastest technology to find an efficient and less time-consuming recognition algorithm.

By considering limitations that is the speed of computation and accuracy for recognizing a face. we have proposed a Robust Home Security System Using PIR Sensor and LBPH Face Recognition which requires less computation cost and provides good accuracy is explained further in the next section.

4 Proposed Methodology

This is interdisciplinary research to address the remote monitoring of the dear ones using smart ubiquitous environment. The different components and the technology required to address the problem are microcontroller, sensors, signal processing, communication, image processing, database management, log keeping and soft computing techniques to build a ubiquitous environment.

The flow of the proposed research is as shown in Figure 2 followed by the explanation of different important technology/components.

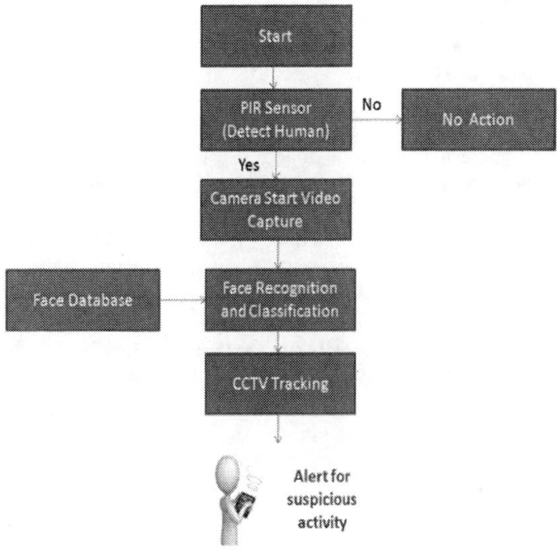

Figure 2: System Design

Passive Infrared (PIR) Sensor

A PIR detector is a motion sensor that detects a living body's heat. The sensor is passive because it is sensitive to the infrared energy generated by every living thing, instead of transmitting a light beam or microwave energy disturbed for a passing person to detect another person. When an intruder enters the field of vision of the detector, it senses the sudden increase in energy level.

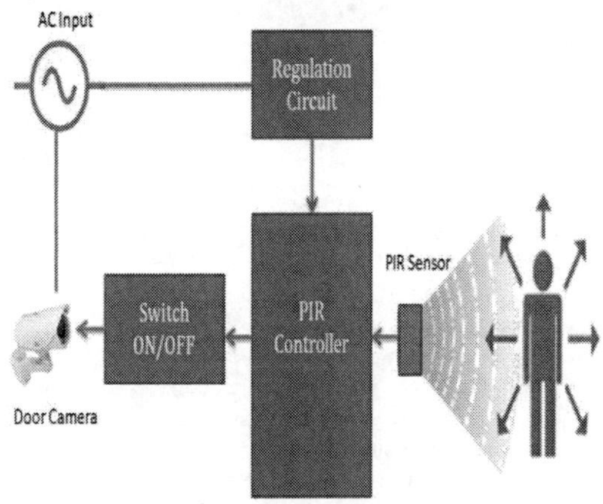

Figure 3: Passive Infrared (PIR) controller diagram

The machine is installed at the front door, as shown in Figure 3. The sensors are designed to distinguish between intrusion movement and vegetation oscillation. The program focuses on the smallest of creatures like dogs, wolves, leopards, and tigers compared to humans. The machine can distinguish between humans and animals by measuring the intruders ' height. To do this, the system would conduct two classifications for each signal detected: firstly, the distinction between vegetation and non-vegetation, and then human and animal classifications. Once people are spotted, the camera turned on and start recording. This saves the energy of the system.

Face Recognition and Classification

A pattern recognition function primarily performed on faces is called facial recognition. It can be described as classifying a face as Family Member, Relative, Friend or unknown, after matching it with stored known individuals as a database. Upon confirmation that the family member or relative is the individual, the door is opened immediately. It is also important to have a program that can learn to recognize unfamiliar faces. Five key functional blocks have their roles in Figure 4, as shown below.

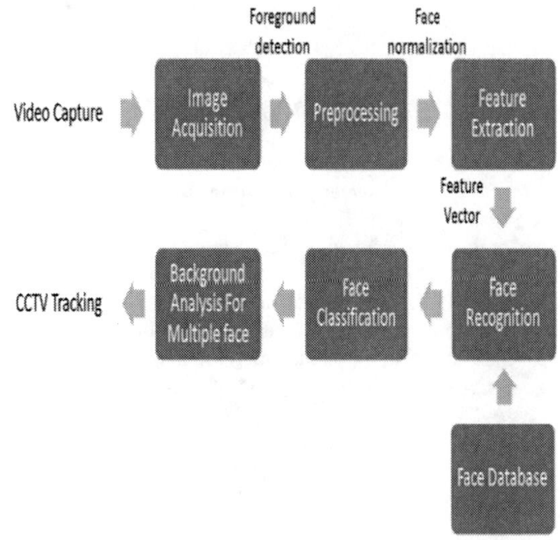

Figure 4: Flow chart of Face Recognition and Classification

A. The acquisition module

The user sets the face image as the input for the face recognition device within this module. This is the entry point of the face recognition process. The face is captured from the real-time input stream.

B. Pre-Processing

The images are normalized in this module to improve machine recognition. Pre-processing steps are normalization of image size, background elimination, translation and rotational normalization, normalization of illumination.

C. Face Detection

In our system, we only need the frontal faces that are normalized in scale from the input images. To reduce the computation for feature extraction, it is important to localize and extract the facial region from an image. We are using Haar cascade for face detection.

The haar features are used for face detection and are of a rectangular type which is determined by an integral image. Figure 5 shows different types of haar features that are similar to a few properties common to human faces. The eye region is the one which is darker than the upper cheeks so the second type of haar feature in figure 6 is used to detect that facial region and haar feature for the nose bridge region which is brighter than the cheeks as shown in the figure 6.

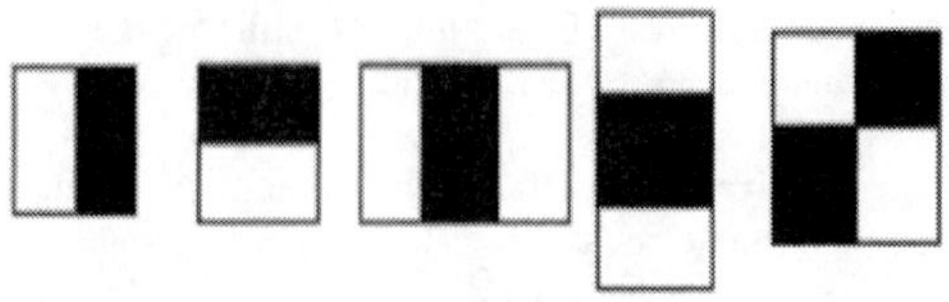

Figure 5: Haar Features

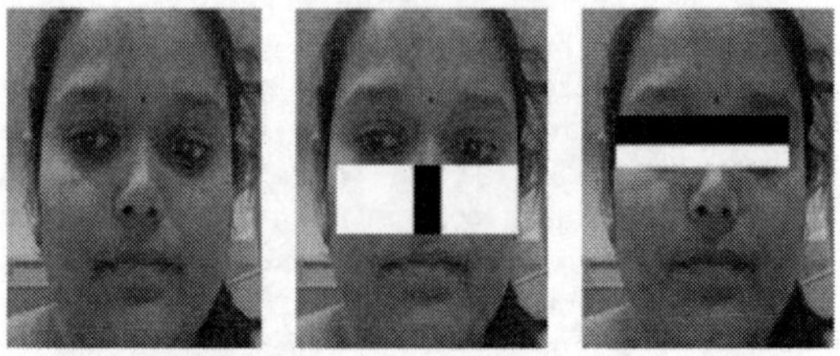

Figure 6: Haar Features used to recognize eyes and the bridge of nose regions of face

Here, using these features we can find the locations of eyes, bridge of nose and mouth by calculating
Feature Value = $\sum$ the sum of pixels in the black area - $\sum$ the sum of pixels in the white area

It is used for facial edge detection and hence the output is a horizontal high-value line. In haar, we use 24 X 24-pixel sub-window on the image window to find the edge therefore many possible features can be extracted which are further used for facial region detection. We use this window size 24 X 24 as we are ignoring the face which are smaller than window size. We create a kernel using haar features to extract this line. Then apply the kernel to the whole image and it has a high output only where the image value matches the kernel that is our expected output.

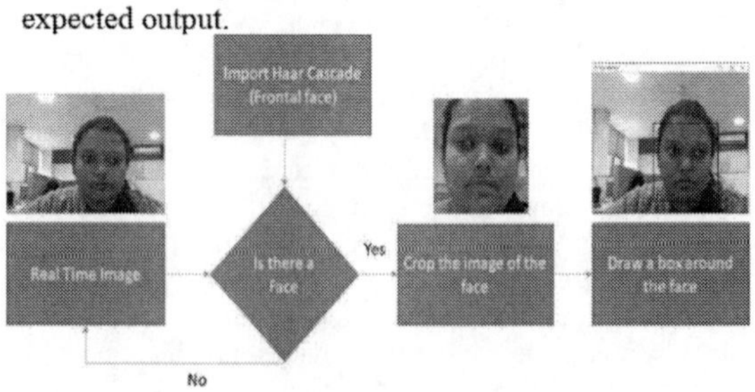

Figure 7: Face detection using Haar Features

In Cascade classifier, the term cascade means several filters on the resultant, that's why cascade is used to combine many features efficiently. It on discarding non-faces images to avoid the unnecessary work and spend more time on images with probable face regions. Therefore, a cascade classifier issued which is composed of stages containing strong classifiers. So that with the output from each we can discard non-facial images. Although training to create new Haar-cascade is important, OpenCV has a robust collection of Haar-cascades that were used for the project. Figure 7 displays the flow diagram of the detection system.

D. Recognition and Classification

To recognize a person, we have to train our machine for certain images of that person. Here training is done using Haar cascades and Local Binary Pattern Histograms

(LBPH). Haar features are used to detect faces in video frames, now each detected face is treated as a dataset to train our machine and LBPH features corresponding to each face are saved in a file. The extracted features of the face image are compared with the ones stored in the face database. It is used to match the test face image is then classified as either Family Member, Relative, Friend or unknown. There are three stages for face recognition as follows:

Stage 1: Feature encoding

The image is split into 3x 3 pixels (cells). The centers, as shown in Figure 8, are compared with the centers in a clockwise or anti-clockwise direction. The neighbor is equal to the middle pixel with the frequency or luminosity. A 1 or 0 is allocated to the position based on the difference being higher or lower than 0. This gives the cell value of 8-bit. The benefit of this approach is even if there is a variation in the luminosity the result will remain the same.

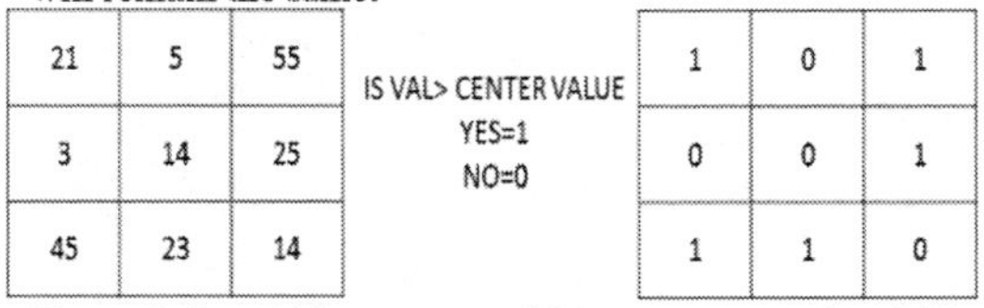

Figure 8: Local binary pattern histogram generating 8-bit number

Stage 2: Generation of Feature Vector

We can generate a histogram for this image with the LBP result and create a vector function to describe the original image pattern as shown in Figure 9. The histogram defines the frequency of LBP results occurring for every pixel. The value for each pixel is between 0 and 255 from the last part after the LBP operator. So there are only 256 values in the histogram.

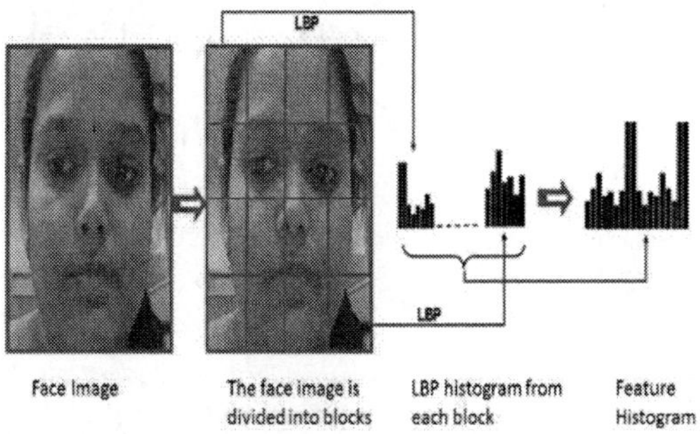

Figure 9: Face description with LBP

Stage 3: LBPH Classification

The ID is attached to each person's face feature in the database. The image is classified based on the ID attached to faces. Initially, the Input Image face is extracted and its feature is compared with all the face features in the dataset. Then it finds the difference in the feature vector of all the images in the dataset to the input image. The ID of a face which has the minimum distance with the input face vector is attached with the input face. By setting up a threshold, it can be identified if it is a known or unknown face. Eigenface and Fisherface compute the dominant features of the whole training set while LBPH analyzes them individually as shown in Figure 10. Histogram intersection is used to measure the similarity between two histograms. It is a distance matching algorithm that measures the dissimilarity strength between two entities using Chi-square as a dissimilarity measure.

$$x^2(S,M) = \sum_i \frac{(S_i - M_i)^2}{S_i + M_i}$$

The same can be used for weighted histograms generated using LBPH. The less the distance, the more similar are the two histograms. The extracted feature histogram of the new sample images is compared with the one stored in the Template Database and check for the extent of match between the two.

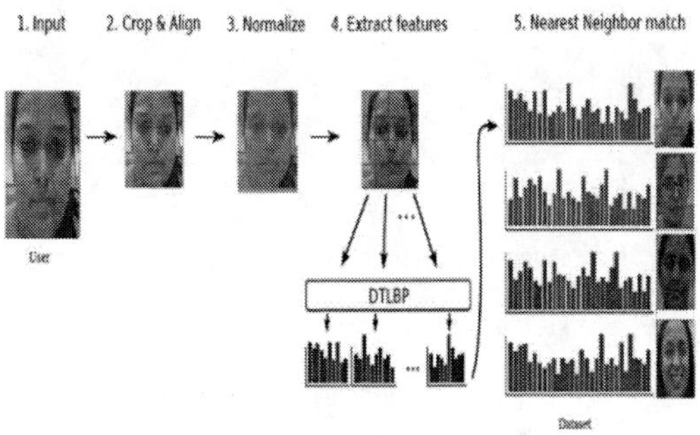

Figure 10: LBPH Classification

E. Face Database

Face database is a collection of train images of 65 users stored in a database. The database consists of 12 images per user from a different angle which is further classified as family, Friend, and Relative.

F. Alarm Generation

After detecting the face of the person, he will classify as Family Member, Relative, friend or unknown. If an unknown person is detected an alert message through the wireless signal is sent to the Owner mobile.

5 Results

The proposed methodology was implemented on a system that had an Intel i3 processor running with 4GB RAM, on windows 10.

For facial recognition, It only takes face as an entry, if the camera doesn't find a face then it stays paused for any face. Our research has been checked on 75 people in which 5 faces of Family members, 30 faces of Friends, 30 of relative and 10 unknown faces are taken. LBPH is trained for 65 faces, if a face is not from the training set then it will be labeled as unknown.

We used the confusion matrix table for calculating the accuracy of our techniques as shown in Table 2.

Table 2: Confusion Matrix

S.No	Category	Family	Friend	Relative	unknown
1	Family	4	1	0	0
2	Friend	2	23	4	1
3	Relative	2	2	25	1
4	unknown	0	1	2	7

A live implementation of the project on a real-time scenario is depicted below. Figure 11 shows the result of face detection and recognition which shows that the system is used to recognize the real-time face and classified as either Family, Relative, Friend or unknown.

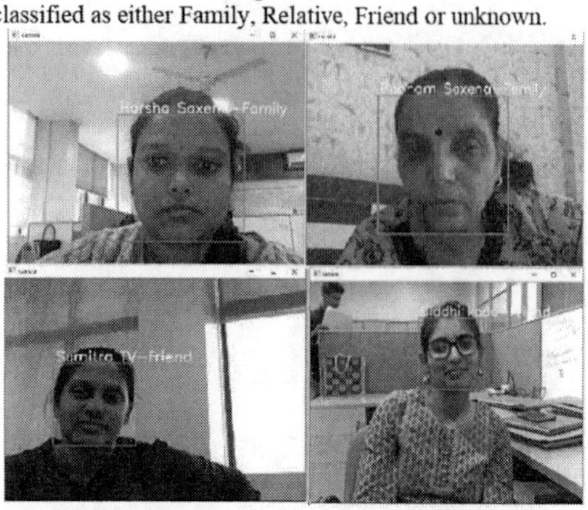

Figure 11: Face Recognition and classification

It can recognize a face with or without spectacles and it can also recognize multiple faces in real-time as shown in Figure 12. If f unknown person is detected an Alert message through wireless signal is sent to the Owner mobile through email as shown in Figure 13

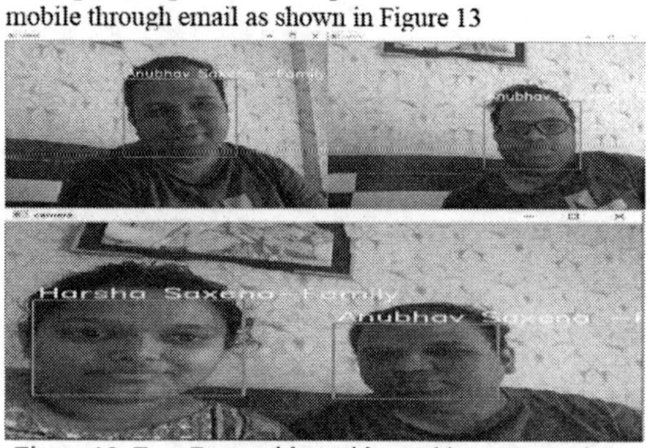

Figure 12. Face Recognition with or without Spectacles and Multiple Face Recognition

We achieved 78.67% accuracy in Real-Time scenarios under many variations and unconstrained environments like change in illumination, low-resolution camera, occlusion due to objects like eye-glasses, hair styling, and makeup. The accuracy of the system is less as compared to the accuracy achieved by Roshni Singh[23] that is 96.8% by using MTCNN for face detection and Inception-Res Net network for Face Recognition, but the system need more computational time for training and testing. The proposed system takes 112.95 seconds time for training and 0.1259 seconds time for testing the model. The system can detect, recognize and classify a face as Family, Relative, Friend or unknown and generate alert accordingly in Real-Time.

Figure 13: Alarm Generation when Unknown person detected

6 Conclusion

This is a research problem with huge scope to improve the performance of the system with respect to sensing, tracking, and controlling the home access by the visitors. In this paper we discussed the advantages and limitation of various techniques. To overcome the limitation, we develop the system that senses the humans and activates the camera as long as the visitors are in the house. It works on Real-Time and identify the visitors as family, friend, relative and an unknown with good accuracy of 78.67 percent and will raise an alarm in case of an emergency.

References

[1] Dommaraju, Premchand. (2015). One-Person Households in India. Demographic Research. 32. 1239-1266. 10.4054/DemRes.2015.32.45.

[2] Sanjeev Kumar1 and Harpreet Kaur "Face Recognition Techniques: Classification and Comparisons "International Journal Of Information Technology And Knowledge Management July-December 2012, Volume 5, No. 2, Pp. 361-363

[3] Akshada Deshmukh, Harshalata Wadaskar "Webcam Based Intelligent Surveillance System "Research Inventy: International Journal of Engineering and Science Vol.2, Issue 8 (March 2013), Pp 38-42 Issn(E): 2278-4721.

[4] M. G. F. HusniTejaSukmana, Prototype Utilization of Pir Motion Sensor for Real Time Surveillance System and Web-Enabled Lamp Automation. Ieee Asia Pacific Conference on Wireless and Mobile, 2015.

[5] Dinalankara, Lahiru. Face Detection Face Recognition Using Open Computer Vision Classifies, (2017)

[6] M.Gowsalya, M.Sangeetha, K. Sri Dhivya Krishnan, N.Divya, T.Devika "A Novel Approach Automatic Digital Door Opening And Closing Security System", International Journal Of Innovative Research In Electrical , Electronics, Instrumentation And Control Engineering , Vol. 2 Issue 2, Feb 2014.

[7] RabailShafiqueSatti, Sidra Ejaz, Madiha Arshad, "A Smart Visitors Notification System with Automatic Secure Door Lock Using Mobile Communication Technology", International Journal Of Computer And Communication System Engineering, Vol. 02 No.01 February 2015

[8] KawserWazedNafi, TonnyShekhaKar, SayedAnisulHoque, "An Advanced Door Lock Security System Using Palmtop Recognition System", International Journal of Computer Applications (0975 – 8887), Volume 56– No.17, October 2012

[9] S.Ramesh, SoundaryaHariharan And Shruti Arora "Monitoring And Controlling Of Bank Security System", International Journal Of Advanced Research In Computer Science And Software Engineering, Volume 2, Issue 10, October 2012.

[10] Seung-Soo Shin, Kun-Hee Han, Kwang-Yoon Jin, "Digital Door Lock on The Access Control System Using Otp-Based User International Journal Of Computer Applications (0975 –

 https://doi.org/10.1051/itmconf/20224403070

8887)
Volume 153 – No2, November 2016
[11] Human Infrared Signal Recognition Using Single Pir Detector. Linhong Wang Chongqing College Of Electrical Engineering Chongqing, China.
[12] I.Yugashini, S.Vidhyasri, K.Gayathri Devi, "Design And Implementation Of Automated Door Accessing System With Face Recognition", International Journal Of Science And Modern Engineering (Ijisme), Volume-1, Issue12, November 2013
[13] Annie P. Oommen, Rahul A P, Pranav V, Ponni S, RenjithNadeshan, "Design and Implementation of A Digital Code Lock", International Journal Of Advanced Research In Electrical, Electronics And Instrumentation Engineering, Vol. 3, Issue 2, February 2014.
[14] Harshada B. More1, Anjali R. Bodkhe," Survey Paper on Door Level Security Using Face Recognition" International Journal of Advanced Research In Computer And Communication Engineering.
[15] Davis Matt and Ferat Sahin" HOG feature human detection system." Systems, Man, and Cybernetics (SMC), International Conference IEEE, 2016.
[16] Hoang, Van-Dung, My-Ha Le, and Kang-Hyun Jo. "Hybrid cascade boosting machine using variant scale blocks based HOG features for pedestrian detection." Neurocomputing (2014)
[17] Dalal, Navneet, and Bill Triggs. "Histograms of oriented gradients for human detection." Computer Vision and Pattern Recognition. IEEE Computer Society Conference on. Vol. 1. IEEE, 2005.
[18] Wang, Xiaoyu, Tony X. Han, and Shuicheng Yan. "An HOG- LBP human detector with partial occlusion handling." Computer Vision, 2009 IEEE 12th International Conference, 2009.
[19] Viola, Paul, and Michael Jones. "Rapid object detection using a boosted cascade of simple features." Computer Vision and Pattern Recognition. Computer Society Conference on. Vol. 1. IEEE, 2001.
[20] Jones, Michael, and Paul Viola. "Fast multi-view face detection", (2003)
[21] Viola, Paul, and Michael J. Jones. "Robust real-time face detection." International journal of computer vision (2004)
[22] Hwang, Jae Jeong, Young Min Kim, and Kang Hyeon Rhee. "Faces Recognition Using HAARCASCADE, LBPH, HOG and Linear SVM Object Detector." International Conference on Green and Human Information Technology. Springer, Singapore, 2018.
[23] Roshni Singh, Manas Singh, Leena Ragha "Real-Time Face Recognition Under Different Environment", 2nd International Conference on Advances in Science & Technology (ICAST) 2019 on 8th, 9th April 2019.

Regional language Speech Emotion Detection using Deep Neural Network

Sweta Padman[1,*], *Dhiraj* Magare[2]

[1]Ramrao Adik college of Engineering, Electronics & Telecommunication Department, Nerul, Navi Mumbai, India
[2] Ramrao Adik college of Engineering, Electronics Department, Nerul, Navi Mumbai, India

Abstract. Speaking is the most basic and efficient mode of human contact. Emotions assist people in communicating and understanding others' viewpoints by transmitting sentiments and providing feedback.The basic objective of speech emotion recognition is to enable computers to comprehend human emotional states such as happiness, fury, and disdain through voice cues. Extensive Effective Method Coefficients of Mel cepstral frequency have been proposed for this problem. The characteristics of Mel frequency ceptral coefficients(MFCC) and the audio based textual characteristics are extracted from the audio characteristics and the hybrid textural framework characteristics of the video are extracted. Voice emotion recognition is used in a variety of applications such as voice monitoring, online learning, clinical investigations, deception detection, entertainment, computer games, and call centres.

1 Introduction

The purpose of emotional speech recognition is to use a person's voice to automatically assess their emotional or physical state. During speech, air moves from the lungs to the larynx through the trachea, vibrating the vocal cords and producing speech signals [9][7]. People transmit their underlying intention through paralinguistic features such as emotions, intonation, and style through the interaction of human speech. The purpose of emotional speech recognition is to use a person's voice to automatically assess their emotional or physical state. This technology has a bright future and is critical for natural language comprehension [4]. Empathic and natural human–computer interactions necessitate the ability to perceive emotions [11][12]. Speech emotion recognition (SER) [12] has garnered a lot of academic interest in recent years, thanks to the rapid growth of conversational agents like Siri, Alexa, and Cortana. Emotions aid communication and understanding by transmitting sentiments and providing feedback to others [13]. Human voice provides a natural and instinctive interface for robot communication, and it is thus commonly used in robots that interact with humans [2]. The ability of computers to understand human emotional states such as joyful, angry, and disgust from speech signals is a fundamental goal of speech emotion recognition [12]. In recent years, a variety of viable solutions to this problem have been offered [14]. [15][6]. Speech emotion recognition is utilised in many applications, including voice surveillance, e-learning, clinical studies, lie detection, entertainment, computer games, and call centres [7].

1.1 Autonomous speech emotion recognition

In essence, autonomous speech emotion recognition systems employ a computer to mimic human emotions, including traits like accentuation, intonation, and pause, and match them to the target emotions using spectrum-based properties. To match their desired emotions, pause uses spectrum-based characteristics. Then, for accentuation, intonation, and pause, spectrum-based traits are used to match them to the intended emotions. A voice emotion recognition system is made up of three phases at its core: speech data preprocessing, emotion feature extraction, and emotion categorization [16][7][28]. As a result, two critical components of emotion detection are a sophisticated categorization architecture and speech emotion characteristics incorporating crucial information [7]. There are now numerous models for audio emotion identification that involve machine learning and deep learning [17][18][19][7]. The categorization process begins with the extraction of features. The quality and amount of characteristics employed determine how well a categorization system performs. Feature engineering is a key stage in categorization in this regard [8]. Speech emotions have been classified using hidden Markov models, support vector machines, deep belief networks, convolutional neural networks (CNN), and long shortterm memory networks (LSTM) [12] [1]. Acoustic characteristics of speech are extracted to identify emotions in speech. For understanding the relationship between retrieved speech data and preset emotion tags, many types of machine learning approaches are used [7]. The act of recognising emotions between people is called recognition. The effort of trying to design classifiers that generalise across application situations and acoustic

* Corresponding author: sweta.parkhedkar@gmail.com

settings is particularly crucial for the building of successful and practical systems of speech emotion recognition. [20][2].

Emotion detection has been used in a variety of industries, including smart homes, travel suggestion systems, and health monitoring, thanks to the rapid growth of artificial intelligence. Externally, by sight, verbal, and gestures; internally, through heart rate, respiration, blood pressure, body temperature, EEG signals, and so on. Because building speech and visual datasets is straightforward and intuitive, speech and visual characteristics are commonly employed in emotion identification. [3]. Intelligent services such as chatbots, psychological diagnosis aid, intelligent healthcare, sales advertising, and intelligent entertainment are examples of intelligent services that address not only the fulfilment of services but also the humanization of the human-computer link. [5]. It's difficult to model human emotions in words. The following are the key reasons: 1) Human emotions may be seen as noise and rejected by many existing speech recognition algorithms due to their abstraction. 2) Throughout general, human emotion can only be recognised at certain points in a protracted speech [21][4]. Nonverbal noises can effectively assist the brain in determining the difference in emotion expression when the human brain analyses emotional speech [22][5]. The automated identification and appraisal of human emotions is one of the most current research areas in fields spanning from biomedical engineering and psychophysiology to computer engineering and artificial intelligence [23][7].

2 Literature Review

Weighted Fusion and Consistent & Random fusion algorithms were suggested by Sheng Zhang et al.[1]. Adaptable and appropriate for activities requiring several modalities. Wisha Zehra et al.[2] investigated the Ensemble learning approach, which proved to be highly useful in the development of an emotion identification system for robots that deal with consumers from all over the world. Chen Guanghui et al.[3] adopt a Multi-modal emotion identification approach that successfully distinguishes similar classes and fuses speech and visual information, resulting in improved multi-modal emotion recognition performance. Chenghao Zhang et al.[4] employed an emotion embedding autoencoder capable of learning strong emotional information from labels. Jia-Hao Hsu et al.[5] Speech emotion detection in affective discussions with nonverbal vocalisation. Not only is it useful for recognising pleasant emotions, but it's also useful for recognising bad emotions. Transfer subspace learning was utilised by Na Liu et al.[6] to solve the unsupervised cross-corpus speech emotion recognition (SER) issue. Mehmet Bilal Er and his colleagues [7] The use of a novel hybrid architecture based on acoustic and deep features improves classification accuracy. Sofia Kanwal et al.[8] utilised a clustering-based genetic algorithm that can distinguish between different emotions.

3 Emotional Speech Databases

The naturalness of the information determines the success of speech emotion recognition. The Danish Emotional Speech corpus (DES) and Berlin Emotional info (EMO-DB) are two public databases, and four databases are available in Spanish, Slovenian, French, and English emotional speech. Only a few databases are authentic, and the majority of them include performed emotional speech. There appear to be three types of databases used in the SER research in terms of relevancy credibility: type one is performed emotional speech by human labeling. This information is gathered by having an actor talk with a predetermined emotion. Recently, strong challenges to the use of performed emotions have surfaced. The read of alternatives and accuracies differs between performed and spontaneous samples [15], and type two is genuine emotional speech with human labeling. Type 3 involves induced emotional speech using self-report instead of labeling, and this database comes from real-world systems (for example, contact centers). Anger and self-report are employed for labeling management whenever emotions are a unit.

4 Speech Emotion Recognition

The methods for emotion identification from Marathi speech, as well as the databases used to evaluate it, are presented in this chapter. The objective is to uncover acoustic emotion units that are suited for real-time applications first, and then to identify potential acoustic characteristics for emotion detection that can be extracted fast and automatically second. as well as evaluating a reasonable technique for picking the most relevant characteristics for a certain goal, and lastly selecting a quick yet accurate classification algorithm. As a result, for both the training and test phases, the approach utilized for all three processes as outlined in the overview in this chapter is detailed. Evaluation tests on the Marathi Language database with performed emotions are undertaken in order to be able to make as broad claims as feasible. The outcomes of the experiments will be discussed in the next chapter.

A fundamental challenge in speech emotion detection is defining a set of core emotions that can be classified by an automatic emotion recognizer. Languages have compiled a list of the most common emotional states we face in our everyday lives. There are 300 emotional states in a typical set. Classifying such a large variety of emotions, on the other hand, is incredibly difficult. Emotion is commonly broken down into core emotions, much as any colour may be broken down into a few basic hues. The basic emotions are anger, contempt, fear, pleasure, sadness, and surprise [2]. These are the most visible and recognisable feelings in our life. Table 1 depicts a strong relationship between mood and a few speech features.

Table 1. A Summary of the most common spoken emotion correlations.

	Anger	Happiness	Sadness	Fear	Disgust
Speech Rate	Slightly Faster	Faster/ slower	Slightly slower	Much Faster	Very much slower
Pitch Average	Very much higher	Much higher	Slightly slower	Very much higher	Very much lower
Pitch Range	*Much wider*	*Much wider*	Slightly narrower	*Much wider*	Slightly narrower
Intensity	*Higher*	*Higher*	lower	normal	Lower
Voice quality	*Breathy chest tone*	*Breathy blaring*	Resonant	Irregular voking	*Grumbled chest tone*
Pitch change	*Abdrupt on stressed*	Smooth, Upward inflection	Downward inflections	Normal	Wide, downward terminal inflections
Articulation	*Tense*	Normal	Slurring	Precise	Normal

4.1. Automatic Speech Emotion Recognition

Speech emotion recognition systems use a person's speech to automatically detect his or her emotional state. Analysis of the speech signal's generating process, as well as the extraction of some aspects that include emotional identification methods for identifying emotional states. The components of the speech emotion system are shown in Figure 1. The pattern recognition technology is similar to spoken emotion recognition. This demonstrates that the steps seen in the pattern recognition system are also found in the Speech emotion recognition system. The speech emotion recognition system has five main modules for training and testing: emotional speech input, preprocessing, feature extraction, feature normalization, classification, and recognised emotional output [2].

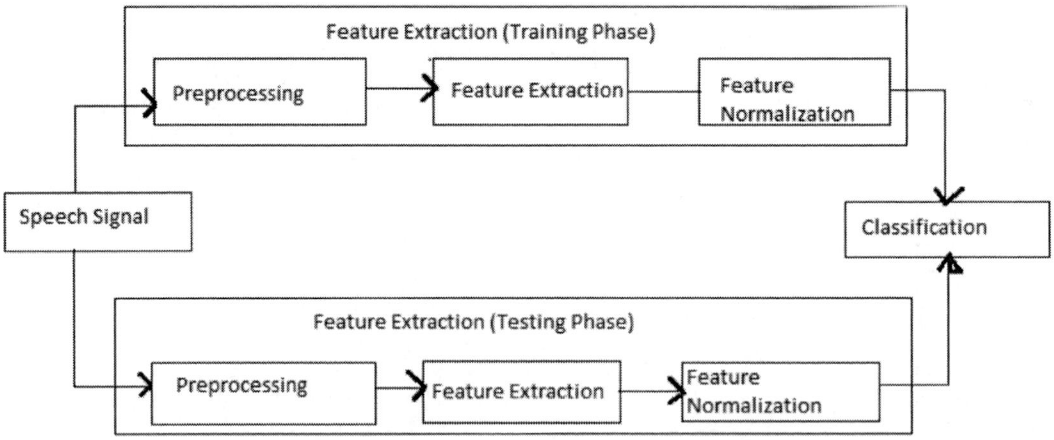

Fig. 1. Basic Emotion Recognition system.

4.1.1 Database Creation

The naturalness of the database used as an input is used to evaluate the speech emotion recognition system. If the system is given a bad database as an input, the system may come to incorrect conclusions. The database used as an input to the voice emotion detection system might comprise real-life or staged emotions. It is more feasible to employ a database that has been compiled from real-world scenarios [15]. The first step is to Emotion identification from speech is derived from acoustic measurements of those units, which are then used to determine the real characteristics from the audio input signal. The units are frequently phrases or utterances, which are linguistically motivated medium-length time periods. Despite the fact that choosing whatever type of unit to join is clearly vital, it has gotten little attention. Neither the division into utterances nor the expectation of a continuous feeling over each speech are straightforward. A good emotion unit, in general, must meet a set of conditions.

The voice emotion recognition system is estimated based on the naturalness of the database that is utilised as an input. If a bad database is used as an input to the method, it is possible that incorrect conclusions will be made. The database, which is used as an input to the spoken emotion detection system, may contain actual or acted emotions. It is more feasible to employ a database that has been compiled from real-world scenarios [15]. The first step in speech emotion recognition is to make the audio input signal meaningful, after which acoustic measurements of those units are used to extract the real features. The units are generally medium-length linguistically inspired units. phrases or utterances are examples of temporal intervals. Despite the fact that choosing whatever type of unit to join is clearly vital, it has gotten little attention. Neither the division into utterances nor the expectation of a consistent feeling over a speech are correct. In general, a good emotion unit must adhere to a set of guidelines. It should be, in particular:

1. To be consistently extracted, it must be well-defined.

2. Long enough that statistical functions may be used to calculate characteristics convincingly.

3. Short enough to maintain consistent emotional acoustic characteristics throughout the Segment.

4. Consistent with the training database's labeling.

Speech samples used in training, testing, and applications should all follow the same set of criteria, i.e., they must have the same properties. The Marathi Database is created using these guidelines.

4.2. Feature Evaluation

A multi-algorithm technique for detecting emotion from audio signals is given. The suggested MFCC and Discrete Wavelet Transform-based algorithms will be utilised to extract emotional information from speech data, using characteristics derived from pitch and formant frequency to support them. Pitch contour features such as local maxima, local minima, frequency distance, temporal distance, and slope between nearby local extrema are determined for each frame of speech sample. In addition, the first four formant frequencies are determined. The MFCC technique is a time-honored way of analysing speech signals. It is based on a linear cosine transform of a log power spectrum on a nonlinear Mel frequency scale of frequency and depicts the short-term power spectrum of a sound. DWT is used to breakdown the input speech signal and provide approximation and detail coefficients as an alternate approach. 4th level decomposition using db4 wavelets will be used to derive wavelet characteristics for each input spoken sound. The SVM classifier will be used to determine the similarity between the recovered features and a set of reference features. The database will have Marathi speech samples for each of the six emotions.

5 Database:

The Emotional Prosody Speech corpus provided us with our data. This corpus comprises Marathi continuous utterances created by 6 speakers in 6 emotions: happy, rage, neutral, fear, sorrow, and boredom (3 female, 3 male). For the examination of distinct emotions, a corpus of 180 utterances of phrases was recorded. For each of the six emotions, each speaker is given a set of five sentences to utter.

Recording: The recording was done using an electric microphone in a partly sound 16 kHz/16 bit format, with the distance between the lips and the microphone set to about 30 cm.

Listening Test: We initially randomized all of one speaker's continuous sentence files, which were then shown to ten naive listeners who were asked to rate the process was repeated for all 10 speakers, and the feelings were divided into six categories: neutral, pleased, angry, grief, fear, and surprise. All of the listeners were educated and aged 18 to 28 years old. For this study, only those statements were picked that had at least 80% of all listeners recognising their sentiments.

6 Acoustic Analysis of Emotions:

The acoustic characteristics of the voice, such as intensity, pitch, and length, are also influenced by emotion. Sentence acoustic analysis is performed. The spectrograms of one of the sentences are shown in Figures (4.1-4.6). Both prosody-related and spectral variables were taken into account in our research of emotions.

Emotional Class	Neutral	Anger	Boredom	Fear	Happiness	Sadness
Neutral	20	20	60	0	0	0
Anger	0	100	0	0	0	0
Boredom	0	0	40	0	60	0
Fear	0	20	20	20	40	0
Happiness	0	20	20	0	60	0
Sadness	0	0	40	40	20	0

Table 2. Confusion Matrix for recognition using pitch based features

Emotional Class	Neutral	Anger	Boredom	Fear	Happiness	Sadness
Neutral	80	0	20	0	0	0
Anger	0	100	0	0	0	0
Boredom	0	0	100	0	0	0
Fear	0	0	0	80	20	0
Happiness	0	0	0	0	100	0
Sadness	0	0	0	0	0	100

Table 3. Confusion Matrix for recognition using MFCC based features

4.3.1 **Pitch:**

 https://doi.org/10.1051/itmconf/20224403071

Tone height is the acoustic equivalent of pitch, which is a fundamental frequency. Pitch estimate by machine is a difficult problem. Effects of vocal tract resonance and short-term disruptions in the spoken stream can obscure pitch detection [17]. Figures shown below indicate pitch contour curves for specific emotion for one of Marathi sentence "KOKILACHA AWAZ KHUP MADHUR ASTO".

Both prosody-related and spectral variables were taken into account while studying emotions. Pitch contour curves in utterances of angry feeling increase and decrease towards the beginning of the phrase and descend towards the conclusion of the sentence, according to the findings (fig 2)[28]. For dread, the pitch increases at the start of the phrase and then stays the same before falling at the end (fig 3). Pitch contour curves of utterances in a happy mood exhibit a hold pattern at the start of the phrase and rise and fall at the conclusion of the sentence (fig 4)[28]. Pitch lowers at the end of the sentence and rises and falls at the beginning of the sentence in a neutral mood (fig 5.). The intensity of the sad emotion's F0 curve falls and rises in the start of sentences (fig. 6), whereas the intensity of the boring emotion's F0 curve falls and rises in the commencement of sentences and declines at the end position (fig. 7). With the exception of surprise, it has been determined that the intensity curve varies in proportion to the pitch in most emotions. The length of phrases pronounced in different emotions has varied values, according to the pitch contours of the sentences. As can be seen from the pitch contour (fig. 6), boredom emotions have the longest length of 2.8 seconds, while rage emotions have the shortest duration of 1.6 seconds. To evaluate single features, the information gain for all characteristics with relation to emotion classes was estimated in a database[28].

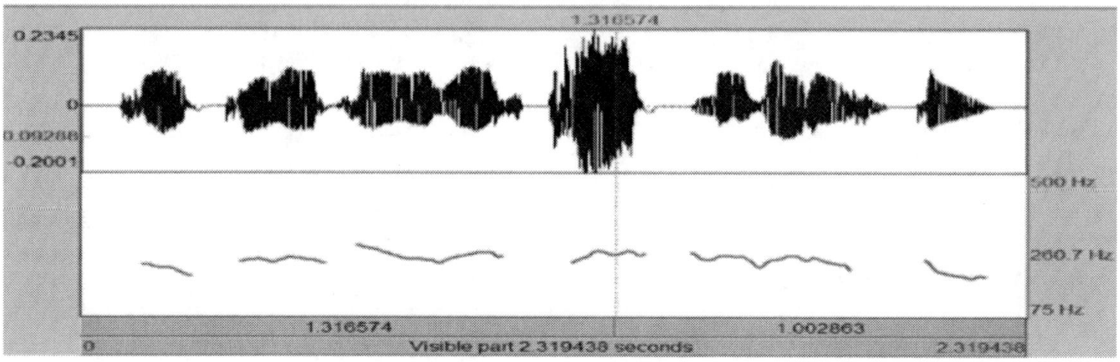

fig.2 Pitch contour for Fear

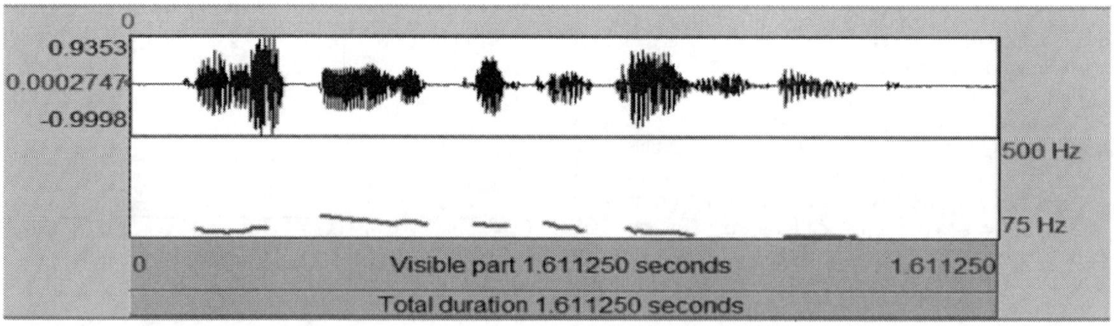

Fig.3 Pitch contour for Neutral

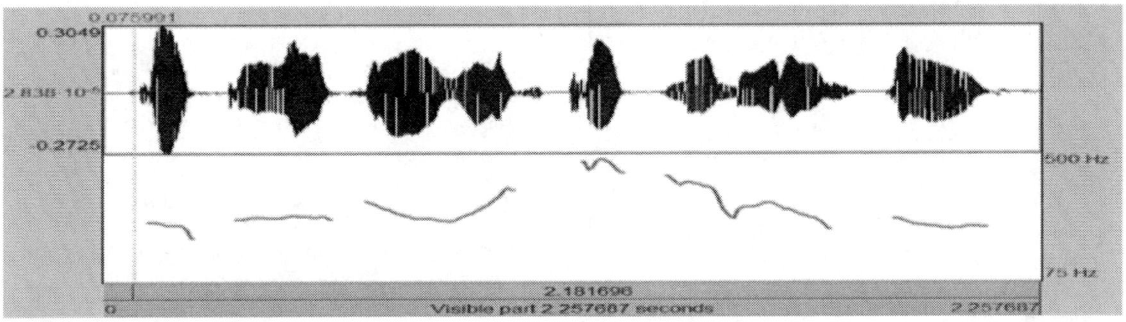

Fig.4 Pitch contour for Happiness

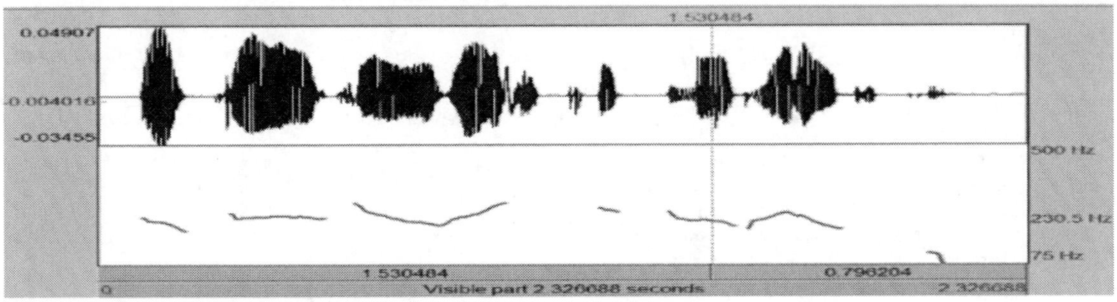

Fig.5 Pitch contour for Boredom

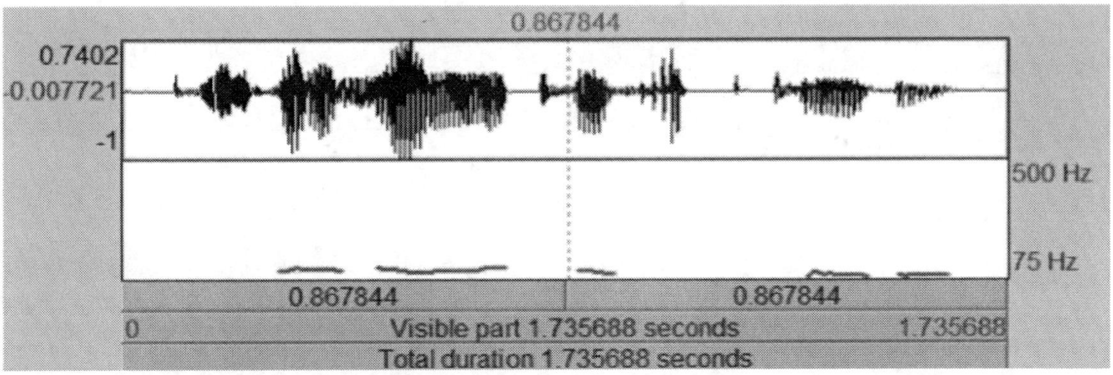

Fig.6 Pitch contour for Sadness

Emotio nal Class	Neu tral	Anger	Bore dom	Fear	Happ iness	Sad ness
Neutral	80	0	20	0	0	0
Anger	0	100	0	0	0	0
Boredo m	0	0	100	0	0	0
Fear	0	0	0	80	20	0
Happin ess	0	0	0	0	100	0
Sadness	0	0	0	0	0	100

Table 4. Confusion Matrix for recognition using Fusion features

7 Results

The multi- algorithm approach is proposed for emotion recognition from speech signals. The system is designed with specifications as: The Database created of one of Indian regional languages, the Marathi speech samples database created for the six emotions. The prosodic analysis on the Marathi speech samples shows that linguistic changes do not affect the prosody and emotion correlation. Feature extraction performed using the Acoustic features like Pitch and Formant frequency along with MFCC and Discrete Wavelet Transform based algorithms. Evaluation carried out for recognition of emotions by using individual features first. Results show that recognition is better when pitch and MFCC algorithms are employed individually.

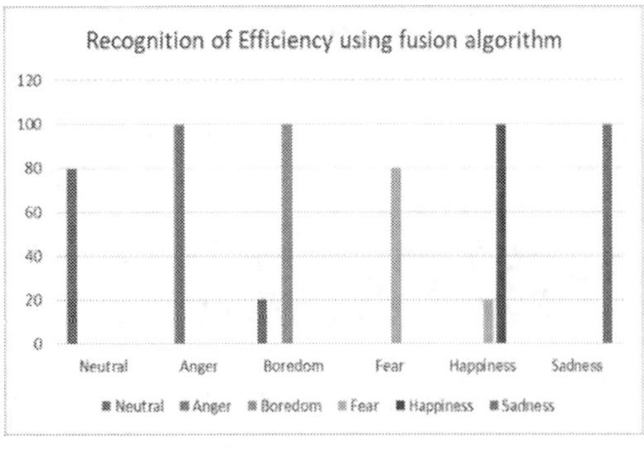

Fig 7. Recognition efficiency using fusion based features

8 Conclusion

When all of the characteristics are combined, the system's performance is deemed to be good. This suggests that the problem of speech emotion recognition is better handled when using a group of multiple descriptors. Moreover, our methodology significantly outperforms the ability of a human listener to classify the respective signals which is equal to 100 percent. The Classifier shows better results when fusion of all the features are used as shown in table 5 and 6. In future work, we aim to test system performance on spontaneously expressed (Non acted) emotions of Marathi language. To enhance system performance

rate. further introduction of one more classifier can be proposed which can reduce confusion rate. classifier can be proposed which can reduce confusion

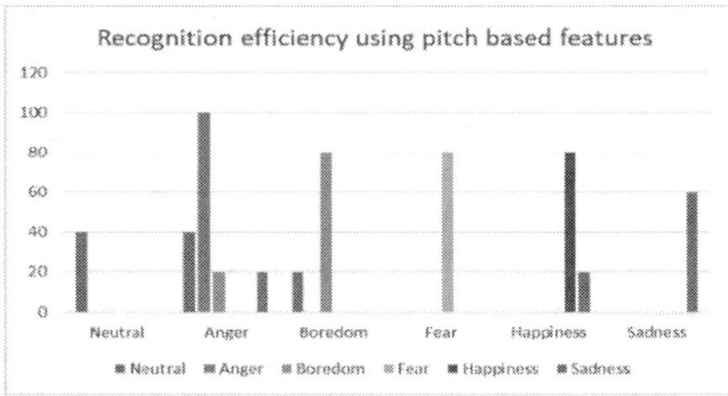

Fig. 8 Recognition efficiency using pitch based features

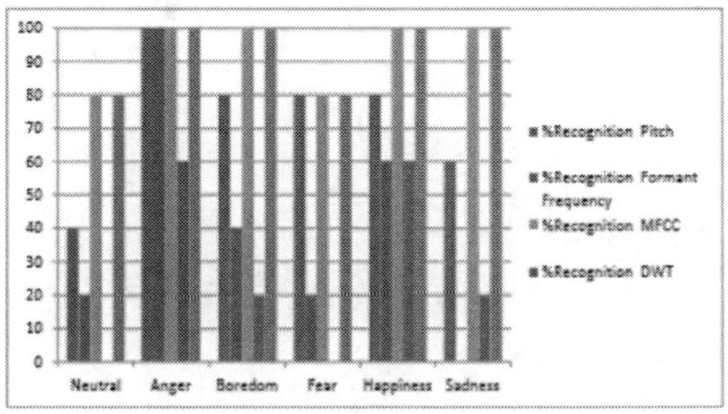

Fig 9. Comparison of % recognition for feature extraction algorithm

References

[1] Zhang, Sheng, Min Chen, Jincai Chen, Yuan-Fang Li, Yiling Wu, Minglei Li, and Chuanbo Zhu, "Combining cross-modal knowledge transfer and semi-supervised learning for speech emotion recognition," in Knowledge-Based Systems, Vol.229, pp.107340, 2021.

[2] Zehra, Wisha, Abdul Rehman Javed, Zunera Jalil, Habib Ullah Khan, and Thippa Reddy Gadekallu, "Cross corpus multi-lingual speech emotion recognition using ensemble learning," in Complex & Intelligent Systems, Vol.7, no.4, pp.1845-1854, 2021.

[3] Guanghui, Chen, and Zeng Xiaoping, "Multi-modal emotion recognition by fusing correlation features of speech-visual," in IEEE Signal Processing Letters, Vol.28, pp.533-537, 2021.

[4] C. Zhang and L. Xue, "Autoencoder With Emotion Embedding for Speech Emotion Recognition," in IEEE Access, vol.9, pp.51231-51241, 2021.

[5] Hsu, Jia-Hao, Ming-Hsiang Su, Chung-Hsien Wu, and Yi-Hsuan Chen, "Speech emotion recognition considering nonverbal vocalization in affective conversations," in IEEE/ACM Transactions on Audio, Speech, and Language Processing, Vol. 29, pp.1675-1686, 2021.

[6] N. Liu, Baofeng Zhang, Bin Liu, Jingang Shi, Lei Yang, Zhiwei Li and Junchao Zhu, "Transfer Subspace Learning for Unsupervised Cross-Corpus Speech Emotion Recognition," in IEEE Access, vol. 9, pp. 95925-95937, 2021.

[7] M. B. Er, "A Novel Approach for Classification of Speech Emotions Based on Deep and Acoustic Features," in IEEE Access, vol. 8, pp. 221640-221653, 2020.

[8] S. Kanwal and S. Asghar, "Speech Emotion Recognition Using Clustering Based GA-Optimized Feature Set," in IEEE Access, vol.9, pp.125830-125842, 2021.

[9] Schlegel, Patrick, Stefan Kniesburges, Stephan Dürr, Anne Schützenberger, and Michael Döllinger, "Machine learning based identification of relevant parameters for functional voice disorders derived from endoscopic high-speed recordings," in scientific reports, Vol.10, no.1, pp.1-14, 2020.

[10] E. Cambria, S. Poria, A. Hussain and B. Liu, "Computational Intelligence for Affective Computing and Sentiment Analysis [Guest Editorial]," in IEEE Computational Intelligence Magazine, vol. 14, no. 2, pp. 16-17, 2019.

[11] Chen, Min, and Yixue Hao, "Label-less learning for emotion cognition," in IEEE transactions on neural networks and learning systems, Vol.31, no.7, pp.2430-2440, 2019.

[12] El Ayadi, Moataz, Mohamed S. Kamel, and Fakhri Karray, "Survey on speech emotion recognition: Features, classification schemes, and databases," in Pattern recognition, Vol.44, no. 3, pp.572-587, 2011.

[13] Zvarevashe, Kudakwashe, and Oludayo Olugbara, "Ensemble learning of hybrid acoustic features for speech emotion recognition," in Algorithms, Vol.13, no.3, pp.70, 2020.

[14] Liu, Zhen-Tao, Qiao Xie, Min Wu, Wei-Hua Cao, Ying Mei, and Jun-Wei Mao, "Speech emotion recognition based on an improved brain emotion learning model," in Neurocomputing, Vol.309, pp.145-156, 2018.

[15] Gideon, John, Melvin G. McInnis, and Emily Mower Provost, "Improving cross-corpus speech emotion recognition with adversarial discriminative domain generalization (ADDoG)," in IEEE Transactions on Affective Computing, Vol.12, no.4, pp.1055-1068, 2019.

[16] Lu, Guanming, Liang Yuan, Wenjuan Yang, Jingjie Yan, and Haibo Li, "Speech emotion recognition based on long short-term memory and convolutional neural networks," in Journal of Nanjing University of Posts and Telecommunications, Vol.38, no.5, pp.63-69, 2018.

[17] Liu, Zhen-Tao, Qiao Xie, Min Wu, Wei-Hua Cao, Ying Mei, and Jun-Wei Mao, "Speech emotion recognition based on an improved brain emotion learning model," in Neurocomputing, Vol.309, pp.145-156, 2018.

[18] Bhavan, Anjali, Pankaj Chauhan, and Rajiv Ratn Shah, "Bagged support vector machines for emotion recognition from speech," in Knowledge-Based Systems, Vol.184, pp.104886, 2019.

[19] Issa, Dias, M. Fatih Demirci, and Adnan Yazici, "Speech emotion recognition with deep convolutional neural networks," in Biomedical Signal Processing and Control, Vol.59, pp.101894, 2020.

[20] Schuller, Björn, Zixing Zhang, Felix Weninger, and Gerhard Rigoll, "Using multiple databases for training in emotion recognition: To unite or to vote?," in Twelfth Annual Conference of the International Speech Communication Association, 2011.

[21] Atmaja, Bagus Tris, and Masato Akagi, "Speech emotion recognition based on speech segment using LSTM with attention model," in 2019 IEEE International Conference on Signals and Systems (ICSigSys), pp.40-44, 2019.

[22] Schirmer, Annett, and Thomas C. Gunter, "Temporal signatures of processing voiceness and emotion in sound," in Social cognitive and affective neuroscience, Vol.12, no.6, pp.902-909, 2017.

[23] Nardelli, Mimma, Gaetano Valenza, Alberto Greco, Antonio Lanata, and Enzo Pasquale Scilingo, "Recognizing emotions induced by affective sounds through heart rate variability," in IEEE Transactions on Affective Computing, Vol.6, no.4, pp.385-394, 2015.

[24] F. Dellaert, T. Polzin and A. Waibel, "Recognizing emotion in speech," Proceeding of Fourth International Conference on Spoken Language Processing. ICSLP, vol.3, pp. 1970-1973 1996.

[25] https://ieeexplore.ieee.org/document/1623803?reload=true

[26] Askarzadeh, Alireza, "A novel metaheuristic method for solving constrained engineering optimization problems: crow search algorithm," in Computers & Structures, Vol.169, pp.1-12, 2016.

[27] Rao, R. Venkata, "Teaching-learning-based optimization algorithm," in Teaching learning based optimization algorithm, pp. 9-39, 2016.
[28] Agrawal, Shyam Sunder. "Emotions in Hindi speech- analysis, perception and recognition." *2011 International Conference on Speech Database and Assessments (Oriental COCOSDA)* (2011): 7-13.

 https://doi.org/10.1051/itmconf/20224403072

Offline Transaction System

Shitanshu Kochar[1,*], *Hrishikesh* Nikam[1,**], *Rahul* Tripathi[1,***], and *Amarsinh* Vidhate[1]

[1]Department of Computer Engineering, Ramrao Adik Institute of Technology, D Y Patil Deemed to be University, Navi Mumbai, India

Abstract. The usage of online payment systems is increasing multifold these days due to their transaction reliability and ease of completion in real-time. The most preferred service is UPI service, which is majorly available in Metropolitan cities, in some semi-urban and rural areas which have a proper communication infrastructure. However, there is a major issue to expand these services in rural areas, which lack proper communication infrastructure. Even after the internet revolution, it is ensured that all Indians are connected to data at all times but there might be situations where the user could not able to connect to the Internet owing to many reasons like High Altitude, Low Network availability, in-flight, etc. As an individual, such situations cannot be avoided, and there it comes to the applicability of offline transaction systems. We plan to change that by introducing an offline transaction system. We plan to expand on the idea of an offline payment system that will ensure that a reliable & secured digital payment system is available to all individuals.

1 Introduction

Creating an *off*line payment system to democratize digital payment technology and ensure that the facility is made available to everyone. We have used react for the Front end, MySQL for maintaining databases, and JavaScript for the backend. We plan to eliminate various problems with digital systems too.

1.1 Objective

The major objective of our system is to make the current digital payment process *off*line and make it more flexible for both entities. We aim to make digital payment available to every segment of the population and not just metropolitan cities as is the case largely now. The current digital payment system proved to be a great revolution but now thereis a need to democratize it further and make it accessible to all. We believe that introduction of such a system can bring about a lot of positive changes. We want to address limitations in the current system which include:

1. Creation of DUAL NETWORK SERVER (DNS) which will process payment *off*line by using SMS Service.
2. Using encrypted sound waves using a specific high-frequency sound, which stores transaction data that will process payment later.
3. Using Batch-processing (mesh technology) in which case payments will be processed after a designated amount of time and will require an internet connection.
4. Bluetooth / NFC-based *off*line transaction systems which will be wallet-to-wallet transfers which will also not be reflected immediately.
5. Launching these features like an inbuilt function wherephones will be inbuilt to act as an independent server (DNS) to give call the API of the particular bank.

*e-mail: shi.koc.rt18@rait.ac.in, **e-mail: hri.nik.rt18@rait.ac.in, ***e-mail: rah.tri.rt18@rait.ac.in

2 Literature Survey

Existing Online payment solutions are popular and we more or less know how this works. Let us take a lookat the available *off*line transaction services.

1. Ahamad, S.S at.el [1] has expressed the flaws in existing solutions like end-to-end communication, information privacy as well as client's anonymity. The authors have proposed a secure and Privacy-preserving mobile commerce for NFC proximity-based. The author claims about the low cost computations and communication.
2. The proposal based on the recent Covid -19 pandemic, where a huge increase in online transactions have taken place. However many times the internet connectivity is not available due to many reasons. Yash Kumar Gupta at. El [2] has proposed a cheaper offline solution without involving higher online transaction costs.
3. General overall observations of the survey are that existing OFFLINE solutions are either heavily reliant on Infrastructure or unable to process payments to OFFLINE[3,4,5].

 https://doi.org/10.1051/itmconf/20224403072

4. Major problems with these cards were that the majority number of times *offl*ine transactions did not process, absence of Infrastructure like availability of the card hence were unsuccessful [4,5].

5. Pay TM 2019 launched a Tap-to-Tap pay card, in which users loaded money and could pay with or without the internet, provided that the receiver had the Tap-to-Tap card as well[6]

6. Googlepay partnered with pine labs in July 2019 to allow merchants to initiate payment requests by punching in a customer's mobile number on their PoS service which was later unsuccessful [7].

Addressing a few issues which could potentially contribute to solving this problem are:

1. Creation of DUAL NETWORK SERVER (DNS) which will process payment *offl*ine by using SMS service (mobile network). Using SMS service will ensure convenience and secure transactions.

2. Using encrypted sound waves using a specific high-frequency sound, which stores transaction data that will process payment later.

3. Using Batch-processing (mesh technology) in which case payments will be processed after a designated amount of time and will require an internet connection.

4. Bluetooth / NFC-based *offl*ine transaction systems which will be wallet-to-wallet transfers which will also not be reflected immediately.

5. Launching these features like an inbuilt function wherephones will be inbuilt to act as an independent server (DNS) to give call the API of the particular bank.

6. To conclude, the Majority of the existing payment services make use of the internet as infrastructure, and *offl*ine transaction systems are have been tried by various organizations but have not succeeded in creating this system.

3 Proposed Methodologies

1. For the web app/ web portal mentioned in the system design section, where users will register on the website, will be made using the React.js library(JavaScript UI library).

2. For the server mentioned in the system design section, a Node.js server (Asynchronous JavaScript Runtime server framework) will be used.

3. The Database used to store the user's data will be implemented using the MySQL Database (the most popular relational database).

4. The *Offl*ine Pay app will be made using the React Native framework (a React Framework for making cross-platform Mobile Applications) all technologies used in this project will be Open-source.

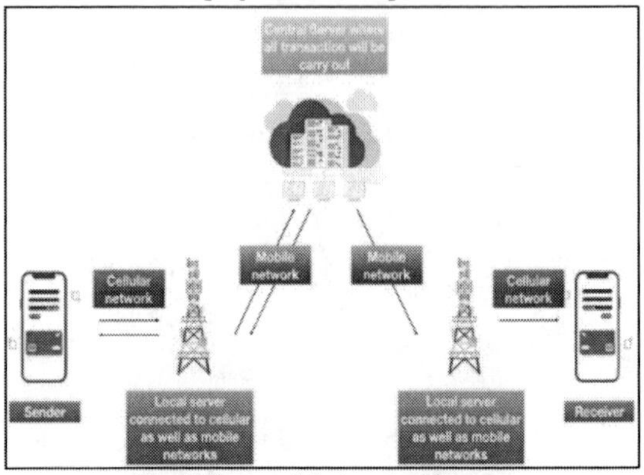

Fig. 1. Proposed System Architecture

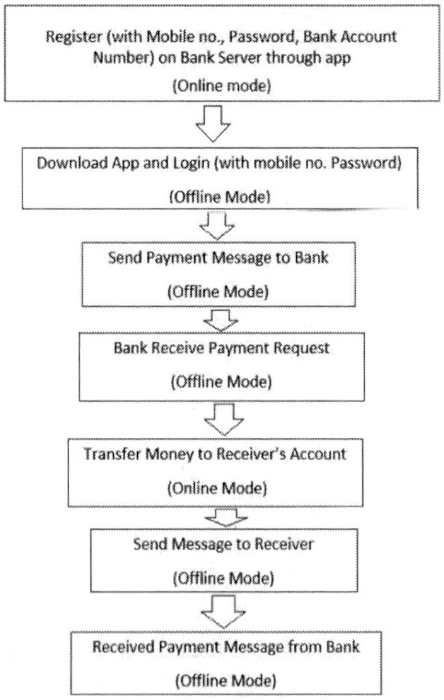

Fig. 2. System Workflow

1. Bank Database
   - Customer ID
   - Account Details
   - Transaction Details

2. Bank Website
   - Add User: This is a function used to register any new user. Using this user is added to the bank database. All the information of the user like name, communication details, bank details, etc.
   - User Database: Large database which is usedto store all the information provided by the customer.
   - Transaction list: This list is used to store transaction details and make them available on-demand.

 https://doi.org/10.1051/itmconf/20224403072

- Transaction database: Used to store all the transaction details. Can be used by the user to retrieve details of all the transactions done by him as and when the need arises.

3. Working of System
   - Register on the app using mobile no., password, account no., etc. In online mode.
   - Download the app and login in *offline* mode.
   - Payment requests to the bank will be sent in *offline* mode by the user using SMS and OTP services which ensure safe and efficient transaction requests.
   - Bank will receive the payment request in *offline* mode through SMS gateway and it will start processing this request after authenticating the user and its request.
   - Money will be transferred to the receiver account by the bank after authenticating the request and credentials.
   - Message will be sent to the receiver and the senderwill get a payment message from the bank.
   - Detailed working of the external system:The process comprises two users for the transaction to take place – the sender and the receiver.

1. Both the users first register on the website.
2. Both the users have to download the app to be able to carry out transactions.
3. Users log in to the app using their credentials.
4. The sender might be faced with two scenarios, if an internet connection is available then the sender can directly click on send now and do the transaction and if the internet is not there then the sender makes use of the SMS to do the transaction.
5. Both the users get SMS of the transaction.

- Detailed working of the internal system:

1. In cases of the device does not have the internet, the app triggers it to send an SMS using an API called React-Native-get-SMS. This app sends SMS to the device whose number is registered on the website and is present in the database.
2. MySQL database stores the details of the registered users.
3. The SMS is received and a record of the transaction is completed.
4. The offline transaction is later logged in the history whenever the device has an active internet connection and is connected to the database.

4. Technologies Used
   - We have used Dual Network Server.
   - Database is synced using MongoDB ATLAS.
   - Interface for API is restful.

Response time of Node Server is recorded in an offline transaction.

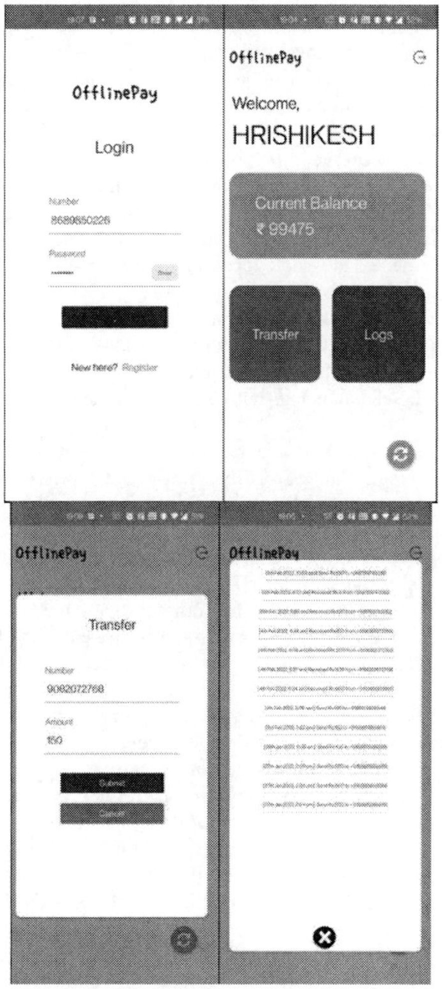

Figure 3: Login and subsequent pages

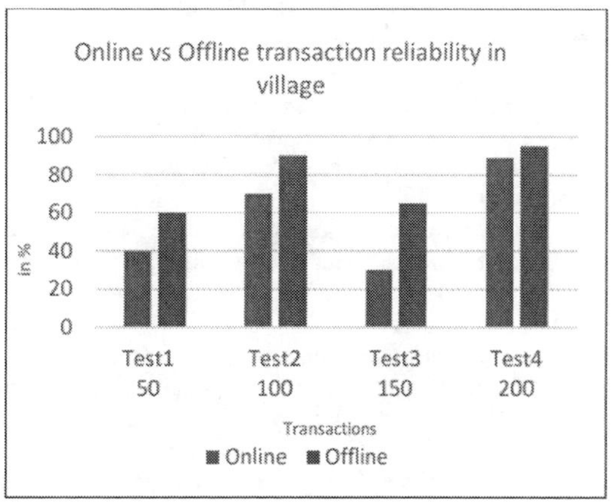

Fig. 4. Average Success Ratio

5. Results and Analysis

The system developed works fine for offline transactions and it can be conveniently used by people even in remote areas and even by a feature phone. SMS service provides a great way to implement offline transaction systems in terms of both ease of use and security.

We observed numerous advantages of offline transaction systems over online transaction systems while implementing our model. The most prominent advantages are no data expenditure, no network hassles, easily implemented at remote locations, and accessible by lower range mobile phones.

Offline transactions require more time as compared to online transactions in areas with good internet connectivity. There are various reasons responsible for this. Load balancing, network interference, throughput, etc.

On average, the speed of online transactions is faster than that of offline transactions provided the internet connectivity is sufficient. This is a result of higher throughput provided via the distributed nature of the internet and lower rate-limiting resulting in greater speeds of transaction.

Considering successful transactions, the chances of a transaction completing is more in the case of offline transactions. There are many reasons responsible for it. Majorly in rural and semi-urban areas, the absence or scarcity of internet results in a lower rate of success of online transactions. The reasons responsible for it are higher rate-limiting and lower throughput.

The offline success ratio is higher because the availability of the cellular network is more common in various rural and semi-urban areas. The offline transactions are dependent on sending and receiving of SMS. This process is dependent on the availability of the cellular network.

Due to various reasons like the sim card revolution and the increase in the number of cellular network consumers, offline transactions are more plausible. The security of offline transactions depends on the security of the bank server. The bank server is the one processing the transactions and logging these changes in the history/logs.

Offline vs online transaction speeds concerning successful transactions completed. Urban, semi-urban and rural areas all have different connection speeds and network availability.

Hence, they result in lower throughput, higher network interference, and lesser load balancing. This affects online transactions adversely and reduces the speed of transactions severely.

In areas with perfect connectivity and decent internet access, online transactions have the upper hand as they tend to be faster as compared to offline transactions.

Also, in areas with good connectivity, the step of sending and receiving an SMS is reduced; the load balancing works in the favor of online transactions and we can assume less to no network interference. So in areas with good Internet connections, online transactions were faster.

In semi-urban areas with decent cellular network range but lesser internet/ data speeds, offline transactions are faster. This is because the speed and success rate of transactions depend on the availability and speed of the internet.

Semi-urban areas perfectly describe the need for offline transaction systems. This is where the majority of existing online transaction systems fail due to the scarcity of the internet.

Areas where people can afford mobile devices and can pay for goods and services, but are stopped by the incapability of online transactions/ requiring internet. This is where offline transactions can make a difference.

6. Conclusion and Future Scope

We performed various tests and recorded success ratios of online and offline transaction systems under various circumstances. In conditions where there was really good network availability, the performance of both the systems was somewhat similar (Test4).

In cases with low network availability like Test3, offline transactions were far ahead of online transactions in terms of success ratio.

In cases with average network availability, like in Test1 and 2, the performance of offline transaction systems was still better owning to uncertainties in online transaction systems on account of network fluctuations, data availability, etc.

UPI services are currently limited to Metropolitan cities; whereas these services could be used in semi-urban and even rural areas which have proper infrastructure. We plan to expand on the idea of having an *off*line payment systemthat will ensure the availability of digital payment systems to everyone.

We conclude that such a system will enhance living standards and will open door to a lot of opportunities for people and will ensure ease of living for every individual.

There is scope for future developmental work too on this system in the field of increasing the easiness and making it more convenient to be used anywhere. Various features such as speed dial themed speed and instant payment system can be added to this and hence can be made much more hassle-free and contribute actively towards increasing ease of living.

References

1. Ahamad, S.S., Pathan, AS.K. Trusted service manager (TSM) based privacy preserving and secure mobile commerce framework with formal verification. *Complex Adapt Syst Model* **7**, 3 (2019).

2. J. Luo and M. Yang, "O*ff*line Transferable E-Cash Mechanism," IEEE Conference on Dependable and Secure Computing (DSC), (2018)

3. Rahat, Abdullah Alam, at.el., " An *off*ine electronic payment system based on an untraceable blind signature scheme," KSII Transactionson Internet and Information Systems. **11**. 2628- 2645,(2017)

4. Siddiqui, Faria Ashraf, H. Ullah, Ata., "Dual server-based security system for multimedia Services inNext

Generation Networks. Multimedia Tools and Applications, **79**, (2020).

5. N. C. Kiran and G. N. Kumar, "Building robust m-commerce payment system on the *offline* wireless network," Fifth IEEE International Conference on Advanced Telecommunication Systems and Networks (ANTS), pp 1-3, (2011)

6. https://paytm.com/offer/taptopay (2019)

7. https://www.pinelabs.com/media-analyst/pine-labs-collaborates-with-google-pay-in-india (2019)

AUTHOR INDEX

AUTHOR INDEX

AUTHOR INDEX